MICROBIAL FOODBORNE DISEASES

HOW TO ORDER THIS BOOK

BY PHONE: 800-233-9936 or 717-291-5609, 8AM–5PM Eastern Time

BY FAX: 717-295-4538

BY MAIL: Order Department
Technomic Publishing Company, Inc.
851 New Holland Avenue, Box 3535
Lancaster, PA 17604, U.S.A.

BY CREDIT CARD: American Express, VISA, MasterCard

BY WWW SITE: http://www.techpub.com

MICROBIAL FOODBORNE DISEASES

Mechanisms of Pathogenesis and Toxin Synthesis

EDITED BY

Jeffrey W. Cary, Ph.D.
John E. Linz, Ph.D.
Deepak Bhatnagar, Ph.D.

CRC Press is an imprint of the
Taylor & Francis Group, an **informa** business

CRC Press
Taylor & Francis Group
6000 Broken Sound Parkway NW, Suite 300
Boca Raton, FL 33487-2742

First issued in paperback 2019

ISBN-13: 978-1-56676-787-3 (hbk)
ISBN-13: 978-0-367-39920-7 (pbk)

Library of Congress Catalog Card No. 99-66881

Visit the Taylor & Francis Web site at
http://www.taylorandfrancis.com

and the CRC Press Web site at
http://www.crcpress.com

Table of Contents

Part III: Fungal and Marine Toxins

Part IV: Parasitic Protozoa

Foreword

THIS very useful and timely volume recognizes that food safety hazards are not confined to *Salmonella, Campylobacter* and the pathogenic *E. coli.* But rather these hazards result from a wide variety of toxins produced in crops and food by several agents, including pathogenic gram positive and gram negative bacteria, fungi and algae, parasitic protozoa and viruses, and related agents.

The scientific knowledge regarding these zoonotic pathogens is the basis of our food safety system today. It is science that develops the principles and the technology necessary for methodology that can detect pathogens in amounts that could cause disease. And it is science that develops the interventions that are used by animal and plant based food producers to prevent the survival and growth of pathogens in their food products. Knowledge of the molecular basis of pathogenesis will help point the way toward effective, affordable interventions in the production of both plant and animal based food products. A food safety system based on science also supports quantitative risk assessment, which leads to placing the greatest priority for action against the risks calculated to have the greatest potential impact on assuring a safe food supply.

An understanding of the mechanisms of pathogenesis of foodborne pathogens is particularly timely because of the current recognition of the importance of science in ensuring food safety. The recent report from the National Academy of Science (NAS), "Ensuring Safe Food from Production to Consumption," stated that "The mission of an effective food safety system is to protect and improve the public health by ensuring that foods meet science-based safety standards through the integrated activities of the public and private sectors." This volume will further help to achieve the goals of the NAS Report to achieve a "science based" food safety system. The work discussed in Chapter 8, "Mechanisms of Pathogenesis and Toxin Synthesis in *Clostridium botulinum,*" will help to further assure the effectiveness of low acid canned food technology—which was mentioned in the NAS Report. In addition, several of the chapters in Section I, "Gram Negative Foodborne Bacterial Pathogens," are particularly applicable to the HACCP systems now being developed by the meat and poultry slaughter and processing industries. Verification of the effectiveness of HACCP requires understanding the genetic heterogeneity of pathogens such as variants of *E. coli* in order to develop tests to detect the pertinent virulent pathotypes of food safety concern. Other very relevant discussions regarding the role of science in food safety include (1) the genetics of the virulence of the various *Campylobacter,* which is crucial as the basis of molecular diagnostic techniques to better define the role of the various *Campylobacter* serotypes as etiologic agents of human enteric infection, and (2) the pathogenic determinants of *Listeria monocytogenes.* Extensive recent outbreaks of *Listeria* in prepared meat products have emphasized the importance of learning more about the nature of this bacterial pathogen and the necessity of devising effective control procedures.

Two other contemporary concerns are dependent on a science-based food safety system. The first is international trade where "science-based" recommendations are becoming the standard for international trade agreements such as the export of mycotoxin-contaminated commodities. No longer can a nation exclude a food product for strictly arbitrary reasons. Even though the implementation of this principle is still excruciatingly slow at times, the scientific understanding that is accumulating regarding the nature of food safety hazards will eventually bring rationality to this area of international trade. Secondly, bioterrorism, with respect to both food production and food safety, is a rapidly escalating concern in many developed countries. The wide availability of the basic type of biological information provided in this volume regarding the pathogenic causative agents of hazards to our food supply is an essential component of protection against bioterrorism. This protection requires methodology to detect toxic and infectious agents and track them to their sources; and it requires the availability of effective intervention and control measures to prevent and deal with food supply disasters.

This volume will be an excellent reference for researchers as they seek to provide the food industry and the government agencies with the necessary scientific advances to further assure food safety and meet the expectations of the consuming public.

JANE F. ROBENS
U.S. Department of Agriculture

Preface

FOODBORNE diseases are major causes of illness and death throughout the world. Though often considered a problem associated with developing, Third World countries, there are millions of cases of illness reported each year in developed countries such as those of the European Union and the United States. It is estimated that in the United States alone, some 9000 people a year, mostly the very young and elderly, die as a result of infections and intoxications from ingestion of contaminated food and water. Ever increasing reports of illness and deaths from ingestion of tainted meats, vegetables and juices, not to mention the anxiety raised worldwide over the possible transmission of "Mad Cow" disease to humans has caused many governments to reassess their policies and industry practices on safeguarding their food supplies. This reassessment was most evident in the United States where beginning in 1997, in an effort to restore American's trust in its own food production system as well as its imports, President Clinton announced the National Food Safety Initiative that pledged 46 million dollars to develop a comprehensive plan to improve the safety of the food supply. The President has requested funding in excess of 100 million dollars for FY 1999 and 2000. Similar food safety initiatives are being undertaken in other countries in response to the public awareness and demands for a safer food supply. Though much of this funding is earmarked for improved surveillance, inspections, education, and training, funds have also been designated for expansion of food-safety research.

Development of efficacious strategies for prevention and treatment of foodborne diseases has relied and will continue to rely to a significant extent on the tools of molecular biology. Through the use of molecular and cellular biological techniques, numerous advances have been made in understanding the molecular basis of virulence mechanisms and toxin biosynthesis in organisms that routinely contaminate food and feed. *Microbial Foodborne Diseases: Mechanisms of Pathogenesis and Toxin Synthesis* is intended to serve as an advanced text providing useful, up-to-date information by recognized authorities on the molecular mechanisms of pathogenicity and toxin production of what we felt were some of the most significant foodborne pathogens. Selection of the pathogens included in this book depended on the body of molecular biological information available as well as the organism's impact on economics and human health. The main focus of this book, therefore, has been on the molecular and cellular processes that govern pathogenicity and toxin production in foodborne pathogens, be they viral, bacterial, fungal, or protozoan. Additional information presented includes the latest information related to the association of the pathogen with particular foods (water will be considered a food), epidemiology, methods of early detection, toxicology, and economic impact of the pathogen. Topics such as spoilage, preservation methods, fermentation, and food processing techniques that are well documented in other food microbiology texts were not included. In general, this book is designed for scientists involved in food microbiology and food safety, as well as human and veterinary medicine, both at the graduate and postgraduate level. In addition to serving as an invaluable reference resource, this book should be of particular interest to molecular biologists desiring current information emphasizing the molecular mechanisms governing the disease process. This text should also serve as a valuable tool in the rational design of preventative controls and therapeutic approaches to the disease process.

The text is divided into five major sections covering bacterial foodborne pathogens, toxigenic fungi and marine dinoflagellates, protozoan pathogens, and viral and virus-like foodborne pathogens. We feel that the first nine chapters provide in depth, authoritative coverage of the molecular mechanisms of pathogenicity and toxin production in bacterial foodborne pathogens. Chapter 10 represents the most comprehensive review to date on the molecular biology of aflatoxin biosynthesis by *Aspergillus,* whereas Chapters 11 and 12 provide up-to-date information on the molecular biology of trichothecene and fumonisin production by *Fusarium* and the biosynthesis of PSP toxins by marine dinoflagellates, respectively. Two detailed reviews on the factors governing pathogenicity in the protozoan parasites *Toxoplasma gondii* and *Entamoeba histolytica* and *Cryptosporidium parvum* are presented in Chapters 13 and 14. Chapter 15 provides a thorough discussion of the epidemiology and pathogenicity of Norwalk and other human caliciviruses. The book concludes with what we think is a very informative chapter on the molecular biology of prion diseases that are associated with bovine spongiform encephalopathy, "Mad Cow" disease, and Creutzfeldt-Jakob disease in humans.

We are extremely grateful to all of our contributors for taking the time and effort required to produce a comprehensive examination of their respective subject areas. The editors also want to thank the many reviewers who took the time to provide critical comments and suggestions to the authors in the final preparation of their chapters. We are grateful to the staff at Technomic Publishing Co., Inc., for their careful editing which allowed this book to come together as a cohesive unit. J. W. C. would like to especially thank Eleanor Riemer at Technomic Publishing Company for her guidance, encouragement, and patience throughout the course of preparation of this book.

JEFFREY W. CARY
JOHN E. LINZ
DEEPAK BHATNAGAR

List of Contributors

Sheila R. Abner
Michigan State University
College of Veterinary Medicine
Animal Health Diagnostic Laboratory
P.O. Box 30076
Lansing, MI 48909-7576

Robert L. Atmar
Division of Molecular Virology
Baylor College of Medicine
One Baylor Plaza
Houston, TX 77030

Deepak Bhatnagar
USDA, ARS
Southern Regional Research Center
1100 Robert E. Lee Blvd.
New Orleans, LA 70124

Jeffrey W. Cary
USDA, ARS
Southern Regional Research Center
1100 Robert E. Lee Blvd.
New Orleans, LA 70124

Guy R. Cornelis
Microbial Pathogenesis Unit
Christian de Duve Institute of Cellular Pathology
and Université Catholique de Louvain
Av. Hippocrate, 74 P. Box 74.49
B-1200 Brussels, Belgium

Michael S. Donnenberg
University of Maryland
School of Medicine
Department of Medicine
Division of Infectious Diseases
10 S. Pine Street
Baltimore, MD 21201-1192

Mary K. Estes
Division of Molecular Virology
Baylor College of Medicine
One Baylor Plaza
Houston, TX 77030

Yukako Fujinaga
Department of Bacteriology
Okayama University Medical School
2-5-1 Shikata-cho, Okayama 700, Japan

Francisco García-del Portillo
Centro de Biología Molecular "Severo Ochoa"
Universidad Autónoma de Madrid-CSIC
Campus de Cantoblanco
28049 Madrid, Spain

Kaoru Inoue
Department of Bacteriology
Okayama University Medical School
2-5-1 Shikata-cho, Okayama 700, Japan

Yuzaburo Ishida
Fukayama University
Department of Marine Biotechnology
Faculty of Engineering
Fukayama, Hiroshima Prefecture 729-02, Japan

Sophia Kathariou
University of Hawaii-Manoa
Department of Microbiology
207 Snyder Hall
2538 The Mall
Honolulu, HI 96822

Corinne Ida Lasmézas
Service de Neurovirologie
CEA/DSV/DRM/SSA 60-68
av. du Général Leclerc
BP. 6
92 265 Fontenay-aux-Roses Cédex, France

John E. Linz
Michigan State University
Department of Food Science and Human Nutrition
234 GM Trout FSHN Bldg.
East Lansing, MI 48824-1224

Barbara J. Mann
University of Virginia
Health Sciences Center
Department of Internal Medicine
Division of Infectious Diseases
MR-4 Building, Room 2115
Charlottesville, VA 22908

Linda S. Mansfield
Michigan State University
Department of Microbiology
B43 Food Safety Toxicology Center
East Lansing, MI 48824

Bruce A. McClane
University of Pittsburgh
School of Medicine
Department of Molecular Genetics and Biochemistry
E1240 Biomedical Science Tower
Pittsburgh, PA 15261

James P. Nataro
University of Maryland
School of Medicine
Center for Vaccine Development
10 S. Pine Street
Baltimore, MD 21201-1192

Keiji Oguma
Department of Bacteriology
Okayama University Medical School
2-5-1 Shikata-cho, Okayama 700, Japan

Kenneth M. Peterson
Louisiana State University Medical Center
Department of Microbiology and Immunology
1501 Kings Highway
P.O. Box 33932
Shreveport, LA 71130-3932

Robert H. Proctor
USDA, ARS
National Center for Agricultural Utilization Research
1815 North University Street
Peoria, IL 61604-3999

Naomi Rebuck
Leeds University
Department of Biology
Leeds LS2 9JT, UK

Yoshihiko Sako
Kyoto University
Division of Applied Biosciences
Graduate School of Agriculture
Kyoto 606, Japan

Philippe J. Sansonetti
Institut Pasteur
Unité de Pathogénie Microbienne Moléculaire
INSERM U 389
28, rue du Dr. Roux
75724 Paris, Cédex 15
France

Kellogg J. Schwab
Division of Molecular Virology
Baylor College of Medicine
One Baylor Plaza
Houston, TX 77030

Upinder Singh
Stanford University
Department of Microbiology and Immunology
Fairchild D305
299 Campus Drive
Stanford, CA 94305-5124

Judith E. Smith
Leeds University
Department of Biology
Leeds LS2 9JT, UK

Marie-Paule Sory
Microbial Pathogenesis Unit
Christian de Duve Institute of Cellular Pathology
and Université Catholique de Louvain
Av. Hippocrate, 74 P. Box 74.49
B-1200 Brussels, Belgium

Theodore S. Steiner
University of Virginia
Health Sciences Center
Department of Internal Medicine
Division of Geographic Medicine
Box 485
Charlottesville, VA 22908

Stefan Weiss
Laboratorium fur Molekulare Biologie
Genzentrum-Institut für Biochemie der LMU München
Feodor-Lynen Str. 25
D-81377 Munich, Germany

Christine Wennerås
Department Medical Microbiology and Immunology
Göteborg University
Guldhedsgatan 10
S-413 46 Göteborg
Sweden

Kenji Yokota
Department of Bacteriology
Okayama University Medical School
2-5-1 Shikata-cho, Okayama 700, Japan

PART I

GRAM NEGATIVE FOODBORNE BACTERIAL PATHOGENS

CHAPTER 1

Molecular and Cellular Biology of *Salmonella* Pathogenesis

FRANCISCO GARCÍA-DEL PORTILLO

1. INTRODUCTION

SALMONELLAE are pathogens with an enormous impact on public health as they infect all types of domestic animals used in the human food chain. These pathogens are one of the most prevalent agents causing foodborne diseases in both developing and developed countries.

Our understanding of *Salmonella* infections has registered a tremendous advance in the last two decades. Novel molecular approaches have permitted the creation of new models on how *Salmonella* engages basic functions of the eucaryotic cell. In this chapter, the most recent contributions in cell biology and molecular genetics of *Salmonella* pathogenesis are presented. Interactions of bacteria with host eucaryotic cells are described, followed by the current knowledge of specific DNA regions and regulatory mechanisms acting on virulence. All this molecular information has led to a further understanding of many new aspects of *Salmonella* pathogenesis and evolution (Finlay and Cossart, 1997; Groisman and Ochman, 1997). This knowledge should enable us to increase the chances of preventing diseases caused by these successful pathogens.

2. THE GENUS *SALMONELLA*

The genus *Salmonella* comprises more than 2,600 serovars of gram-negative facultative anaerobic bacilli. Classification and detection of these bacteria is based in serology and phage susceptibility assays (Rubin and Weinstein, 1977). In the last two decades, new DNA-based typing methods, such as random amplified polymorphic DNA (RAPD) technique, ribotyping, pulsed field gel electrophoresis (PFGE), and multilocus enzyme electrophoresis (MLEE), have contributed to reclassification of serovars of *Salmonella* in a new subspecies groups scheme (Reeves et al., 1989; Bäumler et al., 1998). The genus *Salmonella* contains two lineages that diverged early in its evolution: the species *S. enterica* and *S. bongori* (Le Minor and Popoff, 1987; Reeves et al., 1989; Boyd et al., 1996). Members of *S. enterica* species are divided into seven subspecies groups (I, II, IIIa, IIIb, IV, VI, and VII). Group I includes serovars causing diseases in human and other warm-blooded animals, such as *S. enterica* sv. *Typhi, Paratyphi, Sendai, Typhimurium, Enteritidis, Choleraesuis, Dublin, Gallinarum/Pullorum,* and *Abortusovis* (Bäumler et al., 1998). Groups II to VII include *S. enterica* serovars frequently isolated from cold-blooded vertebrates (Bäumler et al., 1998). Group V includes *S. bongori* (Boyd et al., 1996).

3. DISEASES CAUSED BY *SALMONELLA* INFECTIONS

Salmonella infections lead to a variety of diseases known as salmonellosis. Infection is initiated by consumption of raw or undercooked contaminated animal food or water containing fecal material (Rubin and Weinstein, 1977; Pang et al., 1995). The outcome of this infection largely depends on the serovar and type of host (Bäumler et al., 1998). Thus, certain serovars of *S. enterica* are host-restricted pathogens: *Typhi, Paratyphi* a, b, c, and *sendai* cause diseases only in humans; *Pullorum/Gallinarum* in poultry; *Dublin* in cattle; *Choleraesuis,* in pigs; *Abortusovis* in sheep; and *Abortusequi* in horses (Ekperigin and Nagaraja, 1998; Bäumler et al., 1998). *Dublin* and *Choleraesuis* serovars can also infect and cause severe disease in humans, although on rare occasions. Serovars *Typhimurium* and *Enteritidis* the major incidence, cause disease in humans, cattle, poultry, sheep, pigs, horses and wild rodents (Bäumler et al., 1998).

Diseases caused by salmonellae are divided into two major groups: (1) a localized, self-limiting bacterial infection of the intestinal epithelium, known as ''non-typhoid salmonellosis'' or gastroenteritis; and (2) a systemic infection known as ''typhoid salmonellosis'' or ''enteric fever'' (Miller et al., 1995). While non-typhoid salmonellosis is characterized by diarrhea, abdominal pain, and in some cases vomiting and fever, typhoid salmonellosis is defined by malaise, headache, a non-productive cough, abdominal pain, constipation and increasing fever (Goldberg and Rubin, 1988; Miller et al., 1995). Bacteremia has been associated with highly invasive serovars such as *Choleraesuis* or *Dublin.* Other important clinical sequels of *Salmonella* infections can be erythrema nodosum, meningitis, osteomelitis, septic arthritis, pneumonia, cholecystitis, endocarditis, pericarditis, and cystitis (Rubin and Weinstein, 1977; Miller et al., 1995; Brodov et al., 1996).

A complication linked to infection by certain *Salmonella* serovars is the carrier state (Rubin and Weinstein, 1977; Goldberg and Rubin, 1988). The infected asymptomatic host carries viable bacteria for long periods of time up to several months or years (Buchwald and Blaser, 1984). The carrier state does not require a previous disease period, and seems to be related more to infections by low doses of bacteria. An estimate of 5% of humans infected with serovar *Typhi* develop a further carrier state. Cattle that recover from serovar *Dublin* infection or swine from serovar *Choleraesuis* infection also become carriers for five or more months afterwards (Ekperigin and Nagaraja, 1998). The carrier state is an important cause of foodborne salmonellosis as it leads to spreading the pathogen to the farm, food, food-handlers and consumers.

4. EPIDEMIOLOGY OF *SALMONELLA* INFECTIONS

Typhoid and non-typhoid salmonellosis remain major public health problems, and are clearly the most economically important foodborne diseases. The incidence of typhoid salmonellosis is stable, with very low numbers of cases in developed countries, but cases of non-typhoid salmonellosis are increasing worldwide. Non-typhoid cases account for 1.3 billion cases of acute gastroenteritis/diarrhea with 3 million deaths, and for 16 million cases of typhoid fever with nearly 600,000 deaths (Pang et al., 1995).

Newborns, infants, the elderly, and immunocompromised individuals are more susceptible to *Salmonella* infections. In developed countries more than 80% of all non-typhoid salmonellosis cases may occur individually rather than in outbreaks (WHO, 1997b). In recent years, a notable increase in cases related to a multi-drug resistant *S. enterica* sv. *Typhimurium* strain, named DT104, has been reported in several European countries (WHO, 1997b). Emergence of *S. enterica* sv. *Typhimurium* DT104 occurred in 1988 in England and Wales from infected cattle. Case-fatality and hospitalization rates due to this strain are twice that of other foodborne associated non-typhoid salmonellosis. This strain has spread to other European countries such as Germany, as well as to the United States. In Asiatic countries, such as Japan, non-typhoid salmonellosis is now more frequent as a result of an increased consumption of eggs and egg products (WHO, 1997a). In Australia, *Salmonella* is the predominant cause of morbidity and mortality caused by microbial foodborne diseases. Annual cases of non-typhoid salmonellosis in the United States and Canada are estimated at 40,000 and 9,000 respectively. Data on salmonellosis are scarce in many countries of Africa, Asia and South and Central America, where only 1–10% of cases is reported. The few foodborne surveillance studies performed in these countries show that the disease is often associated to 20–30% mortality rates.

Typhoid fever is still a major public health problem in many developing countries. Humans are the only reservoir and host of serovars *Typhi* and *Paratyphi* a and b, which are transmitted by contaminated water and food in endemic areas, and carriers handling food in developed countries. Higher incidence of human typhoid salmonellosis range from 1:650/year in South America to 1:100/year in most affected countries of other continents such as Indonesia and Papua New Guinea (Pang et al., 1995; WHO, 1998). From 1950, the progressive resistance of *S. enterica* sv. *Typhi* to first-line antibiotics has been reported (Pang et al., 1995). However, multidrug-resistant strains are now emerging in India and Southeast Asia (Pang et al., 1995; Rowe, 1997.) In India 50–70% of strains are resistant to chloramphenicol and other antibiotics.

5. DETECTION OF *SALMONELLA* IN FOOD AND WATER

Microbiological enrichment media containing inhibitors such as bile, tetrathionate, selenite, nitrofurantoin, and brilliant green and malachite green dyes, are widely used for detection of *Salmonella* (Varman and Evans, 1991; de Boer, 1998). These procedures are being replaced by modern molecular methods, including DNA-based and immunological techniques (Gouws et al., 1998).

DNA-based methods include polymerase chain reactions (PCR) with primers for amplifying *Salmonella*-specific genes (Bäumler et al., 1997a; Bennet et al., 1998; Brasher et al., 1998; Manzano et al., 1998; Cocolin et al., 1998). PCR tests have been shown as sensitive in detecting as few as 50 CFUs (Gouws et al., 1998). Amplified PCR products can also be confirmed by microplate-capture hybridization assay (Manzano et al., 1998). Immunological assays use magnetic beads containing anti-*Salmonella* antibodies, permitting rapid and specific capture of *Salmonella* from food and environmental samples (Shaw et al., 1998). A macroporous polyester cloth that provides a high-affinity absorbent for *Salmonella* LPS has also been used (Blais et al., 1998). Food tests have also been done with satisfactory results by applying chemiluminiscent (CL) endpoints to PCR-detection systems (Kricka, 1998).

6. PREVALENCE OF *SALMONELLA* IN FOODS

Salmonellae grow at an optimum at 37°C, although growth at temperatures as low as 10°C has been reported. Heat resistance has been shown to be higher at low pH (5.5) and low water activity (a_w), and in food with a high fat content. Viability of salmonellae declines during frozen storage, being greater at temperatures close to the freezing point (−2 to–5°C). Optimum pH for salmonellae is 6.5 to 7.5, although growth has been reported over the range of 4.5 to 9.0.

Acidulants such as acetic acid exert a protective effect (Varman and Evans, 1991). An a_w of 0.93 and an oxidation-reduction potential of 30 mV have been reported as lower limits for *Salmonella* growth when other inhibitory factors are absent. In optimal growth conditions, salmonellae are able to grow in c.a. 4% NaCl and 350 mg/L of $NaNO_2$. Presence of other inhibitory factors, such as acidic pH, reduces these levels significantly. Organic acids that reduce persistence of *Salmonella* in foods are succinic acid and formic acid (Humphrey and Lanning, 1988).

Ubiquity of salmonellae in the environment coupled with intensive husbandry practices used in meat, fish and shellfish industries, and the lack of microbial control on animals feeds favor the continued prevalence of this pathogen in the global food chain (D'Aoust, 1997). Poultry products such as chicken, turkey and waterfowl remain the main reservoir of *Salmonella* in many countries. Eggs and egg-containing products are of particular concern since there is transovarian transmission of the pathogen into the interior of the egg before deposition. Foods associated with large salmonellosis outbreaks include milk powder, raw milk, cheddar cheese, egg salad, egg drinks, and liver paté (D'Aoust, 1997). Some fruits and vegetables linked to outbreaks are cantaloupes, chocolate, mustard dressing, and paprika chips (D'Aoust, 1997).

Large variations in the infectious dose present in food have been shown to be responsible for salmonellosis outbreaks (D'Aoust, 1997). As few as 1 to 10 cells can constitute an infectious human dose (Kapperud et al., 1990). Foods implicated in outbreaks with low infectious doses have a high fat content. Entrapment of salmonellae within lipid micelles may afford protection to gastric acidic pH.

7. ANALYSIS OF *SALMONELLA* VIRULENCE FACTORS: IN VIVO AND IN VITRO MODELS

The most frequently used in vivo study is the murine typhoid fever model. Bacteria are administrated by oral or intraperitoneal routes to such susceptible animal hosts as BALB/c mice. *Salmonella* serovars normally causing gastroenteritis in humans, e.g., *Typhimurium* or *Enteritidis,* provoke in these animals a typhoid-like disease. The in vitro models utilize established phagocytic or non-phagocytic cell lines of different origin.

Salmonella induces its own uptake by non-phagocytic cells cultured in vitro (Kihlstrom and Edebo, 1976; Kihlstrom and Nilsson, 1977; Jones et al., 1981; Yokoyama et al., 1987). This remarkable process, often referred as "invasion," illustrated an essential virulence trait of *Salmonella,* its capacity to cross the intestinal barrier. These observations were in agreement with the pioneering microscopical description of *Salmonella* infection in guinea pig ileal epithelial cells (Takeuchi, 1967). In this study, evidence was provided for *Salmonella* invasion of columnar epithelial cells. The analogy observed in the *Salmonella* internalization process, both in cultured epithelial cells and in animals models, validated the in vitro model (reviewed in Galán and Bliska, 1996; Finlay, 1994; Finlay and Falkow, 1997). Further studies demonstrated that *Salmonella* mutants, selected in cultured cells for decreased survival in macrophages or intracellular proliferation in epithelial cells, were attenuated in BALB/c mice (Fields et al., 1986; Leung and Finlay, 1991).

These observations led to the assumption that *Salmonella* mutants with a discernible phenotype in the in vitro model should be impaired for virulence in the murine typhoid model and vice versa. However, while this is the case for many essential virulence factors of *Salmonella,* others do not follow this rule. Thus, mutants that do not invade epithelial cells are not highly attenuated in the in vivo model (Galán and Curtiss, 1989). In contrast, certain mutants unable to cause disease in mice do not have any phenotype in the tissue culture model (García-del Portillo et al., 1993a; Farrant et al., 1997; de Groote et al., 1997). It is also known that mutations in certain genes induced during the in vivo infection do not impair *Salmonella* interaction with cultured eucaryotic cells (reviewed in Heithoff et al., 1997b). Virulence-related phenotypes observed in a specific host animal may not be reproduced in others and the extent of bacterial interaction with intestinal epithelial cells is dependent on the serovar used (Watson et al., 1995; Watson et al., 1998; Weinstein et al., 1998). Furthermore, mutations in certain DNA regions harbored exclusively by *Salmonella* do not lead to any phenotype in the in vivo and in vitro models (Wong et al., 1998; Blanc-Potard et al., 1999). All these observations suggest that *Salmonella* infection is a multifactorial process with different outcomes depending on which host animal and *Salmonella* serovar interact. These pathogens may also harbor redundant functions for colonizing such environmental niches as soil, food, water, and host eucaryotic cells of different susceptible animals (Galán and Bliska, 1996; Slauch et al., 1997; Finlay and Falkow, 1997). Despite these observations, comparative analysis in in vivo and in vitro models continues to be an essential tool for studying *Salmonella* pathogenesis.

8. CELL BIOLOGY OF THE *SALMONELLA*-EUCARYOTIC CELL INTERACTION

8.1. PASSAGE THROUGH THE STOMACH AND THE INTESTINAL BARRIER

Salmonella infection is initiated after consumption of contaminated food or water. Previous to their arrival in the intestine, bacteria pass through the stomach. *Salmonella* is rather sensitive to the low acidic pH of the stomach and, unlike *Shigella* or enteroinvasive *Escherichia coli,* a large infective dose (10^4–10^6 organisms) is required to cause disease. *Salmonella* can elicit an acid tolerance response (ATR) which enables organisms to survive at a decreased pH after exposure to low pH (reviewed in Foster and Spector, 1995; Bearson et al., 1997). This response seems to play a role in vivo during bacterial transit through the stomach (García-del Portillo et al., 1993a; Riesenberg-Wilmes et al., 1996).

Once *Salmonella* has successfully passed through the stomach, it reaches and colonizes the distal ileum. The preferred portals of entry are specialized cells, named M cells, located in the Peyer's patches (Siebers and Finlay, 1996; Jones, 1997; Jensen et al., 1998; Raupach et al., 1999). M cells play an essential role in sampling the contents of the intestinal lumen (Siebers and Finlay, 1996; Hamzaoui and Pringault, 1998; Neutra, 1999). M cells are also used as portals of entry by other enteric pathogens such as *Shigella, Yersinia,* enteropathogenic *E. coli* and *Listeria* (Jones et al., 1995; Siebers and Finlay, 1996; Jones, 1997; Jensen et al., 1998; Raupach et al., 1999). Penetration and destruction of M cells by *Salmonella* is followed by recruitment of lymphoid cells and polymorphonuclear phagocytic cells (PMNs), leading to inflammation of the area (reviewed in Siebers and Finlay, 1996; Jones, 1997; Phalipon and Sansonetti, 1999). Inflammatory response is accompanied by secretion of large amounts of fluid to the intestinal lumen. Colonization of Peyer's patches seems to be the critical stage of the infection for eliciting a proper immune response (Jones, 1997; Dunstan et al., 1998). Scanning and transmission electron microscopy studies have shown that internalization of *Salmonella* by the M cell is associated with the apical formation of large membrane ruffles that surround the bacteria (Jones et al., 1995). Within a short period of time, 30 to 60 min, the *Salmonella*-infected M cells are destroyed.

Binding of *Salmonella* to M cells is mediated by specialized fimbriae, named *lpf* (Bäumler and Heffron, 1995a; Bäumler et al., 1996b; Bäumler et al., 1997b). Expression of *lpf* fimbriae occurs by a heritable phase variation mechanism, probably triggered by an intestinal-derived signal (Norris et al., 1998b). In vivo studies have shown that *lpf* defective mutants are partially impaired from causing disease in the murine typhoid model (Bäumler et al., 1997c). This attenuation is more drastic when the *lpf* mutation is combined with mutations in invasion determinants (Bäumler et al., 1997c). Other fimbriae operons, such as the *pef,* mediate bacterial binding to the villous small intestine (Bäumler et al., 1996). Therefore, *Salmonella* has a wide repertoire of adhesins that play a crucial role in vivo on deciding the portal of entry.

8.2. INVASION OF NON-PHAGOCYTIC CELLS

8.2.1. Morphological Description

As mentioned above, *Salmonella* preferentially utilize the M cell as the main port of entry (Jones et al., 1995). Cultures of epithelial cells infected with *Salmonella* show identical morphological membrane alterations as M cells (Francis et al., 1993; Galán and Bliska, 1996; Finlay and Cossart, 1997; Dramsi and Cossart, 1998). Microscopical analyses revealed that *Salmonella* induces transient cytoskeletal rearrangements that culminate in formation of membrane ruffles (Finlay et al., 1991; Francis et al., 1993; Finlay and Cossart, 1997). Localized accumulation of cytoskeletal proteins such as actin, vinculin, vimentin, and ezrin occurs at the site of bacterial entry (Finlay et al., 1991; Dramsi and Cossart, 1998). Increased macropinocytosis is also observed in the same area (García-del Portillo and Finlay, 1994). *Salmonella* mutants, unable to trigger these events, are impaired from invading non-phagocytic cells, supporting the absolute requirement of these morphological changes for bacterial invasion (Galán, 1996; Collazo and Galán, 1997b). Membrane ruffling induced by *Salmonella* in the host cell has been observed in every study model used to date, including M cells (Jones et al., 1995), columnar intestinal epithelial cells (Takeuchi, 1967), and cultured cell lines (Francis et al., 1993; Jones et al., 1993; Alpuche-Aranda et al., 1994). In epithelial cells harboring a developed brush border, such as columnar intestinal cells or in vitro-polarized epithelial cells, degeneration of microvilli in the site of bacterial contact precedes membrane ruffling (Finlay et al., 1988a; Finlay and Falkow, 1990). All these processes take place extremely rapidly, within minutes, and in a transient manner (Francis et al., 1992). Concomitant to these cytoskeletal changes, redistribution of host cell surface membrane proteins has been observed (García-del Portillo et al., 1994). The *Salmonella* invasion process ends up with the enclosure of bacteria in a membrane-bound vacuole.

8.2.2. Role of the Type III Secretion System

In 1989, in *S. enterica* sv. *Typhimurium*, the first invasion locus, named *invA-C*, was reported (Galán and Curtiss, 1989). The *invA-C* locus is part of a large 40 kb region of the chromosome mapping at centisome 63, which is absent in phylogenetic related bacteria such as *E. coli* (Mills et al., 1995; Galán, 1996; Collazo and Galán, 1997b). An estimate of 20 genes form this region, known as *Salmonella* pathogenicity island 1 (SPI1) (Galán, 1996). A similar gene organization was shown to be conserved in other animal and plant pathogens (reviewed in Hueck, 1998). Moreover, the invasion loci shared homologies to genes involved in flagellum synthesis (Grolsman and Ochman, 1993; Van Gijsegem et al., 1993; Hueck, 1998). The function of the invasion gene products also seems to be related to flagellar proteins (Hueck, 1998). Thus, filamentous appendages resembling flagella, named "invasomes," appear in *Salmonella* upon contact with epithelial cells (Ginocchio et al., 1994). Transient formation of these appendages is linked to bacterial invasion and requires a functional SPI1-encoded type III system. However, this system is not an absolute requirement for formation of the appendages (Reed et al., 1998). The type III secretion apparatus has been isolated from the bacterial envelope and appears as a syringe-like structure resembling the membrane-anchor part of flagella (Kubori et al., 1998). Altogether, these results suggest that *Salmonella* uses a supramacromolecular apparatus in the cell envelope to signal the host cell for bacterial uptake.

SPI-1 encodes the specialized type III protein secretion apparatus (Collazo and Galán, 1997b; Suárez and Rüssmann, 1998). Type III secretion systems have unique features that differentiate them from other protein-secretion systems (Hueck, 1998). Thus, in this system, more than 15 proteins are inserted in the inner or outer membrane. Specific cytosolic chaperones prevent miss-folding and degradation of proteins to be secreted. Moreover, secreted proteins are not cleaved by specific proteases during their transport through the envelope (Hueck, 1998). Secretion-related motifs are located in the *N*-terminal amino acid sequence and/or the 5′ leader mRNA of the secreted proteins (Hueck, 1998). Type III secretion systems are also activated upon contact of bacteria with the host cell surface, as shown for *Yersinia* spp., and later proposed for *Salmonella* (Zierler and Galán, 1995). However, activation of a *Salmonella* type III secretion system can also be obtained by a shift from acidic to alkaline pH in the absence of eucaryotic cells (Daefler, 1999).

Several classes of proteins form the *Salmonella* SPI-1 type III secretion system: (i) inner and outer membrane proteins InvA, SpaP, SpaQ, SpaR, SpaS, InvG, InvH, PrgH, and PrgK; (ii) an energizer of the system, the inner membrane ATPase InvC; (iii) secreted proteins required for the secretion process InvJ, SpaO, SipD; (iv) secreted proteins with putative targets in the host cell, also known as effector proteins SipB, SipC, SptP, AvrA; (v) other secreted proteins SipA; (vi) cytosolic chaperonins which prevent premature degradation of secreted proteins SicA, SicP (Fu and Galán, 1998a); and (vii) transcriptional regulatory proteins InvF, HilA, InvE. Excellent reviews on *Salmonella* SPI-1-encoded type III system proteins have been published elsewhere (Collazo and Galán, 1997b; Suárez and Rüssmann, 1998). Current research is focused on the functional role played by the *Salmonella* secreted proteins, since they are expected to play key roles in the cross-talk between the bacteria and host cell.

Ssps. essential for *Salmonella* invasion of non-phagocytic cells include SipB, SipC, SipD, SpaO, InvJ. Other SPI-1-encoded proteins are secreted by the type III system, but apparently not required for bacterial invasion, such as SipA, SptP, and AvrA (Kaniga et al., 1996; Hardt and Galán, 1997; Hardt et al., 1998b). SptP is a tyrosine phosphatase that affects cytoskeleton structure and, like SipB and SipC, is translocated into eucaryotic cells (Collazo and Galán, 1997a; Fu and Galán, 1998b). AvrA is encoded within SPI-1, and is similar to the avirulence proteins of plant pathogens (Hardt and Galán, 1997). Environmental conditions known to induce secretion by the SPI-1 type III system include high osmolarity, low oxygen tension, DNA supercoiling, and alkaline pH (reviewed in Hueck, 1998).

SPI-1-encoded type III system can also secrete proteins not encoded by SPI-1. Examples are the SopE protein of *S. enterica* sv. *Typhimurium* (Hardt et al., 1998b), and the SopB, and SopD proteins of *S. enterica* sv. *Dublin* (Wood et al., 1996; Galyov et al., 1997). These proteins are translocated into the host cell cytosol and play an essential role in signal transduction to epithelial cells (Hardt et al., 1998a), and enteropathogenesis (Jones et al., 1988). Moreover, a second type III secretion system has been described in *S. enterica* sv. *Typhimurium* at centisome 30.5 (Ochman et al., 1996; Shea et al., 1996). Unlike the SPI-1-encoded type III system required for invasion of the intestinal epithelium, the SPI-2 type III system plays an essential role in bacterial survival and intracellular replication within macrophages and epithelial cells (Ochman et al., 1996; Hensel et al., 1997b; Cirillo et al., 1998; Hensel et al., 1998).

8.2.3. Other *Salmonella* Invasion Factors

Additional surface components are also required for *Salmonella* invasion. Loss of invasiveness is observed in non-motile *S. enterica* sv. *Typhi* mutants (Liu et al., 1988). Intact LPS structure is also required for invasion of serovars *Typhi* and *Choleraesuis* (Finlay et al., 1988b; Mroczenski et al., 1989). Fimbrial adhesins, such as *lpf, fim, agf,* and *pef,* also play a role in *Salmonella* adherence and invasion of intestinal epithelium (Bäumler et al., 1996b; Baümler et al., 1996; Baümler et al., 1997c; Bäumler et al., 1997b; Bäumler et al., 1998; van der Velden et al., 1998). In vitro, loss of specific fimbriae affects the capacity of *S. typhimurium* to adhere and invade certain epithelial cells (Bäumler et al., 1996a). No mutant defective in a single fimbriae operon is attenuated for virulence, so *Salmonella* may compensate for loss of a specific fimbrial adhesin by alternative fimbria faciliting colonization of the intestinal barrier. Indeed, simultaneous inactivaction of *fim, lpf, pef,* and *agf* fimbriae operons markedly attenuates virulence (van der Velden et al., 1998). *S. enterica* sv. *Typhimurium* has an additional invasion locus not encoded by SPI-1, and named *sigDE* (Hong and Miller, 1998). SigD of serovar *Typhimurium* is secreted by the SPI-1 type III system, and corresponds to SopB protein of serovar *Dublin* (Galyov et al., 1997; Hong and Miller, 1998). Whether these proteins are essential in some specific stage of the infection awaits further investigation.

8.2.4. Host Factors Involved In *Salmonella* Invasion

Salmonella induces internalization by non-phagocytic cells by subverting essential functions of the host cell (reviewed in Galán and Bliska, 1996; Finlay and Cossart, 1997; Finlay and Falkow, 1997; Dramsi and Cossart, 1998).

Actin cytoskeleton reorganization and membrane ruffling are the two major changes occurring in the physiology of the host cell. Both processes are triggered in eucaryotic cells by linfoquines and activation of certain growth factor receptors (Hotchin and Hall, 1996). It is known that upon *Salmonella* invasion, an increase in intracellular free Ca^{2+} occurs (Ginocchio et al., 1992; Ruschkowski et al., 1992). This Ca^{2+} flux, essential for the cytoskeletal reorganization, is linked to the release of such mediators as arachidonic acid derived metabolites (Pace et al., 1993). Recent findings have provided new insights on the signal transduction pathways triggered by *Salmonella* in the host cell. Thus, SopE of *S. enterica* sv. *Typhimurium,* a substrate of the SPI-1 encoded type III system (Hardt et al., 1998b), stimulates actin reorganization by increasing GDP/GTP nucleotide exchange in several GTPases, such as Rac-1 and CDC-42 (Hardt et al., 1998a). This action is exerted by direct binding of SopE to these GTPases (Hardt et al., 1998a). In concordance, bacterial invasion requires activation of CDC-42, to a lesser extent Rac-1, but not of Rac (Jones et al., 1993; Chen et al., 1996a). *Salmonella* invasion also stimulates production of such proinflammatory cytokines as IL-8 (Eckmann et al., 1993; McCormick et al., 1993; McCormick et al., 1995; Hobbie et al., 1997; Wilson et al., 1998); and nuclear responses, including activation of transcriptional factors NF-κB, AP-1 due to stimulation of MAP-kinases Erk, JNK, and p38 (Hobbie et al., 1997). Beside IL-8, upregulation of other proinflammatory cytokines, such as monocyte chemotatic protein-1 (MCP-1), GM-CSF, and TNF-α, has been correlated to *Salmonella* invasion of human colon epithelial cells (Jung et al., 1995). In conclusion, current data indicate that membrane ruffling is the result of elaborate host-pathogen cross-talk, initiated by translocation of type III effector proteins.

8.3. INTRACELLULAR PROLIFERATION WITHIN NON-PHAGOCYTIC CELLS

In vitro studies have shown that 4–6 h after bacterial internalization by epithelial cells, *Salmonella* proliferates in membrane-bound vacuoles (reviewed in García-del Portillo and Finlay, 1995b; Finlay and Falkow, 1997). During this lag period, the *Salmonella*-containing vacuole (SCV) fuses with host compartments containing lysosomal-membrane glycoproteins (lgps) and bypasses compartments of the endocytic route (García-del Portillo and Finlay, 1995a). Filamentous structures containing lgps, named Sifs (*Salmonella*-induced-filaments), are observed at the time bacteria initiates intracellular growth (García-del Portillo et al., 1993b; Stein et al., 1996). Several *Salmonella* proteins have been linked to induction of these structures, such as SifA and the two component regulatory system OmpR/EnvZ (Stein et al., 1996; Mills et al., 1998). *S. enterica* sv. *Typhimurium* mutants, unable to replicate within epithelial cells and attenuated for virulence in the mouse typhoid model, are also defective for Sif induction (Leung and Finlay, 1991; García-del Portillo et al., 1993b). The role played by Sifs is still a matter of debate. A putative role in the acquisition of nutrients by intracellular bacteria has been proposed. However, neither *sifA* nor *ompR* or *envZ* mutants show any defect for intracellular proliferation (Stein et al., 1996; Mills et al., 1998). Elucidation of the biological function of Sifs deserves further investigation.

The microenvironment of the SCV in epithelial cells is unique, with limiting amounts of Mg^{2+}, Fe^{2+}, and a mild acidic pH (García-del Portillo et al., 1992). Limiting Mg^{2+} is an inducing signal for the *Salmonella* two-component regulatory system PhoP-PhoQ essential for virulence (García Vescovi et al., 1996). To date no gene ''exclusively'' linked to *Salmonella* intracellular proliferation has been identified. As shown by recent findings, SPI-2 is required for intracellular proliferation within epithelial cells (Cirillo et al., 1998; Hensel et al., 1998). A detailed functional characterization of SPI-2 proteins may provide new clues on how *Salmonella* proliferates within cultured non-phagocytic cells.

Salmonella intracellular proliferation also occurs during the in vivo infection, as salmonellae can increase in number in animals treated with gentamicin, an antibiotic that does not affect viability of intracellular bacteria. Different in vivo studies showed contradictory results on the specific host cell type where *Salmonella* proliferates. Thus, it has been claimed that non-phagocytic cells, polymorphonuclear (PMNs), or macrophages are the main cell types where *Salmonella* proliferates (Dunlap et al., 1992; Conlan and North, 1992; Nnalue et al., 1992; Conlan, 1996). However, recent findings provide strong evidence that macrophages are the only cell type where *Salmonella* proliferates intracellularly (Richter-Dahlfors et al., 1997; Gulig et al., 1998).

8.4. INTERACTION OF *SALMONELLA* WITH PHAGOCYTIC CELLS

Macrophages and PMNs play an essential role in controlling *Salmonella* infections (reviewed in Jones and Falkow, 1996; Jones, 1997). When stimulated, these specialized cells are able to ingest and kill microbes. Macrophages kill microbes by generating reactive oxygen and nitrogen compounds, antibacterial peptides, and lysosomal degradative enzymes (Baümler and Heffron, 1995b). The acidic pH of the phagolysosome also contributes to prevention of bacterial survival or growth. *Salmonella* is an example of a pathogen able to resist the attack of macrophages and survive within.

As in non-phagocytic cells, active membrane ruffling and macropinocytosis occurs upon ingestion of *Salmonella* by phagocytic cells (Alpuche-Aranda et al., 1994; Chen et al., 1996b). However, subsequent events markedly differ. Thus, targeting of intracellular bacteria to lgp-containing compartments occurs more rapidly in macrophages. The trafficking route used by *Salmonella* within cultured macrophages is different than within epithelial cells, as fusion of the SCV with mature lysosomes seems to occur (Oh et al., 1996). Results from other groups suggest that the trafficking route could resemble that reported for epithelial cells (Rathman et al., 1997). Isolation of *Salmonella*-containing phagosomes in macrophages has recently been described (Mills and Finlay, 1998). Host cell markers, such as Igps, mannose-6-phosphate receptor (M6P-R), and lysosomal enzymes, are present in this phagosome (Mills and Finlay, 1998). Nonetheless, this compartment does not behave as a conventional mature lysosome since there is no processing of lysosomal enzymes contained within (Mills and Finlay, 1998). Contradictory results have been obtained in studies analyzing fusion of *Salmonella* phagosomes with macrophage lysosomes. In some cases fusion exists while in others the evidence suggests that it does not occur (reviewed in García-del Portillo, 1996). Regarding phagosome acidification, conflicting reports also exist. While the first studies indicated that phagosome acidification is retarded by *Salmonella* (Alpuche-Aranda et al., 1992), a recent report shows that the phagosome rapidly acidifies to a final pH value of 4.0 (Rathman et al., 1996). Acidification triggers bacterial responses for survival and replication within these cells (Rathman et al., 1996).

One explanation for these contradictory results is the apparent heterogeneity existing in the population of intracellular bacteria (Abshire and Neidhardt, 1993b; Buchmeier and Libby, 1997). Intracellular salmonellae residing within macrophages may exist in two populations, one actively growing, and other static but viable (Abshire and Neidhardt, 1993b). A recent study indicates that most of the intracellular bacteria are killed by the macrophage (Buchmeier and Libby, 1997). Molecular analysis of this phenomenon has not been performed, and it is not known whether it also occurs in the in vivo infection.

Salmonella is also able to trigger apoptosis of the infected macrophage. This process has been observed in both in vitro and in vivo models (Chen et al., 1996b; Monack et al., 1996; Baran et al., 1996; Lindgren et al., 1996; Richter-Dahlfors et al., 1997). Bacterial products involved in triggering macrophage apoptosis include OrgA an oxygen-regulated protein encoded in SPI-1 (Monack et al., 1996); OmpR/EnvZ (Lindgren et al., 1996); and SPI-1 type III secreted proteins such as InvJ, SpaO, SipB, SipC, and SipD (Chen et al., 1996b). *Salmonella*-induced apoptosis may have a role in destroying phagocytic cells before they activate lymphoid cells, or alternately in facilitating the bacteria spreading to neighboring cells. It also may be possible that the macrophage responds to *Salmonella* infection by programming its own death to avoid excessive release of immune mediators.

Specific bacterial factors enable *Salmonella* to survive and/or proliferate within phagocytic cells (García Vescovi et al., 1994; Miller et al., 1995; Groisman and Aspedon, 1997). The two-component regulatory system PhoP-PhoQ is an essential regulator in the bacterial defense against the attack of phagocytes (Fields et al., 1989; Miller et al., 1989; Groisman and Saier, 1990; Miller, 1991; García Vescovi et al., 1994; Groisman, 1996; Groisman and Ochman, 1997). This system regulates expression of over 40 genes. Products of some PhoPQ-regulated genes are involved in resistance to antimicrobial peptides (Groisman et al., 1992; Gunn and Miller, 1996; Soncini and Groisman, 1996; Groisman et al., 1997; Groisman and Aspedon, 1997; Gunn et al., 1998b). Specific PhoPQ-regulated genes involved in *Salmonella* survival within macrophages are *pagC* (Gunn et al., 1995), *mgtC* (Blanc-Potard and Groisman, 1997), and *pqaB,* a *S. typhi* gene under control of *pmrAB* (Baker et al., 1999). How products of these genes mediate bacterial intramacrophage survival is still unknown, although a role for *mgtC* in adapting to limiting Mg^{2+} present in the macrophage phagosome has been proposed (Blanc-Potard and Groisman, 1997).

Other bacterial factors required for *Salmonella* survival within phagocytes are RpoE and HtrA (Chatfield et al., 1992; Humphreys et al., 1999). RpoE is an alternative sigma factor, σ^E that controls expression of HtrA, a stress-induced periplasmic serine protease. A *rpoE* mutant is more sensitive to paraquat (a superoxide generator) H_2O_2, and polymixin than a *htrA* mutant (Humphreys et al., 1999), suggesting that additional RpoE-regulated genes may play a role in bacterial survival within macrophages. RpoE could mediate periplasmic processing of stress-damaged proteins produced by oxygen-reactive intermediates. Oxygen-derived DNA damage may also occur in the phagosomal environment, as proteins involved in DNA repair, e.g., RecA and RecBC, are required for *Salmonella* survival within macrophages (Buchmeier et al., 1993; Buchmeier et al., 1995). Increased levels of SodA, a Mn-dependent superoxide dismutase that cleaves superoxide anion, result in increased survival within cultured macrophages (Tsolis et al., 1995). However, *Salmonella sodA* mutants are virulent in animal models. A reverse situation occurs with mutants in SodC, a Zn^{2+}-Cu^{2+}-cofactored superoxide dismutase. This protective enzyme is required for virulence in the mouse typhoid model but not for survival in cultured macrophages (Farrant et al., 1997; de Groote et al., 1997). SlyA is another transcriptional regulator involved in *Salmonella* survival within macrophages and resistance to toxic oxidative products (Buchmeier et al., 1997). All these data suggest that *Salmonella* must mount a defense against oxygen-dependent mechanisms triggered by phagocytic cells. Additional proteins have been linked to *Salmonella* survival within macrophages, including Prc, PurD, FliD, NagA, SmpB (Bäumler et al., 1994). Other sets of *Salmonella* proteins, named Sap, are required for antimicrobial peptide resistance (Groisman et al., 1992, Parra-Lopez et al., 1993; Parra-Lopez et al., 1994). Secreted proteins encoded by the SPI-2 type III system could also have an important role in mediating *Salmonella* survival within macrophages (Ochman et al., 1996; Shea et al., 1996; Hensel et al., 1998; Cirillo et al., 1998; Shea et al., 1999). The function of many of these proteins remains to be determined.

Macrophage attack to *Salmonella* infection has been linked to a protein, Nramp-1 (natural resistance-associated macrophage protein). This protein is macrophage-restricted and encoded by the *Lsh/Ity/Bcg* gene in mice. BALB/c mice have a *lty*^{s} allele that renders them hypersensitive to *Salmonella* infection. *lty*^{r} congenic mice show increased resistance to infection by such intracellular pathogens as *Salmonella, Mycobacterium,* and *Leishmania* (reviewed in Blackwell and Searle, 1999). Nramp-1 is an integral membrane protein with characteristics of membrane transporters and ion channels (Cellier et al., 1996). It is upregulated by mediators, such as interferon-γ and LPS, and localizes to host vacuolar compartments that fuse with pathogen-containing vacuoles (Gruenheid et al., 1997; Searle et al., 1998). Nramp-1 could impair pathogen intracellular proliferation by modifying the phagosomal environment and/or promoting phagosome-lysosome fusion (Searle et al., 1998).

PMNs also play a crucial role against *Salmonella* infections (Vassiloyanakopoulos et al., 1998). In fact, PMNs are cells recruited to the initial infection site upon *Salmonella* invasion of the intestinal barrier. Defense made by PMN has been demonstrated as essential by comparing the infection outcome in neutropenic *lty*^{r} or *lty*^{s} mice (Vassiloyanakopoulos et al., 1998). PMNs efficiently kill *Salmonella* and their role is crucial in a *lty*^{s} host genetic background. It remains to be shown whether *Salmonella* harbors unique determinants to defense from the PMN attack.

8.5. PERSISTENCE WITHIN NON-PHAGOCYTIC CELLS

Macrophages are the only host cell type in which *Salmonella* proliferates intracellularly in vivo (Richter-Dahlfors et al., 1997; Gulig et al., 1998). It is therefore tempting to postulate that interaction of *Salmonella* with non-phagocytic cells may not automatically lead to bacterial intracellular proliferation. This hypothesis seems to contradict all the evidence accumulated in cell-cultured models, in which *Salmonella* has shown an ability to proliferate actively within non-phagocytic cells (reviewed by Finlay and Falkow, 1997). However, recent findings have shown that certain non-phagocytic cells prevent *Salmonella* intracellular proliferation. Thus, the human intestinal epithelial cell line Henle-407 and NRK or 3T3 fibroblasts effectively block *Salmonella* intracellular proliferation (Saarinen et al., 1996; Martínez-Moya et al., 1998). Intracellular bacteria, although non-growing, remain viable for periods of 10–14 days (Saarinen et al., 1996; Martínez-Moya, M. and García-del Portillo, F., unpublished results). Intracellular bacteria rescued from NRK fibroblasts after a long persistence period grow slowly under laboratory conditions, suggesting that *Salmonella* may decrease its metabolic rate to perpetuate within the host cell (García-del Portillo, F., Casadesús, J., and Cano, D., 1999a). Interestingly, the PhoPQ system is involved in preventing growth of intracellular *Salmonella,* as *phoP* mutants grow actively in non-permissive cells (Figure 1.1) (García-del Portillo, F., Groisman, E.A., and Martínez-Moya, M., 1999b). This is the first example of a bacterial-mediated mechanism that prevents its own proliferation in a specific host cell. Long persistence of *Salmonella* within

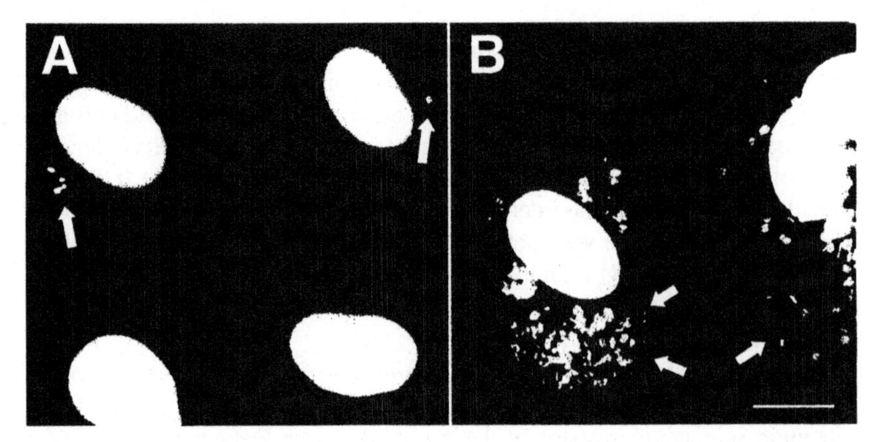

Figure 1.1 The two-component regulatory system PhoPQ attenuates intracellular proliferation of *Salmonella typhimurium* in NRK fibroblasts. Cells were infected with the following bacterial strains: (a) wild type or (b) *phoP* isogenic mutant, and stained at 24 h post infection with Dapi, a DNA-binding dye. Arrows point to intracellular bacteria. Bar: 10 μm.

non-phagocytic cells could be linked to salmonellosis-derived reactive arthritis (Saarinen et al., 1996), chronic infections, or the carrier asymptomatic state.

8.6. INFLAMMATORY RESPONSE TO *SALMONELLA* INFECTION OF INTESTINAL EPITHELIUM: A MOLECULAR MODEL FOR GASTROENTERITIS

Colonization and penetration of the intestinal epithelium by *Salmonella* elicits a profound inflammatory response (reviewed in Jones and Falkow, 1996; Jones, 1997; Phalipon and Sansonetti, 1999). This response consists largely of an early recruitment of PMNs to the infected area of the intestinal epithelium (Jones and Falkow, 1996). These immune cells are recruited in response to pro-inflammatory cytokines released by bacteria-infected epithelial cells. Epithelial cells infected with *Salmonella* basolaterally secrete several chemokines, including IL-8, MCP-1, GM-CSF, and TNF-α (McCormick et al., 1995; Jung et al., 1995; Hobble et al., 1997). A novel pathogen-elicited epithelial chemoattractant (PEEC), secreted apically by the infected cell, has been linked to the transepithelial migration of PMNs observed in *Salmonella* infections (McCormick et al., 1998). All these responses are key factors responsible for subsequent gastroenteritis and fluid secretion. A recent report has shown that both IL-8 and PEEC secretion may not depend on bacterial internalization (Gewirtz et al., 1999). γδ T-cells, present in mucosal sites, also play a role in control of *Salmonella* infection (Phalipon and Sansonetti, 1999). These cells represent a first line of defense against bacterial infections (Haas et al., 1993). IL-15, an interleukin secreted by *Salmonella*-infected macrophages, has been shown to stimulate γδ T-cells

(Nishimura et al., 1996). Bacterial factors responsible for triggering the inflammatory response have been recently identified. SopB from *S. enterica* sv. *Dublin,* a protein secreted by the SPI-1-encoded type III system, is translocated to the host cell eliciting fluid secretion and inflammatory responses in infected ileum (Galyov et al., 1997). *Salmonella* invasion of epithelial cells also triggers formation of D-*myo*-inositol 1,4,5,6-tetrakisphosphate[-Ins(1,4,5,6)P_4] (Eckmann et al., 1997). This compound antagonizes the inhibitory action of phosphatidylinositol 3,4,5-triphosphate[PtdIns(3,4,5)P_3] on Cl^- flux, and therefore increases fluid secretion (Eckmann et al., 1997). Coincident with this model is the fact that SopB is a phosphatase that cleaves several types of inositol phosphates (Norris et al., 1998a). SopB transforms ins(1,3,4,5,6)P_5 to Ins(1,4,5,6)P_4 which, as mentioned, triggers fluid secretion (Eckmann et al., 1997). SopD from serovar *Dublin,* also secreted by the SPI-1 type III system and translocated into the host cell, seems to act in concert with SopB to elicit gastroenteritis and fluid secretion (Jones et al., 1998). Therefore, *Salmonella* triggers gastroenteritis by subverting host signaling pathways in which inositol phosphates are essential mediators. These cascades could be also linked to cytokine and nuclear responses, since signaling through small GTPases and inositol phosphates are interconnected. Indeed, SopB of serovar *Typhimurium* can also stimulate cytokine release and cytoskeleton reorganization (mentioned in Galán, 1998). LPS could also play a role in host signaling upon invasion as it has been shown to be released by *Salmonella* from intracellular locations (García-del Portillo et al., 1997).

Finally, it should be noted that the immune host system responds to *Salmonella* infections by activating both humoral and cell-mediated immunity (Mastroeni et al., 1993). Type-1 cytokines, particularly INF-γ, play a crucial role in controlling *Salmonella* infection (Ottenhoff et al., 1998; de Jong et al., 1998). INF-γ is synthetized by type 1 helper T-cells (T_H1) in response to IL-12 released by *Salmonella*-infected macrophages or dendritic cells (reviewed in Ottenhoff et al., 1998). Beside INF-γ, other immune mediators such as TNF-α, secreted by *Salmonella*-infected macrophages, can play a role in resistance to infection (Nauciel and Espinasse-Maes, 1992; Phalipon and Sansonetti, 1999).

9. MOLECULAR GENETICS OF *SALMONELLA* VIRULENCE

9.1. TRANSCRIPTIONAL REGULATORY SYSTEMS INVOLVED IN VIRULENCE

Salmonella, as well as other enterobacterial pathogens, possesses a series of transcriptional regulators sensing diverse environmental cues (Slauch et al., 1997). Tight regulation of virulence-related factors is essential for the pathogen to express, in a proper temporal and spatial manner, specific virulence determinants. *Salmonella* is an excellent example of bacteria utilizing a variety of transcriptional regulators during its interaction with eucaryotic cells (Table 1.1).

9.1.1. The Two-Component Regulatory System PhoP-PhoQ

PhoP and PhoQ proteins are encoded by genes forming a single operon (*phoPQ*). These proteins constitute a two-component regulatory system sensing extracellular signals such as limiting amounts of Mg^{2+} and mild acidic pH (Alpuche-Aranda et al., 1992; García Vescovi et al., 1996). PhoQ is an inner membrane integral protein capable of binding Mg^{2+} and Ca^{2+} (García Vescovi et al., 1997). Upon this binding, PhoQ auto-phosphorylates and transfers the phosphate to PhoP, a cytosolic protein (Gunn et al., 1996). Phosphorylated PhoP acts then as a transcriptional regulator of at least 40 genes, some of them involved in virulence (Miller and Mekalanos, 1990; García Vescovi et al., 1994; Miller et al., 1995). Some of these genes are activated by PhoP (*pag*), while others are repressed (*prg*).

Direct proof of the role of the PhoP-Q system in *Salmonella* virulence was found when *phoP* and *phoQ* mutants were highly attenuated in the mouse typhoid model (Fields et al., 1989; Miller et al., 1989; Miller and Mekalanos, 1990). Virulence traits linked to the function of PhoPQ include: (1) bacterial survival within macrophages (Groisman and Saier, 1990; Miller, 1991); (2) resistance to cationic antimicrobial peptides (Gunn and Miller, 1996; Groisman et al., 1997); (3) invasion of epithelial cells (Behlau and Miller, 1993); (4) control of antigen presentation by bacteria-infected macrophages (Wick et al., 1995); and (5) resistance to bile acids (van Velkinburgh and Gunn, 1999).

The fact that *Salmonella phoPQ* mutants do not withstand the attack of phagocytic cells indicates that these regulatory proteins are functional in intracellularly localized bacteria. Indeed, several *pag* genes are induced at low pH, a condition shown to occur in *Salmonella*-containing phagosomes (Alpuche-Aranda et al., 1992). A low amount of Mg^{2+} is also responsible for PhoP-PhoQ induction (García Vescovi et al., 1996). Growth of *Salmonella* in laboratory media containing limiting (10 μM), but not high (10 mM), Mg^{2+} concentration leads to *pag* induction (García Vescovi et al., 1996; Soncini and Groisman, 1996; Soncini et al., 1996). Among these genes are such Mg^{2+}-transporters as MgtA and MgtB (García Vescovi et al., 1996). Although magnesium seems to be the signal that controls transcription of the *phoPQ* regulon, there are some exceptions. *pcgJ,* a PhoPQ-activated gene, is not induced in low Mg^{2+} media (Soncini et al., 1996). Certain PhoPQ-repressed genes, such as *psgB* and *psgD,* are not activated in media with millimolar amounts of Mg^{2+} (Soncini et al., 1996). Therefore, Mg^{2+} plays a crucial role in triggering the PhoPQ-mediated response, but is not the only inducer signal. In fact, acidic pH can also mediate activation of PhoPQ-regulated genes

(Alpuche-Aranda et al., 1992; Soncini and Groisman, 1996). The relative role played by each of these inducing signals has been studied with the two-component system *pmrAB,* a gene locus also positively regulated by PhoPQ (Soncini and Groisman, 1996; Gunn and Miller, 1996; Groisman et al., 1997). The *pmrAB*-regulated genes control bacterial resistance to polymixin and other LPS-binding antimicrobial peptides (reviewed in Groisman and Aspedon, 1997), but not to defensins (Gunn and Miller, 1996). The *pmrAB* operon is autoregulated as has also been shown for *phoPQ* (Soncini et al., 1995; Soncini and Groisman, 1996). Expression of *pmrAB*-regulated genes can be induced by either low pH or Mg^{2+} limitation (Soncini and Groisman, 1996; Gunn and Miller, 1996; Groisman et al., 1997). In contrast, expression of *pmrAB*-independent genes regulated by PhoPQ are only activated by low Mg^{2+} (Soncini and Groisman, 1996).

The biological role of products encoded by PhoP-regulated genes has been elucidated. Included are enzymes that modify the lipid A structure of LPS by adding 4-aminoarabinose and palmitic acid (PagC and PagP respectively) (Guo et al., 1997; Gunn et al., 1998a; Gunn et al., 1998b); nonspecific acid phosphatase (PhoN) (Kasahara et al., 1991); UDP-glucose-dehydrogenase (Ugd-PagA) (Soncini et al., 1996; Gunn et al., 1998a); and PcgL, a D-Ala-D-Ala dipeptidase homologous to VanX of *Enteroccocus* (Hilbert et al., 1999). Genetic identification of other *pag* genes has evidenced homologies to membrane protein-encoding genes of other enterobacteria (Gunn et al., 1998a). This study also showed the existence of duplicated genes in the *Salmonella* chromosome which are PhoPQ-activated (Gunn et al., 1998a).

9.1.2. The Two Component System OmpR/EnvZ

OmpR/EnvZ is a two-component regulatory system sensing changes in external medium osmolarity. OmpR is the transcriptional regulator and EnvZ its cognate membrane sensor. This system controls expression of such major outer membrane proteins as OmpC and OmpF. High osmolarity favors expression of OmpC while low osmolarity shifts the expression to OmpF. *Salmonella ompR* mutants are avirulent in the mouse typhoid model (Dorman et al., 1989), and mutants defective in both OmpC and OmpF are attenuated in the in vivo model when administrated orally but not intraperitoneally (Chatfield et al., 1991). Therefore involvement of OmpR/EnvZ in *Salmonella* virulence may be related to expression of specific outer membrane proteins required for colonization of intestinal epithelium.

9.1.3. The Alternative RpoS (KatF) Sigma Factor

RpoS(KatF) is an alternative sigma factor, known as σ^{30} or σ^{s}, that has an important regulatory function in starved bacteria (Lange and Hengge-Aronis,

TABLE 1.1. **Regulatory Proteins Involved in *Salmonella* Virulence.**

Regulator	Type	Signals Involved	Target Genes	Biological Function(s)	Reference(s)
PhoPQ	transducer/sensor two-components system	Mg^{2+}, acidic pH	*phoN, pags, prgs, pmrAB, hilA, phoPQ,* others	Intramacrophage survival, invasion epithelial cells, modification lipid A, antimicrobial peptide resistance, bile resistance, etc.	Garcia Vescovi et al., 1994; Groisman, 1996, Guo et al., 1997 Gunn et al., 1998; van Velkinburgh and Gunn, 1999
OmpR/EnvZ	transducer/sensor two-components system	osmolarity	*ompC, ompF,* others	colonization intestinal epithelium, Sif formation, macrophage apoptosis	Chatfield et al., 1991
SirA	transducer of two-components system	?	*hilA* (SPI-1), SPI-4 and SPI-5 genes	bacterial invasion, enterophatogenesis?	Johnston et al., 1996; Ahmer et al., 1999
HilA	transducer of two-components system	alkaline pH, low oxygen, SirA-mediated activation	*invF* (SPI-1), SPI-4 and SPI-5 genes	bacterial invasion, enterophatogenesis?	Bajaj et al., 1996; Ahmer et al., 1999
InvF	AraC-like regulator	HilA-mediated activation	*sip/ssp* and *inv-spa-org*	bacterial invasion	Galán, 1996
SsrAB	transducer/sensor two-components system	acidic pH?	SPI-2 genes, *hilA, sipC,* and *prgK* (SPI-1)	intramacrophage survival, intracellular proliferation, systemic disease	Hensel et al., 1998; Cirillo et al., 1998
RpoS	alternative σ^{30} factor	starvation, stress	*spvR,* acid tolerance response (ATR) genes, many others	starvation stress response (SSR), acid resistance (ATR), Peyer's patches colonization, intramacrophage survival, systemic disease	Foster and Spector, 1995; Lee et al., 1995; Hengge-Aronis, 1996; Spector, 1998

TABLE 1.1. (continued)

Regulator	Type	Signals Involved	Target Genes	Biological Function(s)	Reference(s)
RpoE	alternative σ^E factor	heat, damaged proteins	*htrA*, pcr, others	intramacrophage survival, intracellular proliferation, systemic disease	Humphreys et al., 1999
CRP/Cya	—	cAMP levels	*spvR*, many others	SSR, systemic disease	Kelly et al., 1992; Foster and Spector, 1995; Spector, 1998
FurA	CRP-like regulator	iron-starvation	many	acid resistance	Garcia-del Portillo et al., 1993; Foster and Spector, 1995; Riesenberg-Wilmes et al., 1996
SpvR	LysR-like regulator	RpoS-mediated activation	*SpvR*, *spvABCD*	systemic disease	Libby et al., 1997
SlyA	transcriptional regulator	?	?	destruction M cells, intramacrophage survival, resistance to oxydative stress	Daniels et al., 1996; Buchmeier et al., 1997
MviA	response regulator family	?	*rpoS*	acid protection, virulence in *Ity*^{s} mice	Bearson et al., 1996; Swords et al., 1997
Dam	methylation of adenines in GATC sites	?	*finP* (virulence plasmid), others unidentified	systemic disease	Garcia-del Portillo et al., 1999c

1991; Tanaka et al., 1993; Mulvey and Loewen, 1993). High levels of RpoS are measured in cells reaching the stationary phase upon exhaustive growth in rich media. RpoS-dependent proteins are essential for bacterial survival in the stationary phase (Tanaka et al., 1993). RpoS is also required to withstand specific nutrient stresses in *Salmonella,* such as carbon (C)-, nitrogen (N)-, or phosphate (P)-starvations (reviewed in Spector, 1998), and adaptation to acidic pH environments (Lee et al., 1994; Lee et al., 1995; Bearson et al., 1996; Foster and Spector, 1995).

Synthesis of RpoS is increased in intracellular bacteria (Chen et al., 1996), and in vivo analysis has shown that RpoS is required for virulence (Coynault et al., 1996; Swords et al., 1997a; Wilmes-Riesenberg et al., 1997). Its role in virulence could be related to bacterial colonization of Peyer's patches (Nickerson and Curtiss, 1997), and an increase in-SpvR levels (Gulig et al., 1993; Kowarz et al., 1994; Guiney et al., 1995b). SpvR is a regulator required for expression of genes located in the virulence plasmid and linked to systemic disease (Libby et al., 1997).

9.1.4. Other Regulators of *Salmonella* Virulence

SsrAB is a two-component regulatory system encoded in SPI-2 (Hensel et al., 1998; Cirillo et al., 1998). SsrAB may regulate intracellular bacteria expression of structural and secreted type III proteins encoded in SPI-2. *sirA* is a *Salmonella* gene homologous to an *E. coli* gene mapping upstream of *uvrC* (Johnston et al., 1996). SirA is a transcriptional regulatory protein coupled to an as yet unidentified sensor membrane protein. SirA control's the expression of *Salmonella* virulence genes located in SPI-1, SPI-4 and SPI-5, and is required for virulence (Johnston et al., 1996; Ahmer et al., 1999). HilA, as SirA, has homology to transcriptional regulators of two-component system. HilA is encoded in SP-1, and is required for expression of components of the SPI-1-encoded type III secretion system (Bajaj et al., 1996).

Other regulators with a role in *Salmonella* virulence are FurA, which mediates control of iron uptake (García-del Portillo et al., 1993a; Riesenberg-Wilmes et al., 1996); MviA, which mediates acid protection via the RpoS protein and virulence in *Ity*^{s} mice (Bearson et al., 1996; Swords et al., 1997b); SlyA, required for macrophage survival (Daniels et al., 1996; Buchmeier et al., 1997); cyclic AMP receptor protein (CRP) and adenylate cyclase (cya) (Kelly et al., 1992); and DNA adenine methylase (Dam) (García-del Portillo, F., Pucciarelli, M.G., and Casadesús, J., 1994c). Part of the role exerted by Dam could be linked to its control over the expression of a gene located in the *Salmonella* virulence plasmid (Torreblanca and Casadesús, 1996; Torreblanca et al., 1999).

9.2. NOVEL METHODS FOR IDENTIFICATION OF *SALMONELLA* GENES INDUCED IN THE HOST

New strategies have been developed to select and characterize *Salmonella* genes induced during infection of an animal host or cultured eucaryotic cells. These novel approaches led to the identification of the SPI-2, and other previously unknown DNA regions (Ochman et al., 1996; Shea et al., 1996). Metabolic and housekeeping genes were also selected, providing new insights into the complex adaptation mechanisms employed by *Salmonella* during colonization of an animal host (reviewed in Heithoff et al., 1997b).

9.2.1. The in vivo-Induced Expression Technology (IVET)

In 1993, Mahan et al. reported a method, named in vivo-induced expression technology (IVET), for selecting *Salmonella* genes expressed during infection of BALB/c mice (Mahan et al., 1993). The method was based on the in vivo complementation of purine (*pur*) auxotrophy. Fragments of the *Salmonella* chromosome were cloned upstream of a promoterless *purA-lac ZY* gene fusion and clones rescued from the mice having no detectable β-galactosidase activity in laboratory media were characterized (Mahan et al., 1993). The IVET strategy was also used with an antibiotic resistance reporter gene (Mahan et al., 1995). The IVET method has permitted identification of more than 100 in vivo-induced (*ivi*) genes, including: (1) previously known virulence genes, *phoP, spvRAD, spvB,* and *pmrB*; (2) genes named *iviVI-A, ivi-VI-B* encoding proteins homologous to adherence and invasion factors; (3) genes encoding proteins required for acquisition of iron, magnesium and copper; (4) genes required for the synthesis of heme group (*hemD, hemA*), vitamin B_{12} (*cobl, iviXVII*), pyrimidine synthesis (*carAB*), nucleotide balance (*ndk*), and mRNA-tRNA processing (*vacB, vacC*); and (5) genes related to functions such as DNA repair (*recB, recD, mutL*), synthesis of osmoprotectant (*otsA*), cadaverine synthesis (*cadC*), and membrane modifications (reviewed in Heithoff et al., 1997a). The function of approximately 25% of the *ivi* genes is still unknown. Induction of *ivi* genes is observed in intracellular bacteria in cultured macrophages and epithelial cells (Heithoff et al., 1999). Some of the *ivi* genes show regulation by low Mg^{2+} and acidic pH in a PhoPQ-dependent manner. Others, not dependent on PhoPQ, are regulated by iron concentration. Therefore, expression of *ivi* loci varies between different host cell types and environmental conditions (Heithoff et al., 1999), supporting the capacity of *Salmonella* to modulate gene expression during the in vivo infection.

9.2.2. Signature-Tagged Mutagenesis (STM)

In 1995, subsequent to the IVET strategy, Hensel et al. reported an alternative method to identify *Salmonella* genes induced in host animals (Hensel et al., 1995). A negative selection is made on pools of tagged transposon insertion mutants. Those mutants with insertions in essential virulence genes are not rescued upon infection (Hensel et al., 1995). This strategy, known as signature-tagged mutagenesis (STM), led to the identification in centisome 30.5 of a new pathogenicity island, resembling SPI-1 in size and gene organization, and therefore named SPI-2 (Shea et al., 1996). SPI-2, like SPI-1, encodes components of a type III secretion system, and may exert control on functionality of SPI-1 (Hensel et al., 1997b; Deiwick et al., 1998). Mutations in SPI-2 genes have a severe effect an *Salmonella* pathogenicity, and many SPI-2 mutants are highly attenuated when administrated by oral or intraperitoneal routes (Shea et al., 1996; Ochman et al., 1996; Cirillo et al., 1998; Shea et al., 1999).

Other genes selected by the STM strategy include: (1) regulatory genes, such as *ompR/envZ* and *spvR*; (2) *rfb* genes, involved in LPS biosynthesis; and (3) genes required for purine synthesis (*purD, purL*) (Hensel et al., 1995). As with the IVET strategy, the STM selected several as yet uncharacterized genes with no similarity to the database entries (Hensel et al., 1995).

9.2.3. Differential Fluorescence Induction (DFI)

The gene encoding the green fluorescent protein (*gfp*) has also been used to identify *Salmonella* genes induced in bacteria residing in intracellular locations of eucaryotic cells (Valdivia et al., 1996). Specific *gfp* fusions can be selected in specific environmental conditions by fluorescent-activated cell sorting (FACS) (Valdivia and Falkow, 1996). The DFI system was first used for the identification of *Salmonella* genes induced in laboratory media at pH 4.5 (Valdivia and Falkow, 1996). This acidic pH mimics the environment of the *Salmonella*-containing phagosomes in macrophages. Acid-inducible fusions were mapped in genes encoding cell surface maintenance enzymes, stress proteins, and efflux pumps (Valdivia and Falkow, 1996). DFI was later used for selection of *Salmonella* genes induced within macrophages (Valdivia and Falkow, 1997). Genes selected in macrophages included: (1) *phoPQ*-regulated genes; (2) genes of the SPI-2 type III secretion system; (3) *aas*, an *ompR/envZ*-regulated gene required for phospholipid recycling; and (4) genes with no homologies in databases (Valdivia and Falkow, 1997).

9.3. *SALMONELLA* PATHOGENICITY ISLANDS (SPIs)

Pathogenicity islands (Pais) are defined as large regions of DNA encoding for virulence determinants. These regions are present in pathogenic bacteria but absent in phylogenetically related non-phatogenic bacteria (reviewed in Groisman and Ochman, 1996; Hacker et al., 1997). Pais were first reported in uropathogenic *E. coli* (Hacker et al., 1997). Pais have in many cases different G + C content than that of the host organism, and are often flanked by specific DNA sequences such as direct repeats, or insertion sequences of transposons or phages. These features indicate genetic horizontal transfer mechanisms as the basis for Pais formation. Insertion sites for many Pais are tRNA loci, which may contain sequences representing target sites for extrachromosomal elements. In most cases Pais are stable regions, although some show unusually high excision frequencies (Hacker et al., 1997). Acquisition of Pais provides the receptor organism with traits to colonize new environmental niches. Expression of the acquired DNA necessitates that it be under regulatory control to be stabilized (Groisman and Ochman, 1996). *Salmonella* represents an excellent example of how a pathogen has evolved by acquisition of different Pais during its evolution, and how these Pais are tightly regulated in a spatial and temporal manner.

9.3.1. Organization of SPIs

Most genetic studies made in *Salmonella* have been performed with the serovar *Salmonella enterica* sv. *Typhimurium.* This serovar is known to have five large SPIs ranging in size from 17 kb to 40 kb. SPI-1 and SPI-2, mapping at centisomes 63 and 30.5 respectively, are 40 kb long DNA regions encoding for components of type III secretion systems (reviewed in Galán, 1996; Suárez and Rüssmann, 1998; Hueck, 1998). SPI-1 also encodes an iron transport system (Zhuou et al., 1999). Unlike SPI-1, which does not have any specific flanking sequences in its boundaries (Mills et al., 1995), SPI-2 maps downstream of a $tRNA^{Val}$ locus (Hensel et al., 1997a). SPI-1 seems to have spontaneous partial excisions in certain *Salmonella* environmental, but not clinical, isolates (Ginocchio et al., 1997). SPI-1 is required to cross the intestinal epithelium while SPI-2 mediates systemic stages of the infection (Cirillo et al., 1998; Hensel et al., 1998). The SPI-2 role on systemic disease is related to bacterial growth in the animal host (Shea et al., 1999). In concordance with their biological function, SPI-1 and SPI-2 genes are mainly expressed in extracellular and intracellular bacteria respectively. Nonetheless, SPI-1 proteins, such as SipC and SptP, are also apparently synthesized by intracellular bacteria (Collazo and Galán, 1997a; Fu and Galán, 1998b; Rüssmann et al., 1998). Switching from SPI-1 to SPI-2 function may not be an all-or-none response.

Genetic and functional characterization of *Salmonella* SPI-3 has been recently reported (Blanc-Potard et al., 1999). SPI-3 is 17 kb long, and contains genes such as *mgtB* and *mgtC,* encoding for a high-affinity Mg^{2+} transporter and a protein required for macrophage survival respectively (Blanc-Potard and Groisman, 1997; Smith and Maguire, 1998). SPI-3 is inserted in the *selC* locus at centisome 82 of the *Salmonella* chromosome. The G + C content of

SPI-3 is not uniform along the 17 kb, varying from 37% to 47% G + C, compared to the genomic average of 52–53% G + C (Blanc-Potard et al., 1999). This ''modular'' structure of SPI-3 has also been described for SPI-1 and SPI-2 (Zhuou et al., 1999; Hensel et al., 1999).

Other SPIs known in *Salmonella* are SPI-4 and SPI-5. SPI-4 is a 27-kb chromosomal fragment located at centisome 92 of the *Salmonella* chromosome (Wong et al., 1998; Wood et al., 1998). SPI-4 also has a modular structure and is not yet fully characterized at a functional level, although it contains a gene required for *Salmonella* survival within macrophages and a putative Type I secretion apparatus (Wong et al., 1998). As in SPI-1, sequence analysis of boundaries of SPI-4 did not reveal repeated sequences, insertion elements (IS) or phage attachment sites. SPI-5, located at centisome 20, contains 6 genes, *pipA, pipB, pipC,* and *pipD, orfX* and *sopB* (Wood et al., 1998). Mutations in these genes affect *Salmonella* capacity for eliciting intestinal secretory and inflammatory responses (Wood et al., 1998). SopB protein is secreted by the SPI-1-encoded type III system.

Besides SPIs, other small regions of the *Salmonella* chromosome are required for virulence and not present in *E. coli* (Table 1.2). These smaller DNA regions have been named ''pathogenicity islets'' (Groisman and Ochman, 1997). These regions comprise the *iviVI, iviXVII, iviXXII,* and *iviXXV* loci (Conner et al., 1998). These *ivi* loci encode for adhesins, invasin-like functions, and metabolism-related functions such as catabolism of 1,2-propanediol or the enzyme aldehyde dehygrogenase. Most of these regions have characteristic flanking repeated sequences or insertion sites for phages (Conner et al., 1998). Finally, other pathogenicity islets found in *Salmonella* include the *sifA* gene, required for induction of Sifs in epithelial cells (Stein et al., 1996), fimbrial operons (Bäumler et al., 1998), and PhoPQ-regulated loci such as *pagC* (Gunn et al., 1995) and *pagK* (Gunn et al., 1998a).

Therefore, numerous regions of DNA in the *S. enterica* chromosome exist which fit within the definition of pathogenicity islands and are the result of frequent horizontal gene transfer events (Groisman et al., 1993; Ochman and Groisman, 1994; Groisman and Ochman, 1996; Groisman and Ochman, 1997). More importantly, some of these regions, such as SPI-1, SP-2, SP-3 and SPI-4, have a modular organization. Another example is the recently characterized intergenic 45 kb region between the *smpB-nrdE* genes (Bäumler and Heffron, 1998). The *smpB-nrdE* region contains genes absent in *E. coli,* and required for expression of FljB, the phase 2 flagellin, uptake of tricarboxylates, and utilization of catecholate-type siderophores (Bäumler and Heffron, 1998). Clearly, this interchange of genetic information has increased *Salmonella* fitness in its interaction with the environment and animal hosts.

Finally, it should be noted that the present scenario of SPIs has been obtained from genetic information obtained mostly from *S. enterica* sv. *Typhimurium.* Additional SPIs may be present in other serovars and absent in serovar *Typhi-*

TABLE 1.2. *Salmonella*-Specific DNA Regions Involved in Virulence.

Locus	Map Position (Centisome)	Insertion Site	Role in Virulence	Reference(s)
SPI-1	63.0	*fhlA-mutS*	invasion	Galán, 1996
SPI-2	30.5	*ydhE-pykF*	survival in macrophages, intracellular proliferation	Ochman et al., 1996; Shea et al., 1996; Hensel et al., 1997a; 1997b; Cirillo et al., 1998
SPI-3	82.0	*selC*	survival in macrophages	Blanc-Potard et al., 1999
SPI-4	92.0	*ssb-yjcB*	survival in macrophages?	Wong et al., 1998
SPI-5	20.0	*serT*	enteropathogenesis	Wood et al., 1998
rfc	35.7	*prdB-stiA*	LPS biosynthesis	Collins et al., 1991
pacC-msgA	25.0		survival in macrophages	Gunn et al., 1995
spvRABCD	virulence plasmid	—	survival in macrophages, systemic disease	Gulig, 1996
lpfC	80.0	—	bacterial binding to M cells	Bäumler et al., 1996
iviVI-A	7.0	*yafH*	adhesion/invasion?	Conner et al., 1998
iviXVII	*44.0*	*his*	utilization of alternative carbon sources?	Conner et al., 1998
iviXXV				
iviXXII	60.0	—	?	Conner et al., 1998
sifA	27.3	*potBC*	intracellular proliferation?	Stein et al., 1996
SPI (serovar *typhi*)	94.0	—	?	Liu and Sanderson, 1995; Zhang et al., 1997

murium. In fact, when genomes of serovars *Typhimurium* and *Typhi* were compared by subtractive hybridization, it was shown that 20% of the *Typhimurium* genome is not present in *Typhi* and vice versa (Lan and Reeves, 1996). Serovar *Typhi* is known to contain several SPIs not present in *Typhimurium* (Liu and Sanderson, 1995). A major SPI, 118 kb in size, maps at centisome 94, and contains genes encoding a site-specific recombinase and putative pilus-tip adhesin protein (Zhang et al., 1997). A chromosome of serovar *Typhi* has also been shown to contain an unexpectedly high number of rearrangements mediated by recombination of genes encoding ribosomal RNA (Liu and Sanderson, 1996).

9.3.2. Regulation of SPIs

SPIs represent foreign DNA acquired by *Salmonella* after its divergence from the common ancestor to *E. coli* some 100 to 160 million years ago (Ochman and Groisman, 1994). This DNA has been perpetuated in *Salmonella* since it provided advantages to colonizing new environmental niches. However, it is essential to control expression of foreign (SPI)-encoded genes and use them only when beneficial for the pathogen. This control has been undertaken by "housekeeping" regulators previously present in the host chromosome, and thus, also shared by *E. coli.* An example is PhoPQ, which exerts negative control over the SPI-1 type III secretion system via HilA, one regulator encoded within SPI-1 (Bajaj et al., 1996). HilA is required to activate another SPI-1 encoded regulator, InvF (reviewed by Galán, 1996; Hueck, 1998). The invasion capacity linked to SPI-1 is impaired when PhoPQ is active (Pegues et al., 1995). A second regulator of *hilA* expression is SirA, a protein having features of two-component transcriptional regulators (Johnston et al., 1996). Recent work has demonstrated that in addition to SPI-1, SirA regulates other SPIs, such as SPI-4 and SPI-5 (Ahmer et al., 1999). Genes under control of SirA include: (1) *hilA* and *spaS* of SPI-1; (2) genes of unknown function in SPI-4; and (3) *sopB (sigD)* gene of SPI-5, involved in bacterial invasion and eliciting of gastroentiritis. Relevant is the fact that SirA control over SPI-4 and SPI-5 genes requires a functional HilA protein encoded by SPI-1 (Ahmer et al., 1999). Therefore, the SirA/HilA regulatory cascade could be a major regulatory circuit in *Salmonella* pathogenesis.

SPI-1 and SPI-2 are another example of corregulation between SPIs. Mutations in SPI-2 genes, such as *ssaT* and *ssrB,* reduce the capacity of *Salmonella* for secreting SipC, a effector protein encoded by SPI-1 (Deiwick et al., 1998). Expression of *sipC* and *prgK* genes, forming part of two different SPI-1 operons, is reduced in SPI-2 mutants (Deiwick et al., 1998). Although *hilA* expression was also shown to be reduced in SPI-2 mutants HilA production in *trans* does not restore invasion. Therefore, more studies are needed to define whether SPI-2 exerts control over SPI-1 via the HilA regulator.

Definition of environmental signals sensed by regulatory proteins is a major challenge in *Salmonella* pathogenesis. Regulation of virulence gene expression in intracellular bacteria is mostly governed by the PhoPQ system, sensing signals as limitation of Mg^{2+} and/or acidic pH, characteristics of the phagosomal environment (García-del Portillo et al., 1992; Alpuche-Aranda et al., 1992; Soncini and Groisman, 1996; García Vescovi et al., 1996; Groisman et al., 1997). In contrast, little is known about the signals sensed by regulators such as SirA and HilA, which drive expression of invasion-related genes. Some environmental factors that could mediate activation of HilA are high osmolarity, alkaline pH, and low oxygen tension (Bajaj et al., 1996). Factors activating SirA, and as the as yet unidentified cognate membrane sensor, remain to be determined.

9.3.3. Presence of SPIs in Different *Salmonella* Serovars: Evolution of *Salmonella* Pathogenicity

SPI-1 was the first SPI shown to be distributed in all *Salmonella* subspecies groups (I to VII) (Li et al., 1995; Ochman and Groisman, 1996). Therefore, SPI-1 probably was acquired by *Salmonella* shortly after their divergence from *E. coli,* enabling salmonellae to colonize the intestinal epithelium of animal hosts (Groisman and Ochman, 1997; Bäumler et al., 1998). Unlike SPI-1, SPI-2 is present in all *S. enterica* serovars, but not in the species *S. bongori,* suggesting that SPI-2 was acquired by *S. enterica* after it split from *S. bongori* (Ochman and Groisman, 1996; Groisman and Ochman, 1997). Acquisition of SPI-2 may have driven adaptation of all *S. enterica* serovars to colonize deeper tissues (Groisman and Ochman, 1997; Bäumler et al., 1998).

Some SPI-3 genes are present in all *S. enterica* subspecies and *S. bongori,* while others are specific to certain subspecies groups (Blanc-Potard et al., 1999). Furthermore, other SPI-3 genes were present in *S. bongori* but absent in certain *S. enterica* subspecies (Blanc-Potard et al., 1999). These results emphasize the numerous horizontal gene transfer events occurring in this specific SPI, and the fact that some of those events may have led not only to gain, but also to loss of concrete regions or modules. Similar results have been described for certain ORFs of SPI-2 located in the boundary of centisome 30.5 (Hensel et al., 1999). SPI-4 and SPI-5 have only been tested for serovars of *S. enterica* subspecies group I, and all of them harbor both SPIs (Wood et al., 1998; Wong et al., 1998).

Distribution of *ivi* loci has also been tested for serovars of *S. enterica* subspecies I having different grades of host adaptation (Conner et al., 1998). Three classes of *ivi* sequences were differentiated: (1) class I, present in broad host-range and host-adapted serovars; (2) class II, present in all but the host-adapted serovar; and (3) class III, present in broad host-adapted serovars and absent from serovars *Typhi* and *Choleraesusis* (Conner et al., 1998). These observations denote a high heterogeneity in the arsenal of virulence factors harbored by different salmonellae. These differences may explain the diverse capacity of *Salmonella* serovars to infect and cause disease in a broad or restricted range of animal hosts.

9.4. ROLE OF THE VIRULENCE PLASMID

Certain *Salmonella* serovars contain a plasmid ranging in size from 60 to 100 kb implicated in virulence (Gulig, 1990; Rexach et al., 1994; Guiney et al., 1995a). Plasmid-cured derivatives are unable to grow and cause systemic disease in the mouse typhoid model, but retain capacity for colonizing the intestinal epithelium and spreading to target tissues and organs (Gulig and Doyle, 1993; Wallis et al., 1995). A minimum region of the virulence plasmid is required for causing disease and is known as the *spv* (*Salmonella* virulence plasmid)*RABCD* locus (Gulig et al., 1993). SpvR is a transcriptional positive regulator for expression of *spvABCD* genes (Caldwell and Gulig, 1991; Coynault et al., 1992). SpvR is under control of the stationary-phase sigma factor RpoS (Fang et al., 1992; Norel, 1992; Kowarz et al., 1994). Spv proteins are not expressed in the logarithmic growth phase, but appear in the stationary phase (Coynault et al., 1992; El-Gedaily et al., 1997; Wilson and Gulig, 1998). SpvA and SpvB are located in the membrane, SpvC in the cytosol, and SpvD is secreted to the medium (El-Gedaily et al., 1997). The function of Spv proteins is still unknown, although they could play a role in signaling between intracellular *Salmonella* and the host cell. In fact, Spv proteins are synthesized in intracellular bacteria (Rhen et al., 1993; Chen et al., 1996), and promote intracellular growth within macrophages in vivo (Gulig et al., 1998). This last observation could explain the plasmid-mediated lysis of infected macrophages (Guilloteau et al., 1996b). All these events seem to occur late in the infection, since no differences in the intestinal inflammatory response are observed in mice infected with wild type and plasmid-cured strains (Guilloteau et al., 1996a).

Several observations indicate that additional bacterial factors are required for systemic disease. First, introduction of the *spv* operon in serovar *Typhi* does not confer mouse virulence and, second, *S. enterica* sv. *Gallinarum Pullorum* and *S. bongori* do not cause systemic disease in mice although both contain the *spv* operon.

10. CONCLUSIONS AND FUTURE PERSPECTIVES

Salmonellae are extremely successful pathogens infecting humans and animals. Their ubiquity in the environment and ability to colonize animals used in the human food chain are factors which certainly make diseases caused by these bacteria one of the most difficult and important to eradicate. Molecular studies have demonstrated the sophisticated strategies used by salmonellae to engage host cell functions to promote bacterial uptake, inflammation, and fluid secretion. For the first time, we have a reasonable model of how enteropathogenesis is elicited by salmonellae. Bacterial proteins responsible for these events as well as their host targets are being characterized.

Another area of tremendous advance in the last years is the molecular genetics of *Salmonella* virulence. Diverse *Salmonella*-specific DNA regions contain essential virulence functions for adherence, invasion, elicitation of enteropathogenesis, intramacrophage survival, intracellular proliferation, etc. Genetic analysis of these regions is providing essential clues to how *Salmonella* arised as a pathogen and how the large number of serovars we know today may have diverged during evolution (Bäumler et al., 1998). A detailed functional characterization of these regions will provide new insights on how *Salmonella* interacts with animal hosts, and copes with different environmental stresses encountered in a variety of biological niches. Certainly, the complete genome sequence of representative serovars will facilitate these types of studies, which are clearly an important challenge in the understanding of *Salmonella* pathogenesis.

Another relevant issue is the extraordinary variety of interactions existing between salmonellae and host eucaryotic cells. It is obvious that within the animal there are many cell types that stimulate different responses in the invading bacteria. Indeed, different studies have confirmed that *Salmonella* synthesizes unique proteins when inhabiting different host eucaryotic cells (Abshire and Neidhardt, 1993a; Burns-Keliher et al., 1997; Burns-Keliher et al., 1998). More studies are needed to characterize in depth the interaction of *Salmonella* with potential new host target cells. Adaptation to the intracellular lifestyle is another aspect that is being intensively studied. Future efforts in dissecting all these adaptation mechanisms are mandatory if we want to combat this pathogen.

Finally, two aspects related to *Salmonella* infection may deserve more interest in future investigations. One is the carrier state, a physiological condition that highly influences transmission of the pathogen in the community and can lead to chronic infections such as reactive arthritis (Mäki-Ikola and Granfors, 1993; Miller et al., 1995; Wuorela and Granfors, 1998). Virtually nothing is known about the molecular mechanisms underlying coexistence of *Salmonella* with the host animal for long periods of time, although a putative role of surface fiber Agf has been proposed (Sukupolvi et al., 1997). Recent findings indicate that intracellular persistence of *Salmonella* with host cells could be relevant, and that this phenomenon is tightly regulated. Further studies on this topic will unravel which bacterial and host factors are involved. The second aspect, also having a high impact on transmissibility of salmonellae, is the viable but not culturable state (VBNC). *Salmonella* has been shown capable of entering this state and remaining viable for long periods of time in soil and water (Turpin et al., 1993; Cho and Kim, 1999). In fact, it has been proposed that some *Salmonella* genes are involved in this process (Romanova et al., 1996). Future work on defining how *Salmonella* enters into a VBNC state is also necessary.

I apologize that, owing to space limitations, and the extensive area covered in this chapter, some original publications may have been omitted or referred to in other review articles referenced herein.

11. REFERENCES

Abshire, K. Z. and Neidhart, F. C. 1993a. "Analysis of proteins synthesized by *Salmonella typhimurium* during growth within a host macrophage." *J. Bacteriol.*, 175 (12):3734–3743.

Abshire, K. Z. and Neidhardt, F. C. 1993b. "Growth rate paradox of *Salmonella typhimurium* within host macrophages." *J. Bacteriol*, 175 (12):3744–3748.

Ahmer, B. M. M., van Reeuwijk, J., Watson, P. R., Wallis, T. S. and Heffron, F. 1999. "*Salmonella* SirA is a global regulator of genes mediating enteropathogenesis." *Mol. Microbiol.*, 31 (3):971–982.

Alpuche-Aranda, C. M., Racoosin, E. L., Swanson, J. A. and Miller, S. I. 1994. "*Salmonella* stimulate macrophage macropinocytosis and persist within spacious phagosomes." *J. Exp. Med.*, 179 (2): 601–608.

Alpuche-Aranda, C. M., Swanson, J. A., Loomis, W. P. and Miller, S. I. 1992. "*Salmonella typhimurium* activates virulence gene transcription within acidified macrophage phagosomes." *Proc. Natl. Acad. Sci. USA*, 89 (21):10079–10083.

Bajaj, V., Lucas, R. L., Hwang, C. and Lee, C. A. 1996. "Co-ordinate regulation of *Salmonella typhimurium* invasion genes by environmental and regulatory factors is mediated by control of *hilA* expression." *Mol. Microbiol.*, 22 (4): 703–714.

Baker, S. J., Gunn, J. S. and Morona, R. 1999. "The *Salmonella typhi* melittin resistance gene *pqaB* affects intracellular growth in PMA-differentiated U937 cells, polymyxin B resistance and lipopolysaccharide." *Microbiology*, 145:367–378.

Baran, J., Guzik, K., Hryniewicz, W., Ernst, M., Flad, H. D. and Pryjma, J. 1996. "Apoptosis of monocytes and prolonged survival of granulocytes as a result of phagocytosis of bacteria." *Infect. Immun.*, 64 (10):4242–4248.

Bäumler, A. J. and Heffron, F. 1995a. "Identification and sequence analysis of *lpfABCDE*, a putative fimbrial operon of *Salmonella typhimurium.*" *J. Bacteriol.*, 177 (8): 2087–2097.

Bäumler, A. J. and Heffron, F. 1995b. "Microbial resistance to macrophage effector functions: strategies for evading microbicidal mechanisms and scavenging nutrients within mononuclear phagocytes." In *Virulence Mechanisms of Bacterial Pathogens.* eds. J. A. Roth, C. A. Bolin, K. A. Brodgen, F. C. Minion and M. J. Wannenmuehler. Washington, D. C. ASM Press, pp. 115–131.

Bäumler, A. J. and Heffron, F. 1998. "Mosaic structure of the *smpB-nrdE* intergenic region of *Salmonella enterica.*" *J. Bacteriol.*, 180 (8):2220–2223.

Bäumler, A. J., Heffron, F. and Reissbrodt, R. 1997a. "Rapid detection of *Salmonella enterica* with primers specific for *iroB.*" *J. Clin. Microbiol.*, 35 (5):1224–1230.

Bäumler, A. J., Kusters, J. G., Stojiljkovic, I. and Heffron, F. 1994. "*Salmonella typhimurium* loci involved in survival within macrophages." *Infect. Immun.*, 62 (5):1623–1630.

Bäumler, A. J., Tsolis, R. M., Bowe, F., Kusters, J. G., Hoffman, S. and Heffron, F. 1996. "The *pef* fimbrial operon mediates adhesion to murine small intestine and is necessary for fluid accumulation in infant mice." *Infect. Immun.*, 64:61–68.

Bäumler, A. J., Tsolis, R. M., Ficht, T. A. and Adams, L. G. 1998. "Evolution of host adaptation in *Salmonella enterica.*" *Infect. Immun.*, 66 (10):4579–4587.

Bäumler, A. J., Tsolis, R. M. and Heffron, F. 1996a. "Contribution of fimbrial operons to attachment to and invasion of epithelial cells by *Salmonella typhimurium.*" *Infect. Immun.*, 64: 1862–1865.

Bäumler, A. J., Tsolis, R. M. and Heffron, F. 1996b. "The *lpf* fimbrial operon mediates adhesion of *Salmonella typhmurium* to murine Peyer's patches." *Proc. Natl. Acad. Sci. USA*, 93:279–283.

Bäumler, A. J., Tsolis, R. M. and Heffron, F. 1997b. "Fimbrial adhesins of *Salmonella typhimurium.* Role in bacterial interactions with epithelial cells." *Adv. Exp. Med. Biol.*, 412: 149–158.

Bäumler, A. J., Tsolis, R. M., Valentine, P. J., Ficht, T. A. and Heffron, F. 1997c. "Synergistic effect of mutations in *invA* and *lpfC* on the ability of *Salmonella typhimurium* to cause murine typhoid." *Infect. Immun.*, 65 (6):2254–2259.

Bearson, S., Bearson, B. and Foster, J. W. 1997. "Acid stress responses in enterobacteria." *FEMS Microbiol. Lett.*, 147 (2): 173–180.

Bearson, S. M., Benjamin, W. H., Jr., Swords, W. E. and Foster, J. W. 1996. "Acid shock induction of RpoS is mediated by the mouse virulence gene *mviA* of *Salmonella typhimurium.*" *J. Bacteriol.*, 178 (9):2572–2579.

Behlau, I. and Miller, S. I. 1993. "A PhoP-repressed gene promotes *Salmonella typhimurium* invasion of epithelial cells." *J. Bacteriol.*, 175 (14):4475–4484.

Bennet, A. R., Greenwood, D., Tennant, C., Banks, J. G. and Betts, R. P. 1998. "Rapid and definitive detection of *Salmonella* in foods by PCR." *Lett. Appl. Micobiol.*, 26 (6):437–441.

Blackwell, J. M. and Searle, S. 1999. "Genetic regulation of macrophage activation: understanding the function of *Nramp1* (= *Ity/Lsh/Bcg*)." *Immunol. Lett.*, 65:73–80.

Blais, B. W., Pietrzak, E., Oudit, D., Wilson, C., Philippe, L. M. and Howlett, J. 1998. "Polymacron enzyme immunoassay system for detection of naturally contaminating *Salmonella* in foods, feeds, and environmental samples." *J. Food Prot.*, 61 (9):1187–1190.

Blanc-Potard, A.-B., Solomon, F., Kayser, J. and Groisman, E. A. 1999. "The SPI-3 pathogenecity island of *Salmonella enterica.*" *Infect. Immun.*, 181 (3):998–1004.

Blanc-Potard, A. B. and Groisman, E. A. 1997. "The *Salmonella selC* locus contains a pathogenicity island mediating intramacrophage survival." *EMBO J.*, 16 (17):5376–5385.

Boyd, E. F., Wang, F.-S., Whittman, T. S. and Selander, R. K. 1996. "Molecular genetic relationships of the salmonellae." *Appl. Environ. Microbiol.*, 62:804–808.

Brasher, C. W., DePaola, A., Jones, D. D. and Bej, A. K. 1998. "Detection of microbial pathogens in shellfish with multiplex PCR." *Curr. Microbiol.*, 37 (2):101–107.

Buchmeier, N., Bossle, S., Chen, C. Y., Fang, F. C., Guiney, D. G. and Libby, S. J. 1997. "SlyA, a transcriptional regulator of *Salmonella typhimurium*, is required for resistance to oxidative stress and is expressed in the intracellular environment of macrophages." *Infect. Immun.*, 65 (9):3725–3730.

Buchmeier, N. A. and Libby, S. J. 1997. "Dynamics of growth and death within a *Salmonella typhimurium* population during infection of macrophages." *Can. J. Microbiol.*, 43 (1):29–34.

Buchmeier, N. A., Libby S. J., Xu, Y., Loewen, P. C., Switala, J., Guiney, D. G. and Fang, F. C. 1995. "DNA repair is more important than catalase for *Salmonella* virulence in mice." *J. Clin. Invest.*, 95:1047–1053.

Buchmeier, N. A., Lipps, C. J., So, M. Y. and Heffron, F. 1993. "Recombination-deficient mutants of *Salmonella typhimurium* are avirulent and sensitive to the oxidative burst of macrophages." *Mol. Microbiol.*, 7 (6):933–936.

Buchwald, D. S. and Blaser, M. J. 1984. "A review of human salmonellosis. II. Duration of excretion following interaction with non*typhi Salmonella.*" *Rev. Infect. Dis.*, 6: 345–356.

Burns-Keliher, L., Nickerson, C. A., Morrow, B. J. and Curtis, R. I. 1998. "Cell-specific proteins synthetized by *Salmonella typhimurium.*" *Infect. Immun.*, 66 (2):856–861.

Burns-Keliher, L. L., Portteus, A. and Curtiss, R. I. 1997. "Specific detection of *Salmonella typhimurium* proteins synthesized intracellularly." *J. Bacteriol.,* 179 (11):3604–3612.

Caldwell, A. L. and Gulig, P. A. 1991. "The *Salmonella typhimurium* virulence plasmid encodes a positive regulator of a plasmid-encoded virulence gene." *J. Bacteriol.,* 173 (22): 7176–7185.

Cellier, M., Belouchi, A. and Gros, P. 1996. "Resistance to intracellular infections: comparative genomic analysis of *Nramp 1.*" *Trends Genet.,* 12:201–204.

Chatfield, S., Dorman, C. J., Hayward, C. and Dougan, G. 1991. "Role of *ompR*-dependent genes in *Salmonella typhimurium* virulence: mutants deficient in both OmpC and OmpF are attenuated in vivo." *Infect. Immun.,* 59:449–452.

Chatfield, S. N., Strahan, K., Pickard, D., Charles, I., Hormaeche, C. and Dougan, G. 1992. "Evaluation of *Salmonella typhimurium* strains harboring defined mutations in *htrA* and *aroA* in the murine salmonellosis mode." *Microb. Pathogen.,* 12: 145–151.

Chen, C. Y., Eckmann, L., Libby, S. J., Fang, F. C., Okamoto, S., Kagnoff, M. F., Fierer, J. and Guiney, D. G. 1996. "Expression of *Salmonella typhimurium rpoS* and *rpoS*-dependent genes in the intracellular environment of eukaryotic cells." *Infect. Immun.,* 64 (11):4739–4743.

Chen, L. M., Hobbie, S. and Galán, J. E. 1996a. "Requirement of CDC42 for *Salmonella*-induced cytoskeletal and nuclear responses." *Science,* 274:2115–2118.

Chen, L. M., Kaniga, K. and Galán, J. E. 1996b. "*Salmonella* spp. are cytotoxic for cultured macrophages," *Mol Microbiol,* 21 (5):1101–1115.

Cho, J. C. and Kim, S. J. 1999. "Viable, but non-culturable, state of a green fluorescence protein-tagged environmental isolate of *Salmonella typhi* in groundwater and pond water." *FEMS Microbiol. Lett.,* 170:257–264.

Cirillo, D. M., Valdivia, R. H., Monack, D. M. and Falkow, S. 1998. "Macrophage-dependent induction of the *Salmonella* pathogenicity island 2 type III secretion system and its role in macrophage survival." *Mol. Microbiol.,* 30 (1):175–188.

Cocolin, L., Manzano, M., Cantoni, C. and Comi, G. 1998. "Use of polymerase chain reaction and restriccion enzyme analysis to directly detect and identify *Salmonella typhimurium* in food." *J. Appl. Microbiol.,* 85 (4):673–677.

Collazo, C. M. and Galán, J. E. 1997a. "The invasion-associated type III system of *Salmonella typhimurium* directs the translocation of Sip proteins into the host cell." *Mol. Microbiol.,* 24 (4):747–756.

Collazo, C. M. and Galán, J. E. 1997b. "The invasion-associated type-III protein secretion system in *Salmonella*-a review." *Gene,* 192 (1):51–59.

Conlan, J. W. 1996. "Neutrophiles prevent extracellular colonization of the liver microvasculature by *Salmonella typhimurium.*" *Infect. Immun.,* 64:1043–1047.

Conlan, J. W. and North, R. J. 1992. "Early pathogenesis of infection in the liver with the facultative intracelular bacteria *Listeria monocytogenes, Francisella turalensis,* and *Salmonella typhimurium* involves lysis of infected hepatocytes by leukocytes." *Infect. Immun.,* 60:5164–5171.

Conner, C. P., Heithoff, D. M., Julio, S. M., Sinsheimer, R. L. and Mahan, M. J. 1998. "Differential patterns of acquired virulence genes distinguish *Salmonella* strains." *Proc. Natl. Acad. Sci. USA,* 95:4641–4645.

Coynault, C., Robbe-Saule, V. and Norel, F. 1996. "Virulence and vaccine potential of *Salmonella typhimurium* mutants deficient in the expression of the RpoS (sigma S) regulon." *Mol. Microbiol.,* 22 (1):149–160.

Coynault, C., Robbe-Saule, V., Popoff, M. Y. and Norel, F. 1992. "Growth phase and SpvR regulation of transcription of *Salmonella typhimurium spvABC* virulence genes." *Microb. Pathog.,* 13 (2):133–143.

D'Aoust, J.-V. 1997. "*Salmonella* species." In *Food Microbiology: Fundamentals and Frontiers*, eds. M. P. Doyle, L. R. Beuchat and T. J. Montville. Washington, D.C.: American Society for Microbiology, pp. 129–158.

Daefler, S. 1999. "Type III secretion by *Salmonella typhimurium* does not require contact with a eukaryotic host." *Mol. Microbiol.*, 31 (1):45–51.

Daniels, J. J., Autenrieth, I. B., Ludwig, A. and Goebel, W. 1996. "The gene *slyA* of *Salmonella typhimurium* is required for destruction of M cells and intracellular survival but not for invasion or colonization of the murine small intestine." *Infect. Immun.*, 64 (12):5075–5084.

de Boer, E. 1998. "Update of media for isolation of Enterobacteriaceae from foods." *Int. J. Food. Microbiol.*, 45 (1):43–53.

de Groote, M. A., Ochsner, U. A., Shiloh, M. U., Nathan, C., McCord, J. M., Dinauer, M. C., Libby, S. J., Vazquez-Torres, A., Xu, Y. and Fang, F. C. 1997. "Periplasmic superoxide dismutase protects *Salmonella* from products of phagocyte NADPH-oxidase and nitric oxide synthase." *Proc. Natl. Acad. Sci. USA*, 94: 13997–14001.

de Jong, R., Altare, F., Haagen, I.-A., Elfernik, D. G., de Boer, T., vand Breda-Vriesman, P. J. C., Kabel, P. J., Draaisma, J. M. T., van Dissel, J. T., Kroon, F. P., Casanova, J.-L. and Ottenhoff, T. H. M. 1998. "Severe Mycobacterial and *Salmonella* infections in interleukin-12 receptor-deficient patients." *Science*, 280: 1435–1438.

Deiwick, J., Nikolaus, T., Shea, J. E., Gleeson, C., Holden, D. W. and Hensel, M. 1998. "Mutations in the *Salmonella* pathogenicity island 2 (SPI2) genes affecting transcription of SPI1 genes and resistance to antimicrobial agents." *J. Bacteriol.*, 180 (18): 4775–4780.

Dorman, C. J., Chatfield, S., Higgins, C. F., Hayward, C. and Dougan, G. 1989. "Characterization of porin and *ompR* mutants of a virulent strain of *Salmonella typhimurium: ompR* mutants are attenuated in vivo." *Infect. Immun.*, 57:2136–2140.

Dramsi, S. and Cossart, P. 1998. "Intracellular pathogens and the actin cytoskeleton." *Annu. Rev. Cell Dev. Biol.*, 14: 137–166.

Dunlap, N. E., Benjamin, W. H. J., Berry, A. K., Eldrigde, J. H. and Briles, D. E. 1992. "A 'safe-site' for *Salmonella typhimurium* is within splenic polymorphonuclear cells." *Microb. Pathog.*, 13: 181–190.

Dunstan, S. J., Simmons, C. P. and Strugnell, R. A. 1998. "Comparison of the abilities of different attenuated *Salmonella typhimurium* strains to elicit humoral immune responses against a heterologus antigen." *Infect. Immun.*, 66 (2):732–740.

Eckmann, L., Kagnoff, M. F. and Fiere, J. 1993. "Epithelial cells secrete the chemokine interleukin-8 in response to bacterial entry." *Infect. Immun.*, 61 (11):4569–4574.

Eckmann, L., Rudolf, M. T., Ptasznik, A., Schultz, C., Jiang, T., Wolfson, N., Tsien, R., Fierer, J., Shears, S. B., Kagnoff, M. F. and Traynor-Kaplan, A. E. 1997. "d-*myo*-Inositol 1,4,5,6-tetrakisphosphate produced in human intestinal epithelial cells in response to *Salmonella* invasion inhibits phosphoinositide 3-kinase signaling pathways." *Proc. Natl. Acad. Sci. USA*, 94: 14456–14460.

Ekperigin, H. E. and Nagaraja, K. V. 1998. "Microbial food borne pathogens. *Salmonella*." *Vet. Clin. North Am. Food. Anim. Pract.*, 14 (1):17–29.

El-Gedaily, A., Paesold, G. and Krause, M. 1997. "Expression profile and subcellular location of the plasmid-encoded virulence (Spv) proteins in wild-type *Salmonella dublin*." *Infect. Immun.*, 65 (8): 3406–3411.

Fang, F. C., Libby, S. J., Buchmeier, N. A., Loewen, P. C., Switala, J., Harwood, J. and Guiney, D. G. 1992. "The alternative sigma factor *katF (rpoS)* regulates *Salmonella* virulence." *Proc. Natl. Acad. Sci. USA*, 89 (24):11978–11982.

Farrant, J. L., Sansone, A., Canvin, J. R., Pallen, M. J., Langford, P. R., Wallis, T. S., Dougan,

G. and Kroll, J. S. 1997. "Bacterial copper- and zinc-cofactored superoxide dismutase contributes to the pathogenesis of systemic salmonellosis." *Mol. Microbiol.*, 25 (4):785–796.

Fields, P. I., Groisman, E. A. and Heffron, F. 1989. "A *Salmonella* locus that controls resistance to microbicidal proteins from phagocytic cells." *Science*, 243:1059–1062.

Fields, P. I., Swanson, R. V., Haidaris, C. G. and Heffron, F. 1986. "Mutants of *Salmonella typhimurium* that cannot survive within the macrophage are avirulent." *Proc. Natl. Acad. Sci. USA*, 83 (14):5189–5193.

Finlay, B. B. 1994. "Molecular and cellular mechanisms of *Salmonella* pathogenesis." *Curr. Top. Microbiol. Immunol.*, 192:163–185.

Finlay, B. B. and Cossart, P. 1997. "Exploitation of mammalian host cell functions by bacterial pathogens." *Science*, 276: 718–725.

Finlay, B. B. and Falkow, S. 1990. "*Salmonella* interactions with polarized human intestinal Caco-2 epithelial cells." *J. Infect. Dis.*, 162 (5):1096–1106.

Finlay, B. B. and Falkow, S. 1997. "Common themes in microbial pathogenicity revisited." *Microbiol. Mol. Biol. Rev.*, 61 (2): 136–169.

Finlay, B. B., Gumbiner, B. and Falkow, S. 1988a. "Penetration of *Salmonella* through a polarized Madin-Darby canine kidney epithelial cell monolayer." *J. Cell Biol.*, 107 (1):221–230.

Finlay, B. B., Ruschkowski, S. and Dedhar, S. 1991. "Cytoskeletal rearrangements accompanying *Salmonella* entry into epithelial cells." *J. Cell Sci.*, 99:283–296.

Finlay, B. B., Starnbach, M. N., Francis, C. L., Stocker, B. A., Chatfield, S., Dougan, G. and Falkow, S. 1988b. "Identification and characterization of TnphoA mutants of *Salmonella* that are unable to pass through a polarized MDCK epithelial cell monolayer." *Mol. Microbiol.*, 2 (6):757–766.

Foster, J. W. and Spector, M. P. 1995. "How *Salmonella* survive against the odds." *Annu. Rev. Microbiol.*, 49: 145–174.

Francis, C. L., Ryan, T. A., Jones, B. D., Smith, S. J. and Falkow, S. 1993. "Ruffles induced by *Salmonella* and other stimuli direct macropinocytosis of bacteria." *Nature*, 364:639–642.

Francis, C. L., Stambach, M. N. and Falkow, S. 1992. "Morphological and cytoskeletal changes in epithelial cells occur immediately upon interaction with *Salmonella typhimurium* grown under low-oxygen conditions." *Mol. Microbiol.*, 6 (21):3077–3087.

Fu, Y. and Galán, J. E. 1998a. "Identification of a specific chaperone for SptP, a substrate of the centisome 63 type III secretion system of *Salmonella typhimurium*." *J. Bacteriol.*, 180 (13):3393–3399.

Fu, Y. and Galán, J. E. 1998b. "The *Salmonella typhimurium* tyrosine phosphatase SptP is translocated into host cells and disrupts the actin cytoskeleton." *Mol. Microbiol.*, 27 (2): 359–368.

Galán, J. E. 1996. "Molecular genetic bases of *Salmonella* entry into host cells." *Mol. Microbiol.*, 20 (2):263–271.

Galán, J. E. 1998. "Interactions of *Salmonella* with host cells: encounters of the closest kind." *Proc. Natl. Acad. Sci. USA*, 95:14006–11408.

Galán, J. E. and Bliska, J. B. 1996. "Cross-talk between bacterial pathogens and their host cells." *Annu. Rev. Cell Dev. Biol.*, 12: 221–255.

Galán, J. E. and Curtiss, R. D. 1989. "Cloning and molecular characterization of genes whose products allow *Salmonella typhimurium* to penetrate tissue culture cells." *Proc. Natl. Acad. Sci USA*, 86 (16):6383–6387.

Galyov, E. E., Wood, M. W., Rosqvist, R., Mullan, P. B., Watson, P. R., Hedges, S. and Wallis, T. S. 1997. "A secreted effector protein of *Salmonella dublin* is translocated into eukaryotic cells and mediates inflammation and fluid secretion in infected ileal mucosa." *Mol. Microbiol.*, 25 (5):903–912.

García Vescovi, E., Ayala, Y. M., Di Cera, E. and Groisman, E. A. 1997. "Characterization of the bacterial sensor protein PhoQ. Evidence for distinct binding sites for Mg^{2+} and Ca^{2+}." *J. Biol. Chem.*, 272 (3):1440–1443.

García Vescovi, E., Soncini, F. C. and Groisman, E. A. 1994. "The role of the PhoP/PhoQ regulon in *Salmonella* virulence." *Res. Microbiol.*, 145:473–480.

García Vescovi, E., Soncini, F. C. and Groisman, E. A. 1996. "Mg^{2+} as an extracellular signal: environmental regulation of *Salmonella* virulence." *Cell*, 84 (1):165–174.

García-del Portillo, F. 1996. "Interaction of *Salmonella* with lysosomes of eukaryotic cells." *Microbiologia*, 12: 259–266.

García-del Portillo, F. and Finlay, B. B. 1994. "*Salmonella* invasion of nonphagocytic cells induces formation of macropinosomes in the host cell." *Infect. Immun.*, 62 (10): 4641–4645.

García-del Portillo, F. and Finlay, B. B. 1995a. "Targeting of *Salmonella typhimurium* to vesicles containing lysosomal membrane glycoproteins bypasses compartments with mannose-6-phosphate receptors." *J. Cell Biol.*, 129:81–97.

García-del Portillo, F. and Finlay, B. B. 1995b. "The varied lifestyles of intracellular pathogens within eukaryotic vacuolar compartments." *Trends Microbiol.*, 3:373–380.

García-del Portillo, F., Casadesús, J. and Cano, D. 1999a. Unpublished data.

García-del Portillo, F., Foster, J. W. and Finlay, B. B. 1993a. "Role of acid tolerance response genes in *Salmonella typhimurium* virulence." *Infect. Immun.*, 61 (10):4489–4492.

García-del Portillo, F., Groisman, E. A. and Martínez-Moya, M. 1999b. Unpublished data.

García-del Portillo, F., Pucciarelli, M. G. and Casadesús, J. 1999c. "Essential role of DNA adenine methylation in *Salmonella typhimurium* virulence." Submitted.

García-del Portillo, F., Foster, J. W., Maguire, M. E. and Finlay, B. B. 1992. "Characterization of the micro-environment of *Salmonella typhimurium*-containing vacuoles within MDCK epithelial cells." *Mol. Microbiol.*, 6 (22):3289–3297.

García-del Portillo, F., Pucciarelli, M. G., Jefferies, W. A. and Finlay, B. B. 1994. "*Salmonella typhimurium* induces selective aggregation and internalization of host cell surface proteins during invasion of epithelial cells." *J. Cell Sci.*, 107:2005–2020.

García-del Portillo, F., Stein, M. A. and Finlay, B. B. 1997. "Release of lipopolysaccharide from intracellular compartments containing *Salmonella typhimurium* to vesicles of the host epithelial cell." *Infect. Immun.*, 65 (1):24–34.

García-del Portillo, F., Zwick, M. B., Leung, K. Y. and Finlay, B. B. 1993b. "*Salmonella* induces the formation of filamentous structures containing lysosomal membrane glycoproteins in epithelial cells." *Proc. Natl. Acad. Sci. USA*, 90 (22):10544–10548.

Gewirtz, A. T., Siber, A. M., Madara, J. L. and McCormick, B. A. 1999. "Orchestration of neutrophil movement by intestinal epithelial cells in response to *Salmonella typhimurium* can be uncoupled from bacterial internalization." *Infect. Immun.*, 67 (2):608–617.

Ginocchio, C., Pace, J. and Galán, J. E. 1992. "Identification and molecular characterization of a *Salmonella typhimurium* gene involved in triggering the internalization of salmonellae into cultured epithelial cells." *Proc. Natl. Acad. Sci. USA.*, 89: 5976–5980.

Ginocchio, C. C., Olmsted, S. B., Wells, C. L. and Galán, J. E. 1994. "Contact with epithelial cells induces the formation of surface appendages on *Salmonella typhimurium*." *Cell*, 76 (4): 717–724.

Ginocchio, C. C., Rahn, K., Clarke, R. C. and Galán, J. E. 1997. "Naturally occurring deletions in the centisome 63 pathogenicity island of environmental isolates of *Salmonella* spp." *Infect. Immun.* 65 (4):1267–1272.

Goldberg, M. B. and Rubin, R. H. 1988. "The spectrum of *Salmonella* infection." *Infect. Dis. Clin. North Am.*, 2 (3):571–598.

Gouws, P. A., Visser, M. and Brozel, V. S. 1998. "A polymerase chain reaction procedure for the detection of *Salmonella* spp. within 24 hours." *J. Food Prot.*, 61 (8):1039–1042.

Groisman, E. A. 1996. "Bacterial responses to host-defense peptides." *Trends Microbiol.*, 4 (4):127–128.

Groisman, E. A. and Aspedon, A. 1997. "The genetic basis of microbial resistance to antimicrobial peptides." *Methods Mol. Biol.*, 78: 205–215.

Groisman, E. A., Kayser, J. and Soncini, F. C. 1997. "Regulation of polymyxin resistance and adaptation to low-Mg^{2+} environments." *J. Bacteriol.*, 179 (22):7040–7045.

Groisman, E. A. and Ochman, H. 1993. "Cognate gene clusters govern invasion of host epithelial cells by *Salmonella typhimurium* and *Shigella flexneri.*" *EMBO J.*, 12 (10):3779–3787.

Groisman, E. A. and Ochman, H. 1996. "Pathogenicity islands: bacterial evolution in quantum leaps." *Cell*, 87:791–794.

Groisman, E. A. and Ochman, H. 1997. "How *Salmonella* became a pathogen." *Trends Microbiol.*, 5 (9):343–349.

Groisman, E. A., Parra-Lopez, C., Salcedo, M., Lipps, C. J. and Heffron, F. 1992. "Resistance to host antimicrobial peptides is necessary for *Salmonella* virulence." *Proc. Natl. Acad. Sci. USA*, 89 (24):11939–11943.

Groisman, E. A. and Saier, M. H., Jr. 1990. "*Salmonella* virulence: new clues to intramacrophage survival." *Trends Biochem. Sci.*, 15 (1):30–33.

Groisman, E. A., Sturmoski, M. A., Solomon, F. R., Lin, R. and Ochman, H. 1993. "Molecular, functional, and evolutionary analysis of sequences specific to *Salmonella.*" *Proc. Natl. Acad. Sci. USA*, 90 (3):1033–1037.

Gruenheid, S., Pinner, E., Desjardins, M. and Gros, P. 1997. "Natural resistance to infection with intracellular pathogens: the Nramp 1 protein is recruited to the membrane of the phagosome." *J. Exp. Med.*, 185 (4):717–730.

Guilloteau, L. A., Lax, A. J., MacIntyre, S. and Wallis, T. S. 1996a. "The *Salmonella dublin* virulence plasmid does not modulate early T-cell responses in mice." *Infect. Immun.*, 64 (1):222–229.

Guilloteau, L. A., Wallis, T. S., Gautier, A. V., MacIntyre, S., Platt, D. J. and Lax, A. J. 1996b. "The *Salmonella* virulence plasmid enhances *Salmonella*-induced lysis of macrophages and influences inflammatory responses." *Infect. Immun.*, 64 (8): 3385–3393.

Guiney, D. G., Fang, F. C., Krause, M., Libby, S., Buchmeier, N. A. and Fierer, J. 1995a. "Biology and clinical significance of virulence plasmids in *Salmonella* serovars." *Clin. Infect. Dis.*, 21:146–151.

Guiney, D. G., Libby, S., Fang, F. C., Krause, M. and Fierer, J. 1995b. "Growth-phase regulation of plasmid virulence genes in *Salmonella.*" *Trends Microbiol.*, 3 (7):275–279.

Gulig, P. A. 1990. "Virulence plasmids of *Salmonella typhimurium* and other salmonellae." *Microb. Pathog.*, 8 (1): 3–11.

Gulig, P. A., Danbara, H., Guiney, D. G., Lax, A. J., Norel, F. and Rhen, M. 1993. "Molecular analysis of spv virulence genes of the *Salmonella* virulence plasmids." *Mol. Microbiol.*, 7 (6):825–830.

Gulig, P. A. and Doyle, T. J. 1993. "The *Salmonella typhimurium* virulence plasmid increases the growth rate of salmonellae in mice." *Infect. Immun.*, 61 (2):504–511.

Gulig, P. A., Doyle, T. J., Hughes, J. A. and Matsui, H. 1998. "Analysis of host cells associated with the spv-mediated increased intracellular growth rate of *Salmonella typhimurium* in mice." *Infect. Immun.*, 66 (6):2471–2485.

Gunn, J. S., Alpuche-Aranda, C. M., Loomis, W. P., Belden, W. J. and Miller, S. I. 1995. "Characterization of the *Salmonella typhimurium* pagC/pagD chromosomal region." *J. Bacteriol.*, 177 (17):5040–5047.

Gunn, J. S., Belden, W. J. and Miller, S. I. 1998a. "Identification of PhoP-PhoQ activated genes within a duplicated region of the *Salmonella typhimurium* chromosome." *Microb. Pathog.*, 25:77–90.

Gunn, J. S., Hohmann, E. L. and Miller, S. I. 1996. "Transcriptional regulation of *Salmonella* virulence: a PhoQ periplasmic domain mutation results in increased net phosphotransfer to PhoP." *J. Bacteriol.*, 178 (21):6369–6373.

Gunn, J. S., Lim, K. B., Krueger, J., Kim, K., Guo, L., Hackett, M. and Miller, S. I. 1998b. "PmrA-PmrB-regulated genes necessary for 4-aminoarabinose lipid A modification and polymyxin resistance." *Mol. Microbiol.*, 27 (6):1171–1182.

Gunn, J. S. and Miller, S. I. 1996. "PhoP-PhoQ activates transcription of *pmrAB*, encoding a two-component regulatory system involved in *Salmonella typhimurium* antimicrobial peptide resistance." *J. Bacteriol.*, 178 (23):6857–6864.

Guo, L., Lim, K. B., Gunn, J. S., Bainbridge, B., Darveau, R. P., Hackett, M. and Miller, S. I. 1997. "Regulation of lipid A modifications by *Salmonella typhimurium* virulence genes *phoP-PhoQ*." *Science*, 276 (5310):250–253.

Haas, W., Pereira, P. and Tonegawa, S. 1993. "γ/δ T cells." *Annu. Rev. Immunol.*, 11:637–656

Hacker, J., Blum-Oehler, G., Mohldorfer, I. and Tschäpe, H. 1997. "Pathogenicity islands of virulent bacteria: structure, function and impact on microbial evolution." *Mol. Microbiol.*, 23 (6): 1089–1097.

Hamzaoui, N. and Pringault, E. 1998. "Interaction of microorganisms, epithelium, and lymphoid cells of the mucosa-associated lymphoid tissue." *Ann. N. Y. Acad. Sci.*, 859:65–74.

Hardt, W.-D., Chen, L.-M., Schuebel, K. E., Bustelo, X. and Galán, J. E. 1998a. "*S. typhimurium* encodes an activator of Rho GTPases that induces membrane ruffling and nuclear responses in host cells." *Cell*, 93:815–826.

Hardt, W.-D. and Galán, J. E. 1997. "A secreted *Salmonella* protein with homology to an avirulence determinant of plant pathogenic bacteria." *Proc. Natl. Acad. Sci. USA*, 94: 9887–9892.

Hardt, W.-D., Urlaub, H. and Galán, J. E. 1998b. "A substrate of the centisome 63 type III secretion system of *Salmonella typhimurium* is encoded by a cryptic bacteriophage." *Proc. Natl. Acad. Sci. USA*, 95:2574–2579.

Heithoff, D. M., Conner, C. P., Hanna, P. C., Julio, S. M., Hentschel, U. and Mahan, M. J. 1997a. "Bacterial infection as assessed by in vivo gene expression." *Proc. Natl. Acad. Sci. USA*, 94 (3): 934–939.

Heithoff, D. M., Conner, C. P., Hentschel, U., Govantes, F., Hanna, P. C. and Mahan, M. J. 1999. "Coordinate intracellular expression of *Salmonella* genes induced during infection." *J. Bacteriol.*, 181 (3):799–807.

Heithoff, D. M., Conner, C. P. and Mahan, M. J. 1997b. "Dissecting the biology of a pathogen during infection." *Trends Microbiol.*, 5 (12):509–513.

Hengge-Aronis, R. 1996. "Regulation of gene expression during entry into stationary phase." In Escherichia coli *and* Salmonella typhimurium: *Cellular and Molecular Biology*, eds. F. C. Neidhardt, R. Curtis III, J. L. Ingraham, E. C. C. Lim, K. B. Low, B. Magasanik, W. S. Reznikoff, M. Riley, M. Schaechter and H. E. Umbarger. Washington, D.C.: American Society for Microbiology, pp. 1497–1512.

Hensel, M., Nikolaus, T. and Egelseer, C. 1999. "Molecular and functional analysis indicates a mosaic structure of *Salmonella* pathogenicity island 2." *Mol. Microbiol.*, 31 (2):489–498.

Hensel, M., Shea, J. E., Baumler, A. J., Gleeson, C., Blattner, F. and Holden, D. W. 1997a. "Analysis of the boundaries of *Salmonella* pathogenicity island 2 and the corresponding chromosomal region of *Escherichia coli* K-12." *J. Bacteriol.*, 179 (4): 1105–1111.

Hensel, M., Shea, J. E., Gleeson, C., Jones, M. D., Dalton, E. and Holden, D. W. 1995. "Simultaneous identification of bacterial virulence genes by negative selection." *Science*, 269:400–403.

Hensel, M., Shea, J. E., Raupach, B., Monack, D., Falkow, S., Gleeson, C., Kubo, T. and Holden, D. W. 1997b. "Functional analysis of *ssaj* and the *ssaK/U* operon, 13 genes encoding components of the type III secretion apparatus of *Salmonella* pathogenicity island 2." *Mol. Microbiol.*, 24 (1):155–167.

Hensel, M., Shea, J. E., Waterman, S. R. Mundy, R., Nikolaus, T., Banks, G., Vazquez-Torres, A., Gleeson, C., Fang, F. C. and Holden, D. W. 1998. "Genes encoding putative effector proteins of the type III secretion system of *Salmonella* pathogenicity island 2 are required for bacterial virulence and proliferation in macrophages." *Mol. Microbiol.*, 30 (1):163–174.

Hilbert, F., García-del Portillo, F. and Groisman, E. A. 1999. "A periplasmic D-alanyl-D-alanine dipeptidase in the gram negative bacterium *Salmonella enterica*." *J. Bacteriol.*, 181 (7): 2158–2165.

Hobbie, S., Chen, L. M., Davis, R. and Galán, J. E. 1997. "Involvement of the mitogen-activated protein kinase pathways in the nuclear responses and cytokine production induced by *Salmonella typhimurium* in cultured intestinal cells." *J. Immunol.*, 159:5550–5559.

Hong, K. H. and Miller, V. L. 1998. "Identification of a novel *Salmonella* invasion locus homologous to *Shigella* ipgDE." *J. Bacteriol.*, 180 (7):1793–1802.

Hotchin, N. A. and Hall, A. 1996. "Regulation of the actin cytoskeleton, integrins and cell growth by the Rho family of small GTPases." *Cancer Surv.*, 27:311–322.

Hueck, C. J. 1998. "Type III protein secretion systems in bacterial pathogens of animals and plants." *Microbiol. Mol. Biol. Rev.*, 62 (2):379–433.

Humphrey, T. J. and Lanning, D. G. 1988. "The vertical transmission of salmonellas and formic acid treatment of chicken feed. A possible strategy for control." *Epidemiol. Infect.*, 100:43–49.

Humphreys, S., Stevenson, A., Bacon, A., Weinhardt, A. B. and Roberts, M. 1999. "The alternative sigma factor, σ^E is critically important for the virulence of *Salmonella typhimurium*." *Infect. Immun.*, 67 (4):1560–1568.

Jensen, V. B., Harty, J. and Jones, B. D. 1998. "Interactions of the invasive pathogens *Salmonella typhimurium, Listeria monocytogenes,* and *Shigella flexneri* with M cells and murine Peyer's patches," *Infect. Immun.*, 66 (8):3758–3766.

Johnston, C., Pegues, D. A., Hueck, C. J., Lee, A. and Miller, S. I. 1996. "Transcriptional activation of *Salmonella typhimurium* invasion genes by a member of the phosphorylated response-regulator superfamily." *Mol. Microbiol.*, 22 (4):715–727.

Jones, B., Pascopella, L. and Falkow, S. 1995. "Entry of microbes into the host: using M cells to break the mucosal barrier." *Curr. Opin. Immunol.*, 7:474–478.

Jones, B. D. 1997. "Host responses to pathogenic *Salmonella* infection." *Genes Dev.*, 11:679–687.

Jones, B. D. and Falkow, S. 1996. "Salmonellosis: host immune responses and bacterial virulence determinants." *Annu. Rev. Immunol.*, 14:533–561.

Jones, B. D., Paterson, H. F., Hall, A. and Falkow, S. 1993. "*Salmonella typhimurium* induces membrane ruffling by a growth factor- receptor-independent mechanism." *Proc. Natl. Acad. Sci. USA*, 90 (21):10390–10394.

Jones, G. W., Richardson, L. A. and Uhlman, D. 1981. "The invasion of HeLa cells by *Salmonella typhimurium:* reversible and irreversible bacterial attachment and the role of bacterial motility." *J. Gen. Microbiol.*, 127:351–360.

Jones, M. A., Wood, M. W., Mullan, P. B., Watson, P. R., Wallis, and Galyov, E. E. 1998.

"Secreted effector proteins of *Salmonella dublin* act in concert to induce enteritis." *Infect. Immun.,* 66 (12):5799–5804.

Jones, M. A., Wood, M. W., Mullan, P. B., Watson, P. R., Wallis, T. S. and Galyov, E. E. 1988. "Secreted effector proteins of *Salmonella dublin* act in concert to induce enteritis." *Infect. Immun.,* 66 (12):5799–5804.

Jung, H. C., Eckmann, L., Yang, S. K., Panja, A., Fierer, J., Morzycka-Wroblewska, E. and Kagnoff, M. F. 1995. "A distinct array of proinflammatory cytokines is expressed in human colon epithelial cells in response to bacterial invasion." *J. Clin. Invest.,* 95 (1): 55–65.

Kaniga, K., Uralil, J., Bliska, J. B. and Galán, J. E. 1996. "A secreted protein tyrosine phosphatase with modular effector domains in the bacterial pathogen *Salmonella typhimurium.*" *Mol. Microbiol.* 21 (3):633–641.

Kapperud, G., Gustavsen, S., Hellesnes, I., Hansen, A. H., Lassen, J., Him, J., Jahkola, M., Montenegro, M. A. and Helmuth, R. 1990. "Outbreak of *Salmonella typhimurium* infection traced to contaminated chocolate and caused by a strain lacking the 60-megadalton virulence plasmid." *J. Clin. Microbiol.,* 28:2597–2601.

Kasahara, M. A., Nakata, A. and Shinagawa, H. 1991. "Molecular analysis of the *Salmonella typhimurium phoN* gene, which encodes nonspecific acid phosphatase." *J. Bacteriol.,* 173: 6760–6765.

Kelly, S. M., Bosecker, B. A. and Curtiss, R. I. 1992. "Characterization and protective properties of attenuated mutants of *Salmonella choleraesuis.*" *Infect. Immun.,* 60 (11): 4881–4890.

Kihistrom, E. and Edebo, L. 1976. "Association of viable and inactivated *Salmonella typhimurium* 395 MS and MR 10 with HeLa cells." *Infect. Immun.,* 14 (4):851–857.

Kihistrom, E. and Nilsson, L. 1977. "Endocytosis of *Salmonella typhimurium* 395 MS and MR10 by HeLa cells." *Acta Pathol. Microbiol. Scand.,* 85B (5):322–328.

Kowarz, L., Coynault, C., Robbe-Saule, V. and Norel, F. 1994. "The *Salmonella typhimurium katF (rpoS)* gene: cloning, nucleotide sequence, and regulation of *spvR* and *spvABCD* virulence plasmid genes." *J. Bacteriol.,* 176 (22):6852–6860.

Kricka, L. J. 1998. "Prospects for chemiluminescent and bioluminescent immunoassay and nucleic acid assays in food testing and the pharmaceutical industry." *J. Biolumin. Chemilumin.,* 13 (4): 189–193.

Kubori, T., Matsushima, Y., Nakamura, D., Uralil, J., Lara-Tejero, M., Sukhan, A., Galán, J. E. and Aizawa, S.-I. 1998. "Supramolecular structure of the *Salmonella typhimurium* type III protein secretion system." *Science,* 280:602–605.

Lan, R. T. and Reeves, P. R. 1996. "Gene transfer is a major factor in bacterial evolution." *Mol. Biol. Evol.,* 13:47–55.

Lange, R. and Hengge-Aronis, R. 1991. "Identification of a central regulator of stationary phase gene expression in *Escherichia coli.*" *Mol. Microbiol.,* 5:49–59.

Le Minor, L. and Popoff, M. Y. 1987. "Designation of *Salmonella enterica* sp. nov., nom. rev., as the type and only species of the genus *Salmonella.*" *Int. J. Syst. Bacteriol.,* 37: 465–468.

Lee, I. S., Slonczewski, J. L. and Foster, J. W. 1994. "A low-pH-inducible, stationary-phase acid tolerance response in *Salmonella typhimurium.*" *J. Bacteriol.,* 176 (5): 1422–1426.

Lee, S., Lin, J., Hall, H. K., Bearson, B. and Foster, J. W. 1995. "The stationary phase sigma factor σ^s (RpoS) is required for a sustained acid tolerance response in virulent *Salmonella typhimurium.*" *Mol. Microbiol.,* 17 (1):155–167.

Leung, K. Y. and Finlay, B. B. 1991. "Intracellular replication is essential for the virulence of *Salmonella typhimurium.*" *Proc. Natl. Acad. Sci. USA,* 88 (24):11470–11474.

Li, J., Ochman, H., Groisman, E. A., Boyd, E. F., Solomon, F., Nelson, K. and Selander, R. K.

1995. "Relationship between evolutionary rate and cellular location among the Inv/Spa invasion proteins of *Salmonella enterica.*" *Proc. Natl. Acad. Sci. USA,* 92 (16): 7252–7258.

Libby, S. J., Adams, L. G., Ficht, T. A., Allen, C., Whitford, H. A., Buchmeier, N. A., Bossie, S. and Guiney, D. G. 1997. "The spv genes on the *Salmonella dublin* virulence plasmid are required for severe enteritis and systemic infection in the natural host." *Infect. Immun.,* 65 (5):1786–1792.

Lindgren, S. W., Stojiljkovic, I. and Heffron, F. 1996. "Macrophage killing is an essential virulence mechanism of *Salmonella typhimurium.*" *Proc. Natl. Acad. Sci. USA,* 93 (9): 4197–4201.

Liu, S.-L. and Sanderson, K. E. 1996. "Highly plastic chromosomal organization in *Salmonella typhi.*" *Proc. Natl. Acad. Sci. USA,* 93:10303–10308.

Liu, S. L., Ezaki, T., Miura, H., Matsul, K. and Yabuuchi, E. 1988. "Intact motility as a *Salmonella typhi* invasion-related factor." *Infect. Immun.,* 56 (8):1967–1973.

Liu, S. L. and Sanderson, K. E. 1995. "Rearrangements in the genome of the bacterium *Salmonella typhi.*" *Proc. Natl. Acad. Sci. USA,* 92:1018–1022.

Mahan, M. J., Slauch, J. M. and Mekalanos, J. J. 1993. "Selection of bacterial virulence genes that are specifically induced in host tissues." *Science,* 259:686–688.

Mahan, M. J., Tobias, J. W., Slauch, J. M., Hanna, P. C., Collier, R. J. and Mekalanos, J. J. 1995. "Antibiotic-based selection for bacterial genes that are specifically induced during infection of a host." *Proc. Natl. Acad. Sci. USA,* 92 (3):669–673.

Mäki-Ikola, O. and Granfors, K. 1993. "The bacteriology of reactive arthritis." *Rev. Med. Microbiol.,* 4:144–150.

Manzano, M., Cocolin, L., Astori, G., Pipan, C., Botta, G. A., Cantoni, C. and Comi, G. 1998. "Development of a PCR microplate-capture hybridization method for simple, fast and sensitive detection of *Salmonella* serovars in food." *Mol. Cell Probes,* 12 (4):227–234.

Martínez-Moya, M. and García-del Portillo, F. 1999. Unpublished data.

Martínez-Moya, M., de Pedro, M. A., Schwarz, H. and García-del Portillo, F. 1998. "Inhibition of *Salmonella* intracellular proliferation by non-phagocytic cells." *Res. Microbiol.,* 149: 309–318.

Mastroeni, P., Villarreal-Ramos, B. and Hormaeche, C. E. 1993. "Adoptive transfer of immunity to oral challenge with virulent Salmonellae in innately susceptible BALB/c mice requires both immune serum and T cells." *Infect. Immun.,* 61:3981–3984.

McCormick, B. A., Colgan, S. P., Delp-Archer, C., Miller, S. I. and Madara, J. L. 1993. "*Salmonella typhimurium* attachment to human intestinal epithelial monolayers: transcellular signalling to subepithelial neutrophils." *J. Cell Biol.,* 123 (4):895–907.

McCormick, B. A., Hofman, P. M., Kim, J., Cames, D. K., Miller, S. I. and Madara, J. L. 1995. "Surface attachment of *Salmonella typhimurium* to intestinal epithelia imprints the subepithelial matrix with gradients chemotactic for neutrophils." *J. Cell Biol.,* 131 (6):1599–1608.

McCormick, B. A., Parkos, C. A., Colgan, S. P., Carnes, D. K. and Madara, J. L. 1998. "Apical secretion of a pathogen-elicited epithelial chemoattractant activity in response to surface colonization of intestinal epithelia by *Salmonella typhimurium.*" *J. Immunol.,* 160:455–466.

Miller, S. I. 1991. "PhoP/PhoQ: macrophage-specific modulators of *Salmonella* virulence?" *Mol. Microbiol.,* 5 (9): 2073–2078.

Miller, S. I., Hohmann, E. and Pegues, D. 1995. "*Salmonella* (including *Salmonella typhi*)." In *Principles and Practice of Infectious Diseases,* eds. G. Mandell, J. Bennett and R. Dolin. New York, N.Y.: Churchill Livingstone, pp. 2013–2033.

Miller, S. I., Kukral, A. M. and Mekalanos, J. J. 1989. "A two-component regulatory system *(phoP phoQ)* controls *Salmonella typhimurium* virulence." *Proc. Natl. Acad. Sci. USA,* 86 (13):5054–5058.

Miller, S. I. and Mekalanos, J. J. 1990. "Constitutive expression of the PhoP regulon attenuates *Salmonella* virulence and survival within macrophages." *J. Bacteriol.*, 172 (5):2485–2490.

Mills, D. M., Bajaj, V. and Lee, C. A. 1995. "A 40 kb chromosomal fragment encoding *Salmonella typhimurium* invasion genes is absent from the corresponding region of the *Escherichia coli* K-12 chromosome." *Mol. Microbiol.*, 15 (4):749–759.

Mills, S. D. and Finlay, B. B. 1998. "Isolation and characterization of *Salmonella typhimurium* and *Yersinia pseudotuberculosis*-containing phagosomes from infected mouse macrophages: *Y. pseudotuberculosis* traffics to terminal lysosomes where they are degraded." *Eur. J. Cell Biol.*, 77:35–47.

Mills, S. D., Ruschkowski, S., Stein, M. A. and Finlay, B. B. 1998. "Trafficking of porin-deficient *Salmonella typhimurium* mutants inside HeLa cells: *ompR* and *envZ* mutants are defective for the formation of *Salmonella*-induced filaments." *Infect. Immun.*, 66 (4):1806–1811.

Monack, D. M., Raupach, B., Hromockyj, A. E. and Falkow, S. 1996. "*Salmonella typhimurium* invasion induces apoptosis in infected macrophages." *Proc. Natl. Acad. Sci. USA*, 93 (18): 9833–9838.

Mroczenski, M. J., Di, F. J. L. and Cabello, F. C. 1989. "Invasion and lysis of HeLa cell monolayers by *Salmonella typhi:* the role of lipopolysaccharide." *Microb. Pathog.*, 6:143–152.

Mulvey, M. R. and Loewen, P. C. 1993. "Nucleotide sequence of *katF* of *Escherichia coli* suggest KatF protein is a novel σ transcription factor." *Nucleic Acid Res.*, 17: 9979–9991.

Nauciel, C. and Espinasse-Maes, F. 1992. "Role of gamma interferon and tumor necrosis factor alpha in resistance to *Salmonella typhimurium* infection." *Infect. Immun.*, 60:450–454.

Neutra, M. 1999. "M cells in antigen sampling in mucosal tissues." *Curr. Top. Microbiol. Immunol.*, 236:17–32.

Nickerson, C. A. and Curtiss, R. R. 1997. "Role of sigma factor RpoS in initial stages of *Salmonella typhimurium* infection." *Infect. Immun.*, 65 (5):1814–1823.

Nishimura, H., Hiromatsu, K., Kobayashi, N., Grabstein, K. H., Paxton, R., Sugamura, K., Bluestone, J. A. and Yoshikai, Y. 1996. "IL-15 is a novel growth factor for murine gamma delta T cells induced by *Salmonella* infection." *J. Immunol.*, 156 (2): 663–669.

Nnalue, N. A., Shnyra, A., Hultenby, K. and Lindberg, A. A. 1992. "*Salmonella choleraesuis* and *Salmonella typhimurium* associated with liver cells after intravenous inoculation of rats are mainly localized mainly in Kupffer cells and multiply intracellularly." *Infect. Immun.*, 60 (7):2758–2768.

Norel, F., Robbe-Saule, V., Popoff, M. Y. and Coynault, C. 1992. "The putative sigma factor KatF (RpoS) is required the transcription of the *Salmonella typhimurium* virulence gene SpvB in *Escherichia coli.*" *FEMS Microbiol. Lett.*, 78:271–276.

Norris, F. A., Wilson, M. P., Wallis, T. S., Galyov, E. E. and Majerus, P. W. 1998a. "SopB, a protein required for virulence of *Salmonella dublin,* is an inositol phosphate phosphatase." *Proc. Natl. Acad. Sci. USA*, 95:14057–14059.

Norris, T. L., Kingsley, R. A. and Bäumler, A. J. 1998b. "Expression and transcriptional control of the *Salmonella typhimurium lpf* fimbrial operon by phase variation." *Mol. Microbiol.*, 29 (1):311–320.

Ochman, H. and Groisman, E. A. 1994. "The origin and evolution of species differences in *Escherichia coli* and *Salmonella typhimurium.*" *Exs.*, 69:479–493.

Ochman, H. and Groisman, E. A. 1996. "Distribution of pathogenicity islands in *Salmonella* spp." *Infect. Immun.*, 64 (12): 5410–5412.

Ochman, H., Soncini, F. C., Solomon, F. and Groisman, E. A. 1996. "Identification of a pathogenicity island required for *Salmonella* survival in host cells." *Proc. Natl. Acad. Sci. USA*, 93 (15):7800–7804.

Oh, Y. K., Alpuche-Aranda, C., Berthiaume, E., Jinks, T., Miller, S. I. and Swanson, J. A. 1996. "Rapid and complete fusion of macrophage lysosomes with phagosomes containing *Salmonella typhimurium.*" *Infect. Immun.*, 64 (9):3877–3883.

Ottenhoff, T. H. M., Kumararatne, D. and Casanova, J.-L. 1998. "Novel human immunodeficiencies reveal the essential role of type-I cytokines in immunity to intracellular bacteria." *Trends Immunol.*, 19 (11): 491–494.

Pace, J., Hayman, M. J. and Galán, J. E. 1993. "Signal transduction and invasion of epithelial cells by *S. typhimurium.*" *Cell,* 72 (4):505–514.

Pang, T., Bhutta, Z. A., Finlay, B. B. and Altwegg, M. 1995. "Typhoid fever and other salmonellosis: a continuing challenge." *Trends Microbiol.*, 3 (7):253–255.

Parra-Lopez, C., Baer, M. T. and Groisman, E. A. 1993. "Molecular genetic analysis of a locus required for resistance to antimicrobial peptides in *Salmonella typhimurium.*" *EMBO J.*, 12 (11): 4053–4062.

Parra-Lopez, C., Lin, R., Aspedon, A. and Groisman, E. A. 1994. "A *Salmonella* protein that is required for resistance to antimicrobial peptides and transport of potassium." *EMBO J.*, 13 (17):3964–3972.

Pegues, D. A., Hantman, M. J., Behlau, I. and Miller, S. I. 1995. "PhoP/PhoQ transcriptional repression of *Salmonella typhimurium* invasion genes: evidence for a role in protein secretion." *Mol. Microbiol.*, 17 (1):169–181.

Phalipon, A. and Sansonetti, P. J. 1999. "Microbial-host interactions at mucosal sites. Host response to pathogenic bacteria at mucosal sites." *Curr. Top. Microbiol. Immunol.*, 236:163–189.

Rathman, M., Barker, L. P. and Falkow, S. 1997. "The unique trafficking pattern of *Salmonella typhimurium*-containing phagosomes in murine macrophages is independent of the mechanism of bacterial entry." *Infect. Immun.*, 65 (4):1475–1485.

Rathman, M., Sjaastad, M. D. and Falkow, S. 1996. "Acidification of phagosomes containing *Salmonella typhimurium* in murine macrophages." *Infect. Immun.*, 64 (7):2765–2773.

Raupach, B., Mecsas, J., Heczko, U., Falkow, S. and Finlay, B. B. 1999. "Bacterial epithelial cells cross talk." *Curr. Top. Microbiol. Immunol.*, 236:137–161.

Reed, K. A., Clark, M. A., Booth, T. A., Hueck, C. J., Miller, S. I., Hirst, B. H., and Jepson, M. A. 1998. "Cell-contact-stimulated formation of filamentous appendages by *Salmonella typhimurium* does not depend on the type III secretion system encoded by *Salmonella* pathogenicity island 1". *Infect. Immun.*, 66 (5):2007–2017.

Reeves, M. W., Evins, G. M., Heiba, A. A., Plikaytis, D. and Farmer III, J. J. 1989. "Clonal nature of *Salmonella typhi* and its genetic relatedness to other salmonellae as shown by multilocus enzyme electrophoresis, and proposal of *Salmonella bongori* comb. nov." *J. Clin. Microbiol.*, 27:311–320.

Rexach, L., Dilasser, F. and Fach, P. 1994. "Polymerase chain reaction for *Salmonella* virulence-associated plasmid genes detection: a new tool in *Salmonella* epidemiology." *Epidemiol. Infect.*, 112:33–43.

Rhen, M., Riikonen, P. and Taira, S. 1993. "Transcriptional regulation of *Salmonella enterica* virulence plasmid genes in cultured macrophages." *Mol. Microbiol.*, 10 (1):45–56.

Richter-Dahlfors, A., Buchan, A. M. J. and Finlay, B. B. 1997. "Murine salmonellosis studied by confocal microscopy: *Salmonella typhimurium* resides intracellularly inside macrophages and exerts a cytotoxic effect on phagocytes in vivo." *J. Exp. Med.*, 186 (4): 569–580.

Riesenberg-Wilmes, M. R., Bearson, B., Foster, J. W. and Curtis, R. I. 1996. "Role of the acid tolerance response in virulence of *Salmonella typhimurium.*" *Infect. Immun.*, 64 (4):1085–1092.

Romanova, Y. M., Kirillov, M. Y., Terekhov, A. A. and Gintsburg, A. L. 1996. "Identification of genes controlling the transition of *Salmonella typhimurium* bacteria to a nonculturable state." *Genetika,* 32 (9):1184–1190.

Rowe, B., Ward, L. R. and Threlfall, E. J. 1997. "Multidrug-resistant *Salmonella typhi:* a worldwide epidemic." *Clin. Infect. Dis.,* 24:106–109.

Rubin, R. H. and Weinstein, L. 1977. *Salmonellosis: Microbiologic, Pathogenic, and Clinical Features.* New York: Stratton Intercontinental Medical Book Corp.

Ruschkowski, S., Rosenshine, I. and Finlay, B. B. 1992. "*Salmonella typhimurium* induces an inositol phosphate flux in infected epithelial cells." *FEMS Microbiol. Lett.,* 74 (2–3): 121–126.

Rossmann, H., Shams, H., Poblete, F., Fu, Y., Galán, J. E. and Donis, R. O. 1998. "Delivery of epitopes by the *Salmonella* type III secretion system for vaccine development." *Science,* 281: 565–568.

Saarinen, M., Pelliniemi, L. J. and Granfors, K. 1996. "Survival and degradation of *Salmonella enterica* serotype Enteritidis in intestinal epithelial cells in vitro." *J. Med. Microbiol.,* 45 (6):463–471.

Searle, S., Bright, N. A., Roach, T. I. A., Atkinson, P. G. P., Barton, H., Meloen, R. H. and Blackwell, J. M. 1998. "Localisation of Nramp1 in macrophages: modulation with activation and infection." *J. Cell Sci.,* 111:2855–2866.

Shaw, S. J., Blais, B. W. and Nundy, D.C. 1998. "Performance of the Dynabeads anti-*Salmonella* system in the detection of *Salmonella* species in foods, animal feeds, and environmental samples." *J. Food Prot.,* 61 (11):1507–1510.

Shea, J. E., Beuzón, C. R., Gleeson, C., Mundy, R. and Holden, D. W. 1999. "Influence of the *Salmonella typhimurium* pathogenicity island 2 type III secretion system on bacterial growth in the mouse." *Infect. Immun.,* 67 (1):213–219.

Shea, J. E., Hensel, M., Gleeson, C. and Holden, D. W. 1996. "Identification of a virulence locus encoding a second type III secretion system in *Salmonella typhimurium.*" *Proc. Natl. Acad. Sci. USA,* 93 (6):2593–2597.

Siebers, A. and Finlay, B. B. 1996. "M cells and the pathogenesis of mucosal and systemic infections." *Trends Microbiol.,* 4 (1): 22–29.

Slauch, J., Taylor, R. and Maloy, S. 1997. "Survival in a cruel world: how *Vibrio cholerae* and *Salmonella* respond to an unwilling host." *Genes Dev.,* 11 (14):1761–1774.

Smith, R. L. and Maguire, M. E. 1998. "Microbial magnesium transport: unusual transporters searching for identity." *Mol. Microbiol.,* 28 (2):217–226.

Soncini, F. C., García Vescovi, E., Solomon, F. and Groisman, E. A. 1996. "Molecular basis of the magnesium deprivation response in *Salmonella typhimurium:* identification of PhoP-regulated genes." *J. Bacteriol.,* 178 (17):5092–5099.

Soncini, F. C. and Groisman, E. A. 1996. "Two-component regulatory systems can interact to process multiple environmental signals." *J. Bacteriol.,* 178 (23):6796–6801.

Soncini, F. C., Vescovi, E. G. and Groisman, E. A. 1995. "Transcriptional autoregulation of the *Salmonella typhimurium* phoPQ operon." *J. Bacteriol.,* 177 (15):4364–4371.

Spector, M. P. 1998. "The starvation-stress response (SSR) of *Salmonella.*" *Avd. Microb. Physiol.,* 40:233–279.

Stein, M. A., Leung, K. Y., Zwick, M., García-del Portillo, F. and Finlay, B. B. 1996. "Identification of a *Salmonella* virulence gene required for formation of filamentous structures containing lysosomal membrane glycoproteins within epithelial cells," *Mol. Microbiol.,* 20 (1):151–164.

Suárez, M. and Rüssmann, H. 1998. "Molecular mechanisms of *Salmonella* invasion: the type III secretion system of the pathogenicity island 1." *Internatl. Microbiol.,* 1:197–204.

Sukupolvi, S., Edelstein, A., Rhen, M., Normark, S. J. and Pfeifer, J. D. 1997. "Development of a murine model of chronic *Salmonella* infection." *Infect. Immun.,* 65 (2):838–842.

Swords, W. E., Cannon, B. M. and Benjamin, W. H., Jr. 1997a. "Avirulence of LT2 strains of *Salmonella typhimurium* results from a defective *rpoS* gene." *Infect. Immun.*, 65 (6): 2451–2453.

Swords, W. E., Giddings, A. and Benjamin, W. H., Jr. 1997b. "Bacterial phenotypes mediated by *mviA* and their relationship to the mouse virulence of *Salmonella typhimurium.*" *Microb. Pathog.*, 22 (6):353–362.

Takeuchi, A. 1967. "Electron microscope studies of experimental *Salmonella* infection. I. Penetration into the intestinal epithelium by *Salmonella typhimurium.*" *Am. J. Pathol.* 50:109–136.

Tanaka, K., Takayanagi, N., Fujita, N., Ishihama, A. and Takahashi, H. 1993. "Heterogeneity of the principal *s* factor in *Escherichia coli:* the *rpoS* gene product, σ^{38}, is second *s* factor of RNA polymerase in stationary-phase *Escherichia coli.*" *Proc. Natl. Acad. Sci. USA,* 90: 3511–3515.

Torreblanca, J. and Casadesús, J. 1996. "DNA adenine methylase mutants of *Salmonella typhimurium* and a novel Dam-regulated locus." *Genetics,* 144:15–26.

Torreblanca, J., Marqués, S. and Casadesús, J. 1999. "Synthesis of FinP RNA by plasmids F and pSTL is regulated by DNA adenine methylation." *Genetics* (in press).

Tsolis, R. M., Baumler, A. J. and Heffron, F. 1995. "Role of *Salmonella typhimurium* M_n-superoxide dismutase (SodA) in protection against early killing by J774 macrophages." *Infect. Immun.,* 63 (5):1739–1744.

Turpin, P. E., Maycroft, K. A., Rowlands, C. L. and Wellington, E. M. 1993. "Viable but non-culturable salmonellas in soil." *J. Appl. Bacteriol.,* 74 (4):421–427.

Valdivia, R. H. and Falkow, S. 1996. "Bacterial genetics by flow cytometry: rapid isolation of *Salmonella typhimurium* acid-inducible promoters by differential fluorescence induction." *Mol. Microbiol.,* 22 (2):367–378.

Valdivia, R. H. and Falkow, S. 1997. "Fluorescence-based isolation of bacterial genes expressed within host cells." *Science,* 277:2007–2011.

Valdivia, R. H., Hromockyj, A. E., Monack, D., Ramakrishnan, L. and Falkow, S. 1996. "Applications for green fluorescent protein (GFP) in the study of host-pathogen interactions." *Gene,* 173:47–52.

van der Velden, A. W. M., Bäumler, A. J., Tsolis, R. M. and Heffron, F. 1998. "Multiple fimbrial adhesins are required for full virulence of *Salmonella typhimurium* in mice." *Infect. Immun.,* 66 (6):2803–2808.

Van Gijsegem, F., Genin, S. and Boucher, C. 1993. "Conservation of secretion pathways for pathogenicity determinants of plant and animal bacteria." *Trends Microbiol.,* 1:157–162.

van Velkinburgh, J. C. and Gunn, J. S. 1999. "PhoP-PhoQ-regulated loci are required for enhanced bile resistance in *Salmonella* spp." *Infect. Immun.,* 67 (4):1614–1622.

Varman, A. H. and Evans, M. G. 1991. "*Salmonella.*" In *Foodborne Pathogens:* an *Illustrated Text,* eds. A. H. Varman and M. G. Evans. Aylesbury, England: Wolfe Publishing Ltd., pp. 51–85.

Vassiloyanakopoulos, A. P., Okamoto, S. and Fierer, J. 1998. "The crucial role of polymorphonuclear leukocytes in resistance to *Salmonella dublin* infections in genetically susceptible and resistant mice." *Proc. Natl. Acad. Sci. USA,* 95 (13): 7676–7681.

Wallis, T. S., Paulin, S. M., Plested, J. S., Watson, P. R. and Jones, P. W. 1995. "The *Salmonella dublin* virulence plasmid mediates systemic but not enteric phases of salmonellosis in cattle." *Infect. Immun.,* 63 (7):2755–2761.

Watson, P. R., Galyov, E. E., Paulin, S. M., Jones, P. W. and Wallis, T. S. 1998. "Mutation of *invH,* but not *stn,* reduces *Salmonella*-induced enteritis in cattle." *Infect. Immun.,* 66 (4):1432–1438.

Watson, P. R., Paulin, S. M., Bland, A. P., Jones, P. W. and Wallis, T. S. 1995. "Characterization of intestinal invasion by *Salmonella typhimurium* and *Salmonella dublin* and effect of a mutation in the *invH* gene.'' *Infect. Immun.*, 63 (7): 2743–2754.

Weinstein, D. L., O'Neili, B. L., and Metcalf, E. S. 1998. ''Differential early interactions between *Salmonella enterica* serovar *typhi* and two other pathogenic *Salmonella* serovars with intestinal epithelial cells.'' *Infect. Immun.*, 66 (5):2310–2318.

WHO, World Health Organization. 1997a. ''Foodborne diseases-possibly 350 times more frequent than reported.'' http://www.who.int/dsa/justpub/food.htm.

WHO, World Health Organization. 1997b. ''Multidrug resistant *Salmonella typhimurium,*'' http://www.who.int/inf-fs/en/fact139.html.

WHO, World Health Organization. 1998. ''Typhoid fever.'' http://www.who.int/gpv-dvacc/diseases/typhoid_fever.htm.

Wick, M. J., Harding, C. V., Twesten, N. J., Normark, S. J. and Pfeifer, J. D. 1995. ''The *phoP* locus influences processing and presentation of *Salmonella typhimurium*-antigens by activated macrophages.'' *Mol. Microbiol.*, 16 (3):465–476.

Wilmes-Riesenberg, M. R., Foster, J. W. and Curtiss, R. r. 1997. ''An altered *rpoS* allele contributes to the avirulence of *Salmonella typhimurium* LT2.'' *Infect. Immun.*, 65 (1): 203–210.

Wilson, J. A. and Gulig, P. A. 1998. ''Regulation of the *spvR* gene of the *Salmonella typhimurium* virulence plasmid during exponential-phase growth in intracellular salts medium and at stationary phase in L broth.'' *Microbiology*, 144 (7): 1823–1833.

Wilson, M., Seymour, R., Henderson, B. 1998. ''Bacterial perturbation of cytokine networks.'' *Infect. Immun.*, 66 (6):2401–2409.

Wong, K.-K., McClelland, M., Stillwell, L. C., Sisk, E. C., Thurston, S. J. and Saffer, J. D. 1998. ''Identification and sequence analysis of a 27-kilobase chromosomal fragment containing a *Salmonella* pathogenicity island located at 92 minutes on the chromosome map of *Salmonella enterica* serovar *typhimurium* LTZ.'' *Infect. Immun.*, 66 (7):3365–3371.

Wood, M. W., Jones, M. A., Watson, P. R., Hedges, S., Wallis, T. S. and Galyov, E. E. 1998. ''Identification of a pathogenicity island required for *Salmonella* enteropathogenicity.'' *Mol. Microbiol.*, 29 (3):883–891.

Wood, M. W., Rosqvist, R., Mullan, P. B., Edwards, M. H. and Galyov, E. E. 1996. ''SopE, a secreted protein of *Salmonella dublin*, is translocated into the target eukaryotic cell via a sip-dependent mechanism and promotes bacterial entry.'' *Mol. Microbiol.*, 22 (2): 327–338.

Wuorela, M. and Granfors, K. 1998. ''Infectious agents as triggers of reactive arthritis.'' *Am. J. Med. Sci.*, 316 (4):264–270.

Yokoyama, H., Ikedo, M., Kohbata, S., Ezaki, T. and Yabuuchi, E. 1987. ''An ultrastructural study of HeLa cell invasion with *Salmonella typhi* GIFU 10007.'' *Microbiol. Immunol.*, 31 (1):1–11.

Zhang, X.-L., Morris, C. and Hackett, J. 1997. ''Molecular cloning, nucleotide sequence, and function of a site-specific recombinase encoded in the major 'pathogencity island' of *Salmonella typhi*.'' *Gene*, 202:139–146.

Zhuou, D., Hardt, W.-D. and Galán, J. E. 1999. ''*Salmonella typhimurium* encodes a putative iron transport system within centisome 63 pathogenicity island.'' *Infect. Immun.*, 67 (4):1974–1981.

Zierler, M. K. and Galán, J. E. 1995. ''Contact with cultured epithelial cells stimulates secretion of *Salmonella typhimurium* invasion protein InvJ.'' *Infect. Immun.*, 63 (10):4024–4028.

CHAPTER 2

Shigella Infections: Epidemiology, Pathogenesis and Host Immune Response

CHRISTINE WENNERÅS
PHILIPPE SANSONETTI

BACTERIAL diarrheal pathogens may be divided into two groups, based on their pathogenic mechanisms: (1) The noninvasive group exemplified by on the one hand *Vibrio cholerae* and enterotoxigenic *Escherichia coli* (ETEC), which cause disease by adhering to the small intestinal mucosa and secreting potent enterotoxins, and enteropathogenic *E. coli* (EPEC), which colonize and partially destroy the intestinal epithelium without invading it. And (2) The invasive group, typified by *Salmonella* and *Shigella,* has invasion of the intestinal mucosa as a central feature, which in the case of *Salmonella* may proceed to systemic dissemination (typhoid fever), which is rarely seen during *Shigella* infections.

Shigella infections are localized in the colon, where they may cause dysentery, a disease characterized by fever, severe abdominal cramps and diarrhea containing blood and mucus. Shigellosis is a disease of poverty, primarily affecting children in developing countries. However, epidemic outbreaks do occur in developed countries as well. *Shigella* has developed a sophisticated process of invasion, by which it induces its own uptake by colonic epithelial cells, makes use of the host cell's cytoskeletal proteins to travel through the cytoplasm, and manages to pass from cell to cell without exiting into the external environment. The infection is as a rule restricted to the outermost layer of the intestinal wall, where it elicits a strong inflammatory reaction, which most probably clears the infection.

Hippocrates coined the term *dysentery,* and further noted the seasonal pattern of dysenteric infections. In Bangladesh, the incidence of *Shigella* infections has two peaks: the winter season, characterized by low temperatures and little rainfall, and the summer season, which is hot and dry (Stoll et al., 1982).

Although we now know dysentery may be caused by *Shigella* or *Entamoeba histolytica,* it seems as if Hippocrates in fact described bacillary dysentery, as he never mentioned liver complications such as jaundice, a feature of amebic, but not of bacillary dysentery. *Shigella* is named after the Japanese microbiologist Shiga, who isolated one of the four species of *Shigella, Shigella dysenteriae,* in 1898.

1. BACTERIOLOGY

Shigella are gram-negative, nonsporulating, nonencapsulated, straight rods that belong to the family *Enterobacteriaceae.* The genus *Shigella* comprises four different species, a division based on the capacity of the various strains to ferment different sugars and on the type of LPS O-chains they express on their surface (Table 2.1).

Shigella may be regarded as a variant of *Escherichia coli* since the relatedness of *Shigella* and *E. coli* is of the same magnitude as that between the various species of *Shigella,* i.e., more than 80% of their nucleotide sequences are homologous (Brenner et al., 1973). It is not surprising, then, that *Shigella* can mate with *E. coli* in vitro (Luria and Burrous, 1957). Actually, if the taxonomic division were to be made today, *Shigella* and *E. coli* would probably constitute a single genus, or perhaps even a single species with five subgroups (Brenner et al., 1984).

Shigella are facultatively anaerobic and can thus utilize both respiratory and fermentative metabolic pathways. However, compared to other *E. coli,* they are less metabolically active. With a few exceptions, *Shigella* ferment sugars without gas production. Traditionally, *Shigella* have been defined as non-motile and devoid of flagella. However, the genes coding for flagellin, the building block of flagella, have been identified in *Shigella* (Tominaga et al., 1994). Furthermore, electronmicroscopic studies have revealed that all four species of *Shigella,* including fresh clinical isolates, may under certain conditions express one to three polar flagella (Girón, 1995). The lack of motility of *Shigella* is only evident in the extracellular environment: once

TABLE 2.1. *Shigella* Taxonomy.

Species	Subgroup	Serotypes
Shigella dysenteriae	A	1, 2, 3, 4, 5, 6, 7, 8, 9, 10, 11, 12, 13
Shigella flexneri	B	1a, 1b, 2a, 2b, 3a, 3b, 4a, 4b, 5a, 5b, 6, X, Y
Shigella boydii	C	1, 2, 3, 4, 5, 6, 7, 8, 9, 10, 11, 12, 13, 14, 15, 16, 17, 18
Shigella sonnei	D	Single serotype

inside an eukaryotic cell, they are highly motile, employing a non-flagellar mechanism of locomotion.

2. LIPOPOLYSACCHARIDE

LPS, lipopolysaccharide, is one of the main constituents of the bacterial outer membrane and is composed of three covalently linked components: lipid A, core and O-specific side chain. The lipid A part mediates the toxic effects of the molecule and is identical to that of *E. coli.* The core region consists of an oligosaccharide, which is similar to, although not identical with that of *Salmonella* and *E. coli.* The O-side chain is composed of repeating sugar units, and constitutes the basis for the various *Shigella* serotypes (Table 2.1). The tetrasaccharide 3β-*N*-acetyl glucosamine-α1→2-rhamnose-α1→2-rhamnose-α1→3-rhamnose is the repeating unit of the *S. flexneri* serotype Y, and slight alterations of this basic motif yields all the other *S. flexneri* serotypes, with the exception of serotype 6 (Simmons and Romanowska, 1987). Hence, there is extensive cross-reactivity between these serotypes. The relatedness of *Shigella* with *E. coli* is also apparent at the O-chain level. Thus, the O-side chains of several *Shigella* serotypes cross-react immunologically with certain O-chains of *E. coli.* In fact, *Shigella* and *E. coli* may share identical O-chains. *S. sonnei,* however, shares no O-chain similarity with *E. coli* (Brahmbhatt et al., 1992).

In *S. flexneri,* all of the genes involved in the synthesis of LPS are chromosomal, with the exception of a gene encoded on a 3-kb plasmid, distinct from the large virulence plasmid. Some of the chromosomal genes that determine the O-chain specificity of certain *S. flexneri* strains are derived from lysogenic bacteriophages (Hong and Payne, 1997). The LPS biosynthesis genes of *S. dysenteriae* are located on the chromosome and on a 9-kb plasmid. In contrast, all of the genes coding for the O-chain of *S. sonnei* are located on the large virulence plasmid. This plasmid is easily lost upon culture, yielding rough, avirulent variants named *S. sonnei* Form II (Table 2.1) (Brahmbhatt, et al., 1992).

3. TRANSMISSION

The only natural hosts of *Shigella* are humans and monkeys. One study revealed that 18% of monkeys maintained in captivity were carriers of *S. flexneri* (Labrec et al., 1964). Transmission of *Shigella* from pet monkeys to children has been reported frequently (Robinson et al., 1965). Although monkeys and humans appear to be susceptible to the same *Shigella* species, monkeys are more often infected by *S. flexneri* serotype Y than humans.

Transmission of *Shigella* occurs through the fecal-oral route. The most common means of spread is probably person-to-person contact. The bacteria can survive on skin. *Shigella* is also frequently transmitted by the ingestion of contaminated foods and water. Examples of types of food that have been described in the literature as spreading *Shigella* include lettuce, onions, coconutmilk dessert, hamburgers, raw oysters, meat soup and pasta dishes. These microorganisms can survive for more than six months in water at room temperature. Flies may transmit bacteria from human excrement to foods (Levine and Levine, 1991). Sexual practices such as oral-anal and oral-genital sex may also be a means of transmission. This may in part explain why shigellosis in San Francisco in 1985 was largely a disease restricted to homosexual men (Tauxe et al., 1988).

The occasionally awesome rate of *Shigella* infections may be exemplified by a recent epidemic in the U.S., in which more than half of 12,700 people attending a mass gathering fell ill with *S. sonnei* infection. In this particular case, sanitation at the campsite was poor and the routes of transmission were suspected to be foodborne (sick persons were not excluded from preparation of food), contaminated water (same water sources used for drinking and bathing) and person-to-person contact (Wharton et al., 1990). Another major outbreak occurred in Israel, where 8000 cases of *S. sonnei* infection arose during a two-week period because of massive fecal contamination of drinking water by a leak from a sewage pipe (Egoz et al., 1991).

Studies of American prisoners experimentally infected with *Shigella* reveal that a few hundred bacteria given by the oral route give rise to disease in 25–50% of the cases. Ingestion of as few as 10 bacteria may cause symptomatic infection (DuPont et al., 1989). By increasing the bacterial dose from 5000 to more than 100,000, there is no change with respect to severity of disease or increased incidence of illness (DuPont et al., 1972). It is unknown why *Shigella* has such an outstandingly low infectious dose. Its resistance to gastric juice is only a partial explanation, since enteroinvasive *E. coli* (EIEC) are as acid-resistant as *Shigella,* but their infectious dose is one thousand times higher than that of *Shigella* (Gorden and Small, 1993).

4. EPIDEMIOLOGY

Worldwide, there are more than 200 million cases of *Shigella* dysentery every year, killing about 650,000 people, of which the majority are children. It was estimated that children living in a poor area of Santiago, Chile, had a 67% risk of developing shigellosis during the first five years of life (Ferreccio et al., 1991). In Bangladesh, where many epidemiological studies of shigellosis have been performed, *Shigella* infections may account for up to 20% of the total mortality among children between ages one and four (Bennish and

Wojtyniak, 1991). During the 1980s, the mortality rate for hospitalized Bangladeshi children with *Shigella* dysentery was 10%, with very similar mortality figures for all four *Shigella* species (Bennish and Wojtyniak, 1991). During *Shigella* epidemics, attack rates have been calculated to range from 1% to 50%, and case fatality rates have been estimated to vary from six per thousand to seven per hundred (Bennish and Wojtyniak, 1991).

The most common *Shigella* species in developing countries are *S. flexneri* and *S. dysenteriae* 1 (66% and 16% of patients hospitalized with shigellosis in Bangladesh, respectively) (Stoll et al., 1982). *S. boydii* is rarely encountered and seems mainly to exist on the Indian subcontinent. *S. dysenteriae* 1 is the most frequent cause of the bacillary dysentery epidemics that often ravage refugee camps in poor countries. In industrialized countries, epidemics are usually caused by *S. sonnei.* The majority (60–80%) of dysentery cases in the U.S. are the result of *S. sonnei* infections, except among homosexual men, where *S. flexneri* is more common (Dritz and Back, 1974).

It appears that when the economy of a developing country improves, a shift occurs, such that *S. flexneri* ceases to be the predominant *Shigella* species and is replaced by *S. sonnei.* An interesting theory for this observation is that persons living in non-industrialized countries may be naturally immunized against *S. sonnei* because of exposure to poor-quality drinking water contaminated with *Plesiomonas shigelloides.* This gram-negative bacillus is often found in surface water and has one serotype, which is identical to the single *S. sonnei* serotype (Sack et al., 1994).

5. DISEASE

5.1. CLINICAL SYMPTOMS

There are two different patterns of disease caused by *Shigella.* The classic picture of bacillary dysentery, i.e., fever, abdominal cramps and frequent purging of stool containing mucus and blood, is typical of *S. flexneri* and *S. dysenteriae* infections. It is common for the dysenteric symptoms to be preceded by watery diarrhea lasting one to two days. The other clinical syndrome is mostly associated with *S. sonnei* infections, and may resemble enterotoxigenic *E. coli* diarrhea, i.e., the patients frequently vomit and have large volumes of watery diarrhea, and consequently, often present with dehydration. It should be emphasized that up to 50% of all *Shigella* infections may be asymptomatic (Guerrero et al., 1994).

In the studies of American prisoners experimentally infected with *Shigella*, it was possible to follow the progression of symptoms arising due to the disease. The first symptom was fever, which occured on average 1.6 days after oral ingestion of the bacteria, but was only experienced by 30% of the persons. Almost all the volunteers suffered from abdominal pain, which arose on average 3.6 days after oral challenge. About 60% developed diarrhea, on average four days after bacterial ingestion. The mean duration of symptoms was seven days, which is quite long compared to most other enteric infections. Positive stool cultures were detected on average 2.6 days after intake of bacteria, and remained positive several weeks after clinical recovery. The "classic" triad of shigellosis, fever, abdominal pain and blood/mucus-containing diarrheal stool was seen in fewer than half of those falling ill (DuPont et al., 1969).

Although *S. dysenteriae* 1 is the *Shigella* species most often associated with severe shigellosis and complications, all *Shigella* species may give rise to fatal disease. In one study from Bangladesh, *S. sonnei* was the species associated with the highest mortality in young children (Bennish, 1991). The most frequent causes of death among hospitalized children are sepsis and hypoglycemia. Septicemia is mainly seen among young malnourished children (Struelens et al., 1985). It is not always the *Shigella* strain itself that causes septicemia—one study reported that in about half of the cases, another gram-negative strain was encountered in the blood of the patients (Bennish, 1991). *Shigella* septicemia is also seen among adult patients with impaired immunity, i.e., patients with AIDS, diabetes mellitus, cirrhosis, sickle cell anemia, or patients who are on immunosuppressive therapy (Baskin et al., 1987).

Advanced hypoglycemia, with blood glucose concentrations below 1 millimole per liter, is most common in children infected with *S. flexneri.* These children are generally not malnourished. The main reason for this metabolic abnormality seems to be a block in gluconeogenesis, since the concentration of insulin is usually low in these patients, whereas the concentration of glucose-releasing hormones such as glucagon, epinephrine and norepinephrine is high. To avoid death from cerebral edema and anoxia, the patients must be given glucose intravenously (Bennish et al., 1990).

A well-known complication of *S. dysenteriae* 1 infection is the hemolytic uremic syndrome, which may cause death by renal failure and/or anemia. The underlying mechanism is only incompletely understood, but Shiga toxin is clearly implicated (see Toxins section below). Another complication that sometimes occurs in patients infected with *S. dysenteriae* 1 is toxic megacolon, an enormous widening of the colon, which may lead to intestinal perforation. However, intestinal perforation may also occur with other *Shigella* species. The exact mechanism behind these intestinal complications is unknown, but it is believed that severe intestinal inflammation predisposes for both complications.

Other features of *Shigella* infections are the leukemoid reaction, i.e., white blood cell counts in excess of 40×10^9/L, which is a negative prognostic factor, but in itself not dangerous; seizures near-identical to common febrile seizures except the *Shigella* patients are usually older; and in adults, Reiter's syndrome (sterile arthritis, uveitis and urethritis), especially after *S. flexneri* infection.

5.2. MALNUTRITION

Malnutrition not only predisposes for severe shigellosis, but shigellosis itself may give rise to malnourishment. Several processes during *Shigella* infection may elicit a negative caloric balance: the lack of appetite, which is a frequent symptom, as well as the intestinal inflammation with concomitant protein loss. Food malabsorption is of minor importance, which is to be expected since the small intestine is rarely affected by *Shigella* infection. A serious consequence of dysentery in children is that it may cause growth retardation and that catch-up growth after resolution of the infection does not always occur (Bennish and Wojtyniak, 1991). In fact, *Shigella* infections are frequently seen among children with persistent diarrhea (lasting more than two weeks) in developing countries, a condition that is often associated with malnutrition and stunted growth.

Monkeys with folic acid deficiency or vitamin-deficient diets (A and B_2) have been shown to be more susceptible to acute dysentery (Janota and Dack, 1939; Watkins, 1960). It is plausible that *Shigella* infection itself may also precipitate vitamin deficiency, as suggested by the finding that more than half of the prisoners of war in Japanese camps who contracted dysentery also developed beriberi (Scrimshaw et al., 1959).

5.3. INTESTINAL PHYSIOLOGY

Shigella infection leads to colonic dysfunction, characterized by diminished net absorption of water, increased secretion of potassium ions and decreased absorption of chloride ions (Butler et al., 1986).

5.4. MACROSCOPIC APPEARANCE

Colonoscopy of patients suffering from shigellosis has shown that the disease is always concentrated in the rectosigmoid area (100% of patients). More proximal parts of the large intestine are less frequently affected. Thus, in one study, half of the patients had disease engaging the descending colon and distal transverse colon, one third of the patients also had disease in the proximal transverse section of the colon, and 15% had infection encompassing the entire colon, i.e., including the cecum (Speelman et al., 1984). The intestinal lesions are continuous and diffuse and may comprise edema, erythema, focal hemorrhages and occasionally a white mucopurulent layer, which adheres to the intestinal wall (Speelman et al., 1984).

5.5. HISTOPATHOLOGICAL FINDINGS

The intestinal wall layer most affected by *Shigella* infection is the mucosa, whereas the submucosa is relatively spared. Bacteria are usually confined to the epithelium of the surface and upper thirds of the colonic crypts. The relationship between the degree of bacterial invasion of colonocytes and colonocyte damage is not obvious, as evidenced by one study in which bacteria were found in the upper part of the crypts, although the most damaged colonocytes were localized in the bottom of the crypts. Colonocytes may lack microvilli altogether, have swollen nuclei and mitochondria, and exhibit increased numbers of cytoplasmic glycogen granules (Mathan and Mathan, 1986). It appears as if the epithelium overlying the lymphoid follicles in the colon is especially damaged by *Shigella* infection (Mathan and Mathan, 1991).

An exudate consisting of colonocytes, polymorphonuclear granulocytes with whole or partly digested bacteria, and erythrocytes in a fibrin mesh may be found in the intestinal lumen of dysenteric patients (Mathan and Mathan, 1991). Damaged epithelial cells are shed and goblet cells reduced in numbers. The epithelial monolayer is frequently infiltrated by polymorphonuclear cells, lymphocytes and mast cells (Mathan and Mathan, 1986). In the crypt epithelium, the most prevalent leukocytes are activated intraepithelial lymphocytes exhibiting prominent pseudopods and mitotic figures. Macrophages often arrive at a later stage of disease, probably to clean up the debris of dead polymorphonuclear granulocytes and epithelial cells. Eosinophilic cell invasion is also a late phenomenon.

The lamina propria is often infiltrated by neutrophilic granulocytes, macrophages and plasma cells. Edema and focal hemorrhages are also seen. Widespread vascular lesions in the lamina propria have been reported ranging from swollen or pyknotic vascular endothelial cells to a complete destruction of the endothelium. Thrombi consisting of platelets, polymorphonuclear cells, eosinophils, red blood cells and fibrin can be seen in larger vessels (Mathan and Mathan, 1986).

While dilated colonic crypts are common during the early stages of disease, elongated crypts are only seen when the duration of disease exceeds one week (Mathan and Mathan, 1991). In about one third of patients, the crypts become so heavily infiltrated with polymorphonuclear granulocytes that crypt abscesses develop (Speelman et al., 1984).

Round, superficial, mucosal erosions, a few millimeters in diameter, named *aphthoid lesions,* were in one study observed in the intestine of 15% of patients with shigellosis (Speelman et al., 1984). These lesions are typical of Crohn's inflammatory bowel disease. Mathan and co-workers noted such lesions overlying the intestinal lymphoid follicles of patients infected with *Shigella* (Mathan and Mathan, 1991). However, *Shigella* infections of the colon are usually compared to another inflammatory bowel disease of unknown etiology, namely, ulcerous colitis: Both conditions are characterized by a continuous, superficial inflammation of the colonic mucosa, centered around the rectosigmoid area.

5.6. CARRIER STATE

In *Shigella*-endemic areas, the carrier rate is 1% among healthy children (Bennish and Wojtyniak, 1991), and around 6% among children with other illnesses (Hossain et al., 1994). Long-term carriage of *Shigella* strains (17 months) has been described in children living in institutions in the U.S. during the 1970s, and in one of the American prisoners who volunteered for *Shigella* infection studies (see above) (Levine et al., 1973).

6. DIAGNOSIS

Freshly collected stool should be immediately streaked on MacConkey agar, on which non-lactose-fermenting colonies will appear in positive cases. Such colonies may be subcultured on more selective medium, such as Salmonella-Shigella agar. Agglutination using *Shigella* species-specific antisera and biochemical tests is needed for confirmation of the diagnosis. *S. sonnei* is the only *Shigella* species that can decarboxylate ornithine. Microscopic examination of the patient's stool specimen is necessary to exclude the presence of *Entamoeba histolytica,* the other causative agent of dysentery. The presence of large numbers of polymorphonuclear granulocytes in stained fecal smears is also a typical feature of *Shigella* dysentery.

Shigella is present in the stool of patients in concentrations ranging form 10^3 to 10^9 viable bacteria per gram feces during the first days of illness. It is more difficult to get positive cultures at a later stage of illness. The fact that *Shigella* are rather fragile organisms and that there is no enrichment medium available for *Shigella* (unlike *Salmonella*) results in a lot of cases being missed. This is not only a problem for the clinician, but also for epidemiological surveys.

7. ANIMAL MODELS

The capacity of a whole range of animals to be infected by *Shigella* has been tested. Arthropods and amphibians are completely resistant, as are the majority of mammals. Out of the 18 different species of mammals examined in one study, only monkeys, young bears and kittens developed disease when given 10^{10} bacteria perorally (Watkins, 1960).

Nevertheless, there are many animal models of experimental shigellosis, although none mimics human disease completely. Experimental infection of monkeys is the model that most closely resembles human shigellosis, but the infectious dose required to elicit disease is about a million times higher than in humans (Labrec et al., 1964). Guinea pigs are only infectable if they are starved for four days, their colonic flora disturbed by administration of streptomycin and the colon paralyzed by opium. Rabbits can be infected by *Shigella* via intraintestinal injection of large numbers of bacteria into ligated loops of the small intestine. The oldest animal model is somewhat inadequately named the *Serény test,* as it was first described by Zoeller and Manoussakis in 1924 (Zoeller and Manoussakis, 1924). However, Serény developed a protocol that was both simpler and more reproducible than the original one (Serény, 1955). This assay consists of the deposition of bacteria into the conjunctival sac of guinea pigs or rabbits. Invasive *Shigella* will cause severe inflammation of the cornea and to a lesser extent of the conjunctiva, i.e., keratoconjunctivitis. Histopathologically, bacterial invasion and destruction of the conjunctival cells and a massive infiltration of the cornea by polymorphonuclear granulocytes is seen (Piéchaud et al., 1958).

Mice and rats are not susceptible to intestinal *Shigella* infection. Watkins and Eisler reported that they tried to infect 23 inbred mice strains orally without success (Watkins, 1960). The only known murine model of shigellosis is one consisting of intranasal administration of bacteria, which results in infection and inflammation of the respiratory tract (Voino-Yasenetsky and Voino-Yasenetskaya, 1961).

8. PATHOGENIC MECHANISMS

8.1. COLONIZATION INCLUDING ADHERENCE

Shigella exhibit tropism for the colon. Besredka (1919) showed in experimental shigellosis in rabbits that even when bacteria were injected by the intravenous or subcutaneous routes, the bacteria homed to the intestine.

Like all members of *Enterobacteriaceae, Shigella* are able to express the mannose-specific type 1 fimbriae (Duguid and Gillies, 1957). Initially it was believed that only certain serotypes of *S. flexneri* could express these fimbriae. However, a recent study has demonstrated that *S. boydii* and *S. dysenteriae* may also harbor type 1 fimbriae (Snellings et al., 1997). The type-1 fimbriated phenotype is induced by serial aerobic cultivation of bacteria in liquid media. As in *E. coli,* the type 1 fimbrial operon of *Shigella* is chromosomally located. The type 1 fimbriae of *Shigella* and *E. coli* have shared as well as unique epitopes (Snellings et al., 1997). *Shigella* strains may switch between fimbriated and non-fimbriated states, a phenomenon known as phase variation. The type 1 fimbriae-expressing *Shigella* strains have the capacity to agglutinate guinea pig erythrocytes, an agglutination that is inhibited by the addition of mannose. Furthermore, type 1-fimbriated strains bind strongly to isolated human colonic epithelial cells, which is reversed by the addition of mannose (Duguid and Gillies, 1957).

Type 1 fimbriae are not believed to be essential for the virulence of *Shigella,* as *S. sonnei* and certain *S. flexneri* serotypes (1b and 6) seem unable to express these fimbriae. It has been suggested that type 1 fimbriae may allow *Shigella* to survive in water for long periods of time, since type 1-fimbriated bacteria typically form pellicles in the air-water interface, which promotes bacterial survival (Snellings et al., 1997). Another theory is that these fimbriae may play a role in the ability of bacterial strains to survive in the large intestine and thereby persist in patients and asymptomatic carriers (Duguid and Gillies, 1957).

In addition to type 1 fimbriae, other adhesins are undoubtedly present in *Shigella.* Recently, an electron microscopic study documented that clinical isolates of *S. flexneri* can express at least two additional types of fimbriae (Utsunomiya et al., 1992). Another study demonstrated that binding of *S. flexneri* to colonic epithelial cells of guinea pigs could partly be inhibited by fucose, *N*-acetyl neuraminic acid and *N*-acetyl mannosamine, indicating the presence of various carbohydrate-specific adhesins on the bacteria (Guhathakurta and Datta, 1992).

In vitro studies have revealed that *S. flexneri* carrying the virulence plasmid are more hydrophobic and adhere 10 times better to epithelial cells than isogenic mutants lacking this plasmid (Pál and Hale, 1989). Recent studies have pinpointed that mutations in the *Mxi-spa* locus result in decreased bacterial adherence to epithelial cells, and conversely, that overexpression of these genes enhances bacterial adhesion (Ménard et al., 1996a) (see below).

8.2. HOST CELL ENTRY

A relationship between virulence and the capacity of *Shigella* to grow inside epithelial cells in vitro was first reported by Watkins in 1960. She found that only strains that attained high numbers inside tissue-cultured cells caused disease in monkeys fed the same strain. Conversely, strains that only reached modest intracellular numbers did not cause disease (Watkins, 1960).

In 1964, Labrec et al. confirmed that bacterial penetration of epithelial cells was an essential step in the pathogenesis of bacillary dysentery. Thus, virulent *Shigella* strains were found to possess the capacity of entering epithelial cells in vitro, a trait not shared by avirulent strains (Labrec et al., 1964). A few years later, Ogawa and co-workers (1968) demonstrated in cinemicrographic studies of tissue-derived cell cultures infected with *Shigella* that upon attachment of the bacteria to epithelial cells, the epithelial cell membrane became ruffled and the bacteria were taken up. These researchers also observed that

Shigella moves within the epithelial cell at a speed of 2–7 μm per minute, that it is always the same pole of a particular bacterium that precedes movement and that the bacteria are often found inside epithelial cell protrusions.

Shigella induce their own uptake by the epithelial cell, a process termed *induced phagocytosis* (Hale et al., 1979). Spike-like projections and folds of the epithelial cell membrane surround the bacterium, creating a blossom-like structure (Figure 2.1 and 2.2). Bundles of filamentous actin, cross-linked by plasmin and myosin, accumulate underneath the epithelial cell membrane adjacent to the bacterium. Other host cell components that agglomerate at the site of bacterial entry are cortactin, vinculin, paxillin, talin and α-actinin; all these proteins are normal constituents of focal adhesion plaques, the structures that ''glue'' cells together or to their extracellular matrix (Ménard et al., 1996a).

8.3. M CELLS

In experimental shigellosis of the small intestine of rabbits, it has been found that invasive *Shigella,* non-invasive isogenic mutants as well as heat-killed bacteria, are all taken up by M cells. This cell type is found overlying

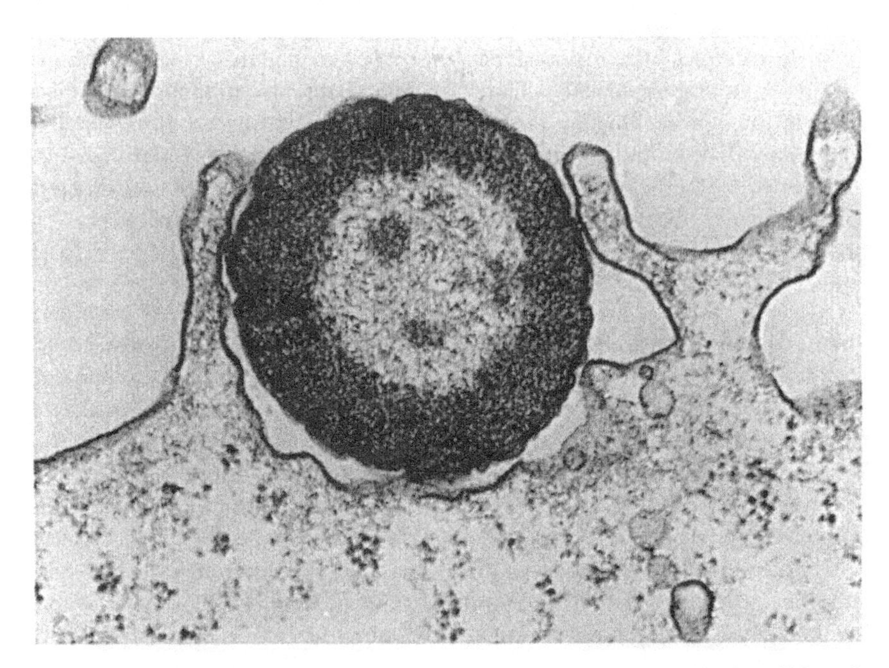

Figure 2.1 Electron micrograph of a *Shigella flexneri* bacterium being engulfed by a HeLa cell (transverse section) (P. Gounon and P. Sansonetti, Station Centrale de Microscopie Electronique, Institut Pasteur).

Figure 2.2 Schematic illustration of the invasion and intercellular dissemination of *Shigella*. The bacteria are taken up by the epithelial cell on the left, lyse the host cell phagosome, move intracellularly by the elaboration of an actin-tail, push the host cell membrane into the epithelial cell on the right, creating a host cell protrusion, the protrusion containing the bacteria buds off into the cell on the right, the bacteria lyse this double-layer phagosome and are once again free in the host cell cytoplasm, ready to repeat the same procedure.

the lymphoid follicles in the intestine, and its function is to transport antigen across the epithelial layer to leukocytes for the initiation of local immune responses against antigens located in the intestinal lumen. Electron microscopy of rabbit intestinal sections reveals that *Shigella* are located inside vesicles in the M cells and that the bacteria do not seem to lyse these vesicles. Although both virulent and avirulent *Shigella* strains are taken up by the M cells, it is only invasive strains that cause ulceration of the epithelium and dome of the lymphoid follicles (Wassef et al., 1989). Whether the "M cell pathway" constitutes a portal of entry for invasive *Shigella* into the intestinal mucosa or merely reflects the normal sampling of antigens by the intestinal immune system is unknown.

8.4. BACTERIAL GENES INVOLVED IN ENTRY

For extensive reviews, please see Parsot (1994) and Sasakawa (1995). The genes associated with virulence in *Shigella* are found on large plasmids, often referred to as *virulence plasmids* or *invasion plasmids.* The size of the virulence plasmids varies between 180 kb (*S. sonnei*) and 220 to 240 kb (*S. flexneri* 5). In 1982, it was shown that only *S. flexneri* strains harboring such plasmids have the capacity to invade HeLa cells and cause keratoconjunctivitis (Sansonetti et al., 1982). It has been revealed that more than 30 genes encoded on a 30 kb fragment of the virulence plasmid are required for bacterial invasiveness (Ménard et al., 1996a). Although the virulence plasmids of the various *Shigella* species have different restriction sites, they share extensive sequence homologies (Hale et al., 1983). These large plasmids are a metabolic burden to the bacteria, testified by their slow growth compared to plasmid-cured strains. The following genes are required for invasiveness: genes coding for the invasion plasmid antigens IpaB, IpaC and IpaD, the chaperone IpgC (invasion plasmid gene) and most of the 20 *mxi-spa* genes (membrane expression of invasion plasmid antigens-surface presentation of antigens) (Ménard et al., 1996a) (Table 2.2).

Bacteria lacking either of the three Ipas (IpaB, C or D) are able to attach to epithelial cells, but no actin accumulates at their attachment site and hence the bacteria fail to induce host cell phagocytosis. It appears that the secretion of the Ipa proteins is necessary for invasion; bacterial mutants that only express Ipas on their surface are not invasive. Conversely, the addition of pure IpaC to culture medium results in the uptake of a normally noninvasive *Shigella* strain by epithelial cells (Marquart et al., 1996). However, latex beads coated with Ipa proteins are also taken up by epithelial cells, which may seem to contradict the notion that Ipas must be secreted, unless a fraction of the Ipas detach from the same beads (Ménard et al., 1996b).

Bacteria grown in laboratory media do not spontaneously secrete the Ipa proteins. Secretion of these proteins may be induced by the addition of serum factors, Congo red, extracellular matrix glycoproteins such as fibronectin,

TABLE 2.2. Virulence Factors of *Shigella* Essential for Disease.

Virulence Factor	Location on Bacterium	Function in vivo
Invasion plasmid antigens B, C, and D (IpaB, IpaC, and IpaD)	Cytoplasm Secreted	Induced phagocytosis Lysis 1-layer phagosome Macrophage apoptosis
Invasion plasmid gene C (IpgC)	Cytoplasm	Chaperone Secretion of IpaB and IpaC
Intracellular/intercellular spreading A (IcsA)	Unipolar surface Secreted	Creation of actin tail Bacterial intracellular movement and intercellular spread
Intracellular/intercellular spreading B (IcsB)	Periplasm	Lysis of 2-layer phagosome
Membrane-expression of Ipas-surface presentation of Ipas (Mxi-Spa)	Inner and outer membrane	Type III secretion system Secretion of Ipas
Lipopolysaccharide (LPS)	Outer membrane	Protection against complement and antibodies Unipolar localization of IcsA

laminin and collagen type IV (Watarai et al., 1995) or by contact with epithelial cells (Ménard et al., 1994a). *Shigella* preferentially invade the basolateral side of tissue-cultured intestinal cells (Mounier et al., 1992). Integrins may constitute the cellular targets of Ipas as it has recently been reported that the $\alpha_5\beta_1$ integrin could serve as the binding site for IpaB, C and D, in experiments utilizing Chinese hamster ovary cells (Watarai et al., 1996).

Invasion plasmid gene C (IpgC) is a chaperone of 17 kDa, and its main function is to inhibit the binding of IpaB to IpaC in the bacterial cytoplasm, thus preventing the premature proteolytic degradation of these Ipas. Mutants lacking IpgC have impaired secretion of IpaB, and thus reduced invasive capacity, but secrete near-normal levels of IpaC (Ménard et al., 1994b). The *mxi-spa* genes code for the Mxi-spa secretion system, one of the three protein secretion systems of gram-negative bacteria. What characterizes the proteins secreted by this system is that they lack signal sequences, are not processed during translocation and do not have periplasmic intermediates. Until now, 15 of the *mxi-spa* genes have been mutated, and it has been shown that 14 of these are necessary for bacterial invasiveness.

8.6. INTRACELLULAR LOCOMOTION

Bacteria taken up by an epithelial cell are enclosed in a phagocytic vesicle (Figure 2.2). *Shigella* rapidly (within 15 minutes) lyse this vesicle with the help of each of the three Ipa proteins B, C and D (Ménard et al., 1993). Once the bacteria are free in the cytoplasm, a bacterial protein named IcsA (Intra/intercellular spreading, previously designated VirG) begins to accumulate at one of the bacterial poles. The unipolarly localized IcsA associates itself with host proteins such as polymerized actin, vinculin, plastin, creating an "actin tail," which may be likened to the tail of a comet. The formation of this tail enables the bacteria to move forward in the host cell cytoplasm in rapid spurts. Bacterial mutants devoid of IcsA lack these actin-tails and are incapable of invading adjacent cells (Bernardini et al., 1989). Furthermore, such mutants only give rise to small infectious foci and cause limited mucosal destruction in the intestine of monkeys, compared to wild-type strains.

IcsA is synthesized as a 120 kDa protein that binds and hydrolyzes ATP (Goldberg et al., 1993). The N-terminal domain (signal sequence) consists of 52 amino acids and is cleaved off during export to the periplasm (not via the type III secretion system). The C-terminal domain of the protein translocates the central portion of IcsA across the outer membrane, leaving this moiety exposed on the bacterial surface. In vitro-grown bacteria secrete a truncated form of IcsA of 95 kDa into the medium, which corresponds to the surface-exposed central moiety of the protein. Recently, it has been reported that IcsA cleavage is necessary for the unipolar localization of IcsA, i.e., a requirement for bacterial intracellular movement. The protein responsible for this cleavage is named

SopA, *Shigella* outer membrane protease. It is a serine protease, which is homologous to OmpT, the outer membrane protease of *E. coli* (Egile et al., 1997).

8.7. DISSEMINATION INTO ADJACENT CELLS

Intracellular *Shigella* tend to move towards the intercellular junctions, where they push the host cell membrane into the adjacent cell, thereby creating protrusions that are up to 20 μm long and wide enough to encompass one bacterium (Figure 2.2). In a majority of cases, such protrusions contain a dividing bacterium, which is why it is thought that there is a relationship between bacterial division and movement. *Shigella* are among the few intracellular bacteria that are able to replicate freely inside the host cell cytoplasm. Their generation time inside the host cell is rapid—about 40 minutes (Sansonetti et al., 1986).

Host cell proteins such as L-CAM, N-cadherin, α- and β-catenins, vinculin and α-actinins are involved in the creation of the epithelial cell protrusions (Sansonetti et al., 1994). Once the individual protrusion has invaded the adjacent cell, it buds off into the cytoplasm. The bacteria are thus enclosed by a double membrane bilayer whose inner leaf is derived from the previous cell and outer leaflet is derived from the host cell in which the bacteria are at that moment (Figure 2.2). Remarkably, *Shigella* are able to escape this double phagosome, although not by the same mechanism previously used to escape from the single-membrane phagosome. The product of *icsB,* another plasmid-encoded gene is required at this stage (Allaoui et al., 1992), as well as the chromosomal *vacJ* gene product (Suzuki et al., 1994). IcsB is a periplasmic protein of 54 kDa and VacJ a surface-expressed lipoprotein of 28 kDa; their exact functions have not been elucidated.

8.9. LPS

Smooth *Shigella* strains, i.e., *Shigella* with a complete O-side chain, are considerably more virulent than rough strains, which terminate with the core. In one study, all people fed 1–3× 10^9 smooth *S. sonnei* bacteria developed disease, whereas the isogenic rough strain gave rise to disease in only one of the 18 volunteers (Watkins, 1960).

Although rough strains are able to invade epithelial cells in vitro, they do not retain the capacity of spreading to neighboring cells (Okamura and Nakaya, 1977). This may be explained by the inability of IcsA to aggregate at one of the bacterial poles when the O-chain is missing (Sandlin et al., 1995). Furthermore, it seems as if the secretion of the truncated form of IcsA is impaired in rough strains, and it is possible that the conformation of SopA in the bacterial outer membrane is altered. All these changes interfere with IcsA-actin-driven bacterial movement and thus intercellular dissemination.

Rough strains also consistently yield negative Serény tests, i.e., do not induce keratoconjunctivitis in the eyes of rabbits or guinea pigs (Okamura and Nakaya, 1977). This is only partly explained by their impaired capacity of intercellular spread. An equally important explanation is that rough strains are much more susceptible to complement-mediated killing (Rowley, 1968). The O-chain protects the bacteria by limiting or blocking the access of lytic antibodies and complement factors to their binding sites on the outer bacterial membrane (Rowley, 1968). In fact, a recent study describes a *S. flexneri* mutant with short O-side chains, with normal capacity to disseminate in an epithelial cell monolayer in vitro, but exhibiting increased susceptibility to serum-mediated killing, which was probably the reason why it only gave a very weak Serény test (Hong and Payne, 1997).

8.10 TOXINS

Shigella produce various toxins, of which Shiga toxin is the most well known (O'Brien et al., 1992). Only *S. dysenteriae* 1 is able to produce this toxin, and it synthesizes large quantities of it. Shiga toxin is a holotoxin composed of five B-subunits of 7 kDa each, and one A-subunit of 32 kDa. The B-subunits bind to the disaccharide Galα1-4Galβ, found in glycolipids such as globotriaosylceramide, Gb_3. It is not clear whether this glycolipid is found on colonic epithelium. A study of the distribution of glycolipids in the human colon revealed that Gb_3 was mainly detected in the non-epithelial fraction of colonic sections (Holgersson et al., 1991). Another study demonstrated that the expression of this glycolipid on human colonic cancer cell lines was upregulated when the colonic cells were differentiated into enterocytes of small intestinal phenotype (Jacewicz et al., 1995). Since Shiga toxin induces fluid secretion in the small intestine of rabbits, and watery diarrhea is often seen initially during *Shigella* infections, it has been thought that Shiga toxin may elicit this symptom by binding to the human small intestinal mucosa. However, there is no evidence of increased fluid secretion in the small intestine of patients with ongoing *Shigella* infection (Butler, et al., 1986).

The A-subunit is the toxic part of Shiga toxin, which becomes enzymatically active upon proteolytic cleavage: Two fragments are produced, the A1-fragment of 27 kDa, which is the toxic moiety, and the carboxyterminal A2-portion of 4 kDa (O'Brien et al., 1992). The A1-subunit exerts its toxic effect by inhibiting protein synthesis. More precisely, it is an *N*-glycosidase, which cleaves off adenine from a specific adenosine of the 28S component of the 60S ribosomal subunit, thus irreversibly hindering ribosomal function (Endo et al., 1988). Shiga toxin-mediated inhibition of protein synthesis has been described both in epithelial cells (Thompson et al., 1976) and in bacteria such as *E. coli* (Olenick and Wolfe, 1980), although eukaryotic cell ribosomes are much more susceptible to the action of Shiga toxin than are bacterial ribosomes.

Shiga toxin is cytotoxic to intestinally derived epithelial cells in vitro (Moyer et al., 1987), which is why it has been speculated that it might damage the intestinal epithelium in vivo. In experimental infection of monkeys, toxin-positive strains differed from toxin-negative ones by causing hemorrhages in the intestinal mucosa (Fontaine et al., 1988). There was no difference in the severity of disease, however, since toxin-positive and toxin-deficient strains equally often gave rise to fatal disease. Epidemiological studies confirm these findings, revealing that patients infected with *S. dysenteriae* 1 more often have blood in their stool, compared to patients infected with other *Shigella* species (Stoll et al., 1982). Taken together, these findings suggest that during natural infection, Shiga toxin targets blood vessels rather than the epithelium, thereby giving rise to mucosal hemorrhages and blood in the feces. As an aside, it may be mentioned that Shiga toxin has been attributed neurotoxic activity, since it causes limb paralysis when injected intravenously into animals. However, it has been revealed that this neurotoxic effect was secondary to Shiga toxin's destruction of blood vessels in the spinal cord and other organs (Bridgwater et al., 1955).

Shiga toxin plays a central role in eliciting the hemolytic uremic syndrome, although how this occurs is unknown. It is likely that the toxin penetrates the blood vessels in the intestinal wall and then reaches the kidneys via the circulation. The glycolipid Gb_3 is found in high concentrations on human kidney endothelial cells, which are very sensitive to the cytotoxic action of Shiga toxin (Obrig et al., 1993). Interestingly, it has been found that Shiga toxin has particular affinity for the glomeruli of infants of less than two years of age, who are most often afflicted by the hemolytic uremic syndrome (Lingwood, 1994). Autopsies of children who died from this complication show thrombosis and destruction of the glomerular capillaries and renal cortical necrosis (Koster et al., 1978). Whether LPS (endotoxin) acts in synergy with Shiga toxin in destroying the renal blood vessels is under debate (Heyderman et al., 1994). However, it has been found that persons developing diarrhea-associated hemolytic uremic syndrome more often (50%) have circulating endotoxin compared to controls with uncomplicated shigellosis (5%) (Koster et al., 1978).

Shigella also produce other toxins. To date, two Shiga-like toxins have been described: Shiga-like toxin 1, which is near-identical to Shiga toxin, except for one amino acid (Strockbine et al., 1988), and Shiga-like toxin 2, which has the same biological activity as Shiga toxin, but is different with respect to physicochemical and immunological characteristics as it is only 56% homologous to Shiga toxin (Jackson et al., 1987). Two more toxins have been described: *Shigella* enterotoxin 1, a chromosomally encoded, iron-dependent toxin of 55 kDa, mainly expressed by *S. flexneri* 2a (Noriega et al., 1995), and *Shigella* enterotoxin 2, a plasmid-encoded protein of 63 kDa (Nataro et al., 1995). It is unknown what role these toxins play in *Shigella* disease.

8.11. OTHER VIRULENCE ANTIGENS

SepA, *Shigella*-extracellular protein, is yet another plasmid-encoded protein of importance for *Shigella* pathogenicity. This protein is the major protein secreted by *Shigella* grown in laboratory media. Even though one portion of SepA is very homologous to the N-terminal region of IgA, proteases of gonococci and *Haemophilus influenzae,* SepA is unable to cleave IgA1. As of yet, the function of this protein is unknown. It probably constitutes a virulence determinant as SepA-negative mutants are less virulent than wild-type strains in the rabbit model of shigellosis (Benjelloun-Touimi et al., 1995).

OmpC is a porin, a pore-producing protein. In *E. coli,* OmpC is expressed under conditions of high osmolarity, high temperature and anaerobic conditions, a milieu resembling the colon. The protein creates pores in the bacterium itself, which is believed to be an adaptive mechanism to an osmotically hostile environment. In *Shigella,* however, OmpC is constitutively expressed, independently of the molarity of its habitat. Hence, the function of OmpC in *Shigella* is less well understood. Nonetheless, it is of importance for virulence, since a *Shigella* strain lacking this porin has diminished capacity of intercellular dissemination (Bernardini et al., 1993).

All *Shigella* strains express siderophores and their corresponding outer membrane protein receptors. Siderophores are low-molecular proteins with extremely high affinity for iron, capable of removing iron from host iron-binding proteins such as transferrin and lactoferrin. Most strains of *S. flexneri* and *S. boydii* express a siderophore named aerobactin, whereas *S. dysenteriae* 1 and some *S. sonnei* strains preferentially express the siderophore enterobactin. Some *Shigella* strains express both siderophores (Lawlor et al., 1987). In vivo assays reveal that aerobactin-negative mutants have an infectious dose that is one to two log lower than wild-type strains. It is assumed that there is enough iron in the cytoplasm of epithelial cells to meet the metabolic demands of invasive *Shigella,* but that siderophores such as aerobactin optimize the utilization of iron in less hospitable environments such as tissues.

8.12. CHROMOSOMAL VIRULENCE GENES

Although the transfer of the invasion plasmid of *Shigella* to *E. coli* K12 renders *E. coli* invasive in vitro, such a bacterial strain is not invasive in vivo. Thus, it neither gives a positive Serény test nor induces fluid accumulation in rabbit small intestinal loops. Hence, *Shigella* has chromosomal genes that are necessary for the tissue-invasion process (Sansonetti et al., 1983). Examples of such genes are those coding for aerobactin, LPS and the transcriptional regulator VirR (see below).

8.13. REGULATION OF VIRULENCE GENES

Shigella are invasive at 37°C, but not at 30°C (Maurelli et al., 1984). This is because the transcription of the genes of the *ipa* and *mxi-spa* operons are under the control of the proteins VirB, VirF and VirR. The first two of these are transcriptional activators encoded by the invasion plasmid, whereas VirR is chromosomal. VirB is believed to bind to the promoters of the *ipa* and *mxi-spa* operons, but this has not been shown yet. VirF is a positive regulator of the expression of VirB and exerts it function by binding to the *virB* promoter. In the absence of VirF, VirB is not expressed, and if VirB is not expressed, the *ipa* genes are not transcribed. VirR thermoregulates *ipa* expression by binding to the promoter of *virB* at temperatures lower than 37°C, thus inactivating *virB*. The gene product of *virR* is a histone-like protein, named H-NS. There is another regulatory gene of importance, which is also located on the chromosome, named *vacB*. Mutants lacking this gene have diminished expression of both *icsA* and *ipaB* gene products, despite unaltered transcription of the genes. It is possible that the downregulatory effect exerted by *vacB* occurs at the post-transcriptional level (Sasakawa, 1995). Another reflection of the effect temperature has on *Shigella* is that the invasion plasmid is tightly supercoiled at 37°C, the temperature at which *Shigella* is most invasive. The Ipa proteins themselves also regulate their own secretion, demonstrated by the fact that mutants lacking either IpaB or IpaD constitute "leaky phenotypes," in which the non-mutated Ipas are spontaneously secreted into the medium (Ménard et al., 1993).

9. *SHIGELLA* AND THE IMMUNE SYSTEM

As *Shigella* are intracellular parasites, the host immune response must be able to detect and eliminate host cells harboring *Shigella*. It is unknown how frequently *Shigella* are found outside the hosts cells during infection, but presumably this occurs at least when infected cells are lysed, thus making it necessary for the immune system to be capable of identifying and killing bacteria in the extracellular environment as well.

9.1. INTERACTIONS OF *SHIGELLA* WITH COMPONENTS OF THE INNATE IMMUNE SYSTEM

9.1.1. Polymorphonuclear Granulocytes

Shigella are phagocytosed by monocytes/macrophages and polymorphonuclear granulocytes in vivo, as evident from histochemical stains of intestinal tissue derived from humans and the large numbers of white blood cells found in the stool of patients with shigellosis.

Polymorphonuclear granulocytes are believed to be very important in clearing *Shigella* infections. However, it has also been suggested that they may enhance bacterial invasion of the intestine. In vitro experiments showed that the addition of polymorphonuclear granulocytes to an epithelial cell monolayer resulted in the disruption of the epithelium, and secondarily, enhanced bacterial invasion. Increased bacterial invasion was also seen when the polymorphonuclear granulocytes were replaced by a calcium chelator, a substance known to open intercellular spaces by disruption of the tight junctions (Perdomo et al., 1994a). Finally, *Shigella* infection of rabbits was considerably attenuated, reflected by decreased bacterial invasion and inflammation of the intestinal mucosa, after the injection of a monoclonal antibody directed against CD18, the integrin β2 subunit expressed by leukocytes. This protective effect was interpreted to be the result of blockage of migration of leukocytes across the intestinal epithelium (Perdomo et al., 1994b).

Shigella carry the gene *sodB* encoding an iron-containing superoxide dismutase. This enzyme converts superoxide radicals into less toxic substances, and is believed to thwart the bactericidal effects of oxygen radicals produced by polymorphonuclear granulocytes during the phagocytosis of bacteria. Mutants lacking this enzyme are rapidly killed off by oxidative stress in human polymorphonuclear granulocytes in vitro and cause little tissue damage in animal models (Franzon et al., 1990). *Shigella* are also catalase positive with the exception of *S. dysenteriae* 1. This enzyme catalyzes the liberation of water and oxygen from hydrogen peroxide. However, catalase-negative mutants are as virulent as wild-type strains in vivo, despite being killed to a higher extent than catalase-positive strains by human polymorphonuclear cells in vitro. Thus, it seems as if catalase plays a less important role than superoxide dismutase in *Shigella* virulence (Franzon et al., 1990).

9.1.2. Macrophages

As early as 1970, Yee and Buffenmeyer, showed that murine peritoneal macrophages are able to phagocytose virulent as well as avirulent *Shigella* species, but that the virulent strains are often lethal for the macrophages. Peritoneal macrophages are able to kill virulent strains in vitro when presented with low numbers of virulent bacteria, but are themselves killed by the virulent *Shigella* when severely outnumbered by the bacteria (Yee and Buffenmeyer, 1970).

More recent studies have revealed that invasive *Shigella* kill macrophages by the induction of apoptosis (Zychlinsky et al., 1992). It seems that IpaB induces the apoptotic death of the macrophage by binding to interleukin-1β-converting enzyme (ICE) (Chen et al., 1996). ICE is part of a family of cysteine proteases that have the capacity to initiate apoptotic death programs in eukaryotic cells. One important function of ICE is to cleave the proinflammatory cytokine IL-1β to its biologically active form. Prior to being killed by *Shigella,* macrophages that have been prestimulated in vitro with LPS release large quantities of IL-1α and IL-1β from cytoplasmic pools (Zychlinsky et al., 1994). It has been shown that large numbers of apoptotic macrophages are found in the lymphoid follicles of rabbit small intestinal loops incubated with virulent *Shigella,* but not in loops exposed to avirulent bacteria (Zychlinsky et al., 1996).

9.1.3. Complement

The complement system protects the host against invading microorganisms by generating opsonic, chemotactic and lytic factors. The complement factors are located in the blood vessels, but may seep out into inflamed tissues when the blood vessels are dilated or destroyed, as happens during shigellosis. There are two pathways of complement activation: the alternative pathway, considered to be an early warning system, as it can be activated by the bacterial surface in the absence of antibody, and the classical pathway, which is typically triggered by the presence of specific antibody (Joiner, 1988). A recent study of a number of *S. flexneri* strains with various mutations in the genes coding for LPS revealed that strains with low amounts of O-side chains and/or incomplete core were highly susceptible to complement-mediated killing, whereas the wild-type strain with complete LPS was not (Hong and Payne, 1997).

9.1.4. Opsonization

It is generally easier for phagocytic cells such as polymorphonuclear granulocytes and macrophages to bind to and destroy bacteria coated with either complement factors or antibodies. Phagocytic cells have receptors both for the Fc-portion of antibodies and for complement factors. Polymorphonuclear granulocytes are able to kill non-opsonized *Shigella,* but their killing efficiency is enhanced when the bacteria are opsonized by specific antibodies, or by a mixture of IgM and complement factors of the alternative pathway of complement activation (Lowell et al., 1980; Reed, 1975).

9.1.5. Cytotoxic Lymphocytes

Large granular lymphocytes isolated from peripheral blood of normal donors are able to kill *Shigella*-infected HeLa cells, but not uninfected HeLa cells (Klimpel et al., 1986). It is feasible that cytotoxic lymphocytes present in the intestinal crypt epithelium may exert similar natural cytotoxic effector activity against *Shigella*-infected colonocytes.

9.1.6. Cytokines

The first study of cytokines elicited by shigellosis was performed on Sri Lankan children with *S. dysenteriae* 1 infection: they had elevated levels of IL-6 and TNF-α in serum and these cytokines were often found in the stool as well (Harendra de Silva et al., 1993). A very extensive mapping of cytokines in rectal biopsies of *Shigella* patients from Bangladesh could demonstrate the presence of all cytokines tested for in that study: interleukin-1 alpha and beta (IL-1α and IL-1β), interleukin-one receptor antagonist (IL-IRA), tumor necrosis factor-α (TNF-α), IL-4, IL-6, IL-8, IL-10, interferon-γ (IFN-γ), tumor necrosis factor-β (TNF-β) and transforming growth factor-β (TGF-β_{1-3}). The intestinal levels of these cytokines, some of which are proinflammatory (IL-1β, IL-8, TNF-α) and others are mainly anti-inflammatory (IL-10, IL-1RA), were elevated compared to healthy controls. Further, in the more severe clinical pictures, higher levels of IL-1β, IL-6, TNF-α and IFN-γ were detected in the rectal mucosa (Raqib et al., 1995a). The same authors also studied cytokine secretion in stool and blood of patients with shigellosis. They found that cytokine levels were 100 times higher in stool than in plasma, and that the levels of cytokines detected in stool, e.g., TNF-α, IL-1β, IL-1RA, IL-6, IL-8 and granulocyte-macrophage colony-stimulating factor (GM-CSF), were highest at the onset of disease and decreased to control levels within two weeks. One notable exception was interferon-γ: the levels of this cytokine in stool were severely depressed compared to controls and remained so for over a month (Raqib et al., 1995b). The significance of this finding is unclear.

In vitro experiments have indicated that epithelial cells pretreated with either interferon-γ or interferon-α are more resistant to invasion by *Shigella* than untreated cells (Niesel et al., 1986). Furthermore, fibroblasts infected with *S. flexneri* in vitro produce interferon, a production that requires not only bacterial invasion, but also intracellular bacterial metabolism (Hess et al., 1990). *Shigella*-infected fibroblasts are also more sensitive to the cytotoxic action of TNF-α (Klimpel et al., 1990). In addition, TNF-α is able to bind to high-affinity TNF-α receptors present on the bacterial surface, which results in enhanced bacterial uptake by a human monocytic cell line. Preliminary evidence suggests that TNF-α-coated bacteria are more readily killed by monocytes. However, *Shigella* coated with TNF-α also more readily invade HeLa cells in vitro; the significance of this finding is not clear (Luo et al., 1993).

The importance of proinflammatory cytokines in the pathogenesis of *Shigella* infections has been demonstrated in experimental *Shigella* infection of rabbits. Animals treated with IL-1RA, the natural antagonist of the proinflammatory cytokine IL-1, prior to and during infection, have much less tissue destruction and bacterial invasion of the intestine than sham-treated animals (Sansonetti et al., 1995).

10. ADAPTIVE HOST IMMUNE RESPONSES

10.1. PROTECTIVE IMMUNITY

Evidence for the dogma that one *Shigella* infection may protect against renewed infection has been gained from studies of experimental *Shigella* infection of American volunteers: Only 20% of the volunteers who had previously been infected with *Shigella* became ill upon reinfection with *S. flexneri*, compared with 56% of a previously unexposed group of volunteers (DuPont et al., 1972). Protective immunity against *Shigella* infection is serotype-specific. The protective efficacy of an initial natural infection with a particular *Shigella* serotype against reinfection with the same serotype has been estimated to be 72% (Ferreccio et al., 1991).

10.2. LOCAL MUCOSAL IMMUNITY

As early as 1919, Besredka claimed that local immunity in the gut was of paramount importance for clearance of *Shigella* infection, since rabbits that had received killed *Shigella* by the oral route were protected against intravenous challenge with virulent *Shigella* despite the absence of agglutinating antibodies in the serum (Besredka, 1919). A few years later, Davies (1922) found that early during natural infection, agglutinating antibodies could be detected in the stool of patients but not in the serum, and concluded that diarrheal fluid constituted the fluid "par excellence" for the examination of antibodies to *Shigella*. Indirect proof of the protective efficacy of local antibodies directed against *Shigella* comes from a study in which human volunteers fed cow milk containing high levels of bovine immunoglobulins directed against *Shigella* LPS were protected from challenge with a virulent *Shigella* strain expressing the same LPS (Tacket et al., 1992).

Volunteers orally immunized with a live *Shigella* vaccine developed fecal antibodies to the LPS of the vaccine strain, and significantly lower antibody titers to heterologous LPS (Mel et al., 1965). Antibodies of the IgA, IgM and IgG isotypes directed against *Shigella* may be recovered from the stool of patients with ongoing infection. It seems as if the quantities of total IgA in the diarrheal fluid of *Shigella* patients is higher than in patients with diarrhea caused by other pathogens (Reed and Williams, 1971). A recent study of Bangladeshi patients with shigellosis revealed that the patients developed IgA antibodies in feces not only against LPS, but also against the Ipas and Shiga toxin, and that these responses peaked five days after onset of disease (Islam et al., 1995b). *Shigella*-specific IgA has also been detected in the saliva of children with dysentery (Schultsz et al., 1992).

10.3. MILK ANTIBODIES

Breast-fed children under the age of two are less often affected by symptomatic *Shigella* infection and have less severe illness than non-breast-fed infants of the same age. Since this protective effect could not be ascribed to superior nutritional status of the breast-fed infants, it seems likely that antibacterial substances in the milk are involved (Stoll et al., 1982). Ipa-specific IgA antibodies have been detected in human milk samples derived from poor Mexican women (Cleary et al., 1989).

10.4. ANTIBODY-SECRETING CELLS

High numbers of antibody-secreting cells of the IgA isotype specific for the LPS of the infecting *Shigella* strains were detected in the blood of Israeli patients with shigellosis. These antibody-secreting cells could be detected five to 10 days after the onset of disease and probably constituted stimulated B cells that had migrated from the intestinal mucosa into the circulation (Orr et al., 1992). These same patients had very low antibody-secreting cell numbers directed against LPS derived from *Shigella* species other than the infecting strain, confirming the serotype-specific nature of immunity to *Shigella.*

10.5. SERUM ANTIBODY RESPONSES

Antibodies against *Shigella* appear later in the serum than in the stool, peaking between 1–2 weeks after onset of disease (Keusch et al., 1976; Reed and Williams, 1971). The LPS of *Shigella* elicits specific serum antibodies of the IgA, IgG and IgM isotypes. It appears as if different IgG subclasses are induced by different *Shigella* serotypes, such that IgG_2 dominates in *S. flexneri* 2a infection, while IgG_1 is the dominant subclass induced by *S. sonnei* LPS (Robin et al., 1997). IgG_1 is superior to IgG_2 both in the capacity to fix complement and to bind to Fc-receptors on phagocytes. The presence of serum antibodies directed against the LPS of *S. sonnei* was in a prospective study shown to correlate with decreased risk of acquiring symptomatic *S. sonnei* infection (Cohen et al., 1991). It is likely that these antibodies constituted a marker of existent local protective immunity to *S. sonnei* infection, rather than being protective on their own. However, a parenteral vaccine directed against the LPS of *S. sonnei* was shown to confer considerable protection against natural disease. The mean serum anti-LPS IgG and IgA titer was about 2.5 times higher among protected vaccinees than unprotected vaccinees. It is unknown whether these serum-derived antibodies exerted their function by penetrating into the intestinal mucosa or if other immune mechanisms were at work (Cohen et al., 1997). Thus, whether serum antibodies may protect against shigellosis is still an open question.

The serum antibodies directed against Shiga toxin are mainly of the IgM isotype (Keusch et al., 1976). Both humans and monkeys infected with *Shigella* respond with serum antibodies directed against IcsA, IpaA, IpaB, IpaC and IpaD (Oaks et al., 1986). Antibodies elicited by infection with one *S. flexneri* serotype are able to react with the Ipas produced by other *S. flexneri* serotypes as well as the other *Shigella* species. The fact that protective immunity to *Shigella* is serotype-specific suggests that these plasmid-encoded antigens are not protective antigens.

10.6. T CELL RESPONSES

An analysis of the T cells found in the blood of patients with ongoing *Shigella* infection revealed that a fraction of these express activation markers, such as CD25, the IL-2 receptor. Furthermore, the degree of peripheral T cell activation seems related to the severity of disease (Islam et al., 1995a). A *Shigella*-specific T cell clone reacting with an unspecified component of the bacterial outer membrane has been produced from the blood of a patient who developed reactive arthritis subsequent to *Shigella* infection (Zwillich et al., 1989).

10.7. PATIENTS WITH COMPLICATED SHIGELLOSIS

It is possible that dysenteric patients who develop complications have depressed immunity. Azim and co-workers (1996) found alterations in both the phenotype and function of peripheral blood lymphocytes in children with *Shigella* infection who later developed complications.

11. TREATMENT

There is no effective treatment for shigellosis at present. Since patients suffering from dysentery rarely are dehydrated, oral or intravenous rehydration therapy is of no avail. The only currently used therapeutics are antibiotics: nalidixic acid, ampicillin, tetracycline, chloramphenicol, trimetoprim-sulfonamide may be used, although resistance is frequently seen. In such cases, fluoroquinolones or third-generation cephalosporins may be employed. Multiresistant *Shigella* strains are unfortunately common. Antibiotics shorten the duration of the illness and abbreviate the duration of excretion of bacteria in the stool. However, due to the rapid emergence of resistant strains and the fact that *Shigella* infections are generally self-limited, it has been proposed that antibiotics should be reserved for severe cases of shigellosis only (Weissman et al., 1973). An additional reason to limit the usage of antibiotics is that they have been implicated in the development of the hemolytic uremic syndrome (Al-Qarawi et al., 1995; Butler et al., 1987). Thus, the risk ratios of developing this complication were elevated among patients given certain antibiotics, some of which were considered to be ineffective against *Shigella*. A possible explanation for this epidemiological observation is that subinhibitory concentrations of antibiotics lead to increased production of Shiga toxin by *Shigella* strains in vitro (Walterspiel et al., 1992). Alternately, antibiotics that alter the normal enteric flora might allow the infecting *Shigella* strain to proliferate with less restraint. As an addendum, a study has shown that anti-diarrheal drugs such as Lomotil, which function by decreasing intestinal motility, may worsen *Shigella* disease (DuPont and Hornick, 1973).

At the community-level, important measures to combat *Shigella* infections include chlorination of water to impede fecal contamination of drinking water, efficient sewage systems, fly control, and availability of clean food.

12. VACCINATION

Besredka proposed that an efficient vaccine against *Shigella* should be given by the oral route, in view of his findings that rabbits orally immunized with killed *Shigella* were protected against reinfection. Since he found no correlation between the presence of agglutinating antibodies in the serum and protection against *Shigella* challenge, he even suggested vaccinologists should aim at developing vaccines that did not elicit serum antibodies (Besredka, 1919). Although work on vaccine development has been ongoing for 100 years, no commercially available vaccine exists. A large array of vaccine candidates have been tested in humans, but these have either elicited poor protective immunity or side-effects, or both. The following vaccine formulations have been assayed: heat-killed bacteria in milligram doses given orally, parenteral vaccines, oral attenuated vaccines with deletions in either the chromosome and/or virulence plasmid, as well as hybrid vaccines, consisting of non-pathogenic *E. coli* harboring cloned *Shigella* virulence factors. One problem with attenuated vaccines is that the virulence phenotype is often unstable in *Shigella* (Hale, 1995). A clinical trial of a parenteral *S. sonnei* O-specific polysaccharide-protein conjugate vaccine has yielded promising results: a single injection elicited a protective efficacy of 74% against natural disease (Cohen et al., 1997).

13. CONCLUSION

Whereas a lot of information has emerged about the pathogenic mechanisms of *Shigella* infection, the role played by the host, in particular by the immune system, is very poorly understood. It is generally accepted that immunity against LPS is protective, but how to measure such immunity is less certain—whether mucosal secretions or serum should be employed. Further, it is not known to what extent the immune response engendered by *Shigella* infection contributes to tissue destruction, and to what extent it is protective. Although blocking inflammatory mediators during shigellosis in animal models has been successful, it is questionable whether this approach should be attempted in human patients. The most appealing alternative is, of course, to prevent disease altogether—the creation of a safe, cheap and efficacious vaccine is therefore of high priority. Continuing epidemiological surveys of which *Shigella* species and serotypes account for shigellosis in various populations in the world are much needed. Existing studies suggest that a vaccine inducing protective immunity against the three strains that dominate worldwide, i.e., *S. flexneri* 2a, *S. sonnei* and *S. dysenteriae* 1, could reduce *Shigella*-associated morbidity by 70–80%.

14. REFERENCES

Allaoui, A., Mounier, J., Prévost, M. -C., Sansonetti, P. J., and Parsot, C. 1992. "*icsB:* a *Shigella flexneri* virulence gene necessary for the lysis of protrusions during intercellular spread," *Mol. Microbiol.* 6:1605–1616.

Al-Qarawi, S., Fontaine, R. E., and Al-Qahtani, M. -S. 1995. "An outbreak of hemolytic uremic syndrome associated with antibiotic treatment of hospital inpatients for dysentery," *Emerging Infectious Diseases* 1:138–140.

Azim, T., Sarker, M. S., Hamadani, J., Khanum, N., Halder, R. C., Salam, M. A., and Albert, J. A. 1996. "Alterations in lymphocyte phenotype and function in children with shigellosis who develop complications," *Clin. Diagn. Lab. Immunol.* 38:91–196.

Baskin, D. H., Lax, J. D., and Barenberg, D. 1987. "Shigella bacteremia in patients with the acquired immune deficiency syndrome," *Am. J. Gastroenterol.* 82:338–341.

Benjelloun-Touimi, Z., Sansonetti, P. J., and Parsot, C. 1995. "SepA, the major extracellular protein of *Shigella flexneri:* autonomous secretion and involvement in tissue invasion," *Mol. Microbiol.* 17:123–135.

Bennish, M. L. 1991. "Potentially lethal complications of shigellosis," *Rev. Infect. Dis.* 13:S319–324.

Bennish, M. L. and Wojtyniak, B. J. 1991. "Mortality due to shigellosis: community and hospital data," *Rev. Infect. Dis.* 13:S245–251.

Bennish, M. L., Azad, A. K., Rahman, O., and Phillips, R. E. 1990. "Hypoglycemia during diarrhea in childhood: prevalence, pathophysiology and outcome," *N. Engl. J. Med.* 322:1357–1363.

Bernardini, M. L., Sanna, M. G., Fontaine, A., and Sansonetti, P. J. 1993. "OmpC is involved in invasion of epithelial cells by *Shigella flexneri,*" *Infect. Immun.* 61:3625–3635.

Bernardini, M. L., Mounier, J., d'Hauteville, H., Coquis-Rondon, M., and Sansonetti, P. J. 1989. "Identification of *icsA,* a plasmid locus of *Shigella flexneri* that governs bacterial intra- and intercellular spread through interactions with F-actin," *Proc. Natl. Acad. Sci.* 86:3867–3871.

Besredka, A. 1919. "Du mécanisme de l'infection dysentérique, de la vaccination contre la dysenterie par la voie buccale et de la nature de l'immunité antidysentérique," *Ann. Inst. Past.* 33:301–317.

Brahmbhatt, H. N., Lindberg, A. A., and Timmis, K. N. 1992. "*Shigella* lipopolysaccharide: structure, genetics, and vaccine development," *Curr. Top. Microbiol. Immunol.* 180, 45–64.

Brenner, D. J., Rowe, B., and Cross, R. J. 1984. "Facultatively anaerobic gram negative rods." In *Bergey's Manual of Systematic Bacteriology,* ed., N. R. Kreig, Baltimore, MD: Williams and Wilkins, pp. 411–427.

Brenner, D. J., Fanning, G. R., Miklos, G. V., and Steigerwalt, A. G. 1973. ''Polynucleotide sequence relatedness among *Shigella* species,'' *Int. J. Syst. Bacteriol.* 23:1–7.

Bridgwater, F. A. J., Morgan, R. S., Rowson, K. E. K., and Payling Wright, G. 1955. ''The neurotoxin of *Shigella shigae.* Morphological and functional lesions produced in the central nervous system of rabbits,'' *Br. J. Exp. Pathol.* 36:447–453.

Butler, T., Islam, M. R., Azad, M. A. K., and Jones, P. K. 1987. ''Risk factors for development of hemolytic uremic syndrome during shigellosis,'' *J. Pediatr.* 110:894–897.

Butler, T., Speelman, P., Kabir, I., and Banwell, J. 1986. ''Colonic dysfunction during shigellosis,'' *J. Infect. Dis.* 154:17–824.

Chen, Y., Smith, M. R., Thirumalai, K., and Zychlinsky, A. 1996. ''A bacterial invasin induces macrophage apoptosis by binding directly to ICE,'' *Embo J.* 15:3853–3860.

Cleary, T. G., Winsor, D. K., Reich, D., Ruiz-Palacios, G., and Calva, J. J. 1989. ''Human milk immunoglobulin A antibodies to *Shigella* virulence determinants,'' *Infect. Immun.* 57:1675–1679.

Cohen, D., Green, M. S., Block, C., Slepon, R., and Ofek, I. 1991. ''Prospective study of the association between serum antibodies to lipopolysaccharide O antigen and the attack rate of shigellosis,'' *J. Clin. Microbiol.* 29:386–389.

Cohen, D., Ashkenazi, S., Green, M. S., Gdalevich, M., Robin, G., Slepon, R., Yavzori, M., Orr, N., Block, C., Ashkenazi, I., Shemer, J., Taylor, D. N., Hale, T. L., Sadoff, J. C., Pavliakova, D., Schneerson, R., and Robbins, J. B. 1997. ''Double-blind vaccine-controlled randomised efficacy trial of an investigational *Shigella sonnei* conjugate vaccine in young adults,'' *Lancet* 349:155–159.

Davies, A. 1922. ''An investigation into the serological properties of dysentery stools,'' *Lancet* 2:1009–1012.

Dritz, S. K. and Back, A. F. 1974. ''Shigella enteritis venereally transmitted,'' *N. Engl. J. Med.* 291:1194.

Duguid, J. P. and Gillies, R. R. 1957. ''Fimbriae and adhesive properties in dysentery bacilli,'' *J. Path. Bact.* 74:397–411.

DuPont, H. and Hornick, R. 1973. ''Adverse effects of Lomotil therapy in shigellosis,'' *JAMA* 226:1525.

DuPont, H. L., Levine, M. M., Hornick, R. B., and Formal, S. B. 1989. ''Inoculum size in shigellosis and implications for expected mode of transmission,'' *J. Infect. Dis.* 159:1126–1128.

DuPont, H. L., Hornick, R. B., Dawkins, A. T., Snyder, M. J., and Formal, S. B. 1969. ''The response of man to virulent *Shigella flexneri* 2a,'' *J. Infect. Dis.* 119:296–299.

DuPont, H. L., Hornick, R. B., Snyder, M. J., Libonati, J. P., Formal, S. B., and Gangarosa, E. J. 1972. ''Immunity in shigellosis. II. Protection induced by oral live vaccine or primary infection,'' *J. Infect. Dis.* 125:12–16.

Egile, C., d'Hauteville, H., Parsot, C., and Sansonetti, P. J. 1997. ''SopA, the outer membrane protease responsible for polar localization of IcsA in *Shigella flexneri,*'' *Mol. Microbiol.* 23: 1063–1073.

Egoz, N., Shmilovitz, M., Kretzer, B., Lucian, M., Porat, V., and Raz, R. 1991. ''An outbreak of *Shigella sonnei* infection due to contamination of a municipal water supply in Northern Israel,'' *J. Infect.* 22:87–93.

Endo, Y., Tsurugi, K., Yutsudo, T., Takeda, Y., Ogasawara, T., and Igarashi, K. 1988. ''Site of action of a Vero toxin (VT2) from *Escherichia coli* O157:H7 and of Shiga toxin on eukaryotic ribosomes,'' *Eur. J. Biochem.* 171:45–50.

Ferreccio, C., Prado, V., Ojeda, A., Cayyazo, M., Abrego, P., Guers, L., and Levine, M. M.

1991. "Epidemiologic patterns of acute diarrhea and endemic *Shigella* infections in children in a poor periurban setting in Santiago, Chile," *Am. J. Epidemiol.* 134:614–627.

Fontaine, A., Arondel, J., and Sansonetti, P. J. 1988. "Role of shiga toxin in the pathogenesis of bacillary dysentery, studied by using a tox mutant of *Shigella dysenteriae* 1," *Infect. Immun.* 56:3099–3109.

Franzon, V. L., Arondel, J., and Sansonetii, P. J. 1990. "Contribution of superoxide dismutase and catalase activities to *Shigella flexneri* pathogenesis," *Infect. Immun.* 58:529–535.

Girón, J. A. 1995. "Expression of flagella and motility by *Shigella*," *Mol. Microbiol.* 18:63–75.

Goldberg, M. B., Barzu, O., Parsot, C., and Sansonetti, P. J. 1993. "Unipolar localization and ATPase activity of IcsA, a *Shigella flexneri* protein involved in intracellular movement," *J. Bacteriol.* 175:2189–2196.

Gorden, J. and Small, P. L. C. 1993. "Acid resistance in enteric bacteria," *Infect. Immun.* 61:364–367.

Guerrero, L., Calva, J. J., Morrow, A. L., Velazquez, F. R., Tuz-Dzib, F., Lopez-Vidal, Y., Ortega, H., Arroyo, H., Cleary, T. G., Pickering, L. K., and Ruiz-Palacios, G. M. 1994. "Asymptomatic *Shigella* infections in a cohort of Mexican children younger than two years of age," *Pediatr. Infect. Dis. J.* 13:597–602.

Guhathakurta, D. S. and Datta, A. 1992. "Adhesion of *Shigella dysenteriae* type 1 and *Shigella flexneri* to guinea-pig colonic epithelial cells in vitro, *Med. Microbiol.* 36:403–405.

Hale, T. L. 1995. "*Shigella* vaccines." In *Molecular and Clinical Aspects of Bacterial Vaccine Development,* eds., Ala'Aldeen and Hormaeche, John Wiley & Sons Ltd. pp 179–204.

Hale, T. L., Morris, R. E., and Bonventre, P. F. 1979. "*Shigella* infection of Henle intestinal epithelial cells: role of the host cell." *Infect. Immun.* 24:887–894.

Hale, T. L., Sansonetti, P. J., Schad, P. A., Austin, S., and Formal, S. B. 1983. "Characterization of virulence plasmids and plasmid-associated outer membrane proteins in *Shigella flexneri, Shigella sonnei,* and *Escherichia coli.*" *Infect. Immun.* 40:340–350.

Harendra de Silva, D. G., Mendis, L. N., Sheron, N., Alexander, G. J. M., Candy, D. C. A., Chart, H., and Rowe, B. 1993. "Concentrations of interleukin 6 and tumor necrosis factor in serum and stools of children with *Shigella dysenteriae* 1 infection," *Gut* 34:194–198.

Hess, C. B., Niesel, D. W., Holmgren, J., Jonson, G., and Klimpel, G. R. 1990. "Interferon production by *Shigella flexneri*-infected fibroblasts depends upon intracellular bacterial metabolism," *Infect. Immun.* 58:399–405.

Heyderman, R. S., Fitzpatrick, M. M., and Robin Barclay, G. 1994. "Haemolytic-uraemic syndrome, *Lancet* 343:1042.

Holgersson, J., Jovall, P.-Å., and Breimer, M. E. 1991. "Glycosphingolipids of human large intestine: detailed structural characterization with special reference to blood group compounds and bacterial receptor structures," *J. Biochem.* 110:120–131.

Hong, M. and Payne, S. M. 1997. "Effect of mutations in *Shigella flexneri* chromosomal and plasmid-encoded lipopolysaccharide genes on invasion and serum resistance," *Mol. Microbiol.* 24:779–791.

Hossain, M. A., Hasan, K. Z., and Albert, M. J. 1994. "*Shigella* carriers among non-diarrhoeal children in an endemic area of shigellosis in Bangladesh," *Trop. Geograph. Med.* 46:40–42.

Islam, D., Bardhan, P. K., Lindberg, A. A., and Christensson, B. 1995a. "*Shigella* infection induces cellular activation of T and B cells and distinct species-related changes in peripheral blood lymohocyte subsets during the course of the disease," *Infect. Immun.* 63:2941–2949.

Islam, D., Wretlind, B., Ryd, M., Lindberg, A. A., and Christensson, B. 1995b. "Immunoglobulin subclass distribution and dynamics of *Shigella*-specific antibody responses in serum and stool samples in shigellosis," *Infect. Immun.* 63:2054–2061.

Jacewicz, M. S., Acheson, D. W. K., Mobassaleh, M., Donohue-Rolfe, A., Balasubramanian, K. A., and Keusch, G. T. 1995. "Maturational regulation of globotriaosylceramide, the Shiga-like toxin 1 receptor, in cultured human gut epithelial cells," *J. Clin. Invest.* 96:1328–1335.

Jackson, M. P., Neil, R. J., O'Brien, A. D., Holmes, R. K., and Newland, J. W. 1987. "Nucleotide sequence analysis and comparison of the structural genes for Shiga-like toxin I and Shiga-like toxin II encoded by bacteriophages from *Escherichia coli* 933," *FEMS Microbiol. Lett.* 44:109–114.

Janota, M. and Dack, G. M. 1939. "Bacillary dysentery developing in monkeys on a "vitamin M" deficient diet," *J. Infect. Dis.* 65:219–224.

Joiner, K. A. 1988. "Complement evasion by bacteria and parasites," *Ann. Rev. Microbiol.* 42:201–230.

Keusch, G. T., Jacewicz, M., Levine, M. M., Hornick, R. B., and Kochwa, S. 1976. "Pathogenesis of Shigella diarrhea: serum anticytoxin antibody response produced by toxigenic and nontoxigenic *Shigella dysenteriae* 1," *J. Clin. Invest.* 57:194–202.

Klimpel, G. R., Niesel, D. W., and Klimpel, K. D. 1986. "Natural cytotoxic effector cell activity against *Shigella flexneri*-infected HeLa cells," *J. Immunol.* 136:1081–1086.

Klimpel, G. R., Shaban, R., and Niesel, D. W. 1990. "Bacteria-infected fibroblasts have enhanced susceptibility to the cytotoxic action of tumor necrosis factor," *J. Immunol.* 145:711–717.

Koster, F., Levin, J., Walker, L., Tung, K. S. K., Gilman, R. H., Rahaman, M. M., Majid, M. A., Islam, S., and Williams, R. C., Jr. 1978. "Hemolytic-uremic syndrome after shigellosis: relation to endotoxemia and circulating immune complexes," *N. Engl. J. Med.* 298:927–933.

Labrec, E. H., Schneider, H., Magnani, T. J., and Formal, S. B. 1964. "Epithelial cell penetration as an essential step in the pathogenesis of bacillary dysentery," *J. Bacteriol.* 88:1503–1518.

Lawlor, K. M., Daskaleros, P. A., Robinson, R. E., and Payne, S. M. 1987. "Virulence of iron transport mutants of *Shigella flexneri* and utilization of host iron compounds," *Infect. Immun.* 55:594–599.

Levine, M. M., DuPont, H. L., Khodabandelou, M., and Hornick, R. B. 1973. "Long-term shigella-carrier state," *N. Engl. J. Med.* 288:1169–1171.

Levine, O. S. and Levine, M. M. 1991. "Houseflies (*Musca domestica*) as mechanical vectors of shigellosis," *Rev. Infect. Dis.* 13:688–696.

Lingwood, C. A. 1994. "Verotoxin-binding in human renal sections," *Nephron* 66:21–28.

Lowell, G. H., MacDermott, R. P., Summers, P. L., Reeder, A. A., Bertovich, M. J., and Formal, S. B. 1980. "Antibody-dependent cell-mediated antibacterial activity: K lymphocytes, monocytes, and granulocytes are effective against *Shigella,*" *J. Immunol.* 125:2778–2784.

Luo, G., Niesel, D. W., Shaban, R. A., Grimm, E. A., and Klimpel, G. R. 1993. "Tumor necrosis factor alpha binding to bacteria: evidence for a high-affinity receptor and alteration of bacterial virulence properties," *Infect. Immun.* 61:830–835.

Luria, S. E. and Burrous, J. W. 1957. "Hybridization between *Escherichia coli* and Shigella," *J. Bacteriol.* 74:461–476.

Marquart, M. E., Picking, W. L., and Picking, W. D. 1996. "Soluble invasion plasmid antigen C (IpaC) from *Shigella flexneri* elicits epithelial cell responses related to pathogen invasion," *Infect. Immun.* 64:4182–4187.

Mathan, M. M. and Mathan, V. I. 1986. "Ultrastructural pathology of the rectal mucosa in *Shigella* dysentery," *Am. J. Pathol.* 123:25–38.

Mathan, M. M. and Mathan, V. I. 1991. "Morphology of rectal mucosa of patients with shigellosis," *Rev. Infect. Dis.* 13:S314–S318.

Maurelli, A. T., Blackmon, B., and Curtiss, R., III. 1984. "Temperature-dependent expression of virulence genes in *Shigella* species," *Infect. Immun.* 43:195–201.

Mel, D. M., Papo, R. G., Terzin, A. L., and Vuksic, L. 1965. "Studies on vaccination against bacillary dysentery: 2. Safety tests and reactogenicity studies on a live dysentery vaccine intended for use in field trials," *Bull. Wld Hlth Org.* 32:637–645.

Ménard, R., Dehio, C., and Sansonetti, P. J. 1996a. "Bacterial entry into epithelial cells: the paradigm of *Shigella,*" *Trends Microbiol.* 4:220–226.

Ménard, R., Sansonetti, P. J., and Parsot, C. 1993. "Nonpolar mutagenesis of the *Ipa* genes defines IpaB, IpaC, and IpaD as effectors of *Shigella flexneri* entry into epithelial cells," *J. Bacteriol.* 175:5899–5906.

Ménard, R., Sansonetti, P., and Parsot, C. 1994a. "The secretion of the *Shigella flexneri* Ipa invasins is activated by epithelial cells and controlled by IpaB and IpaD," *EMBO J.* 13:5293–5302.

Ménard, R., Sansonetti, P., Parsot, C., and Vasselon, T. 1994b. "Extracellular association and cytoplasmic partitioning of the IpaB and IpaC invasins of *S. flexneri,*" *Cell* 79:515–525.

Ménard, R., Prévost, M. C., Gounon, P., Sansonetti, P., and Dehio, C. 1996b. "The secreted Ipa complex of *Shigella flexneri* promotes entry into mammalian cells," *Proc. Natl. Acad. Sci. USA* 93:1254–1258.

Mounier, J., Vasselon, T., Hellio, R., Lesourd, M., and Sansonetti, P. J. 1992. "*Shigella flexneri* enters human colonic Caco-2 epithelial cells through the basolateral pole," *Infect. Immun.* 60:237–248.

Moyer, M. P., Dixon, P. S., Rothman, S. W., and Brown, J. E. 1987. "Cytotoxicity of shiga toxin for primary cultures of human colonic and ileal epithelial cells," *Infect. Immun.* 55:1533–1535.

Nataro, J. P., Seriwatana, J., Fasano, A., Maneval, D. R., Guers, L. D., Noriega, F., Dubovksy, F., Levine, M. M., and Morris, J. G., Jr. 1995. "Identification and cloning of a novel plasmid-encoded enterotoxin of enteroinvasive *Escherichia coli* and *Shigella* strains," *Infect. Immun.* 63:4721–4728.

Niesel, D. W., Hess, C. B., Cho, Y. J., Klimpel, K. D., and Klimpel, G. R. 1986. "Natural and recombinant interferons inhibit epithelial cell invasion by *Shigella* spp," *Infect. Immun.* 52:828–833.

Noriega, F. R., Liao, F. M., Formal, S. B., Fasano, A., and Levine, M. M. 1995. "Prevalence of *Shigella* enterotoxin 1 among *Shigella* clinical isolates of diverse serotypes," *J. Infect. Dis.* 172:1408–1410.

Oaks, E. V., Hale, T. L., and Formal, S. B. 1986. "Serum immune response to *Shigella* protein antigens in Rhesus monkeys and humans infected with *Shigella* spp," *Infect. Immun.* 53:57–63.

O'Brien, A. D., Tesh, V. L., Donohue-Rolfe, A., Jackson, M. P., Olsnes, S., Sandvig, K., Lindberg, A. A., and Keusch, G. T. 1992. "Shiga toxin: biochemistry, genetics, mode of action, and role in pathogenesis," *Curr. Top. Microbiol. Immunol.* 180:65–94.

Obrig, T. G., Louise, C. B., Lingwood, C. A., Boyd, B., Barley-Maloney, L., and Daniel, T. O. 1993. "Endothelial heterogeneity in Shiga toxin receptors and responses," *J. Biol. Chem.* 268:15484–15488.

Ogawa, H., Nakamura, A., and Nakaya, R. 1968. "Cinemicrographic study of tissue cell cultures infected with *Shigella flexneri,*" *Japan J. Med. Sci. Biol.* 21:259–273.

Okamura, N. and Nakaya, R. 1977. "Rough mutant of *Shigella flexneri* 2a that penetrates tissue culture cells but does not evoke keratoconjunctivitis in guinea pigs." *Infect. Immun.* 17:4–8.

Olenick, J. G. and Wolfe, A. D. 1980. "*Shigella* toxin inhibition of binding and translation of polyuridylic acid by ribosomes of *Escherichia coli.*" *J. Bacteriol.* 141:1246–1250.

Orr, N., Robin, G., Lowell, G., and Cohen, D. 1992. "Presence of specific immunoglobulin A-secreting cells in peripheral blood after natural infection with *Shigella sonnei*" *J. Clin. Microbiol.* 30:2165–2168.

Pál, T. and Hale, T. L. 1989. "Plasmid-associated adherence of *Shigella flexneri* in a HeLa cell model," *Infect. Immun.* 57:2580–2582.

Parsot, C. 1994. "*Shigella flexneri*: genetics of entry and intercellular dissemination in epithelial cells," *Curr. Top. Microbiol. Immunol.* 192:217–241.

Perdomo, J. J., Gounon, P., and Sansonetti, P. J. 1994a. "Polymorphonuclear leukocyte transmigration promotes invasion of colonic epithelial monolayer by *Shigella flexneri.*" *J. Clin. Invest.* 93:633–643.

Perdomo, O. J. J., Cavaillon, J. M., Huerre, M., Ohayon, H., Gounon, P., and Sansonetti, P. J. 1994b. "Acute inflammation causes epithelial invasion and mucosal destruction in experimental shigellosis," *J. Exp. Med.* 180:1307–1319.

Piéchaud, M., Szturm-Rubinsten, S., and Piéchaud, D. 1958. "Evolution histologique de la kérato-conjonctivite a bacilles dysentériques du cobaye," *Ann. Inst. Pasteur* 94:298–309.

Raqib, R., Wretlind, B., Andersson, J., and Lindberg, A. A. 1995b. "Cytokine secretion in acute shigellosis is correlated to disease activity and directed more to stool than to plasma," *J. Infect. Dis.* 171:376–384.

Raqib, R., Lindberg, A. A., Wretlind, B., Bardhan, P. K., Andersson, U., and Andersson, J. 1995a. "Persistence of local cytokine production in shigellosis in acute and convalescent satges," *Infect. Immun.* 63:289–296.

Reed, W. P. 1975. "Serum factors capable of opsonizing *Shigella* for phagocytosis by polymorphonuclear neutrophils," *Immunology* 28:1051–1059.

Reed, W. P. and Williams, R. C., Jr. 1971. "Intestinal immunoglobulins in shigellosis," *Gastroenterology* 61:35–45.

Robin, G., Cohen, D., Orr, N., Markus, I., Slepon, R., Ashkenazi, S., and Keisari, Y. 1997. "Characterization and quantitative analysis of serum IgG class and subclass response to *Shigella sonnei* and *Shigella flexneri* 2a lipopolysaccharide following natural *Shigella* infection," *J. Infect. Dis.* 175:1128–1133.

Robinson, D. T., Armstrong, E. C., and Carpenter, K. P. 1965. "Outbreak of dysentery due to contact with a pet monkey," *Brit. Med. J.* 1:903–905.

Rowley, D. 1968. "Sensitivity of rough Gram-negative bacteria to the bactericidal action of serum," *J. Bacteriol.* 95:1647–1650.

Sack, D. A., Hoque, A. T. M. S., Huq, A., and Etheridge, M. 1994. "Is protection against shigellosis induced by natural infection with *Plesiomonas shigelloides*?" *Lancet* 343:1413–1415.

Sandlin, R. C., Lampel, K. A., Keasler, S. P., Goldberg, M. B., Stolzer, A. L., and Maurelli, A. T. 1995. "Avirulence of rough mutants of *Shigella flexneri*: requirement of O antigen for correct unipolar localization of IcsA in the bacterial outer membrane," *Infect. Immun.* 63:229–237.

Sansonetti, P. J., Kopecko, D. J., and Formal, S. B. 1982. "Involvement of a plasmid in the invasive ability of *Shigella flexneri,*" *Infect. Immun.* 35:852–860.

Sansonetti, P. J., Arondel, J., Cavaillon, J.-M., and Huerre, M. 1995. "Role of interleukin-1 in the pathogenesis of experimental shigellosis," *J. Clin. Invest.* 96:884–892.

Sansonetti, P. J., Mounier, J., Prévost, M. C., and Mège, R.-M. 1994. "Cadherin expression is required for the spread of *Shigella flexneri* between epithelial cells," *Cell* 76:829–839.

Sansonetti, P. J., Ryter, A., Clerc, P., Maurelli, A. T., and Mounier, J. 1986. "Multiplication of *Shigella flexneri* within HeLa cells: lysis of the phagocytic vacuole and plasmid-mediated contact hemolysis," *Infect. Immun.* 51:461–469.

Sansonetti, P. J., Hale, T. L., Dammin, G. J., Kapfer, C., Collins, H. H., Jr., and Formal, S. B. 1983. "Alterations in the pathogenicity of *Escherichia coli* K-12 after transfer of plasmid and chromosomal genes from *Shigella flexneri,*" *Infect. Immun.* 39:1392–1402.

Sasakawa, C. 1995. "Molecular basis of pathogenicity of *Shigella*," *Rev. Med. Microbiol.* 6:257–266.

Schultsz, C., Qadri, F., Hossain, S. A., Ahmed, F., and Ciznar, I. 1992. "*Shigella*-specific IgA in saliva of children with bacillary dysentery," *FEMS Microbiol. Immunol.* 89:65–72.

Scrimshaw, N. S., Taylor, C. E., and Gordon, J. E. 1959. "Interactions of nutrition and infection," *Am. J. Med. Sci.* 23:367–403.

Serény, B. 1955. "Experimental *Shigella* keratoconjunctivitis. A preliminary report," *Acta Microbiol. Acad. Sci. Hung.* 2:293–295.

Simmons, D. A. R., and Romanowska, E. 1987. "Structure and biology of *Shigella flexneri* O antigens," *J. Med. Microbiol.* 23:289–302.

Snellings, N. J., Tall, B. T., and Venkatesan, M. M. 1997. "Characterization of *Shigella* type 1 fimbriae: expression, FimA sequence, and phase variation," *Infect. Immun.* 65:2462–2467.

Speelman, P., Kabir, I., and Islam, M. 1984. "Distribution and spread of colonic lesions in shigellosis: a colonoscopic study," *J. Infect. Dis.* 150:899–903.

Stoll, B. J., Glass, R. I., Huq, M. I., Khan, M. U., Banu, H., and Holt, J. 1982. "Epidemiologic and clinical features of patients infected with *Shigella* who attended a diarrheal disease hospital in Bangladesh," *J. Infect. Dis.* 146:177–183.

Strockbine, N. A., Jackson, M. P., Sung, L. M., Holmes, R. K., and O'Brien, A. D. 1988. "Cloning and sequencing of the genes from Shiga toxin from *Shigella dysenteriae* type 1," *J. Bacteriol.* 170:1116–11122.

Struelens, M. J., Patte, D., Kabir, I., Salam, A., Nath, S. K., and Butler, T. 1985. "Shigella septicemia: prevalence, presentation, risk factors, and outcome," *J. Infect. Dis* 152:784–790.

Suzuki, T., Murai, T., Fukuda, I., Tobe, T., Yoshikawa, M., and Sasakawa, C. 1994. "Identification and characterization of a chromosomal virulence gene, *vacJ*, required for intercellular spreading of *Shigella flexneri*," *Mol. Microbiol.* 11:31–41.

Tacket, C. O., Binion, S. B., Bostwick, E., Losonsky, G., Roy, M. J., and Edelman, R. 1992. "Efficacy of bovine milk immunoglobulin concentrate in preventing illness after *Shigella flexneri* challenge," *Am. J. Trop. Med. Hyg.* 47:276–283.

Tauxe, R. V., McDonald, R. C., Hargrett-Bean, N., and Blake, P. A. 1988. "The persistence of *Shigella flexneri* in the United States: increasing role of adult males," *Am. J. Public Health* 78:1432–1435.

Thompson, M. R., Steinberg, M. S., Gemski, P., Formal, S. B., and Doctor, B. P. 1976. "Inhibition of *in vitro* protein synthesis by *Shigella dysenteriae* 1 toxin," *Biochem. Biophys. Res. Commun.* 71:783–788.

Tominaga, A., Mahmoud, M. A. H., Mukaihara, T., and Enomoto, M. 1994. "Molecular characterization of intact, but cryptic, flagellin genes in the genus *Shigella*," *Mol. Microbiol.* 12:277–285.

Utsunomiya, A., Naito, T., Ehara, M., Ichinose, Y., and Hamamoto, A. 1992. "Studies on novel pili from *Shigella flexneri*. I. Detection of pili and hemagglutination activity," *Microbiol. Immunol.* 36:803–813.

Voino-Yasenetsky, M. V., and Voino-Yasenetsky, M. K. 1961. "Experimental pneumonia caused by bacteria of the shigella group," *Acta Morphol.* XI:440–454.

Walterspiel, J. N., Ashkenazi, S., Morrow, A. L., and Cleary, T. G. 1992. "Effect of subinhibitory concentrations of antibiotics on extracellular shiga-like toxin I," *Infection* 20:25–29.

Wassef, J. S., Keren, D. F., and Mailloux, J. L. 1989. "Role of M cells in initial antigen uptake and in ulcer formation in the rabbit intestinal loop model of shigellosis," *Infect. Immun.* 57:858–863.

Watarai, M., Funato, S., and Sasakawa, C. 1996. "Interaction of Ipa proteins of *Shigella flexneri* with α5β1 integrin promotes entry of the bacteria into mammalian cells,'' *J. Exp. Med.* 183:991–999.

Watarai, M., Tobe, T., Yoshikawa, M., and Sasakawa, C. 1995. ''Contact of *Shigella* with host cells triggers release of Ipa invasins and is an essential function of invasiveness,'' *Embo J.* 14:2461–2470.

Watkins, H. M. S. 1960. ''Some attributes of virulence in *Shigella,*'' *Ann. N. Y. Acad. Sci.* 88:1167–1187.

Weissman, J. B., Gangarosa, E. J., and DuPont, H. L. 1973. ''Changing needs in the antimicrobial therapy of shigellosis,'' *J. Infect. Dis.* 127:611–613.

Wharton, M., Spiegel, R. A., Horan, J. M., Tauxe, R. V., Wells, J. G., Barg, N., Herndon, J., Meriwether, R. A., MacCormack, J. N., and Levine, R. H. 1990. ''A large outbreak of antibiotic-resistant shigellosis at a mass gathering,'' *J. Infect. Dis.* 162:1324–1328.

Yee, R. B. and Buffenmeyer, C. L. 1970. ''Infection of cultured mouse macrophages with *Shigella flexneri,*'' *Infect. Immun* 1:459–463.

Zoeller, C. and Manoussakis, 1924. ''De la kérato-conjonctivite expérimentale dysentérique,'' *Compt. Rend. Soc. Biol.* 91:257–258.

Zwillich, S. H., Duby, A. D., and Lipsky, P. E. 1989. ''T-lymphocyte clones responsive to *Shigella flexneri,*'' *J. Clin. Microbiol.* 27:417–421.

Zychlinsky, A., Prevost, M. C., and Sansonetti, P. J. 1992. ''*Shigella flexneri* induces apoptosis in infected macrophages,'' *Nature* 358:167–169.

Zychlinsky, A., Fitting, C., Cavaillon, J.-M., and Sansonetti, P. J. 1994. ''Interleukin 1 is released by murine macrophages during apoptosis induced by *Shigella flexneri,*'' *J. Clin. Invest.* 94:1328–1332.

Zychlinsky, A., Thirumalai, K., Arondel, J., Cantey, J. R., Aliprantis, A. O., and Sansonetti, P. J. 1996. ''In vivo apoptosis in *Shigella flexneri* infections,'' *Infect. Immun.* 64:5357–5365.

CHAPTER 3

The Molecular Pathogenesis of *Escherichia coli* Infections

MICHAEL S. DONNENBERG
JAMES P. NATARO

1. INTRODUCTION

ESCHERICHIA *coli* may be the most versatile of human pathogens. This organism is not only the dominant gram-negative facultative anaerobe in the human gastrointestinal tract, it is also a potent pathogen capable of causing a variety of diseases by a dizzying array of mechanisms. *E. coli* can cause no less than six clinical syndromes of diarrhea with overlapping, but distinct symptoms and epidemiology; it can cause urinary tract infections ranging from asymptomatic bacteriuria to urosepsis; it can cause neonatal meningitis, pneumonia, cholecystitis, and wound infections.

The ability of a single species to co-exist with its host as a symbiote and to cause so many distinct diseases is a property of the genetic heterogeneity of the organism. While much of the *E. coli* genome is common to all strains of the species, the various clinical syndromes ascribed to *E. coli* are caused by pathotypes that differ from each other and from commensal strains because they have acquired distinct sets of virulence genes. These genes are carried on plasmids, on lysogenic bacteriophages, or on large chromosomal insertions known as pathogenicity islands. The distinctive clinical syndromes recognized in patients infected with different pathotypes of *E. coli* are a direct result of the interactions with the host encoded by the various combinations of gene sets. Presumably, these combinations of virulence genes were not selected by virtue of their ability to harm humans, rather these genes endow the organism with the ability to occupy distinctive niches within (or between) hosts. Unfortunately, little is known about the precise selective advantage provided to the organism by the various sets of virulence genes.

This chapter is organized by pathotype, with sections on each of six categories of diarrheagenic *E. coli* and two on *E. coli* that cause extraintestinal infections. Within each section, we will attempt to describe the current state of knowledge of the clinical and epidemiological features of the infection, the mechanisms by which the pathotype causes disease, and the prospects for preventing and controlling the infection. In keeping with the goals of the text, we will include available information on transmission of the infection through food and water. Despite these attempts at organization, the reader should bear in mind that the free flow of genetic information in nature is a force that complicates our efforts to identify individual *E. coli* strains as members of one or another pathotype. Strains that share virulence features with more than one group have been identified and will likely be recognized with increasing frequency. The species is not static, but is constantly evolving in leaps and bounds by the acquisition of new genetic determinants.

2. ENTEROPATHOGENIC *E. COLI* (EPEC)

2.1. CLINICAL AND EPIDEMIOLOGICAL FEATURES

The first *E. coli* implicated as a cause of diarrhea was the pathotype now known as EPEC (Bray, 1945). Once a feared cause of devastating outbreaks of nosocomial and community-acquired infant diarrhea, EPEC disease is now rare in developed countries. However, EPEC continue to be a leading cause of endemic diarrhea in infants in developing countries throughout the world (Donnenberg, 1995). Disease due to EPEC is distinctive in that it occurs almost exclusively in very young infants, it is characterized by profuse watery diarrhea with vomiting and low-grade fever, and it can be protracted and resistant to oral rehydration therapy (Donnenberg, 1995). EPEC is transmitted primarily by person-to-person spread (Bower et al., 1989; Wu and Peng, 1992), but outbreaks of EPEC disease have occasionally been linked to contaminated food and water (Viljanen et al., 1990; Schroeder et al., 1968).

2.2. PATHOGENESIS

The pathogenesis of EPEC infection is yielding to molecular and cellular analysis. EPEC can be defined as strains of *E. coli* that are capable of inflicting upon host cells a particular type of damage known as the attaching and effacing lesion, but which do not produce Shiga toxins (Kaper, 1996). Typical EPEC also possess a large plasmid that confers upon them the ability to form three-dimensional microcolonies on the surface of cells in a pattern termed localized adherence (Baldini et al., 1983; Scaletsky et al., 1984). Localized adherence is highly specific to EPEC, as strains that exhibit this feature almost invariably cause attaching and effacing lesions and lack Shiga toxins. A three-stage

EPEC

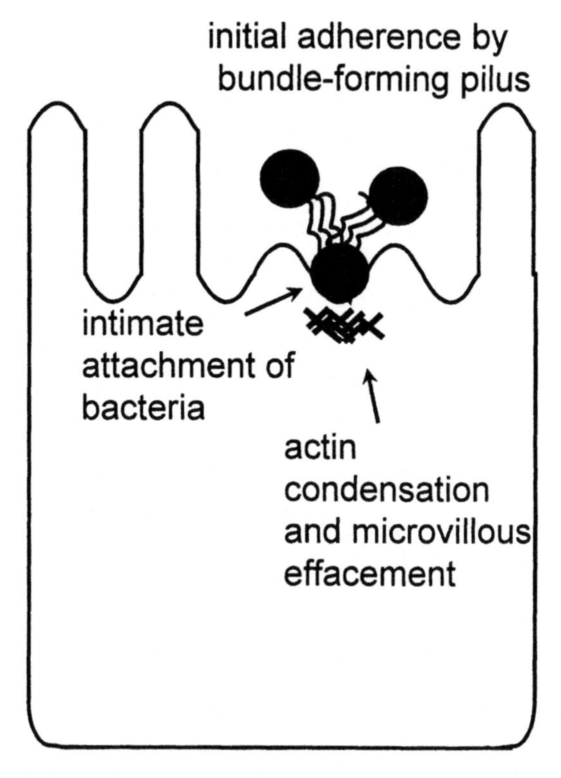

Figure 3.1 Interactions between enteropathogenic *E. coli* (EPEC) and epithelial cells. Three stages of EPEC interactions with host cells have been proposed, initial localized adherence mediated by the plasmid-encoded bundle-forming pilus, signal transduction mediated by secreted Esp proteins, and intimate attachment mediated by the outer membrane protein intimin. The result is loss of microvilli and the formation a cup-like pedestal composed of actin and other cytoskeletal proteins upon which the bacteria rest.

model has been put forward that describes many of the interactions between EPEC and epithelial cells (Figure 3.1) (Kaper, 1996; Donnenberg and Kaper, 1992). In the first stage, EPEC adhere to cells in the localized adherence pattern. In the second stage, EPEC signal cells to activate a receptor. In the third stage, EPEC attaches intimately to the cells via this receptor and the cell responds by dramatically altering its cytoskeleton.

2.2.1. Stage One: Localized Adherence and the Bundle-Forming Pilus

Typical EPEC strains produce a type IV fimbria known as the bundle-forming pilus (BFP) (Girón et al., 1991a). Fourteen contiguous *bfp* genes on the large EPEC plasmid are sufficient for biogenesis of the pili in a recombinant *E. coli* strain (Stone et al., 1996; Sohel et al., 1996). Mutations in several of these genes block expression of BFP and eliminate localized adherence (Ramer et al., 1996; Donnenberg et al., 1992). An anti-serum raised against BFP also reduces localized adherence (Girón et al., 1991a). Expression of the *bfp* gene cluster by a recombinant *E. coli* strain endows that strain with the ability to perform localized adherence, although at lower levels than EPEC (Stone et al., 1996; Sohel et al., 1996). Thus, it appears that BFP are responsible at least in part for the localized adherence phenotype of EPEC. However, a direct role for these pili in attachment to a host cell receptor has not been confirmed. Since these pili aggregate into rope-like bundles, it is also possible that their primary role is to tether bacteria to each other, rather than to host cells.

2.2.2. Stage Two: Signal Transduction Through Secreted Proteins

In addition to the large plasmid encoding BFP and localized adherence, EPEC possess a 35,637 bp pathogenicity island termed the locus of enterocyte effacement (LEE) (McDaniel et al., 1995). In many EPEC strains, the LEE is inserted into the chromosome 14 bp downstream of the *selC* locus, the precise site at which an entirely different pathogenicity island is inserted in a uropathogenic *E. coli* isolate (see uropathogenic *E. coli*). The base composition of the LEE is radically different from that of the rest of the *E. coli* genome, suggesting an origin in another species. The LEE is necessary and sufficient to confer upon recombinant *E. coli* the ability to perform the attaching and effacing effect (McDaniel and Kaper, 1997). The attaching and effacing effect consists of the localized destruction of microvilli and the formation of a cup-like pedestal composed of cytoskeletal proteins at the site at which bacteria bind intimately to host cells (Finlay et al., 1992; Knutton et al., 1989a; Moon et al., 1983).

Much of the LEE is composed of *sep* (secretion of *E. coli* proteins) genes encoding a type III secretion apparatus (Jarvis et al., 1995). Type III secretion systems, found in a variety of plant and animal pathogens, are dedicated to the export and, in at least some cases, to the delivery into the host cell cytoplasm, of factors critical to the pathogenesis of infection (Lee, 1997; Mecsas and Strauss, 1996). In the case of EPEC, at least three proteins, EspA, EspB, and EspD, rely on the Sep apparatus for export out of the bacterium, as mutations in *sep* genes block export of all three Esp proteins (Lai et al., 1997; Kenny et al., 1996; Kenny and Finlay, 1995). The Esp proteins are encoded by three contiguous *esp* genes located near one end of the LEE. Mutations in each of the *esp* genes and in *sepB* completely eliminate the ability of EPEC to form attaching and effacing lesions (Donnenberg et al., 1993c; Lai et al., 1997; Kenny et al., 1996). Thus, it appears that export of the Esp proteins by the Sep apparatus is necessary for this effect.

The functions of the Esp proteins are not yet defined, but it appears that their major role is in transducing signals to host cells. EPEC strains with *esp* mutations lose the ability to lower the resting membrane potential of epithelial cells and to stimulate a change in short circuit current in epithelial cell monolayers, attributes that may be correlated with fluid secretion in the intestine (Knutton et al., 1996; Stein et al., 1996). They also lose the ability to induce tyrosine phosphorylation of Hp90, the putative receptor for intimate attachment (Lai et al., 1997; Kenny et al., 1996; Rosenshine et al., 1992; Foubister et al., 1994).

2.2.3. Stage Three: Intimate Attachment and Maturation of the Attaching and Effacing Lesion

The *eae* gene, located in the LEE between the *esp* genes and most of the *sep* genes, encodes an outer membrane protein known as intimin (Jerse et al., 1990). Intimin is closely related to the invasins produced by enteric *Yersinia,* but unlike invasin, is not sufficient to promote cellular invasion. EPEC strains with mutations in *eae* lose the ability to attach intimately to host cells. The ability of *eae* mutants to cause diarrhea in experimental human infections is attenuated (Donnenberg et al., 1993a). However, such mutants remain capable of signaling cells to induce tyrosine phosphorylation of Hp90 and to cause changes in membrane polarization and shortcircuit current (Knutton et al., 1996; Rosenshine et al., 1992; Stein et al., 1996). Moreover, when tissue culture cells are co-infected with an *eae* mutant along with a *sep* or *esp* mutant, the *sep* or *esp* mutant regains the ability to form attaching and effacing lesions (Foubister et al., 1994; Rosenshine et al., 1992). Thus, the *eae* mutant is able to signal the cells, presumably through the Esp proteins, to complement the defect in the signaling mutant. Recently Hp90 has been shown to bind to intimin and has been proposed to be the intimin receptor (Rosenshine et al., 1996). Thus, the purpose of EPEC signaling to cells seems to be to deliver or activate the Hp90 receptor to allow EPEC to bind intimately to cells via intimin.

2.3. PREVENTION AND CONTROL

EPEC infections are most common in poor urban areas in developing countries. There is no known animal reservoir for human EPEC strains and EPEC infection has rarely been linked to contaminated food or water. EPEC disease is much more common in bottle-fed than breast-fed infants and human milk has been shown to block EPEC adherence to host cells (Blake et al., 1993; Silva and Giampaglia, 1992; Cravioto et al., 1991b). Therefore, the most practical program for reducing EPEC infection may be educational efforts to increase breast feeding in poor urban areas of developing countries. The possibility of vaccines against EPEC targeting BFP, intimin or the Esp proteins has not received much attention, perhaps because of the formidable barriers to delivery of such a vaccine to the population most in need.

3. ENTEROHEMORRHAGIC *E. COLI* (EHEC)

3.1. CLINICAL AND EPIDEMIOLOGICAL FEATURES

The enterohemorrhagic *E. coli* include strains of serotype O157:H7, 0157:NM, O111:NM, O111:H8 and O26:H11 and others that produce Shiga toxins and cause attaching and effacing lesions in vitro (Caprioli et al., 1994; Ørskov and Ørskov, 1992; Scotland et al., 1990) (Figure 3.2). EHEC are the most important members of a larger group of Shiga toxin-producing *E. coli,* which includes strains that do not cause attaching and effacing lesions. However, all large outbreaks of infection due to Shiga toxin-producing *E. coli* reported thus far have been due to EHEC strains, with O157:H7 strains by far the most important cause in developed countries (Griffin and Tauxe, 1991).

Disease due to EHEC infection is characterized by diarrhea, which is usually watery at onset, but often progresses to grossly bloody stools (hemorrhagic colitis), and by severe crampy abdominal pain (Tarr, 1995; Griffin and Tauxe, 1991; Griffin et al., 1988). Fever is not a prominent feature of the infection. The disease can mimic non-infectious illnesses such as inflammatory bowel disease, ischemic colitis, diverticulosis, or appendicitis. Approximately 5–10% of individuals infected with *E. coli* O157:H7 go on to develop a thrombotic microangiopathy manifest either as hemolytic-uremic syndrome or thrombotic thrombocytopenic purpura. These severe complications are characterized by damage to the endothelium lining small blood vessels leading to microthrombi, fragmentation and destruction of red blood cells, consumption of platelets, and tissue ischemia. Severe end-organ damage ensues. The kidneys are especially susceptible, but the brain, intestines and other organs can also be affected. Mortality from HUS is about 3%.

EHEC can be isolated from healthy cattle, but may cause disease in calves (Dean-Nystrom et al., 1997; Moxley and Francis, 1986). EHEC have also been isolated from deer and caribou (Orr et al., 1994; Keene et al., 1997). The most common source of epidemic EHEC infection is contaminated ground beef (Bell et al., 1994; Griffin and Tauxe, 1991). However, large outbreaks have been linked to salami, sausage, milk, apple cider, and unpasteurized juice (Tarr, 1995; Besser et al., 1993; Griffin and Tauxe, 1991; CDC, 1995). The largest outbreak to date has been linked to radish sprouts (Watanabe and Guerrant, 1997). At the time of this writing an outbreak linked to alfalfa sprouts is under investigation (CDC, 1997a). While contamination of implicated products with bovine feces has been suggested for some of the non-beef-associated outbreaks, the source of the bacteria in the sprout outbreaks

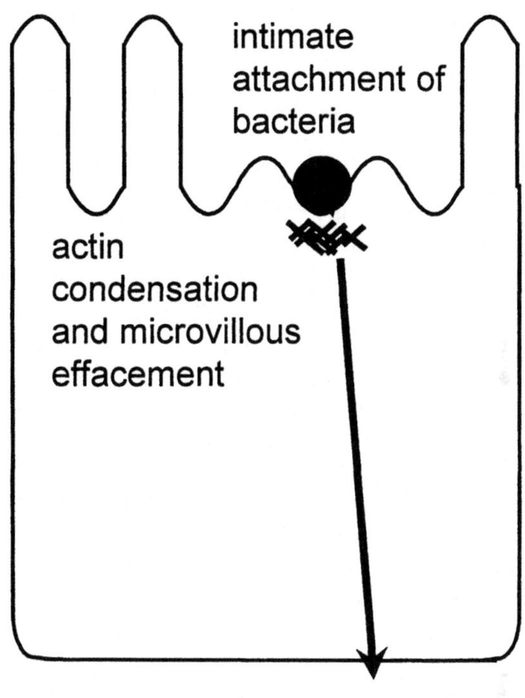

Figure 3.2 Interactions between enterohemorrhagic *E. coli* (EHEC) and epithelial cells. EHEC cause attaching and effacing lesions with loss of microvilli and formation of adhesion pedestals similar to those caused by EPEC. In addition, EHEC deliver Shiga toxins, which presumably leave the bowel lumen and enter the body to cause systemic effects.

remains obscure. EHEC can also be spread by person-to-person contact, by contaminated drinking water, and by exposure during swimming (CDC, 1996; Keene et al., 1994; Belongia et al., 1993; Swerdlow et al., 1992; Griffin and Tauxe, 1991; Carter et al., 1987).

3.2. PATHOGENESIS

It is now clear from several lines of investigation that EHEC serotype O157:H7 and EPEC serotype O55:H7 share a recent common ancestor (Whittam et al., 1993). Interestingly, like EHEC, many strains of EPEC O55:H7 lack the adherence plasmid found in typical EPEC strains (Scotland et al., 1991). The evidence indicates that O157:H7 evolved from an EPEC strain by the acquisition of at least three genetic elements, perhaps in stepwise fashion: bacteriophages encoding Shiga toxins, a plasmid encoding a hemolysin, and the genes encoding the lipopolysaccharide sides chains specifying the O157 antigen (Bilge et al., 1996).

3.2.1. Shiga Toxins

The Shiga toxins (also known as Shiga-like toxins, verotoxins, and verocytotoxins) are a family of molecules composed of a single catalytic A subunit and five receptor-binding B subunits (Tesh and O'Brien, 1991). The B subunits form a pentamer with a central pore through which the carboxyl terminus of the A subunit protrudes (Fraser et al., 1994). Upon binding to cells expressing globo series glycolipids containing the minimal receptor Galactose-$\alpha(1 \rightarrow 4)$-Galactose (Jacewicz et al., 1994), the A subunit enters the cell by endocytosis, transits to and beyond the Golgi (Sandvig et al., 1992), and enters the cytoplasm where it catalyses the *N*-glycosidation of adenine 4324 of 28S ribosomal RNA (Tesh and O'Brien, 1991). This leads to a failure of peptide chain elongation, a block in protein synthesis and cell death. Two members of the Shiga toxin family, Stx1 and Stx2, are produced by EHEC strains pathogenic for humans. Each is encoded by a bacteriophage related to lambda. Strains may produce either or both toxins.

Toxin production is thought to be crucial to the severe complications of EHEC infection. It has been proposed, but not demonstrated, that systemic absorption of Shiga toxins leads to intoxication of endothelial cells (Acheson et al., 1996). Receptors for Shiga toxins have been demonstrated on endothelial cells, and human endothelial cells of the kidney are especially sensitive to the toxins (Louise and Obrig, 1995). Dying endothelial cells are thought to provide the surface for activation of fibrin and coagulation factors that lead to the microangiopathy.

3.2.2. Attaching and Effacing Lesions

Like EPEC, EHEC possess the entire LEE and cause attaching and effacing lesions in animals (McDaniel et al., 1995; Tzipori et al., 1989). However, such lesions have not been reported in biopsies from patients with EHEC infection. Like EPEC, EHEC secrete EspA and EspB proteins into culture supernatants (Ebel et al., 1996; Jarvis and Kaper, 1996). As in EPEC, mutations in the *eae* gene of EHEC lead to inability to bind intimately to host cells and to attenuation of diarrhea in vivo, in this case in a gnotobiotic piglet model (Donnenberg et al., 1993b). However, an EHEC *eae* mutant is still capable of causing neurologic dysfunction and death in piglets, presumably the results of Shiga toxin production (Tzipori et al., 1995).

3.2.3. Heat and Acid Stability

The infectious dose required to cause EHEC infection has been calculated to be less than 1000 organisms (CDC, 1997b). This low inoculum suggests that EHEC are able to resist killing by the acidic pH normally found in the stomach. Indeed, outbreaks of EHEC infection caused by contaminated salami and sausage also indicate that the organism is able to withstand acidic conditions, since fermentation of these products causes the pH to fall below 4.5. In vitro EHEC is able to survive at pH 2.5 for at least 5 h (Benjamin and Datta, 1995). EHEC is also able to resist killing by temperatures as high as 62.8°C and dehydration during preparation of jerky (Keene et al., 1997).

3.2.4. Plasmid Factors

EHEC strains of O157:H7 and other serotypes often possess a large plasmid (Levine et al., 1987). This plasmid encodes a hemolysin (Schmidt et al., 1996), a serine protease (Brunder et al., 1997) and may encode an outer membrane protein (Fratamico et al., 1993). However a role for plasmid encoded factors in pathogenesis has not been demonstrated in animal models (Tzipori et al., 1987).

3.3. PREVENTION AND CONTROL

One of the most important factors in preventing illness due to EHEC infection is the prompt recognition and interruption of outbreaks (Watanabe and Guerrant, 1997; Bell et al., 1994). Although the Centers for Disease Control recommend culturing all bloody stools for EHEC, many laboratories do not. Lack of detection undoubtedly contributes to under-appreciation of common source outbreaks and sporadic cases of EHEC disease (CDC, 1995a). The more stringent policies recently approved by the United States Department of Agriculture may help to identify critical control points that lead to contamination of meat. There is no current vaccine against EHEC, but the notion of immunizing cattle to reduce the reservoir of colonized animals may have merit.

4. ENTEROTOXIGENIC *E. COLI*

4.1. CLINICAL AND EPIDEMIOLOGICAL FEATURES

Enterotoxigenic *E. coli* (ETEC) cause watery diarrhea, which can range from mild, self-limiting disease to severe purging (Levine et al., 1977). The organism is a major cause of weanling diarrhea in the developing world and the major cause of travelers' diarrhea (Tornieporth et al., 1995; Levine, et al.,

1993; Albert et al., 1995; Hoque et al., 1994). The epidemiologic pattern of ETEC infection is in large part determined by three features: (1) mucosal immunity to ETEC infection develops in exposed individuals; (2) even immune asymptomatic individuals may shed high numbers of virulent ETEC in the stool; and (3) the infection requires a relatively high infectious dose (DuPont et al., 1971). These three features create a situation in which ETEC contamination of the environment in endemic areas is extremely prevalent and most infants in such areas will encounter ETEC soon after weaning. The percentage of cases of sporadic endemic infant diarrhea that are due to ETEC in developing areas usually varies from 10-30% (Mangia et al., 1993; Hoque et al., 1994; Flores-Abuxapqui et al., 1994; Schultsz, 1994; Tornieporth et al., 1995; Levine et al., 1993; Albert et al., 1995). School-age children and adults in these areas typically have a very low incidence of symptomatic ETEC infection. Travelers from the developed world who visit developing areas have no immunity to ETEC and, therefore, ETEC represent the number one cause of travelers' diarrhea.

Epidemiologic investigations have implicated contaminated food and water as the most common vehicles for ETEC infection (Long et al., 1994; Wood et al., 1996; Black et al., 1982; Black et al., 1981) and transmission is highest during the warm seasons. Since the infectious dose required to initiate infection is relatively high (10^8 with bicarbonate buffer), ETEC contamination rates are predictably high. ETEC infections tend to be clustered in warm, wet months in endemic areas, when multiplication of ETEC in food and water is most efficient (Levine, 1987). Person-to-person transmission was not found to occur in ETEC volunteer studies (Levine et al., 1980).

4.2. PATHOGENESIS

ETEC elicit diarrhea by colonization of the small bowel mucosa, followed by elaboration of the heat labile (LT) and/or heat stable (ST) enterotoxins (Figure 3.3) (O'Brien and Holmes, 1996; Sears and Kaper, 1996; Levine, 1987). Colonization of the mucosal surface is mediated by fimbriae known as colonization factor antigens (CFAs) (Gaastra and Svennerholm, 1996; Cassels and Wolf, 1995; de Graaf and Gaastra, 1994).

4.2.1. Colonization Factors

There are a very large number of CFAs from human ETEC (and a similarly large number from animal ETEC). ETEC fimbriae confer the species-specificity of the pathogen; for example, ETEC expressing K99 are pathogenic for calves, lambs and pigs, whereas K88-expressing organisms cause disease only in pigs (Cassels and Wolf, 1995).

The CFAs can be subdivided based upon their morphologic characteristics. CFA/I is a rigid rod-shaped fimbria of 7 nm diameter (Jann and Hoschutsky,

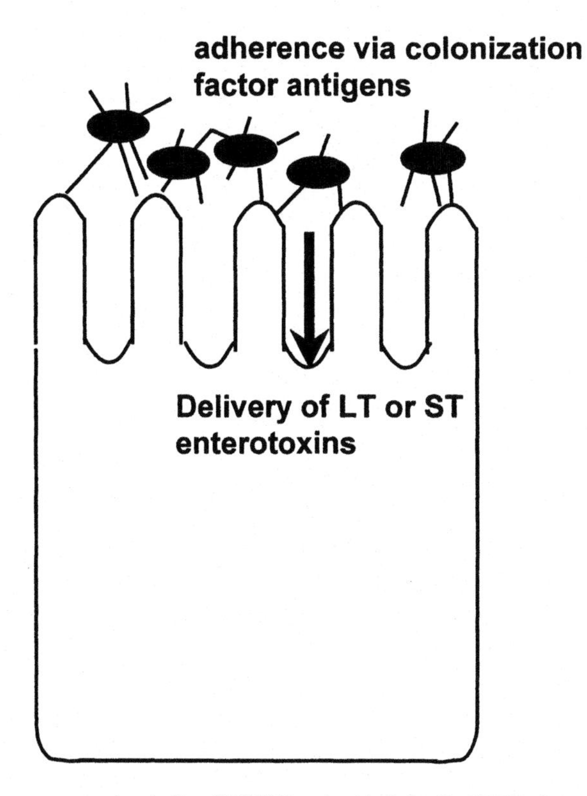

Figure 3.3 Interactions between enterotoxigenic *E. coli* (ETEC) and epithelial cells. ETEC adhere to epithelial cells via fimbriae known as colonization factor antigens and deliver heat-labile (LT) and heat-stable (ST) enterotoxins that lead to fluid secretion.

1991). CFA/III is a bundle-forming pilus with homology to the type 4 fimbrial family (Taniguchi et al., 1994; Taniguchi et al., 1995). CFA/II and CFA/IV are composed of multiple distinct fimbrial structures: CFA/II producers express the flexible CS3 structure either alone or in association with the rod-shaped CS1 or CS2 (Sjoberg et al., 1988; Levine et al., 1984); CFA/IV producers express CS6 in conjunction with CS4 or CS5 (Cassels and Wolf, 1995; Knutton et al., 1989). A large number of other, less common adhesins have also been described in ETEC (de Graaf and Gaastra, 1994), yet epidemiologic studies suggest that together CFA/I, CFA/II and CFA/IV are expressed by approximately 75% of human ETEC worldwide (Wolf, 1997). A newly described ETEC type IV fimbriae, designated Longus, has been found on a large proportion of human ETEC (Girón et al., 1994).

The genetics of CFAs have been studied extensively and such studies have served to illuminate models for fimbrial expression, protein secretion and translocation, and the assembly of bacterial organelles. CFA genes are usually encoded upon plasmids, which typically also encode the enterotoxins ST and/or LT (de Graaf and Gaastra, 1994). Typical fimbrial gene clusters consist of a series of genes encoding a primary fimbrial subunit protein and accessory proteins, which are required for processing, secretion and assembly of the fimbrial structure itself (Jones et al., 1996; Jann and Hoschutsky, 1991; Kusters and Gaastra, 1994; de Graaf and Gaastra, 1994).

4.2.2. Enterotoxins

ETEC cause diarrhea through the action of two unrelated groups of toxins, LT and ST. ETEC strains may express an LT only, an ST only, or both an LT and an ST.

4.2.3. Heat-Labile Enterotoxins

LTs produced by ETEC are toxins that are closely related in structure and function to the cholera enterotoxin (CT) expressed by *Vibrio cholerae* (Sixma et al., 1993). The LT found predominantly in human isolates (called LT-I) is approximately 75% identical at the amino acid level with CT. The two toxins are virtually identical in crystal structure and share several phenotypes, including the primary eukaryotic cell receptor (Teneberg et al., 1994) and the principal mechanisms of action (Dickinson and Clements, 1995).

LT-I is an oligomeric toxin of ca. 86 kDa composed of one 28 kDa A subunit and five identical 11.5 kDa B subunits (Spangler, 1992; Sixma et al., 1991). The B subunits are arranged in a ring or "doughnut" and bind strongly to the ganglioside GM1 and weakly to GD1b and some intestinal glycoproteins (Spangler, 1992; Teneberg et al., 1994). The A subunit is responsible for the enzymatic activity of the toxin (Moss et al., 1979) and is cleaved to yield A_1 and A_2 peptides joined by a disulfide bond. Two closely related variants of LT-I (called LTp for porcine and LTh for human) have been described, which exhibit partial antigenic cross-reactivity.

Genes encoding LT-I (called *elt* or *etx*) reside on plasmids, which also may contain genes encoding ST and/or CFAs. The genes are arranged as an A-B cluster with overlap of the B ribosomal binding site with the A coding region (Dallas and Falkow, 1980). The A and B subunits are synthesized individually and secreted through the inner membrane coupled with processing of their signal sequences (Hirst et al., 1984). In the periplasm, the subunits assemble into the mature A_1B_5 configuration, which is held together by non-covalent bonds (Spangler, 1992; Hol et al., 1995; Hirst et al., 1984). The last four residues of the A subunit are required for stability of the holotoxin structure (Hirst, 1995).

After binding to host cell membranes, the LT holotoxin is endocytosed and translocated through the cell in a process involving trans-Golgi retrograde transport (Lencer et al., 1995). The A_1 protein acts by transferring an ADP-ribosyl moiety from NAD to the alpha subunit of G_s a regulatory G protein that serves to regulate adenylate cyclase. ADP-ribosylation of the $G_{s\alpha}$ results in adenylate cyclase being "locked on," thereby leading to increased levels of intracellular cAMP. cAMP, in turn, stimulates the cAMP-dependent protein kinase (A kinase) to phosphorylate and thereby activate chloride channels, including the Cystic Fibrosis Transmembrane Conductance Regulator (CFTR). The net result of CFTR phosphorylation is increased Cl^- secretion from secretory crypt cells. In addition, LT serves to inhibit NaCl absorption by villus tip cells (Sears and Kaper, 1996; Field et al., 1972).

There is increasing evidence that stimulation of secretion and inhibition of absorption by LT involves several accessory effector pathways (McGee et al., 1993; Eklund et al., 1987; Sears and Kaper, 1996). The mediators of these pathways include prostaglandins, platelet-activating factor, vasoactive intestinal peptide and the enteric neural system (ENS), and cytokines. A role for any of these secondary mechanisms in human disease has yet to be demonstrated.

4.2.4. Heat-Stable Enterotoxins

The STs of ETEC are small, monomeric toxins that contain multiple cysteine residues whose disulfide bonds account for the heat stability of these toxins (Hirayama, 1995). There are two unrelated classes of STs (STa and STb), which differ in structure and mechanism of action. Genes for both classes are predominantly found on plasmids and some ST-encoding genes have been found on transposons (So and McCarthy, 1980).

The mature STa toxin is an 18- or 19-amino acid peptide with a molecular mass of ca. 2 kDa. Two STa variants exist, designated STp (porcine) and STh (human) (Hirayama, 1995). Both variants can be found in human ETEC isolates and are presumed to be equally pathogenic. STh and STp are nearly identical in the 13 residues that are necessary and sufficient for enterotoxic activity. Of these 13 residues, six are cysteines, which form three intramolecular disulfide bridges. STa is synthesized as a 72 amino acid precursor ("pre-pro form"), which is cleaved by signal peptidase to a 53 amino-acid peptide "pro-form" in the periplasm (Rasheed et al., 1990). An undefined periplasmic protease then processes pro-STa to the final 18 or 19 residue-mature toxin, which is released by diffusion across the outer membrane.

The major receptor for STa is a membrane-spanning guanylate cyclase (GC-C) (de Sauvage et al., 1991). GC-C belongs to a family of receptor cyclases that includes the atrial natriuretic peptide receptors GC-A and GC-B. Binding of STa to GC-C stimulates guanylate cyclase activity, leading to increased intracellular cyclic GMP levels, which in turn produce stimulation of chloride secretion and inhibition of sodium chloride absorption (Dreyfus et al., 1984; deJonge and Lohman, 1985; Crane et al., 1992). Ultimately, STa is believed to phosphorylate and thereby activate the CFTR chloride channel, as does LT. However, phosphorylation of CFTR by STa apparently involves cGMP-dependent kinases instead of or in addition to A kinase (Sears and Kaper, 1996). Alternative mechanisms of action for STa involving prostaglandins and the ENS have been proposed, but the evidence for the involvement of these factors is inconsistent (Nzegwu and Levin, 1996; Nzegwu and Levin, 1994). The secretory response to STa may also involve phosphatidyl inositol and diacyl glycerol release, activation of protein kinase C, elevation of intracellular calcium, and/or microfilament (F-actin) rearrangement (reviewed in Sears and Kaper, 1996).

STb is synthesized as a 71-amino-acid precursor protein, which is processed to a mature 48-amino-acid protein with a molecular weight of 5.1 kDa (Fujii et al., 1991). The STb protein sequence has no homology with that of STa although it does contain four cysteine residues, which form disulfide bonds (Arriaga et al., 1995). Unlike STa, STb induces histologic damage in the intestinal epithelium, consisting of loss of villus epithelial cells and partial villus atrophy. The mechanism by which STb causes intestinal secretion is not well characterized (Sears and Kaper, 1996), but it apparently does not stimulate increases in intracellular cAMP or cGMP. STb also stimulates the release of PGE_2 and serotonin, suggesting that the ENS may also be involved in the secretory response to this toxin (Fujii et al., 1995; Hitotsubashi et al., 1992).

4.3. PREVENTION AND CONTROL

The cornerstone of prevention for ETEC diarrhea is the provision of uncontaminated food and water sources. For infants and children in the developing world, the major victims of endemic disease, this entails breaking the chain of fecal-oral contamination of weaning foods and water. Such contamination may come from cases of ETEC diarrhea or from individuals who excrete ETEC asymptomatically. In turn, providing clean water depends on the maintenance of clean water sources, adequate sewage disposal and personal fecal-oral hygiene. Hygienic food practices entail minimizing the length of time during which foods are maintained at ambient temperatures, especially in warm seasons.

Similar recommendations are also relevant to the prevention of travelers' diarrhea. Travelers should seek uncontaminated food and water sources while visiting endemic areas.

5. ENTEROAGGREGATIVE *E. COLI*

5.1. CLINICAL AND EPIDEMIOLOGICAL FEATURES

Enteroaggregative *E. coli* (EAEC) are pathogens associated with pediatric diarrhea in the developing world and are particularly associated with cases that persist for more than 14 days (Mayer and Wanke, 1995; Fang et al., 1995; Bhatnagar et al., 1993; Bhan et al., 1989; Cravioto et al., 1991a; Cobeljic et al., 1996). Huppertz et al. have suggested that EAEC may be a cause of sporadic infant diarrhea in Germany (Huppertz et al., 1997) and the role of EAEC in developed countries has not been studied systematically. Clinically, EAEC infection is manifested by the presence of watery mucoid diarrhea, but blood may be found in a significant proportion of patients' stools (Cravioto et al., 1991a; Cobeljic et al., 1996).

Although most reports have implicated EAEC in sporadic endemic diarrhea, a growing number of reports have described EAEC outbreaks (Cobeljic et al., 1996; Eslava et al., 1993; Smith et al., 1994). Cobeljic et al. have described an outbreak affecting 19 infants in the nursery of a hospital in Belgrade, Serbia, occurring over a nine-day period in 1995 (Cobeljic et al., 1996). The illness lasted 3–9 days (mean 5.2 days) in 16 babies, but in three infants, persistent diarrhea developed, lasting 18–20 days. Infants with diarrhea typically manifested liquid green stools; in three, mucus was visibly apparent. There was no gross blood. The source of infection was unclear. Cravioto and Eslava have described outbreaks of EAEC diarrhea in the malnutrition ward of a Mexico City hospital (Eslava et al., 1993). A massive outbreak of EAEC diarrhea occurred in Gifu Prefecture, Japan, in 1993 (Itoh et al., 1997). A total of 2697 children at 16 schools became ill with abdominal pain, nausea and severe diarrhea after consuming contaminated school lunches. A single EAEC strain of serotype O?:H10 was implicated, but the organism was not found in any of the foods served in the implicated luncheon.

Perhaps even more significant than the association of EAEC with persistent diarrhea are the recent data from Brazil and Australia that link EAEC with growth retardation in infants (Steiner et al., 1997). In each of these studies, the isolation of EAEC from the stools of infants was associated with a low *z*-score for height and/or weight, irrespective of the presence of diarrheal symptoms. Given the high prevalence of asymptomatic EAEC excretion in many areas, such an observation may imply that the contribution of EAEC to the human disease burden is significantly greater than is currently appreciated.

5.2. PATHOGENESIS

EAEC are currently defined as *E. coli,* which do not secrete enterotoxins

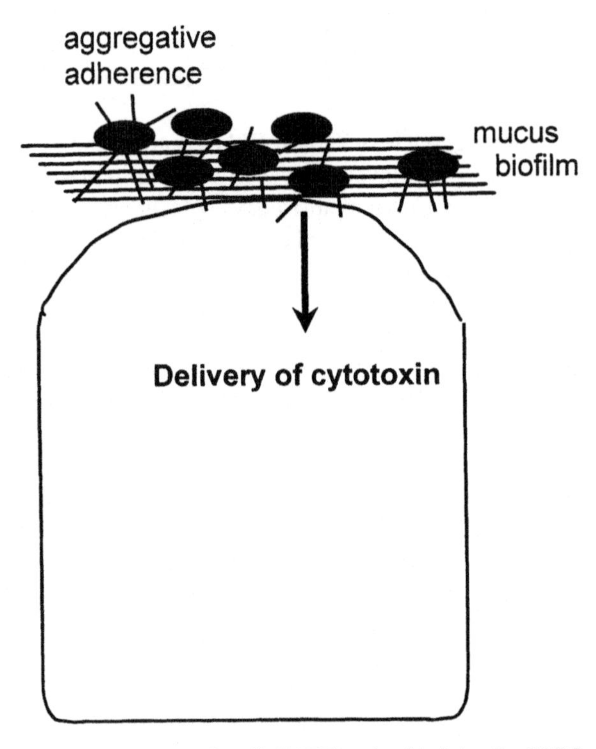

Figure 3.4 Interactions between enteroaggregative *E. coli* (EAEC) and epithelial cells. EAEC adhere initially by virtue of Aggregative Adherence Fimbriae. This is followed by stimulation of mucosal mucus secretion leading to the development of a mucus/bacteria biofilm. EAEC have been shown to secrete enterotoxins (EAST1 and/or the 108 kDa toxin), which stimulate fluid secretion and cause damage to epithelial cells.

LT or ST and which adhere to HEp-2 cells in an aggregative adherence (AA) pattern. Heterogeneous pathogenicity of EAEC has been confirmed in volunteer studies. (Nataro et al., 1995b).

EAEC pathogenesis is not thoroughly understood; however, a characteristic histopathologic lesion and several candidate virulence factors have been described (Figure 3.4). EAEC characteristically enhance mucus secretion from the mucosa, with trapping of the bacteria in a bacterium-mucus biofilm. Whereas only a minority of EAEC caused diarrhea in a gnotobiotic piglet model (Tzipori et al., 1992), all strains induced an unusual mucoid gel packed with aggregating bacteria. Hicks et al. reported that EAEC strains adhered to sections of pediatric small bowel mucosa in an in vitro organ culture (IVOC) model (Hicks et al., 1996). These investigators also observed EAEC to be embedded within a mucus-containing biofilm. In addition, EAEC have been shown to bind mucus in vitro (Wanke et al., 1990), and volunteers fed EAEC developed diarrhea which is predominantly mucoid in character (Nataro et al., 1995b). The role of excess mucus production in EAEC pathogenesis is unclear; however, the formation of a heavy biofilm may be related to the organism's diarrheagenicity and, perhaps, to its ability to cause persistent colonization and diarrhea.

Adherence to the intestinal mucosa requires the presence of surface fimbriae (Czeczulin et al., 1997; Nataro et al., 1992). Nataro et al. have identified a flexible, bundle-forming fimbrial structure of 2–3 nm diameter, designated Aggregative Adherence Fimbriae I (AAF/I) (Nataro et al., 1992), which mediates HEp-2 adherence and human erythrocyte hemagglutination in EAEC strain 17–2. The AAF/I fimbriae are bundle-forming fimbriae but do not show homology to members of the type 4 class of fimbriae (Tennant and Mattick, 1994). AAF/I is encoded as two separate gene clusters on the 60 MDa plasmid of 17–2, separated by 9 kb of intervening DNA (Savarino et al., 1994; Nataro et al., 1994; Nataro et al., 1993). Region 1 encodes a cluster of genes required for fimbrial synthesis and assembly, including the structural subunit of the fimbria itself. Region 1 is highly homologous to members of the Dr family of adhesins (Savarino et al., 1994). Region 2 encodes a transcriptional activator of AAF/I expression, which shows homology to members of the AraC family of DNA-binding proteins (Nataro et al., 1994). These investigators have also identified a second fimbriae, designated AAF/II, which is distinct morphologically and genetically from AAF/I (Nataro et al., 1995b). Still more AA adhesins are likely to exist.

5.2.1. EAEC ST-Like Toxin

Savarino et al. have described a 4100 Da homolog of the heat-stable enterotoxin (ST), designated Enteroaggregative ST1 (EAST1) (Savarino et al., 1993; Savarino et al., 1991). EAST1 is a 38-amino acid protein, which features four cysteine residues, unlike the six residues characteristic of *E. coli* ST. EAST1 clones yield net increases in shortcircuit current in the rabbit mucosal Ussing Chamber model (Savarino et al., 1991) but a role for this toxin in vivo has not been shown. Interestingly, other *E. coli* categories, notably the EHEC, have been shown to elaborate EAST1 with higher frequency than EAEC (Savarino et al., 1996).

5.2.2. Invasiveness

Benjamin et al. have suggested that some EAEC strains may invade intestinal epithelial cells in vitro (Benjamin et al., 1995). However, human intestinal explants do not reveal internalization of EAEC (Hicks et al., 1996), and clinical evidence for invasiveness is as yet lacking.

5.2.3. Cytotoxins

EAEC infection has been shown to induce mucosal damage in both animal models and infected patients. In rabbit and rat ileal loop models (Vial et al., 1988), EAEC induce shortening of the villi, hemorrhagic necrosis of the villous tips and a mild inflammatory response with edema and mononuclear infiltration of the submucosa.

Cravioto et al. demonstrated mucosal destruction in ileal sections of patients who succumbed to EAEC persistent diarrhea during an outbreak in a Mexico City hospital (Eslava et al., 1993) and Hicks et al. have shown that EAEC cytotoxicity can be demonstrated in IVOC using pediatric intestinal biopsies (Hicks et al., 1996). Nataro and Sears have shown that EAEC strain 042 elicits cytotoxic effects on T84 cells (human intestinal carcinoma) in vitro (Nataro et al., 1996). In the T84 cell model, EAEC give rise to a unique phenotype, characterized by vesiculation of the microvillar membrane, increased vacuole formation and separation of the nucleus from the surrounding cytoplasm. Ultimately, this effect leads to cell death and exfoliation of cells from the monolayer. In both the T84 and IVOC systems, the toxicity requires the presence of genes encoded on the 65 MDa plasmid in addition to those encoding the adherence fimbriae.

Eslava et al. have identified a 108 kDa cytotoxin which elicits destructive lesions in the rat ileal loop (Eslava et al., 1993). Navarro-Garcia et al. have also demonstrated enterotoxic activity of this protein in the Ussing chamber (Navarro-Garcia et al., 1997). This 108 kDa protein was recognized by serum from patients infected with EAEC in the Mexican outbreak described below. Eslava and Nataro have determined the sequence of the 108 kDa toxin gene and have found it to be homologous to member of the auto-transporter family of proteins (Jose et al., 1995).

5.2.4. A Model of EAEC Pathogenesis

EAEC pathogenesis apparently comprises a three-stage model. Stage I involves initial adherence to the intestinal mucosa and/or the mucus layer. AAF/I and AAF/II are the leading candidates for factors that may facilitate initial colonization. Stage II includes enhanced mucus production, apparently leading to deposition of a thick mucus-containing biofilm encrusted with EAEC. The mucus blanket may promote persistent colonization and perhaps nutrient malabsorption. Stage III, suggested from histopathologic and molecular evidence, includes the elaboration of an EAEC cytotoxin(s) and enterotoxin, which result in mucosal damage and fluid secretion. Malnourished hosts may be particularly impaired in their ability to repair this damage, leading to the persistent diarrhea syndrome.

5.3. PREVENTION AND CONTROL

The mode of transmission of EAEC is not known: however, the relatively high infectious dose in volunteer studies and the relative paucity of reported outbreaks suggest that food- and waterborne transmission is likely more important than person-to-person spread. Thus, as in ETEC control, the maintenance of sanitation and hygienic food and water is essential to controlling the infection. No EAEC vaccine is currently available.

6. ENTEROINVASIVE *E. COLI*

6.1. CLINICAL AND EPIDEMIOLOGICAL FEATURES

Enteroinvasive *E. coli* (EIEC) are biochemically, genetically and pathogenetically closely related to *Shigella spp.* Both characteristically cause an invasive inflammatory colitis, but either may also elicit a watery diarrhea syndrome indistinguishable from that due to other *E. coli* pathogens.

Most epidemiologic studies of EIEC describe outbreaks. EIEC outbreaks are usually food- or waterborne (Tornieporth et al., 1995; Snyder et al., 1984; Lanyi et al., 1959; Marier et al., 1973), although person-to-person transmission occurs (Harris et al., 1985). The infectious dose for EIEC in volunteers is higher than that for *Shigella* (DuPont et al., 1971) and thus the potential for person-to-person transmission is lower. Endemic sporadic disease occurs in some areas, generally where *Shigella* is also prevalent, but epidemiologic features are not well characterized (Ketyi, 1989; Taylor et al., 1988).

6.2. PATHOGENESIS

The pathogenesis of both EIEC and *Shigella* involves cellular invasion and spread, and requires specific chromosomal and plasmid-borne virulence genes (Figure 3.5) (Levine, 1987).

Both EIEC and *Shigella* have been shown to invade the colonic epithelium, a phenotype mediated by both plasmid and chromosomal loci. In addition, both EIEC and *Shigella* elaborate one or more secretory enterotoxins which may play roles in diarrheal pathogenesis.

6.2.1. Invasiveness

The current model of *Shigella* and EIEC pathogenesis comprises (1) epithelial cell penetration, followed by (2) lysis of the endocytic vacuole, (3) intracel-

EIEC

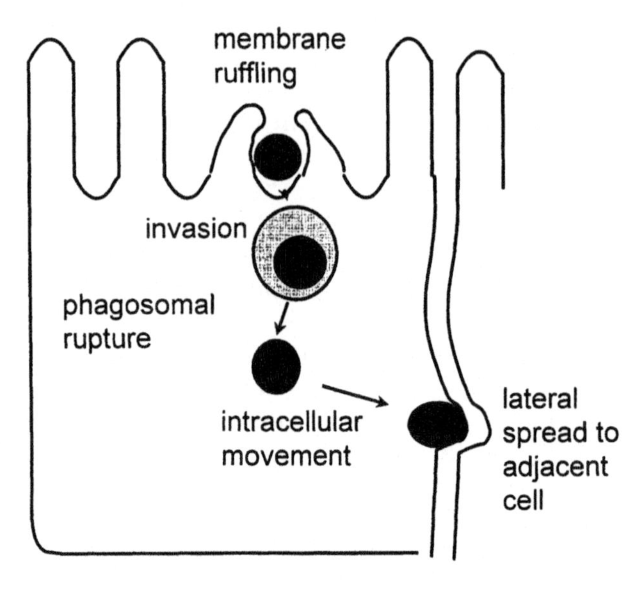

Figure 3.5 Interactions between enteroinvasive *E. coli* (EIEC) and epithelial cells. Like *Shigella*, EIEC invade intestinal epithelial cells and lyse the phagosomal vacuole. The bacteria then mediate the nucleation of cellular F-actin in such a way as to propel the bacterium directionally through the cytoplasm. This movement results in lateral spread of the bacterium from cell to cell.

lular multiplication, (4) movement through the cytoplasm and (5) extension into adjacent epithelial cells (Nataro and Deng, 1997; Sansonetti, 1992b; Goldberg and Sansonetti, 1993). When severe, this sequence of events elicits a strong inflammatory reaction which is manifested grossly as mucosal ulceration, predominantly in the colon (Sansonetti, 1992a; Sansonetti, 1992b).

Invasion-related genes are carried on a 120 MDa plasmid (pInv) in *S. sonnei* and a 140 MDa plasmid in other *Shigella* serotypes and in EIEC (Baudry et al., 1987; Sasakawa et al., 1992; Small and Falkow, 1988). The *mxi* and *spa* gene clusters, which encode a plasmid-borne type III secretion apparatus (Maurelli, 1994; Venkatesan et al., 1992; Allaoui et al., 1995; Andrews et al., 1991), are required for the secretion of multiple virulence effector proteins called Ipas (invasion plasmid antigens) (Hale et al., 1985; Ménard et al., 1993; Ménard et al., 1996; Baudry et al., 1987). IpaC has been shown to promote uptake of *Shigella* into the eukaryotic cell (Marquart et al., 1996), whereas IpaB is thought to function in lysis of the phagocytic vacuole (High et al., 1992) and in the induction of apoptosis in macrophages (Zychlinsky et al.,

Shigella and EIEC movement within the cytoplasm is mediated by nucleation of cellular actin into a "tail," which extends from one pole of the bacterium (Sansonetti, 1992b; Adam et al., 1995; Vasselon et al., 1992). As additional actin is added to this structure, the bacterium is propelled through the cytoplasm. VirG (IcsA) is a surface protein that is essential for the nucleation of actin filaments, movement through the cytoplasm and spread into adjacent cells (Goldberg et al., 1993; Sansonetti, 1992a).

Regulation of *Shigella* virulence is complex and features at least one regulatory cascade. VirR is a chromosomally encoded histone-like protein related to *drdX* (Hromockyj et al., 1992; Maurelli et al., 1992). VirR apparently acts in concert with VirF, a transcriptional activator encoded on pInv (Dorman, 1992). VirF exerts pleiotropic effects, some of which are through the intermediate transcriptional activator VirB (Tobe et al., 1993).

6.2.2. Enterotoxin Production

Both *Shigella* and EIEC infections are characterized by a period of watery diarrhea that precedes the onset of scanty dysenteric stools containing blood and mucus. Indeed, in the majority of EIEC cases and many *Shigella* cases, only watery diarrhea occurs. Nataro et al. (Nataro et al., 1995a) have cloned and sequenced a plasmid-borne gene from EIEC (designated *sen*), which encodes a novel protein of predicted size 63 kDa. These investigators have shown that a mutation of the *sen* gene causes a significant diminution of the enterotoxic activity of the parent strain and that the purified Sen protein elicits rises in shortcircuit current without a significant effect on tissue conductance.

6.3. PREVENTION AND CONTROL

Control of EIEC infection requires the provision of clean water and uncontaminated food. There is little evidence that person-to-person transmission is relevant to the transmission of the disease. There is no EIEC vaccine currently available.

7. DIFFUSELY ADHERENT *E. COLI*

7.1. CLINICAL AND EPIDEMIOLOGICAL FEATURES

Diffusely adherent *E. coli* (DAEC) cause a watery diarrhea syndrome in adults and children outside of infancy (Levine et al., 1993). It is still not clear that all DAEC strains elicit diarrhea, as their link in diarrhea is strictly one of epidemiologic association.

An age-dependent association of DAEC with diarrhea has been reported

(Baqui et al., 1992; Girón et al., 1991b; Levine et al., 1993; Gunzberg et al., 1993). Unlike most other diarrheal pathogens, DAEC seems to be associated with diarrhea only outside of infancy. Levine et al. showed that the relative risk of DAEC in association with diarrhea increased with age from 1 year to age 4–5 years in Santiago, Chile (Levine et al., 1993). Other epidemiologic features, such as the mode of acquisition of DAEC infection, are undetermined.

DAEC were found to account for a large proportion of diarrheal cases among hospitalized patients in France who had no other identified enteropathogen (Jallat et al., 1993). However, a subsequent case-control study failed to implicate DAEC as a cause of diarrhea among children in France (Forestier et al., 1996).

7.2. PATHOGENESIS

The pathogenesis of DAEC diarrhea is not as yet elucidated, but several virulence-related characteristics have been identified. Prototype DAEC strains have been shown to induce finger-like projections that extend from the surface of infected Caco-2 or HEp-2 cells (Cookson and Nataro, 1996; Yamamoto et al., 1994) (Figure 3.6). The projections "wrap around" the bacteria, but do not effect complete internalization. A role for this phenotype in pathogenesis has yet to be demonstrated.

Bilge et al. have described the cloning and characterization of a surface fimbria in a DAEC strain, which mediates the DA phenotype (Bilge et al., 1989; Bilge et al., 1993a; Bilge et al., 1993b). The genes encoding the fimbria (designated F1845) can be found on either the bacterial chromosome or on a plasmid. The fimbrial genes show homology to members of the Dr group of bacterial adhesins, so called because they mediate adherence to the Dr blood group antigen.

Benz et al. (Benz and Schmidt, 1992a; Benz and Schmidt, 1992b) described a 100 kDa outer membrane protein, which is associated with the DA phenotype in one strain of serotype O126:H27. The gene encoding this factor (designated AIDA-I) has been sequenced. Use of a DNA probe specific for AIDA-I suggests that this factor is expressed by a small percentage of DAEC isolates (Benz and Schmidt, 1989).

7.3. PREVENTION AND CONTROL

Proper control measures to prevent the spread of DAEC infection are not elucidated, but provision of adequate sanitation and fecal-oral hygiene, as in the other *E. coli* pathotypes, would be advisable. No vaccine is available and none is currently undergoing testing. The public health value of a vaccine for DAEC infection is questionable.

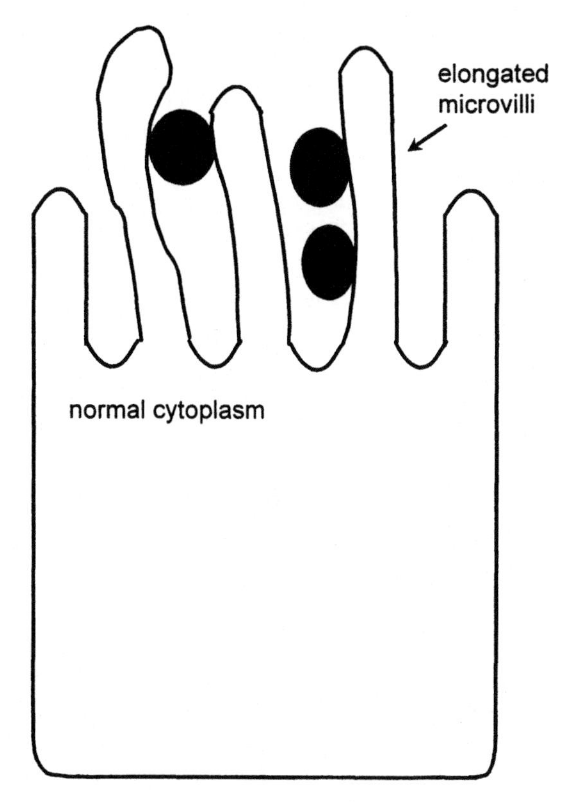

Figure 3.6 Interactions between diffusely adherent *E. coli* (DAEC) and epithelial cells. Little is known of the interaction of DAEC with intestinal epithelium. Initial adherence is thought to be by virtue of the fimbrial antigen F1845. Some adhering DAEC then elicit the formation of long cellular processes that extend from the surface of the cell and partially envelop the attached bacteria without leading to invasion.

8. UROPATHOGENIC *E. COLI*

In addition to causing gastroenteritis and colitis, *E. coli* is responsible for a wide range of extraintestinal infections including sepsis without known source, pneumonia, skin and soft tissue infections, cholecystitis, meningitis and infections of the urinary tract. Very little is known about the virulence factors that contribute to all but the last two of these infections.

8.1. CLINICAL AND EPIDEMIOLOGICAL FEATURES

It is estimated that 40% of women will have a symptomatic urinary tract infection (UTI) during their lifetime (Warren, 1996). *E. coli* is by far the most common cause of all types of UTIs, covering the spectrum from asymptomatic bacteriuria to urosepsis. Asymptomatic bacteriuria is extremely common, especially in the elderly, but is of uncertain clinical significance except during pregnancy. Cystitis, infection of the bladder characterized by dysuria, urgency and frequency in the absence of fever, accounts for the vast majority of the eight million annual physician visits for UTIs in the United States. Pyelonephritis, infection of the kidney parenchyma, is characterized by fever and flank pain with or without the symptoms of cystitis. Pyelonephritis is responsible for most of the 100,000 annual hospitalizations and almost all of the mortality due to UTIs.

Numerous studies have demonstrated that a strain of *E. coli* isolated from the urine of a patient with UTI arises from the fecal flora of that individual (Hooton and Stamm, 1996). How an individual acquires the strain is not clear. There is some evidence to support transmission between couples (Foxman et al., 1997). In addition, there have been rare outbreaks of community-acquired and nosocomial urinary tract infections (Tullus et al., 1984; Phillips et al., 1988). The sources of these infections have not been identified.

8.2. PATHOGENESIS

There is a strong relationship between sexual activity and UTI in women (Strom et al., 1987; Hooton et al., 1996). *E. coli* that may be colonizing the periurethral area can be inoculated into the bladder during sexual intercourse. In addition, the use of spermicidal cream or jelly increases the numbers of *E. coli* that colonize the vagina and periurethral area (Hooton et al., 1991).

It is clear that *E. coli* that cause urinary tract infections, especially those that cause infections in normal hosts, differ from fecal and diarrheagenic strains (Donnenberg and Welch, 1996). These uropathogenic *E. coli* are more likely than are fecal strains to have the genes for a number of fimbrial and afimbrial adhesins, for hemolysin and cytotoxic necrotizing factor, and are more likely than fecal strains to belong to certain O serogroups. It is now apparent that many of these factors are linked, either physically or clonally.

8.2.1. P Fimbriae

Of the factors that have been implicated in the pathogenesis of urinary tract infections, the evidence is perhaps strongest for P fimbriae. These structures are composite fibers made up of a rigid, rod-like shaft joined end-to-end with a flexible fibrillar tip (Kuehn et al., 1992). At the very end of the fiber is the PapG adhesin, which mediates binding to globoseries glycolipids present on a variety of host cell types including renal epithelial cells (Svensson et al., 1994; Korhonen et al., 1986). An operon of nine genes is required for synthesis and export of the major and minor subunits and for the ordered assembly of these proteins to form the final structure (Jones et al., 1996).

An analysis of many published studies has revealed that P fimbriae are present in 81% of *E. coli* strains isolated from the urine of women with normal urinary tracts and acute pyelonephritis (Donnenberg and Welch, 1996). Such strains are six times more likely to have P fimbriae than are strains isolated from the feces of healthy individuals. P fimbriae are also more common in strains from patients with cystitis and in strains from patients with asymptomatic bacteriuria than in fecal isolates. P fimbriae are significantly more common in strains isolated from the patients whose episodes of pyelonephritis were complicated by bacteremia than in other patients and in patients with pyelonephritis who have normal urinary tracts than in patients with abnormal urinary tracts.

P fimbriae have been demonstrated to contribute to the pathology and the duration of renal colonization in a non-human primate model of urinary tract infection (Roberts et al., 1994), but they have not been shown to be necessary for colonization or pathology in a murine model (Mobley et al., 1993).

8.2.2. Other Adhesins

E. coli strains that cause urinary tract infection may produce a plethora of other adhesins. Most *E. coli* produce type I fimbriae, which are similar in structure to P fimbriae, but mediate binding to mannose-containing glycoproteins and glycolipids (Gbarah et al., 1991; Wu et al., 1996; Ofek et al., 1977). There is strong evidence to implicate type I fimbriae in the pathogenesis of urinary tract infections (Keith et al., 1986; Connell et al., 1996; Hultgren et al., 1985). The roles of other adhesins in pathogenesis have not been as well studied. Among these adhesins are S and F1C fimbriae, which are closely related to P fimbriae, but do not share the same receptor (Ott et al., 1988). F1C fimbriae have been epidemiologically associated with pyelonephritis (Pere et al., 1985). A group of fimbrial and afimbrial adhesins known as the Dr family, which share similar operon structures and each binds to decay-accelerating factor (Nowicki et al., 1990), has been convincingly linked to cystitis (Zhang et al., 1997), but not as convincingly to pyelonephritis (Nowicki et al., 1989). As noted above, similar fimbriae may be found in DAEC.

8.2.3. Toxins

E. coli strains isolated from the urine of patients with pyelonephritis and strains from patients with cystitis are more likely to produce hemolysin than are fecal strains (Donnenberg and Welch, 1996). Hemolysin is a pore-forming toxin capable of lysing not only erythrocytes but renal tubular epithelial cells as well (Mobley et al., 1990). There is evidence from animal models that hemolysin, when overexpressed from a plasmid gene, contributes to renal damage (O'Hanley et al., 1991). Another toxin, cytotoxic necrotizing factor (CNF), is also found more commonly in isolates from UTIs than in fecal strains, almost always in association with hemolysin (Foxman et al., 1995; Blanco et al., 1990). CNF causes changes in the distribution of actin in cells by modifying the small GTP-binding signaling protein Rho (Flatau et al., 1997; Schmidt et al., 1997). The role of CNF in the pathogenesis of urinary tract infections in unknown.

8.2.4. Pathogenicity Islands

The genes for P fimbriae, hemolysin, and CNF are often found on large chromosomal insertions known as pathogenicity islands (High et al., 1988; Blum et al., 1995; Swenson et al., 1996). As most of the sequences of these large genetic determinants (some are over 100 kb in length) have not yet been reported, it is possible that these islands contain many more virulence determinants that have yet to be recognized. Thus, the epidemiologic data cited above, linking the genes for various putative virulence factors to disease, must be interpreted with the caveat that these genes may merely be markers for other genes, linked in the same pathogenicity islands, that may be the more relevant determinants of pathogenicity.

8.3. PREVENTION AND CONTROL

In women with histories of recurrent urinary tract infection, the incidence of UTI can be reduced by using prophylactic antibiotics after sexual intercourse (Stapleton et al., 1990). There is some evidence that voiding after intercourse is also protective (Strom et al., 1987). Progress is being made on a vaccine, based on the tip adhesin of type I fimbriae, for the prevention of UTI (Langermann et al., 1997).

9. *E. COLI* STRAINS THAT CAUSE NEONATAL MENINGITIS

9.1. CLINICAL AND EPIDEMIOLOGICAL FEATURES

E. coli is one of the leading causes of sepsis of unknown origin and meningitis in neonates and young infants. The organisms that cause these serious illnesses arise from the fecal flora of the mother (Sarff et al., 1975).

9.2. PATHOGENESIS

E. coli strains isolated from neonates with meningitis and sepsis differ from control strains isolated from the feces of healthy children (Korhonen et al.,

1985; Siitonen et al., 1993). Strains isolated from such patients are far more likely to produce the K1 capsule than are control strains. The K1 capsule, a homopolymer of sialic acid similar to the capsule of group B *Neisseria meningitidis,* may help the bacteria resist activating the alternative pathway of complement (Pluschke et al., 1983). Evidence from animal models suggests that the K1 capsule may augment the ability of *E. coli* to cause bacteremia (Silver and Vimr, 1990). *E. coli* strains from infants with meningitis are also more likely than control strains to produce S fimbriae (Korhonen et al., 1985; Siitonen et al., 1993). S fimbriae, similar in structure to P fimbriae (Ott et al., 1988), mediate binding to sialyl(alpha-2–3)galactosyl glycoproteins and bind to the endothelium lining the choroid plexus of the brain (Korhonen et al., 1990) and to sialoglycoproteins isolated from brain microvascular cells (Prasadarao et al., 1997).

One of the most common serotypes isolated from neonates and infants with sepsis or meningitis is O18:K1:H7 (Siitonen et al., 1993). Strains from this serotype usually make S fimbriae and, by definition, K1 capsule. In addition, two groups have demonstrated that such strains are capable of invading into cultured epithelial and endothelial cells (Bloch et al., 1996; Meier et al., 1996). Mutants incapable of cellular invasion have been isolated. An analysis of the locations of the mutations in these mutants reveals that such strains may contain as yet uncharacterized pathogenicity islands involved in the invasion process (Bloch et al., 1996).

9.3. PREVENTION AND CONTROL

Efforts to prevent disease in neonates and infants due to *E. coli* center around detection and prevention of colonization, either of the mother or of the neonate. Strategies under consideration include detecting colonization of the mother with pathogenic strains during pregnancy and selecting such individuals for treatment or vaccination to eliminate the organism.

10. SUMMARY

The ability of *E. coli* to cause a wide variety of diseases is dependent upon the presence of particular virulence genes that define pathotypes of the species. Common themes of *E. coli* pathogenesis include the presence of mobile genetic elements such as plasmids, pathogenicity islands and bacteriophages, which encode virulence factors; the ability to produce pili, which are often involved in adherence to host tissues; and the expression of extracellular effector molecules such as enterotoxins, hemolysin, or proteins secreted through type III secretion systems, which affect host cell functions (Table 3.1). Prevention of *E. coli* infections will require several approaches, principally a guarantee of non-

TABLE 3.1. **Summary of the Clinical Syndromes, Mode of Transmission and Major Virulence Factors of Pathogenic *E. coli*.**

Pathotype[a]	Clinical Syndrome	Transmission[b]	Proposed Virulence Factors
EPEC	Infantile gastroenteritis	Person-to-person	Bundle-forming pili mediating localized adherence; type III secretion system mediating attaching and effacing lesions
EHEC	Hemorrhagic colitis	Food	Shiga toxins; type III secretion system mediating attaching and effacing lesions
ETEC	Watery diarrhea in young children and travelers	Food and water	Colonization factor antigen fimbriae; heat-labile and heat-stable enterotoxins
EAEC	Childhood mucoid diarrhea	Unknown	Fimbriae mediating aggregative adherence; mucous production; cytotoxin
EIEC	Watery diarrhea and dysentery	Food	Enterotoxin; type III secretion system mediating cellular invasion
DAEC	Diarrhea in older children	Unknown	Fimbrial and afimbrial adhesins; elongation of microvilli
UPEC	Urinary tract infections	Unknown	P and other fimbriae; hemolysin
NMEC	Sepsis and meningitis in neonates and infants	Person-to-person	Capsules; S-fimbriae; cellular invasion

[a] EPEC, enteropathogenic *E. coli*; EHEC, enterohemorrhagic *E. coli*; ETEC, enterotoxigenic *E. coli*; EAEC, enteroaggregative *E. coli*; EIEC, enteroinvasive *E. coli*; DAEC, diffusely adherent *E. coli*; UPEC, uropathogenic *E. coli*; NMEC, *E. coli* that cause neonatal meningitis.

[b] Predominant route of transmission.

contaminated food and water sources, as well as the development of new vaccines.

11. ACKNOWLEDGMENTS

The authors' efforts were supported by Public Health Service Awards AI37606, DK49720, and AI32074 (to MSD) and AI33096 (to JPN) from the National Institutes of Health.

12. REFERENCES

Acheson, D. W. K., Moore, R., De Breucker, S., Lincicome, L., Jacewicz, M., and Keusch, G. T. 1996. "Translocation of Shiga toxin across polarized intestinal cells in tissue culture," *Infect. Immun.* 64:3294–3300.

Adam, T., Arpin, M., Prévost, M. C., Gounon, P., and Sansonetti, P.J. 1995. "Cytoskeletal rearrangements and the functional role of T-plastin during entry of *Shigella flexneri* into HeLa cells," *J. Cell Biol.* 129:367–381.

Albert, M. J., Faruque, S. M., Faruque, A. S., Neogi, P. K., Ansaruzzaman, M., Bhuiyan, N. A., Alam, K., and Akbar, M. S. 1995. "Controlled study of *Escherichia coli* diarrheal infections in Bangladeshi children," *J. Clin. Microbiol.* 33:973–977.

Allaoui, A., Sansonetti, P. J., Ménard, R., Barzu, S., Mounier, J., Phalipon, A., and Parsot, C. 1995. "MxiG, a membrane protein required for secretion of *Shigella* spp. Ipa invasins: Involvement in entry into epithelial cells and in intercellular dissemination," *Mol. Microbiol.* 17:461–470.

Andrews, G. P., Hromockyj, A. E., Coker, C., and Maurelli, A. T. 1991. "Two novel virulence loci, *mxiA* and *mxiB,* in *Shigella flexneri* 2a facilitate excretion of invasion plasmid antigens," *Infect. Immun.* 59:1997–2005.

Arriaga, Y. L., Harville, B. A., and Dreyfus, L. A. 1995. "Contribution of individual disulfide bonds to biological action of *Escherichia coli* heat-stable enterotoxin B," *Infect. Immun.* 63:4715–4720.

Baldini, M. M., Kaper, J. B., Levine, M. M., Candy, D. C., and Moon, H. W. 1983. "Plasmid-mediated adhesion in enteropathogenic *Escherichia coli,*" *J. Pediatr. Gastroenterol. Nutr.* 2:534–538.

Baqui, A. H., Sack, R. B., and Black, R. E. 1992. "Enteroapthogens asociated with acute and persistent diarrhea in Bagladeshi children <5 years of age," *J. Infect. Dis.* 166:792–796.

Baudry, B., Maurelli, A. T., Clerc, P., Sadoff, J. C., and Sansonetti, P. J. 1987. "Localization of plasmid loci necessary for the entry of *Shigella flexneri* into HeLa cells, and characterization of one locus encoding four immunogenic polypeptides," *J. Gen. Microbiol.* 133:3403–3413.

Bell, B. P., Goldoft, M., Griffin, P. M., Davis, M. A., Gordon, D. C., Tarr, P. I., Bartleson, C. A., Lewis, J. H., Barrett, T. J., Wells, J. G., Baron, R., and Kobayashi, J. 1994. "A multistate outbreak of *Escherichia coli* O157:H7-associated bloody diarrhea and hemolytic uremic syndrome from hamburgers: The Washington experience," *JAMA* 272:1349–1353.

Belongia, E. A., Osterholm, M. T., Soler, J. T., Ammend, D. A., Braun, J. E., and Macdonald, K. L. 1993. "Transmission of *Escherichia coli* 0157:H7 infection in Minnesota child day-care facilities," *JAMA,* 269:883–888.

Benjamin, M. M. and Datta, A. R. 1995. "Acid tolerance of enterohemorrhagic *Escherichia coli,*" *Appl. Environ. Microbiol.* 61:1669–1672.

Benjamin, P., Federman, M., and Wanke, C. A. 1995. ''Characterization of an invasive phenotype associated with enteroaggregative *Escherichia coli,*'' *Infect. Immun.* 63:3417–3421.

Benz, I. and Schmidt, M. A. 1989. ''Cloning and expression of an adhesin (AIDA-I) involved in diffuse adherence of enteropathogenic *Escherichia coli,*'' *Infect. Immun.* 57:1506–1511.

Benz, I. and Schmidt, M. A. 1992a. ''AIDA-I, the adhesin involved in diffuse adherence of the diarrhoeagenic *Escherichia coli* strain 2787 (O126:H27), is synthesized via a precursor molecule,'' *Mol. Microbiol.* 6:1539–1546.

Benz, I. and Schmidt, M. A. 1992b. ''Isolation and serologic characterization of AIDA-I, the adhesin mediating the diffuse adherence phenotype of the diarrhea-associated *Escherichia coli* strain 2787 (O126: H27),'' *Infect. Immun.* 60:13–18.

Besser, R. E., Lett, S. M., Weber, J. T., Doyle, M. P., Barrett, T. J., Wells, J. G., and Griffin, P. M. 1993. ''An outbreak of diarrhea and hemolytic uremic syndrome from *Escherichia coli* O157:H7 in fresh-pressed apple cider,'' *JAMA,* 269:2217–2220.

Bhan, M. K., Raj, P., Levine, M. M., Kaper, J. B., Bhandari, N., Srivastava, R., Kumar, R., and Sazawal, S. 1989. ''Enteroaggregative *Escherichia coli* associated with persistent diarrhea in a cohort of rural children in India,'' *J. Infect. Dis.* 159:1061–1064.

Bhatnagar, S., Bhan, M. K., Sommerfelt, H., Sazawal, S., Kumar, R., and Saini, S. 1993. ''Enteroaggregative *Escherichia coli* may be a new pathogen causing acute and persistent diarrhea,'' *Scand. J. Infect. Dis.* 25:579–583.

Bilge, S. S., Apostol, J. M., Jr., Fullner, K. J., and Moseley, S. L. 1993b. ''Transcriptional organization of the F1845 fimbrial adhesin determinant of *Escherichia coli,*'' *Mol. Microbiol.* 7:993–1006.

Bilge, S. S., Apostol, J. M., Jr., Aldape, M. A., and Moseley, S. L. 1993a. ''mRNA processing independent of RNase III and RNase E in the expression of the F1845 fimbrial adhesin of *Escherichia coli,*'' *Proc. Natl. Acad. Sci. USA,* 90:1455–1459.

Bilge, S. S., Clausen, C. R., Lau, W., and Moseley, S. L. 1989. ''Molecular characterization of a fimbrial adhesin, F1845, mediating diffuse adherence of diarrhea-associated *Escherichia coli* to HEp-2 cells,'' *J. Bacteriol,* 171:4281–4289.

Bilge, S. S., Vary, J. C., Jr., Dowell, S. F., and Tarr, P. I. 1996. ''Role of the *Escherichia coli* O157:H7 O side chain in adherence and analysis of an *rfb* locus,'' *Infect. Immun.* 64:4795–4801.

Black, R. E., Brown, K. H., Becker, S., Abdul Alim, A. R. M., and Merson, M. H. 1982. ''Contamination of weaning foods and transmission of enterotoxigenic *Escherichia coli* diarrhoea in children in rural Bangladesh,'' *Trans. Roy. Soc. Trop. Med. Hyg.* 76:259–264.

Black, R. E., Merson, M. H., Rowe, B., Taylor, P. R., Abdul Alim, A. R. M., Gross, R. J., and Sack, D. A. 1981. ''Enterotoxigenic *Escherichia coli* diarrhoea: acquired immunity and transmission in an endemic area,'' *Bull. WHO,* 59:263–268.

Blake, P. A., Ramos, S., Macdonald, K. L., Rassi, V., Gomes, T. A. T., Ivey, C., Bean, N. H., and Trabulsi, L. R. 1993. ''Pathogen-specific risk factors and protective factors for acute diarrheal disease in urban Brazilian infants,'' *J. Infect. Dis.* 167:627–632.

Blanco, J., Alonso, M. P., Gonzalez, E. A., Blanco, M., and Garabal, J. I. 1990. ''Virulence factors of bacteraemic *Escherichia coli* with particular reference to production of cytotoxic necrotising factor (CNF) by P-fimbriate strains.'' *J. Med. Microbiol.* 31:175–183.

Bloch, C. A., Huang, S. H., Rode, C. K., and Kim, K. S. 1996. ''Mapping of noninvasion Tn*phoA* mutations on the *Escherichia coli* O18:K1:H7 chromosome,'' *FEMS Microbiol. Lett.* 144:171–176.

Blum, G., Falbo, V., Caprioli, A., and Hacker, J. 1995. ''Gene clusters encoding the cytotoxic necrotizing factor type 1, Prs-fimbriae and α-hemolysin form the pathogenicity island II of the uropathogenic *Escherichia coli* strain J96,'' *FEMS Microbiol. Lett.* 126:189–196.

Bower, J. R., Congeni, B. L., Cleary, T. G., Stone, R. T., Wanger, A., Murray, B. E., Mathewson, J. J., and Pickering, L. K. 1989. "*Escherichia coli* O114:nonmotile as a pathogen in an outbreak of severe diarrhea associated with a day care center," *J. Infect. Dis.* 160:243–247.

Bray, J. 1945. "Isolation of antigenically homogeneous strains of *Bact. coli neapolitanum* from summer diarrhoea of infants," *J. Pathol. Bacteriol.* 57:239–247.

Brunder, W., Schmidt, H., and Karch, H. 1997. "EspP, a novel extracellular serine protease of enterohaemorrhagic *Escherichia coli* O157:H7 cleaves human coagulation factor V," *Mol. Microbiol.* 24:767–778.

Caprioli, A., Luzzi, I., Rosmini, F., Resti, C., Edefonti, A., Perfumo, F., Farina, C., Goglio, A., Gianviti, A., and Rizzoni, G. 1994. "Communitywide outbreak of hemolytic-uremic syndrome associated with non-O157 verocytotoxin-producing *Escherichia coli*," *J. Infect. Dis.* 169:208–211.

Carter, A. O., Borczyk, A. A., Carlson, J. A. K., Harvey, B., Hockin, J. C., Karmali, M. A., Krishnan, C., Korn, D. A., and Lior, H. 1987. "A severe outbreak of *Escherichia coli* O157:H7-associated hemorrhagic colitis in a nursing home," *N. Engl. J. Med.* 317:1496–1500.

Cassels, F. J. and Wolf, M. K. 1995. "Colonization factors of diarrheagenic *E. coli* and their intestinal receptors," *J. Indust. Microbiol.* 15:214–226.

CDC. 1995a. "Enhanced detection of sporadic *Escherichia coli* O157:H7 infections—New Jersey, July 1994," *MMWR* 44:417–418.

CDC. 1995b. "*Escherichia coli* O157:H7 outbreak linked to commercially distributed dry-cured salami—Washington and California, 1994," *MMWR* 44:157–160.

CDC. 1996. "Lake-associated outbreak of *Escherichia coli* O157:H7—Illinois, 1995," *MMWR* 45:437–439.

CDC. 1997a. "Outbreaks of *Escherichia coli* O157:H7 infection associated with eating alfalfa sprouts—Michigan and Virginia, June–July 1997," *MMWR* 46:741–744.

CDC. 1997b. "Outbreaks of *Escherichia coli* O157:H7 infections and cryptosporidiosis associated with drinking unpasteurized apple cider—Connecticut and New York, October 1996," *MMWR* 46:4–8.

Cobeljic, M., Miljkovic-Selimovic, B., Paunovic-Todosijevic, D., Velickovic, Z., Lepsanovic, Z., Savic, D., Ilic, R., Konstantinovic, S., Jovanovic, B., and Kostic, V. 1996. "Enteroaggregative *Escherichia coli* associated with an outbreak of diarrhoea in a neonatal nursery ward," *Epidemiol. Infect.* 117:11–16.

Connell, H., Agace, W., Klemm, P., Schembri, M., Mårild, S., and Svanborg, C. 1996. "Type 1 fimbrial expression enhances *Escherichia coli* virulence for the urinary tract," *Proc. Natl. Acad. Sci. USA,* 93:9827–9832.

Cookson, S. T. and Nataro, J. P. 1996. "Characterization of HEp-2 cell projection formation induced by diffusely adherent *Escherichia coli*," *Microb. Pathog.* 21:421–434.

Crane, J. K., Wehner, M. S. Bolen, E. J., Sando, J. J., Linden, J., Guerrant, R. L., and Sears, C. L. 1992. "Regulation of intestinal guanylate cyclase by the heat-stable enterotoxin of *Escherichia coli* (STa) and protein kinase C," *Infect. Immun.* 60:5004–5012.

Cravioto, A., Tello, A., Villafán, H., Ruiz, J., Del Vedovo, S., and Neeser, J.-R. 1991b. "Inhibition of localized adhesion of enteropathogenic *Escherichia coli* to HEp-2 cells by immunoglobulin and oligosaccharide fractions of human colostrum and breast milk," *J. Infect. Dis.* 163:1247–1255.

Cravioto, A., Tello, A., Navarro, A., Ruiz, J., Villafán, H., Uribe, F., and Eslava, C. 1991a. "Association of *Escherichia coli* HEp-2 adherence patterns with type and duration of diarrhoea," *Lancet,* 337:262–264.

Czeczulin, J. R., Balepur, S., Hicks, S., Phillips, A., Hall, R., Kothary, M. H., Navarro-Garcia,

F., and Nataro, J. P. 1997. "Aggregative Adherence Fimbria II, a second fimbrial antigen mediating aggregative adherence in enteroaggregative *Escherichia coli,*" *Infect. Immun.* 65:4135–4145.

Dallas, W. S. and Falkow, S. 1980. "Amino acid sequence homology between cholera toxin and *Escherichia coli* heat-labile toxin," *Nature,* 288:499–501.

de Graaf, F. K. and Gaastra, W. 1994. "Fimbriae of enterotoxigenic *Escherichia coli,*" in *Fimbriae: Adhesion, Genetics, Biogenesis, and Vaccines,* ed., Klemm, P. Boca Raton, FL: CRC Press. pp. 58–83.

de Sauvage, F. J., Camerats, T. R., and Goeddel, D. V. 1991. "Primary structure and functional expression of the human receptor for *Escherichia coli* heat-stable enterotoxin," *J. Biol. Chem.* 266:17912–17918.

Dean-Nystrom, E. A., Bosworth, B. T., Cray, W. C., Jr., and Moon, H. W. 1997. "Pathogenicity of *Escherichia coli* O157:H7 in the intestines of neonatal calves," *Infect Immun.* 65:1842–1848.

deJonge, H. R. and Lohman, S. M. 1985. "Mechanism by which cyclic nucleotides and other intracellular mediators regulate secretion," *Ciba Foundation Symp.* 112:116–138.

Dickinson, B. L. and Clements, J. D. 1995. "Dissociation of *Escherichia coli* heat-labile enterotoxin adjuvanticity from ADP-ribosyltransferase activity." *Infect. Immun.* 63:1617–1623.

Donnenberg, M. S. 1995. "Enteropathogenic *Escherichia coli,*" in *Infections of the Gastrointestinal Tract,* eds., Blaser, M. J., Smith, P. D., Ravdin, J. I., Greenberg, H. B., and Guerrant, R. L. New York: Raven Press, Ltd. pp. 709–726.

Donnenberg, M. S. and Kaper, J. B. 1992. "Minireview: Enteropathogenic *Escherichia coli,*" *Infect. Immun.* 60:3953–3961.

Donnenberg, M. S. and Welch, R. A. 1996. "Virulence determinants of uropathogenic *Escherichia coli,*" in *Urinary Tract Infections:* Molecular Pathogenesis and Clinical Management, eds., Mobley, H. L. T. and Warren, J. W. Washington, D.C.: ASM Press. pp. 135–174.

Donnenberg, M. S., Yu, J., and Kaper, J. B. 1993c. "A second chromosomal gene necessary for intimate attachment of enteropathogenic *Escherichia coli* to epithelial cells." *J. Bacteriol.* 175:4670–4680.

Donnenberg, M. S., Girón, J. A., Nataro, J. P., and Kaper, J. B. 1992. "A plasmid-encoded type IV fimbrial gene of enteropathogenic *Escherichia coli* associated with localized adherence," *Mol. Microbiol.* 6:3427–3437.

Donnenberg, M. S., Tacket, C. O., James, S. P., Losonsky, G., Nataro, J. P., Wasserman, S. S., Kaper, J. B., and Levine, M. M. 1993a. "The role of the *eoeA* gene in experimental enteropathogenic *Escherichia coli* infection," *J. Clin. Invest.* 92:1412–1417.

Donnenberg, M. S., Tzipori, S., McKee, M., O'Brien, A. D., Alroy, J., and Kaper, J. B. 1993b. "The role of the *eae* gene of enterohemorrhagic *Escherichia coli* in intimate attachment in vitro and in a porcine model," *J. Clin. Invest.* 92:1418–1424.

Dorman, C. J. 1992. "The VirF protein from *Shigella flexneri* is a member of the AraC transcription factor superfamily and is highly homologous to Rns, a positive regulator of virulence genes in enterotoxigenic *Escherichia coli,*" *Mol. Microbiol.* 6:1575.

Dreyfus, L. A., Jaso-Friedman, L., and Robertson, D. C. 1984. "Characterization of the mechanism of action of *Escherichia coli* heat-stable enterotoxin," *Infect. Immun.* 44:493–501.

DuPont, H. L., Formal, S. B., Hornick, R. B., Snyder, M. J., Libonati, J. P., Sheahan, D. G., LaBrec, E. H., and Kalas, J. P. 1971. "Pathogenesis of *Escherichia coli* diarrhea," *N. Eng. J. Med.* 285:1–9.

Ebel, F., Deibel, C., Kresse, A. U., Guzmán, C. A., and Chakraborty, T. 1996. "Temperature- and medium-dependent secretion of proteins by Shiga toxin-producing *Escherichia coli,*" *Infect. Immun.* 64:4472–4479.

Eklund, S., Brunsson, I., Jodal, M., and Lundgren, O. 1987. "Changes in cyclic 3′5′-adenosine monophosphate tissue concentration and net fluid transport in the cat's small intestine elicited by cholera toxin, arachidonic acid, vasoactive intestinal polypeptide, and 5-hydroxytryptamine," *Acta Physiol. Scand.* 129:115–125.

Eslava, C., Villaseca, J., Morales, R., Navarro, A., and Cravioto, A. 1993. "Identification of a protein with toxigenic activity produced by enteroaggregative *Escherichia coli.*" *Abstr. Gen. Meet. Amer. Soc. Microbiol.* B105:44(Abstract).

Fang, G. D., Lima, A. A. M., Martins, C. V., Nataro, J. P., and Guerrant, R. L. 1995. "Etiology and epidemiology of Persistent Diarrhea in northeastern Brazil: a hospital-based, prospective, case-control study." *J. Pediatr. Gastroenterol. Nutr.* 21:137–144.

Field, M., Fromm, D., Al-Awqati, Q., and Greenough, W. B. 1972. "Effect of cholera enterotoxin on ion transport across isolated ileal mucosa," *J. Clin. Invest.* 51:796–804.

Finlay, B. B., Rosenshine, I., Donnenberg, M. S., and Kaper, J. B. 1992. "Cytoskeletal composition of attaching and effacing lesions associated with enteropathogenic *Escherichia coli* adherence to HeLa cells," *Infect. Immun.* 60:2541–2543.

Flatau, G., Lemichez, E., Gauthier, M., Chardin, P., Paris, S., Fiorentini, C., and Boquet, P. 1997. "Toxin-induced activation of the G protein p21 Rho by deamidation of glutamine," *Nature* 387:729–733.

Flores-Abuxapqui, J. J., Suarez-Itoil, G. J., Heredia-Navarrete, M. R., Puc-Franco, M. A., and Franco-Monsreal, J. 1994. "Frequency of enterotoxigenic *Escherichia coli* in infants during the first three months of life," *Arch. Med. Res.* 25:303–307.

Forestier, C., Meyer, M., Favre-Bonte, S., Rich, C., Malpuech, G., Le Bouguenec, C., Sirot, J., Joly, B., and De Champs, C. 1996. "Enteroadherent *Escherichia coli* and diarrhea in children: A prospective case-control study," *J. Clin. Microbiol.* 34:2897–2903.

Foubister, V., Rosenshine, I., Donnenberg, M. S., and Finlay, B. B. 1994. "The *eaeB* gene of enteropathogenic *Escherichia coli* is necessary for signal transduction in epithelial cells," *Infect. Immun.* 62:3038–3040.

Foxman, B., Zhang, L., Palin, K., Tallman, P., and Marrs, C. F. 1995. "Bacterial virulence characteristics of *Escherichia coli* isolates from first-time urinary tract infection," *J. Infect. Dis.* 171:1514–1521.

Foxman, B., Zhang, L. X., Tallman, P., Andree, B. C., Geiger, A. M., L. T., Gillespie, B. W., Palin, K. A., Sobel, J. D., Rode, C. K., Bloch, C. A., and Marrs, C. F. 1997. "Transmission of uropathogens between sex partners," *J. Infect. Dis.* 175:989–992.

Fraser, M. E., Chernaia, M. M., Kozlov, Y. V., and James, M. N. 1994. "Crystal structure of the holotoxin from *Shigella dysenteriae* at 2.5 A resolution," *Nat. Struct. Biol.* 1:59–64.

Fratamico, P. M., Bhaduri, S., and Buchanan, R. L. 1993. "Studies on *Escherichia coli* serotype O157:H7 strains containing a 60-MDa plasmid and on 60-MDa plasmid-cured derivatives," *J. Med. Microbiol.* 39:371–381.

Fujii, Y., Kondo, Y., and Okamoto, K. 1995. "Involvement of prostaglandin E2 synthesis in the intestinal secretory action of *Escherichia coli* heat-stable enterotoxin II," *FEMS Microbiol. Lett.* 130:259–266.

Fujii, Y., Hayashi, M., Hitotsubashi, S., Fuke, Y., Yamanaka, H., and Okamoto, K. 1991. "Purification and characterization of *Escherichia coli* heat-stable enterotoxin II," *J. Bacteriol.* 173:5516–5522.

Gaastra, W. and Svennerholm, A.-M. 1996. "Colonization factors of human enterotoxigenic *Escherichia coli* (ETEC)," *Trends Microbiol.* 4:444–452.

Gbarah, A., Gahmberg, C. G., Ofek, I., Jacobi, U., and Sharon, N. 1991. "Identification of the leukocyte adhesion molecules CD11 and CD18 as receptors for Type 1-fimbriated (mannose-specific) *Escherichia coli,*" *Infect. Immun.* 59:4524–4530.

Girón, J. A., Ho, A. S. Y., and Schoolnik, G. K. 1991a. "An inducible bundle-forming pilus of enteropathogenic *Escherichia coli*," *Science* 254:710–713.

Girón, J. A., Levine, M. M., and Kaper, J. B. 1994. "Longus: A long pilus ultrastructure produced by human enterotoxigenic *Escherichia coli*," *Mol. Microbiol.* 12:71–82.

Girón, J. A., Jones, T., Millan Velasco, F., Castro Munoz, E., Zarate, L., Fry, J., Frankel, G., Moseley, S. L., Baudry, B., Kaper, J. B., et al. 1991b. "Diffuse-adhering *Escherichia coli* (DAEC) as a putative cause of diarrhea in Mayan children in Mexico," *J. Infect. Dis.* 163:507–513.

Goldberg, M. B. and Sansonetti, P. J. 1993. "*Shigella* subversion of the cellular cytoskeleton: a strategy for epithelial colonization," *Infect. Immun.* 61:4941–4946.

Goldberg, M. B., Bârzu, O., Parsot, C., and Sansonetti, P. J. 1993. "Unipolar localization and ATPase activity of IcsA, a *Shigella flexneri* protein involved in intracellular movement," *J. Bacteriol.* 175:2189–2196.

Griffin, P. M. and Tauxe, R. V. 1991. "The epidemiology of infections caused by *Escherichia coli* O157:H7, other enterohemorrhagic *E. coli*, and the associated hemolytic uremic syndrome," *Epidemiol. Rev.* 13:60–98.

Griffin, P. M., Ostroff, S. M., Tauxe, R. V., Greene, K. D., Wells, J. G., Lewis, J. H., and Blake, P. A. 1988. "Illnesses associated with *Escherichia coli* O157:H7 infections: a broad clinical spectrum," *Ann. Intern. Med.* 109:705–712.

Gunzberg, S. T., Chang, B. J., Elliott, S. J., Burke, V., and Gracey, M. 1993. "Diffuse and enteroaggregative patterns of adherence of enteric *Escherichia coli* isolated from aboriginal children from the Kimberley region of western Australia," *J. Infect. Dis.* 167:755–758.

Hale, T. L., Oaks, E. V., and Formal, S. B. 1985. "Identification and antigenic characterization of virulence-associated, plasmid-coded proteins of *Shigella spp.* and enteroinvasive *Escherichia coli*," *Infect. Immun.* 50:620–629.

Harris, J. R., Mariano, J., Wells, J. G., Payne, B. S., Donnel, H. D., and Cohen, M. L. 1985. "Person-to-person transmission in an outbreak of enteroinvasive *Escherichia coli*," *Amer. J. Epidemiol.* 122:245–252.

Hicks, S., Candy, D. C. A., and Phillips, A. D. 1996. "Adhesion of enteroaggregative *Escherichia coli* to pediatric intestinal mucosa in vitro," *Infect. Immun.* 64:4751–4760.

High, N., Mounier, J., Prevost, M.-C., and Sansonetti, P. J. 1992. "IpaB of *Shigella flexneri* causes entry into epithelial cells and escape from the phagocyic vacuole," *EMBO J.* 11:1991–1999.

High, N. J., Hales, B. A., Jann, K., and Boulnois, G. J. 1988. "A block of urovirulence genes encoding multiple fimbriae and hemolysin in *Escherichia coli* O4:K12:H-," *Infect. Immun.* 56:513–517.

Hirayama, T. 1995. "Heat-stable enterotoxin of *Escherichia coli*," in *Bacterial Toxins and Virulence Factors in Disease*, eds., Moss, J., Iglewski, B., Vaughn, M., and Tu, A. T. New York: Marcel Dekker, Inc. pp. 281–296.

Hirst, R., Sanchez, J., Kaper, J. B., Hardy, S. J., and Holmgren, J. 1984. "Mechanism of toxin secretion by *Vibrio cholerae* investigated in strains harboring plasmids that encode heat-labile enterotoxins of *Escherichia coli*," *Proc. Natl. Acad. Sci. USA* 81:7752–7756.

Hirst, T. R. 1995. "Biogenesis of cholera toxin and related oligomeric enterotoxins," in *Bacterial Toxins and Virulence Factors in Disease*, eds., Moss, J., Iglewski, B., Vaughan, M., and Tu, A. T. New York: Marcel Dekker, Inc. pp. 123–184.

Hitotsubashi, S., Fujii, Y., Yamanaka, H., and Okamoto, K. 1992. "Some properties of purified *Escherichia coli* heat-stable enterotoxin II," *Infect. Immun.* 60:4468–4474.

Hol, W. G. J., Sixma, T. K., and Merritt, E. A. 1995. "Structure and function of *E. coli* heat-labile enterotoxin and cholera toxin B pentamer," in Bacterial Toxins and Virulence Factors in Disease, eds., Moss, J., Iglewski, B., Vaughan, M., and Tu, A. T. New York: Marcel Dekker Inc. pp. 185–223.

Hooton, T. M. and Stamm, W. E. 1996. "The vaginal flora and urinary tract infections," in *Urinary* Tract Infections: Molecular Pathogenesis and Clinical Management, eds., Mobley, H. L. T. and Warren, J. W. Washington, D. C.: Asm Press. pp. 67–94.

Hooton, T. M., Hillier, S., Johnson, C., Roberts, P. L., and Stamm, W. E. 1991. "Escherichia coli bacteriuria and contraceptive method," *JAMA* 265:64–69.

Hooton, T. M., Scholes, D., Hughes, J. P., Winter, C., Roberts, P. L., Stapleton, A. E., Stergachis, A., and Stamm, W. E. 1996. "A prospective study of risk factors for symptomatic urinary tract infection in young women," *N. Engl. J. Med.* 335:468–474.

Hoque, S. S., Faruque, A. S., Mahalanabis, D., and Hasnat, A. 1994. "Infectious agents causing acute watery diarrhoea in infants and young children in Bangladesh and their public health implications," *J. Trop. Peds.* 40:351–354.

Hromockyj, A. E., Tucker, S. C., and Maurelli, A. T. 1992. "Temperature regulation of *Shigella* virulence: Identification of the repressor gene *virR,* an analogue of *hns,* and partial complementation by tyrosyl transfer RNA (tRNA1Tyr)," *Mol. Microbiol.* 6:2113–2124.

Hultgren, S. J., Porter, T. N., Schaeffer, A. J., and Duncan, J. L. 1985. "Role of type 1 pili and effects of phase variation on lower urinary tract infections produced by *Escherichia coli,*" *Infect. Immun.* 50:370–377.

Huppertz, H. I., Rutkowski, S., Aleksic, S., and Karch, H. 1997. "Acute and chronic diarrhoea and abdominal colic associated with enteroaggregative *Escherichia coli* in young children living in western Europe," *Lancet* 349:1660–1662.

Itoh, Y., Nagano, I., Kunishima, M., and Ezaki, T. 1997. "Laboratory investigation of enteroaggregative *Escherichia coli* O untypeable:H10 associated with a massive outbreak of gastrointestinal illness," *J. Clin. Microbiol.* 35:2546–2550.

Jacewicz, M. S., Mobassaleh, M., Gross, S. K., Balasubramanian, K. A., Daniel, P. F., Raghavan, S., McCluer, R. H., and Keusch, G. T. 1994. "Pathogenesis of *Shigella* diarrhea: XVII. A mammalian cell membrane glycolipid, Gb3, is required but not sufficient to confer sensitivity to Shiga toxin," *J. Infect. Dis.* 169:538–546.

Jallat, C., Livrelli, V., Darfeuille-Michaud, A., Rich, C., and Joly, B. 1993. "*Escherichia coli* strains involved in diarrhea in France: High prevalence and heterogeneity of diffusely adhering strains," *J. Clin. Microbiol.* 31:2031–2037.

Jann, K. and Hoschutsky, H. 1991. "Nature and organization of adhesins," *Curr. Top. Microbiol. Immunol.* 151:55–70.

Jarvis, K. G. and Kaper, J. B. 1996. "Secretion of extracellular proteins by enterohemorrhagic *Escherichia coli* via a putative type III secretion system," *Infect. Immun.* 64:4826–4829.

Jarvis, K. G., Girón, J. A., Jerse, A. E., McDaniel, T. K., Donnenberg, M. S., and Kaper, J. B. 1995. "Enteropathogenic *Escherichia coli* contains a putative type III secretion system necessary for the export of proteins involved in attaching and effacing lesion formation," *Proc. Natl. Acad. Sci. USA* 92:7996–8000.

Jerse, A. E., Yu, J., Tall, B. D., and Kaper, J. B. 1990. "A genetic locus of enteropathogenic *Escherichia coli* necessary for the production of attaching and effacing lesions on tissue culture cells," *Proc. Natl. Acad. Sci. USA* 87:7839–7843.

Jones, C. H., Dodson, K., and Hultgren, S. J. 1996. "Structure, function, and assembly of adhesive P pili," in *Urinary Tract Infections: Molecular Pathogenesis and Clinical Management.* eds., Mobley, H. L. T. and Warren, J. W. Washington, D.C.: American Society for Microbiology. pp. 175–219.

Jose, J., Jahnig, F., and Meyer, T. F. 1995. "Common structural features of IgA1 protease-like outer membrane protein autotransporters," *Mol. Microbiol.* 18:377–382.

Kaper, J. B. 1996. "Defining EPEC," *Rev. Microbiol.,* Sao Paulo, 27 (Suppl. 1):130–133.

Keene, W. E., Sazie, E., Kok, J., Rice, D. H., Hancock, D. D., Balan, V. K., Zhao, T., and Doyle, M. P. 1997. "An outbreak of *Escherichia coli* O157:H7 infections traced to jerky made from deer meat," *JAMA* 277:1229–1231.

Keene, W. E., McAnulty, J. M., Hoesly, F. C., Williams, L. P., Jr., Hedberg, K., Oxman, G. L., Barrett, T. J., Pfaller, M. A., and Fleming, D. W. 1994. "A swimming-associated outbreak of hemorrhagic colitis caused by *Escherichia coli* O157:H7 and *Shigella sonnei,*" *N. Engl. J. Med.* 331:579–584.

Keith, B. R., Maurer, L., Spears, P. A., and Orndorff, P. E. 1986. "Receptor-binding function of type I pili effects bladder colonization by a clinical isolate of *Escherichia coli,*" *Infect. Immun.* 53:693–696.

Kenny, B. and Finlay, B. B. 1995. "Protein secretion by enteropathogenic *Escherichia coli* is essential for transducing signals to epithelial cells," *Proc. Natl. Acad. Sci. USA* 92:7991–7995.

Kenny, B., Lai, L.-C., Finlay, B. B., and Donnenberg, M. S. 1996. "EspA, a protein secreted by enteropathogenic *Escherichia coli* (EPEC), is required to induce signals in epithelial cells," *Mol. Microbiol.* 20:313–323.

Ketyi, I. 1989. "Epidemiology of the enteroinvasive *Escherichia coli.* Observations in Hungary," *J. Hyg. Epidemiol. Microbiol. Immunol.* 33:261–267.

Knutton, S., Baldwin, T., Williams, P. H., and McNeish, A. S. 1989a. "Actin accumulation at sites of bacterial adhesion to tissue culture cells: basis of a new diagnostic test for enteropathogenic and enterohemorrhagic *Escherichia coli,*" *Infect. Immun.* 57:1290–1298.

Knutton, S., McConnell, M. M., Rowe, B., and McNeish, A. S. 1989b. "Adhesion and ultrastructural properties of human enterotoxigenic *Escherichia coli* producing colonization factor antigens III and IV," *Infect. Immun.* 57:3364–3371.

Knutton, S., Collington, G. K., Baldwin, T. J., Haigh, R. D., and Williams, P. H. 1996. "Cellular responses to EPEC infection," *Rev. Microbiol.,* Sao Paulo, 27:89–94.

Korhonen, T. K., Valtonen, M. V., Parkkinen, J., Väisänen-Rhen, V., Finne, J., Ørskov, F., Ørskov, I., Svenson, S. B., and Mäkelä, P. H. 1985. "Serotypes, hemolysin production, and receptor recognition of *Escherichia coli* strains associated with neonatal sepsis and meningitis," *Infect. Immun.* 48:486–491.

Korhonen, T. K., Virkola, R., and Holthofer, H. 1986. "Localization of binding sites for purified *Escherichia coli* P fimbriae in the human kidney," *Infect. Immun.* 54:328–332.

Korhonen, T. K., Virkola, R., Westurlund, B., Holthofer, H., and Parkkinen, J. 1990. "Tissue tropism of *Escherichia coli* adhesins in human extraintestinal infections," *Curr. Top. Microbiol. Immunol.* 151:115–127.

Kuehn, M. J., Heuser, J., Normark, S., and Hultgren, S. J. 1992. "P pili in uropathogenic *E. coli* are composite fibres with distinct fibrillar adhesive tips," *Nature* 356:252–255.

Kusters, J. G. and Gaastra, W. 1994. "Fimbrial operons and evolution," in *Fimbriae: Adhesion, Genetics, Biogenesis and Vaccines,* ed., Klemm, P. Boca Raton: CRC Press. pp. 179–196.

Lai, L. C., Wainwright, L. A., Stone, K. D., and Donnenberg, M. S. 1997. "A third secreted protein that is encoded by the enteropathogenic *Escherichia coli* pathogenicity island is required for transduction of signals and for attaching and effacing activities in host cells," *Infect. Immun.* 65:2211–2217.

Langermann, S., Palaszynski, S., Barnhart, M., Auguste, G., Pinkner, J. S., Burlein, J., Barren, P., Koenig, S., Leath, S., Jones, C. H., and Hultgren, S. J. 1997. "Prevention of mucosal *Escherichia coli* infection by FimH-adhesin-based systemic vaccination," *Science* 276:607–611.

Lanyi, B., Szita, J., Ringelhann, A., and Kovach, K. 1959. "A waterborne outbreak of enteritis associated with *Escherichia coli* serotype 124:72:32," *Acta. Microbiol. Hung.* 6:77–78.

Lee, C. A. 1997. "Type III secretion systems: Machines to deliver bacterial proteins into eukaryotic cells?" *Trends Microbiol.* 5:148–156.

Lencer, W. I., Constable, C., Moe, S., Jobling, M. G., Webb, H. M., Ruston, S., Madara, J. L., Hirst, T. R., and Holmes, R. K. 1995. "Targeting of cholera toxin and *Escherichia coli* heat labile toxin in polarized epithelia: role of COOH-terminal KDEL," *J. Cell Biol.* 131:951–962.

Levine, M. M. 1987. "*Escherichia coli* that cause diarrhea: enterotoxigenic, enteropathogenic, enteroinvasive, enterohemmorrhagic, and enteroadherent," *J. Infect. Dis.* 155:377–389.

Levine, M. M., Caplan, E. S., Watermann, D., Cash, R. A., Homick, R. B., and Snyder, M. J. 1977. "Diarrhea caused by *Escherichia coli* that produce only heat-stable enterotoxin," *Infect. Immun.* 17:78–82.

Levine, M. M., Rennels, M. B., Cisneros, L., Highes, T. P., Nalin, D. R., and Young, C. R. 1980. "Lack of person-to-person transmission of enterotoxigenic *Escherichia coli* despite close contact," *Am. J. Epidemiol.* 111:347–355.

Levine, M. M., Xu, J.-G., Kaper, J. B., Lior, H., Prado, V., Tall, B., Nataro, J., Karch, H., and Wachsmuth, K. 1987. "A DNA probe to identify enterohemorrhagic *Escherichia coli* of O157:H7 and other serotypes that cause hemorrhagic colitis and hemolytic uremic syndrome," *J. Infect. Dis.* 156:175–182.

Levine, M. M., Ristaino, P., Marley, G., Smyth, C., Knutton, S., Boedeker, E., Black, R., Young, C., Clements, M. L., Cheney, C., and Patnaik, R. 1984. "Coli surface antigens 1 and 3 of colonization factor antigen II-positive enterotoxigenic *Escherichia coli:* Morphology, purification, and immune responses in humans," *Infect. Immun.* 44:409–420.

Levine, M. M., Ferreccio, C., Prado, V., Cayazzo, M., Abrego, P., Martinez, J., Maggi, L., Baldini, M. M., Martin, W., Maneval, D., Kay, B., Guers, L., Lior, H., Wasserman, S. S., and Nataro, J. P. 1993b. "Epidemiologic studies of *Escherichia coli* diarrheal infections in a low socioeconomic level peri-urban community in Santiago, Chile," *Am. J. Epidemiol.* 138:849–869.

Long, K. Z., Wood, J. W., Vasquez Gariby, E., Weiss, K. M., Mathewson, J. J., de la Cabada, F. J., Du Pont, H. L., and Wilson, R.A. 1994. "Proportional hazards analysis of diarrhea due to enterotoxigenic *Escherichia coli* and breast feeding in a cohort of urban Mexican children," *Amer. J. Epidemiol.* 139:193–205.

Louise, C. B. and Obrig, T. G. 1995. "Specific interaction of *Escherichia coli* O157:H7-derived Shiga-like toxin II with human renal endothelial cells," *J. Infect. Dis.* 172:1397–1401.

Mangia, A. H., Duarte, A. N., Duarte, R., Silva, L. A., Bravo, V. L., and Leal, M. C. 1993. "Aetiology of acute diarrhoea in hospitalized children in Rio de Janeiro City, Brazil," *J. Trop. Peds.* 39:365–367.

Marier, R., Wells, J. G., Swanson, R. C., Dallhan, W., and Mehlman, I. J. 1973. "An outbreak of enteropathogenic *Escherichia coli* foodborne disease traced to imported french cheese," *Lancet* 2:1376–1378.

Marquart, M. E., Picking, W. L., and Picking, W. D. 1996. "Soluble invasion plasmid antigen C (IpaC) from *Shigella flexneri* elicits epithelial cell responses related to pathogen invasion," *Infect. Immun.* 64:4182–4187.

Maurelli, A. T. 1994. "Virulence protein export systems in *Salmonella* and *Shigella:* A new family or lost relatives?" *Trends Cell Biol.* 4:240–242.

Maurelli, A. T., Hromockyj, A. E., and Bernardini, M. L. 1992. "Environmental regulation of *Shigella* virulence," *Curr. Top. Microbiol. Immunol.* 180:95–116.

Mayer, H. B. and Wanke, C. A. 1995. "Enteroaggregative *Escherichia coli* as a possible cause of diarrhea in an HIV-infected patient," *N. Engl. J. Med.* 332:273–274.

McDaniel, T. K., Jarvis, K. G., Donnenberg, M. S., and Kaper, J. B. 1995. "A genetic locus of enterocyte effacement conserved among diverse enterobacterial pathogens,'' *Proc. Natl. Acad. Sci. USA* 92:1664–1668.

McDaniel, T. K. and Kaper, J. B. 1997. ''A cloned pathogenicity island from enteropathogenic *Escherichia coli* confers the attaching and effacing phenotype on K-12 *E. coli*,'' *Mol. Microbiol.* 23:399–407.

McGee, D. W., Elson, C. O., and McGhee, J. R. 1993. ''Enhancing effect of cholera toxin on interleukin-6 intestinal epithelial cells: mode of action and augmenting effet of inflammatory cytokines,'' *Infect. Immun.* 61:4637–4644.

Mecsas, J. J. and Strauss, E. J. 1996. ''Molecular mechanisms of bacterial virulence: type III secretion and pathogenicity islands,'' *Emerg. Infect. Dis.* 2:270–288.

Meier, C., Oelschlaeger, T. A., Merkert, H., Korhonen, T. K., and Hacker, J. 1996. ''Ability of *Escherichia coli* isolates that cause meningitis in newborns to invade epithelial and endothelial cells,'' *Infect. Immun.* 64:2391–2399.

Ménard, R., Sansonetti, P. J., and Parsot, C. 1993. ''Nonpolar mutagenesis of the *ipa* genes defines IpaB, IpaC and IpaD as effectors of *Shigella flexneri* entry into epithelial cells,'' *J. Bacteriol.* 175:5899–5906.

Ménard, R., Prévost, M. C., Gounon, P., Sansonetti, P., and Dehio, C. 1996. ''The secreted Ipa complex of *Shigella flexneri* promotes entry into mammalian cells,'' *Proc. Natl. Acad. Sci. USA* 93:1254–1258.

Mobley, H. L. T., Green, D. M., Trifillis, A. L., Johnson, D. E., Chippendale, G. R., Lockatell, C. V., Jones, B. D., and Warren, J. W. 1990. ''Pyelonephritogenic *Escherichia coli* and killing of cultured human renal proximal tubular epithelial cells: role of hemolysin in some strains,'' *Infect. Immun.* 58:1281–1289.

Mobley, H. L. T., Jarvis, K. G., Elwood, J. P., Whittle, D. I., Lockatell, C. V., Russell, R. G., Johnson, D. E., Donnenberg, M. S., and Warren, J. W. 1993. ''Isogenic P-fimbrial deletion mutants of pyelonephritogenic *Escherichia coli:* The role of αGal(1–4)βGal binding in virulence of a wild-type strain,'' *Mol. Microbiol.* 10:143–155.

Moon, H. W., Whipp, S. C., Argenzio, R. A., Levine, M. M., and Giannella, R. A. 1983. ''Attaching and effacing activities of rabbit and human enteropathogenic *Escherichia coli* in pig and rabbit intestines,'' *Infect. Immun.* 41:1340–1351.

Moss, J., Garrison, S., Oppenheimer, N. J., and Richardson, S. H. 1979. ''NAD-dependent ADP-ribosylation of arginine and proteins by *Escherichia coli* heat-labile enterotoxin,'' *J. Biol. Chem.* 254:6270–6272.

Moxley, R. A. and Francis, D. H. 1986. ''Natural and experimental infection with an attaching and effacing strain of *Escherichia coli* in calves,'' *Infect. Immun.* 53:339–346.

Nataro, J. P. and Deng, Y. 1997. Unpublished data.

Nataro, J. P., Yikang, D., Yingkang, D., and Walker, K. 1994. ''AggR, a transcriptional activator of Aggregative Adherence Fimbria I expression in enteroaggregative *Escherichia coli*,'' *J. Bacteriol.* 176:4691–4699.

Nataro, J. P., Hicks, S., Phillips, A. D., Vial, P. A., and Sears, C. L. 1996. ''T84 cells in culture as a model for enteroaggregative *Escherichia coli* pathogenesis,'' *Infect. Immun.* 64:4761–4768.

Nataro, J. P., Deng, Y., Maneval, D. R., German, A. L., Martin, W. C., and Levine, M. M. 1992. ''Aggregative adherence fimbriae I of enteroaggregative *Escherichia coli* mediate adherence to HEp-2 cells and hemagglutination of human erythrocytes,'' *Infect. Immun.* 60:2297–2304.

Nataro, J. P., Yikang, D., Girón, J. A., Savarino, S. J., Kothary, M. H., and Hall, R. 1993. ''Aggregative adherence fimbria I expression in enteroaggregative *Escherichia coli* requires two unlinked plasmid regions,'' *Infect. Immun.* 61:1126–1131.

Nataro, J. P., Yikang, D., Cookson, S., Cravioto, A., Savarino, S. J., Guers, L. D., Levine, M.

M., and Tacket, C. O. 1995b. "Heterogeneity of enteroaggregative *Escherichia coli* virulence demonstrated in volunteers," *J. Infect. Dis.* 171:465–468.

Nataro, J. P., Seriwatana, J., Fasano, A., Maneval, D. R., Guers, L. D., Noriega, F., Dubovsky, F., Levine, M. M., and Morris, J. G., Jr. 1995a. "Identification and cloning of a novel plasmid-encoded enterotoxin of enteroinvasive *Escherichia coli* and *Shigella* strains," *Infect. Immun.* 63:4721–4728.

Navarro-Garcia, F., Eslava, C., Nataro, J. P., and Cravioto, A. 1997. Unpublished data.

Nowicki, B., Svanborg-Eden, C., Hull, R., and Hull, S. 1989. "Molecular analysis and epidemiology of the Dr hemagglutinin of uropathogenic *Escherichia coli,*" *Infect. Immun.* 57:446–451.

Nowicki, B., Labigne, A., Moseley, S. L., Hull, R., Hull, S., and Moulds, J. 1990. "The Dr hemagglutinin, afimbrial adhesins AFA-I and AGA-III, and F1845 fimbriae of uropathogenic and diarrhea-associated *Escherichia coli* belong to a family of hemagglutinins with Dr receptor recognition," *Infect. Immun.* 58:279–281.

Nzegwu, H. C. and Levin, R. J. 1994. "Neurally maintained hypersecretion in undernourished rat intestine activated by *E. coli* STa enterotoxin and cyclic nucleotides in vitro," *J. Physiol.* (Lond.) 479:159–169.

Nzegwu, H. C. and Levin, R. J. 1996. "Luminal capsaicin inhibits fluid secretion induced by enterotoxin *E-coli* STa, but not by carbachol, in vivo in rat small and large intestine," *Exp. Physiol.* 81:313–315.

O'Brien, A. D. and Holmes, R. K. 1996. "Protein toxins of *Escherichia coli* and *Salmonella,*" in *Escherichia coli and Salmonella,* ed., Neidhardt, F. C. Washington, D.C.: ASM Press. pp. 2788–2802.

O'Hanley, P., Lalonde, G., and Ji, G. 1991. "Alpha-hemolysin contributes to the pathogenicity of piliated digalactoside-binding *Escherichia coli* in the kidney—efficacy of an alpha-hemolysin vaccine in preventing renal injury in the BALB/c mouse model of pyelonephritis," *Infect. Immun.* 59:1153–1161.

Ørskov, F. and Ørskov, I. 1992. "*Escherichia coli* serotyping and disease in man and animals," *Can. J. Microbiol.* 38:699–674.

Ofek, I., Mirelman, D., and Sharon, N. 1977. "Adherence of *Escherichia coli* to human mucosal cells mediated by mannose receptors," *Nature* 265:623–625.

Orr, P., Lorencz, B., Brown, R., Kielly, R., Tan, B., Holton, D., Clugstone, H., Lugtig, L., Pim, C., Macdonald, S., Hammond, G., Moffatt, M., Spika, J., Manuel, D., Winther, W., Milley, D., Lior, H., and Sinuff, N. 1994. "An outbreak of diarrhea due to verotoxin-producing *Escherichia coli* in the Canadian Northwest Territories," *Scand. J. Infect. Dis.* 26:675–684.

Ott, M., Hoschutzky, H., Jann, K., van Die, I., and Hacker, J. 1988. "Gene clusters for S fimbrial adhesin (*sfa*) and F1C fimbriae (*foc*) of *Escherichia coli*: comparative aspects of structure and function," *J. Bacteriol.* 170:3983–3990.

Pere, A., Leinonen, M., Väisänen-Rhen, V., Rhen, M., and Korhonen, T. K. 1985. "Occurrence of type-1C fimbriae on *Escherichia coli* strains isolated from human extraintestinal infections," *J. Gen. Microbiol.* 131:1705–1711.

Phillips, I., Eykyn, S., King, A., Gransden, W. R., Rowe, B., Frost, J. A., and Gross, R. J. 1988. "Epidemic multiresistant *Escherichia coli* infection in West Lambeth health district," *Lancet* 1:1038–1041.

Pluschke, G., Mayden, J., Achtman, M., and Levine, R. P. 1983. "Role of the capsule and the O antigen in resistance of O18:K1 *Escherichia coli* to complement-mediated killing," *Infect. Immun.* 42:907–913.

Prasadarao, N. V., Wass, C. A., and Kim, K. S. 1997. "Identification and characterization of S fimbria-binding sialoglycoproteins on brain microvascular endothelial cells," *Infect. Immun.* 65:2852–2860.

Ramer, S. W., Bieber, D., and Schoolnik, G. K. 1996. "BfpB, an outer membrane lipoprotein required for the biogenesis of bundle-forming pili in enteropathogenic *Escherichia coli,*" *J. Bacteriol.* 178:6555–6563.

Rasheed, J. K., Guzman-Versezco, L.-M., and Kupersztoch, Y. M. 1990. "Two precursors of the heat-stable enterotoxin of *Escherichia coli:* evidence of extracellular processing," *Molec. Microbiol.* 4:265–273.

Roberts, J. A., Marklund, B.-I., Ilver, D., Haslam, D., Kaack, M. B., Baskin, G., Louis, M., Möllby, R., Winberg, J., and Normark, S. 1994. "The Gal(α1-4)Gal-specific tip adhesin of *Escherichia coli* P-fimbriae is needed for pyelonephritis to occur in the normal urinary tract," *Proc. Natl. Acad. Sci. USA* 91:11889–11893.

Rosenshine, I., Donnenberg, M. S., Kaper, J. B., and Finlay, B. B. 1992. "Signal exchange between enteropathogenic *Escherichia coli* (EPEC) and epithelial cells: EPEC induce tyrosine phosphorylation of host cell protein to initiate cytoskeletal rearrangement and bacterial uptake," *EMBO J.* 11:3551–3560.

Rosenshine, I., Ruschkowski, S., Stein, M., Reinscheid, D. J., Mills, S. D., and Finlay, B. B. 1996. "A pathogenic bacterium triggers epithelial signals to form a functional bacterial receptor that mediates actin pseudopod formation," *EMBO J.* 15:2613–2624.

Sandvig, K., Garred, O., Prydz, K., Kozlov, J. V., Hansen, S. H., and Van Deurs, B. 1992. "Retrograde transport of endocytosed Shiga toxin to the endoplasmic reticulum," *Nature* 358:510–512.

Sansonetti, P. J. 1992a. "*Escherichia coli, Shigella,* antibiotic-associated diarrhea, and prevention and treatment of gastroenteritis," *Curr. Opin. Infect. Dis.* 5:66–73.

Sansonetti, P. J. 1992b. "Molecular and cellular biology of *Shigella flexneri* invasiveness: From cell assay systems to shigellosis," *Curr. Top. Microbiol. Immunol.* 180:1–19.

Sarff, L. D., McCracken, G. H., Schiffer, M. S., Glode, M. P., Robbins, J. B., Orskov, I., and Orskov, F. 1975. "Epidemiology of *Escherichia coli* K1 in healthy and diseased newborns," *Lancet* 1:1099–1104.

Sasakawa, C., Buysse, J. M., and Watanabe, H. 1992. "The large virulence plasmid of *Shigella,*" *Curr. Top. Microbiol. Immunol.* 180:21–44.

Savarino, S. J., Fasano, A., Robertson, D. C., and Levine, M. M. 1991. "Enteroaggregative *Escherichia coli* elaborate a heat-stable enterotoxin demonstrable in an in vitro rabbit intestinal model," *J. Clin. Invest.* 87:1450–1455.

Savarino, S. J., Fox, P., Yikang, D., and Nataro, J. P. 1994. "Identification and characterization of a gene cluster mediating enteroaggregative *Escherichia coli* aggregative adherence fimbria I biogenesis," *J. Bacteriol.* 176:4949–4957.

Savarino, S. J., Fasano, A., Watson, J., Martin, B. M., Levine, M. M., Guandalini, S., and Guerry, P. 1993. "Enteroaggregative *Escherichia coli* heat-stable enterotoxin 1 represents another subfamily of E. coli heat-stable toxin," *Proc. Natl. Acad. Sci. USA* 90:3093–3097.

Savarino, S. J., McVeigh, A., Watson, J., Cravioto, A., Molina, J., Echeverria, P., Bhan, M. K., Levine, M. M., and Fasano, A. 1996. "Enteroaggregative *Escherichia coli* heat-stable enterotoxin is not restricted to enetroaggregative *E. coli,*" *J. Infect. Dis.* 173:1019–1022.

Scaletsky, I. C. A., Silva, M. L. M., and Trabulsi, L. R. 1984. "Distinctive patterns of adherence of enteropathogenic *Escherichia coli* to HeLa cells," *Infect. Immun.* 45:534–536.

Schmidt, G., Sehr, P., Wilm, M., Selzer, J., Mann, M., and Aktories, K. 1997. "Gln 63 of Rho is deamidated by *Escherichia coli* cytotoxic necrotizing factor-1," *Nature* 387:725–729.

Schmidt, H., Kernbach, C., and Karch, H. 1996. "Analysis of the EHEC *hly* operon and its location in the physical map of the large plasmid of enterohaemorrhagic *Escherichia coli* O157:H7," *Microbiology* 142:907–914.

Schroeder, S. A., Caldwell, J. R., Vernon, T. M., White, P. S., Granger, S. I., and Bennett, J. V. 1968. "A waterborne outbreak of gastroenteritis in adults associated with enteropathogenic *Escherichia coli,*" *Lancet* 1:737–740.

Schultsz, C. 1994. "Detection of enterotoxigenic *Escherichia coli* in stool samples by using nonradioactively labeled oligonucleotide DNA probes and PCR," *J. Clin. Microbiol.* 1994:2393–2397.

Scotland, S. M., Willshaw, G. A., Smith, H. R., and Rowe, B. 1990. "Properties of strains of *Escherichia coli* O26:H11 in relation to their enteropathogenic or enterohemorrhagic classification," *J. Infect. Dis.* 162:1069–1074.

Scotland, S. M., Smith, H. R., Said, B., Willshaw, G. A., Cheasty, T., and Rowe, B. 1991. "Identification of enteropathogenic *Escherichia coli* isolated in Britain as enteroaggregative or as members of a subclass of attaching-and-effacing *E. coli* not hybridising with the EPEC adherence-factor probe," *J. Med. Microbiol.* 35:278–283.

Sears, C. L. and Kaper, J. B. 1996. "Enteric bacterial toxins: Mechanisms of action and linkage to intestinal secretion," *Microbiol. Rev.* 60:167–215.

Siitonen, A., Takala, A., Ratiner, Y. A., Pere, A., and Mäkelä, P. H. 1993. "Invasive *Escherichia coli* infections in children: Bacterial characteristics in different age groups and clinical entities," *Pediatr. Infect. Dis. J.* 12:606–612.

Silva, M. L. M. and Giampaglia, C. M. S. 1992. "Colostrum and human milk inhibit localized adherence of enteropathogenic *Escherichia coli* to HeLa cells," *Acta Paediatr. Scand.* 81:266–267.

Silver, R. P. and Vimr, E. R. 1990. "Polysialic acid capsule of *Escherichia coli* K1," in *Molecular Basis of Bacterial Pathogenesis,* eds., Iglewski, B. H. and Clark, V. L. San Diego: Academic Press, Inc. pp. 39–60.

Sixma, T. K., Kalk, K. H., van Zanten, B. A., Dauter, Z., Kingma, J., Witholt, B., and Hol, W. G. 1993. "Refined structure of *Escherichia coli* heat-labile enterotoxin, a close relative of cholera toxin," *J. Mol. Biol.* 230:890–918.

Sixma, T. K., Pronk, S. E., Kalk, K. H., Wartna, E. S., van Zanten, B. A., Witholt, B., and Hol, W. H. J. 1991. "Crystal structure of a cholera toxin-related heat-labile enterotoxin from *E. coli, Nature* 351:371–378.

Sjoberg, P. O., Lindahl, M., Porath, J., and Wadstrom, T. 1988. "Purification and characterization of CS2, a sialic acid-specific haemagglutinin of enterotoxigenic *Escherichia coli,*" *Biochem. J.* 255:105–111.

Small, P. L. C. and Falkow, S. 1988. "Identification of regions on a 230-kilobase plasmid from enteroinvasive *Escherichia coli* that are required for entry into HEp-2 cells," *Infect. Immun.* 56:225–229.

Smith, H. R., Scotland, S. M., Willshaw, G. A., Rowe, B., Cravioto, A., and Eslava, C. 1994. "Isolates of *Escherichia coli* O44:H18 of diverse origin are enteroaggregative," *J. Infect. Dis.* 170:1610–1613.

Snyder, J. D., Wells, J. G., Yashuk, J., Puhr, N., and Blake, P. A. 1984. "Outbreak of invasive *Escherichia coli* gastroenteritis on a cruise ship," *Am. J. Trop. Med. Hyg.* 33:281–284.

So, M. and McCarthy, B. J. 1980. "Nucleotide sequence of the bacterial transposon Tn1681 encoding a heat-stable (ST) toxin and its identification in enterotoxigenic *Escherichia coli* strains," *Proc. Natl. Acad. Sci. USA* 77:4011–4015.

Sohel, I., Puente, J. L., Ramer, S. W., Bieber, D., Wu, C.-Y., and Schoolnik, G. K. 1996. "Enteropathogenic *Escherichia coli:* identification of a gene cluster coding for bundle-forming pilus morphogenesis," *J. Bacteriol.* 178:2613–2628.

Spangler, B. D. 1992. "Structure and function of cholera toxin and the related *Escherichia coli* heat-labile enterotoxin," *Microbiol. Rev.* 56:622–647.

Stapleton, A., Latham, R. H., Johnson, C., and Stamm, W. E. 1990. "Postcoital antimicrobial prophylaxis for recurrent urinary tract infection. A randomized, double-blind, placebo-controlled trial," *JAMA* 264:703–706.

Stein, M. A., Mathers, D. A., Yan, H., Baimbridge, K. G., and Finlay, B. B. 1996. "Enteropathogenic *Escherichia coli* (EPEC) markedly decreases the resting membrane potential of Caco-2 and HeLa human epithelial cells," Infect. Immun. 64:4820–4825.

Steiner, T. S., Lima, A. A. M., Nataro, J. P., and Guerrant, R. L. 1997. Enteroaggregative *Escherichia coli* produce intestinal inflammation and growth impairment and cause interleukin-8 release from intestinal epithelial cells," J. Infect. Dis. 177:88–96.

Stone, K. D., Zhang, H.-Z., Carlson, L. K., and Donnenberg, M. S. 1996. "A cluster of fourteen genes from enteropathogenic *Escherichia coli* is sufficient for biogenesis of a type IV pilus," *Mol. Microbiol.* 20:325–337.

Strom, B. L., Collins, M., West, S. L., Kreisberg, J., and Weller, S. 1987. "Sexual activity, contraceptive use, and other risk factors for symptomatic and asymptomatic bacteriuria. A case-control study," *Ann. Intern. Med.* 107:816–823.

Svensson, M. Lindstedt, R., Radin, N. S., and Svanborg, C. 1994. "Epithelial glucosphingolipid expression as a determinant of bacterial adherence and cytokine production," *Infect. Immun.* 62:4404–4410.

Swenson, D. L., Bukanov, N. O., Berg, D. E., and Welch, R. A. 1996. "Two pathogenicity islands in uropathogenic Escherichia coli J96:Cosmid cloning and sample sequencing," *Infect. Immun.* 64:3736–3743.

Swerdlow, D. L., Woodruff, B. A., Brady, R. C., Griffin, P. M., Tippen, S., Donnell, H. D., Jr., Geldreich, E., Payne, B. J., Meyer, A., Jr., Wells, J. G., Greene, K. D., Bright, M., Bean, N. H., and Blake, P. A. 1992. "A waterborne outbreak in Missouri of *Escherichia coli* O157:H7 associated with bloody diarrhea and death," *Ann. Intern. Med.* 117:812–819.

Taniguchi, T., Fujino, Y., Yamamoto, K., Miwatani, T., and Honda, T. 1995. "Sequencing of the gene encoding the major pilin of pilus colonization factor antigen III (CFA/III) of human enterotoxigenic *Escherichia coli* and evidence that CFA/III is related to type IV pili," *Infect. Immun.* 63:724–728.

Taniguchi, T., Arita, M., Sato, M., Yamamoto, K., Miwatani, T., and Honda, T. 1994. "Evidence that the N-terminal amino acid sequence of pilus colonization factor antigen III produced by human enterotoxigenic *Escherichia coli* is similar to that of TcpA pilin of Vibrio cholerae," *J. Infect. Dis.* 170:1049–1050.

Tarr, P. I. 1995. "*Escherichia coli* O157:H7: Clinical, diagnostic, and epidemiological aspects of human infection," *Clin. Infect. Dis.* 20:1–10.

Taylor, D. N., Echeverria, P., Sethabutr, O., Pitarangi, C., Leksomboon, U., Blacklow, N. R., Rowe, B., Gross, R., and Cross, J. 1988. "Clinical and microbiologic features of *Shigella* and enteroinvasive *Escherichia coli* infections detected by DNA hybridization," *J. Clin. Microbiol.* 26:1362–1366.

Teneberg, S., Hirst, T. R., Ångström, J., and Karlsson, K.-A. 1994. "Comparison of the glycolipid-binding specificities of cholera toxin and porcine *Escherichia coli* heat-labile enterotoxin: Identification of a receptor-active non-ganglioside glycolipid for the heat-labile toxin in infant rabbit small intestine," Glycoconjugate J. 11:533–540.

Tennant, J. M. and Mattick, J. S. 1994. "Type 4 fimbriae," in *Fimbriae: Adhesion, Genetics, Biogenesis and Vaccines,* ed., Klemm, P. Boca Raton, IL: CRC Press. pp. 127–146.

Tesh, V. L. and O'Brien, A. D. 1991. "The pathogenic mechanisms of Shiga toxin and the Shiga-like toxins," *Mol. Microbiol.* 5:1817–1822.

Tobe, T., Yoshikawa, M., Mizuno, T., and Sasakawa, C. 1993. "Transcriptional control of the invasion regulatory gene *virB* of *Shigella flexneri:* Activation by VirF and repression by H-NS,'' *J. Bacteriol.* 175:6142–6149.

Tornieporth, N. G., John, J., Salgado, K., de Jesus, P., Latham, E., Melo, M. C., Gunzburg, S. T., and Riley, L. W. 1995. ''Differentiation of pathogenic *Escherichia coli* strains in Brazilian children by PCR,'' *J. Clin. Microbiol.* 33:1371–1374.

Tullus, K., Hörlin, K., Svenson, S. B., and Kallenius, G. 1984. ''Epidemic outbreaks of acute pyelonephritis caused by nosocomial spread of P fimbriated *Escherichia coli* in children,'' *J. Infect. Dis.* 150:728–736.

Tzipori, S., Gibson, R., and Montanaro, J. 1989. ''Nature and distribution of mucosal lesions associated with enteropathogenic and enterohemorrhagic *Escherichia coli* in piglets and the role of plasmid-mediated factors,'' *Infect. Immun.* 57:1142–1150.

Tzipori, S., Gunzer, F., Donnenberg, M. S., De Montigny, L., Kaper, J. B., and Donohue-Rolfe, A. 1995. ''The role of the *eaeA* gene in diarrhea and neurological complications in a gnotobiotic piglet model of enterohemorrhagic *Escherichia coli* infection,'' *Infect. Immun.* 63:3621–3627.

Tzipori, S., Montanaro, J., Robins-Browne, R. M., Vial, P., Gibson, R., and Levine, M. M. 1992. ''Studies with enteroaggregative *Escherichia coli* in the gnotobiotic piglet gastroenteritis model,'' *Infect. Immun.* 60:5302–5306.

Tzipori, S., Karch, H., Wachsmuth, K. I., Robins-Browne, R. M., O'Brien, A. D., Lior, H., Cohen, M. L., Smithers, J., and Levine, M. M. 1987. ''Role of a 60-megadalton plasmid and Shiga-like toxins in the pathogenesis of infection caused by enterohemorrhagic *Escherichia coli* O157:H7 in gnotobiotic piglets,'' *Infect. Immun.* 55:3117–3125.

Vasselon, T., Mounier, J., Hellio, R., and Sansonetti, P. J. 1992. ''Movement along actin filaments of the perijunctional area and de novo polymerization of cellular actin are required for *Shigella flexneri* colonization of epithelial Caco- 2 cell monolayers,'' *Infect. Immun.* 60:1031–1040.

Venkatesan, M. M., Buysse, J. M., and Oaks, E. V. 1992. ''Surface presentation of *Shigella flexneri* invasion plasmid antigens requires the products of the *spa* locus,'' *J. Bacteriol.* 174:1990–2001.

Vial, P. A., Robins Browne, R., Lior, H., Prado, V., Kaper, J. B., Nataro, J. P., Maneval, D., Elsayed, A., and Levine, M. M. 1988. ''Characterization of enteroadherent-aggregative *Escherichia coli,* a putative agent of diarrheal disease,'' *J. Infect. Dis.* 158:70–79.

Viljanen, M. K., Peltola, T., Junnila, S. Y. T., Olkkonen, L., Järvinen, H., Kuistila, M., and Huovinen, P. 1990. ''Outbreak of diarrhoea due to *Escherichia coli* O111:B4 in schoolchildren and adults: association of Vi antigen-like reactivity,'' *Lancet* 336:831–834.

Wanke, C. A., Cronan, S., Goss, C., Chadee, K., and Guerrant, R. L. 1990. ''Characterization of binding of *Escherichia coli* strains which are enteropathogens to small-bowel mucin,'' *Infect. Immun.* 58:794–800.

Warren, J. W. 1996. ''Clinical presentations and epidemiology of urinary tract infections,'' in *Urinary Tract Infections: Molecular Pathogenesis and Clinical Management,* eds., Mobley, H. L. T. and Warren, J. W. Washington, D.C: ASM Press. pp. 3–27.

Watanabe, H. and Guerrant, R. L. 1997. ''Summary: Nagasaki enterohemorrhagic *Escherichia coli* meeting and workshop,'' *J. Infect. Dis.* 176:247–249.

Whittam, T. S., Wolfe, M. L., Wachsmuth, I. K., Orskov, F., Orskov, I., and Wilson, R. A. 1993. ''Clonal relationships among *Escherichia coli* strains that cause hemorrhagic colitis and infantile diarrhea,'' *Infect. Immun.* 61:1619–1629.

Wolf, M. K. 1997. ''Occurrence, distribution, and associations of O and H serogroups, colonization factor antigens, and toxins of enterotoxigenic *Escherichia coli,*'' *Clin. Microbiol. Rev.* 10:569–584.

Wood, L. V., Ferguson, L. E., Hogan, P., Thurman, D., Morgan, D., DuPont, H. L., and Ericsson, C. D. 1983. ''Incidence of bacterial enteropathogens in foods from Mexico,'' *Appl. Environ. Microbiol.* 46:328–332.

Wu, S.-X. and Peng, R.-Q. 1992. "Studies on an outbreak of neonatal diarrhea caused by EPEC O127:H6 with plasmid analysis restriction analysis and outer membrane protein determination," *Acta Paediatr. Scand.* 81:217–221.

Wu, X. R., Sun, T. T., and Medina, J. J. 1996. "*In vitro* binding of type 1-fimbriated *Escherichia coli* to uroplakins Ia and Ib: Relation to urinary tract infections," *Proc. Natl. Acad. Sci. USA* 93:9630–9635.

Yamamoto, T., Kaneko, M., Changchawalit, S., Serichantalergs, O., Ijuin, S., and Echeverria, P. 1994. "Actin accumulation associated with clustered and localized adherence in *Escherichia coli* isolated from patients with diarrhea," *Infect. Immun.* 62:2917–2929.

Zhang, L. X., Foxman, B. Tallman, P., Cladera, E., Le Bouguenec, C., and Marrs, C. F. 1997. "Distribution of *drb* genes coding for Dr binding adhesins among uropathogenic and fecal *Escherichia coli* isolates and identification of new subtypes," *Infect. Immun.* 65:2011–2018.

Zychlinsky, A., Kenny, B., Ménard, R, Prevost, M. C., Holland, I. B., and Sansonetti, P. J. 1994. "IpaB mediates macrophage apoptosis induced by *Shigella flexneri*," *Mol. Microbiol.* 11:619–627.

CHAPTER 4

Virulence Determinants of the Bacterial Pathogen *Yersinia enterocolitica:* Mode of Action and Global Regulation

MARIE-PAULE SORY
GUY R. CORNELIS

1. INTRODUCTION

THREE *Yersinia* species are pathogenic for humans or animals: *Y. pestis,* propagating plague, a nowadays resurgent disease, *Y. pseudotuberculosis,* causing diarrhea and death in rodents as a result of septicaemia, and *Y. enterocolitica,* an etiological agent of gastroenteritis in humans. *Y. enterocolitica* is widespread across the globe and is a common foodborne pathogen. Infection mainly occurs by ingestion of contaminated food or water, pork meat being a current contamination source (Tauxe et al., 1987).

The severity of the diseases produced by *Y. enterocolitica* is variable. Yersiniosis is often a self-limiting infection with ileitis, ulcerative colitis and mesenteric lymphadenitis as main clinical symptoms. Infection may, however, be followed by severe immunological sequellae such as erythema nodosum, Reiter's syndrome or polyarthritis, and exceptionally, *Y. enterocolitica* can provoke overwhelming infections such as septicaemia and meningitis (Cover and Aber, 1989; Attwood et al., 1989). *Y. enterocolitica* is an heterogenous species, divided in serogroups according to the reactivity of the O-chain of the lipopolysaccharide. Pathogenic strains are restricted to a limited number of serogroups. The virulence of these pathogenic strains also varies according to the serogroups. Strains of serotypes O:3, O:9 or O:5,27 are of lower natural virulence than those of serotypes O:8 or O:21.

Most of the cases due to the former serogroups are asymptomatic, undiagnosed or rather benign. However, because of its psychrophilic nature, *Y. enterocolitica* can multiply in stored blood bags, which leads to some rare but severe septic post-transfusional shocks (Prentice, 1992).

All three pathogenic *Yersinia* spp. harbor a common 70 kb virulence plasmid (pYV) encoding a constellation of tightly regulated anti-host factors. All yersiniae share a common strategy to escape the immune response of the host. A number of other virulence factors are encoded by the chromosome or other plasmids, some of them being responsible for the highest virulence of *Y. pestis* (for complete reviews on *Y. pestis,* see Perry and Fetherston, 1997; Brubaker, 1991). This review will essentially be devoted to the mechanisms involved in *Y. enterocolitica* evasion of the host immune response, but attention will also be given to studies on the pYV-dependent virulence factors of *Y. pseudotuberculosis* and *Y. pestis.*

2. PATHOLOGY

Y. enterocolitica is an invasive pathogen able to cross the gut epithelium of the host and to proliferate locally in the underlying tissue. *Y. enterocolitica* enter via the M (microfold) cells and reach the Peyer's patches (Grützkau et al., 1990). From the Peyer's patches, they can disseminate to the mesenteric lymph nodes and eventually to the liver and the spleen. They multiply mainly extracellularly (Hanski et al., 1989; Simonet et al., 1990). However, one cannot exclude an intracellular state of *Y. enterocolitica* at some stage of the infection, perhaps allowing transmigration of bacteria from Peyer's patches to deeper tissues.

Histological studies of mice infected by parenteral inoculation show the formation of pyogenic lesions and abscesses in the spleen and liver (Carter, 1975). Small granuloma-like lesions have also been described in the liver of mice having received sublethal doses of *Y. enterocolitica.* Abscesses contained *Y. enterocolitica* while no bacteria were detected in the granuloma-like lesions. Rather, the latter contained CD11b/18 positive cells (monocytes or granulocytes) as well as CD4 (helper) and CD8 (cytotoxic) T lymphocytes (Autenrieth et al., 1993).

In mice or rabbits orally inoculated with a high amount of bacteria, heavily colonized Peyer's patches were infiltrated with polymorphonuclear leukocytes (PMN) (Pai et al., 1980; Hanski et al., 1989). However, only little phagocytosis was observed. Monocytes also infiltrated the Peyer's patches and matured into inflammatory macrophages to produce interleukins (Beuscher et al., 1995).

All the histological studies confirmed the affinity of *Y. enterocolitica* for lymphoid tissue and the reticuloendothelial system. The variability sometimes observed between the different studies may be due to the amount of bacteria inoculated and the susceptibility of the animal used. Indeed, the morphological tissue alterations after bacterial infection result from both the virulence properties of the bacteria and the defense features of the immune system of the host and hence reflect the fragile equilibrium between the *Yersinia* invasion and the defense of the host.

3. INTERACTION WITH EUKARYOTIC HOST CELLS

3.1. OVERVIEW ON THE VIRULENCE FUNCTIONS

Mice undergoing bacterial infection with *Yersinia* initiate their defense by producing Interleukin-12, gamma interferon and tumor necrosis factor alpha (TNFa) (Autenrieth et al., 1994; 1996; Bohn and Autenrieth, 1996). The macrophages are presumably the final effector cells, killing *Y. enterocolitica* after activation by T cells. Indeed, evidence suggests that a specific T-cell-mediated host response is required for resolution of *Y. enterocolitica* infections (Autenrieth et al., 1992). Both CD4 Th1 and CD8 T cells confer resistance to *Y. enterocolitica* when transferred into T-cell-deficient mice. However, during the first steps of the infection, professional phagocytes also severely restrict the rate at which *Yersinia* multiply in the host tissues, thereby allowing the host to develop a specific protective immunity (Conlan, 1997).

The challenge of a pathogen is to cope with and subvert these host immune responses. Microscopic studies have provided evidence for a predominantly extracellular multiplication of *Yersinia* in lymphatic organs of infected mice. According to the literature, the strategies used by *Yersinia* to establish an infection is to avoid lysis by complement, to resist phagocytosis by macrophages and neutrophils and to inhibit cytokine production. To achieve this, *Y. enterocolitica* specifically targets host-cell receptors to adhere to the surface of the hosts cells in order to invade them or to inject some toxic proteins inside their cytosol, thereby inducing dysregulation of their signal transduction pathways.

Chromosomal as well as plasmidic factors are responsible for the virulence of *Y. enterocolitica* (Figure 4.1). The invasin InvA, the factor MyfA, the yersinial enterotoxin Yst and the proteins involved in iron acquisition are chromosome-encoded. The virulence plasmid, referred to as the pYV plasmid, encodes an integrated virulence apparatus called the Yop virulon and an outer membrane protein called YadA. The Yop virulon consists of a panoply of secreted proteins—the Yop proteins—, a dedicated secretion system and a regulatory network. A subset of Yop proteins—the Yop effectors—are addressed to the cytoplasm of eukaryotic cells by bacteria adhering at their surface. The other subset of Yop proteins—the translocators—form an apparatus that translocates the Yop effectors across the membrane of eukaryotic cells. The genes encoding the effector Yop proteins are scattered around the pYV plasmid, while those encoding the Yop translocators, the components

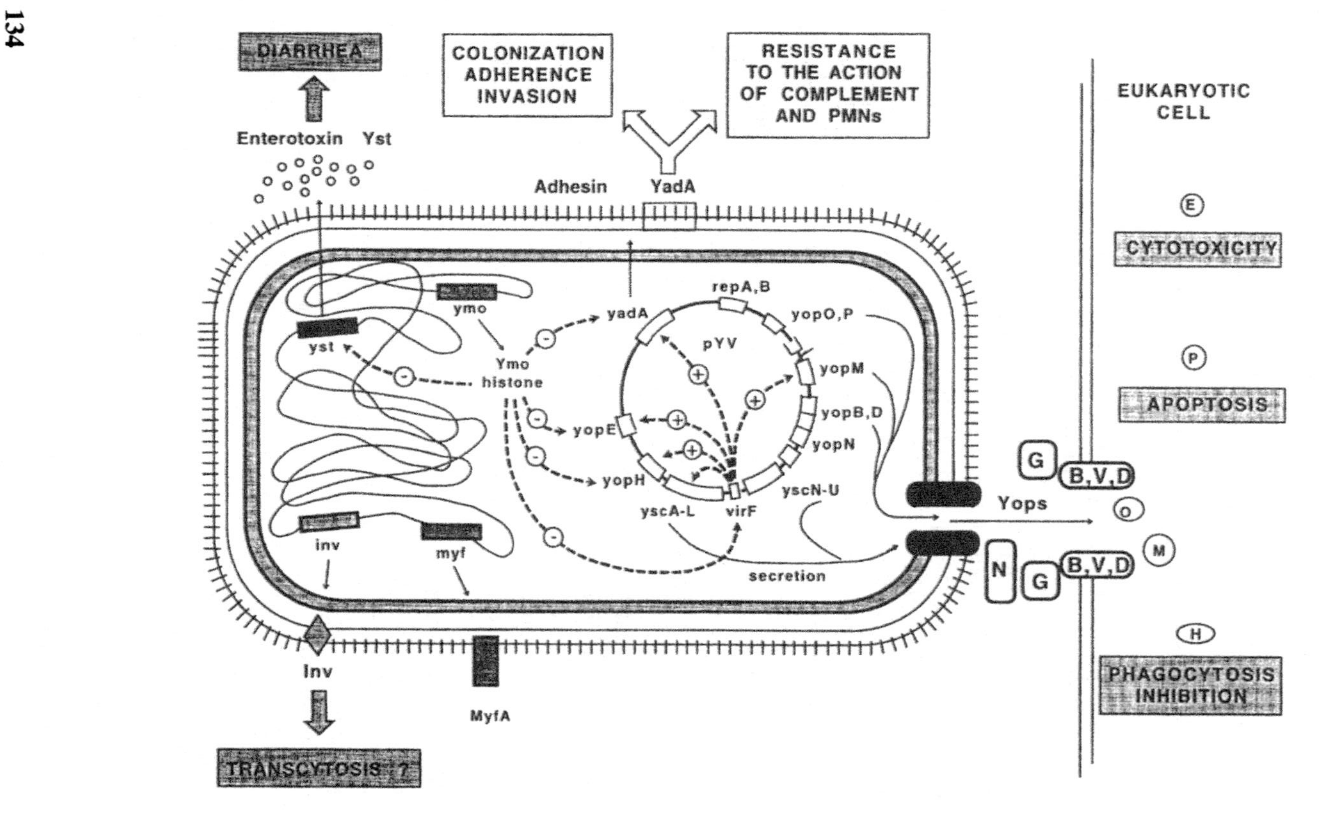

Figure 4.1 Overview of the main proven and putative virulence factors of *Y. enterocolitica.*

of the secretion system and the regulators are encoded by at least 27 genes mainly arranged in operons (Figure 4.2) (for recent reviews in the field, see also Cornelis, 1994; Forsberg et al., 1994; Cornelis and Wolf-Watz, 1997; Fällman et al., 1997). Besides virulence factors, the pYV plasmid also encodes resistance to arsenite (Neyt et al., 1997).

3.2. CELL ADHESION

Adherence to host cells and to the extracellular matrix is essential, especially during the first step of a disease. Two major surface proteins, YadA, encoded by the virulence pYV plasmid, and the chromosome-encoded MyfA, fulfill this function in *Y. enterocolitica.*

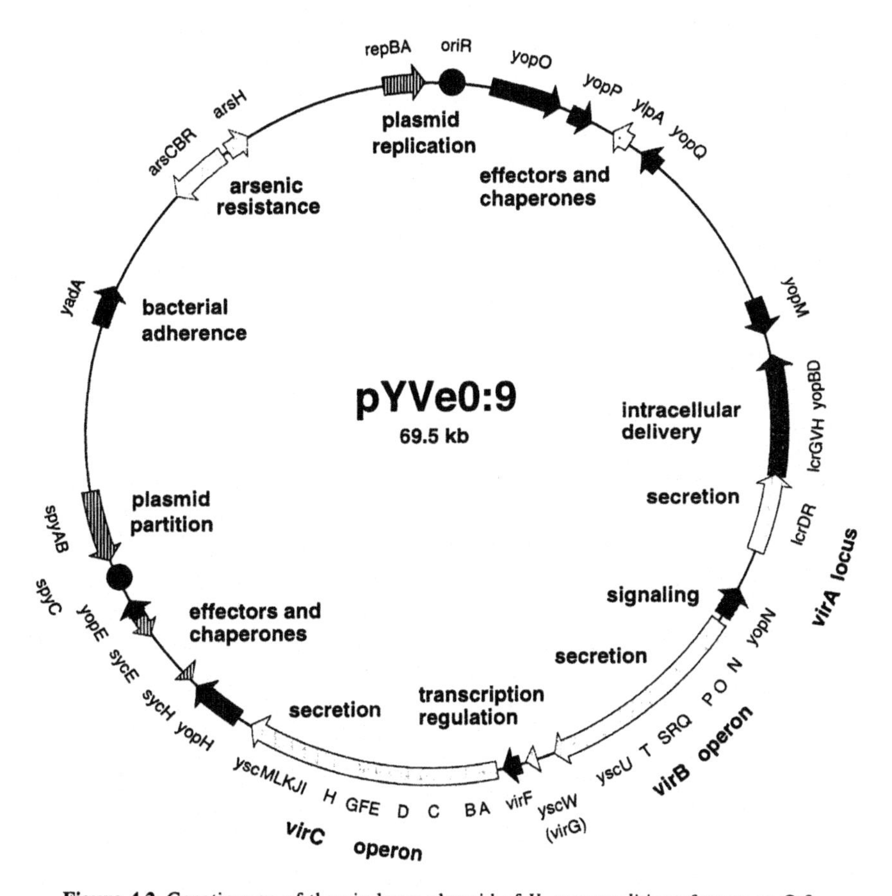

Figure 4.2 Genetic map of the virulence plasmid of *Y. enterocolitica* of serotype O:9.

3.2.1. YadA

YadA (*Yersinia* adhesin) is a prominent outer membrane protein of about 50 kDa forming undefined polymers (200 to 240 kDa) at the bacterial surface. It is a multifunctional protein mediating bacterial auto-agglutination, adhesion to epithelial cells and granulocytes and interaction with several extracellular matrix (ECM) proteins. In *Y. enterocolitica* but not in *Y. pseudotuberculosis,* YadA also confers resistance to the bactericidal action of complement and to phagocytosis and killing by PMNs. A YadA mutant is avirulent for mice infected either orally or intraperitoneally. The YadA protein is essential for persistence of *Y. enterocolitica* in Peyer's patches (Pepe et al., 1995) as well as for dissemination to mesenteric lymph nodes, spleen and liver (Roggenkamp et al., 1995).

Plasmid-bearing strains of *Y. enterocolitica* inhibit the oxidative burst of granulocytes and resist phagocytosis and killing. YadA is involved in these phenomena by binding the complement-inhibitory factor H, which reduces the C3b deposition on the bacterial surface by a rapid degradation of C3b into C3bi (China et al., 1993, 1994). By promoting close contact between *Yersinia* and hosts target cells, YadA also contributes to the delivery of Yops inside PMNs (Visser et al., 1995; Roggenkamp et al., 1996), thereby mediating resistance to phagocytosis and killing (see the role of Yop effectors below).

YadA mediates bacterial adherence to soluble and immobilized type I, II, III, IV, V and XI collagens (Emody et al., 1989; Schulze-Koops et al., 1992, 1995; Tamm et al., 1993). YadA-positive bacteria adhere to isolated collagen chain as well as to denaturated collagen, which suggests that a linear epitope in the collagen is involved in the binding. YadA also binds to laminin but the binding affinity is lower than that for collagen.

The YadA monomer of *Y. enterocolitica* contains a N-terminal signal sequence and two hydrophobic domains. The amino-terminal hydrophobic sequence (80–101 aa of *Y. enterocolitica* serotype O:8) is responsible for the auto-agglutination potential of *Y. enterocolitica* and for its collagen-binding property. More precisely, substitution of His-156 and His-159 results in abrogation of binding to collagens and adherence to HEp-2 cells (Roggenkamp et al., 1995). This loss of collagen binding is associated with the reduction of mouse virulence but has no effect on the ability of *Yersinia* to survive in human serum. The collagen-binding function of YadA is not required for crossing from the intestinal lumen to Peyer's patches but it appears to be crucial for transport from Peyer's patches to spleen and for survival and multiplication in extraintestinal tissues. Deletion of the amino-terminal hydrophobic sequence does not affect the binding to laminin. The hydrophobic carboxy-terminal sequence of YadA is involved in polymerization of the YadA subunits and in YadA expression on the cell surface (Tamm et al., 1993).

YadA elicits in mice a protective humoral immunity against *Yersinia* infection. Protective antibodies most likely function by opsonization of extracellularly multiplying bacteria. This then favors phagocytosis and killing by macrophages but could also inhibit the adherence of *Yersinia* to eukaryotic cells and therefore hinder internalization of the anti-host Yop proteins inside eukaryotic cells. This protective ability is, however, serotype specific, since a monospecific serum against YadA of *Y. enterocolitica* O:8 confers immunity on mice challenged with the autologous strain but not with a *Y. enterocolitica* O:9 strain.

3.2.2. MyfA

At 37°C and in an acidic medium, *Y. enterocolitica* synthesizes a fibrillar structure resembling the CS3 pili of enterotoxigenic *E. coli* (Iriarte et al., 1993). A chromosomal locus is devoted to the production and exportation of MyfA (mucoid yersinia factor), the subunit of the fibrillae. The counterpart of the *myf* operon in *Y. pestis* is responsible for the synthesis of pH6 antigen, a structure known since the mid-1950s (Lindler and Tall, 1993). The role of MyfA in *Y. enterocolitica* is unknown, but Myf could fufill the role of an adherence factor, as recently shown for pH6 Ag (Yang et al., 1996), reinforcing the action of either the yersinial enterotoxins or the anti-host Yop proteins (see below).

3.3. CELL INVASION

Y. enterocolitica induces its own endocytosis in a variety of epithelial cell lines (HEp-2, HEC1B, CHO, MDCK). Unlike *Salmonella* and *Shigella,* which induce extensive cell membrane ruffling to trigger their uptake, *Yersinia* induces localized cytoskeletal reorganization involving accumulation of polymerized actin, filamin and talin around entering bacteria (Young et al., 1992). This quiescent invasion, described as ''zippering,'' is induced by the bacterial outer membrane protein invasin, Inv. This protein binds multiple beta1 integrins (alpha3beta1, alpha4beta1, alpha5beta1, alpha6beta1 and alphavbeta1), the natural receptors for fibronectin, collagen, laminin, VCAM-1, LFA-1 or CR3 (Isberg and Leong, 1990; for review, Isberg and Tra Van Nhieu, 1994). The Invasin-integrin interaction then activates signal transduction events that promotes the entry of *Yersinia* within an endocytic vacuole. These events include tyrosine phosphorylation since specific tyrosine protein kinase inhibitors block the internalization but not the binding of bacteria to eukaryotic cells (Rosenshine et al., 1992).

The N-terminal region of invasin mediates export to the bacterial surface and the carboxy-terminal end is sufficient for binding and invasion of eukaryotic cells. In spite of the absence of any RGD domain, binding to alpha5beta1 integrin can be inhibited by RGD-containing peptides. The aspartic acid at position 911 (D911) could thus play the same role as the aspartate residue of the RGD motif. D911 is indeed critical for promoting binding of invasin to integrin (Leong et al., 1995).

Invasin seems to play a role in the pathogenesis of *Y. enterocolitica* since only pathogenic *Y. enterocolitica* strains harbor functional *inv*-homologous sequences. In support of this hypothesis, Inv promotes an efficient penetration of the intestinal tissue during the initial stages of mice infection by *Y. enterocolitica.* However, during the subsequent establishment of a systemic infection, invasin is of secondary importance since an *inv* mutant can also cause a fatal infection (Pepe and Miller, 1993). On the other hand, *Y. enterocolitica* still expresses invasin in Peyer's patches two days after being introduced intragastrically into mice (Pepe et al., 1994). A recent study also showed that invasin promotes binding between *Yersinia* and resting peripheral B cells via beta1 integrins and initiates the activation of these cells by other unknown factors. The exact role of this interaction in the infection is, however, unknown and could be a source for prolonged immunogenic stimulation, leading to post-infection complications (Lundgren et al., 1996). The question of the real function of invasin in a naturally occurring *Y. enterocolitica* infection is thus still a matter of debate. However, one favors the hypothesis that invasin functions mainly as an adhesin that attaches to eukaryotic cells to initiate the contact-dependent secretion and internalization of the anti-host Yop proteins of *Yersinia* into eukaryotic cells (see below).

3.4. CELL INTOXICATION

3.4.1. The Heat-Stable Enterotoxins Yst

The chromosome of *Y. enterocolitica* O:9 encodes a heat-stable enterotoxin called Yst, detectable in the culture medium by the infant mouse test (Pal and Mors, 1978). Yst is a polypeptide of 30 amino acids, which resembles other methanol-soluble STs and guanylin, an activator of the guanylate cyclase in the eukaryotic intestinal cells (Currie et al., 1992). The enterotoxin is specifically associated with most of the pathogenic serotypes (Delor et al., 1990; Yoshino et al., 1995). A study conducted in the young rabbit concluded that, at least in this model, Yst was responsible for the diarrhea (Delor and Cornelis, 1992).

Related enterotoxins, called Y-STb and Y-STc (Yoshino et al., 1995), are produced by *Y. enterocolitica* strains that belong to serotypes generally considered as nonpathogenic. Y-STc, the largest peptide (5638 Da), is also the most potent member of the ST family.

3.4.2. The Toxic Effectors of the Yop Virulon

3.4.2.1. The Antiphagocytic Activity of YopH

The phagocytic activity and oxidative burst of macrophages and PMN are rapidly paralyzed by strains of *Yersinia* harboring the virulence pYV plasmid (Rosqvist et al., 1988; Bliska and Black, 1995; China et al., 1994; Visser et al., 1995; Fällman et al., 1995). Anti-phagocytosis is mediated primarily by YopH, a protein tyrosine phosphatase (PTPase) related to eukaryotic PTPases. The C-terminal domain is catalytically active and substitution of C-403 results in abrogation of both PTPase activity and mouse virulence (Guan and Dixon, 1990).

In *Yersinia*-infected HeLa cells (cervical epithelial cells), YopH causes dephosphorylation of p130^{Cas} and FAK, two focal adhesion (FA) components with molecular weights of 120–135 kDa (Persson et al., 1997; Black and Bliska, 1997). FAs are multi-molecular complexes where integrins, in contact with the extracellular matrix, are associated with cytoplasmic derived proteins such as alpha-actinin, paxillin or focal adhesion kinase (FAK). During in vitro infection of HeLa cells by plasmid-cured yersiniae, binding of invasin to beta1-integrin leads to the phosphorylation of FAK and to the recruitment and phosphorylation of the cytoplasmic p130^{Cas} protein. Upon infection with a YopH-producing strain, these proteins are rapidly dephosphorylated, the peripheral FAs are disrupted and uptake of *Yersinia* by these cells is inhibited. The YopH protein is thus sufficient to impair beta1-integrin-mediated bacterial uptake. It is likely, but not shown yet, that similar events also occur when YopH abolishes the phagocytosis of *Yersinia* by PMN, mediated by complement receptors (beta2-integrin receptor) (Ruckdeschel et al., 1996) and by macrophages when phagocytosis is mediated by Fc receptors (Fällman et al., 1995).

3.4.2.2. The Cytotoxic YopE Protein

YopE (25 kDa) is one of the major Yop effectors of the Yop virulon. The mutational loss of YopE results in avirulence in mice (Straley and Bowmer, 1986; Rosqvist et al., 1990). The mutated strain can still invade the Peyer's patches but is unable to reach deeper tissues such as the liver and the spleen (Sory and Cornelis, 1988).

The YopE protein is addressed to the cytosol of cultured eukaryotic cells where it exerts a strong cytotoxic effect on cultured cells by depolymerizing the F-actin (Rosqvist et al., 1991). Its mode of action is unknown. It probably acts on some cytoskeleton-regulating proteins since it does not directly affect the actin filaments. The C-terminal region of YopE, necessary for its toxic activity, is homologous to the N-terminal domain of ExoS of *Pseudomonas aeruginosa.* This toxin, also secreted by a type III secretion system, ADP-ribosylates small G proteins and has the same cytotoxic effect as YopE. No ADP-ribosylating function has been ascribed to YopE but both proteins could have similar substrates.

3.4.2.3. Apoptosis Induced by YopP

In vitro, *Y. enterocolitica* induces apoptosis in the murine macrophage-like cell lines J774A.1 and PU5-1.8 (Mills et al., 1997). Cells undergoing apoptosis have also been observed during infection of hepatocytes and macrophages by *Listeria* and *Salmonella/Shigella,* respectively. The phenomenon is strictly dependent on the integrity of the Yop secretion system and internalization apparatus. The 30 kDa (288 aa) YopP protein of *Y. enterocolitica* (called YopJ in *Y. Pseudotubercolosis*) is the effector responsible for this programmed cell death (pcd). Strikingly, this protein shares homologies with the protein AvrRxv of the phytopathogen *Xanthomonas campestris P. vesicatoria,* an avirulence protein involved in the hypersensitive response of tomato towards *Xanthomonas.* The way YopP induces apoptosis is unknown. Further studies are necessary to establish the real impact of apoptosis in the pathogenesis of yersiniosis.

3.4.2.4. Suppression of Immunity Mediators

In addition to their role in food absorption or ion transport, cells lining the intestinal surface appear to function as an integral component of the mucosal immune system. Among others, they produce IL-8, a potent neutrophil chemo-attractant and activator involved in pro-inflammatory events during bacterial infections. Upon contact with the human intestinal epithelial cell line T_{84} or with HeLa cells, non-pathogenic pYV^- *Y. enterocolitica* strains trigger IL-8 production. In contrast, pYV-harboring bacteria prevent IL-8 release. This effect is dependent on the YopB and YopD translocators (Schutte et al., 1996), which indicates that one or several so-far-unidentified Yop effectors abrogate the release of IL-8 by epithelial cells, thereby counteracting the immune response of the host. In the same way, *Y. enterocolitica* prevents production of TNFa by the macrophage-like cell line J774A-1. The effector involved in this inhibition is the YopP protein (YopJ in *Y. pseudotuberculosis*) (Boland et al., 1998). Suppression of TNFa involves a reduction of ERK1/2, p38 and JNK kinase activities as well as of Raf-1 kinase activity (Ruckdeschel et al., 1997). It is likely that YopP acts upstream of this pathway to inhibit TNFa production.

3.4.2.5. YopO and YopM

Two other Yop effectors are part of the Yop virulon but their role remains elusive.

The YopO protein of *Y. enterocolitica* (YpkA in *Y. pseudotuberculosis*) is an 84 kDa-protein showing a high degree of homology with eukaryotic serine/threonine kinases. It can autophosphorylate and is essential for full virulence of *Yersinia* (Galyov et al., 1993). It is targeted to the inner leaflet of the eukaryotic cell membrane by adherent *Yersinia* and induces a cytotoxic effect (Hakansson et al., 1996a). The signal transduction pathway sabotaged by YopO/YpkA remains to be determined.

YopM is a protein of 48 kDa containing repeated sequences rich in leucine residues. It binds human thrombin and has been suggested to play an anti-inflammatory role by inhibition of platelet aggregation (Leung et al., 1990). However, contradicting this hypothesis, YopM has recently been shown to be delivered into the cytosol of eukaryotic cells by bacteria adhering at their surface (Boland et al., 1996). YopM is thus an intracellular effector but its role is unknown.

3.5. BACTERIAL IRON ACQUISITION

In mammalian hosts, iron is tightly sequestered by the high-affinity, iron-binding proteins transferrin and lactoferrin. Free iron, essential for bacterial growth, is thus found in very low concentrations. To survive, high-virulence *Y. enterocolitica* strains have acquired specialized mechanisms to pull iron from the host chelators. These strains produce a siderophore—yersiniabactin (Ybt)—containing aromatic and nonaromatic iron-chelating groups (Haag et al., 1993). Its structure resembles that of pyochelin and anguibactin, two siderophores produced by *Pseudomonas* and *Vibrio,* respectively. Transport of the iron-yersiniabactin complex back into the bacterial cell requires the surface receptor FyuA that also confers pesticin sensitivity to the bacterium. Genes for production of Ybt and FyuA are assembled with genes *irp1,* encoding a large iron-regulated protein, and *irp2,* necessary for Ybt synthesis, in a pathogenicity island of about 40 kb (Carniel et al., 1996).

Other pathogenic *Y. enterocolitica* strains lack this capacity to synthesize and release a siderophore, which probably explains their low animal virulence. However, these strains have the capacity to use a number of other siderophores, including desferrioxamine B, a siderophore produced by *Streptomyces* and used to treat patients with iron overload (Robins-Browne and Prpic, 1985). These patients are thus at risk for *Yersinia* infections.

3.6. THE ROLE OF UREASE

Y. enterocolitica tolerates the acidic pH (<3) prevailing in the normal human fasting stomach. This resistance depends on the bacterial growth phase and the concentration of urea in the medium. Urease-negative mutants have lost this acid resistance and have a reduced viability after passage through the stomach of mice (De Koning-Ward and Robins-Browne, 1995). This suggests that urease contributes to the virulence of *Y. enterocolitica* by enhancing the likelihood of bacterial survival during passage through the stomach.

4. PRODUCTION AND REGULATION OF *Y. ENTEROCOLITICA* VIRULENCE FACTORS

4.1. SECRETION AND INTERNALIZATION OF YOP PROTEINS INTO EUKARYOTIC CELLS

4.1.1. The Type III Secretion System

Secretion of the Yop proteins is realized by a devoted type III secretion system called Ysc (for Yop Secretion). An increasing number of pathogens, zoo- and phytopathogens, such as *Shigella, Salmonella, E. coli, Pseudomonas, Erwinia,* and *Xanthomonas,* are now recognized to express versions of this system.

The secretion apparatus of *Y. enterocolitica* is encoded by about 20 genes clustered in three operons, *virA, virB* and *virC* (Michiels et al., 1991; Plano and Straley, 1993; Bergman et al., 1994; Fields et al., 1994; Allaoul et al., 1994, 1995; for a review, see Cornelis and Wolf-Watz, 1997). The *virA* (containing at least IcrD) and *virB* (genes *yscN* to *yscU*) loci seem mainly devoted to encoding inner membrane proteins. VirB also encodes an ATP-binding protein called YscN (Woestyn et al., 1994). The *virC* locus (genes *yscB* to *yscM*) encodes an outer membrane protein called YscC, that polymerizes to form a pore in the bacterial membrane (Koster et al., 1997). YscC shares significant homology with PulD and XcpQ, two outer-membrane proteins also assembled in homopolymers to form a pore. Nonpolar mutations in genes *yscB* to *yscG* and *yscI* to *yscL* abolish the secretion of the Yop proteins. Surprisingly, the *yscH* gene encodes a protein, YopR, which is secreted, and is not necessary for the secretion of the Yop proteins (Allaoui et al., 1995).

Secretion of several Yop proteins requires individual chaperones called Syc (specific Yop chaperone). SycE specifically assists YopE secretion, SycH is the chaperone of YopH, and SycD is required for both secretion of YopB and YopD (for a review, see Wattiau et al., 1996). The Syc proteins bind their cognate Yop partner in the bacterial cytoplasm. The Syc-binding domain on YopE is localized between aa 15 and 50. Similarly, YopH contains a unique SycH-binding domain localized between residue 20 and 70. The Syc chaperones are, however, dispensable for secretion of the respective Yop proteins if the Syc-binding domain is deleted (Woestyn et al., 1996). The N-terminal regions of YopE (1–15 aa) and YopH

(1–17 aa) are also involved in the secretion of YopE and YopH and sufficient for the secretion of hybrid proteins (Sory et al., 1995; Schesser et al., 1996). A YopE protein deleted of codons 2–15 is, however, still secreted, albeit at a lower level, suggesting the existence of two secretory pathways (Cheng et al., 1997). Both regions for YopE secretion could also act synergistically to promote high secretion of the protein.

The mechanism of the type III secretion system is not known in great detail. The peculiarity of this secretion system seems to be its link with subsequent internalization of some of the secreted proteins into eukaryotic cells.

4.1.2. Internalization of Yops Inside Eukaryotic Cells

When *Y. enterocolitica* adheres to the surface of eukaryotic cells, the toxic Yop effectors such as YopE, YopH, YopO/YpkA and YopM are addressed directly into the cytosol (Rosqvist et al., 1994; Sory and Cornelis, 1994; Sory et al., 1995; Persson et al., 1995; Hakansson et al., 1996a; Boland et al., 1996).

Analysis by confocal microscopy and by an enzyme reporter strategy demonstrated that this translocation of Yop effectors is dependent upon at least four other secreted proteins, YopB, YopD, LcrV and LcrG. YopB presents two hydrophobic regions and is responsible for a hemolytic phenotype of *Yersinia* on sheep erythrocytes. The hypothesis is that YopB polymerizes and forms a pore in the eukaryotic cell membrane, allowing the Yop proteins to cross this membrane (Hakansson et al., 1996b). YopD is also required for internalization of the Yop effector proteins but its mode of action is unknown (Rosqvist et al., 1994; Sory et al., 1994; Hartland et al., 1994). YopB and YopD need to be secreted by the bacterium to achieve their function. This secretion requires not only the complete Ysc secretion apparatus but also LcrV, which is secreted simultaneously (Sarker et al., in preparation). LcrG is a small protein that binds to YopD in an in vitro system and that also recognizes an as of yet unidentified component on the eukaryotic cell surface (Sarker et al., in preparation). LcrG is thought to sense the eukaryotic cell, promoting the opening as well as the assembly of the channel for secretion and internalization of Yop proteins inside eukaryotic cells. LcrG, LcrV, YopB and YopD are encoded by the large *lcrGVHyopBD* operon. All these proteins could act in concert to form a complex structure between the adherent bacteria and the eukaryotic cell. This structure could be physically linked to the secretion apparatus, which could allow a direct injection of YopE from the bacteria to the eukaryotic cell.

Delivery of Yop proteins into eukaryotic cells is also influenced by YopK (from *Y. pseudotuberculosis,* known as YopQ) in *Y. enterocolitica* (Mulder et al., 1969). YopK could control the pore size formed in the eukaryotic cell by the YopB protein (Holmstrom et al., 1997).

Mutants of *Y. enterocolitica* unable to secrete YopN, a protein of about 35 kDa, have an abnormal phenotype with regard to the internalization process. Instead of delivering most of their YopE or YopH effectors inside the eukaryotic target cell, they waste large amounts in the eukaryotic cell culture medium (Rosqvist et al., 1994; Persson et al., 1995; Boland et al., 1996). YopN is thus involved in the control of the internalization event: It blocks Yop release unless there is close contact with a eukaryotic cell. YopN could thus be viewed as a stop valve, anchored at the more external part of the secretion apparatus.

The N-terminal 15–50 aa region of YopE and the 20–70 aa domain of YopH have been shown to be necessary for their internalization inside the cells (Sory et al., 1995; Schesser et al., 1996). These regions coincide with those necessary for Syc binding and were suggested to hinder a forward association of YopE and YopH with the components of the internalization apparatus, such as YopD or YopB, in the cytosol of the bacteria (Woestyn et al., 1996).

4.2. REGULATION OF THE VIRULENCE FUNCTIONS

4.2.1. Regulation of Inv, Yst and MyfA Production

4.2.1.1. Growth-Phase Regulation

Expression of the three chromosome-encoded virulence proteins Inv, Yst and MyfA is growth-phase regulated, in the sense that none of them is expressed during exponential growth. Expression of *inv* and *yst* begins when cells enter late-exponential phase to early stationary phase and continues throughout the stationary phase (Pepe et al., 1994; Mikulskis et al., 1994). By contrast, *myfA* transcription occurs during transition between the late-exponential phase and early stationary phase and then decreases gradually throughout the stationary phase (Iriarte et al., 1995).

In *E. coli,* expression of many stationary-phase genes requires the alternative sigma factor RpoS. In a *rpoS* mutant of *Y. enterocolitica, yst* expression is reduced though not completely abolished (Iriarte et al., 1995). The *Y. enterocolitica* homolog of RpoS is thus involved in the expression of *yst* but, as for many *E. coli rpoS*-regulated genes, additional control mechanisms responding to growth phase must also be involved. In agreement with this, *yst* expression is modulated by osmolarity and by the histone-like protein YmoA (see below) (Mikulskis et al., 1994). The mechanism of action of RpoS in *yst* regulation appears thus to be quite complex. So far, it is impossible to conclude whether RpoS regulates *yst* directly by binding to the promoter region, or indirectly through other regulatory proteins.

The role of *rpoS* in the expression of *inv* has not been established yet. Although expression of *myf* occurs at the onset of the stationary phase, it is not significantly affected in a *rpoS* mutant (Iriarte et al., 1995).

4.2.1.2. Silencing During Storage

The level of *yst* expression depends on the bacterial strain analyzed: It gradually decreases in some strains during storage, suggesting the existence of a mechanism switching the expression of *yst* to a silent state. The silencing of *yst* is not due to modifications in the gene itself but rather reflects the status of bacterial host factors (Mikulskis et al., 1994). The production of Yst can be restored in a silent strain by a *ymoA* mutation (Cornelis et al., 1991), but the typical growth phase-dependent expression pattern is lost. Taking into account the histone-like properties of Ymo (see below), chromatin structure could be involved in this gene reactivation phenomenon.

Another chromosomal gene restores the production of Yst in silent strains (Nakao et al., 1995). The gene, called *yrp* (*Yersinia* regulator for pleiotropic phenotype), encodes a small protein of 101 aa showing similarity to *E. coli* factor 1 gene. Mutations in the *Irp* gene have been shown to occur in several strains silent for Yst production. Irp is involved in regulating negative supercoiling of plasmids and regulates not only transcription of the *yst* gene but also influences other phenotypes such as colony morphology, growth rate, carbon fermentation and ornithine decarboxylase activity.

4.2.1.3. Temperature Regulation

All the yersinial pathogenicity functions are thermoregulated, though not always in the same manner.

The host temperature of 37°C reduces transcription of *inv* compared to ambient temperature (Isberg et al., 1988; Plerson and Falkow, 1990). Maximal expression at low temperature may prime *Y. enterocolitica* for cellular penetration once inside the host intestinal tract. After ingestion, *Y. enterocolitica* encounters a temperature of 37°C, which could down regulate the production of Inv. However, temperature upshift is not the only stimulus that triggers *Y. enterocolitica* once they enter the host. Several environmental cues modulate thermoregulation and thus affect expression of *Inv*. The most significant parameters in the intestinal tract are osmolarity, pH, oxygen and ion concentrations. In vitro, *Inv* expression in *Y. enterocolitica* increases significantly at 37°C as pH of the growth medium goes below 7. *Inv* expression at 37°C also increases with increasing concentration of Na^+(Pepe et al., 1994). Thus, in vivo, the acidity of the stomach, and the high Na^+ concentrations close to the enterocytes brush border could counteract the negative effect of temperature and allow continous *inv* expression. In agreement with this hypothesis, Pepe et al. (1994) showed that the level of *inv* expression by *Y. enterocolitica* after two days in the mouse intestine is comparable to expression levels at 23°C in laboratory cultures.

Production of Yst, like that of Inv, is also very low at 37°C in standard growth media (Delor et al., 1990; Mikulskis et al., 1994). This temperature downregulation of *yst* expression is puzzling because the enterotoxin needs to be secreted in vivo to induce diarrhea. Moreover, as in the case of *inv*, thermoregulation of *yst* expression is modulated by additional factors such as osmolarity and pH: In vitro, *yst* transcription can be induced at 37°C by increasing Na^+ and K^+ ion concentrations as well as pH up to values normally present in the ileum lumen (Mikulskis et al., 1994). This observation reconciles the in vitro regulation studies with the known role of Yst in inducing diarrhea.

Myf fibrillae expression is regulated at the transcriptional level and only occurs at 37°C (Iriarte et al., 1993; Iriarte and Cornelis, 1995). Myf production is thus downregulated when *Y. enterocolitica* are in the environment and it is stimulated when they enter their host. In agreement with this, we observed that mice infected with *Y. enterocolitica* grown at room temperature develop antibodies against Myf (Iriarte, unpublished observation).

4.2.1.4. Regulation of Myf Synthesis by pH

Transcription of *myfA* requires a pH below 6 (Iriarte and Cornelis, 1995). The same applies to pH6 Ag, the Myf counterpart in *Y. pestis* (Lindler and Tall, 1993). After entering the host, Myf synthesis could be turned on by the acidic pH of the stomach. Nondividing bacteria might then keep their fibrillae in the intestine where alkaline pH prevails.

Tight regulation by acidity suggests that the regulatory network controlling *myf* expression is more specific and probably more sophisticated that those controlling the expression of *yst* and *inv*. Transcription of *myfA* requires at least two genes, *myfF* and *myfE*, situated immediately upstream from *myfA* (Iriarte and Cornelis, 1995). Their products do not show similarity to any known regulatory protein. MyfF is a 18.5-kDa protein with no typical helix-turn-helix motif and a unique hydrophobic domain in the NH_2 terminal part. MyfF is associated with the inner membrane by means of its hydrophobic domain while the hydrophilic part protrudes in the periplasm. Gene *myfE*, situated immediately upstream from *myfA*, has not been characterized yet. In a *Y. enterocolitica myfE* mutant, transcription of *myfA* is completely abolished (Iriarte and Cornelis, 1995).

Genes *myfF* and *myfE* are homologous to genes *psaF* and *psaE* of *Y. pseudotuberculosis*. Both genes regulate *psaA* (encoding pH6 Ag) at the transcriptional level. PsaF and PsaE are inner membrane-associated proteins. The N-terminal cytoplasmic domain of PsaE contain sequence similarity to transcriptional regulators found in two component systems (Yang and Isberg, 1997). PsaF and PsaE as well as MyfF and MyfE are thus probably parts of a two-component system.

4.2.2. Regulation of pYV-Encoded Virulence Functions

4.2.2.1. VirF and the virF *Regulon*

Transcription of many genes among which are the *yop* genes, *sycE, ylpA, yadA* and the *virC* operon requires the product of *virF,* a gene located between *virB* and *virC* (Cornelis et al., 1989). These genes and operons constitute what we call the *virF* regulon. By contrast, VirF seems to be dispensible for the transcription of the *virA* locus, of the *virB* operon and of some genes like *sycH* (Lambert de Rouvroit et al., 1992; Wattiau et al., 1994). VirF is a 30 kDa transcriptional activator of the AraC family of regulators (Cornelis et al., 1989), a very large family including regulators of degradative pathways in *E. coli* and *P. putida* as well as regulators involved in the control of virulence of *Shigella,* enterotoxigenic *E. coli* and the phytopathogen *P. solanacearum.* VirF acts as a DNA-binding protein. DNaseI footprinting experiments on the VirF binding sites of four genes identified a protected region spanning about 40 bp immediately upstream from the RNA polymerase binding site. This VirF binding sequence is located in an AT-rich region and comprises two sites, each containing a 13 bp consensus sequence, either alone or as inverted repeats (Wattiau and Cornelis, 1994).

4.2.2.2. Thermoregulation of the yop Stimulon

The *yop, yadA* and *yipA* genes as well as the *virA, virB* and *virC* loci are silent at low temperature and strongly expressed at 37°C. They constitute what we call the *yop* stimulon. (Note that the *yop* stimulon is larger than the *virF* regulon.)

In *Y. enterocolitica,* gene *virF* itself is strongly thermoregulated (Cornelis et al., 1989). This thermoinduction still occurs in *E. coli* containing an isolated *virF* gene. This indicated that *virF* must be thermoregulated by a chromosomal gene rather than by the pYV plasmid. The fact that *virF* is thermoregulated can explain why the *virF* regulon is only expressed at 37°C. However, it does not demonstrate that the temperature regulation of the *virF* regulon only involves the temperature regulation of *virF.* Indeed, when *virF* is transcribed at low temperature from a *tac* promoter, the *yop* and *yadA* genes are only poorly transcribed (Lambert de Rouvroit et al., 1992). By contrast, at 37°C, the response to IPTG mimicks the normal response to thermal induction. In conclusion, the expression of the *yop* stimulon is first controlled by temperature but the expression of some genes is reinforced by the action of VirF, the synthesis of which is also temperature controlled.

In *Y. pestis* the situation is slightly different: Transcription of *lcrF,* the homolog of *virF,* is insensitive to temperature. However, comparison of the amount of LcrF protein produced per unit of message at low and high temperatures indicates that the efficiency of translation of *lcrF* mRNA increases with temperature (Hoe and Goguen, 1993). The model is that destabilization of a mRNA secondary structure sequestering the *lcrF* Shine-Dalgamo sequence regulates LcrF synthesis.

4.2.2.3. The YmoA Regulator

By transposon mutagenesis, Cornelis and co-workers (1991) identified a chromosomal regulator of the *yop* stimulon of *Y. enterocolitica.* In such a mutant, transcription of genes such as *yopE* and *yadA* is strongly increased at 28°C. The Yop and YadA proteins are nevertheless not secreted at this temperature. Transcription of the regulatory gene *virF* itself is increased at 28°C, which may account for the increased transcription of *yopH, yopE* and *yadA.* Although these genes are overexpressed at low temperature in the mutants, there is still an increase of transcription upon transfer to 37°C. Hence, the thermal response is not abolished but rather "modulated." The phenotype is thus not that of a classical repressor minus mutant. The gene was called *ymoA* for "*Yersinia* modulator."

The *ymoA* gene encodes a 8 kDa protein extremely rich in positively and negatively charged residues. Although there is no similarity between YmoA and HU, IHF or H-NS (H1), it is very likely that YmoA is a histone-like protein. Chromatin structure is thus involved in the increasing susceptibility at 37°C of *yop* gene promoters to VirF activation. Temperature could somehow modify the structure of chromatin, making the promoters more accessible to VirF. Rohde et al. (1994) confirmed that temperature alters DNA supercoiling and DNA bending in *Y. enterocolitica* and they hypothesized that temperature dislodges a repressor, perhaps YmoA, bound on promoter regions of *virF* and of some other thermoregulated genes.

4.2.2.4. Contact Regulation of Yop Synthesis and Secretion

It has been known since the mid-fifties that *Yersinia* do not grow at 37°C in Ca^{2+}-deprived media. Moreover, in the same conditions, *Yersinia* govern the massive release of a set of at least 12 Yop proteins (Heesemann et al., 1985; Michiels et al., 1990, 1991). The Ca^{2+} was thus viewed as an external signal regulating Yop protein synthesis. At present, it is believed that the real signal mediating Yop protein production is a physical contact between *Yersinia* and eukaryotic cells. Indeed, the internalization phenomenon of Yop proteins inside eukaryotic cells requires an intact secretion apparatus and is only realized by bacteria firmly attached to the surface of eukaryotic cells by Inv or YadA (Rosqvist et al., 1994: Sory and Cornelis, 1994). Bacteria that are not in close contact with eukaryotic cells do not secrete Yop proteins. Internalization of Yop proteins into cells also occurs in a polarized manner: While a wild-type strain mainly targets the Yop proteins into the cytosol of eukaryotic cells, a mutated *yopN* strain releases high amount of proteins in the eukaryotic cell growth medium (Rosqvist et al., 1994; Boland et al., 1996). Hence, secretion and internalization of Yop proteins in mammalian cells occur in a medium that is not permissive of Yop secretion and only occur if bacteria are in direct contact with eukaryotic cells.

4.2.2.5. Feedback Inhibition of Yop Protein Synthesis in the Absence of Secretion

Polar mutations in any of the *virA, B* and *C* loci encoding the secretion apparatus prevent the expression of the *yop* genes themselves (Cornelis et al., 1989). This effect is maximal for mutations affecting the *virB* operon. Nonpolar mutations inhibiting synthesis of the ATPase YscN or of the inner membrane protein YscU drastically reduce synthesis of the Yops (Woestyn et al., 1994). This reveals the existence of a feedback inhibition mechanism of Yop synthesis when export is compromised.

LcrQ could act as a negative element in this regulatory pathway. This protein, known as YscM in *Y. enterocolitica,* is encoded by the last gene of the *virC* operon (Rimpilainen et al., 1992; Stainier et al., 1997). Recently, LcrQ was suggested to be a secreted regulator (Petterson et al., 1996). The protein is indeed secreted in the surrounding medium by the type III secretion apparatus (Pettersson et al., 1996; Stainier et al., 1997). Without contact with eukaryotic cells, the secretion apparatus is blocked and the high intrabacterial concentration of LcrQ represses *yop* transcription. Upon contact with eukaryotic cells, the secretion channel is opened and LcrQ is secreted. Consequently, the intracellular concentration of LcrQ is lowered, leading to derepression of *yop* expression. The type III secretion system thus seems to play a key role in the coordination between secretion and expression of virulence factors after physical contact with the eukaryotic cells.

5. REFERENCES

Allaoui, A., Schulte, R., and Cornelis, G. R. 1995. "Mutational analysis of the *Yersinia enterocoliticia virC* operon: characterization of *yscE, F, G, I, J, K* required for Yop secretion and *yscH* encoding YopR. *Mol. Microbiol.*" 18:343–355.

Allaoui, A., Woestyn, S., Sluiters, C., and Cornelis, G. R. 1994. "YscU, a *Yersinia enterocolitica* inner membrane protein involved in Yop secretion," *J. Bacteriol,* 176:4534–4542.

Attwood, S. E. A., Cafferkey, M. T., and Keane, F. B. V. 1989. "Yersinia infections in surgical practice," *Br. J. Surg.* 76:499–504.

Autenrieth, I. B., Tingle, A., Reske Kunz, A., and Heesemann, J. 1992. "T lymphocytes mediate protection against *Yersinia enterocolitica* in mice: characterization of murine T-cell clones specific for *Y. enterocolitica,*" *Infect. Immun.* 60:1140–1149.

Autenrieth, I. B., Beer, M., Bohn, E., Kaufmann, S. H., and Heesemann, J. 1994. "Immune responses to *Yersinia enterocolitica* in susceptible BALB/c and resistant C57BL/6 mice: an essential role for gamma interferon,'' *Infect. Immun.* 62:2590–2599.

Autenrieth, I. B., Kempf, V., Sprinz, T., Preger, S., and Schneil, A. 1996. ''Defense mechanisms in Peyer's patches and mesenteric lymph nodes against *Yersinia enterocolitica* involve integrins and cytokines,'' *Infect. Immun.* 64(4):1357–1368.

Autenrieth, I. B., Vogel, U., Preger, S., Heymer, B., and Heesemann, J. 1993. ''Experimental *Yersinia enterocolitica* infection in euthymic and T-cell-deficient athymic nude C57BL/6 mice: comparison of time course, histomorphology, and immune response,'' Infect. Immun. 61:2585–2595.

Bergman, T., Erickson, K., Galyov, E., Persson, C., and Wolf Watz, H. 1994. ''The *lcrB* (*yscN/U*) gene cluster of *Yersinia pseudotuberculosis* is involved in Yop secretion and shows high homology to the *spa* gene clusters of *Shigella flexneri* and Salmonella typhimurium,'' *J. Bacteriol.* 176:2619–2626.

Beuscher, H. U., Rödel, F., Forsberg, A., and Röllinghoff, M. 1995. ''Bacterial evasion of host immune defense: *Yersinia enterocolitica* encodes a suppressor for tumor necrosis factor alpha expression,'' *Infect. Immun.* 63(4):1270–1277.

Black, D. S. and Bliska, J. B. 1997. ''Identification of p130^{Cas} as a substrate of *Yersinia* YopH (Yop51), a bacterial protein tyrosine phosphatase that translocates into mammalian cells and targets focal adhesions,'' *EMBO J.* 16:2730–2744.

Bliska, J. B. and Black, D. S. 1995. ''Inhibition of the Fc receptor-mediated oxidative burst in macrophages by the *Yersinia pseudotuberculosis* tyrosine phosphatase,'' *Infect. Immun.* 63:681–685.

Bohn, E., and Autenrieth, I. B. 1996. ''IL-12 is essential for resistance against *Yersinia enterocolitica* by triggering IFN-production in NK cells and CD4$^+$ T cells,'' *J. Immunol.* 156:1458–1468.

Boland, A. and Cornelis, G. R. 1998. ''Role of YopP in suppression of tumor necrosis factor alpha release by macrophages during *Yersinia* infection,'' *Infection and Immunity* 66:1878–1884.

Boland A., Sory, M. P., Iriarte, M., Kerbourch, C., Wattlau, P., and Cornelis, G. R. 1996. ''Status of YopM and YopN in the *Yersinia* Yop virulon: YopM of *Y. enterocolitica* is internalized inside the cytosol of PU5-1.8 macrophages by the YopB, D, N delivery apparatus,'' *EMBO J.* 15:5191–5201.

Brubaker, R. R. 1991. Factors promoting acute and chronic diseases caused by Yersiniae,'' *Cl. Microbiol. Rev.* 4:309–324.

Carniel, E., Gullvout, I., and Prentice, M. 1996. ''Characterization of a large chromosomal ''high pathogenicity island'' in biotype 1B *Yersinia enterocolitica,*'' *J. Bacteriol.* 178:6743–6751.

Carter, P. B. 1975. ''Pathogenecity of *Yersinia enterocolitica* for mice,'' *Infect. Immun.* 11:164–170.

Cheng, L. W., Anderson, D., and Schneewind, O. 1997. ''Two independent type III secretion mechanisms for YopE in *Yersinia enterocolitica.*'' *Mol. Microbiol.* 24:757–765.

China, B., N'Guyen, B. T., de Bruyere, M., and Cornelis, G. R. 1994. ''Role of YadA in resistance of *Yersinia enterocolitica* to phagocytosis by human polymorphonuclear leukocytes,'' *Infect. Immun.* 62:1275–1281.

China, B., Sory, M. P., N'Guyen, B. T., de Bruyere, M., and Cornelis, G. R. 1993. ''Role of the YadA protein in prevention of opsonization of *Yersinia enterocolitica* by C3b molecules,'' *Infect. Immun.* 61:3129–3136.

Conlan, J. W. 1997. ''Critical roles of neutrophils in host defense against experimental systemic infections of mice by *Listeria monocytogenes, Salmonella typhimurium* and *Yersinia enterocolitica,*'' *Infection and Immunity* 65:630–635.

Cornelis, G. R. 1994. ''*Yersinia* pathogenicity factors,'' *Curr. Top. Microbiol. Immunol.* 192:243–263.

Cornelis, G. R. and Wolf-Watz, H. 1997. "The *Yersinia* Yop virulon: a bacterial system for subverting eukaryotic cells," *Mol. Microbiol.* 23:861–867.

Cornelis, G., Sluiters, C., de Rouvroit, C. L., and Michiels, T. 1989. "Homology between *virF*, the transcriptional activator of the *Yersinia* virulence regulon, and AraC, the *Escherichia coli* arabinose operon regulator," *J. Bacteriol.* 171:254–262.

Cornelis, G. R., Sluiters, C., Delor, I., Gelb, D., Kaniga, K., Lambert de Rouvroit, C., Sory, M. P., Vanooteghem, J. C., and Michiels, T. 1991. "*ymoA*, a *Yersinia enterocolitica* chromosomal gene modulating the expression of virulence functions," *Mol. Microbiol.* 5:1023–1034.

Cover, T. L. and Aber, R. C. 1989. "*Yersinia enterocolitica*," *N. Engl. J. Med.* 321:16–24.

Currie, M. G., Fok, K. F., Kato, J., Moore, R. J., Hamra, F. K., Duffin, K. L., and Smith, C. E. 1992. "Guanylin: an endogenous activator of intestinal guanylate cyclase," *Proc. Natl. Acad. Sci. USA* 89:947–951.

De Koning-Ward, T. F. and Robins-Browne, R. M. 1995. "Contribution of urease to acid tolerance in *Yersinia enterocolitica*," *Infect. Immun.* 63:3790–3795.

Delor, I. and Cornelis, G. R. 1992. "Role of *Yersinia enterocolitica* Yst toxin in experimental infection of young rabbits," *Infect. Immun.* 60:4269–4277.

Delor, I., Kaeckenbeeck, A., Wauters, G., and Cornells, G. R. 1990. "Nucleotide sequence of *yst*, the *Yersinia enterocolitica* gene encoding the heat-stable enterotoxin, and prevalence of the gene among pathogenic and nonpathogenic yersiniae," *Infect. Immun.* 58:2983–2988.

Emody, L., Heesemann, J., Wolf Watz, H., Skurnik, M., Kapperud, G., O'Toole, P., and Wadstrom, T. 1989. "Binding to collagen by *Yersinia enterocolitica* and *Yersinia pseudotuberculosis:* evidence for *yopA*-mediated and chromosomally encoded mechanisms," *J. Bacteriol.* 171:6674–6679.

Fällman, M., Persson, C., and Wolf-Watz, H. 1997. "Yersinia proteins that target host cell signaling pathways," *J. Clin. Invest.* 99:1153–1157.

Fällman, M., Andersson, K., Hakansson, S., Magnusson, K. E., Stendahl, O., and Wolf Watz, H. 1995. "*Yersinia pseudotuberculosis* inhibits Fc receptor-mediated phagocytosis in J774 cells," *Infect. Immun.* 63:3117–3124.

Fields, K. A., Plano, G. V., and Straley, S. C. 1994. "A low-Ca2+ response (LCR) secretion (*ysc*) locus lies within the *lcrB* region of the LCR plasmid in *Yersinia pestis*," *J. Bacteriol.* 176:569–579.

Forsberg, A., Rosqvist, R., and Wolf Watz, H. 1994. "Regulation and polarized transfer of the *Yersinia* outer proteins (Yops) involved in antiphagocytosis," *Trends Microbiol.* 2:14–19.

Galyov, E. E., Hakansson, S., Forsberg, A., and Wolf Watz, H. 1993. "A secreted protein kinase of *Yersinia pseudotuberculosis* is an indispensable virulence determinant," *Nature* 361:730–732.

Grützkau, A., Hanski, C., Hahn, H., and Riecken, E. O. 1990. "Involvement of M cells in the bacterial invasion of Peyer's patches: a common mechanism shared by *Yersinia enterocolitica* and other enteroinvasive bacteria," *GUT* 31:1011–1015.

Guan, K. L. and Dixon, J. E. 1990. "Protein tyrosine phosphatase activity of an essential virulence determinant in *Yersinia*," *Science* 249:553–556.

Haag, H., Hantke, K., Drechsel, H., Stojiljkovic, I., Jung, G., and Zahner, H. 1993. "Purification of yersiniabactin: a siderophore and possible virulence factor of *Yersinia enterocolitica*," *J. Gen. Microbiol.* 139:2159–2165.

Hakansson, S., Galyov, E. E., Rosqvist, R., and Wolf-Watz, H. 1996a. "The *Yersinia* YpkA Ser/Thr kinase is translocated and subsequently targeted to the inner surface of the HeLa cell plasma membrane," *Mol. Microbiol.* 20:593–603.

Hakansson, S., Schesser, K., Persson, C., Galyov, E. E., Rosqvist, R., Homble, F., and Wolf-

Watz, H. 1996b. "The YopB protein of *Yersinia pseudotuberculosis* is essential for translocation of Yop effector proteins across the target plasma membrane and displays a contact-dependent membrane disrupting activity," *EMBO J.* 15:5812–5823.

Hanski, C., Kutschka, U., Schmoranzer, H. P., Naumann, M., Stallmach, A., Hahn, H., Menge, H., and Riecken, E. O. 1989. "Immunohistochemical and electron microscopic study of interaction of *Yersinia enterocolitica* serotype O8 with intestinal mucosa during experimental enteritis," *Infect. Immun.* 57:673–678.

Hartland, E. L., Green, S. P., Phillips, W. A., and Robins Browne, R. M. 1994. "Essential role of YopD in inhibition of the respiratory burst of macrophages by *Yersinia enterocolitica*," *Infect. Immun.* 62:4445–4453.

Heesemann, J., Gross, U., Schmidt, N., and Laufs, R. 1986. "Immunochemical analysis of plasmid-encoded proteins released by enteropathogenic *Yersinia* sp. grown in calcium-deficient media," *Infect. Immun.* 54:561–567.

Hoe, N. P. and Goguen, J. D. 1993. "Temperature sensing in *Yersinia pestis:* translation of the LcrF activator protein is thermally regulated, *J. Bacteriol.* 175:7901–7909.

Holmstrom, A., Pettersson, J., Rosqvist, R., Hakansson, S, Tafazoli, F, Fällamn, M., Magnusson, K. E, Wolf-Watz, H., and Forsberg, A. 1997. "YopK of *Yersinia pseudotuberculosis* controls translocation of Yop effectors across the eukaryotic cell membrane," *Mol. Microbiol.* 24(1):73–91.

Iriarte, M. and Cornelis, G. R. 1995. "MyIF, an element of the network regulating the synthesis of fibrillae in *Yersinia enterocolitica*," *J. Bacteriol.* 177:738–744.

Iriarte, M., Stainier, I., and Cornelis, G. R. 1995. "The *rpoS* gene from *Yersinia enterocolitica* and its influence on expression of virulence factors," *Infect. Immun.* 63:1840–1847.

Iriarte, M., Vanooteghem, J. C., Delor, I., Diaz, R., Knutton, S., and Cornelis, G. R. 1993. "The Myf fibrillae of *Yersinia enterocolitica*." *Mol. Microbiol.* 9:507–520.

Isberg, R. R. and Leong, J. M. 1988. "Cultured mammalian cells attach to the invasin protein of *Yersinia pseudotuberculosis*," *Proc. Natl. Acad. Sci. USA* 85:6682–6686.

Isberg, R. R. and Leong, J. M. 1990. "Multiple beta1 chain integrins are receptors for invasin, a protein that promotes bacterial penetration into mammalian cells," *Cell* 60:861–871.

Isberg, R. R. and Van Nhieu, G. T. 1994. "Two mammalian cell internalization strategies used by pathogenic bacteria," *Annu. Rev. Genet.* 28:395–422.

Isberg, R. R., Swain, A., and Falkow, S. 1988. "Analysis of expression and thermoregulation of the *Yersinia pseudotuberculosis inv* gene with hybrid proteins," *Infect. Immun.* 56:2133–2138.

Koster, M., Bitter, W., de Cock, H., Allaoui, A., Cornelis, G. R., and Tommassen, J. 1997. "The outer membrane component, YscC, of the Yop secretion machinery of *Yersinia enterocolitica* forms a ring-shaped multimeric complex," *Molecular Microbiology* 26:789–797.

Lambert de Rouvroit, C., Slulters, C., and Cornelis, G. R. 1992. "Role of the transcriptional activator, VirF, and temperature in the expression of the pYV plasmid genes of *Yersinia enterocolitica*," *Mol. Microbiol.* 6:395–409.

Leong, J. M., Morrissey, P. E., Marra, A., and Isberg, R. R. 1995. "An aspartate residue of the *Yersinia pseudotuberculosis* invasin protein that is critical for integrin binding," *EMBO J.* 14(3):422–431.

Leung, K. Y., Reisner, B. S., and Straley, S. C. 1990. "YopM inhibits platelet aggregation and is necessary for virulence of *Yersinia pestis* in mice," *Infect. Immun.* 58:3262–3271.

Lindler, L. E. and Tall, B. D. 1993. "*Yersinia pestis* pH 6 antigen forms fimbriae and is unduced by intracellular association with macrophages," *Mol. Microbiol.* 8:311–324.

Lundgren, E., Carballeira, N., Vasquez, R., Dubinina, E., Bränden, H., persson, H., and Wolf-

Watz, H. 1996. "Invasin of *Yersinia pseudotuberculosis* activates human peripheral B cells," *Infect. Immun.* 64(3):829–835.

Michiels, T., Wattlau, P., Brasseur, R., Ruysschaert, J. M., and Cornelis, G. R. 1990. "Secretion of Yop proteins by *Yersiniae*," *Infect. Immun.* 58:2840–2849.

Michiels, T., Vanooteghem, J. C., Lambert de Rouvroit, C., China, B., Gustin, A., Boudry, P., and Cornelis, G. R. 1991. "Analysis of *virC*, an operon involved in the secretion of Yop proteins by *Yersinia enterocolitica*." *J. Bacteriol.* 173:4994–5009.

Mikulskis, A. V., Delor, I., Thi, V. H., and Cornelis, G. R. 1994. "Regulation of the *Yersinia enterocolitica* enterotoxin Yst gene. Influence of growth phase, temperature, osmolarity, pH and bacterial host factors," *Mol. Microbiol.* 14:905–915.

Mills, S. D., Boland, A., Sory, M. P., Vandermissen, P., Finlay, B. B., and Cornelis, G. R. 1997. "Yop, a new *Yersinia* effector protein, is delivered into macrophages to induce apoptosis," *Proc. Natl. Acad. Sci. USA* 94:12638–12643.

Mulder, B., Michiels, T., Simonet, M., Sory, M. P., and Cornelis, G. 1989. "Identification of additional virulence determinants on the pYV plasmid of *Yersinia enterocolitica* W227," *Infect. Immun.* 57:2534–2541.

Nakao, H., Watanabe, H., Nakayama, S.-I., and Takeda, T. 1995. "*yst* gene expression in *Yersinia enterocolitica* is positively regulated by a chromosomal region that is highly homologous to *Escherichia coli* host factor 1 gene (*hfq*), *Mol. Microbiol.* 18:859–865.

Neyt, C., Iriarte, M., Ha Ti, V., and Cornelis, G. R. 1997. "Virulene and arsenic resistance in Yersiniae," *J. Bacteriol.* 179(3):612–619.

Pai, C. H. and Mors, V. 1978. "Production of enterotoxin by *Yersinia enterocolitica*," *Infect. Immun.* 19:908–911.

Pal, C. H., Mors, V., and Seemayer, T. A. 1980. "Experimental *Yersinia enterocolitica* enteritis in rabbits," *Infect. Immun.* 28:238–244.

Pepe, J. C. and Miller, V. L. 1993. "*Yersinia enterocolitica* invasin: a primary role in the initiation of infection," *Proc. Natl. Acad. Sci.* USA 90:6473–6477.

Pepe, J. C., Badger, J. L., and Miller, V. L. 1994. "Growth phase and low pH affect the thermal regulation of the *Yersinia enterocolitica inv* gene," *Mol. Microbiol.* 11:123–135.

Pepe, J. C., Wachtel, M. R., Wagar, E., and Miller, V. L. 1995. "Pathogenesis of defined invasion mutants of *Yersinia enterocolitica* in a BALB/c mouse model of infection," *Infect. Immun.* 63:4837–4848.

Pefry, R. D. and Fetherston, J. D. 1997. "*Yersinia pestis*-etiologic agent of plague," *Cl. Microbiol. Rev.* 10:35–66.

Persson, C., Carbalieira, N., Wolf-Watz, H., and Fällman, M. 1997. "The PTPase YopH inhibits uptake of *Yersinia*, tyrosine phosphorylation of p130^{Cas} and FAK, and the associated accumulation of these proteins in peripheral focal adhesions." *EMBO J.* 16:2307–2318.

Persson, C., Nordfeith, R., Holmström, A., Hakansson, S., Rosqvist, R., and Wolf-Watz, H. 1995. "Cell-surface-bound *Yersinia* translocate the protein tyrosine phosphatase YopH by a polarized mechanism into the target cell," *Mol. Microbiol.* 18:135–150.

Petterson, J., Nordfeith, R., Dubinina, E., Bergman, T., Gustafsson, M., Magnusson, K. E., and Wolf-Watz, H. 1996. "Modulation of virulence factor expression by pathogen target cell contact," *Science* 273:1231–1233.

Pierson, D. E., and Falkow, S. 1990. "Nonpathogenic isolates of *Yersinia enterocolitica* do not contain functional *inv*-homologous sequences," *Infect. Immun.* 58:1059–1064.

Piano, G. V. and Straley, S. C. 1993. "Multiple effects of *lcrD* mutations in *Yersinia pestis*," *J. Bacteriol.* 175:3536–3545.

Prentice, M. 1992. "Transfusing *Yersinia enterocolitica*. Rare but deadly, *BMJ* 305:664–665.

Rimpilainen, M., Forsberg, A., and Wolf-Watz, H. 1992. "A novel protein, LcrQ, involved in the low-calcium response of *Yersinia pseudotuberculosis* shows extensive homology to YopH," *J. Bacteriol.* 174:3355–3363.

Robins Browne, R. M. and Prpic, J. K. 1985. "Effects of iron and desferrioxamine on infections with *Yersinia enterocolitica*," *Infect. Immun.* 47:774–779.

Roggenkamp, A., Neuberger, H.-R., Flügel, A., Schmoll, T., and Heesemann, J. 1995. "Substitution of two histidine residues in YadA protein of *Yersinia enterocolitica* abrogates collagen binding, cell adherence and mouse virulence," *Mol. Microbiol.* 16:1207–1219.

Roggenkamp, A., Ruckdeschel, K., Leitritz, L., Schmitt, R., and Heesemann, J. 1996. "Deletion of amino acids 29 to 81 in adhesion protein YadA of *Yersinia enterocolitica* serotype O:8 results in selective abrogation of adherence to neutrophils," *Infect. Immun.* 64(7):2506–2514.

Rohde, J. R., Fox, J. M., and Minnich, S. A. 1994. "Thermoregulation in *Yersinia enterocolitica* is coincident with changes in DNA supercoiling," *Mol. Microbiol.* 12:187–199.

Rosenshine, I., Duronio, V., and Finlay, B. B. 1992. "Tyrosine protein kinase inhibitors block invasin-promoted bacterial uptake by epithelial cells," *Infect. Immun.* 60:2211–2217.

Rosqvist, R., Bolin, I., and Woll-Watz, H. 1988. "Inhibition of phagocytosis in *Yersinia pseudotuberculosis:* a virulence plasmid-encoded ability involving the Yop2b protein," *Infect. Immun.* 56:2139–2143.

Rosqvist, R., Forsberg, A., and Wolf-Watz, H. 1991. "Intracellular targeting of the *Yersinia* YopE cytotoxin in mammalian cells induces actin microfilament disruption," *Infect. Immun.* 59:4562–4569.

Rosqvist, R., Magnusson, K. E., and Wolf-Watz, H. 1994. "Target cell contact triggers expression and polarized transfer of *Yersinia* YopE cytotoxin into mammalian cells," *EMBO J.* 13:964–972.

Rosqvist, R., Forsberg, A., Rimpilainen, M., Bergman, T., and Wolf-Watz, H. 1990. "The cytotoxic protein YopE of *Yersinia* obstructs the primary host defence," *Mol. Microbiol.* 4:657–667.

Ruckdeschel, K., Machold, J., K., Roggenkamp, A., Schubert, S., Pierre, J., Zumbihi, R., Liautard, J.-P., Heesemann, J., and Rouot, B. 1997. "*Yersinia enterocolitica* promotes deactivation of macrophage mitogen-activated protein kinases extracellular signal-regulated kinase-1/2, p38, and c-Jun NH_2-terminal kinase," *J. Biol. Chem.* 272(25):15920–15927.

Ruckdeschel, K., Roggenkamp, A., Schubert, S., and Heesemann, J. 1996. "Differential contribution of *Yersinia enterocolitica* virulence factors to evasion of microbicidal action of neutrophils," *Infect. Immun.* 64:724–733.

Schesser, K., Frithz-Lindsten, E., and Wolf-Watz, H. 1996. "Delineation and mutational analysis of the *Yersinia pseudotuberculosis* YopE domains which mediate translocation across bacterial and eukaryotic cellular membranes," *J. Bacteriol.* 178:7227–7223.

Schulte, R., Wattlau, P., Hartland, E. L., Robins-Browne, R. M., and Cornells, G. R. 1996. "Differential secretion of interleukin-8 by human epithelial cell lines upon entry of virulent or nonvirulent *Yersinia enterocolitica*," *Infect. Immun.* 64:2106–2113.

Schulze-Koops, H., Burkhardt, H., Heesemann, J., von der Mark, K., and Emmrich, F. 1992. Plasmid-encoded outer membrane protein YadA mediates specific binding of enteropathogenic *Yersiniae* to various types of collagen," *Infect. Immun.* 60:2153–2159.

Schulze-Koops, H., Burkhardt, H., Heesemann, J., von der Mark, K., and Emmrich, F. 1995. "Characterization of the binding region for the *Yersinia enterocolitica* adhesin YadA on types I and II collagen," *Arthritis Rheum.* 38:1283–1289.

Simonet, M., Richard, S., and Berche, P. 1990. "Electron microscopic evidence for in vivo extracellular localization of *Yersinia pseudotuberculosis* harboring the pYV plasmid," *Infect. Immun.* 58:841–845.

Sory, M. P. and Cornells, G. 1988. "*Yersinia enterocolitica* O:9 as a potential live oral carrier for protective antigens," *Microb. Pathog.* 4:431–442.

Sory, M. P. and Cornelis, G. R. 1994. "Translocation of a hybrid YopE-adenylate cyclase from *Yersinia enterocolitica* into HeLa cells," *Mol. Microbiol.* 14:583–594.

Sory, M. P., Boland, A., Lambermont, I., and Cornelis, G. R. 1995. "Identification of the YopE and YopH domains required for secretion and internalization into the cytosol of macrophages, using the *cyaA* gene fusion approach," *Proc. Natl. Acad. Sci. USA* 92:11998–12002.

Stainier, I., Iriarte, M., and Cornelis, G. R. 1997. "YscM1 and YscM2, two proteins causing downregulation of *yop* transcription," *Molecular Microbiology* 26:833–843.

Straley, S. C. and Bowner, W. S. 1986. "Virulence genes regulated at the transcriptional level by Ca^{2+} in *Yersinia pestis* include structural genes for outer membrane proteins," *Infection and Immunity* 5:445–454.

Tamm, A., Tarkkanen, A. M., Korthonen, T. K., Kuusela, P., Toivanen, P., and Skurnik, M. 1993. "Hydrophobic domains affect the collagen-binding specificity and surface polymerization as well as the virulence potential of the YadA protein of *Yersinia enterocolitica,*" *Mol. Microbiol.* 10:995–1011.

Tauxe, R. V., Vandepitte, J., Wauters, G., Martin, S. M., Goossens V., De Mol, P., Van Noyen, R., and Thiers, G., 1987. "*Yersinia enterocolitica* infections and pork: the missing link," *Lancet* II:1129–1132.

Visser, L. G., Annema, A., and van Furth, R. 1995. "Role of Yops in inhibition of phagocytosis and killing of opsonized *Yersinia enterocolitica* by human granulocytes," *Infect. Immun.* 63:2570–2575.

Wattiau, P. and Cornelis, G. R. 1994. "Identification of DNA sequences recognized by VirF the transcriptional activator of the *Yersinia* yop regulon," *J. Bacteriol.* 176:3878–3884.

Wattiau, P., Woestyn, S., and Cornelis, G. R. 1996. "Customized secretion chaperones in pathogenic bacteria," *Mol. Microbiol.* 20:255–262.

Wattiau, P., Bernier, B., Deslee, P., Michiels, T., and Cornelis, G. R. 1994. "Individual chaperones required for Yop secretion by *Yersinia,*" *Proc. Natl. Acad. Sci. USA* 91:10493–10497.

Woeslyn, S., Allaoui, A., Wattiau, P., and Cornelis, G. R. 1994. "YscN, the putative energizer of the *Yersinia* Yop secretion machinery," *J. Bacteriol.* 176:1561–1569.

Woeslyn, S., Sory, M. P., Boland, A., Lequenne, O., and Cornelis, G. R. 1996. "The cytosolic SycE and SycH chaperones of *Yersinia* protect the region of YopE and YopH involved in translocation across eukaryotic cell membranes," *Mol. Microbiol.* 20:1261–1271.

Yang, Y. and Isberg, R. R. 1997. "Transcriptional regulation of the *Yersinia pseudotuberculosis* pH6 antigen adhesin by two envelope-associated components," *Mol. Microbiol.* 24:499–510.

Yang, Y., Merriam, J. J., Mueller, J. P., and Isberg, R. R. 1996. "The *psa* locus is responsible for thermoinducible binding of *Yersinia pseudotuberculosis* to cultured cells," *Infect. Immun.* 64:2483–2489.

Yoshino, K., Takao, T., Huang, X., Murata, H., Nakao, H., Takeda, T., and Shimonishi, Y. 1995. "Characterization of a highly toxic, large molecular size heat-stable enterotoxin produced by a clinical isolate of *Yersinia enterocolitica,*" *FEBS Lett.* 362:319–322.

Young, V. B., Falkow, S., and Schoolnik, G. K. 1992. "The invasin protein of *Yersinia enterocolitica:* Internalization of invasin-bearing bacteria by eukaryotic cells is associated with reorganization of the cytoskeleton," *J. Cell. Biol.* 116:197–207.

CHAPTER 5

Molecular Pathogenesis of *Vibrio* Infections

KENNETH M. PETERSON

1. INTRODUCTION

MEMBERS of the genus *Vibrio* are motile, gram-negative rods which are among the most common of bacteria native to marine and estuarine environments. This genus includes 35 currently recognized species, 12 of which are considered to be human pathogens. Vibrio-associated human infections typically result from the ingestion of raw, undercooked or recontaminated seafood. Fresh, frozen, or iced fish/shellfish have been found to harbor various species of vibrios, in some markets these organisms are isolated at rates approaching 100%. Not surprisingly, of the various bacteria associated with seafood-related human disease, *Vibrio* spp. cause the majority of illnesses. Human infections can also follow exposure of wounds to saltwater and/or shellfish. Epidemiological studies show that the rates of many vibrio infections increase during the warmer months paralleling the numbers of vibrios isolated from environmental sources. Besides temperature and salinity, other factors that affect the occurrence and distribution of vibrios in the environment are not well understood.

Pathogenic vibrios are capable of producing three clinical manifestations of infection: gastroenteritis, soft tissue infections and systemic infections including bacteremia. In compromised individuals, a rapidly fatal septicemia is a common manifestation of infection with *Vibrio* organisms. Although the overall incidence of vibrio infections appears to be low, it is likely that many mild infections that do occur are not reported, making an estimate of the true incidence of vibrio infections difficult to establish. Adequate sewage treatment, proper water sanitation, thorough cooking of seafood and public warnings regarding the potential hazards of eating raw shellfish appear to be the most effective means of preventing vibrio infections.

The 12 *Vibrio* spp. that are considered pathogenic for humans possess a variety of virulence determinants ranging from the well-characterized cholera enterotoxin of noninvasive *V. cholerae* O1 to the polysaccharide capsule of invasive *V. vulnificus*. Our current understanding of the molecular mechanisms by which *Vibrio* spp. cause disease will be reviewed below.

2. *VIBRIO CHOLERAE*

Vibrio cholerae is the etiologic agent of Asiatic cholera, an acute diarrheal disease that is acquired by oral ingestion of toxigenic vibrios found in contaminated water or food sources. Infection with *V. cholerae* O1 may be the result of endemic, epidemic or pandemic disease. The O1 designation refers to the *V. cholerae* serogroup antigen that is synthesized by El Tor and classical biotypes that include Inaba and Ogawa serotypes. Non-O1 *V. cholerae* serogroups can cause diarrheal disease indistinguishable from cholera, but until recently have not caused epidemic or pandemic disease. During the last decade we have witnessed an increase in the global incidence of cholera infections typified by the unexpected return of epidemic cholera to South America following a century of absence (Tauxe and Blake, 1992). Another disturbing event was the appearance of a variant of *V. cholerae* (O139) on the Indian subcontinent capable of causing widespread cholera infections indistinguishable from the typical clinical features of cholera (Albert et al., 1993; Shimada et al., 1993; Ramamurthy et al., 1993).

A long-held perception regarding *V. cholerae* ecology was that this organism was highly host-adapted and incapable of long-term survival outside of infected humans. Evidence collected over the last 30 years demonstrates that *V. cholerae* is a normal inhabitant of brackish water and estuarine systems (Colwell and Spira, 1992). *V. cholerae* is transmitted via the fecal-oral route and is spread mainly through contaminated water and food. During the recent epidemic in Peru, fecal contamination of municipal water supplies in large urban areas resulted in a large number of cases (Swerdlow and Ries, 1992; Ries et al., 1992). River water, well water, spring water, bottled water, street vendor drinks, water stored in the home, municipal water and ice have been incriminated in cholera transmission (Blake, 1993). Case-control investigations of cholera epidemics demonstrate that in certain areas food is a more important vehicle of transmission than water (Estrada-Garcia and Mintz, 1996). Primary contamination of food occurs in situations in which environmental water sources are contaminated by human feces or by naturally occurring *V. cholerae*. Harvesting seafood from contaminated waters and irrigating food crops with *V. cholerae*-infected water are two examples of primary contamination. Secondary contamination occurs by the addition of contaminated products to the food, or by food handlers infected with cholera. Cooked food such as rice or seafood that is moistened with water contaminated with human feces and then consumed without reheating is a common source of cholera outbreaks (Estrada-Garcia and Mintz, 1996).

Vibrio cholerae produces diarrheal disease in humans by colonizing the small bowel and secreting a proteinaceous enterotoxin. Recent studies utilizing a combination of genetic and biochemical methods have begun to identify the vibrio-associated determinants responsible for *V. cholerae* intestinal colonization and survival (Figure 5.1) (Butterton and Calderwood, 1995). *V. cholerae* possesses a large set of coordinately expressed genes that are required for intestinal colonization, toxin production and vibrio survival within the intestine (Peterson and Mekalanos, 1988). Induction of the virulence regulon in vitro can be affected by specific environmental stimuli that act through a complex and incompletely understood regulatory cascade involving at least two transcriptional regulators, ToxR and ToxT (Skorupski and Taylor, 1997a). Such an adaptive response allows *V. cholerae* to adjust gene expression in response to undetermined host signals, maximizing the ability of the vibrios to survive and multiply within the host. Although much is known regarding the major events involved in the pathogenesis of *V. cholerae,* many aspects of the host-parasite interaction remain to be elucidated.

3. *VIBRIO CHOLERAE* VIRULENCE FACTORS

3.1. CHOLERA TOXIN

The diarrhea associated with *V. cholerae* infection follows the release of cholera toxin into the lumen of the small intestine (Finkelstein, 1992). Cholera toxin is a multi-subunit protein that includes five identical B subunits, each with a molecular weight of 11,600 and a single A subunit with a molecular weight of 27,200 (Gill, 1976). The mature A subunit is proteolytically cleaved in the intestine to produce the A1 and A2 peptides with molecular weights of 21,800 and 5400, respectively, which are joined via disulfide bonds (Gill and Rappaport, 1977). The A2 peptide links the A subunit to the B-subunit pentamer (Gill, 1976). The B subunits are responsible for binding of the holotoxin to GM1 gangliosides found on eukaryotic cell surfaces (van Heyningen et al., 1971; Holmgren et al., 1975) while the A1 subunit possesses the enzymatic activity associated with cholera toxin (Gill and King, 1975). Binding of the holotoxin to the cell surface receptor induces translocation of the A1 fragment through the cell membrane. Following reduction, the A1 subunit catalyzes the ADP-ribosylation of an arginine residue on the α subunit

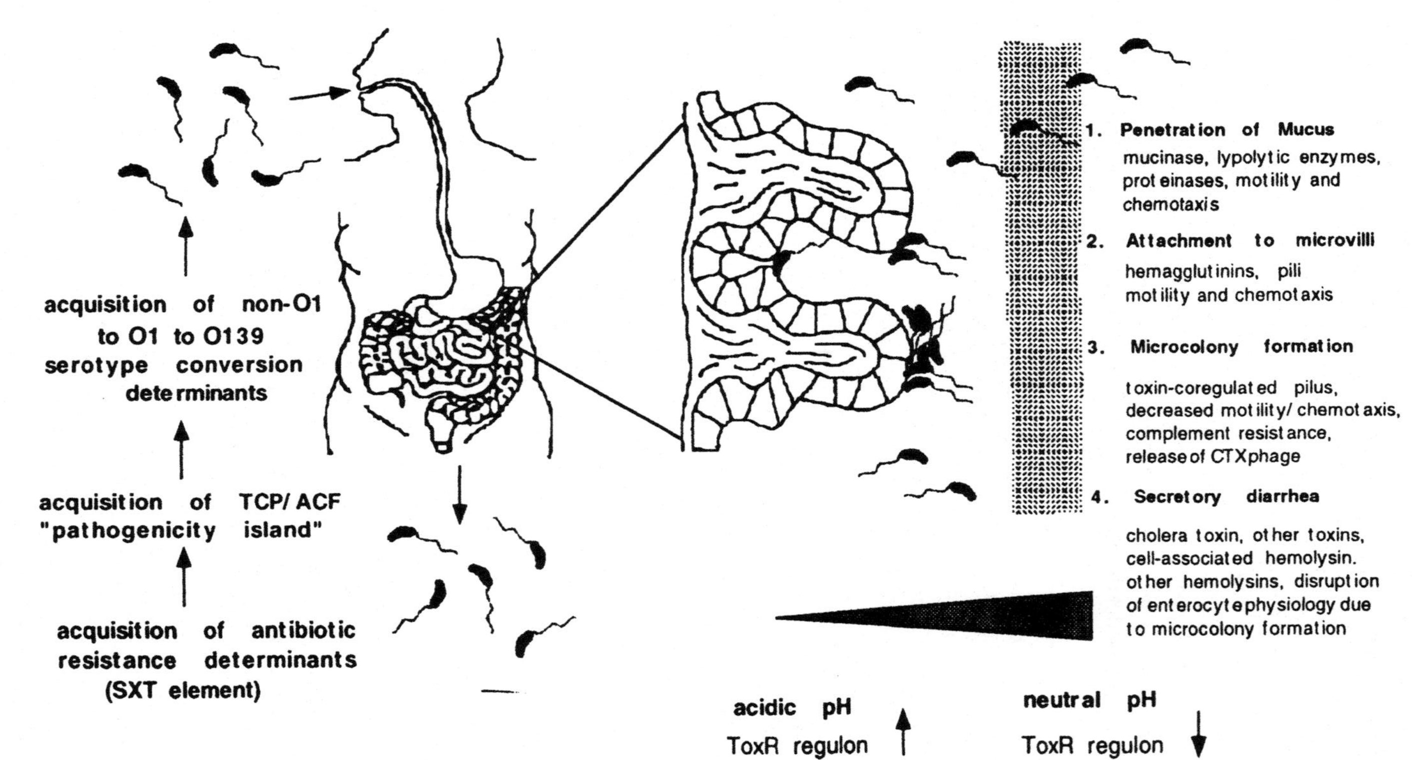

Figure 5.1 Model showing the various steps involved in intestinal colonization by *V. cholerae*.

of Gs, a guanylnucleotide-binding protein involved in the regulation of adenylate cyclase activity (Gill and Mexen, 1978). Modification of Gs results in the activation of adenylate cyclase and the elevation of cAMP levels in the target cells. High levels of cAMP are responsible for alterations in ion transport in villus and crypt cells of the intestinal mucosa (Field, 1980). The net effect is an increase in chloride secretion into the intestinal lumen and an inhibition of sodium absorption, causing the diarrhea associated with Asiatic cholera. Recent studies examining the secretory effects of cholera toxin suggest that prostaglandins and the enteric nervous system are also involved in the cellular response to cholera toxin (Kaper et al., 1994).

V. cholerae secretes a large number of extracellular proteins, including cholera toxin. The A and B subunit proteins of cholera toxin are synthesized as precursor proteins containing typical N-terminal signal peptides. The subunits are translocated through the cytoplasmic membrane followed by processing into the mature subunits via a system analogous to the *E. coli* Sec apparatus (Hirst, 1991). Within the periplasm a disulfide isomerase (*dsbA/tcpG*) catalyzes the formation of disulfide bonds within the A and B subunits (Yu et al., 1992). The mature subunits are assembled into an A:B(5) complex within the periplasm and then translocated across the outer membrane via a large set of accessory proteins encoded by the *eps* (extracellular protein secretion) genes (Sandkvist et al., 1993). The 12 Eps proteins are thought to form a multiprotein complex that promotes selective protein secretion across the enterocyte outer membrane (Sandkvist et al., 1995). Activation of cholera toxin is dependent upon the proteolytic cleavage of the A1 and A2 peptides (Mekalanos et al., 1979). Presently, it is not clear whether vibrio proteases and/or intestinal proteases are responsible for nicking the A subunit (Kaper et al., 1994).

The A and B subunits of cholera toxin are encoded by the *cixAB* operon that is flanked by directly repeated segments called RS1 sequences (Mekalanos, 1983). This segment of DNA was originally termed the CTX genetic element and was thought to be a type of compound transposon associated with toxigenic strains of *V. cholerae* (Pearson et al., 1993). Recent studies demonstrate, however, that the CTX genetic element is in fact the genome of a lysogenic filamentous bacteriophage designated CTXϕ (Waldor and Mekalanos, 1996c). The core of the CTX genetic element carries at least six genes, including *ctxAB*, *zot* (zona accludens toxin), *cep* (core-encoded pilin), *ace* (accessory cholera enterotoxin, and *orfU* (product of unknown function) (Kaper et al., 1994). Mutations within *orfU* and *zot* abolish the ability of CTXϕ to transfer the CTX element (Waldor and Mekalanos, 1996c). The deduced amino acid sequences of Zot and Cep are homologous to the gene I and gene VIII products of coliphage M13 (Waldor and Mekalanos, 1996c). The deduced amino acid sequence of Ace is homologous to filamentous phage Pf1 of *Pseudomonas* (Waldor and Mekalanos, 1996c). The RS-repeated segments encode at least four open reading frames (*rstABCR*) that are required for integration functions (Pearson et al., 1993). The single-stranded CTXϕ infects *V. cholerae* by adsorbing to the toxin-coregulated pilus (TCP), which is essential for *V. cholerae* intestinal colonization (Waldor and Mekalanos, 1996c). Interestingly, TCP is only synthesized under conditions that favor the expression of the cholera toxin genes (Taylor et al., 1987). Lysogenic conversion of cholera toxin-negative strains occurs both in vitro and in vivo. Remarkably, the degree of conversion is much greater in the gastrointestinal tract of mice than under laboratory growth conditions (Waldor and Mekalanos, 1996c), suggesting that horizontal transfer of the CTX genetic element is influenced by in vivo gene expression. The CTXϕ genome can be incorporated into the *V. cholerae* chromosome as an array of tandemly repeated copies via integration into a specific 18-base pair DNA sequence designated *attRS1* (Pearson et al., 1993). The replicative form of CTXϕ can also be transmitted as a plasmid (Waldor and Mekalanos, 1996c). Understanding the biology of CTXϕ acquisition and maintenance will no doubt greatly aid in the understanding of the horizontal transfer of virulence determinants by *V. cholerae* and other pathogens as well as help explain the unusual epidemiology of cholera.

3.1.1. Other Toxins

Human volunteer studies designed to examine the safety and efficacy of live vaccine strain candidates of *V. cholerae* lacking *ctxAB* produced an unexpected result. Some subjects ingesting these strains experienced mild to moderate diarrhea, headache, fever, nausea, intestinal cramps and vomiting (''reactogenicity'') (Levine et al., 1988). These results led to the hypothesis that *V. cholerae* synthesizes additional enterotoxins that may also be important in disease production. In response to these observations, studies that examined the ability of culture supernatants from *ΔctxAB* strains to alter the permeability of rabbit ileal tissue mounted in Ussing chambers led to the discovery of two proteins that can alter electrolyte transport: zona accludens toxin (Zot) and accessory cholera enterotoxin (Ace) (Fasano et al. 1991; Trucksis et al., 1993). Zot increases the permeability of the small intestinal mucosa by affecting the structure of the intercellular tight junctions (Fasano et al., 1991). Ace increases ion transport across rabbit ileal mucosa and causes fluid accumulation in ligated rabbit ileal loops (Trucksis et al., 1993). The mechanism of action of Ace is unknown. Strikingly, the structural genes encoding Zot and Ace are located in tandem directly upstream of *ctxAB* in the core region of the CTXϕ genome (Trucksis et al., 1993). The role of Ace and Zot in the pathogenesis of cholera is unclear since a *V. cholerae* strain deleted for *zot, ace, hlyA* (hemolysin) and *ctxA* still causes mild to moderate diarrhea as well as fever and abdominal cramps in human volunteers (Tacket et al., 1993). The subsequent discovery that the predicted products of *zot* and *ace* are homologous to proteins encoded by filamentous bacteriophage is interesting in light of their biological activity in Ussing chambers. Taken together, these data suggest that the proteins encoded by *ace* and *zot* may be involved in both phage morphogenesis and *V. cholerae* pathogenesis.

V. cholerae produces a hemolysin that is cytolytic for a variety of erythrocytes and mammalian cells and causes death when injected intravenously into mice. The hemolysin also produces a bloody fluid accumulation in ligated rabbit ileal loops (Honda and Hibolstem, 1979). These findings led to the suggestion that the diarrhea produced by *Δctx* strains in human volunteers is due to the production of the hemolysin. However, *V. cholerae* strains deleted for the hemolysin structural gene (*hlyA*) in a *Δctx* background still cause diarrhea in human volunteers (Levine et al., 1988).

Some strains of *V. cholerae* produce a toxin that is homologous to the heat-stable enterotoxins of *E. coli* (Yoshino, 1993). The role of the *V. cholerae* heat-stable-like enterotoxins in human disease is not understood. *V. cholerae* may also produce a toxin that is related to the shiga-like toxin of enterohemorrhagic *E. coli.* This toxin was identified based on the ability of antibodies directed against shiga toxin from *Shigella dysenteriae* to neutralize *V. cholerae* cytotoxicity of tissue culture cells (O'Brien et al., 1985). The genes encoding this shiga-like toxin activity have yet to be isolated.

A toxin that belongs to the RTX family of hemolysins/leukotoxins has recently been indentified in El Tor and O139 strains of *V. cholerae.* The RTX gene cluster is physically linked to the CTX element on the *V. cholerae* genome. The RTX toxin has been implicated in a cell-associated cytotoxicity that causes mammalian cells to detach and round up. Classical strains of *V. cholerae* are defective in the production of the cytotoxic activity due to a DNA deletion within the RTX gene cluster (Lin et al., 1999). It has been suggested that the RTX toxin may be responsible for the inflammatory diarrhea observed in volunteers immunized with RTX-positive vaccine strains.

It is suggested that toxins other than cholera toxin contribute to the pathogenesis of fully virulent *V. cholerae* and the reactogenicity of *ΔctxA* strains in human volunteers. To date, however, there is no convincing evidence proving that any of these other toxins play a role in pathogenesis of human cholera. The uncertainty surrounding the biological relevance of these toxins in human disease complicates the construction of predictably less reactogenic vaccine strains.

3.2. MOTILITY AND CHEMOTAXIS

A key step in the pathogenesis of *V. cholerae* infection is the ability of the vibrios to penetrate the intestinal mucus layer prior to colonization of enterocytes (Guentzel and Berry, 1975). Intestinal mucus provides a chemotactic signal for *V. cholerae* whereby the vibrios direct their movement toward the intestinal surface (Freter et al., 1981). Directed motility, coupled with the vibrios' ability to secrete enzymes (mucinase, lipases, proteinases) capable of degrading the mucus, maximizes the ability of *V. cholerae* to burrow through the mucus to the surface of intestinal cells. *V. cholerae* are motile by the action of a single, polar, sheathed flagellum (Guentzel and Berry, 1975). Numerous studies have shown that motility and chemotaxis are important virulence properties (Guentzel and Berry 1975: Freter et al., 1981: Attridge, and Rowley, 1983). Although the vibrio flagellar structure's primary function in colonization is thought to be due to its contribution to motility and chemotaxis, the flagella may also carry an adhesin that promotes vibrio attachment to the intestinal mucosa (Attridge and Rowley, 1983). A study confirmed that motility itself is important in pathogenesis, while functional flagella are required for optimum virulence (Richardson, 1991).

Interestingly, a recent study examining the relationship between vibrio motility and virulence noted an inverse relationship between motility, as measured by swarm plate activity, and virulence factor expression (Gardel and Mekalanos, 1996). Hyperswarming vibrio cells were defective in TCP expression, cholera toxin synthesis and production of cell-associated hemolysin, whereas non-motile mutants exhibited increased expression of pili, cholera toxin and cell-associated hemolysin (Gardel and Mekalanos, 1996). These data indicate that early in the colonization process when motility is required for penetration of the mucous layer, several key virulence factors including cholera toxin are not synthesized. Upon reaching the underlying epithelial cells where motility is no longer beneficial, however, maximal synthesis of colonization factors (TCP) and cholera toxin occurs. These observations are supported by recent reports demonstrating that methyl-accepting chemotaxis proteins involved in pilus synthesis and accessory colonization factor expression negatively regulate vibrio swarm plate activity (Everiss et al., 1994: Harkey et al., 1994).

Identification and characterization of *V. cholerae* genes whose expression parallels that of cholera toxin and the toxin-coregulated pilus led to the identification of four closely linked accessory colonization factor (ACF) genes that are required for efficient intestinal colonization (Peterson and Mekalanos, 1988). Analysis of the proteins encoded by these genes and identification of altered motility phenotypes due to disruption of three of the four ACF genes suggest that these inducible genes are involved in chemotaxis of the vibrios towards the intestinal surface (Everiss et al., 1994). *acfA* encodes a 23 kDa protein that is homologous to the OmpW outer membrane protein of *V. cholerae* (Hughes et al., 1995). However, the role of AcfA in intestinal colonization is not yet known. *acfB* encodes a 626-amino acid protein that is related to bacterial methyl-accepting chemotaxis proteins. Furthermore, *V. cholerae acfB* mutants display an altered motility phenotype in semisolid agar (Everiss et al., 1994). The relationship between AcfB and vibrio motility and the amino acid similarities between AcfB and chemotaxis signal-transducing proteins suggest that AcfB might interact with the *V. cholerae* chemotaxis machinery. *V. cholerae acfC* mutants lack the ability to swim toward a gradient of sulfate ions (Shaffer et al., 1996). These results are consistent with DNA sequence analysis showing homology between AcfC and sulfate-binding proteins from other bacteria. Taken together, these data suggest that AcfB and AcfC may interact to provide the motile vibrios a mechanism to guide them toward sulfate moieties present in the intestine.

The last gene in the ACF locus (*acfD*) is predicted to encode a lipoprotein that is related to the *fliC* gene product required for motility in *Salmonella* sp. (Hughes et al., 1994). Consistent with the homology between *acfD* and *fliC* is the reduced motility and the altered swarming phenotype of *V. cholerae acfD* mutants (Hughes et al., 1994). As with AcfA, the precise role of AcfD in vibrio motility/chemotaxis is not known. A better understanding of vibrio motility and chemotaxis is important since the reactogenicity of live attenuated, vaccine candidate strains seems to be affected by their motility status, i.e., less motile strains are less reactogenic in humans than are highly motile attenuated vibrio strains (Kenner et al., 1995).

3.3. COLONIZATION FACTORS

3.3.1. Toxin-Coregulated Pilus

The best characterized colonization factor of *V. cholerae* is a type IV pilus designated toxin-coregulated pilus (TCP) because its production relates to cholera toxin synthesis. TCP is composed of 7 nm filaments that form laterally associated bundles composed of the 20.4 kDa TcpA pilin subunit (Taylor et al., 1987). Assembly of TCP on the bacterial cell surface results in vibrio autoagglutination due to the hydrophobic nature of the pilus. The amino-terminal region of TcpA possesses an unusual hydrophilic N-terminal domain that is related to other type IV pilins (Taylor et al., 1987). Following processing of the immature pilin subunit by a specialized prepilin peptidase, TcpJ, a modified N-terminal amino acid in the form of N-methylmethionine is found on the mature pilin subunit (Kaufmann et al., 1991). Construction of defined *tcpA* knockouts in El Tor and classical biotypes and O139 strains of *V. cholerae* demonstrated that TCP is required for intestinal colonization in an infant mouse model (Rhine and Taylor, 1994). TCP is the only *V. cholerae* colonization factor whose importance in human disease has been established. Volunteers ingesting a *ΔtcpA* derivative of a classical strain of *V. cholerae* are not colonized and do not exhibit diarrhea (Herrington et al., 1988). The mechanism by which TCP mediates intestinal colonization is not understood. To date no cellular receptors have been identified that interact specifically with TCP. Monoclonal antibodies employed in passive immunization studies indicate that protective epitopes map to the carboxyl region of TcpA, which contains a disulfide loop (Sun et al., 1990). Further evidence implicating this region in the function of TCP comes from studies demonstrating that mutations in TcpG, a disulfide isomerase, that lead to a failure to modify TcpA do not autoagglutinate and vibrio *tcpG* mutants fail to colonize experimentally infected mice (Peek and Taylor, 1992). Single amino acid substitutions in the N-terminus of TcpA also abolish autoagglutination and vibrios bearing these mutations are colonization defective (Chiang et al., 1995). The correlation between TCP-mediated autoagglutination and *V. cholerae* colonization is also seen with other bacteria that produce type IV pili.

Many of the genes required for TCP biosynthesis have been identified and most lie adjacent to the TcpA subunit gene (Taylor et al., 1988). The nucleotide sequence of the entire contiguous TCP gene cluster has been determined and contains as many as 15 genes, which are involved in TCP biogenesis (Iredell and Manning, 1994; Kaufmann et al., 1994). As with other type IV pili systems, some TCP biogenesis proteins share significant homologies to export-associated proteins from a wide variety of bacteria (Iredell and Manning, 1994; Kaufmann et al., 1994). The presence of several unique TCP biogenesis proteins implies that some of the steps in TCP export/assembly may be unique to *V. cholerae.* The study of TCP biogenesis to date has provided information that is relevant to the development of strategies for prevention of cholera, and has also expanded our fundamental understanding of protein export mechanisms.

3.3.2. Mannose-Sensitive Hemagglutinin (MSHA)

El Tor strains of *V. cholerae* synthesize a thin flexible pilus that is related to type IV pili of other gram-negative organisms. The pilus is composed of 17 kDa subunits encoded by *mshA* (Johnson et al., 1994). Monoclonal antibodies directed against MSHA that inhibit erythrocyte agglutination and that recognize the pilus structure protect experimental animals from cholera caused by El Tor vibrios but not against challenge by classical strains of *V. cholerae* (Osek et al., 1992). These data prompted two independent studies examining the colonization phenotype of El Tor and O139 strains of *V. cholerae* carrying in-frame deletions within the *mshA* structural gene. The results demonstrate that MSHA pili are not necessary for *V. cholerae* intestinal colonization of the infant mouse (Attridge et al., 1996; Thelin and Taylor, 1996).

3.3.3. Mannose-Fucose-Resistant Hemagglutinin (MFRHA)

A gene encoding a predicted protein of 26.9 kDa is associated with the expression of mannose- and fucose-resistant hemagglutination activity by El Tor and classical strains of *V. cholerae* (Franzon et al., 1993). Analysis of the colonization phenotype of an MFRHA mutant revealed a marked defect in intraintestinal survival. The exact nature of the mutation is unknown, however, since plasmids encoding MFRHA activity are unable to complement hemagglutination in the MFRHA mutant (Franzon et al., 1993). Future studies using better defined mutations are needed to clearly establish a role for MFRHA in intestinal colonization.

3.3.4. OmpU

OmpU is a major outer membrane protein of *V. cholerae* and exhibits a molecular weight of 38 kDa during SDS-polyacrylamide gel analysis. N-terminal amino acid sequence analysis reveals a similarity between OmpU and adhesins from *Haemophilus influenzae* and *Bordetella pertussis* (Sperandio et al., 1995). Studies examining the contribution of OmpU to host-cell adherence demonstrate that OmpU binds to fibronectin and that antibodies to OmpU inhibit vibrio-tissue culture cell interactions (Sperandio et al., 1995). Passive immunization studies also show that antibodies to OmpU protect against infection with both El Tor and classical biotype *V. cholerae* in infant mice (Sperandio et al., 1995). A clearer picture regarding the role of OmpU in intestinal colonization will depend on the results of studies designed to examine the colonization profiles of *ΔompU* strains of *V. cholerae.*

3.4. IRON-REGULATED DETERMINANTS

Under low iron conditions, *V. cholerae* synthesizes and secretes an iron-scavenging siderophore called vibriobactin (Griffiths et al., 1984). *V. cholerae* can also acquire iron from heme and hemoglobin in a siderophore-independent manner (Stoebner and Payne, 1988). Virulence testing of heme utilization and vibriobactin uptake mutants reveals a relation between the ability to acquire iron via these systems and the capacity of *V. cholerae* to cause disease (Henderson and Payne, 1994). Mutants defective in only one iron-acquisition mechanism are not drastically reduced in colonization whereas strains carrying mutations in both iron-acquisition systems display a several hundredfold reduction in intestinal colonization (Henderson and Payne, 1994). A 77 kDa iron-regulated outer membrane protein of *V. cholerae* is encoded by the *irgA* gene. A mutation in *irgA* produces a ten fold defect in colonization in the suckling mouse model of cholera infection (Goldberg et al., 1990). The basis for this defect is not known.

4. TOXR REGULON

In *Vibrio cholerae,* the expression of a large number of genes required for intraintestinal survival is activated in response to specific environmental signals. These genes are members of the ToxR regulon and their expression is controlled by the ToxR protein (Figure 5.2).

4.1. ToxR

toxR was identified and isolated based on its ability to activate *ctxAB* transcription in *E. coli* when provided in trans (Miller and Mekalanos, 1984). *toxR* encodes a 32-kDa integral membrane protein, bearing an amino-terminal cytoplasmic domain related to several prokaryotic regulatory proteins that function in DNA binding and transcriptional activation (Miller et al., 1987). Mutations within this domain ablate the ability of ToxR to bind DNA at its target promoters (Ottemann et al., 1992). The function of the periplasmic domain is not known, but is probably involved in sensing specific environmental conditions (Miller and Mekalanos, 1988). In *E. coli,* ToxR increases transcription initiation from the *ctx* promoter and binds a DNA element, TTTTGAT, which is repeated from three to eight times within the *ctxAB* promoter (Mekalanos et al., 1983). *toxR* expression is negatively regulated by *htpG,* a gene that encodes a member of the Hsp90 family of heat shock proteins. *toxR* and *htpG* are divergently transcribed and since their promoters overlap, *htpG* transcription negatively impacts *toxR* expression (Parsot and Mekalanos, 1990). Thus, under conditions of stress, such as those encountered by the vibrios upon entering the stomach and intestine (low pH, anoxia, bile salts), *htpG* transcription inhibits the expression of genes requiring ToxR for transcriptional activation.

4.2. ToxS

The periplasmic carboxy terminal domain of ToxR interacts with ToxS. ToxS enhances the ability of ToxR to function as a transcriptional activator. Although ToxS facilitates the function of ToxR in gene activation, it is unnecessary when ToxR is overexpressed (Miller et al., 1989). The *toxS* open reading frame encodes a 19 kDa protein, which is located in the inner membrane and the periplasm (DiRita and Mekalanos, 1991). ToxS interaction with ToxR appears to stabilize ToxR in a conformation that is optimal for transcriptional activation, possibly as a heterodimer or in combination with a ToxR homodimer (DiRita and Mekalanos, 1991). The oligomerization state of ToxR does not appear to change under different environmental conditions (Otteman and Mekalanos, 1996), making it difficult to predict how environmental stimuli promote activation of ToxR regulated genes.

4.3. ToxT

ToxR directs virulence gene expression in *V. cholerae* via its ability to

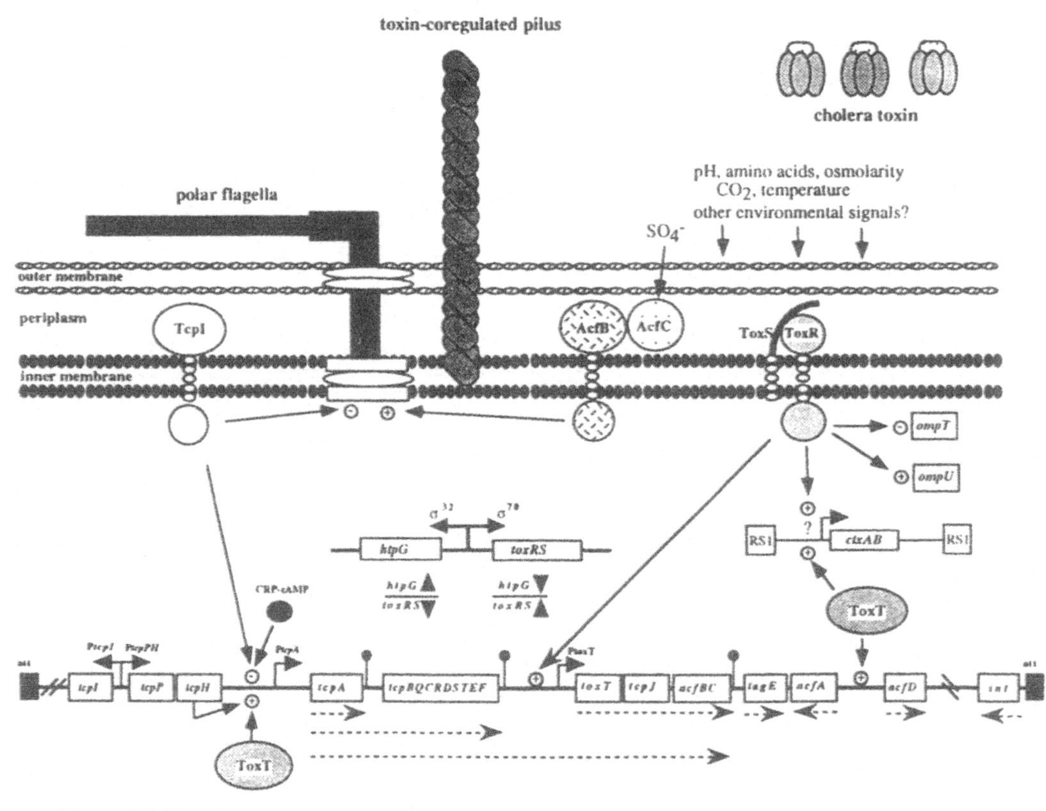

Figure 5.2 Model showing the regulatory pathways that influence the expression of ToxR-regulated genes.

activate the expression of ToxT, a 32-kDa cytoplasmic regulatory protein that is related to the AraC class of transcriptional activators (Higgins et al., 1992). The ToxT carboxy terminal domain possesses a helix-turn-helix DNA binding motif that is found in other AraC-like proteins (Higgins et al., 1992). The ToxT gene lies within the TCP gene cluster (Figure 5.2) and is dependent upon ToxR for expression (DiRita and Mekalanos, 1991). ToxR binds to a promoter upstream of the site of *toxT* transcription initiation. Interestingly, the *toxT* promoter lacks the TTTTGAT heptad repeat found in the *ctxAB* promoter (Higgins et al., 1992). *toxT* expression also is affected by transcriptional readthrough from the ToxT-dependent *tcpA* promoter upstream of *toxT* (Figure 5.2) (Brown and Taylor, 1995). In *E. coli,* both ToxR and ToxT are able to activate *ctxAB* transcription (DiRita and Mekalanos, 1991). Since ToxR null mutants fail to produce cholera toxin, it was assumed that one of the primary functions of ToxR involved the direct transcriptional activation of *ctxAB* in response to specific environmental signals. The biological relevance of ToxR activation of *ctxAB* in *V. cholerae* is unclear since *V. cholerae* strains synthesizing a version of ToxT lacking the helix-turn-helix motif fail to activate *ctxAB* expression (DiRita et al., 1996).

4.4. ENVIRONMENTAL CONTROL OF ToxR REGULON EXPRESSION

In vitro studies examining cholera toxin production have defined several variables that stimulate cholera toxin synthesis. Growth of *V. cholerae* at 30°C, at low pH (6.5) with adequate aeration favors toxin production (Evans and Richardson 1969; Taylor et al., 1987). In addition, the ion concentration of the growth medium and the presence of certain amino acids, carbon dioxide levels and bile salts influence cholera toxin synthesis (Skorupski and Taylor, 1997). Interestingly, the conditions that stimulate high-level expression of the ToxR regulon in classical strains fail to do so for El Tor biotype strains. The reason for this differential expression of ToxR regulon genes by the two biotypes lies in some unknown mechanism of control of *toxT* transcription initiation by ToxR (DiRita et al., 1996). Recent studies suggest that this mechanism may be influenced by TcpP/H and CRP-cAMP (discussed below), two proteins involved in controlling *tcpA* expression (Figure 5.2) (Carroll et al., 1997; Skorupski and Taylor 1997). ToxR stimulates the synthesis of ToxR regulon gene products in a cascade fashion by first activating the expression of *toxT.* ToxT then directly activates the expression of ToxR regulon genes (DiRita et al., 1991). ToxR also controls the expression of *ompU* and *ompT* in addition to *toxT.* Thus, the ToxR regulon is composed of two different branches: a ToxT-dependent branch (*ctxAB, tcp,* and *acf*) and a ToxT-independent branch (*ompU* and *ompT*) (Champion et al., 1997).

The mechanisms by which environmental stimuli control the expression of the ToxR regulon in classical and El Tor strains of *V. cholerae* are poorly understood. Multiple overlapping systems are thought to be involved since all ToxR regulon genes do not respond to the same environmental stimuli (Skorupski and Taylor, 1997). It seems clear that ToxR functions both as a sensor and a signal transducer in controlling the expression of the ToxR regulon. Although temperature, pH, amino acids and osmolarity all influence the expression of ToxR regulon genes, only osmolarity is thought to directly act through ToxR (Miller and Mekalanos, 1988). With the exception of high temperature, signals that influence ToxR regulon expression do not alter the levels of ToxR (Parsol and Mekalanos, 1990). The oligomerization state of ToxR and/or ToxS does not appear to change under environmental conditions that influence ToxR regulon expression (Otteman and Mekalanos, 1996), making it difficult to understand how the environmental stimuli that are perceived by ToxR influence DNA binding and transcriptional activation. Although not much is known about signal transduction and ToxR regulon expression, ToxR activation of *toxT* appears to be a key step in the control of the regulon by specific environmental stimuli (Champion et al., 1997). The results of a recent study involving a search for mutations that derepress toxin expression at the normally repressive temperature of 37°C revealed that the cAMP-CRP system of *V. cholerae* represses the expression of ToxR regulon genes under certain specific conditions (Skorupski and Taylor, 1997). A putative cAMP-CRP binding site overlaps the −35 site of the *tcpA* promoter (Figure 5.2) (Thomas, 1995). Thus, cAMP-CRP binding to this site could prevent activation of the ToxT-dependent branch of the ToxR regulon. It was also noted that expression of the ToxR regulon in response to pH and temperature is altered in the absence of a functional cAMP-CRP system (Skorupski and Taylor, 1997b). These data suggest that intestinal carbon and energy sources impact the way that the ToxR/ToxT regulatory cascade responds to the host environment.

A similar screen for mutations that increase *toxT* transcription revealed a complex relationship between membrane sodium flux regulation of virulence gene expression in *V. cholerae* (Hase and Mekalanos, 1999). These data however, may explain the link between motility/flagellar rotation and virulence gene expression since the *V. cholerae* flagellar motor is driven by a sodium motive force. Consistent with these observations is the finding that inhibition of flagellar rotation by increased media viscosity and/or drugs results in the induction of virulence factor expression (Hase and Mekalanos, 1999). These findings suggest that vibrio sensing of the high-viscosity environment found in intestinal mucus may be a key signal in activating virulence gene expression in the human host.

A protein, TcpI, implicated in repressing TCP synthesis is related to a family of methyl-accepting chemotaxis proteins involved in environmental sensing (Harkey et al., 1994). The ability of *tcpI* mutants to synthesize TCP at elevated pH levels that are normally repressive for TCP production suggests that *tcpI* may negatively regulate *tcpA* and subsequent *toxT* expression in response to pH (see Figure 5.2). The ability of pH to regulate the transcription of *tcpH* is also noteworthy, given that TcpH has been implicated in the positive regulation of TCP production. The influence of pH on TCP synthesis in vitro may be a biologically significant observation, since a pH gradient exists in the intestine (Shiau, 1985). The pH in the lumen of the intestine is relatively neutral, whereas the surface of the microvilli is somewhat acidic. Based on the current model of ToxT activation of *tcpA,* and the TCP phenotype of *tcpI/tcpH* mutants, it is envisioned that the acidic pH surrounding the microvilli acts through TcpI/TcpH to activate *tcpA* expression thereby promoting TCP production.

The in vivo relevance of the individual environmental signals and the overlapping regulatory pathways affecting ToxR regulon gene expression that have been discovered in the laboratory remains to be determined. Nevertheless, it is not difficult to imagine that during human cholera infections recognition of various environmental signals (bile, temperature, low pH, carbon source availability, elevated CO_2) by *V. cholerae* would facilitate within the host microenvironment the appropriate regulation of pili production and toxin synthesis. The presence of overlapping signal transduction pathways would ensure the timely expression of those vibrio genes required for adaptation to intraintestinal growth and multiplication.

5. *V. CHOLERAE* ACQUISITION OF VIRULENCE GENES

As mentioned earlier, cholera toxin is encoded by a single-stranded filamentous phage (CTXϕ). Transduction of cholera toxin genes by CTXϕ occurs most efficiently in the microenvironment of the intestine. The receptor for CTXϕ is TCP, which is essential for vibrio intestinal colonization of humans (Waldor and Mekalanos, 1996a). The large cluster of virulence genes encoding proteins involved in TCP biogenesis and accessory colonization factor (ACF) expression is flanked by sequences that resemble bacteriophage attachment (*att*) half sites. Adjacent to the *att*-like site is a gene that encodes a protein that is related to the integrase family of site-specific recombinases (Kovach et al., 1996). Genomic analysis of *V. cholerae* strains indicates that only vibrios capable of causing epidemic Asiatic cholera possess the TCP-ACF "pathogenicity island" in association with the integrase. This pathogenicity island is located in a region of the *V. cholerae* chromosome that is analogous to the *E. coli* CP4–57 cryptic prophage integration site (Kovach et al., 1996). It is proposed that the evolution of virulent *V. cholerae* began with horizontal acquisition of the TCP-ACF pathogenicity island carrying *toxT.* The ability of ToxR to gain control of *toxT* expression permits TCP production within the microenvironment of the small intestine and thereby provides the receptor for CTXϕ and subsequent elaboration of the potent cholera enterotoxin. Understanding the interaction of chromosomally encoded regulatory proteins with genes acquired by virulence elements represents a emerging and fascinating aspect of bacterial pathogenesis.

Epidemic cholera was associated only with *V. cholerae* O1 until recently. It was, therefore, surprising that a cholera epidemic starting in India in 1992 was caused by a non-O1 strain with a novel serotype (O139) (Ramamurthy et al., 1993). The epidemic *V. cholerae* O139 strains synthesize a novel LPS with shorter O-side chains and a different sugar composition than *V. cholerae* O1 LPS. In addition, *V. cholerae* O139 produces a capsule that is not associated with O1 strains of *V. cholerae* (Johnson et al., 1994). *V. cholerae* O139 also displays a distinct antibiotic resistance profile, i.e., sulfamethoxazole, trimethoprim, streptomycin and furazolidone resistance (Waldor and Mekalanos, 1996a). Microbiological and molecular characterization of *V. cholerae* O139 indicates that this serotype was derived from El Tor *V. cholerae* O1 (Bik et al., 1995). Careful genomic analysis of several *V. cholerae* O139 clinical isolates revealed that the O139 serogroup determinants are encoded by genes that are not found in El Tor O1 strains of *V. cholerae* (Bik et al., 1995: Comstock et al., 1995). Analysis of O139 antigen synthesis genes revealed that a 22-kbp region of DNA necessary for O1 antigen synthesis has been replaced by a 35-kbp DNA fragment that directs the synthesis of both O139 antigen and O139 capsular polysaccharide (Comstock et al., 1995). The finding that the O139 antigen-encoding DNA region contains a mobile element designated IS*1358* that is also found in the O1 antigen-encoding gene cluster (Mooi and Bik, 1997) suggests a mechanism for acquisition of O139 antigen genes by *V. cholerae* O1. It is not known whether IS*1358* can direct transposition of O139 antigen genes from a self-replicating vector or promote acquisition by homologous recombination.

Careful analysis of antibiotic resistance in *V. cholerae* O139 revealed that sulfamethoxazole, trimethoprim and streptomycin resistance is carried on a 62-kbp self-transmissible, chromosomally integrating genetic element (SXT) with properties reminiscent of conjugative transposons (Waldor and Mekalanos, 1996a). It is suggested that the presence of the SXT element may have facilitated the mobilization of O139 antigen genes from an unidentified donor into *V. cholerae* O1 (Waldor and Mekalanos, 1996a). It also is proposed that epidemic strains carrying novel O antigen genes in conjunction with the SXT element may have contributed to the emergence of *V. cholerae* O139 in areas of widespread antibiotic use (Waldor and Mekalanos, 1996a).

6. CONTROL OF CHOLERA

Over the centuries, epidemic cholera has shown a preference for areas of the globe in the process of industrial transformation. Cholera outbreaks in developed countries are limited in nature because of the general availability of safe water, proper food preparation and adequate sewage treatment facilities. On the other hand, cholera outbreaks in developing countries are easily sustained because of secondary transmission due to unsafe community water supplies, inadequate disposal and/or treatment of sewage and improper preparation of food. The most useful public health measures that can take place in these societies are improvements in food and water sanitation. Providing safe drinking water via a combination of point-of-use disinfection and safer water storage are two strategies being employed in Latin America to limit the spread of cholera (Estrada-Garcia and Mintz, 1996; Tauxe et al., 1995). Eliminating transmission of *V. cholerae* via food contaminated in the market or the home, i.e., food and beverages sold by street vendors, leftover rice and unwashed fruits and vegetables, by improved food handling can also lead to dramatic decreases in the incidence of cholera (Estrada-Garcia and Mintz, 1996; Tauxe et al., 1995). Transmission of cholera through seafood can be reduced by eliminating the practice of eating raw and undercooked seafood and by maintaining sewage-free harvest beds and improved sanitation in processing plants. These steps would also reduce the burden of other diseases transmitted by the fecal-oral route.

Compounding the absence of adequate sanitation in large areas of the globe is the lack of an effective cholera vaccine. The killed, parenteral cholera vaccine in current use gives only limited protection of short duration and does not prevent asymptomatic infections (Waldor and Mekalanos, 1996b). This vaccine is no longer recommended for use by the World Health Organization (WHO, 1991). The development of an efficacious cholera vaccine faces several problems. Natural infection does not provide complete long-lasting immunity against reinfection with the El Tor pandemic strain of *V. cholerae* O1 (Clemens et al., 1991). Furthermore, the emergence of epidemic cholera due to *V. cholerae* O139 suggests that the synthesis of variant O-group antigens by fully virulent vibrio strains may also permit the evolution of *V. cholerae* that are resistant to vaccine-induced immunity. Despite these barriers, an active cholera vaccine research program is in the process of developing numerous potential vaccine candidates. Presently, cholera vaccine development is focusing on peroral administration of live attenuated *V. cholerae* strains or inactivated preparations of vibrios and/or subunit preparations (cholera toxin, LPS, TCP) in order to stimulate a secretory immune response to relevant protective antigens. Two oral cholera vaccine candidates (inactivated WC/rBS and live CVD 103 HgR) have recently been marketed in Europe (Steffen, 1994). WC/rBS consists of 1 mg of recombinant cholera toxin B subunit and 2.5×10^{10} each of heat-killed classical Inaba and classical Ogawa vibrios and the same number of formalin-killed El Tor Inaba and classical Ogawa vibrios (Jertborn et al., 1992). A recent field trial of WC/rBS in Peruvian adults indicated that this vaccine preparation is highly immunogenic and safe (Sanchez et al., 1994).

The long-term efficacy of WC/rBS in Peruvian adults and children is currently being evaluated. Wc/rBS is recommended by the WHO for use in cholera prevention or control in emergency situations. CVD103HgR is a derivative of classical strain 569B, which contains a deletion within *ctxA* and carries a mercury-resistance marker introduced into the *hlyA* gene (Levine et al., 1988). Human challenge studies demonstrate that CVD103HgR is highly immunogenic and safe for humans (Levine et al., 1988). This strain provides high levels of protection against reinfection by the classical strains of *V. cholerae* and lower levels of protection against reinfection by El Tor strains (Levine and Kaper, 1993). CVD103HgR is currently undergoing large-scale field trials in Indonesia.

A major concern regarding the development of live-attenuated vaccine strains is the reactogenicity of many recombinant *V. cholerae* derivatives. One possible solution to this problem is the engineering of motility defects into candidate vaccine strains. For example, in human volunteer studies reduced vibrio motility is associated with a lack of reactogenicity. Importantly, these non-motile derivatives retain the ability to colonize the intestine and impart protection from challenge with wild-type *V. cholerae* (Kenner et al., 1995; Waldor and Mekalanos 1996b).

The results obtained from human volunteer studies indicate that safe and effective vaccines for *V. cholerae* O1 and O139 are now available. The new oral cholera vaccines will be more useful in controlling endemic cholera rather than epidemic cholera, since their efficacy against asymptomatic cholera appears limited. As information from field studies using new-generation cholera vaccines becomes available, a more refined approach to the construction of vaccines will likely yield cost-effective products with increased efficacy. Future cholera vaccines may also be useful vectors for the delivery of heterologous antigens to the intestinal mucosal immune system.

7. *VIBRIO PARAHAEMOLYTICUS*

Vibrio parahaemolyticus causes an acute illness characterized by severe cramping, abdominal pain, vomiting and watery to bloody diarrhea following ingestion of raw or incompletely cooked seafood. The disease is usually self-limiting, requiring no treatment other than restoration of water and electrolytes (Blake et al., 1980; Joseph et al., 1982). *V. parahaemolyticus* is found in coastal waters throughout the world and is associated with common-source outbreaks involving shellfish. In countries such as Japan, where raw fish is commonly eaten, *V. parahaemolyticus* accounts for a significant percentage of all diarrheal disease (Janda et al., 1988). *V. parahaemolyticus* gastroenteritis is associated with the synthesis of a thermostable direct hemolysin (TDH), which produces beta hemolysis on agar containing human erythrocytes. This hemolytic reaction is termed the Kanagawa phenomenon and almost all strains isolated from patients with gastroenteritis are Kanagawa-positive (KP+) (Miyamato et al., 1969; Nishibuchi and Kaper, 1995). Although a large amount of information is available regarding TDH, very little is known about other *V. parahaemolyticus* virulence factors.

7.1. *V. PARAHAEMOLYTICUS* IN THE ENVIRONMENT

V. parahaemolyticus is a halophilic microorganism that is widely distributed in marine environments. These vibrios can be isolated from sediment, suspended particles, plankton, fish and shellfish (Oliver and Kaper, 1997). Isolation of *V. parahaemolyticus* from the environment is rare when water temperatures fall below 13–15°C (Doyle, 1990). Freshly harvested seafood from waters containing *V. parahaemolyticus* typically contains 10^2 organisms per gram. It is not uncommon, however, to find 10^3 organisms per gram in market shellfish during the summer (Doyle, 1990). Heating seafood at 60°C for 15 minutes kills *V. parahaemolyticus*. Refrigeration of *V. parahaemolyticus*-contaminated food at 4°C or freezing also results in decreased numbers of vibrios (Doyle, 1990). Thus, refrigeration of seafood before and after cooking is an effective method of preventing *V. parahaemolyticus* gastroenteritis.

Interestingly, most strains of *V. parahaemolyticus* isolated from foods and marine environments are Kanagawa-negative (KP–). Strains isolated from patients with gastroenteritis, however, are usually Kanagawa-positive (KP+). It has been suggested that KP– strains survive better in the environment, while KP+ cells survive better in the gastrointestinal tract (Oliver and Kaper, 1997). The basis for this apparent correlation between the KP phenotype and survival in different environments is presently unknown.

7.2. *VIBRIO PARAHAEMOLYTICUS* VIRULENCE FACTORS

7.2.1. Thermostable Direct Hemolysin

Almost all *V. parahaemolyticus* strains isolated from patients with gastroenteritis synthesize a thermostable direct hemolysin (TDH). Experimental infection of human volunteers with KP+ and KP– strains of *V. parahaemolyticus* reveal that ingestion of 2×10^5 KP+ cells causes gastroenteritis while ingestion of 1.6×10^{10} KP– cells does not result in gastrointestinal illness (Oliver and Kaper, 1997). KP+ strains induce fluid accumulation in rabbit ileal loops while mutant KP– strains do not (Nishibuchi et al., 1992).

The thermostable direct hemolysin is a 23 kDa protein that exhibits hemolytic activity on various red blood cells. This activity is not inactivated by heating at 100°C for 10 minutes. TDH also exhibits enterotoxigenicity, cytotoxicity, cardiotoxicity, as well as lethality in mice and induces increased vascular permeability in rabbit skin (Honda and Tida, 1993). In vitro experiments suggest that TDH alters ion flux in intestinal cells leading to a secretory diarrhea. Studies employing rabbit ileal tissue mounted in Ussing chambers and *V. parahaemolyticus* culture supernatants demonstrate that nanogram levels of TDH can alter ion transport in enterocytes in the absence of histological changes. Purified TDH induces chloride ion secretion using Ca^{2+} as an intracellular second messenger (Raimondi et al., 1995). The cellular receptor for TDH appears to be trisialoganglioside GT_{1b} (Oliver and Kaper, 1997).

V. parahaemolyticus strains that are KP+ typically contain two nonidentical copies of the *tdh* gene (*tdh*1 and *tdh*2). Weakly hemolytic strains usually possess only a single copy of the *tdh* gene and nonhemolytic strains generally lack *tdh* genes. Low-level expression of *tdh* gene(s) is associated with the KP– and/or KP+/– phenotype of *tdh*-bearing *V. parahaemolyticus* isolates. Greater than 90% of the TDH isolated from a KP+ strain bearing *tdh*1 and *tdh*2 is synthesized from the *tdh*2 gene, suggesting that *tdh*2 is preferentially expressed. Expression studies measuring steady-state levels of *tdh* mRNA confirm that *tdh*2 is expressed at higher levels than *tdh*1 (Nishibuchi and Kaper 1995). As is the case with cholera toxin, TDH synthesis is stimulated by culture medium with an acidic pH. *V. parahaemolyticus* possesses ToxRS homologs, which are responsible for the environmentally controlled expression of the *tdh*2 gene. The Vp-*toxRS* operon is structurally and functionally similar to the *V. cholerae toxRS* operon and appears to control the synthesis of proteins other than TDH (Lin et al., 1993). Unlike *tdh* genes, the Vp-*toxRS* operon is found in both clinical and environmental isolates, suggesting that *V. parahaemolyticus* acquired this master regulatory locus prior to acquisition of *tdh* genes.

Hybridization studies demonstrate that genes highly homologous to the *tdh* genes of *V. parahaemolyticus* are also found in non-O1 strains of *V. cholerae, Vibrio hollisae* and *Vibrio mimicus* (Nishibuchi and Kaper, 1995). The presence of *tdh* in other vibrio species is not surprising since the *tdh* genes are flanked by terminal-inverted repeats (ISVs) related to IS*102*/IS*109* (Nishibuchi and Kaper, 1995). It seems likely, therefore, that the acquisition of the *tdh* genes by these other vibrios occurred as the result of horizontal transfer.

7.2.2. Thermostable Related Hemolysin

Some clinical isolates of *V. parahaemolyticus* that are KP– produce a 23 kDa hemolysin (TRH) that is immunologically related to TDH (Honda and Miwatoni, 1988). Although these proteins are closely related, the sensitivities of various erythrocytes to lysis by TDH and TRH are different and TRH is not as heat stable as TDH. The *trh* gene shares 69% identity with the *tdh2* gene. It is thought that the *tdh* and *trh* evolved from a common ancestor. Expression of *trh* appears to be at a much lower level than that of *tdh* from

KP+ strains of *V. parahaemolyticus* and does not appear to be controlled by VP-ToxRS (Nishibuchi and Kaper, 1995). Epidemiological evidence suggests that TRH is associated with gastroenteritis, and thus TRH may be a cause of diarrhea in humans from which only KP– strains of *V. parahaemolyticus* are isolated.

8. *VIBRIO VULNIFICUS*

Vibrio vulnificus causes three types of disease in humans—primary septicemia, gastroenteritis and wound infections. Primary septicemia is the most common of these diseases and usually follows the ingestion of raw oysters. *V. vulnificus*-associated septicemia is rapidly fatal in 50–60% of cases. Nearly all systemic infections occur in persons with chronic diseases, including liver disease and alcoholism. Individuals suffering from immunosuppression and persons with conditions that lead to increased serum iron levels are also susceptible to systemic infections with *V. vulnificus* (Morris and Black, 1985). Wound infections caused by *V. vulnificus* are usually associated with seawater and range from relatively benign infections to a severe and rapidly progressive cellulitis that is often fatal. *V. vulnificus* is a common inhabitant of estuarine environments worldwide, but most of the reported human cases are in the United States (Oliver and Kaper, 1997). Indeed, *V. vulnificus* is responsible for 95% of all seafood-related deaths in the United States, and is the leading cause of foodborne-related deaths in Florida (Hlady et al., 1993). Numerous extracellular products and cell surface structures are thought to be important in the pathogenesis of *V. vulnificus* infections. These include hemolysin, protease, elastase, siderophores, lipopolysaccharide, capsule and outer-membrane proteins (Oliver and Kaper, 1997).

8.1. *VIBRIO VULNIFICUS* IN THE ENVIRONMENT

V. vulnificus is part of the normal microflora of estuarine waters and molluscan shellfish around the world (Oliver et al., 1983). *V. vulnificus* has also been isolated from the intestines of bottom-feeding fish. *V. vulnificus* exhibits a marked seasonal distribution, with the highest levels of these organisms found during the summer months typically between May and October when the water temperature is greater than 20°C and salinity is 5 to 15 parts per thousand (ppt). During this period there is also a higher frequency of isolation of *V. vulnificus* from oysters and crabs (Wright et al., 1996). It is reported that greater than 50% of oyster lots and 11% of crab lots test positive for *V. vulnificus* during certain times of the year (Laughlin and Lavery, 1995). Human infections occur when *V. vulnificus* levels in oysters exceed 10^3/g of meat, indicating that ingestion of 10^5 virulent *V. vulnificus* is sufficient to cause human infection (Oliver and Kaper 1997).

Recent pulsed-field gel electrophoresis studies examining the genetic diversity of *V. vulnificus* in the environment demonstrate that a dozen oysters can contain over 100 different strains and that only certain subsets of these strains appear to be associated with human disease. In fact, restriction fragment length polymorphism analysis indicates that human mortality following ingestion of oysters contaminated with a heterogeneous population of *V. vulnificus* can be the result of infection with a single strain (Jackson et al., 1997).

8.2. *VIBRIO VULNIFICUS* VIRULENCE FACTORS

8.2.1. Iron Acquisition

Clinical isolates of *V. vulnificus* exhibit a high level of virulence in experimental animals. *V. vulnificus* is rapidly lethal in mice when injected intraperitoneally or subcutaneously, and injection of these vibrios into ligated ileal loops leads to bacteremia and death. Iron-loading of experimental animals enhances the virulence of *V. vulnificus* such that the median lethal dose in iron-loaded mice is one vibrio cell whereas in untreated mice the median lethal dose is 10^6 vibrios. These data are consistent with the observation that elevated serum iron levels contribute to the virulence of *V. vulnificus* in human infections (Oliver and Kaper, 1997). *V. vulnificus* is capable of producing two different siderophores, vulnibactin, a phenolate type siderophore, and an unnamed hydroxamate type siderophore. *V. vulnificus* is not able to grow in human serum, suggesting that it is unable to acquire iron from transferrin. A recent study, however, shows that vulnibactin-producing strains of *V. vulnificus* utilize transferrin- and lactoferrin-bound iron following cleavage of these proteins by the *V. vulnificus* exocellular protease (Okujo et al., 1996). It is not known if vibrio-directed degradation of iron-binding proteins represents a biologically relevant mechanism of iron uptake within the host.

8.2.2. Hemolysin/Cytolysin

Both environmental and clinical isolates of *V. vulnificus* produce a heat-stable protein that is lethal in nanogram quantities when injected into mice. This protein exhibits hemolytic activity against a variety of mammalian erythrocytes and cytolytic activity against tissue culture cells, and increases vascular permeability when placed on guinea pig skin (Miyoshi et al., 1993). The role of the cytolysin in the pathogenesis of *V. vulnificus* infection, if any, is unclear since strains lacking the ability to synthesize this protein are as virulent in mice as wild-type *V. vulnificus*.

8.2.3. Capsule

V. vulnificus synthesizes a polysaccharide capsule that enhances its ability to initiate infection. Strains unable to synthesize a capsule due to insertion mutations exhibit an LD_{50} that is 10,000 times higher than wild-type encapsulated strains. Capsule production affords *V. vulnificus* protection from the bactericidal activity of serum complement and also provides resistance to phagocytosis (Oliver and Kaper, 1997). Furthermore, only encapsulated *V. vulnificus* cells are capable of utilizing transferrin-bound iron. Recent studies utilizing *V. vulnificus* capsular polysaccharide conjugated to tetanus toxoid demonstrate the protective efficacy of this vaccine in murine models of *V. vulnificus* infection (Devi et al., 1995).

9. *VIBRIO ALGINOLYTICUS*

V. alginolyticus is a halophilic species that is distributed in coastal waters worldwide and has been isolated from shellfish. *V. alginolyticus* is occasionally isolated from patients with gastroenteritis but is more commonly associated with superficial wounds and otitis media. Persons infected with *V. alginolyticus* typically have had direct contact with water. In immunocompromised individuals, *V. alginolyticus* infection can lead to a fatal bacteremia. *V. alginolyticus* produces numerous extracellular enzymes, none of which has been implicated in the pathogenesis of *V. alginolyticus* infection (Oliver and Kaper, 1997).

10. *VIBRIO FLUVIALIS*

V. fluvialis is a bacterium frequently isolated from brackish and marine waters and is occasionally associated with gastroenteritis following the ingestion of raw oysters. *V. fluvialis* can also be isolated from fish, shellfish and freshwater clams. The clinical symptoms of *V. fluvialis* gastroenteritis are similar to cholera with the additional finding of bloody stools, which is suggestive of an invasive pathogen (Oliver and Kaper, 1997). Oral challenge of infant mice with *V. fluvialis* causes diarrhea and death. *V. fluvialis* produces an uncharacterized heat-labile enterotoxin and several other potential enterotoxins. These putative toxins are identified using Chinese hamster ovary (CHO) cells and characterized as CHO cell cytotoxin, CHO cell-rounding toxin and CHO cell-elongation factor. Partially purified preparations of each of these three proteins elicit fluid accumulation in infant mice (Lockwood et al., 1982). Information is lacking regarding the precise role of these putative virulence factors in the pathogenesis of *V. fluvialis* infection.

11. *VIBRIO HOLLISAE*

V. hollisae is a halophilic vibrio that is infrequently associated with acute gastrointestinal disease and septic infections in immunodeficient individuals. *V. hollisae* has been isolated from oysters and coastal fish. *V. hollisae* infection typically follows the consumption of raw seafood (Oliver and Kaper, 1997). *V. hollisae* is unusual in that it fails to grow on MacConkey and TCBS agar, two media routinely used to culture stool samples for vibrios (Oliver and Kaper, 1997). *V. hollisae* synthesizes a thermostable direct hemolysin that is related to the TDH of *V. parahaemolyticus* (Nishibuchi and Kaper, 1995). It also produces a heat-labile enterotoxin that elongates CHO cells and causes fluid accumulation in the intestines of suckling mice (Kothary and Richardson, 1987). *V. hollisae* has also been reported to invade tissue culture cells (Miliotis et al., 1995).

12. *VIBRIO MIMICUS*

V. mimicus, previously referred to as sucrose-negative *V. cholerae* non-O1, is recognized as a new species (Shandera et al., 1983). *V. mimicus* is mainly associated with gastroenteritis following ingestion of raw seafood and occasionally isolated from patients with ear infections. A recent report from Costa Rica also discovered an association between the ingestion of raw turtle eggs and *V. mimicus* diarrheal illness (Campos et al., 1996). *V. mimicus* is found in both freshwater and brackish water with an average of 4 ppt being optimum for cultivation. This organism is not found in waters with a salinity of greater than 10 ppt. *V. mimicus* has been isolated from fish, oysters, shrimp sediments, plankton and the roots of aquatic plants (Oliver and Kandel, 1997).

Ten to 35% of *V. mimicus* clinical isolates produce a heat-labile protein that is identical to cholera toxin whereas only 1% of environmental *V. mimicus* isolates produce cholera toxin (Chowdhury et al., 1987). The epidemiological relevance of these toxigenic environmental isolates in the pathogenesis of gastrointestinal disease is not known. Some strains of *V. mimicus* are also capable of producing a heat-stable enterotoxin that is similar to those produced by non-O1 *V. cholerae* and *V. fluvialis* (Oliver and Kandel, 1997). The role of these enterotoxins in human gastroenteritis has not been examined in any detail. Furthermore, most virulent strains of *V. mimicus* do not produce any of these enterotoxins.

V. mimicus has also been reported to produce two hemolysins, one heat labile and the other heat stable. The heat-labile hemolysin (VMH) is immunologically related to the *V. cholerae* hemolysin and causes hemolysis via the formation of a transmembrane pore and subsequent disruption of the erythrocyte membrane (Honda et al., 1987). Purified VMH induces bloody fluid accumulation in ligated rabbit ileal loops. Antibodies directed against VMH reduce the enteropathogenicity of *V. mimicus* infection in ligated rabbit ileal loops (Miyoshi et al., 1997). The heat-stable hemolysin produced by *V. mimicus* is nearly identical to the thermostable direct hemolysin of *V. parahaemolyticus.* Most clinical isolates of *V. mimicus* produce thermostable direct hemolysin while environmental isolates do not (Oliver and Kandel, 1997).

A recent study examining the intestinal adhesiveness of environmental isolates of *V. mimicus* reported a correlation between erythrocyte agglutination and vibrio attachment to isolated intestinal tissue. Adherence to human intestinal epithelial cells by *V. mimicus* also correlates with hemagglutination. Elucidation of the precise role of hemagglutination in vibrio attachment to the intestinal mucosa is complicated by the presence of three distinct hemagglutinins—metalloprotease (Vm-HA/protease), lipopolysaccharide (Vm-LPSHA) and 39 kDa outer-membrane protein (Vm-OMPHA), all three of which may play a role in intestinal adherence (Alam et al., 1997).

13. CONCLUSIONS

Human illness caused by foodborne vibrios is a significant public health problem that will not be easily controlled. This is especially true for the halophilic non-cholera *Vibrio* spp., whose presence in the environment is not related to contamination via human waste. Currently, thorough heating of seafood appears to be the only effective means of preventing vibrio infections. Research aimed at improving our understanding of the susceptibility of vibrios to basic food-preservation methods is needed so that more effective methods of preventing or controlling contamination of foodstuffs can be developed. The application of modern genetics to the study of bacterial virulence has led to a tremendous increase in our knowledge of *V. cholerae* virulence mechanisms. Because of this increased understanding of *V. cholerae* virulence determinants, the production of a safe and efficacious cholera vaccine appears to be at hand. Additional advances leading to a more thorough understanding of *V. cholerae*-host interactions will no doubt lead to further improvements in vaccine development. In contrast, the power of bacterial genetics needs to be better utilized with regard to understanding the role of putative toxins and other postulated virulence determinants in the pathogenesis of the non-cholera *Vibrio* infections. The lack of well-characterized genotypic and/or phenotypic markers for the various *Vibrio* spp. has hindered the development of modern tests that can distinguish potentially pathogenic vibrios from non-pathogenic environmental strains. It is obvious that more detailed information is needed regarding the genetic and pathogenic diversity of virulent *Vibrio* strains found in naturally colonized shellfish and their ecological niches before rational policies regarding the risk assessment of vibrio-contaminated food can be developed.

14. REFERENCES

Alam, M., Miyoshi, S., Tomochika, K., and Shinoda, S. 1997. "*Vibrio mimicus* attaches to the intestinal mucosa by outer membrane hemagglutinins specific to polypeptide moieties of glycoproteins," *Infection Immunity* 65:3662–3665.

Albert, M. J., Siddique, A. K., Islam, M. S. et al. 1993. "Large outbreak of clinical cholera due to *Vibrio cholerae* non-O1 in Bangladesh," *Lancet* 341:704.

Attridge, S. R. and Rowley, D. 1983. "The role of the flagellum in the adherence of *Vibrio cholerae,*" *J. Infectious Diseases* 147:864–872.

Attridge, S. R., Manning, P. A., Holmgren, J., and Jonson, G. 1996. "Relative significance of mannose-sensitive hemagglutinin and toxin-coregulated pili in colonization of infant mice by *Vibrio cholerae* El Tor," *Infection and Immunity* 64(8):3369–3373.

Bik, E. M., Bunschoten, A. E., Gouw, R. D., and Mooi, F. R. 1996. "Genesis of the novel epidemic *Vibrio cholerae* O139 strain: Evidence for horizontal transfer of genes involved in polysaccharide synthesis," *EMBO Journal* 20:799–811.

Blake, P. A. 1993. "Epidemiology of cholera in the Americas," *Gastroenterology Clinics of North America,* 22:639–660.

Blake, P. A., Weaver, R. E., and Hollis, D. G. 1980. "Diseases of humans (other than cholera) caused by vibrios," *Annual Review Microbiology* 34:341–367.

Brown, R. C. and Taylor, R. K. 1995. "Organization of *tcp, acf,* and *toxT* genes within a ToxT-dependent operon," *Molecular Microbiology* 16:425–439.

Butterton, J. R. and Calderwood, S. B. 1995. "*Vibrio cholerae* O1," in *Infections of the Gastrointestinal Tract,* eds., M. J. Blaser, P. D. Smith, J. I. Ravdin, H. B. Greenberg and R. L. Guerrant. New York: Raven Press. pp. 649–670.

Campos, E., Bolanos, H., Acuna, M. T., Diaz, G., Matamoros, M. C., Raventos, H., Sanchez, L. M., Sanchez, O., and Barquero, C. 1996. "*Vibrio mimicus* diarrhea following ingestion of raw turtle eggs," *Applied Environmental Microbiology* 62:1141–1144.

Carrol, P. A., Tashima, K. T., Rogers, M. B., DiRita, V. J., and Calderwood, S. B. 1997. "Phase variation in *tcpH* modulates expression of the ToxR Regulon in *Vibrio cholerae,*" *Molecular Microbiology* 25:1099–1111.

Champion, G. A., Neely, M. N., Brennan, M. A., and DiRita, V. J. 1997. "A branch of the ToxR regulatory cascade of *Vibrio cholerae* revealed by characterization of toxT mutant strains," *Molecular Microbiology* 23:323–331.

Chiang, S. L., Taylor, R. K., Koomey, M., and Mekalanos, J. J. 1995. "Single amino acid substitutions in the N-terminus of *Vibrio cholerae* TcpA affect colonization," *Molecular Microbiology* 17(6):133–1142.

Chowdhury, M. A. R., Aziz, K. M. S., Kay, B. A., and Rahim, Z. 1987. "Toxin production by *Vibrio mimicus* strains isolated from human and environmental sources in Bangladesh," *J. Clinical Microbiology* 25:2200–2203.

Clemens, J. D., Van Loon, F., Sack, D. A., Rao, M. R., Ahmed, J., Chakraborty, B. A., Kay, B. A., Khan, M. R., Yunus, M. D., Harris, J. R., Svennerholm, A. M., and Holmgren, J. 1991. "Biotype as a determinant of natural immunizing effect of *V. cholerae* O1: Implications for vaccine development," *J. Infectious Diseases* 23:473–479.

Colwell, R. R. and Spira, W. M. 1992. "The ecology of *Vibrio cholerae,*" in *Cholera,* eds., D. Barua and W. B. Greenough. New York: Plenum Press. pp. 1–36.

Comstock, L. E., Maneva, L. D., Panigrahi, P., Joseph, A., Levine, M. M., Kaper, J. B., Morris, J. G., and Johnson, J. A. 1995. "The capsule and O antigen in *Vibrio cholerae* O139 Bengal are associated with a genetic region not present in *Vibrio cholerae* O1,'' *Infection and Immunity* 63:317–323.

Devi, S. J. N., Hayat, U., Frasch, C. E., Kreger, A. S., and Morris, J. G. 1995. ''Capsular polysaccharide-protein conjugate vaccines of carbotpe 1 *Vibrio vulnificus:* Construction, immunogenicity, and protective efficacy in a murine model,'' *Infection Immunity* 63:2906–2911.

DiRita, V. J. 1992. ''Co-ordinate expression of virulence genes by ToxR in *Vibrio cholerae,'' Molecular Microbiology* 6:451–458.

DiRita, V. J. and Mekalanos, J. J. 1991 ''Periplasmic interaction between two membrane regulatory proteins, ToxR and ToxS, results in signal transduction and transcriptional activation,'' *Cell* 164:29–37.

DiRita, V. J., Neely, M., Taylor, R. K., and Bruss, P. M. 1996. ''Differential expression of the ToxR regulon in classical and El Tor biotypes of *Vibrio cholerae* is due to biotype-specific control over *taxT* expression,'' *Proc. Natl. Acad. Sci. USA* 93:7991–7995.

DiRita, V. J., Parsot, C., Jander, G., and Mekalanos, J. J. 1991. ''Regulatory cascade controls virulence in *Vibrio cholerae,'' Proc. Natl. Acad. Sci. USA* 88:5403–5407.

Doyle, M. P. 1990. ''*Pathogenic Escherichia coli. Yersinia enterocolitica,* and *Vibrio parahaemolyticus,'' Lancet* 336:1111–1115.

Estrada-Garcia, T. and Mintz, E. D. 1996. ''Cholera: Foodborne transmission and its prevention,'' 461–469.

Evans, D. J. and Richardson, S. H. 1968. ''In vitro production of choleragen and vascular permeability factor by *Vibrio cholerae,'' J. Bacteriology,* 96:126–130.

Everiss, K. D., Hughes, K. J., Kovach, M. E., and Peterson, K. M. 1994. ''The *Vibrio cholerae acfB* colonization determinant encodes an inner membrane protein that is related to a family of signal-transducing proteins,'' *Infection and Immunity* 62:3289–3298.

Fasano, A., Baudry, B., Pumplin, D. W., Wasserman, S. S., Tall, B. D., Ketley, J. M., and Kaper, J. B. 1991. ''*Vibrio cholerae* produces a second enterotoxin, which affects intestinal tight junctions,'' *Proc. Natl. Acad. Sci. USA* 88:5242–5246.

Field, M. 1980. ''Regulation of small intestine ion transport by cyclic nucleotides and calcium,'' in *Secretroy Diarrhea,* eds., M. Field, J. S. Fordtran, and S. G. Schultz. Bethesda, MD: American Physiology Society. p. 21.

Finkelstein, R. A. 1992. ''Cholera enterotoxin (choleragen): A historical perspective,'' in *Cholera* eds., D. Barua, W. B. Greenough. New York: Plenum Press. pp. 1–36.

Franzon, V. L., Barker, A., and Manning, P. A. 1993. ''Nucleotide sequence encoding the mannose-fucose-resistant hemagglutinin of *Vibrio cholerae* O1 and construction of a mutant,'' *Infection and Immunity* 61:3031–3037.

Freter, R., Allweiss, B., O'Brien, P. C. M., Halstead, S. A., and Macsai, M. S. 1981. ''Role of chemotaxis in the association of motile bacteria with intestinal mucosa: In vitro studies,'' *Infection and Immunity* 34:241–249.

Gardel, C. L. and Mekalanos, J. J. 1996. ''Alterations in *Vibrio cholerae* motility phenotypes correlate with changes in virulence factor expression,'' *Infection and Immunity* 64:2246–2255.

Gill, D. M. 1976. ''The arrangement of subunits in cholera toxin,'' *Biochemistry* 15:1242–1248.

Gill, D. M. and King, C. A. 1975. ''The mechanism of action of cholera toxin in pigeon erythrocyte lysates,'' *J. Biological Chemistry* 250:6224–6432.

Gill, D. M. and Meren, R. 1978. ''ADP-ribosylation of membrane proteins catalyzed by cholera toxin: Basis of the activation of adenylate cyclase,'' *Proc. Natl. Acad. Sci. USA* 75:3050–3054.

Gill, D. M. and Rappaport, R. S. 1977. ''The origin of A1,'' in *Proceedings of the 12th Joint Conference on Cholera.* US-Japan Cooperative Medical Science Program, Japan, Bethesda, MD: National Institutes of Health.

Goldberg, M. B., DiRita, V. J., and Calderwood, S. B. 1990. "Identification of an iron-regulated virulence determinant in *Vibrio cholerae,* using Tn*phoA* mutagenesis," *Infection and Immunity* 58:55–60.

Griffiths, G. L., Sigel, S. M., Payne, S. M., and Neilands, J. B. 1984. "Vibriobactin, a siderophore from *Vibrio cholerae,*" *J. Biological Chemistry* 259:383–385.

Guentzel, M. N. and Berry, L. J. 1975. "Motility as a virulence factor for *Vibrio cholerae,*" *Infection and Immunity* 11:890–897.

Harkey, C. W., Everiss, K. D., and Peterson, K. M. 1994. "The *Vibrio cholerae* toxin-corregulated pilus gene *tcpI* encodes a homolog of methyl-accepting chemotaxis proteins," *Infection and Immunity* 62:2669–2678.

Hase, C. and Mekalanos, J. J. 1999. "Effects of changes in membrane sodium flux on virulence gene expression in *Vibrio choleras.*" *Proc. Natl. Acad. Sci. USA* 96:3183–3187.

Henderson, D. P. and Payne, S. M. 1994. "*Vibrio cholerae* Iron transport systems: Roles of Heme and siderophore iron transport in virulence and identification of a gene associated with multiple iron transport systems," *Infection and Immunity* 62:5120–5125.

Herrington, D. Q., Hall, R. H., Losnosky, G. A., Mekalanos, J. J., Taylor, R. K., and Levine, M. M. 1988. "Toxin, toxin-coregulated pili and the *toxR* regulon are essential for *Vibrio cholerae* pathogenesis in humans," *J. Experimental Medicine* 168:1487–1492.

Higgins, D. E., Nazareno, E., and DiRita, V. J. 1992. "The virulence gene activator ToxT from *Vibrio cholerae* is a member of the arac family of transcriptional activators," *J. Bacteriology* 174:6974–6980.

Hirst, T. R. 1991. "Assembly and secretion of oligomeric toxins by gram-negative bacteria" in *Sourcebook of Bacterial Protein Toxins,* eds., J. E. Alouf and J. H. Freer. London: Academic Press. pp. 75–100.

Hlady, W. G., Mullen, R. C., and Hopkin, R. S. 1993. "*Vibrio vulnificus* from raw oysters. Leading cause of reported deaths from foodborne illness in Florida," *J. Florida Medical Association* 80:536–538.

Holmgren, J., Lonnroth, I., Mansson, J.-E., and Svennerholm, L. 1975. "Interaction of cholera toxin and membrane G_{MI} ganglioside of small intestine," *Proc. Natl. Acad. Sci. USA* 72:2520–2524.

Honda, T. and Finkelstein, R. A. 1979. "Purification and characterization of a hemolysin produced by *Vibrio cholerae* biotype El Tor: Another toxic substance produced by cholera vibrios," *Infection and Immunity* 26:1020–1027.

Honda, T. and Iida, T. 1993. "The pathogenicity of *Vibrio parahaemolyticus* and the role of the thermostable direct hamolysin and related haemolysins," *Reviews Medical Microbiology* 4:106–113.

Honda, T., Narita, I., Yoh, M., and Miwatani, T. 1987. "Purification and properties of two hemolysins produced by *Vibrio mimicus,*" *Japanese J. Bacteriology* 42:201.

Honda, T., Ni, Y., and Miwatani, T. 1988. "Purification and characterization of a hemolysin produced by a clinical isolate of Kanagawa phennomenon-negative *Vibrio parahaemolyticus* and related to the thermostable direct hemolysin," *Infection Immunity* 56:961–965.

Hughes, K. J., Everiss, K. D., Kovach, M. E., and Peterson, K. M. 1994. "Sequence analysis of the *Vibrio cholerae acfD* gene reveals the presence of an overlapping reading frame, *orfZ* which encodes a protein that shares sequence similarity to the FliA and FliC products of *Salmonella,*" *Gene* 146:79–82.

Hughes, K. J., Everiss, K. D., Kovach, M. E., and Peterson, K. M. 1995. "Isolation and characterization of the *Vibrio cholerae acfA* gene, required for efficient intestinal oblonization," *Gene* 156:59–61.

Iredell, J. R. and Manning, P. A. 1994. "The toxin-co-regulated pilus of *Vibrio cholerae* O1: A model for type 4 pilus biogenesis?" *Trends in Microbiology* 2:187–192.

Jackson, J. K., Murphree, R. L., and Tamplin, M. L. 1997. "Evidence that mortality from *Vibrio vulnificus* infection results from single strains among heterogenous populations in shellfish," *J. Clinical Microbiology* 35:2098–2101.

Janda, J. M., Powers, C. Bryant, R. G., and Abbott, S. L. 1988. "Current perspective on the epidemiology and pathogenesis of clinically significant *Vibrio* spp," *Clinical Microbiology Reviews* 1:245–267.

Jertborn, M., Svennerholm, A. M., and Holmgren, J. 1992. "Safety and immunogenicity of an oral recombinant cholera B subunit-whole cell vaccine in Swedish volunteers," *Vaccine* 10(2):130–132.

Johnson, J. A., Salles, C. A., Panigrahi, P., Albert, M. J., Wright, A. C., Johnson, R. J., and Morris, J. G. 1994. "*Vibrio cholerae* O139 synonym Bengal is closely related to *Vibrio cholerae* El Tor but has important differences," *Infection and Immunity* 62:2108–2110.

Joseph, S. W., Colwell, R. R., and Kaper, J. B. 1982. "*Vibrio parahaemolyticus* and related halophilic vibrios," *Critical Reviews Microbiology* 10:77–124.

Kaper, J. B., Fasano, A., and Trucksis, M. 1994. "Toxins of *Vibrio cholerae,*" in *Vibrio cholerae and cholera,* eds., I. K. Wachsmuth, P. A. Blake, and O. Olsvik. Washington, D.C.: American Society for Microbiology. pp. 145–176.

Kaufman, M. R., Seyer, J. M., and Taylor, R. K. 1991. "Processing of TCP pilin by TcpJ typifies a common step intrinsic to a newly recognized pathway of extracellular protein secretion by gram-negative bacteria," *Genes & Development* 5:1834–1846.

Kaufman, M. R., Shaw, C. E., Jones, I. D., and Taylor, R. K. 1994. "Biogenesis and regulation of the *Vibrio cholerae* toxin-coregulated pilus: analogies to other virulence factor secretory systems," *Gene* 126:43–49.

Kenner, J., Coster, T., Trofa, A., Taylor, D., Barrera-Oro, M., Hyman, T., Adams, J., Beattie, D., Killeen, K., Mekalanos, J. J., and Sadoff, J. C. 1995. "Peru-15, a live, attenuated oral vaccine candidate for *Vibrio cholerae* 01," *J. Infectious Diseases* 172:1126–1129.

Kothary, M. H. and Richardson, S. H. 1987. "Fluid accumulation in infant mice caused by *Vibrio hollisae* and its extracellular enterotoxin," *Infection Immunity* 55:626–630.

Kovach, M. E., Shaffer, M. D., and Peterson, K. M. 1996. "A putative integrase gene defines the distal end of a large cluster of ToxR-regulated colonization genes in *Vibrio cholerae,*" *Microbiology* 142:2165–2174.

Laughlin, T. J. and Lavery, L. A. 1995. "Lower extremity manifestations of *Vibrio vulnificus* infection," *J. Foot Ankle Surgery* 34:354–357.

Levine, M. M. and Kaper, J. B. 1993. "Live oral vaccines against cholera: An update," *Vaccine* 11:207–212.

Levine, M. M., Black, R. E., Clements, M. L., Cisneros, L., Nalin, D. R., and Young, C. R. 1981. "Duration of infection-derived immunity to cholera," *J. Infectious Diseases* 143:818–820.

Levine, M. M., Kaper, J. B., Herrington, D., Losonsky, G., Morris, J. G., Clements, M. L., Black, R. E., Tall, B., and Hall, R. 1988. "Volunteer studies of deletion mutants of *Vibrio cholerae* O1 prepared by recombinant techniques," *Infection and Immunity* 56:161–167.

Lin, W., Fullner, K. J., Clayton, R., Sexton, J. A., Rogers, M. B., Calia, K. E., Calderwood, S. B., Fraser, C., and Mekalanos, J. J. 1999. "Identification of a *Vibrio cholerae* RTX toxin gene cluster that is tightly linked to the cholera toxin prophage," *Proc. Natl. Acad. Sci. USA* 96:1071–1076.

Lin, Z., Kumagai, K., Baba, K., Mekalanos, J. J., and Nishibuchi, M. 1993. "*Vibrio parahaemolyticus* has a homolog of the *Vibrio cholerae toxRS* operon that mediates environmentally induced regulation of the thermostable direct hemolysin gene," *J. Bacteriology* 175:3844–3855.

Lockwood, D. E., Kreger, A. S., and Richardson, S. H. 1982. "Detection of toxins produced by *Vibrio fluvialis*," *Infection Immunity* 35:702–708.

Mekalanos, J. 1983. Duplication and amplification of toxin genes in *Vibrio cholerae*," *Cell* 35:253–263.

Mekalanos, J. J., Collier, R. J., and Romig, W. R. 1979. "The enzymatic activity of cholera toxin. I. Relationships to proteolytic processing, disulfide bond reduction, and subunit composition," *J. Biological Chemistry* 254:5855–5861.

Mekalanos, J. J., Swartz, D. J., Pearson, G. D. N., Harford, N., Groyne, F., and de Wilde, M. 1983. "Cholera toxin genes: Nucleotide sequence, deletion analysis, and vaccine development," *Nature* 306:551–557.

Miliotis, M. D., Tall, B. D., and Gray, R. T. 1995. Adherence to and invasion of tissue culture cells by *Vibrio hollisae*," *Infection Immunity* 63:4959–4963.

Miller, V. L. and Mekalanos, J. J. 1984. "Synthesis of cholera toxin is positively regulated at the transcriptional level by *toxR*," *Proc. Natl. Acad. Sci. USA* 81:3471–3475.

Miller, V. L. and Mekalanos, J. J. 1988. "A novel suicide vector and its use in construction of insertion mutations; Osmoregulation of outer membrane proteins and virulence determinants in *Vibrio cholerae* requires *toxR*," *J. Bacteriology* 170:2575–2583.

Miller, V. L., DiRita, V. J., and Mekalanos, J. J. 1989. "Identification of *toxS*, a regulatory gene whose product enhances ToxR-mediated activation of the cholera toxin promoter," *J. Bacteriology* 171:1288–1293.

Miller, V. L., Taylor, R. K., and Mekalanos, J. J. 1987. "Cholera toxin transcriptional activator ToxR is a transmembrane DNA binding protein," *Cell* 48:271–279.

Miyamoto, Y., Kato, T., Obara, Y., Akiyama, S., Takizawa, K., and Yamai, S. 1969. "In vitro hemolytic characteristic of *Vibrio parahaemolyticus:* Its close correlation with human pathogenicity," *J. Bacteriology* 100:1147–1149.

Miyoshi, S., Oh, E. G., Hirata, K., and Shinoda, S. 1993. "Exocellular toxic factors produced by *Vibrio vulnificus*," *J. Toxicology Toxin Reviews* 12:253–288.

Miyoshi, S., Sasahara, K., Akamatsu, S., Rahman, M. M., Katsu, T., Tomochika, K., and Shinoda, S. 1997. "Purification and characterization of a hemolysin produced by *Vibrio mimicus*," *Infection Immunity* 65:1830–1835.

Mooi, F. R. and Bik, E. M. 1997. "The evolution of epidemic *Vibrio cholerae* strains," *Trends in Microbiology* 5:161–165.

Morris, J. G. and Black, R. E. 1985. "Cholera and other vibrioses in the United States," *New England J. Medicine* 312:343–350.

Nishibuchi, M. and Kaper, J. B. 1995. "Thermostable direct hemolysis gene of *V. parahaemolyticus:* A virulence gene acquired by a marine bacterium," *Infection Immunity* 63:2093–2099.

Nishibuchi, M., Fasano, A., Russel, R. G., and Kaper, J. B. 1992. "Enterotoxicity of *Vibrio parahaemolyticus* with and without genes encoding thermostable direct hemolysin," *Infection Immunity*, 60:3539–3545.

O'Brien, A. D., Chen, M. E., Holmes, R. K., and Kaper, J. B. 1985. "Environmental and human Isolates of *Vibrio cholerae* and *Vibrio parahaemolyticus* produce a *Shigella dysenteriae* 1 (Shiga)-like cytotoxin," *Lancet*, 1:77–78.

Okujo, N., Akiyama, T., Miyoshi, S., Shinoda, S., and Yamamoto, S. 1996. "Involvement of vulnibactin and exocellular protease in utilization of transferrin- and lactoferrin-bound iron by *Vibrio vulnificus*," *Microbiology Immunology* 40:595–598.

Oliver, J. D. and Kaper, J. B. 1997. "*Vibrio* species," in *Food Microbiology*, eds. M. P. Doyle, L. R. Beuchat, and T. J. Montville. Washington, D. C.: ASM Press. pp. 228–264.

Oliver, J. D., Warner, R. A., and Cleland D. R. 1983. "Distribution of *Vibrio vulnificus* and other lactose-fermenting vibrios in the marine environment,'' *Applied Environment Microbiology* 45:985–998.

Osek, J., Svennerholm, A.-M., and Holmgren, J. 1992. ''Protection against *Vibrio cholerae* El Tor infection by specific antibodies against mannose-binding hemagglutinin pili,'' *Infection and Immunity* 60:4961–4964.

Ottemann, K. M. and Mekalanos, J. J. 1996. ''The ToxR protein of *Vibrio cholerae* forms homodimers and heterodimers,'' *J. Bacteriology,* 178:156–162.

Ottemann, K. M., DiRita, V. J., and Mekalanos, J. J. 1992. ''ToxR proteins with substitutions in residues conserved with OmpR fail to activate transcription from the cholera toxin promoter,'' *J. Bacteriology,* 174:6807–6814.

Parsot, C. and Mekalanos, J. J. 1990. ''Expression of ToxR, the transcriptional activator of the virulence factors in *Vibrio cholerae,* is modulated by the heat shock response,'' *Proc. Natl. Acad. Sci. USA* 87:9898–9902.

Pearson, G. D. N., Woods, A., Chiang, S. L., and Mekalanos, J. J. 1993. ''CTX genetic element encodes a site-specific recombinant system and an intestinal colonization factor,'' *Proc. Natl. Acad. Sci. USA* 90:3750–3754; Correction, 1993. *Proc. Natl. Acad. Sci. USA* 90:8302.

Peek, J. A. and Taylor, R. K. 1992 ''Characterization of a periplasmic thiol:disulfide interchange protein required for the functional maturation of secreted virulence factors of *Vibrio cholerae,*'' *Proc. Natl. Acad. Sci. USA* 89:6210–6214.

Peterson, K. M. and Mekalanos, J. J. 1988. ''Characterization of the *Vibrio cholerae toxR* regulon: Identification of novel genes involved in intestinal colonization,'' *Infection and Immunity* 56:288–289.

Raimondi, D., Kao, J. P. Y., Kaper, J. B., Guandalini, S., and Fasano, A. 1995. ''Calcium-dependent intestinal chloride secretion by *Vibrio parahaemolyticus* thermostable direct hemolysin in a rabbit model,'' *Gastroenterology* 109:381–386.

Ramamurthy, T., Garg, S., Sharma, R., Bhattacharya, S. K., Nair, G. B., Shimada, T., Takeda, T., Karasawa, T., Kurazano, H., Pal, A., and Takeda, Y. 1993. ''Emergence of a novel strain of *Vibrio cholerae* with epidemic potential in Southern and Eastern India,'' *Lancet* 341:703–704.

Rhine, J. A. and Taylor, R. K., 1994. ''TcpA pilin sequences and colonization requirements for O1 and O139 *Vibrio cholerae,*'' *Molecular Microbiology* 13(6):1013–1020.

Richardson, K. 1991. ''Role of motility and flagellar structure in pathogenicity of *Vibrio cholerae*: Analysis of motility mutants in three animal models,'' *Infection and Immunity* 59:2727–2736.

Ries, A. A., Vugia, D. J., Beingolea, L., et al. 1992. ''Cholera in Piura, Peru: a modern urban epidemic,'' *J. Infectious Diseases* 166:1429–1433.

Sanchez, J. L., Vasquez, B., Begue, R. E., Meza, R., Castellares, G., Cabezas, C., Watts, C., Svennerholm, A.-M., Sadoff, J. C., and Taylor, D. N. 1994. ''Protective efficacy of oral whole-cell/recombinant B-subunit cholera vaccines in Peruvian military recruits,'' *Lancet* 344:1273–1276.

Sandkvist, M., Morales, V., and Bagdasarian, M. 1993. ''A Protein required for secretion of cholera toxin through the outer memberane of *Vibrio cholerae,*'' *Gene* 123:81–86.

Sandkvist, M., Bagdasarian, M., Howard, S. P., and DiRita, V. J. 1995. ''Interaction between the autokinase EpsE and EpsL in the cytoplasmic membrane is required for extracellular secretion in *Vibrio cholerae,*'' *EMBO Journal* 14(8):1664–1673.

Shaffer, M. D., Kovach, M. E., and Peterson, K. M. 1996. *Proceedings of the American Society for Microbiology,* May 19–23, 1996. New Orleans, LA, p. 171.

Shandera, W. X., Johnston, J. M., Davis, B. R., and Blake, P. A. 1983. ''Disease from infection with *Vibrio mimicus,* a newly recognized *Vibrio* species,'' *Annals Internal Medicine* 99:169–171.

Shiau, Y., Fernandez, P., Jackson, M. J., and McMonagle, S. 1985. ''Mechanisms maintaining a low-pH microclimate in the intestine,'' *American J. Physiology* 248:G608–G617.

Shimada, T., Nair, G. B., Deb, B. C., Albert, M. J., Sack, R. B., and Takeda, Y. 1993. "Outbreak of *Vibrio cholerae* non-O1 in India and Bangladesh," *Lancet* 341:1346.

Skorupski, K. and Taylor, R. K. 1997a. "Control of the ToxR virulence regulon in *Vibrio cholerae* by environmental stimuli," *Molecular Microbiology* 25:1003–1009.

Skorupski, K. and Taylor, R. K. 1997b. "Cyclic AMP and its receptor protein negatively regulate the coordinate expression of cholera toxin and toxin-coregulated pilus in *Vibrio cholerae,*" *Proc. Natl. Acad. Sci. USA* 94:265–270.

Sperandio, V., Giron, J. A., Silveira, W. D., and Kaper, J. B. 1995. "The OmpU outer membrane protein, a potential adherence factor of *Vibrio cholerae,*" *Infection and Immunity* 63(11):4433–4438.

Steffen, R. 1994. "New cholera vaccines—for whom?" *Lancet* 344:1241–1242.

Stoebner, J. A. and Payne, S. M. 1988. "Iron regulated hemolysin production and utilization of heme and hemoglobin by *Vibrio cholerae,*" *Infection and Immunity* 56:2891–2895.

Sun, D., Mekalanos, J. J., and Taylor, R. K. 1990. "Antibodies directed against the toxin-coregulated pilus isolated from *Vibrio cholerae* provide protection in the infant mouse experimental cholera model," *J. Infectious Diseases,* 161:1231–1236.

Swerdlow, D. L. and Ries, A. A. 1992. "Cholera in the Americas," *JAMA* 267:1495–1499.

Tacket, C. O., Losonsky, G., Nataro, J. P., Cryz, S. J., Edelman, R., Fasano, A., Michalski, J., Kaper, J. B., and Levine, M. M. 1993. "Safety, immunogenicity, and transmissibility of live oral cholera vaccine candidate CVD110, a ΔctxA, Δace derivative of El Tor Ogawa *Vibrio cholerae,*" *J. Infectious Disease* 168:1536–1540.

Tauxe, R. V. and Blake, P. A. 1992. "Epidemic cholera in Latin America," *JAMA* 267:1388–1390.

Tauxe, R. V., Mintz, E. D., and Quick, R. E. 1995. "Epidemic cholera in the new world: Translating field epidemiology into new prevention strategies," *Emerging Infectious Diseases* 1:141–145.

Taylor, R. K., Miller, V. L., Furlong, D. B., and Mekalanos, J. J. 1987. "Use of *phoA* gene fusions to identify a pilus colonization factor coordinately regulated with cholera toxin," *Proc. Natl. Acad. Sci. USA* 84:2833–2837.

Taylor, R. K., Shaw, C., Peterson, K. M., Spears, P., and Mekalanos, J. J. 1988. "Safe, live *Vibrio cholerae* vaccines?" *Vaccine* 6:151–154.

Thelin, K. H. and Taylor, R. K. 1996. "Toxin-coregulated pilus, but not mannose-sensitive hemaglutinin, is required for colonization by *Vibrio cholerae* O1 El Tor biotype and O139 strains," *Infection and Immunity* 64:2853–2856.

Thomas, S, Williams, S. G., and Manning, P. A. 1995. "Regulation of *tcp* genes in classical and El Tor strains of *Vibrio cholerae* O1," *Gene* 166:43–48.

Trucksis, M., Galen, J. E., Michalski, J., Fasano, A., and Kaper, J. B. 1993. "Accessory cholera enterotoxin (Ace), the third toxin of a *Vibrio cholerae* virulence cassette," *Proc. Natl. Acad. Sci. USA* 90:5267–5271.

van Heyningen, W. E., Carpenter, C. C., Pierce, N. F., and Greenough, W. B. 1971. "Deactivation of cholera toxin by ganglioside," *J. Infectious Diseases* 124:415–418.

Waldor, M. K. and Mekalanos, J. J. 1996a. "A new type of conjugative transposon encodes resistance to sulfamethoxazole, trimethoprim, and streptomycin in *Vibrio cholerae,*" *J. Bacteriology* 178(14):4157–4165.

Waldor, M. K. and Mekalanos, J. J. 1996b. "Progress toward live-attenuated cholera vaccines," in *Mucosal Vaccines,* eds., H. Kiyono, P. L. Ogra, and J. R. McGhee. New York: Academic Press, Inc. pp. 229–240.

Waldor, M. K. and Mekalanos, J. J. 1996c. "Lysogenic conversion by a filamentous phage encoding cholera toxin," *Nature* 272:1910–1914.

World Health Organization. 1991. "Guidelines for cholera control," in *Program for Control of Diarrhoeal Disease*. Publication WHO/CDD/SER/80.4 rev2.

Wright, A. C., Hill, R. T., Johnson, J. A., Roghman, M., Colwell, R. R., and Morris, J. G. 1996. "Distribution of *Vibrio vulnificus* in the Chesapeake Bay," *Applied Environmental Microbiology* 62:717–724.

Yoshino, K., Miyachi, M., Takao, T., Bag, P. K. Huang, X., Nair, G. B., Takeda, T., and Shimonishi, Y. 1993. "Purification and sequence determination of heat-stable enterotoxin elaborated by a cholera toxin-producing strain of *Vibrio cholerae* O1," FEBS Letters 326:83–86.

Yu, J., Webb, H., and Hirst, T. R. 1992. "A homologue of the *Escherichia coli* DsbA protein involved in disulfide bond formation is required for enterotoxin biogenesis in *Vibrio cholerae*," *Molecular Microbiology* 6:1949–1958.

CHAPTER 6

Molecular Mechanisms Governing *Campylobacter* Pathogenicity

L. S. MANSFIELD
S. R. ABNER

1. INTRODUCTION

1.1. SIGNIFICANCE OF *CAMPYLOBACTER* INFECTION

DISEASE in humans caused by *Campylobacter* spp. is a serious emerging problem in the United States (CDC, 1994; FSIS, 1995; Buzby and Roberts, 1996; MMWR, 1997; ASM, 1997) and the world (WHO, 1995; Gore, 1996). *Campylobacter* has been targeted as one of the four most important foodborne pathogens in the United States based on the number of reported cases and their severity (CDC, 1994; FSIS, 1995; Buzby and Roberts, 1996; ASM, 1997). Currently, the number of cases of *Campylobacter* enteritis is estimated at 1.1–7 million per year, making *C. jejuni* and *C. coli* the most commonly isolated enterobacterial pathogens (Buzby and Roberts, 1996; ASM, 1997). Enteritis with diarrhea is the most common presenting symptom in humans due to *Campylobacter* spp. (Patton and Wachsmuth, 1992). The World Health Organization Surveillance Program for Control of Foodborne Infections and Intoxications in Europe concludes that *C. jejuni* and *C. coli* are responsible for the majority of cases of *Campylobacter* enteritis (PHLS Communicable Disease Surveillance Centre, 1993). Skirrow and Blaser concur and define *Campylobacter* enteritis and Campylobacteriosis as referring to the two species (Skirrow and Blaser, 1992). The majority of studies suggest that *C. jejuni* predominates over *C. coli* in isolates from clinically affected humans. In 1989, in studies from 45 U.S. states 91% of human clinical isolates were typed and 99% of these were *C. jejuni* (Tauxe, 1992). However, the occurrence of *C. coli* as a food-poisoning agent may be underestimated because *C. coli* is more sensitive to short periods of storage at 4°C and tends to be more sensitive to antibiotics used in transport media (Ng et al., 1985; Madden et al., 1996). Recent molecular diagnostic techniques, including the random amplified polymorphic DNA (RAPD) method, will better define the role of each in these infections (Madden et al., 1996). This review will focus on the pathogenesis of enteric infections caused by both *C. jejuni* and *C. coli* emphasizing recent work. An excellent review of *Campylobacter* work up to and including 1992 is available in the book "*Campylobacter jejuni:* Current Status and Future Trends" (Nachamkin et al., 1992). Other recent reviews of enteric *Campylobacter* infection appeared in Microbiology (Ketley, 1997) and Trends in Microbiology, Vol. 96 (Wooldridge and Ketley, 1997).

1.2. *CAMPYLOBACTER* SPP.

Campylobacter spp. are gram-negative, spiral or curved rods, which exhibit a characteristic corkscrew darting motility mediated by a single polar flagellum. These organisms are slow growing with a generation time of approximately 90 minutes (Rollins and Colwell, 1983), fastidious, and require enriched medium and microaerobic conditions with increased CO_2 (3–15% O_2, 3–10% CO_2, 85% N_2) for growth (On, 1996). A variety of antibiotic-supplemented, selective media are suitable for growth of *Campylobacter* including Butzler's, Skirrow's, and Campy-BAP plating medium. *Campylobacter* spp. are unable to utilize sugars either oxidatively or fermentatively and are extremely sensitive to hydrogen peroxide and superoxide anions that appear in the culture medium when it is exposed to air and light (Humphrey, 1988; Hodge and Krieg, 1994). *Campylobacter* spp. are sensitive to low pH (below pH 5.0), particularly organic acids, drying except under refrigeration, NaCl concentrations exceeding 2%, and long periods at temperatures between 10 and 30°C (Doyle and Roman, 1982a, b, c; Gill and Harris, 1983; Abram and Potter, 1984). In unfavorable growth conditions, spiral rods undergo a degenerate conversion to coccoid forms. Thermophilic *Campylobacters, C. jejuni, C. coli* and *C. lari* grow best at 42°C although they are capable of growth at 37°C (Butzler and Skirrow, 1979). The minimal growth temperature was found to be 31°C, but physiological activity is maintained at 4°C despite the absence of detectable cold shock proteins (Hazeleger et al., 1998). Enrichment is required for the majority of clinical sampling unless material can be transported to the laboratory immediately (Hodge and Terro, 1984). Transport media and storage at 4°C produce the best results when sample transport to the laboratory is delayed.

The genus *Campylobacter* currently includes 15 species and six subspecies with new members being routinely described (On, 1996). The non-*jejuni* and non-*coli* species of *Campylobacter* may also be pathogenic for humans. These include *C. fetus* subsp. *fetus, C. fetus* subsp. *venerealis, C. hyointestinalis, C. lari, C. upsaliensis, C. concisus, C. curvus, C. showae, C. gracilis, C. sputorum,*

C. butzleri, C. rectus, and *C. mucosalis* (Butzler, 1984; Gebhart et al., 1989). In general, these species have less importance for humans than *C. jejuni* or *C. coli.* Further information about their clinical manifestations is available in a recent review (On, 1996). However, at least two of these species, including *C. fetus* subsp. *fetus* and *C. upsaliensis,* are acquired through the gastrointestinal tract, but are most often associated with extraintestinal infections (Guerrant et al., 1976; Mishu et al., 1992). Also, several spp. including *C. concisus, C. curvus, C. showae, C. gracilis,* and *C. rectus* can cause periodontal disease (On, 1996). Some closely related species have been removed from the genus. *Campylobacter*-like organisms (CLO) *Helicobacter cinaedi* (CLO1A) and *H. fennelliae* (CLO2) were recently reclassified as helicobacters (Vandamme et al., 1991). *Campylobacter pylori,* an organism associated with gastric ulcers, also has been reclassified as *Helicobacter pylori* due to subtle differences between it and other members of the genus. Ileal symbiont *intracellularis* originally called *Campylobacter*-like was recently reclassified as an Arcobacter, *Campylobacter nitrofragilis* as *Arcobacter nitrofragilis* and *Campylobacter cryaerophilus* as *Arcobacter cryaerophilus.* Finally, subspecies of *C. jejuni* exist, including *C. jejuni* subsp. *jejuni* and *C. jejuni* subsp. *doylei.* Efforts to characterize the most significant species associated with enteritis in humans are currently being encouraged by the Food Safety Inspection Service and the National Institutes of Health (Emerging Diseases Program NIH, 1996; FSIS, 1995).

1.3. DISEASES ASSOCIATED WITH *CAMPYLOBACTER* SPP.

Campylobacter jejuni can cause a spectrum of diseases including gastroenteritis, proctitis, septicemia, meningitis, abortion, arthritis and Guillain-Barré syndrome (GBS) (On, 1996). Commonly, fever, abdominal pain, and diarrhea commence within 2–3 days following ingestion of food or water contaminated with *C. jejuni* (Wallis, 1994). As few as 800 organisms in improperly cooked poultry, beef, pork, or other foods and water can cause disease (Black et al., 1988). *C. jejuni*-induced diarrhea may mimic diarrhea caused by *Vibrio cholerae* with copious amounts of water excreted in response to toxin production, and/or by *Shigella* with mucus and blood present in the stool due to invasion of cells (Black et al., 1988; Wallis, 1994) or may have symptoms of both. Infection with *C. jejuni* is usually confined to the lower gastrointestinal tract (GI) and is self-limiting within a period of 5–8 days (Black et al., 1988). Systemic disease may ensue due to translocation by monocytes and result in persistence, particularly in immunocompromised hosts (Keihlbauch et al., 1985). This is consistent with in vitro experiments using macrophages from peripheral blood of human donors in which a low percentage of donors had macrophages incapable of killing *C. jejuni* (Wassenaar et al., 1997). *Campylobacter* spp. are rarely isolated from healthy individuals in developed countries

(Taylor and Blaser, 1991), but commonly isolated from nonsymptomatic individuals in developing countries (Echeverria et al., 1989). *C. jejuni* infection can result in severe enteric disease and lead to peritonitis, ileitis, and intestinal obstruction (Perkins and Newstead, 1994). Patients with *Campylobacter* enteritis may eventually require laparotomy due to the severity of abdominal symptoms and signs. The transmural inflammatory changes observed grossly and histologically in these patients can be mistaken for Crohn's disease, a form of inflammatory bowel disease.

C. jejuni has been linked to GBS, which is a debilitating inflammatory polyneuritis that is characterized by fever, pain and weakness that progresses to paralysis, and often results in long-term disability (Kuroki et al., 1991; Fujimoto et al., 1992). *C. jejuni* infection is now known to be a precipitating factor in GBS (Rees et al., 1995a; Rees et al., 1995b). Disease due to *C. jejuni* may result in GBS in 1 out of 1000 people (Smith, 1995). Up to 13% of all GBS patients die, 75% have residual symptoms ranging from foot drag to bedridden status, and fewer than 15% fully recover (Smith, 1995). GBS usually occurs 1–3 weeks after the precipitating event (Nachamkin, 1992). Of patients with this disease, 20 to 40% have a history of *C. jejuni* infection and high anti-*C. jejuni* antibody titers (Mishu and Blaser, 1993; Mishu-Allos, 1997). In these patients, antibodies directed against one or more *C. jejuni* antigens cross-react with antigens present on the surface of neural cells (GM1 ganglioside) (Yuki et al., 1997). An unfavorable prognosis is associated with GBS patients who have documented *C. jejuni* infection and anti-GM1 antibodies, which is improved somewhat with intravenous immune globulin administration (Jacobs et al., 1996). There is a strong relationship between GBS and Penner's serotype 19 (Kuroki et al., 1993; Mishu-Allos et al., 1998), Lior 11, Lau 19 and Lau 3/25 *C. jejuni* (Yuki et al., 1994; Bolton, 1995). *Campylobacter jejuni* serotype 019 strains may now be distinguished from non-019 strains using PCR (Misawa et al., 1998).

Reactive arthritis can follow diarrheic episodes of *Campylobacter* (Nickerson et al., 1990; Peterson, 1994) or asymptomatic exposure to the organism (Maki-Ikola, 1991). Reactive arthritis, or Reiter's syndrome, can also occur after infection with *Yersinia, Salmonella, Shigella, Clostridium difficile* and *Brucella* and is generally thought to be caused by autoimmune response in joints (Gerloni and Fantini, 1990; Salyers and Whitt, 1994). IgM, IgG and IgA class antibodies are increased in patients with a history of *Campylobacter* exposure and arthritis (Maki-Ikola, 1991). This arthritis is suspected to be initiated by molecular mimicry between bacterial antigens and the HLA-B27 cell surface antigen (Smith et al., 1993). Antibiotic therapy may decrease symptoms of chronic reactive arthritis. A double-blind study found that a 3-month course of Ciprofloxacin therapy decreased the symptoms reported by patients with chronic reactive arthritis (Toivanen et al., 1993).

Campylobacter spp. are frequently isolated from immunocompromised individuals including children, cancer patients, AIDS patients, organ recipients and the very old (Lastovica et al., 1986; Edmonds et al., 1987; Bernard et al., 1989; Sorvillo et al., 1991). However, *Campylobacter* spp. are capable of causing disease in immunocompetent individuals (Black et al., 1988). Strain differences affect the rate of infection with *Campylobacter* (Black et al., 1988).

1.4. EPIDEMIOLOGY

The incidence of *Campylobacter* reported by clinical laboratories ranges from 6–7/100,000 (Tauxe et al., 1988) to 30–60/100,000 based on active local surveillance (Hopkins and Olmsted, 1985; Johnson and Nolan, 1985). In 1990 in the United Kingdom, clinical laboratories reported a fourfold increase in *Campylobacter* since 1980 to a level of 87/100,000 (Skirrow and Blaser, 1992). This increase is largely due to improvements in means for detection and identification of the organisms.

Foodborne bacteria have been proven as the major source of this disease problem (Skirrow, 1990; Butzler and Oosterom, 1991; Bean and Griffin, 1990). Outbreaks of campylobacteriosis are relatively uncommon, whereas numerous sporadic episodes are most often encountered (Tauxe et al., 1987; Tauxe et al., 1988). Cross-contamination of foods with foods of animal origin lies at the heart of transmission of these bacteria. Foods documented as contaminated with *Campylobacter* include chicken, turkey, beef, pork, fish and milk. (Stern, 1992). Other foods that have been documented to be contaminated with *Campylobacter* spp. include raw seafood such as oysters, and mushrooms (Doyle and Schoeni, 1986; Stelma and McCabe, 1992). Cooking kills *Campylobacter* in foods; adherence to strict food hygiene can prevent infection (Stern, 1992). Raw milk can serve as a culture medium for the organism and inadequate pasteurization has led to documented outbreaks and cases (Potter et al., 1983; Warner et al., 1986). Waterborne transmission of *Campylobacter* to humans has also been documented (Blaser et al., 1983; Hernandez et al., 1996). Although the source of the bacteria-contaminating water has been found in only a few instances, it has been linked in some cases to fecal contamination from domesticated animals (Rollins and Colwell, 1986; Endtz et al., 1991; Stelma and McCabe, 1992; Pearson et al., 1993; Jimenez et al., 1994). *C. jejuni* survived longer in culturable form than *C. coli* in untreated and membrane-filtered lake water at both 4 and 20°C (Korhonen and Martikainen, 1991).

1.5. RESERVOIRS OF INFECTION

Campylobacter spp. are present in warm-blooded animals and their environment throughout most parts of the world (Norcross et al., 1992; Angulo, 1996). The predominant ecological niche for Campylobacteria is the intestinal tract of a wide variety of domesticated and wild vertebrates (Endtz et al., 1991;

Stelma and McCabe, 1992; Jimenez et al., 1994; On, 1996). Chicken is the number one source of *Campylobacter* for humans (Harris et al., 1986; Deming et al., 1987; Fukushima et al., 1987; Adak et al., 1995). Birds appear to have a higher infection rate and carriage of these pathogens than other animals (Montrose et al., 1985; Pearson et al., 1993). Transmission in broilers is extremely rapid which may be due to palantine colonization leading to transmission through communal water troughs and standard fecal-oral spread (Montrose et al., 1985). Fecal shed in cattle mainly leads to contamination of milk, although outbreaks have been linked to contaminated beef (Dilworth et al., 1988). Cattle checked at slaughter harbored the organism in gallbladders, large and small intestines, and liver (Garcia et al., 1985). Swine commonly carry *C. coli* and *C. jejuni* as intestinal commensals (Stern, 1992) or enteric pathogens (Mansfield and Urban, 1996), and studies in the U.S., Netherlands and Germany show that more than half of commercially raised pigs excrete the organisms (Blaser et al., 1983). Reports from western Europe and Canada conclude that biotypes and serotypes from pigs are unlikely to be related to those of human isolates (Banffer, 1985; Munroe et al., 1983), but studies in the U.S. have failed to ascertain this information (Bracewell et al., 1985). Controversy exists concerning the role of pigs in human *C. coli* enteritis (Madden et al., 1996). *Campylobacter* spp. have been isolated from free-living birds including migratory birds and waterfowl, crows, gulls, and domestic pigeons (Norcross et al., 1992). They have also been isolated from flies, rabbits, rodents, dogs and cats. Domesticated animals including dogs and cats are a significant source of the bacterium for the human population (Bruce et al., 1980; Deming et al., 1987). Additionally, antibiotic resistance is contributing to an increased incidence of infection in humans worldwide, which may be affected by large-scale antibiotic usage in animal agriculture (Endtz et al., 1991). The high rate of carriage, close association with epithelium, and chemoattraction by mucin makes control in animals difficult. The mechanisms of transfer from animals to humans are likely to be by the fecal-oral route (Saeed et al., 1993), but few studies exist showing how dynamics and growth characteristics of these species in food animals determine the level and mode of contamination of substrates reaching the human population.

2. PATHOGENESIS AND VIRULENCE FACTORS

2.1. NATURAL INFECTIONS

Human infection with *Campylobacter* is initiated, as in animals, by ingestion of the organism in contaminated foods or water. The bacteria must traverse the stomach to gain access to the appropriate microenvironment of the small and large intestines where it multiples, although a recent report describes infection with *C. jejuni* in the stomach (Sahay et al., 1995). Gastric acid provides a barrier against infection, because ingestion of *Campylobacter* with sodium bicarbonate increases the rate of illness in humans (Black et al., 1988). *Campylobacter* may colonize either the small (King, 1957; Cadranel et al., 1973) or large intestines, but more often results in colitis or cecitis (Lambert et al., 1979). The presence of abdominal pain, fever, diarrhea, frank blood in stools, and inflammatory cells in stools suggests invasion (Lambert et al., 1979; Black et al., 1988). Histological sections of colonic biopsies confirm acute colitis with crypt abscesses, depletion of goblet cells, and inflammatory infiltrates of the lamina propria composed of polymorphonuclear leukocytes, lymphocytes, and plasma cells (Mandal et al., 1984; Lambert et al., 1979). *C. jejuni* enteritis is initiated in the gastrointestinal tract, but can become extraintestinal in severe cases (Wallis, 1994). Clinical reports describing primary infections with *C. jejuni* include infection with systemic spread, infection with mucosal disease, infection without disease with short-term bacterial persistence, and infection with resistance and no bacterial persistence (Karmali and Fleming, 1979; Black et al., 1988; Ketley, 1997). These observations support the notion that this organism is capable of producing a spectrum of disease scenarios depending on the immune status of the host and other factors.

2.2. ANIMAL MODELS

Evidence from studies of animals with natural and experimentally induced infections with *Campylobacter* spp. have provided significant information regarding the pathogenesis of *C. jejuni.* Natural infections with *C. jejuni* resulting in enteritis have been reported in juvenile macaques (Russell, 1992), weaning-age ferrets (Bell and Manning, 1990), dogs (Davies et al., 1984; Fox et al., 1988), cats (Blaser et al., 1982), and swine (Boosinger and Powe, 1988; Vitovec et al., 1989; Babakhani et al., 1993; Mansfield and Urban, 1996). Chickens (Meinersmann et al., 1991), rodents (Humphrey et al., 1985; Bacquar et al., 1993; Bacquar et al., 1996), ferrets (Bell and Manning, 1990), dogs (Prescott et al., 1981), primates (Fitzgeorge et al., 1981; Russell, 1992), rabbits (McSweegan et al., 1987; Walker et al., 1988), and pigs (Boosinger and Powe, 1988; Vitovec et al., 1989; Babakhani et al., 1993; Mansfield et al., in review) have been inoculated experimentally by various routes with *C. jejuni* to mimic the course of infection in humans.

C. jejuni infection in *Macaca nemestrina* (pig-tailed macaques) most closely mimics enteric infection in humans with self-limiting diarrheal disease producing acute colitis (Russell et al., 1989). Bloody diarrhea, fecal leukocytes, vomiting and bacteremia are observed in these animals, but rectal swabs are negative by day 30 after infection. Histological examination early in infection shows acute inflammation with damage to the epithelium, loss of goblet cells, infiltration of neutrophils in the mucosa and lumenal crypts, and replacement of columnar epithelium by flattened epithelium. Electron microscopy (EM) showed epithelial damage with organisms in the submucosa. *M. nemestrina* are susceptible at an early age and may be reinfected with other *C. jejuni* strains with decreased fecal shedding time compared to the initial infection, although maternal nursing correlates with protection (Russell et al., 1989). *M. nemestrina* given challenge infections shed bacteria, but have no clinical signs of disease. Monkeys with primary infections have IgM and IgG in the serum while those with secondary infections have IgA and increases in IgM and IgG. Epidemiological studies in these animal colonies showed that animals were infected and susceptible to multiple strains of *C. jejuni* and *C. coli* serotyped by the Penner and Lior methods (Russell, 1992).

Rabbits have been used extensively to study *C. jejuni* pathogenesis by using a surgically prepared colonic loop termed Removable Intestinal Tie Adult Rabbit Diarrhea (RITARD) (Spira et al., 1981; Caldwell et al., 1983). Bacteria may be inoculated or given orally to mimic host conditions. Infected rabbits show diarrhea, bacteremia, dehydration, acidosis, azotemia and death in some cases. Pathologic lesions in the terminal ileum, cecum and colon include goblet cell hyperplasia, loss of mucus, intense cellular proliferation and a range from mild inflammatory infiltrates to frank epithelial necrosis (Walker et al., 1992). EM showed *Campylobacter* in mucus, in crypts, in close association with brush border epithelial cells, in vacuoles inside of M cells, in lymphoid cells immediately beneath the follicle-associated epithelium and in large groups beneath M cells. In this model, diarrhea is associated with mortality, but acquired immunity with decreased fecal shed does occur mediated by *C. jejuni*-specific mucosal IgA, and serum IgM and IgG (Burr et al., 1988; Pavlovskis et al., 1991). The Lior serotype can be used to predict cross-protection; rabbits challenged with the same organism had no bacteremia whereas those challenged with a heterologous strain were fully susceptible (Abimiku and Dolby, 1988). The pathophysiology of this diarrhea suggests a secretory mechanism mediated by elevations in cyclic AMP, prostaglandin E_2 (PGE_2) and leukotriene B_4 (Everest et al., 1993). This is supported by in vitro work showing that human epithelial cells upregulate PGE_2 and $PGF_{2\alpha}$ production and prostaglandin H synthase (PGHS)-2 after infection with entero-invasive bacteria (Eckmann et al., 1997). PGE_2 can stimulate intestinal epithelial cells to increase chloride secretion and can stimulate enteric nerves to release neurotransmitters that activate epithelial ion transport processes (Eberhart and DuBois, 1995; Eckmann et al., 1997). Flagella are important for colonization of rabbits with *C. jejuni.* Some *Campylobacter* undergo a bidirectional transition between flagellated and nonflagellated phenotypes. However, The Fla^- to Fla^+ shift predominated in infected rabbits (Walker et al., 1992).

Chicken models are used to study mechanisms of colonization. *C. jejuni* containing mutations in flagellin genes were unable to colonize the intestinal tract of 3-day-old chickens (Nackamkin et al., 1993). Similarly, *C. jejuni dnaJ* mutants were unable to colonize newly hatched chickens, suggesting that heat shock proteins are needed to cope with the intestinal environment (Konkel et al., 1998). Congenic *C. jejuni* strains, A74/O, a noncolonizing strain, and A74/C, a colonizing strain, were examined to determine the traits associated with the ability to colonize the gastrointestinal tracts of chickens (Meinersmann et al., 1991). Differences were not observed in plasmid content, restriction digests or morphology using transmission EM. Strain A74/C penetrated to the base of cultured cecal crypts, but A74/O was found in only one case associated with necrotic cells. A74C associated with crypt epithelium without damage and both strains destroyed mature cecal epithelium cells suggesting that resistance may be developmentally regulated in host cells (Meinersmann et al., 1991). Inoculation of chick embryos has been used to ascertain the virulence properties of *C. jejuni.* Changes associated with increased virulence of two strains of *C. jejuni* propagated by this means included alterations in cultural and cellular morphology, loss of flagella, expression of new outer-membrane proteins, alterations in cell-surface carbohydrates and decreases in cell-surface hydrophobicity (Field et al., 1993). Competitive exclusion of *C. jejuni* by cecum colonizing bacteria has been examined also, particularly as a means for practical control (Schoeni and Doyle, 1992). A nine-strain mixture of cecal bacteria provided 41–85% protection from colonization in white leghorn laying chicks challenged with *C. jejuni.*

Mice have been used extensively as models for *C. jejuni* enteritis because of their low cost and ease of maintenance. Mice are not naturally colonized by *Campylobacter* spp., but oral dosing of several inbred and outbred strains resulted in intestinal colonization and in some cases bacteremia without clinical signs of disease (Abimiku et al., 1989; Blaser et al., 1984). However, the site specificity of colonization of the intestinal mucosa in mice is under debate (Fox, 1992). Maternal vaccination protocols have shown efficacies of 57% against heterologous strains of *C. jejuni* as judged by prevention of colonization in mice (Abimiku and Dolby, 1988; Abimiku et al., 1989). Additionally, Kita and others have suggested that repeated exposure to a hepatotoxic activity identified in *C. jejuni* infection in a mouse model may be involved in the pathogenesis of Reiter's or Guillain-Barré syndromes, although these experiments did not report on whether mouse hepatitis virus or other pathogens may have been present in these mice (Kita et al., 1992). Other recent work in germ-free mice exposed to a single bacterial species indicates that commensal lumenal bacteria can influence epithelial gene-cell expression (Bry et al., 1996). Cross-talk between epithelial cells and commensal flora can be studied in these models using wildtype and mutant isogenic bacterial strains to determine how commensals can signal mucosal inflammation (Rath et al., 1996; Kagnoff and Eckmann, 1997).

Gnotobiotic or colostrum-deprived piglets inoculated orally with pathogenic strains of *C. jejuni* develop clinical signs and pathology observed in acute infections in humans with diarrhea (Boosinger and Powe, 1988; Vitovec et al., 1989; Babakhani et al., 1993). Pigs have anorexia, fever, and diarrhea for 1–5 days followed by remission of clinical signs, but continue to shed *C. jejuni* in the feces. However, immunocompetent pigs with a full complement of enteric bacteria and immune system components are refractory to infection with *C. jejuni* (Babakhani et al., 1993). Additionally, colostrum-deprived newborn piglets have provided a source of primary colonic epithelial cells to develop in vitro assays for the study of *C. jejuni* cellular invasion (Babakhani and Joens, 1993).

Young immune-competent pigs exposed to the helminth *Trichuris suis* developed clinical signs and pathology due to naturally acquired *C. jejuni* that closely mimics those seen in some humans with primary infections (Mansfield and Urban, 1996). Pigs develop a self-limiting colitis with watery, bloody diarrhea that correlates with the presence of third-stage larvae in the proximal colon and resolves at 50–60 days of infection when worms are expelled. Gnotobiotic swine infected with *C. jejuni* from a human with enteritis (ATTC 33292) had organisms colonizing the colon with no attachment or invasion (Mansfield et al., in press). *Trichuris suis* exposure converted these pigs with commensal *C. jejuni* to a severe disease status with comparable signs and pathology to that of conventionally reared pigs and humans. Germ-free pigs inoculated with *T. suis* alone or *C. jejuni* alone had no clinical signs and no pathology. Others have shown that concurrent infections with viruses (Konkel and Joens, 1990) and bacteria (Bukholm and Kapperud, 1987) increase the disease and pathology caused by *C. jejuni.* Together these studies suggest that exogenous virulence mechanisms exist that facilitate bacterial invasion.

Other models include ferrets and dogs. Ferrets have been used to study abortion or self-limiting diarrhea associated with *C. jejuni* infection (Fox et al., 1987). Dogs develop bacteremia after mild diarrhea and, thus, provide a good model for examination of systemic progression of *Campylobacter* enteritis (Davies et al., 1984; Fox et al., 1988).

2.3. SUMMARY OF IN VIVO PATHOGENESIS STUDIES

Clinical and experimental data indicate that invasion and progression of disease result from a combination of virulence factors (Fox, 1992; Wallis, 1994) that are influenced by host immune status and co-infection with other enteroinvasive enteric bacteria (Bukholm and Kapperud, 1987; Fox, 1992), viruses (Konkel and Joens, 1990), and parasites (Mansfield and Urban, 1996). The organism must be chemoattracted to the appropriate region of the gastrointestinal tract, penetrate the mucus layer, and associate with the base of the crypt or adhere to the mucosal surface to initiate colonization and infection (Wallis, 1994). Motility is necessary to overcome this physical barrier to permit colonization (Guerry et al., 1992; Nachamkin et al., 1993). Adherence and invasion results in tissue damage, inflammation and immune response on the part of the host. Multiple reports suggest that the ability to associate and invade mucosal epithelial cells may be *Campylobacter*-strain specific (Klipstein et al., 1985; Lindblom et al., 1990; Russell, 1992; Field et al., 1993). However, exposure to an in vivo environment is important for expression of virulence (Field et al., 1993). Roseneau et al., (1987) suggest that pathogenic strains associate with the cell surface, produce cytotoxic effects when high numbers adhere, and are endocytosed in low number when they are not able to survive. Others believe invasion via transcytosis and paracytosis through tight junctions is an important part of *C. jejuni* pathogenesis (Walker et al., 1992). Toxins have been associated with specific disease syndromes; enterotoxin-producing strains produce a cholera-like diarrhea, while cytotoxic strains produce microlesions of enterocytes that lead to bloody diarrhea (Wallis, 1994). Resolution and healing of the intestinal mucosa result, in most cases, in clearance of the organism. Prolonged fecal shedding is most often due to reinfection with different strains (Taylor, 1992; Russell, 1992). *Campylobacter* spp. are opportunistic pathogens that multiply in the gastrointestinal tract of their hosts and sometimes becomes extraintestinal when immune defenses are compromised (Bernard et al., 1989; Sorvillo et al., 1991). Pathogenicity of the organism is dependent on its virulence traits and the host immune response, and susceptibility secondary to other agents probably results from dysregulation of the host immune response (Mansfield, unpublished results).

3. CELL ASSOCIATION AND VIRULENCE MECHANISMS

Many questions regarding pathogenic mechanisms involved in *C. jejuni*-mediated enteritis have been addressed using in vitro model systems. Motility, mucus colonization, toxin production, attachment, internalization, and translocation are among the processes associated with *C. jejuni* virulence that have been investigated in vitro. Cells derived from the intestinal epithelium from multiple species have been used for these studies. A large base of knowledge has been generated by using cells in culture.

3.1. MOTILITY AND MUCUS COLONIZATION

Motility and high-viscosity growth medium mimicking intestinal mucus enhance binding and invasion of *C. jejuni* in Caco-2 cells (Szymanski et al., 1995). *C. jejuni* is chemoattracted to mucin (Hugdahl et al., 1988), which also enhances total cell-associated and internalized *C. jejuni* in HEp-2 cells (DeMelo and Pechere, 1988). Although the intestinal epithelial cell glycocalyx barrier containing mucus and secretory IgA may provide antiadhesive protection in many cases (McSweegan et al., 1987; Mantle et al., 1989; Frey et al., 1996), some bacteria thrive in the mobile mucus biofilm interfacing tissue and digesta (Costerton et al., 1983). There are multiple reports from in vivo model systems that *C. jejuni* is able to colonize mucus, preferentially dwelling within crypts in distal segments of the intestine (Beery et al., 1988; Terzolo et al., 1987; Lee et al., 1986). Spiral cellular morphology and polar flagella provide *C. jejuni* with a distinctive motility well suited to movement in a viscous environment, which might provide an ecological advantage in intestinal mucus (Ferrero and Lee, 1988; Szymanski et al., 1995). Colonization of mucus places *C. jejuni* in close proximity to enterocytes, such that toxins and/or adhesins may reach their cellular targets.

3.2. TOXINS

Numerous reports on *C. jejuni* and *C. coli* describe the incidence of toxins among isolates from various human and animal sources in different geographic locations (Ruiz-Palacios et al., 1983; Wadstrom et al., 1983; Mathan et al., 1984; Klipstein et al., 1985; Johnson and Lior, 1986; Klipstein et al., 1986; Saha et al., 1988; Akhtar and Huq, 1989; Coker and Obi, 1989; Perez-Perez et al., 1989; Lindblom et al., 1989; Coyer et al., 1990; Lindblom et al., 1990; Bok et al., 1991; Prasad et al., 1991; McFarland and Neill, 1992; Florin and Antillon, 1992; Khalil et al., 1993; Misawa et al., 1995; Us et al., 1995; Lindblom and Kaijser, 1995). Most of these studies correlate toxin production with clinical presentation. *C. jejuni* tends to be more toxigenic than *C. coli*, but results vary (Lindblom et al., 1989; Lindblom et al., 1990; Cover et al., 1990; Daikoku et al., 1989; Pickett et al., 1996). Animal isolates are less toxigenic than human isolates (Akhtar and Huq, 1989), and among animal isolates *Campylobacter* strains from pigs are less enterotoxigenic than those from chickens (Lindblom et al., 1990). Distinct cholera-like, cytotonic enterotoxin and cytotoxin activities have been described from *C. jejuni*, sometimes being produced concurrently (Johnson and Lior, 1984; Klipstein et al., 1985; Johnson and Lior, 1986; Taylor et al., 1987; Perez-Perez et al., 1989; Daikoku et al., 1989; Bok et al., 1991; Florin and Antillon, 1992). In general, noninflammatory, acute secretory (watery) diarrhea is associated with the cytotonic enterotoxin in children in developing countries, whereas inflammatory diarrhea and *Campylobacter* invasion are associated with cytotoxin production in sporadic cases in developed countries (Blaser et al., 1979; Ruiz-Palacios et al., 1983; Klipstein et al., 1985, 1986; Walker et al., 1986; Guerrant et al., 1987; Perez-Perez et al., 1989; Bok et al., 1991).

3.2.1. Enterotoxin

The original description of *C. jejuni* enterotoxin (CJT) indicated structural and functional similarity to cholera toxin (CT) based on elongation of CHO cells, heat lability, inactivation by cholera antitoxin, increased intracellular cyclic AMP levels, and fluid secretion in rat ileal loops (Ruiz-Palacios et al., 1983). Other researchers have provided supporting evidence for a cholera-like toxin in *C. jejuni* (McCardell et al., 1984; Klipstein and Engert, 1984a, b; Johnson and Lior, 1984; Collins et al., 1992). In addition to elongation of CHO cells, CJT induces rounding of Y-1 mouse adrenal cells (McCardell et al., 1984; Florin and Antillon, 1992), and a reversible cytotonic response in Vero cells (Johnson and Lior, 1986). Several authors have reported purification and characterization of CJT by various means (Klipstein and Engert, 1984a; Daikoku et al., 1990; Saha and Sanyal, 1990).

Gel immunodiffusion and enzyme-linked immunosorbent assay analyses have demonstrated the relatedness of CJT, in holotoxin and B subunit forms, to CT and *E. coli* heat-labile toxin (LT) (McCardell et al., 1984; Klipstein and Engert, 1984a, b, 1985). Antibodies to CT and LT cross-react with CJT, resulting in inactivation of cytotonic and secretory activities of CJT in cultured cells and ligated ileal loops, respectively (Ruiz-Palacios et al., 1983; McCardell et al., 1984; Klipstein and Engert, 1984a, b, 1985; Goossens et al., 1985a). Although negative results were initially obtained (Olsvik et al., 1984), low-level nucleotide sequence homology was demonstrated between genomic DNA of *C. jejuni* and the B subunits of LT and CT (Baig, et al., 1986; Calva et al., 1989). The B subunits of these toxins specify binding sites for GM1 ganglioside on intestinal epithelial cells (Klipstein and Engert, 1985; Suzuki et al., 1994).

Kanwar and colleagues (1994, 1995) have made substantial contributions toward understanding the mechanisms underlying observed increases in intracellular cAMP (Ruiz-Palacios et al., 1983) and fluid accumulation in ligated ileal loops infected with *C. jejuni* (Ruiz-Palacios et al., 1983; McCardell et al., 1984). The pathophysiology of secretory diarrhea from enterotoxigenic *C. jejuni* infection involves impaired Na^+, K^+-ATPase activity associated with increased Na^+ and Cl^- secretion (Kanwar et al., 1994). Alteration of ion transport is a calcium-dependent process involving protein kinase C activation (Kanwar et al., 1995).

Despite a considerable amount of supporting evidence, several investigators have reported failures to repeat experiments aimed at detecting enterotoxin acitivity from various sources of *C. jejuni* isolates, casting doubt on the existence of a cholera-like enterotoxin in *C. jejuni*. In a study of patients with inflammatory diarrhea from the U.S., antibodies to enterotoxin could not be detected, nor could enterotoxin activity be detected in the *C. jejuni* isolates from these patients (Perez-Perez et al., 1989). This is in contrast to a similar study conducted on children with *Campylobacter* enteritis in Mexico (Ruiz-Palacios et al., 1985). Differences in virulence characteristics between *C. jejuni* isolates from different geographic regions could account for these discrepant results (Perez-Perez et al., 1989). Until the putative CJT is cloned and sequenced, controversy will remain regarding the existence and pathobiological significance of an enterotoxin in *C. jejuni*-mediated enteritis (Wadstrom et al., 1983; Perez-Perez et al., 1989; Perez-Perez et al., 1992; Konkel et al., 1992d).

3.2.2. Cytotoxins

In vitro cytopathic effects from cell-free filtrates of polymxyin B-treated and sonicated *C. jejuni* and stool filtrates from infected individuals and animals have been described repeatedly (Wong et al., 1983; Johnson and Lior, 1984; Goossens et al., 1985b; Pang et al., 1987; Johnson and Lior, 1988; Akhtar and Huq, 1989; Cover et al., 1990; McFarland and Neill, 1992; Florin and Antillon, 1992). Assays for *C. jejuni* cytotoxins are usually based on morphological alterations of a wide variety of cultured cells. Indicators of cytotoxin activity include cell rounding, elongation, distention, loss of adherence, and cell death. Trypan blue staining of nonviable cells is commonly used to assess cytotoxicity. Because interpretation of cytopathic changes may lead to equivocal results (Bag et al., 1993), other methods have been developed to detect cytotoxin activity. A quantitative ^{51}Cr-release assay has been used in conjunction with cytopathic effects to evaluate cytotoxicity (Pang et al., 1987), as well as a recently described dye reduction assay (Coote and Arain, 1996). Cytotoxic effects due to *C. jejuni* toxins are more prominent in freshly seeded fibroblasts (Florin and Antillon, 1992), although HeLa cells are particularly susceptible (McFarland and Neill, 1992; Akhtar and Huq, 1989). The type and concentration of serum used in assays for *C. jejuni* toxins influence cytotoxicity as measured by cell rounding (Misawa et al., 1994), and the ability to elaborate cytotoxin decreases with in vitro passaging of the organism (Mizuno et al., 1994).

The best characterized toxin from *C. jejuni* is the cytolethal-distending toxin (CDT), originally described by Johnson and Lior (1988) as causing progressive cell distention and eventual cytotoxicity of cultured cells. Accompanying morphological changes such as cytoskeletal alterations are associated with interruption of cell proliferation in CHO cells (Aragon et al., 1997). Nucleic acid sequence similarity with the *E. coli* equivalent provided primers for PCR amplification, leading to the cloning of three adjacent *C. jejuni* genes (*cdtA, cdtB,* and *cdtC*) encoding proteins with predicted sizes of 30, 29, and 21 kDa (Pickett et al., 1996). All three genes were required for toxic activity of CDT in a HeLa cell assay (Pickett et al., 1996). It has recently been shown that *C. jejuni* CDT causes HeLa and Caco-2 cells to become arrested in the G_2 cell-cycle stage (Whitehouse et al., 1998). In HeLa cells, this block was associated with failure to dephosphorylate CDC2, which leads to accumulation of the inactive form of this kinase necessary for entry into the M phase (Whitehouse et al., 1998). The discovery that CDT binds to a 59 kDa protein found in both HeLa and CHO cell membranes, as well as a 45 kDa protein in HeLa cell membranes, led to the development of an immunoblot assay as an alternative to conventional tissue culture detection (Bag et al., 1993). Binding of CDT to its receptors was partially inhibited by EDTA (Bag et al., 1993).

The number of distinct cytotoxins produced by *C. jejuni* is unknown, but it is likely that there is more than one (McFarland and Neill, 1992; Florin and Antillon, 1992; Misawa et al., 1994; Misawa et al., 1995; Pickett et al., 1996). Failed attempts to neutralize activity with antisera to toxins from other enteric pathogens indicated that *C. jejuni* cytotoxins were immunologically distinct from *E. coli, Clostridium difficile,* and *Vibrio cholerae* toxins (Guerrant et al., 1987, 1992; Mahajan and Rodgers, 1990), although the CDT genes of *C. jejuni* have homologs in *E. coli* (Pickett et al., 1996). The contribution of cytotoxins to *Campylobacter* enteritis remains to be established.

3.3. ENTEROINVASION

C. jejuni is enteroinvasive and this activity is probably aided by toxin production. *C. jejuni* has been detected in the colonic mucosa and submucosa of experimentally infected infant monkeys and newborn piglets (Russell et al., 1993; Babakhani et al., 1993), and recovered from mesenteric lymph nodes through 23 days post-infection in a gnotobiotic mouse model (Fauchere et al., 1985). In these models, *C. jejuni* has been detected inside intestinal epithelial cells, indicating that transcellular migration accounts for at least some of the invasion into underlying tissues (Russell et al., 1993; Babakhani et al., 1993).

Translocation of *C. jejuni* has been studied in vitro using artificial epithelial barriers of cells grown on microporous membrane filters (Everest et al., 1992; Konkel et al., 1992b; Grant et al., 1993). *C. jejuni* is capable of translocating across polarized Caco-2 cell monolayers not only through but also between cells within 60 minutes following addition to the apical surface (Konkel et al., 1992b). Other investigators have also suggested that *C. jejuni* is capable of penetrating cell monolayers via a paracellular route (Everest et al., 1992), although tight junction integrity is not significantly disrupted in this process (Konkel et al., 1992b; Ketley, 1997). Translocation is reduced when bacterial protein synthesis is inhibited by chloramphenicol treatment (Konkel et al., 1992b), and intact, functional flagella are necessary for this process (Grant et al., 1993).

3.3.1. Adhesins

Regardless of whether *C. jejuni* invasion occurs by transcellular or paracellular pathways, penetration of the intestinal epithelial barrier is dependent on adhesion to host substrates. Surface bacterial structures that bear adhesins include fimbriae, fibrillae, flagella, capsule, outer membrane, and other appendages (Ofek and Doyle, 1994). *C. jejuni* flagella have been considered to be involved in the adhesive process (McSweegan and Walker, 1986; Szymanski et al., 1995), although studies with motility-defective mutants indicate that flagella have an ancillary role in adhesion (Grant et al., 1993; Yao et al., 1994; Russell and Blake, 1994).

Peritrichous pilus-like appendages have recently been identified in *C. jejuni* (Doig et al., 1996). These are induced when the bacterium is grown in the presence of bile salts and confer a highly aggregative phenotype. A site-specific insertional mutation within a gene, termed *pspA* which encodes a predicted protein resembling protease IV of *E. coli*, results in the loss of pilus synthesis (Doig et al., 1996). Although the non-piliated mutant showed no reduction in adherence to or invasion of INT 407 cells in vitro, disease symptoms in a ferret animal model were significantly reduced (Doig et al., 1996).

Adhesins may be anchored in the cytoplasmic membrane, peptidoglycan, S layer, or the outer membrane of gram-negative bacteria (Ofek and Doyle, 1994). Multiple-surface and periplasmic *C. jejuni* proteins have been identified and characterized, some of which are potential adhesins. Although other functions may be inferred based on amino acid sequence analysis, surface-exposed proteins are candidates for adhesins by virtue of their location. The major outer-membrane protein (OMP) is arranged hexagonally as a trimer of 42 kDa subunits and is antigenically related to the *E. coli* OmpC porin (Kervella et al., 1992; Amako et al., 1996; Bolla et al., 1995; Zhuang et al., 1997). Inducible outer membrane proteins of 55, 35 and 20 kDa, not detected when cultivated in artificial media, were expressed when *C. jejuni* was maintained in implants in chicken peritoneal cavities for one week (Chart et al., 1996). Additionally, a highly conserved, immunogenic 18 kDa OMP with significant similarity to other peptidoglycan-associated lipoproteins of gram-negative bacteria has been recently identified in *C. jejuni* (Burnens et al., 1995; Konkel et al., 1996). Based on the homologous proteins, Omp 18 is predicted to form a bridge between the outer membrane and peptidoglycan to stabilize the cell wall.

A 28 kDa surface-exposed protein, cell binding factor 1 (CBF1 or PEB1), possibly involved in binding to host cells and facilitating amino acid transport, has been described in *C. jejuni* (Pei and Blaser, 1993; Kervella et al., 1993). CBF1 is highly conserved among *C. jejuni* strains, but is absent in some nonadherent *C. jejuni* strains (Pei and Blaser, 1993; Kervella et al., 1993). CBF1 plays a role in adhesion and invasion of epithelial cells in culture as well as intestinal colonization in a mouse model (Pei et al., 1998). P29 (HisJ) is a 29 kDa periplasmic protein from *C. jejuni* that functions as a histidine binding protein and has homology to CBF1 (Garvis et al., 1996). However, there is no evidence that P29 promotes binding to eukaryotic cells (Garvis et al., 1996). PEB4 (CBF2) is also a 29 kDa periplasmic protein, which may function as an extracellular chaperone (Kervella et al., 1993; Burucoa et al., 1995). Additional external proteins recently identified include a 36 kDa lipoprotein containing the signature sequence for siderophore-binding proteins, which may function in iron acquisition (Park and Richardson, 1995), and CjaC having significant homology to periplasmic solute-binding proteins of the ABC transport system (Pawelec et al., 1998).

C. jejuni has been subjected to various treatments to assess the nature of adhesins (McSweegan and Walker, 1986; Maruyama and Katsube, 1994). Some of the *C. jejuni* factors mediating agglutination of intestinal epithelial cells and erythrocytes from various species are reported to be formalin-, glutaraldehyde-, and heat-resistant (Maruyama and Katsube, 1994). Adherence of *C. jejuni* to INT 407 cells was inhibited partially by treating the bacterial cells with proteases or fixatives (McSweegan and Walker, 1986). Likewise, pretreatment of *C. jejuni* outer-membrane preparations with proteinase K significantly diminished binding for INT 407 cells (Moser et al., 1992; Moser and Schroder, 1995). At least one adhesin from *C. jejuni* has been identified, CadF, an outer-membrane protein that binds to fibronectin (Konkel et al., 1997).

3.3.2. Lipopolysaccharide

Early work suggested that the oligosaccharides of *C. jejuni* LPS were involved in adhesion, based on reduced cell binding following periodate oxidation of LPS (McSweegan and Walker, 1986). In contrast, Moser and Schroder (1995) reported that the binding of *C. jejuni* outer-membrane preparations to INT 407 cells was not significantly altered following oxidation of LPS by sodium metaperiodate nor by pretreatment with LPS-specific monoclonal antibody. Abundant evidence implicates *C. jejuni* LPS in post-infection pathology although it may not have a primary function in adhesion.

The LPS of *C. jejuni* is a low-molecular weight rough-form, often lacking O polysaccharide antigens, which is unusual for an enteric pathogen (Logan and Trust, 1984; Naess and Hofstad, 1984; Penner and Aspinall, 1997). Considerable work has been done to determine the chemical composition of the core oligosaccharide and lipid A structures of *C. jejuni* (Naess and Hofstad, 1984; Beer et al., 1986; Moran et al., 1991; Aspinall et al., 1992, 1993; Moran, 1995). Lipooligosaccharide (LOS) is used to refer to the glycolipids of gram-negative bacteria lacking O-antigen units with core oligosaccharide structures limited to 10 saccharide units (Preston et al., 1996; Moran et al., 1996). In some cases the LOSs on bacteria are structurally and antigenically similar to LOS moieties on host glycolipids, such that in vivo modification of bacterial structures by the host may occur (Preston et al., 1996). Modification of *C. jejuni* O-antigens by the host was proposed as an explanation for antigenic variation of an isolate from the same patient during the course of an infection

(Mills et al., 1992). Also, LPS was modified following recovery of *C. jejuni* maintained in chicken peritoneal cavities for one week, suggesting that inducible structural variation occurs when the organism is exposed to in vivo stimuli (Chart et al., 1996).

Mimicry of carbohydrate structures on human cells by bacterial LOSs contributes to the development of autoimmune neuropathies such as Guillain-Barré (GBS) and Miller Fisher syndromes (MFS) (Moran et al., 1996; Prendergast et al., 1998). Glycosylated flagellin of *C. jejuni* may also be involved in GBS etiology (Guerry, 1997). LOSs of multiple *C. jejuni* isolates from patients with GBS and MFS have been shown to contain various ganglioside structures (Yuki et al., 1994, 1995; Aspinall et al., 1993; Salloway et al., 1996). Antibodies to the oligosaccharides of *C. jejuni* cross-react with gangliosides on human neurons, possibly leading to autoimmune neuropathy (Wirguin et al., 1994; Yuki et al., 1995; Gregson et al., 1997).

3.3.3. Receptors

Receptors for bacterial adhesins generally contain carbohydrates found on integral, peripheral, and cell-coat components of animal cell membranes (Ofek and Doyle, 1994). Multiple carbohydrates have been considered in several investigations of *C. jejuni* binding to host cells (Cinco et al., 1984; Newell et al., 1985; McSweegan and Walker, 1986; DeMelo and Pechere, 1990; Moser et al., 1992; Russell and Blake, 1994; Szymanski and Armstrong, 1996). Fucose, galactose, mannose, maltose, and glucose may be involved in mediating adhesin-receptor interaction, as these sugars have been reported by different investigators to inhibit adhesion of intact or fractionated *C. jejuni* outer membranes to host cell membranes (Cinco et al., 1984; Moser et al., 1992; Russell and Blake, 1994). Mixed results have been obtained following protease treatment or fixation of host cells in vitro to assess protein involvement (McSweegan and Walker, 1986). *C. jejuni* has an affinity for multiple intact lipid structures, particularly unsaturated fatty acids, suggesting that lipids in host cell membranes are involved in mediating *C. jejuni* adhesion (Szymanski and Armstrong, 1996).

Extracellular matrix components, including fibronectin, laminin, and collagens, are potential anchors for *C. jejuni* adhesion (Kuusela et al., 1989; Moser and Schroeder, 1995; Garvis et al., 1997). In particular, *C. jejuni* binding to fibronectin is mediated by a specific interaction with the 37 kDa outer-membrane protein CadF (Konkel et al., 1997). Flagellin and the major outer-membrane protein of *C. jejuni* have also been reported to bind to fibronectin (Moser et al., 1997).

3.3.4. Internalization

Based on experimentally infected infant monkeys, cell invasion by *C. jejuni* was concluded to be the primary mechanism of colon damage and diarrheal disease (Russell et al., 1993). In in vitro culture, attachment of *C. jejuni* to cells is independent of bacterial protein synthesis, although internalization is contingent upon newly synthesized bacterial proteins (Konkel and Cieplak, 1992; Konkel et al., 1992b; Oelschlaeger et al., 1993). *C. jejuni* synthesizes at least 14 new bacterial proteins within 60 minutes following culture with INT 407 cells, suggesting active engagement in a directed response to facilitate its internalization (Konkel and Cieplak, 1992; Konkel et al., 1993).

The mechanisms underlying *C. jejuni* invasion are clouded by conflicting reports of experiments using various inhibitors of cellular processes. Although de novo protein synthesis by the host cell is not required, internalization does involve active invagination of the target cell membrane (Konkel et al., 1992a). Pretreatment of Caco-2 cell monolayers with filipin III, which disrupts caveolae (plasma membrane invaginations) by chelating cholesterol, significantly reduces the ability of *C. jejuni* to enter these cells (Wooldridge et al., 1996). Internalization of *C. jejuni* by INT 407 cells has been shown to be inhibited by cytochalasin, which disrupts microfilament formation (Konkel et al., 1992). Microfilament depolymerization had no significant effect on entry in another study, although entry was blocked by microtubule depolymerization and inhibitors of coated-pit formation (Oelschlaeger et al., 1993). A separate study argued that coated-pit formation is not likely to be important since there was no inhibition of *C. jejuni* invasion in Caco-2 cells with monodansylcadaverine or g-strophanthin treatment (Russell and Blake, 1994). In contrast, Konkel and colleagues (1992) reported that dansylcadaverine, an inhibitor of receptor cycling, did reduce internalization of *C. jejuni* by INT 407 cells.

C. jejuni is found in membrane-bound vacuoles once internalized by INT 407 and HeLa cells (Konkel et al., 1992; Russell and Blake, 1994; Fauchere et al., 1986). *C. jejuni* has also been detected within enterocytes in vivo (Fauchere et al., 1985; Russell et al., 1993; Babakhani et al., 1993; Mansfield et al., in press). Treatment of INT 407 cells with ammonium chloride and methylamine, two chemicals that inhibit endosomal acidification, did not affect *C. jejuni* internalization, nor did they have a significant impact on their intracellular survival (Konkel et al., 1992e; Oelschlaeger et al., 1993). In the absence of antibiotic in the media, infection of INT 407 cells led to deterioration of monolayers, indicating that *C. jejuni* is able to survive within epithelial cells and elicit a cytotoxic effect (Konkel et al., 1992c). Similar observations have been made in vivo in that surface epithelial cells are damaged and exfoliated into the intestinal lumen in monkey and piglet models (Russell et al., 1993; Babakhani et al., 1993), although the host response contributes to the destruction of mucosal cells.

3.3.5. Pathway to Pathogenicity?

Although mucus may be a barrier for some organisms, it appears to be no obstacle for *C. jejuni* if motility is intact. *C. jejuni* actually thrives in the base of mucus-filled crypts. Exotoxins may be released independent of adhesion (Lee et al., 1986), or *C. jejuni* may become immobilized via adhesin-receptor interactions with cellular or extracellular host substrates. Once physical contact has been established, endotoxins may be placed in close proximity to their cellular targets, and/or the intestinal epithelial barrier may be breached via intra- or intercellular translocation mechanisms. Invasion of intestinal epithelial cells by *C. jejuni* is a means of translocation to the lamina propria and deeper tissues, and these events can occur in lymphoglandular complexes in the distal colon (Mansfield and Urban, 1996).

4. GENETIC BASIS FOR VIRULENCE

Tompkins, Taylor, Tenover et al. and others provide in-depth analysis of the genetic mechanisms of virulence factors in a comprehensive book review (Nackamkin et al., 1992). A more recent review by Ketley (1997) covers many of the same topics, but updates the information with published accounts of recent progress in pathogenesis of enteric *Campylobacter* infection.

4.1. ROLE OF FLAGELLA IN *CAMPYLOBACTER* PATHOGENESIS

Flagella are considered significant virulence determinants for *C. jejuni,* although the full extent of their role(s) in pathogenesis is uncertain. The polar flagellum can occur at one or both ends of the cells and is attached at the flagellar insertion point. Flagella are composed of three structural units: the basal structure anchored in the outer and inner membranes, plus the filament and hook, which are located on the cell surface (Luneberg et al., 1998). The hook, which connects the filament to the basal body, exhibits hypervariability in its central region (Luneberg et al., 1998). The filament is composed of two closely related flagellin proteins with subunit molecular weights of 58 to 62.5 kDa. The flagellins are encoded by two genes *flaA* and *flaB,* which are oriented in tandem (Alm et al., 1993). The flagellar protein is the predominant protein antigen of the *Campylobacter* cell and *flaA* is the major flagellin subunit (Guerry et al., 1992). *FlaA* and *flaB* are independently transcribed; *flaA* is under the control of the σ^{28}-like promoter and flaB the σ^{54} promoter (Alm et al., 1993). *FlaA* is expressed at higher levels than *flaB*. Miller et al. (1993) have identified an additional gene, *flbA,* which may regulate flagellin expression. An insertion mutation in the *flbA* gene produced a *C. jejuni* mutant that did not synthesize flagellin and was nonmotile.

There is a high degree of conservation between *Campylobacter* flagellins and other bacterial species, although the posttranslational phosphorylation of serine residues is unusual (Guerry et al., 1992). *Campylobacter* flagella undergo antigenic variation. The antigenic differences between two antigen types,

T1 and T2, that have been studied extensively, are due to posttranslational modifications to the flagellins rather than changes in primary structure (Alm et al., 1992). In recent studies, mutation of *ptmB* caused a change in apparent M(r) of the flagellin subunit on SDS-PAGE gels (Guerry et al., 1996). Mutation of the genes *ptmA* with homology to alcohol dehydrogenases, and *ptmB* with homology to CMP-N-acetylneuraminic acid synthetase involved in sialic acid capsular biosynthesis suggests that the surface exposed posttranslational modifications play a significant role in protective immune responses against *Campylobacter* (Guerry et al., 1996).

Flagella allow *Campylobacter* to invade mucosal crypts by propelling it through the barrier of viscous mucus in the gut (Nachamkin et al., 1993). *C. jejuni* containing mutations in flagellin genes were unable to colonize INT 407 cells (Wassenaar et al., 1991), and intestines of mice (Morooka et al., 1985), hamsters (Aguero-Rosenfield et al., 1990), rabbits (Pavloskis et al., 1991), and 3-day-old chickens (Nachamkin et al., 1993). The presence of an intact flagella (Aguero-Rosenfield et al., 1990) and motility (Newell et al., 1985) appear to be important factors in the ability of *C. jejuni* to colonize the host. However, flagellar mutants constructed with mutations in the *flaA* structural gene sequence producing bald and stubby mutants and their parent strain (Grant and Tompkins, 1992) were equally unable to adhere to INT 407 and Hep-2 cells, suggesting that other virulence factors are present (Tompkins, 1992). Additional studies using nonflagellated mutants of *C. jejuni* 81116 indicated that either motility or the presence of *flaA* is required for invasion of, but not adherence to, INT407 and Caco-2 cells (Wassenaar et al., 1991; Nuijten et al., 1992; Grant et al., 1993).

Controversy about the role of flagella in adherence and endocytic entry into cells versus active invasion was addressed by the development of defined insertional mutations in flagellar genes. Yao et al. introduced kanamycin-resistance cassettes into various positions in the *C. jejuni* chromosome and isolated eight mutants defective in their ability to invade INT407 cells (Yao et al., 1994). Three of these mutants had motility defects and five were fully motile. Mutant 1, with an insertion into the *flaA* flagellin gene, had greatly reduced motility, a truncated flagellar filament, and was nonadherent and noninvasive. Two mutants with defects in the *pflA* gene were phenotypically paralyzed. They were nonmotile despite the presence of a full-length flagella. All three mutants had defective flagellar structure at the point where the filament attaches to the cell. These experiments showed that motility rather than the *FlaA* protein is necessary for invasion of INT407 cells by *C. jejuni*. They also suggest that flagellin can mediate adherence to eukaryotic cells without internalization, but that other adhesins are involved in a motility-dependent invasion process.

4.2. IRON UTILIZATION AND REGULATION

Bacteria require iron for growth, which may be limiting in their hosts. The iron in mammalian hosts is tightly sequestered by high-affinity, iron-binding proteins such as transferrin or lactoferrin and, thus, are not available to bacteria. Additionally, in aerobic organisms, iron must be assimilated, utilized and stored in ways that sequester iron for essential use, but avoid iron overload. Iron overload can lead to cytotoxicity because iron reacts with the reduced form of O_2 leading to oxidative damage of lipids, proteins and nucleic acids. Bacteria that live on mucosal surfaces, on deeper tissues and inside of cells have evolved different strategies for acquiring iron; some bacteria can directly bind host iron-binding proteins such as transferrin and lactoferrin, and others secrete siderophores, which can bind iron with a higher affinity than the host (Finlay and Falkow, 1997).

Campylobacter spp. require iron for growth (Field et al., 1986), are not thought to produce siderophores (Richardson and Park, 1995), but are able to utilize various exogenous siderophores (Field et al., 1986). Park and Richardson (1995) have cloned a protein from *C. jejuni* with homology to periplasmic siderophore binding proteins of *Vibrio* and *Bacillus* that confers a hemolytic phenotype on the organism. The *ceu* operon encodes a transport system that scavenges siderophores in the intestinal tract (Richardson and Park, 1995). Cawthraw et al. evaluated the effect of the *ceu* operon on the ability of *C. jejuni* to colonize the host. A *C. jejuni ceu* mutant colonized chick epithelium equivalent to wildtype, suggesting that there are additional iron uptake systems in campylobacters (Crawthraw et al., 1996). However, this may not be the best animal model for testing correlates of invasiveness since chickens are colonized without disease.

Ferritin is an iron-storage protein that can allow growth in low-iron environments and can protect against iron overload cytotoxicity. Wai et al. purified and characterized ferritin from *C. jejuni* (Wai et al., 1995). Wai and others also cloned and evaluated the biological relevance of the ferritin-encoding gene (*cft*) of *C. jejuni* by producing a ferritin-deficient mutant (Wai et al., 1996). The growth of the ferritin-deficient strain was inhibited under iron deprivation. The ferritin-deficient mutant was more sensitive to killing by H_2O_2 and paraquat than the isogenic parent strain, indicating that ferritin is active in both iron-storage and protection from intracellular iron overload functions in *C. jejuni.*

In *E. coli* and most other bacteria, iron forms an ferrous complex with the ferric-uptake-regulating protein (Fur), which is a DNA-binding protein, which acts as a repressor by binding to the Fur binding sequence found at the promoter region of iron-regulated genes. The *fur* gene from *C. jejuni* has been cloned and compared to that of other bacteria and has 35% homology to *E. coli fur* (Wooldridge et al., 1994; Chan et al., 1995). The *fur* gene from *C. upsaliensis* has also been cloned, determined to have high homology to *fur* of *C. jejuni* and mapped to a highly conserved region of the genome upstream from *lysS* and *glyA* (Bourke et al., 1996). The Fur-like protein of *C. jejuni* is highly diverged compared to that of other gram-negative bacteria and is potentially the major iron-regulating protein in this organism (Wooldridge et al., 1994). The iron-responsive regulatory system associated with *fur* may regulate a subset of virulence-associated genes (Ketley, 1997). Genes with Fur-operator sequences include *sod*-encoding superoxide dismutase (Pesci et al., 1994; Purdy and Park, 1994) and *katA*-encoding catalase (Grant and Park, 1995). Mutational analysis of the *C. jejuni fur* gene showed that the fur mutant grew slower than the parental strain and had derepressed expression of three iron-regulated outer-membrane proteins, including CfrA (van Vliet et al., 1998). Two iron-repressed proteins, KatA and alkyl hydroperoxide reductase (AhpC), were present in the cytoplasmic fraction of the *fur* mutant, but were still iron-regulated, suggesting the presence of Fur-independent iron regulation. Additionally, as in many other bacteria, *C. jejuni* synthesizes envelope-associated proteins in response to iron stress (Field et al., 1986; Pickett et al., 1992). One of these proteins functions as a component of the high-affinity uptake pathway for hemin and hemoglobin (Pickett et al., 1992). Functional studies into the role of the Fur regulon in *Campylobacter* pathogenesis will soon be possible.

4.3. TWO-COMPONENT REGULATORY SYSTEMS

Gram-negative bacteria possess two-component regulatory systems that regulate many bacterial functions including virulence factors. These systems are composed of a sensor protein located in and spanning the membrane, a histidine kinase intracellular domain, and a regulator protein, which can act as a transcriptional activator or repressor (Finlay and Falkow, 1997). Two-component regulatory systems described to date respond to external stimuli by autophosphorylation of the sensing protein, transfer of a phosphate residue to the amino terminal domain of the regulator protein, change in the ability of the regulatory protein to bind to DNA sequences, and transcription of genes that contain an upstream regulator binding sequence (Finlay and Falkow, 1997). BvgA/BvgS in *Bordetella* spp., which regulates itself, fimbria and several toxins (Uhl and Miller, 1996), and PhoP/PhoQ in *Salmonella typhimurium,* which regulates many proteins needed for survival of the organism within macrophages (Wick et al., 1995), are examples of two-component regulatory systems that regulate virulence factors. Components of this type of regulatory system have been identified in *Campylobacter* spp., including histidine protein kinase (HPK), and response regulator protein genes *cheY* and *regXI,* mutation of which results in loss of chemotactic and invasion abilities, respectively (Ketley, 1997).

4.4. ANTIBIOTIC RESISTANCE

Antibiotic resistance in campylobacters will become increasingly important in the epidemiology of clinical disease as the prevalence of resistance increases and the number of immunosuppressed individuals increases. Taylor and others identified conjugative plasmids in *Campylobacter,* which encode resistance to tetracycline, kanamycin, and chloramphenicol (Taylor et al., 1981; Taylor and Courvalin, 1988). These plasmids are 45 to 50 kb in size and have a host range that is limited to a few closely related *Campylobacter* spp. Subsequently, Tenover et al. identified a 57 kb plasmid-mediating resistance to tetracycline that was conjugally transferred from one *C. jejuni* strain to another during mating (Tenover et al., 1983). Since then, numerous reports have identified plasmids transferring in *Campylobacter* spp. (Bradbury et al., 1983; Tenover et al., 1985). Plasmids have been associated with the occurrence of antibiotic resistance in serologically defined strains of *C. jejuni* and *C. coli* (Bradbury and Monroe, 1985). In this study, 53% (116 of 200) of *Campylobacter* strains contained plasmid DNA. Other conjugative plasmids encoding resistance to nalidixic acid, kanamycin, ampicillin and 16 other antibiotics have been identified (Velazquez et al., 1995). Plasmid profiles for 24 isolates representing six species of *Campylobacter* revealed only two patterns for those containing plasmids. Of these, 37% had only one large 38 Mdal plasmid and 8% had a 38 Mdal and a 1.6 Mdal plasmid (Boosinger et al., 1990). Currently, there are no reports of *Campylobacter* plasmids being transferred to other genera of bacteria. The mechanisms controlling this antibiotic resistance have been explored in detail and described by Taylor in a recent review (Taylor, 1992).

4.4.1. Tetracycline Resistance

Resistance to tetracycline in *C. jejuni* is mediated by the *tetO* gene found on a 45 Kb self-transmissible plasmid (pUA466) with a narrow host range, including *C. lari, C. fetus* subsp. *fetus,* and *C. fetus* subsp. *venerealis* (Taylor et al., 1981; Taylor et al., 1983). This *C. jejuni tetO* encodes resistance to a high level of tetracycline. *TetO* has been identified on a plasmid (pIP1433) in *C. coli* and also integrated into the *C. coli* chromosome (UA703) (Taylor et al., 1987; Ng et al., 1987). The *tetO*s from *C. jejuni* and *C. coli* are highly homologous based on cloning and expression in *E. coli* followed by DNA sequence comparison (Taylor and Courvalin, 1988). The *tetO* ORF is 1911 bp with 300 nucleotides of promotor sequence upstream of the ORF. This ORF encodes a 68 kDa protein when expressed in *E. coli* (Taylor et al., 1987; Manavathu et al., 1990).

Tetracycline is believed to inhibit protein synthesis by binding to a single high-affinity site on the 30S subunit within the 70S ribosome, which blocks entry of amino acyl tRNA into the ribosomal A site. Taylor speculates that TetO and TetM probably bind GDP and GTP and act catalytically as GTPases (Taylor, 1992). It is likely that TetO arose in gram-positive bacteria and spread to *Campylobacter* spp. This hypothesis is supported by the following findings:

(1) TetO is found in *Enterococcus* and *Streptococcus,* (2) the G + C content of *tetO* is 40%, which is close to that of *tetM* but different than that of the *C. jejuni* and *C. coli* chromosome and plasmid DNA, (3) TetO is highly related to TetM based on amino acid sequence analysis, and (4) the Shine-Delgarno sequence in TetO resembles a gram-positive ribosomal binding site (Taylor, 1992). Additionally, recent mutational alteration of Asn-128 of the GTP binding domain of the *tetO* gene showed that substitution resulted in a decrease in tetracycline resistance (Grewal et al., 1993).

4.4.2. Kanamycin Resistance

Kanamycin resistance in *C. coli* was first recognized by Lambert et al. (1985). This isolate had resistance against a wide range of antibiotics. The *aphA*-3 gene controlling this resistance was shown to be very similar to genes previously identified in the gram-positive organisms staphlococcus and streptococcus when nucleotide seqences were compared (Trieu-Cuot and Courvalin, 1983; Trieu-Cuot et al., 1985). This gene is encoded on the plasmid that carries tetracycline resistance (Trieu-Cuot et al., 1985). The *aphA*-3 gene has been isolated from plasmids and chromosome of *C. jejuni* and *C. coli* from Thailand, Vietnam, and Spain, suggesting that the gene is capable of translocation (Papadopoulou and Courvalin, 1988).

4.4.3. Chloramphenicol Resistance

Plasmid-mediated resistance to chloramphenicol in *Campylobacter* spps is rare. Chloramphenicol resistance is mediated by a 24 kDa protein encoded by a 621 bp ORF. The *Campylobacter cat* gene is highly homologous to that of acetyltransferase genes isolated from *Clostridium perfringens* (Steffen and Matzura, 1989) and *Clostridium difficile* (Wren et al., 1989) and may have been acquired from the clostridia (Taylor, 1992).

4.4.4. Streptomycin Resistance

Streptomycin resistance in *C. jejuni* and *C. coli* is mediated by an acetyltransferase (Pinto-Alphandary et al., 1990). The gene encoding this resistance is unlike that of *E. coli.* It is likely that campylobacters have acquired this gene from both gram-positive and gram-negative organisms because DNA probes for the *aadE* gene of gram-positive cocci (encoding a 6-aminoglycoside adenylyltransferase) and probes for the *aadA* gene of gram-negative bacteria hybridized with all streptomycin resistant *Campylobacter* strains (Taylor, 1992).

4.4.5. Erythromycin Resistance

Strains resistant to erythromycin show cross-resistance to spiramycin, lincomycin, clindamycin, azithromycin, and clarithromycin (Burridge et al., 1986; Karmali et al., 1981; Yan, 1990; Taylor and Chang, 1991; Funke et al., 1994). Erythromycin resistance is chromosomally mediated (Taylor and Courvalin, 1988). The gene encoding erythromycin resistance has not been cloned although it has been localized to the 240 kb *Sma*1 fragment 3 of *C. coli* UA417 by pulse-field gel electrophoresis (Taylor, 1992). The gene is constitutively expressed because it is not inducible by pretreatment of Ery *C. jejuni* and *C. coli* strains with sublethal amounts of erythromycin and other macrolides (Yan, 1990; Yan and Taylor, 1991). Probes made from other bacterial erythromycin resistance genes *erm*A, *erm*AM, *erm*C and *erm*E do not hybridize with DNA from Ery^r from campylobacters (Bibb et al., 1985; Murphy, 1985).

4.4.6. Fluoroquinolone Resistance

C. jejuni and *C. coli* have developed resistance to quinolones (Adler-Mosca et al., 1991; Gootz and Martin, 1991; Power et al., 1992; Velazquez et al., 1995; Gaudreau and Gilbert, 1998; Gibreel et al., 1998), which are broad-spectrum antimicrobial agents that inhibit DNA gyrase A subunit activity in DNA supercoiling and arrest DNA replication of bacteria. Comparison of antimicrobial resistance of 158 *C. jejuni* strains isolated from humans in Quebec, Canada, between 1985/1986 and 1995/1997 showed increased resistance to tetracycline, nalidixic acid and ciprofloxacin (Gaudreau and Gilbert, 1998). Resistance to this group of antibiotics is mediated by several different point mutations in *C. jejuni* DNA gyrase that lead to a marked decrease in susceptibility to quinolones (Gootz and Martin, 1991; Wang, 1993; Charvalos et al., 1996). In Sweden, quinolone resistance in *C. jejuni* increased 20-fold associated with chromosomal mutations in codon 86 of the *gyrA* gene and in codon 139 of the *parC* gene (Gibreel et al., 1998). A similar result was found in Spain (Ruiz et al., 1998).

4.4.7. Efflux Pump Systems

Multidrug resistance in *C. jejuni* has been attributed to an efflux pump system with a broad specificity (Charvalos et al., 1995). Mutant *C. jejuni* strains 34PEFr and 34CTXr had cross-resistance to erythromycin, chloramphenicol, tetracycline, beta-lactams, and quinolones. Strain PEFr had a mutation at codon 86 of the *gyrA* gene. However, both strains had overexpression of two outer-membrane proteins of 55 and 39 kDa, and preincubation of these cells with carbonyl cyanide m-chlorophenylhydrazone abolished the accumulation of pefloxacin, ciprofloxacin and moncycline indicating the presence of an efflux system (Charvalos et al., 1995).

5. TOOLS TO EXPLORE THE GENETICS OF VIRULENCE

Recent progress has provided the foundation for exploration of genetic control of virulence through the development of tools to return cloned *Campylobacter* spp. genes to the natural host where they can be assessed individually for their ability to control pathogenicity. Early work to characterize virulence attributes of *Campylobacter* spp. was restricted to biological and biochemical characterization because of the lack of available tools to accomplish this task. Genetic exchange systems were not available for the construction of mutants and the cloning of *Campylobacter* genes. Plasmids could be transferred to other species of *Campylobacter* but not to *E. coli.* Additionally, it was difficult to evaluate the ability of broad-host range plasmids to be maintained in *Campylobacter* because antibiotic resistance markers expressed in *E. coli* were not expressed in *Campylobacter.*

5.1. SHUTTLE VECTORS AND SUICIDE VECTORS

Labigne et al. (1987) were the first to develop a shuttle plasmid that could replicate in both *Campylobacter* and *E. coli* species, contained a kanamycin resistance gene isolated from *C. coli* that expressed resistance in both *E. coli* and *Campylobacter* spp., and contained the *oriT* sequence isolated from a broad-host range plasmid allowing it to be mobilized from *E. coli* donors into *Campylobacter* recipients. This recombinant shuttle plasmid was then successfully mobilized from *E. coli* into *Campylobacter* cells by mating. Labigne proceeded to construct a suicide plasmid based on this product by deleting the sequence encoding the entire cryptic plasmid originally ligated into the *Campylobacter* shuttle plasmid (Labigne et al., 1988). This plasmid could replicate in *E. coli* but not *Campylobacter* cells. This suicide plasmid formed the basis for shuttle mutagenesis when Labigne used the suicide vector to shuttle modified *Campylobacter* genes into *Campylobacter* cells for recombination with chromosomal homologs (Labigne et al., 1992). Labigne et al. also constructed a *Campylobacter* leucine synthetase mutant by shuttle transposon mutagenesis. This technique allowed the construction of a *Campylobacter* mutant by using a recombinant transposon to introduce mutations into a cloned gene in *E. coli.* The advantage of this approach was that the transposon need not transpose in *Campylobacter* spp., which had posed a problem previously. Wang and Taylor (1990) constructed additional shuttle vectors that contain the LacZ' determinant, an origin of replication derived from the *C. coli* plasmid pIP1445, which functions in *Campylobacter* spp, an origin of replication from plasmid pUC13, which functions in *E. coli,* and three antibiotic resistance markers encoding kanamycin resistance specified by an aminoglycoside phosphotransferase, tetracycline resistance specified by the TetO determinant, and chloramphenicol resistance specified by chloramphenicol acetyltransferase (Labigne-Roussel et al., 1987; Vieira and Messing, 1982; Wang and Taylor, 1990). Wang and Taylor have also constructed vectors for the study of transformation in *Campylobacter* spp. (Wang and Taylor, 1990; Taylor, 1992). These have an origin of replication from a *C. coli* plasmid, a resistance determinant that functions in *Campylobacter* spp. and a multiple-cloning site. Yao et al. constructed a *Campylobacter* shuttle vector which is a 6.5–6.8 kb plasmid with *Campylobacter* and *E. coli* replicons, a multiple-cloning site flanked by T7 and T3 late promoters and M13 forward and reverse priming sites, the *lacZa* gene, *oriT* and either a kanamycin or chloramphenicol resistance gene from *Campylobacter,* which functions in both hosts (Yao et al., 1993). All of these vectors provide tools with which to explore the role of specific genes in controlling virulence.

5.2. GENETIC EXCHANGE MECHANISMS

DNA has been successfully introduced into *C. jejuni* by conjugation, natural transformation, and electrotransformation.

5.2.1. Conjugative Plasmids

Taylor was the first to report the transfer of tetracycline resistance encoding plasmids from *C. jejuni* to *C. fetus* subsp. *fetus* (Taylor et al., 1981). Numerous other reports describing transfer of antibiotic resistance encoding plasmids between *C. jejuni* and *C. coli* and other *Campylobacter* recipients have been published (Tenover et al., 1983; Tenover et al., 1985; Kotarski et al., 1986; Sagara et al., 1987). The host range of these plasmids is limited to closely related members of the genera. No sex pili have been identified in plasmid containing *Campylobacter* cells (Taylor, 1992). The similarity of erythromycin resistance genes from *Campylobacter* and *Streptococci* suggests that a genetic pathway between the two exists (Trieu-Cuot et al., 1987).

5.2.2. Natural Transformation

Some strains of *C. jejuni* and *C. coli* will take up DNA in the absence of special treatments (Wang and Taylor, 1990; Wassenaar et al., 1993). Five *C. coli* strains and three of six *C. jejuni* strains were found to be naturally competent when uptake of chromosomal markers for streptomycin resistance and nalidixic acid resistance were tested (Wang and Taylor, 1990). In these studies, transformation of plasmid DNA was much less efficient than for chromosomal DNA. However, plasmid transformation efficiency was greatly increased by the addition of a homologous plasmid, which acted as a rescue plasmid by recombining with the incoming plasmid (Wang and Taylor, 1990). Wassenaar et al. also found that natural transformation was efficient for the uptake of *C. jejuni* chromosomal DNA and not plasmid DNA (Wassenaar et al., 1993).

5.2.3. Electroporation

Campylobacter cells may be transformed using electroporation with plasmids or naked DNA (Miller et al., 1988; Wassenaar et al., 1993). Plasmids replicate in the cytoplasm while naked DNA may be recombined and incorporated into the chromosome. Tompkins observed that *Campylobacter* cells restricted incoming *E. coli* DNA that was not modified by a *Campylobacter*-specific restriction modification system. Shuttle plasmids grown in *Campylobacter* cells could be transformed efficiently while those grown in *E. coli* could not (Labigne-Roussel et al., 1988). Differential uptake of plasmids grown in *Campylobacter* versus those grown in *E. coli* suggested that *Campylobacter* cells restrict incoming *E. coli* DNA that was not modified (Miller et al., 1988). It is likely that *Campylobacter* spp. have specific sequences that allow for binding to and uptake of DNA into cells such as has been identified in *Haemophilus* spp. (Danner et al., 1980) and *Neisseria gonorrhoeae* (Goodman and Scocca, 1988). These sequences have been termed uptake sequences and are 11 bp and 10 bp in size, respectively.

5.2.4. Bacteriophages

Bacteriophages have been reported in *C. coli* and *C. jejuni* (Ritchie et al., 1983; Grajewski et al., 1985; Salama et al., 1989). Currently, bacteriophages have been used solely to type *Campylobacter* spp.; no published reports of bacteriophage transduction of genetic markers exist.

5.2.5. New Approaches

Wren, Henderson and Ketley (1994) have developed a new approach for the production of defined *Campylobacter* deletion mutants. This technique uses amplification of conserved gene sequences with degenerate oligonucleotides primers (PCRDOP) and inverse PCR mutagenesis (IPCRM) to obtain a cloned gene with mutations. A selectable marker or reporter gene is then inserted, the gene cloned into a suicide vector, a wildtype strain transformed and the mutants identified and characterized.

5.3. GENETIC MAPPING OF CAMPYLOBACTER SPP.

5.3.1. Genome

The genome of *C. jejuni* has been sequenced, ordered and is available through the Sanger Centre, U.K. Genome maps for *C. jejuni* and *C. coli* chromosomes are available (Taylor et al., 1992; Taylor 1992) and have recently been updated (Newnham et al., 1996; Karlyshev et al., 1998). Promoter sequences have been analyzed to determine consensus regions (Wosten et al., 1998). The *Campylobacter* genome is approximately 36% the size of the *E. coli* genome (Tompkins, 1992). The *C. jejuni* and *C. coli* genomes were estimated at 1.7 Mb (Taylor, 1992) and reevaluated at 1.8 Mb (Kim et al., 1992). Chromosomal genes encoding the ribosomal RNA operon (rRNA), 16S rRNA, 23S rRNA, ribosomal proteins, flagellar structural proteins, antibiotic resistance genes (streptomycin resistance (StrA), erythromycin resistance), isoleucyl-tRNA synthetase gene, and iron-uptake regulatory (*fur*) gene have been mapped by multiple techniques including restriction endonuclease digestion and pulsed-field gel electrophoresis (Chang and Taylor, 1990; Nuijten et al., 1990; Kim et al., 1993; Trust et al., 1994; Kim et al., 1995; Hong et al., 1995; Bourke et al., 1996; Newnham et al., 1996). Other genes that have been recently cloned but not mapped include *glyA* encoding serine hydroxymethyltransferase (Chan and Bingham, 1992), *Ast* encoding arylsulfatase (Yao and Guerry, 1996), *katA* encoding catalase (Grant and Park, 1995), *proA* encoding gamma-glutamyl phospate reductase (Louie and Chan, 1993), *ftsA* cell division gene (Griffiths et al., 1996), *tig* trigger factor gene (Griffiths et al., 1995), superoxide dismutase gene (Purdy and Park, 1994), *argH* gene encoding argininosuccinate lyase (Hani and Chan, 1994), lysyl-tRNA synthetase gene (Chan and Bingham, 1992), *hup* gene encoding a histone-like protein (Konkel et al., 1994), and *aroA* gene encoding 5-enolpyruvyl-shikimate-3-phosphate synthase (Wosten et al., 1996). Also, a heat-inducible *lon* gene was cloned and sequenced in *C. jejuni* with 56.6% homology to the *lon* gene of *Helicobacter pylori* (Thies et al., 1998).

6. HOST RESPONSE TO *CAMPYLOBACTER* SPP.

Host inflammatory and immune responses affect the pathogenesis of *C. jejuni* infections. Intestinal mucus and secretory antibodies are primary barriers to adherence of *C. jejuni* to intestinal epithelial cells necessary for initiation of disease (McSweegan et al., 1987). In humans, anti-*C. jejuni*-specific serum IgG, IgM, and IgA levels rise during the month after infection (Wallis, 1994). IgG confers lasting immunity that is strain specific. IgA plays a significant role in the clearance of *C. jejuni* from the intestine causing immobilization, aggregation and activation of complement by the alternative pathway. *C. jejuni* can stimulate phagocytosis by polymorphonuclear neutrophils, especially if the bacteria have previously been opsonized (Wallis, 1994). Monocytes that ingest *C. jejuni* convert them to their coccoid forms, although the bacterium may escape prior to degeneration (Wallis, 1994). Monocytes may play a role in the translocation of the organism to the bloodstream, but the role of phagocytosis as a defense mechanism against *C. jejuni* seems to be secondary, with the humoral response having the primary role against infection (Wallis, 1994). Susceptibility to infection depends on previous exposure, and multiple exposure confers homologous protective immunity.

6.1. HOST CYTOKINES INFLUENCE THE COURSE OF INFECTION WITH *C. JEJUNI*

Host cytokines influence the course of infection with *C. jejuni.* Several strains of *C. jejuni* produced secretion of the proinflammatory cytokine IL-8 by INT407 cells (Hickey et al., 1999). Strains causing the highest IL-8 production had the highest amount of adherence and invasion.

Exogenously administered cytokines were used to influence development of specific immune responses and to enhance the immune competency of mice challenged with *C. jejuni* by an oral route (Baquar et al., 1993). A total of 300 units of human recombinant IL-2 (rIL-2), IL-5 (rIL-5), and IL-6 (rIL-6) were given to mice orally before, during and after infection with *C. jejuni* to determine the effect on intestinal colonization by the bacteria and development of *C. jejuni*-specific immune responses. Oral rIL-6 was associated with enhanced intestinal and systemic *C. jejuni*-specific IgA responses and a 3-log-unit reduction in the number of *C. jejuni* found in the feces confirming the importance of mucosal antibody in controlling this bacterium. rIL-5 treatment also reduced *C. jejuni* numbers comparably, but the rate of clearance was slower. rIL-5 probably acted to increase maturation of IgA-positive B cells into IgA secreting cells. rIL-2 had no effect on *C. jejuni* colonization in naive animals given a primary infection. rIL-2 treatment suppressed IgA titers in the intestine of these mice, which may explain this effect. However, *C. jejuni* infection was controlled only in rIL-2-treated mice and not in rIL-6- or rIL-5-treated mice after a homologous challenge suggesting that IL-2 functions to stimulate effective cell-mediated immunity during secondary exposure. From these results it appears that cell-mediated immunity and secretory immunoglobulin are important in protection against *C. jejuni,* but additional work is needed to understand the mechanisms that allow for opportunistic invasion by this bacterium.

6.2. AUTOIMMUNE RESPONSES IN CAMPYLOBACTER INFECTION

C. jejuni colitis can result in primary or secondary inflammatory bowel disease (Lambert et al., 1979; Berberian et al., 1994). BK2+ anti-erythrocyte antibodies were elevated in most sera from patients with Crohn's Disease (CD), Ulcerative Colitis (UC) and in 10 of 38 patients with noninflammatory enterocolitis caused by *C. jejuni* (Berberian et al., 1994). This suggests a common pathogenic factor in CD, UC and *C. jejuni* enterocolitis that stimulates B cell activation for the immunoglobulin heavy chain region VH3-15.

Guillain-Barré is thought to be caused by antibodies elicited by bacterial antigens that cross-react with host antigens, particularly cell surface gangliosides (Aspinall et al., 1994). Lipopolysaccharides from *C. jejuni* serotypes 0:4

and 0:19 isolated from GBS patients had core oligosaccharides with terminal structures similar to human gangliosides GM1 and GD1a (Aspinall et al., 1994). Salloway et al. (1996) found additional evidence for molecular mimicry mechanisms due to *C. jejuni* in a patient with Miller-Fisher syndrome (a variant of GBS). This patient had antibodies directed against a core trisaccharide epitope from *C. jejuni* LPS. This epitope is similar to the terminal region of human ganglioside GD3 and is found in LPS cores of serotype 0:19 *C. jejuni* from Guillain-Barré patients and not from LPS cores of non-neuropathic *C. jejuni.* HLA-class II haplotype can also influence autoimmune reactivity. HLA-class II alleles, particularly HLA-DQB1°03, were significantly associated with Guillain-Barré syndrome, Miller-Fisher syndrome and preceding *C. jejuni* infection (Rees et al., 1995b).

7. FUTURE APPROACHES

Basic and applied research to understand and control *Campylobacter* is urgently needed. Important approaches that are currently underway include mechanisms of pathogenesis studies with in vitro and in vivo models, vaccine-mediated control for animal reservoirs, mutational analysis and mapping of virulence genes of *Campylobacters,* and cloning, mapping and analysis of genes controlling antibiotic resistance. Practical approaches are also needed for application in Hazard Analysis Critical Control Point (HACCP) programs for food safety aimed at decreasing foodborne hazards. Human infection may be controlled with appropriate food-related hygiene and other behavioral measures. However, antibiotic resistance and other factors may spur the need for human *C. jejuni* vaccines, which are currently in commercial testing and development. Advances in understanding the pathogenesis of *Campylobacter* spp. using the numerous molecular tools that have recently been developed should produce significant progress in the control and elimination of this threat in the future.

8. REFERENCES

Abimiku, A. G., and Dolby, J. M. 1988. "Cross-protection of infant mice against intestinal colonization by *Campylobacter jejuni:* importance of heat-labile serotyping (Lior) antigens," *J. Med. Micr.* 26(4):265–268.

Abimiku, A. G., Dolby, J. M., and Borriello, S. P. 1989. "Comparison of different vaccines and induced immune response against *Campylobacter jejuni* colonization in the infant mouse," *Epid. Inf.* 102(2):271–280.

Abram, D. D., and Potter, N. N. 1984. "Survival of *Campylobacter jejuni* at different temperatures in broth, beef, chicken and cod supplemented with sodium chloride," *J. Food Prot.* 47:795–800.

Adak, G. K., Cowden, J. M., Nicholas, S., and Evans, H. S. 1995. "The Public Health Laboratory Service national case-control study of primary indigenous sporadic cases of campylobacter infection," *Epid. Inf.* 115:15–22.

Adler-Mosca, H., Luthy-Hottenstein, J., Martinetti-Lucchini, G., Burnens, A., and Altwegg, M. 1991. "Development of resistance to quinolones in five patients with campylobacteriosis treated with norfloxacin or ciprofloxacin," *Eur. J. Clin. Micro. Infect. Dis.* 10(11):953–957.

Aguero-Rosenfield, M. E., Yang, X. H., and Nachamkin, I. 1990. "Infection of adult Syrian hamsters with flagellar variants of *Campylobacter jejuni,*" *Inf. Imm.* 58:2214–2219.

Akhtar-SQ, Huq-E. 1989. "Effect of *Campylobacter jejuni* extracts and culture supernatants on cell culture," *J-Trop-Med-Hyg.* Apr, 92(2):80–5.

Alm, R. A., Guerry, P., Power, M. E., and Trust, T. J. 1992. "Variation in antigenicity and molecular weight of *Campylobacter coli* VC167 flagellin in different genetic backgrounds," *J. Bact.* 174(13):4230–4238.

Alm, R. A., Guerry, P., and Trust, T. J. 1993. "Distribution and polymorphism of the flagellin genes from isolates of *Campylobacter coli* and *Campylobacter jejuni,*" *J. Bact.* 175(10):3051–3057.

Amako, K., Wai, S. N., Umeda, A., Shigematsu, M., and Takade, A. 1996. "Electron microscopy of the major outer membrane protein of *Campylobacter jejuni,*" *Micr. Immunol.* 40(10):749–754.

American Society for Microbiology Press Release. May, 1997. "Financial impact of foodborne diseases," *Economic Research Service.* USDA, Washington, D.C.:

Angulo, F. J. 1996. "Food safety symposium: responding to the changing epidemiologic characteristics of foodborne diseases," *JAVMA* 208(9):1396–1404.

Aragon, V., Chao, K., and Dreyfus, L. A. 1997. "Effect of cytolethal distending toxin on F-actin assembly and cell division in Chinese hamster ovary cells," *Infect. Immun.* 65(9):3774–3780.

Aspinall, G. O., Fujimoto, S., McDonald, A. G., Pang, H., Kurjanczyk, L. A., and Penner, J. L. 1994. "Lipopolysaccharides from *Campylobacter jejuni* associated with Guillain-Barre syndrome patients mimic human gangliosides in structure," *Inf. Imm.* 62(5):2122–2125.

Aspinall, G. O., McDonald, A. G., Raju, T. S., Pang, H., Moran, A. P., Penner, J. L. 1993. "Chemical structures of the core regions of *Campylobacter jejuni* serotypes O:1, O:4, O:23, and O:36 lipopolysaccharides," *Eur. J. Biochem.* 213(3):1017–1027.

Aspinall, G. O., McDonald, A. G., Raju, T. S., Pang, H., Mills, S. D., Kurjanczyk, L. A., and Penner, J. L. 1992. "Serological diversity and chemical structures of *Campylobacter jejuni* low-molecular-weight lipopolysaccharides," *J. Bact.* 174(4):1324–1332.

Babakhani, F. K., Bradley, G. A., and Joens, L. A. 1993. "Newborn piglet model for campylobacteriosis," *Inf. Imm.* 61(8):3466–3475.

Babakhani, F. K., and Joens, L. A. 1993. "Primary swine intestinal cells as a model for studying *Campylobacter jejuni* invasiveness," *Inf. Imm.* 61:2723–2726.

Baig, B. H. Wachsmuth, J. K., Morris, G. K. Hill, W. E. 1986. "Probing of *Campylobacter jejuni* with DNA coding for *Escherichia coli* heat-labile enterotoxin [letter]," *J-Infect-Dis.* Sep. 154(3):542.

Banffer, J. R. 1985. "Biotypes and serotypes of *Campylobacter jejuni* and *Campylobacter coli* strains isolated from patients, pigs and chickens in the region of Rotterdam," *J. Inf.* 10(3):277–281.

Baqar, S., Pacheco, N. D., Rollwagen, F. M. 1993. "Modulation of mucosal immunity against *Campylobacter jejuni* by orally administered cytokines," *Antimicr. Agents and Chemotherapy* 37:2688–2692.

Baqar, S., Bourgeois, A. L., Applebee, L. A., Mourad, A. S., Kleinosky, M. T., Mohran, Z., and Murphy, J. R. 1996. "Murine intranasal challenge model for the study of *Campylobacter* pathogenesis and immunity," *Inf. Immun.* 64:4933–4939.

Bean, N. H., and Griffin, P. M. 1990. "Foodborne disease outbreaks in the United States, 1973–1987: pathogens, vehicles and trends," *J. Food Prot.* 53(9):804–817.

Beer, W., Adam, M., and Seltmann, G. 1986. "Monosaccharide composition of lipopolysaccharides from *Campylobacter jejuni* and *Campylobacter coli*," J. Basic Micr. 26(4):201–204.

Beery, J. T., Hugdahl, M. B., and Doyle, M. P. 1988. "Colonization of gastrointestinal tracts of chicks by *Campylobacter jejuni*," *Appl. Env. Micr.* 54(10):2365–2370.

Bell, J. A., and Manning, D. D. 1990. "A domestic ferret model of immunity to *Campylobacter jejuni*-induced enteric disease," *Inf. Imm.* 58:1848–1852.

Berberian, L. S., Valles-Ayoub, Y., Gordon, L. K., Targan, S. R., and Braun, J. 1994. "Expression of a novel autoantibody defined by the VH3-15 gene in inflammatory bowel disease and *Campylobacter jejuni* enterocolitis," *J. Immunol.* 153(8):3756–3763.

Bernard, E., Roger, P. M., Carles, D., Bonaldi, V., Fournier, J. P., and Dellamonica, P. 1989. "Diarrhea and *Campylobacter* infections in patients infected with the human immunodeficiency virus," *J. Inf. Dis.* 159:143–144.

Bibb, M. J., Janssen, G. R., and Ward, J. M. 1985. "Cloning and analysis of the promoter region of the erythromycin resistance gene (ermE) of *Streptomyces erythraeus*," *Gene* 38(1–3):215–226.

Black, R. E., Levine, M. M., Clements, M. L., Hughes, T. P., and Blaser, M. J. 1988. "Experimental *Campylobacter jejuni* infection in humans," *J. Inf. Dis.* 157(3):472–479.

Blaser, M. J., Taylor, D. N., and Feldman, R. A. 1983. "Epidemiology of *Campylobacter jejuni* infections," *Epid. Rev.* 5:157–176.

Blaser, M. J., Taylor, D. N., and Feldman, R. A. 1984. "Epidemiology of *Campylobacter* infections," in *Campylobacter Infection in Man and Animals,* ed., Butzler, J. P. Boca Raton, FL:CRC Press. pp. 143–161.

Blaser, M. J., Weiss, S. H., and Barrett, T. J. 1982. "*Campylobacter* enteritis associated with a healthy cat," *JAMA* 247(6):816.

Bok, H. E., Greeff, A. S., and Crewe-Brown H. H. "Incidence of toxigenic Campylobacter strains in South Africa," *J. Clin. Microbiol.* Jun. 29(6):1262–4.

Bolla, J. M., Loret, E., Zalewski, M., and Pages, J. M. 1995. "Conformational analysis of the *Campylobacter jejuni* porin," *J. Bact.* 177(15):4266–4271.

Bolton, C. F. 1995. "The changing concepts of Guillain-Barre syndrome," *N.E.J.M.* 333(21):1415–1416.

Boosinger, T. R., and Powe, T. A. 1988. "*Campylobacter jejuni* infections in gnotobiotic pigs," *Am. J. Vet. Res.* 49:456–458.

Boosinger, T. R., Blevins, W. T., Heron, J. V., and Sunter, J. L. 1990. "Plasmid profiles of six species of *Campylobacter* from human beings, swine and sheep," *Am. J. Vet. Res.* 51:718–722.

Bourke, B., al-Rashid, S. T., Bingham, H. L., and Chan, V. L. 1996. "Characterization of *Campylobacter upsaliensis fur* and its localization in a highly conserved region of the *Campylobacter* genome," *Gene.* 183(1–2):219–224.

Bracewell, A. J., Reagan, J. O., Carpenter, J. A., and Blankenship, L. C. 1985. "Incidence of *Campylobacter jejuni/coli* on pork carcasses in the northeast Georgia area," *J. Food Prot.* 48(9):808–810.

Bradbury, W. C., and Munroe, D. L. G. 1985. "Occurrence of plasmid and antibiotic resistance among *Campylobacter jejuni* and *Campylobacter coli* isolated from healthy and diarrhetic animals," *J. Clin. Micr.* 22:339–346.

Bradbury, W. C., Marko, M. A., Hennessy, J. N., and Penner, J. L. 1983. "Occurrence of plasmid DNA in serologically defined strains of *Campylobacter jejuni* and *Campylobacter coli*," *Inf. Imm.* 40(2):460–463.

Bruce, D., Zochowski, W., and Fleming, G. A. 1980. "*Campylobacter* infections in dogs and cats," *Vet. Rec.* 107(9):200–201.

Bry, L., Falk, P. G., Midtvedt, T., and Gordon, J. I. 1996. "A model of host-microbial interactions in an open mammalian ecosystem," *Science* (Wash. D.C.) 273:1380–1383.

Bukholm, G., and Kapperud, G. 1987. "Expression of *Campylobacter jejuni* invasiveness in cell cultures coinfected with other bacteria," *Inf. and Imm.* 55:2816–2821.

Burnens, A., Stucki, U., Nicolet, J., and Frey, J. 1995. "Identification and characterization of an immunogenic outer membrane protein of *Campylobacter jejuni*," *J. Clin. Micr.* 33(11): 2826–2832.

Burr, D. H., Caldwell, M. B., Bourgeois, A. L., Morgan, H. R., Wistar, R. Jr., and Walker, R. I. 1988. "Mucosal and systemic immunity to *Campylobacter jejuni* in rabbits after gastric inoculation," *Inf. Imm.* 56(1):99–105.

Burridge, R., Warren, C., and Phillips, I. 1986. "Macrolide, lincosamide and streptogramin resistance in *Campylobacter jejuni/coli*," *J. Antimicr. Chemother.* 17(3):315–321.

Burucoa, C., Fremaux, C., Pei, Z., Tummuru, M., Blaser, M. J., Cenatiempo Y., and Fauchere, J. L. 1995. "Nucleotide sequence and characterization of peb4A encoding an antigenic protein in *Campylobacter jejuni*," *Res. Micr.* 146(6):467–476.

Butzler, J. P. 1984. *Campylobacter Infections in Man and Animals.* Boca Raton, FL.

Butzler, J. P. and Oosterom, J. 1991. "*Campylobacter* pathogenicity and significance in foods," *Int. J. Food Micr.* 12(1):1–8.

Butzler, J. P., and Skirrow, M. B. 1979. "*Campylobacter* enteritis," *Clin Gastroenter.* 8:737–765.

Buzby, J. C., and Roberts, T. 1996. "ERS updates US foodborne disease costs for seven pathogens," *Food Rev.* 19(3):20–25.

Cadranel, S., Rodesch, P., Butzler, J. P., and Dekeyser, P. 1973. "Enteritis due to 'related Vibrio' in children," *Am. J. Dis. Child.* 126(2):152–155.

Caldwell, M. B., Walker, R. I., Stewart, S. D., and Rogers, J. E. 1983. "Simple adult rabbit model for *Campylobacter jejuni* enteritis." *Inf. Imm.* 42(3):1176–1182.

Calva, E., Torres, J., Vazquez, M., Angeles, V., de-la-Vega, H., and Ruiz-Palacios-G. M. 1989. *Campylobacter jejuni* chromosomal sequences that hybridize to *Vibrio cholerae* and *Escherichia coli* LT enterotoxin genes." *Gene.* Feb 20 75(2):243–51.

CDC. 1994. "Addressing Emerging Infectious Disease Threats to Health: A Prevention Strategy for the United States, Atlanta, GA: US Dept. of Health and Human Services, Public Health Service.

CDC. 1997. "Foodborne diseases active surveillance network, 1996," *MMWR* 46(12):258261.

Chan, V. L. and Bingham, H. L. 1991. "Complete sequence of the *Campylobacter jejuni glyA* gene encoding serine hydroxymethyltransferase," *Gene* 101(1):51–58.

Chan, V. L. and Bingham, H. L. 1992. "Lysyl-tRNA synthetase gene of *Campylobacter jejuni*." *J. Bact.* 174(3):695–701.

Chan, V. L., Louie, H., and Bingham, H. L. 1995. "Cloning and transcription regulation of the ferric uptake regulatory gene of *Campylobacter jejuni* TGH9011," *Gene,* 164(1):25–31.

Chang, N. and Taylor, D. E. 1990. "Use of pulsed-field agarose gel electrophoresis to size genomes of *Campylobacter* spp. and to construct a SalI map of *Campylobacter jejuni* UA580," *J. Bact.* 172(9):5211–5217.

Chart, H., Conway, D., Frost, J. A., and Rowe, B. 1996. "Outer membrane characteristics of *Campylobacter jejuni* grown in chickens," *FEMS Micr. Lett.* 145(3):469–72.

Charvalos, E., Peteinaki, E., Spyridaki, I., Manetas, S., Tselentis, Y. 1996. "Detection of ciprofloxacin resistance mutations in *Campylobacter jejuni gyrA* by nonradioisotopic single-strand conformation polymorphism and direct DNA sequencing," *J. Clin. Lab. Anal.* 10(3):129–133.

Charvalos, E., Tselentis, Y., Hamzehpour, M. M., Kohler, T., and Pechere, J. C. 1995. "Evidence for an efflux pump in multi drug-resistant *Campylobacter jejuni,*'' *Antimicr. Agents Chemother.* 39(9):2019–2022.

Cinco, M., Banfi, E., Ruaro, E., Crevatin, D., Crotti, D. 1984. ''Evidence for L-fucose (6-deoxy-L-galactopyranose)-mediated adherence of *Campylobacter* spp. to epithelial cells,'' *FEMS Micr. Lett.* 347–351.

Coker, A. O. and Obi, C. I. 1989. ''Endotoxic activity and enterotoxigenicity of human strains of *Campylobacter jejuni* isolated from patients in a Nigerian hospital,'' *Cent Afr J Med.* Nov 35(11):524–527.

Collins, S., O'Loughlin, E., O'Rourle, J., Li, Z., Lee, A., and Howden, M. 1992. ''A cytotonic, cholera toxin-like protein produced by *Campylobacter jejuni,*'' *Comp. Biochem. Physiol.* B. 103(2):299–303.

Coote, J. G., and Arain, T. 1996. ''A rapid, colourimetric assay for cytotoxin activity in *Campylobacter jejuni,*'' *FEMS Immunol. Med. Microbiol.* 13(1):65–70.

Costerton, J. W., Rozee, K. P., and Cheng, K. J. 1983. ''Colonization of particulates, mucus, and intestinal tissue,'' *Prog. Nutr. Food Sci.* 7(3–4):91–105.

Cover, T. L., Perez-Perez, G. I. and Blaser, M. J. 1990. ''Evaluation of cytotoxic activity in fecal filtrates from patients with *Campylobacter jejuni* or *Campylobacter coli* enteritis,'' *FEMS Microbiol Lett.* Aug 58(3):301–4.

Crawthraw, S. A., Wassenaar, T. M., Ayling, R., and Newell, D. G. 1996. ''Increased colonization potential of *Campylobacter jejuni* strain 81116 after passage through chickens and its implication on the rate of transmission within flocks,'' *Epidemiol. Infect.* 117(1):213–5.

Daikoku, T., Kawaguchi, M., Takama, K., Suzuki, S. 1990. ''Partial purification and characterization of the enterotoxin produced by *Campylobacter jejuni,*'' *Infect. Immun.* Aug 58(8): 2414–2419.

Daikoku, T., Suzuki, S., Oka, S., Takama, K. ''Profiles of enterotoxin and cytotoxin production in *Campylobacter jejuni* and *C. coli,*'' FEMS Microbiol Lett. Mar 49(1):33–6.

Danner, D. B., Deich, R. A., Sisco, K. L., Smith, H. O. 1980. ''An eleven-base-pair sequence determines the specificity of DNA uptake in *Haemophilus* transformation,'' *Gene* 11(3–4):311–318.

Davies, A. P., Gebhart, C. J., and Meric, S. A. 1984. ''*Campylobacter*-associated chronic diarrhea in a dog,'' *J. Am. Vet. Med. Assoc.* 184(4):469–471.

deMelo, M. A. and Pechere, J. C. 1988. ''Effect of mucin on *Campylobacter jejuni* association and invasion on HEp-2 cells,'' *Microb. Pathog.* 5(1):71–76.

deMelo, M. A., and Pechere, J. C. 1990. ''Identification of *Campylobacter jejuni* surface proteins that bind to eucaryotic cells in vitro,'' *Inf. Imm.* 58(6):1749–1756.

Deming, M. S., Tauxe, R. V., Blake, P. A., Dixon, S. E., Fowler, B. S., Jones, T. S., Lockamy, E. A., Patton, C. M., and Sikes, R. O. 1987. ''*Campylobacter* enteritis at a university: transmission from eating chicken and from cats'' [published erratum appears in *Am. J. Epid.* 126(6):1220], *Am. J. Epid.* 126(3):526–537.

Dilworth, C. R., Lior, H., and Belliveau, M. A. 1988. ''*Campylobacter* enteritis acquired from cattle,'' *Can. J. Pub. Health* 79:60–62.

Doig, P., Yao, R., Burr, D. H., Guerry, P., and Trust, T. J. 1996. ''An environmentally regulated pilus-like appendage involved in *Campylobacter* pathogenesis,'' *Mol. Micr.* 20(4):885–894.

Doyle, M. P. and Roman, D. J. 1981. ''Growth and survival of *Campylobacter fetus* subsp. *jejuni* as a function of temperature and pH,'' *J. Food Prot.* 44:596–601.

Doyle, M. P. and Roman, D. J. 1982a. ''Prevalence and survival of *Campylobacter jejuni* in unpasteurized milk,'' *Appl. Env. Micr.* 44(5):1154–1158.

Doyle, M. P. and Roman, D. J. 1982b. "Recovery of *Campylobacter jejuni* and *Campylobacter coli* from inoculated foods by selective enrichment," *Appl. Env. Micr.* 43(6):1343–1353.

Doyle, M. P. and Roman, D. J. 1982c. "Sensitivity of *Campylobacter jejuni* to drying," *J. Food Prot.* 45(6):507–510.

Doyle, M. P. and Schoeni, J. L. 1986. "Isolation of *Campylobacter jejuni* from retail mushrooms," *Appl. Env. Micr.* 51(2):449–450.

Eberhart, C. E. and Dubois, R. N. 1995. "Eicosanoids and the gastrointestinal tract," *Gastroenterol.* 109:285–301.

Echeverria, P., Taylor, D. N., Lexsomboon, U., Bhaibulaya, M., Blacklow, N. R., and Sakazaki, R. 1989. "Case-control study of endemic diarrheal disease in Thai children under 5 years old," *J. Inf. Dis.* 159:543–548.

Eckmann, L., Stenson, W. F., Savidge, T. C., Lowe, D. C., Barrett, K. E., Fierer, J., Smith, J. R. and Kagnoff, M. F. 1997. "Role of intestinal epithelial cells in the host secretory response to infection by invasive bacteria: bacterial entry induces epithelial prostaglandin H sythetase-2 expression, and prostaglandin E_2 and F_2 production," *J. Clin. Invest.* 100(2):296–309. Referenced in "Perspectives series: Host/pathogen interactions: Epithelial cells as sensors for microbial infection," M. F. Kagnoff, L. Eckmann, *J. Clin Inv.* 100(1):6–10.

Edmonds, P., Patton, C. M., Griffin, P. M., Barrett, T. J., Schmid, G. P., Baker, C. N., Lambert, M. A., and Brenner, D. J. 1987. "*Campylobacter hyointestinalis* associated with human gastrointestinal disease in the United States," *J. Clin. Micr.* 25:685–691.

Edwards, C. K., Hedegaard, H. B., Zlotnik, A., Gangadharam, P. R., Johnston, R. B., Pabst, M. J. 1986. "Chronic infection due to *Mycobacterium intracellulare* in mice: association with macrophage release of prostaglandin E2 and reversal by injection of indomethacin, muramyl dipeptide, or interferon-gamma." *J. Imm.* 136:1820–1827.

Endtz, H. P., Ruijs, G. J., van Klingeren, B., Jansen, W. H., van der Reyden, T., and Mouton, R. P. 1991. "Quinolone resistance in *Campylobacter* isolated from man and poultry following the introduction of fluoroquinolones in veterinary medicine," *J. Antimicr. Chemother.* 27(2): 199–208.

Everest, P. H., Goossens, H., Butzler, J. P., Lloyd, D., Knutton, S., Ketley, J. M., and Williams, P. H. 1992. "Differentiated Caco-2 cells as a model for enteric invasion by *Campylobacter jejuni* and *C. coli.*" *J. Med. Micr.* 37(5):319–325.

Everest, P. H., Cole, A. T., Hawkey, C. J., Knutton, S., Goossens, H., Butzler, J. P., Ketley, J. M., and Williams P. H. 1993. "Roles of leukotrine B4, prostaglandin E2, and cyclic AMP in *Campylobacter jejuni*-induced intestinal fluid secretion," *Inf. Imm.* 61:4885–4887.

Fauchere, J. L., Veron, M., Lellouch-Tubiana, A., and Pfister, A. 1985. "Experimental infection of gnotobiotic mice with *Campylobacter jejuni*: colonisation of intestine and spread to lymphoid and reticulo-endothelial organs," *J. Med. Micr.* 20(2):215–24

Fauchere, J. L., Rosenau, A., Veron, M., Moyen, E. N., Richard, S., and Pfister, A. 1986. "Association with HeLa cells of *Campylobacter jejuni* and *Campylobacter coli* isolated from human feces," *Inf. Imm.* 54(2):283–287.

Ferrero, R. L. and Lee, A. 1988. "Motility of *Campylobacter jejuni* in a viscous environment: comparison with conventional rod-shaped bacteria," *J. Gen. Micr.* 134(Pt 1):53–59.

Field, L. H., Headley, V. L., Payne, S. M., and Berry, L. J. 1986. "Influence of iron on growth, morphology, outer membrane protein composition, and synthesis of siderophores in *Campylobacter jejuni*," *Inf. Imm.* 54:126–132.

Field, L. H., Underwood, J. L., Payne, S. M., and Berry, L. J. 1993. "Characteristics of an avirulent *Campylobacter jejuni* strain and its virulence-enhanced variants," *J. Med. Micr.* 38(4):293–300.

Finlay, B. B. and Falkow, S. 1997. "Common themes in microbial pathogenicity revisited," *Micr. Mol. Biol. Rev.* 61(2):136–169.

Fitzgeorge, R. B., Baskerville, A., and Lander, K. P. 1981. "Experimental infection of rhesus monkeys with a human strain of *Campylobacter jejuni,*" *J. Hyg.* 86:343–351.

Florin, I. and Antillon, F. 1992. "Production of enterotoxin and cytotoxin in *Campylobacter jejuni* strains isolated in Costa Rica," *J. Med. Micr.* 37(1):22–29.

Fox, J. G 1992. "In vivo Models of enteric campylobacteriosis: Natural and experimental infections" in *Campylobacter jejuni*: Current Status and Future Trends, eds., Nachamkin, I., Blaser, M. J., Tompkins, L. S. Washington, D.C.: American Society for Microbiology. pp. 131–138.

Fox, J. G., Claps, M. C., Taylor, N. S. and Ackermann, J. I. 1988. "*Campylobacter jejuni/coli* in commercially reared beagles: prevalence and serotypes," *Lab. Anim. Sci.* 38(3):262–265.

Fox, J. G., Ackerman, J. I., Taylor N., Claps, M. and Murphy, J. C. 1987. "*Campylobacter jejuni* infection in the ferret: an animal model of human campylobacteriosis," *Am. J. Vet. Res.* 48(1):85–90.

Frey, A., Giannasca, K. T., Weltzin, R., Giannasca, P. J., Reggio, H., Lencer, W. I., and Neutra, M. R. 1996. "Role of the glycocalyx in regulating access of microparticles to apical plasma membranes of intestinal epithelial cells: implications for microbial attachment and oral vaccine targeting," *J. Exp. Med.* 184(3):1045–59.

FSIS. 1995. *Food Safety Research: Current Activities and Future Needs.* Washington, D.C.: USDA.

Fujimoto, S., Yuki, N., Itoh, T., and Amako, K. 1992. "Specific serotype of *Campylobacter jejuni* associated with Guillain-Barre syndrome," [letter] *J. Inf. Dis.* 165:183.

Fukushima, H., Hoshina, K., Nakamura, R., and Ito, Y. 1987. "Raw beef, pork and chicken in Japan contaminated with *Salmonella* sp., *Campylobacter* sp., *Yersinia enterocolitica,* and *Clostridium perfringens*—a comparative study," *Zentralb. Bakteriol. Mikrobiol. Hyg.-B.* 184(1):60–70.

Funke, G., Baumann, R., Penner, J. L., and Altwegg, M. 1994. "Development of resistance to macrolide antibiotics in an AIDS patient treated with clarithromycin for *Campylobacter jejuni* diarrhea," *Eur. J. Clin. Micr. Inf. Dis.* 13(7):612–615.

Garcia, M. M., Lior, H., Stewart, R. B., Ruckerbauer, G. M., Trudel, J. R. R., and Skljarevski, A. 1985. "Isolation, characterization and serotyping of *Campylobacter jejuni* and *Campylobacter coli* from slaughter cattle," *Appl. Environ. Microbiol.* 49(3):667–672.

Garvis, S. G., Puzon, G. J., and Konkel, M. E. 1996. "Molecular characterization of a *Campylobacter jejuni* 29-kiloDalton periplasmic binding protein," *Inf. Imm.* 64(9):3537–3543.

Garvis, S. G., Tipton, S. L., and Konkel, M. E. 1997. "Identification of a functional homolog of the *Escherichia coli* and *Salmonella typhimurium cysM* gene encoding O-acetylserine sulfhydrylase B in *Campylobacter jejuni,*" *Gene* 185(1):63–67.

Gaudreau, C. and Gilbert, H. 1998. "Antimicrobial resistance of clinical strains of *Campylobacter jejuni* subsp. *jejuni* isolated from 1985 to 1997 in Quebec, Canada, *Antimicrobial. Agents and Chemotherapy* 42(8):2106–2108.

Gebhart, C. J., Ward, G. E., and Murtaugh, M. P. 1989. "Species-specific cloned DNA probes for the identification of *Campylobacter hyointestinalis,*" *J. Clin. Micr.* 27:2717–2723.

Gerloni, V. and Fantini, F. 1990. "Reactive arthritis," *Ped. Med. Chir.* (Italian) 12(5):447–451.

Gibreel, A., Sjogren E., Kaijser, B., Wretlind, B., Skold, O. 1998. "Rapid emergence of high-level resistance to quinolones in *Campylobacter jejuni* associated with mutational changes in *gyrA* and *parC, Antimicrobial Agents Chemotherapy* 42(12):3276–3278.

Gill, C. O., and Harris, L. M. 1983. "Limiting conditions of temperatures and pH for growth of 'thermophilic' campylobacters on solid media," *J. Food Prot.* 46:767–768.

Goodman, S. D. and Scocca, J. J. 1988. "Identification and arrangement of the DNA sequence recognized in specific transformation of *Neisseria gonorrhoeae,*" *Proc. Nat. Acad. Sci. USA* 85(18):6982–6986.

Goossens, H., Butler, J. P., and Takeda, Y. 1985a. "Demonstration of cholera-like enterotoxin production by *Campylobacter jejuni,*" *FEMS Microbiology Letters* 29:73–76.

Goossens, H., Rummens, E., Cadranel, S., Butzler, J. P. and Takeda, Y. 1985b. "Cytotoxic activity on Chinese hamster ovary cells in culture filtrates of *Campylobacter jejuni/coli* [letter]," *Lancet* Aug 31, 2(8453):511.

Gootz, T. D. and Martin, B. A. 1991. "Characterization of high-level quinolone resistance in *Campylobacter jejuni,*" *Antimicr. Agents. Chemoth.* 35(5):840–845.

Gore, A. 1996. "Emerging infections threaten national and global security," *ASM News* 62:448–449.

Grajewski, B. A., Kusek, J. W., and Gelfand, H. M. 1985. "Development of a bacteriophage typing system for *Campylobacter jejuni* and *Campylobacter coli,*" *J. Clin. Micr.* 22(1):13–18.

Grant, C. C., Konkel, M. E., Cieplak, W., and Tomkins, L. S. 1993. "Role of flagella in adherence, internalization, and translocation of *Campylobacter jejuni* in nonpolarized and polarized epithelial cultures," *Inf. Imm.* 61(5):1764–1771.

Grant, K. A., and Park S. F. 1995. "Molecular characterization of *katA* from *Campylobacter jejuni* and generation of a catalase-deficient mutant of *Campylobacter coli* by interspecific allelic exchange," *Microbiology* 141(Pt. 6):1369–1376.

Gregson, N. A., Rees, J. H., and Hughes, R. A. 1997. "Reactivity of serum IgG anti-GM1 ganglioside antibodies with the lipopolysaccharide fractions of *Campylobacter jejuni* isolates from patients with Guillain-Barre syndrome (GBS)," *J. Neuroimm.* 73(1–2):28–36.

Grewal, J., Manavathu, E. K., and Taylor, D. E. 1993. "Effect of mutational alteration of Asn-128 in the putative GTP-binding domain of tetracycline resistance determinant Tet(o) from *Campylobacter jejuni,*" *Antimicr. Agents Chemoth.* 37:2645–2649.

Griffiths, P. L., Dougan, G., and Connerton, I. F. 1996. "Transcription of the *Campylobacter jejuni* cell division gene *ftsA,*" *FEMS Micr. Lett.* 143(1):83–87.

Griffiths, P. L., Park, R. W., and Connerton, I. F. 1995. "The gene for *Campylobacter* triggers factor: evidence for multiple transcription start sites and protein products," *Microbiology* 141(Pt. 6):1359–1367.

Guerrant, R. L., Dickens, M. D., Wenzel, R. P., and Kapikian, A. Z. 1976. "Toxigenic bacterial diarrhea: nursery outbreak involving multiple bacterial strains," *J. Ped.* 89(6):885–891.

Guerrant, R. L., Fang, G., Pennie, R. A., and Pearson, R. D. 1992. "In vitro models for studying *Campylobacter jejuni* infections" in *Campylobacter jejuni: Current Status and Future Trends,* eds., Nachamkin, I., Blaser, M. J., and Tompkins, L. S. Washington, D.C.: American Society for Microbiology. pp. 160–167.

Guerrant, R. L., Wanke, C. A., Pennie, R. A., Barrett, L. J., Lima, A. A. and O'Brien, A. D. 1987. "Production of a unique cytotoxin by *Campylobacter jejuni,*" *Infect. Immun.* 1987 Oct; 55(10):2526–30.

Guerry, P. 1997. "Nonlipopolysaccharide surface antigens of *Campylobacter* species," *J. Infect. Dis.* 176 Suppl 2:S122–124.

Guerry, P., Alm, R. A., Power, M. E., and Trust, T. J. 1992. "Molecular and structural analysis of *Campylobacter* flagellin," in Nachamkin, I., Blaser, M. J., and Tompkins, L. S. eds., Campylobacter jejuni: *Current Status and Future Trends,* Washington, D.C.: American Society of Microbiology. pp. 267–281.

Guerry, P., Doig, P., Alm, R. A., Burr, D. H., Kinsella, N., and Trust, T. J. 1996. "Identification and characterization of genes required for post-translational modification of *Campylobacter coli* VC167 flagellin," *Molec. Micr.* 19(2):369–378.

Hani, E. K. and Chan, V. L. 1994. "Cloning, characterization and nucleotide sequence analysis of the *argH* gene from *Campylobacter jejuni* TGH9011 encoding arginosuccinate lyase," *J. Bact.* 176(7):1865–1871.

Harris, N. V., Weiss, N. S., and Nolan, C. M. 1986. "The role of poultry and meats in the etiology of *Campylobacter jejuni/coli* enteritis," *A.J.P.H.* 76(4):407–411.

Harris, N. V., Thompson, D., Martin, D. C., and Nolan, C. M. 1986. "A survey of *Campylobacter* and other bacterial contaminants of pre-market chicken and retail poultry and meats, King County, Washington," *A.J.P.H.* 76(4):401–406.

Hazeleger, W. C., Wouters, J. A., Rombouts, F. M., Abee, T. 1998. "Physiological activity of *Campylobacter jejuni* far below the minimal growth temperature," *Applied Environment Microbiol.* 64(10):3917–3922.

Hernandez, J., Fayos, A., Alonso, J. L., Owen, R. J. 1996. "Ribotypes and AP-PCR fingerprints of thermophilic *Campylobacters* from marine recreational waters," *J. Appl. Bact.* 80(2):157–164.

Hickey, T. E., Shihida, B., Bourgeous, A. L., Ewing, C. P., and Guerry, P. 1999. "*Campylobacter jejuni*-stimulated secretion of IL-8 by INT 407 cells," *Inf. Immun.* 67(1):88–93.

Hodge, D. S., and Terro, R. 1984. "Comparative efficacy of liquid enrichment medium for isolation of *Campylobacter jejuni,*" *J. Clin. Micr.* 19:434.

Hodge, J. P., and Krieg, N. R. 1994. "Oxygen tolerance estimates in *Campylobacter* species depend on the testing medium," *J. Appl. Bact.* 77(6):666–673.

Hong, Y., Wong, T., Bourke, B., and Chan, V. L. 1995. "An isoleucyl-tRNA synthetase gene from *Campylobacter jejuni,*" *Microbiology* 141(Pt. 10):2561–2567.

Hopkins, R. S., and Olmsted, R. N. 1985. "*Campylobacter jejuni* infection in Colorado: unexplained excess of cases in males," *Pub. Health Rep.* 100(3):333–336.

Hugdahl, M. B., Beery, J. T., and Doyle, M. P. 1988. "Chemotactic behavior of *Campylobacter jejuni,*" *Inf. Imm.* 56(6):1560–1566.

Humphrey, C. D., Montag, D. M., and Pittman, F. E. 1985. "Experimental infection of hamsters with *Campylobacter jejuni,*" *J. Inf. Dis.* 151:485–493.

Humphrey, T. J. 1988. "Peroxide sensitivity and catalase activity in *Campylobacter jejuni* following injury and during recovery," *J. Appl. Bact.* 64:337–343.

Jacobs, B. C., Schmitz, P. I. M., and Van der Meche, F. G. A. 1996. "*Campylobacter jejuni* infection and treatment for Guillain-Barre syndrome," *N. E. J. Med.* 335:208–209.

Jimenez, A., Velazquez, J. B., Rodriguez, J., Tinajas, A., and Villa, T. G. 1994. "Prevalence of fluoroquinalone resistance in clinical strains of *Campylobacter jejuni* isolated in Spain," *J. Anti. Micro. Chemother.* 33:188–190.

Johnson, K. E., and Nolan, C. M. 1985. "Community-wide surveillance of *Campylobacter jejuni* infection: evaluation of a laboratory-based method," *Diagn. Micr. Inf. Dis.* 3(5):389–396.

Johnson, W. M., and Lior, H. 1988. "A new heat-labile cytolethal distending toxin (CLDT) produced by *Campylobacter* spp.," *Microb. Pathog.* 4(2):115–126.

Johnson, W. M., and Lior, H. 1984. "Toxins produced by *Campylobacter jejuni* and *Campylobacter coli* [letter]," *Lancet* Jan 28; 1(8370):229–230.

Johnson, W. M., and Lior, H. 1986. "Cytotoxic and cytotonic factors produced by *Campylobacter jejuni, Campylobacter coli,* and *Campylobacter laridis,*" *J. Clin. Microbiol.* Aug; 24(2): 275–281.

Kagnoff, M. F., and Eckmann, L. 1997. "Epithelial cells as sensors for microbial infection," *J. Clin. Inv.* 100(1):6–10.

Kanwar, R. K., Ganguly, N. K., Kanwar, J. R., Kumar, L. and Walia, B. N. 1994. "Impairment of Na+, K(+)-ATPase activity following enterotoxigenic *Campylobacter jejuni* infection:

changes in Na+, Cl+ and 3-O-methyl-D-glucose transport in vitro, in rat ileum," *FEMS Microbiol Lett.* Dec 15; 124(3):381–385.

Kanwar, R. K., Ganguly, N. K., Kumar, L., Rakesh, J. Panigrahi, D. and Walia, B. N. 1995. "Calcium and protein kinase C play an important role in *Campylobacter jejuni*-induced changes in Na+ and Cl-transport in rat ileum in vitro," *Biochim Biophys Acta* Apr 24; 1270(2-3):179–192.

Karlyshev, A. V., Henderson, J., Ketley, J. M., and Wren, B. W. 1998. "An improved physical and genetic man of *Campylobacter jejuni* NCTC 11168 (UA580)," *Microbiology* 144(Pt 2):503–508.

Karmali, M. A., and Fleming, P. C. 1979. "*Campylobacter* enteritis in children," *J. Ped.* 94(4):527–533.

Karmali, M. A., DeGrandis, S., and Fleming, P. C. 1981. "Antimicrobial susceptibility of *Campylobacter jejuni* with special reference to resistance patterns of Canadian isolates," *Antimicr. Agents Chemother.* 19(4):593–597.

Keihlbauch, J. A., Chang, K. P., Albach, R. A., and Baum, L. L. 1985. "Phagocytosis of *Campylobacter jejuni* and its intracellular survival in mononuclear phagocytes," *Inf. Imm.* 48:446–451.

Kervella, M., Fauchere, J. L., Fourel, D., and Pages, J. M. 1992. "Immunological cross-reactivity between outer membrane pore proteins of *Campylobacter jejuni* and *Escherichia coli*," *FEMS Micr. Lett.* 99(2&3):281–285.

Kervella, M., Pages, J. M., Pei, Z., Grollier, G., Blaser, M. J., and Fauchere, J. L. 1993. "Isolation and characterization of two *Campylobacter* glycine-extracted proteins that bind to HeLa cell membranes," *Inf. Imm.* 61(8):3440–3448.

Ketley, J. M. 1997. "Pathogenesis of enteric infection by *Campylobacter jejuni*," *Microbiology* 143(Pt. 1):5–21.

Khalil, K., Lindblom, G. B., Mazhar, K., Sjogren, E., and Kaijser, B. 1993. "Frequency and enterotoxigenicity of *Campylobacter jejuni* and *C. coli* in domestic animals in Pakistan as compared to Sweden," *J. Trop. Med. Hyg.* Feb; 96(1):35–40.

Kim, N. W., Gutell, R. R., and Chan, V. L. 1995. "Complete sequences and organization of the *rrnA* operon from *Campylobacter jejuni* TGH9011 (ATCC43431)," *Gene* 164(1):101–106.

Kim, N. W., Bingham, H., Khawaja, R., Louie, H., Hani, E., Neote, K., and Chan, V. L. 1992. "Physical map of *Campylobacter jejuni* TGH9011 and localization of 10 genetic markers by use of pulsed-field gel electrophoresis," *J. Bact.* 174(11):3494–3498.

Kim, N. W., Lombardi, R., Bingham, H., Hani, E., Louie, H., Ng, D., and Chan, V. L. 1993. "Fine mapping of the three rRNA operons on the updated genomic map of *Campylobacter jejuni* TG9011 (ATCC 43431)," *J. Bact.* 175:7468–7470.

King, E. O. 1957. "Human infection with Vibrio foetus and a closely related Vibrio," *J. Inf. Dis.* 101:119.

Kita, E., Nishikawa, F., Kamikaidou, N., Nakano, A., Katsui, N., and Kashiba, S. 1992. "Mononuclear cell response in the liver of mice infected with hepatotoxigenic *Campylobacter jejuni*," *J. Med. Micr.* 37:326–331.

Klipstein, F. A., and Engert, R. F. 1984b. "Properties of crude *Campylobacter jejuni* heat-labile enterotoxin," *Infect. Immun.* 45(2):314–319.

Klipstein, F. A., and Engert, R. F. 1984a. "Purification of *Campylobacter jejuni* enterotoxin [letter]." *Lancet* May 19; 1(8386):1123–1124.

Klipstein, F. A., Engert, R. F., Short, H., and Schenk, E. A. 1985. "Pathogenic properties of *Campylobacter jejuni:* assay and correlation with clinical manifestations," *Inf. Immun.* 50(1):43–49.

Klipstein, F. A., and Engert, R. F. 1985. "Immunological relationship of the B subunits of

Campylobacter jejuni and *Escherichia coli* heat-labile enterotoxins," *Infect. Immun.* 48(3):629–633.

Klipstein, F. A., Engert, R. F., and Short, H. B. 1986. "Enzyme-linked immunosorbent assays for virulence properties of *Campylobacter jejuni* clinical isolates," *J. Clin. Microbiol.* 23(6):1039–1043.

Konkel, M. E., and Cieplak, W. Jr. 1992. "Altered synthetic response of *Campylobacter jejuni* to cocultivation with human epithelial cells is associated with enhanced internalization," *Inf. Imm.* 60(11):4945–4949.

Konkel, M. E. and Joens, L. A. 1990. "Effect of enteroviruses on adherence to and invasion of HEp-2 cells by *Campylobacter* isolates," *Inf. and Imm.* 58:1101–1105.

Konkel, M. E., Lobet, Y., and Cieplak, W. Jr. 1992d. "Examination of multiple isolates of *Campylobacter jejuni* for evidence of cholera toxin-like activity," in Campylobacter jejuni: *Current Status and Future Trends,* eds., Nachamkin, I., Blaser, M. J., Tompkins, L. S. Washington, D. C.: American Society for Microbiology. pp. 193–200.

Konkel, M. E., Mead, D. J., and Cieplak, W. Jr. 1993. "Kinetic and antigenic characterization of altered protein synthesis by *Campylobacter jejuni* during cultivation with human epithelial cells," *J. Inf. Dis.* 168(4):948–954.

Konkel, M. E., Mead, D. J., and Cieplak, W. 1996. "Cloning, sequencing, and expression of a gene from *Campylobacter jejuni* encoding a protein (Omp 18) with similarity to peptidoglycan-associated lipoproteins," *Inf. Imm.* 64(5):1850–1853.

Konkel, M. E., Corwin, M. D., Joens, L. A., and Cieplak, W. 1992a. "Factors that influence the interaction of *Campylobacter jejuni* with cultured mammalian cells," *J. Med. Micr.* 37(1):30–37.

Konkel, M. E., Hayes, S. F., Joens, L. A., and Cieplak, W. Jr. 1992c. "Characteristics of the internalization and intracellular survival of *Campylobacter jejuni* in human epithelial cell cultures," *Microb. Pathog.* 13(5):357–370.

Konkel, M. E., Marconi, R. T., Mead, D. J., and Cieplak, W. 1994. "Cloning and expression of the *hup* encoding, a histone-like protein of *Campylobacter jejuni,*" *Gene* 146(1):83–86.

Konkel, M. E., Mead, D. J., Hayes, S. F., and Cieplak, W. Jr. 1992b. "Translocation of *Campylobacter jejuni* across human polarized epithelial cell monolayer cultures," *J. Inf. Dis.* 166(2):308–315.

Konkel, M. E., Garvis, S. G., Tipton, S. L., Anderson, D. E. Jr., and Cieplak, W. Jr. 1997. "Identification and molecular cloning of a gene encoding a fibronectin-binding protein (CadF) from *Campylobacter jejuni,*" *Mol. Microbiol.* 24(5):953–963.

Konkel, M. E., Kim, B. J., Klena, J. D., Young, C. R., and Ziprin, R. 1998. "Characterization of the thermal stress response of *Campylobacter jejuni,*" *Infect. Immun.* 66(8):3666–3672.

Korhonen, L. K. and Martikainen, P. J. 1991. "Comparison of the survival of *Campylobacter jejuni* and *Campylobacter coli* in culturable form in surface water," *Can. J. Micr.* 37(7):530–533.

Kotarski, S. F., Merriwether, T. L., Tkalcevic, G. T., and Gemski, P. 1986. "Genetic studies of kanamycin resistance in *Campylobacter jejuni,*" *Antimicr. Agents Chemother.* 30(2):225–230.

Kuroki, S., Haruta, T., Yoshioka, M., Kobayashi, Y., Nukina, M., and Nakanishi, H. 1991. "Guillain-Barré syndrome associated with *Campylobacter* infection," *Ped. Inf. Dis. J.* 10:149–151.

Kuroki, S., Saida, T., Nukina, M., Haruta, T., Yoshioka, M., Kobayashi, Y., and Nakanishi, H. 1993. "*Campylobacter jejuni* strains from patients with Guillain-Barre syndrome belong mostly to Penner serogroup 19 and contain beta-N-acetylglucosamine residues," *Ann. Neurol.* 33(3):243–247.

Kuusela, P., Moran, A. P., Vartio, T., and Kosunen, T. U. 1989. "Interaction of *Campylobacter jejuni* with extracellular matrix components," *Biochem. Biophys. Acta.* 993(2–3):297–300.

Labigne, A., Courcoux, P., and Tompkins, L. 1992. "Cloning of *Campylobacter jejuni* genes required for leucine biosynthesis, and construction of leu-negative mutant of *C. jejuni* by shuttle transposon mutagenesis," *Res. Micr.* 143(1):15–26.

Labigne-Roussel, A., Courcoux, P., and Tompkins, L. S. 1988. "Gene disruption and replacement as a feasible approach for mutagenesis of *Campylobacter jejuni*," *J. Bact.* 170:1704–1708.

Labigne-Roussel, A., Harel, J., and Tompkins, L. 1987. "Gene transfer from *Escherichia coli* to *Campylobacter* species: development of shuttle vectors for genetic analysis of *Campylobacter jejuni*," *J. Bact.* 169(11):5320–5323.

Lambert, M. E., Schofield, P. F., Ironside, A. G., and Mandal, B. K. 1979. "*Campylobacter* colitis," *Brit. Med. J.* 1(6167):857–859.

Lambert, T., Gerbaud, G., Trieu-Cuot, P., and Courvalin, P. 1985. "Structural relationship between the genes encoding 3′-aminoglycoside phosphotransferases in *Campylobacter* and in gram-positive cocci," *Ann. Inst. Pasteur Micr.* 136B(2):135–150.

Lastovica, A. J., LeRoux, E., Congi, R. V., and Penner, J. L. 1986. "Distribution of sero-biotypes of *Campylobacter jejuni* and *C. coli* isolated from paediatric patients," *J. Med. Micr.* 21:1–5.

Lee, A., O'Rourke, J. L., Barrington, P. J., and Trust, T. J. 1986. "Mucus colonization as a determinant of pathogenicity in intestinal infection by *Campylobacter jejuni:* a mouse cecal model," *Inf. Imm.* 1(2):536–546.

Lindblom, G. B. and Kaijser, B. 1995. "In vitro studies of *Campylobacter jejuni/coli* strains from hens and humans regarding adherence, invasiveness, and toxigenicity," *Avian Dis.* 39(4):718–722.

Lindblom, G. B., Kaijser, B., and Sjogren, E. 1989. "Enterotoxin production and serogroups of *Campylobacter jejuni* and *Campylobacter coli* from patients with diarrhea and from healthy laying hens." *J. Clin. Microbiol.* 27(6):1272–6.

Lindblom, G. B., Cervantes, L. E., Sjogren, E., Kaijser, B., and Ruiz-Palacios, G. M. 1990b. "Adherence, enterotoxigenicity, invasiveness and serogroups in *Campylobacter jejuni* and *Campylobacter coli* strains from adult humans with acute enterocolitis," *APMIS* 98(2):179–184.

Lindblom, G. B., Johny, M., Khalil, K., Mazhar, K., Ruiz-Palacios, G. M., and Kaijser, B. 1990a. "Enterotoxigenicity and frequency of *Campylobacter jejuni, C. coli,* and *C. laridis* in human and animal stool isolates from different countries," *FEMS Micr. Letters* 54:163–167.

Logan, S. M. and Trust, T. J. 1984. "Structural and antigenic heterogeneity of lipopolysaccharides of *Campylobacter jejuni* and *Campylobacter coli,*" *Inf. Imm.* 45(1):210–216.

Louie, H. and Chan, V. L. 1993. "Cloning and characterization of the gamma-glutamyl phosphate reductase gene of *Campylobacter jejuni,*" *Molec. Gen. Genet.* 240(1):29–35.

Luneberg, E., Glenn-Calvo, E., Hartmann, M., Bar, W., and Frosch, M. 1998. "The central, surface-exposed region of the flagellar hook protein FlgE of *Campylobacter jejuni* shows hypervariability among strains," *J. Bacteriol.* 180(14):3711–3714.

Madden, R. H., Moran, L., and Scates, P. 1996. "Sub-typing of animal and human *Campylobacter* spp. using RAPD," *Lett. Appl. Micr.* 23(3):167–170.

Mahajan, S. and Rodgers, F. G. 1990. "Isolation, characterization, and host-cell-binding properties of a cytotoxin from *Campylobacter jejuni,*" *J. Clin. Micr* 28(6):1314–1320.

Maki-Ikola, O., Viljanen, M. K., Tiitinen, S., Toivanen, P., and Granfors, K. 1991. "Antibodies to arthritis-associated microbes in inflammatory joint diseases," *Rheumatol. Int.* 10(6):231–234.

Manavathu, E. K., Fernandez, C. L., Cooperman, B. S., and Taylor, D. E. 1990. "Molecular studies on the mechanism of tetracycline resistance mediated by Tet(O)," *Antimicr. Agents Chemother.* 34(1): 71–77.

Mandal, B. K., DeMol, P., and Butzler, J. P. 1984. *Campylobacter Infection in Man and Animals.* Boca Raton, FL: CRC Press. pp. 21–31.

Mansfield, L. S., and Urban, J. F. 1996. ''The pathogenesis of necrotic proliferative colitis in swine is linked to whipworm induced suppression of mucosal immunity to resident bacteria.'' *Vet. Immunol. Immunopath.* 50:1–17.

Mansfield, L. S., Gauthier, D. T., Abner, S. R., Jones, K. M., and Urban, J. F. Submitted 1999. ''Development of a swine animal model where opportunistic invasion of the colon by *Campylobacter jejuni* occurs spontaneously,'' *Inf. Imm.* in press.

Mansfield, L. S., Urban, J. F., Holley-Shanks, R. R., Murtaugh, M. P., Zarlenga, D. S., Foss, D., Canals, A., Gause, W., and Lunney, J. K. 1997. ''Construction of internal cDNA competitors for measuring IL-10 and IL-12 cytokine gene expression in swine,'' *Vet. Immunol. Immunopathol.* 65(1998):63–74.

Mantle, M., Basaraba, L., Peacock, S. C., and Gall, D. G. 1994. ''Binding of *Yersinia enterocolitica* to rabbit intestinal brush border membranes, mucus, and mucin,'' *Inf. Imm.* 57(11):3292–3299.

Maruyama, S., and Katsube, Y. 1994. ''Adhesion activity of *Campylobacter jejuni* for intestinal epithelial cells and mucus and erythrocytes,'' *J. Vet. Med. Sci.* 56(6):1123–1127.

Mathan, V. I., Rajan, D. P., Klipstein, F. A., and Engert, R. E. 1984 Oct 27. ''Enterotoxigenic *Campylobacter jejuni* among children in South India [letter].'' *Lancet,* 2(8409):981.

McCardell, B. A., Madden, J. M., and Lee, E. C. 1984 Feb 25. ''Production of cholera-like toxin by *Campylobacter jejuni/coli* [letter].'' *Lancet* 1(8374):448–449.

McFarland, B. A., and Neill, S. D. 1992. ''Profiles of toxin production by thermophilic *Campylobacter* of animal origin,'' *Vet. Micro.* 30(2–3):257–266.

McSweegan, E. and Walker, R. I. 1986. ''Identification and characterization of two *Campylobacter jejuni* adhesins for cellular and mucous substrates,'' *Inf. Imm.* 53(1):141–148.

McSweegan, E., Burr, D. H., and Walker, R. I. 1987. ''Intestinal mucus gel and secretory antibody are barriers to *Campylobacter jejuni* adherence to INT 407 cells,'' *Inf. Imm.* 55(6):1431–1435.

Meinersmann, R. J., Rigsby, W. E., Stern, N. J., Kelley, L. C., Hill, J. E., and Doyle, M. P. 1991. ''Comparative study of colonizing and noncolonizing *Campylobacter jejuni,*'' *Am. J. Vet. Res.* 52(8):1518–1522.

Miller, J. F., Dower, W. J. and Tompkins, L. S. 1988. ''High-voltage electroporation of bacteria: genetic transformation of *Campylobacter jejuni* with plasmid DNA,'' *Proc. Nat. Acad. Sci. USA* 85(3):856–860.

Miller, S., Pesci E. C., and Pickett, C. L. 1993. ''A *Campylobacter jejuni* homolog of the LcrD/FlbB family of proteins is necessary for flagellar biosynthesis,'' *Inf. Imm.* 61(7):2930–2936.

Miller, S., Pesci, E. C. and Pickett, C. L. 1994. ''Genetic organization of the region upstream from the *Campylobacter jejuni* flagellar gene *flhA,*'' *Gene* 146(1):31–38.

Mills, S. D., Kuzniar, B., Shames, B., Kurjanczyk, L. A. and Penner, J. L. 1992. ''Variation of the O antigen of *Campylobacter jejuni* in vivo,'' *J. Med. Micr.* 36(3):215–219.

Misawa, N., Mishu, B., and Blaser, M. J. 1998. ''Differentiation of *Campylobacter jejuni* serotype 019 strains from non-O19 strains by PCR,'' *J. Clin. Micr.* 36(12):3567–3573.

Misawa, N., Ohnishi, T., Itoh, K., and Takahashi, E. 1994. ''Development of a tissue culture assay system for *Campylobacter jejuni* cytotoxin and the influence of culture conditions on cytotoxin production,'' *J. Med. Micr.* 41(4):224–230.

Misawa, N., Ohnishi, T., Itoh, K., and Takahashi, E. 1995. ''Cytotoxin detection in *Campylobacter jejuni* strains of human and animal origin with three tissue culture assay systems,'' *J. Med. Micr.* 43(5):354–359.

Mishu, B. and Blaser, M. J. 1993. ''Role of infection due to *Campylobacter jejuni* in the initiation of Guillain-Bare syndrome,'' *Clin. Inf. Dis.* 17(1):104–108.

Mishu, B., Patton, C. M., and Tauxe, R. V. 1992. ''Clinical and epidemiological features of non-*jejuni* non-*coli Campylobacter* species,'' in Campylobacter jejuni: Current Status and Future

Trends, eds., Nachamkin, I. Blaser, M. J., Tompkins, L. S., Washington, D.C.: American Society for Microbiology. pp. 31–44.

Mishu-Allos, B. 1997. ''*Campylobacter jejuni* foodborne infection and Guillain-Barre syndrome,'' in *Chronic Impact of Foodborne Disease* [symposium title]. 97th General Meeting, American Society for Microbiology, May 4–8, 1997, Miami Beach, FL.

Mishu-Allos, B., Lippy, F. T., Carlsen, A., Washburn, R. G. and Blaser, M. J. 1998. ''*Campylobacter jejuni* strains from patients with Guillain-Barre syndrome,'' *Emerging Inf. Dis.* 4(2):263–268.

Mizuno, K., Takama, K., and Suzuki, S. 1994. ''Characteristics of cytotoxin produced by *Campylobacter jejuni* strains,'' *Microbiol.* 78(317):215–228.

Montrose, M. S., Shane, S. M., and Harrington, K. S. 1985. ''Role of litter in the transmission of *Campylobacter jejuni,*'' *Avian Dis.* 29(2):392–399.

Moran, A. P. 1995. ''Biological and serological characterization of *Campylobacter jejuni* lipopolysaccharides with deviating core and lipid A structures,'' *FEMS Imm. Med. Micr.* 11(2):121–130.

Moran, A. P., Prendergast, M. M., and Appelmelk, B. J. 1996. ''Molecular mimicry of host structures by bacterial lipopolysaccharides and its contribution to disease,'' *FEMS Imm. Med. Micr.* 16(2):105–115.

Moran, A. P., Rietschel, E. T., Kosunen, T. U., and Zahringer, U. 1991. ''Chemical characterization of *Campylobacter jejuni* lipopolysaccharides containing N-acetylneuraminic acid and 2,3-diamino-2,3-dideoxy-D-glucose,'' *J. Bact.* 173(2):618–626.

Morbidity Mortality Weekly Reports. 1997. Atlanta, GA: Centers for Disease Control.

Morooka, T., Umeda, A., and Amako, K. 1985. ''Motility as an intestinal colonization factor for *Campylobacter jejuni,*'' *J. Gen. Micr.* 131(8):1973–1980.

Moser, I. and Schroder, W. 1995. ''Binding of outer membrane preparations of *Campylobacter jejuni* to INT 457 cell membranes and extracellular matrix proteins,'' *Med. Micr. Imm. Berlin* 184(3):147–153.

Moser, I., Schroeder, W. F., and Hellmann, E. 1992. ''In vitro binding of *Campylobacter jejuni/coli* outer membrane preparations to INT 407 cell membranes,'' *Med. Micr. Imm. Berl.* 180(6):289–303.

Moser, I., Schroeder, W., and Salnikow, J. 1997. ''*Campylobacter jejuni* major outer membrane protein and a 59 kDa protein are involved in binding to fibronectin and INT407 cell membranes,'' *FEMS Microbiol. Lett.* 157(2):233–238.

Munroe, D. L., Prescott, J. F., and Penner, J. L. 1983. ''*Campylobacter jejuni* and *Campylobacter coli* serotypes isolated from chickens, cattle, and pigs,'' *J. Clin. Micr.* 18:877–881.

Murphy, E. 1985. ''Nucleotide sequence of *ermA,* a macrolide-lincosamide-streptogramin B determinant in *Staphylococcus aureus,*'' *J. Bacteriol.* 162:633–640.

Nachamkin, I. 1992. ''Immune responses directed against *Campylobacter jejuni,*'' in Campylobacter jenuni: Current Status and Future Trends, eds., Nachamkin, I., Blaser, M. J., and Thompkins, L. S. Washington, D.C. American Society of Microbiology. pp. 201–206.

Nachamkin, I., Blaser, M. J. and Tompkins, L. S. 1992. Campylobacter jejuni: *Current Status and Future Trends.* Washington, D.C., American Society for Microbiology.

Nachamkin, I., Ung, H., and Patton, C. M. 1996. ''Analysis of HL and O serotypes of *Campylobacter* strains by flagellin gene typing system,'' *J. Clin. Micr.* 34(2):277–281.

Nachamkin, I., Yang, X. H., and Stern, N. J. 1993. ''Role of *Campylobacter jejuni* flagella as colonization factors for three-day-old chicks: analysis with flagellar mutants,'' *Appl. Env. Micr.* 59:1269–1273.

Naess, V. and Hofstad, T. 1984. ''Chemical studies of partially hydrolysed lipopolysaccharides from four strains of *Campylobacter jejuni* and two strains of *Campylobacter coli,*'' *J. Gen. Micr.* 130 (Pt 11):2783–2789.

Nedialkov, Y. A., Motin, V. L., and Brubaker, R. R. 1997. ''Resistance to lipopolysaccharide mediated by *Yersiniae* V antigen-polyhistidine fusion peptide: amplification of interleukin-10,'' *Inf. Imm.* 65(4):1196–1203.

Newell, D. G., McBride, H., and Dolby, J. M. 1985. ''Investigations on the role of flagella in the colonization of infant mice with *Campylobacter jejuni* and attachment of *Campylobacter jejuni* to human epithelial cell lines,'' *J. Hyg. Lond.* 95(2):217–227.

Newnham, E., Chang, N., and Taylor, D. E. 1996. ''Expanded genomic map of *Campylobacter jejuni* UA580 and localization of 23s ribosomal rRNA genes by I-Ceul restriction endonuclease digestion,'' *FEMS Micr. Lett.* 142(2–3):223–229.

Ng, L. K., Stiles, M. E., and Taylor, D. E. 1985. ''Inhibition of *Campylobacter coli* and *Campylobacter jejuni* by antibiotics used in selective growth media,'' *J. Clin. Micr.* 22(4):510–514.

Ng, L. K., Stiles, M. E., and Taylor, D. E. 1987. ''DNA probes for identification of tetracycline resistance genes in *Campylobacter* species isolated from swine and cattle,'' *Antimicr. Agents Chemoth.* 31:1669–1674.

Nickerson, C. L., Luthra, H. S. and David, C. S. 1990. ''Role of enterobacteria and HLA-B27 in spondyloarthropathies: studies with transgenic mice,'' *Ann. Rheum. Dis.* 49(1):426–433.

Norcross, M. A., Johnston, R. W., and Brown, J. L. 1992. ''Importance of *Campylobacter* spp. to the food industry,'' in Campylobacter jejuni: *Current Status and Future Trends,* eds., Nachamkin, I., Blaser, M. J., and Tompkins, L. S. Washington, D.C.: American Society for Microbiology. pp. 61–65.

Nuijten, P. J. M., van Asten, F. J., Gaastra, W., and van der Zeijst, B.A. 1990. ''Structural and functional analysis of two *Campylobacter jejuni* flagellin genes,'' *J. Biol. Chem.* 265(29): 17798–17804.

Nuijten, P. J. M., Wassenaar, T. M., Newell, D. G., and van der Zeijst, B. A. M. 1992. ''Molecular characterization and analysis of *Campylobacter jejuni* flagellin genes and proteins,'' in Campylobacter jejuni: *Current Status and Future Trends,* eds., Nachamkin, I., Blaser, M. J., and Thompkins, L. S. Washington, D.C.: American Society of Microbiology. pp. 282–296.

Oelschlaeger, T. A., Guerry, P., and Kopecki, D. J. 1993. ''Unusual microtubule-dependent endocytosis mechanisms triggered by *Campylobacter jejuni* and *Citrobacter freundii,*'' *Proc. Nat. Acad. Sci. USA* 90(14):6884–6888.

Olek, I. and Doyle, R. J. 1994. *Bacterial adhesion to cells and tissues.* New York: Chapman. pp. 1–578.

Olsvik, O., Wachsmuth, K., Morris, G. and Feeley, J. C. 1984. ''Genetic probing of *Campylobacter jejuni* for cholera toxin and *Escherichia coli* heat-labile enterotoxin [letter],'' *Lancet* Feb 25; 1(8374):449.

On, S. L. W. 1996. ''Identification methods for campylobacters, helicobacters, and related organisms,'' *Clin. Micr. Rev.* 9:405–422.

Pang, T., Wong, P. Y., Puthucheary, S. D., Sihotang, K., and Chang, W. K. 1987. ''In-vitro and in-vivo studies of a cytotoxin from *Campylobacter jejuni,*'' *J. Med. Microbiol.* May; 23(3):193–198.

Papadopoulou, B. and Courvalin, P. 1988. ''Dispersal in *Campylobacter* spp. of *aphA,* a kanamycin resistance determinant from gram positive cocci,'' *Antimicr. Agents Chemother.* 32(6):945–948.

Park, S. F. and Richardson P. T. 1995. ''Molecular characterization of a *Campylobacter jejuni* lipoprotein with homology to periplasmic siderophore-binding proteins,'' *J. Bact.* 177(9): 2259–2264.

Patton, C. M., and Wachsmuth, I. K. 1992. ''Typing schemes: Are current methods useful?'' in

Campylobacter jejuni: *Current Status and Future Trends,* eds., Nachamkin, I., Blaser, M. J., and Thompkins, L. S. Washington, D.C.: American Society of Microbiology. pp. 110–128.

Pavlovskis, O. R., Rollins, D. M., Haberberger Jr., R. L., Green, A. E., Habash, L., Strocko, S., and Walker, R. I. 1991. "Significance of flagella in colonization resistance of rabbits immunized with *Campylobacter* spp." *Inf. Imm.* 59:2259–2264.

Pawelec, D., Jakubowska-Mroz, J., and Jagusztyn-Krynicka, E. K. 1998. "*Campylobacter jejuni* 72 DZ/92 *cjaC* gene coding 28 kDa immunopositive protein, a homologue of the solute-binding components of the ABC transport system," *Lett. Appl. Microbiol.* 26(1):69–76.

Pearson, A. D., Greenwood, M., Healing, T. D., Rollins, D., Shahamat, M., Donaldson, J., and Colwell, R. R. 1993. "Colonization of broiler chickens by waterborne *Campylobacter jejuni,*" *Appl. Env. Micr.* 59(4):987–996.

Pei, Z. and Blaser, M. J. 1993. "PEB1, the major cell-binding factor of *Campylobacter jejuni,* is a homolog of the binding component in gram-negative nutrient transport systems," *J. Biol. Chem.* 268:18717–18725.

Pei, Z., Burucoa, C., Grignon, B., Baqar, S., Huang, X. Z., Kopecko, D. J., Bourgeois, A. L., Fauchere, J. L., and Blaser, M. J. 1998. "Mutation in the peb1A locus of *Campylobacter jejuni* reduces interactions with epithelial cells and intestinal colonization of mice," *Infect. Immun.* 66(3):938–943.

Penner, J. L. and Aspinall, G. O. 1997. "Diversity of lipopolysaccaride structures in *Campylobacter jejuni,*" *J. Infect. Dis.* 176 Suppl. 2:S135–138.

Perez-Perez, G. I., Taylor, D. N., Echeverra, P. D., and Blaser, M. J. 1992. "Lack of evidence of enterotoxin involvement in pathogenesis of *Campylobacter* diarrhea," in Campylobacter jejuni: *Current Status and Future Trends,* eds., Nachamkin, I., Blaser, M. J., Tompkins, L. S. Washington, D.C.: American Society for Microbiology. pp. 184–192.

Perez-Perez, G. I., Cohn, D. L., Guerrant, R. L., Patton, C. M., Reller, L. B., and Blaser, M. J. 1987. "Clinical and immunologic significance of cholera-like toxin and cytotoxin production by *Campylobacter* species in patients with acute inflammatory diarrhea in the USA," *J. Infect. Dis.* Sep; 160(3):460–468.

Perkins, D. J. and Newstead, G. L. 1994. "*Campylobacter jejuni* enterocolitis causing peritonitis, ileitis and intestinal obstruction," *Aust. N. Z. J. Surg.* 64(1):55–58.

Pesci, E. C., Cottle, D. L., and Pickett, C. L. 1994. "Genetic, enzymatic, and pathogenic studies of the iron superoxide dismutase of *Campylobacter jejuni,*" *Inf. Imm.* 62(7):2687–2694.

Peterson, M. C. 1994. "Rheumatic manifestations of *Campylobacter jejuni* and *C. fetus* infections in adults," *Scan. J. Rheumatol.* 23(4):167–170.

PHLS Communicable Disease Surveillance Centre. 1993. "Other gastrointestinal tract infections, England and Wales." Regular summary data. *Commun. Dis. Rep.* 3:10.

Pickett, C. L., Auffenberg, T., Pesci, E. C., Sheen, V. L., and Jusuf, S. S. 1992. "Iron acquisition and hemolysin production by *Campylobacter jejuni,*" *Inf. Imm.* 60(9):3872–3877.

Pickett, C. L., Pesci, E. C., Cottle, D. L., Russell, G., Erdem, A. N., and Zeytin, H. 1996. "Prevalence of cytolethal distending toxin production in *Campylobacter jejuni* and relatedness of *Campylobacter* sp. *cdtB* gene," *Inf. Imm.* 64:2070–2078.

Pinto-Alphandary, H., Mabilat, C., and Courvalin, P. 1990. "Emergence of aminoglycoside resistance genes *aadA* and *aadE* in the genus *Campylobacter,*" *Antimicr. Agents Chemother.* 34:1294–1296.

Potter, M. E., Blaser, M. J., Sikes, R. K., Kaufmann, A. F., and Wells, J. G. 1983. "Human *Campylobacter* infection associated with certified raw milk," *Am. J. Epid.* 117(4):475–483.

Power, E. G., Munoz-Bellido, J. L., and Phillips, I. 1992. "Detection of ciprofloxacin resistance in gram-negative bacteria due to alterations in *gyrA,*" *J. Antimicr. Chemoth.* 29(1):9–17.

Prasad, K. N., Anupurba, S., and Dhole, T. N. 1991. ''Enterotoxigenic *Campylobacter jejuni* & *C. coli* in the etiology of dirrhoea in northern India,'' *Indian J. Med. Res.* Mar; 93:81–86.

Prendergast, M. M., Lastovica, A. J., and Moran, A. P. 1998. ''Lipopolysaccharides from *Campylobacter jejuni* O:41 strains associated with Guillain-Barre syndrome exhibit mimicry of GM1 ganglioside,'' *Infect. Immun.* 66(8):3649–3655.

Prescott, J. F., Barker, I. K., Manninen, K. I., and Miniats, O. P. 1981. ''*Campylobacter jejuni* colitis in gnotobiotic dogs,'' *Can. J. Comp. Med.* 45:37–383.

Preston, A., Mandrell, R. E., Gibson, B. W., and Apicella, M. A. 1996. ''The lipooligosaccharides of pathogenic gram-negative bacteria,'' *Crit. Rev. Micr.* 22(3):139–180.

Purdy, D. and Park, S. F. 1993. ''Heterologous gene expression in *Campylobacter coli:* the use of bacterial luciferase in a promoter probe vector,'' *FEMS Micr. Lett.* 111:233–238.

Purdy, D., and Park, S. F. 1994. ''Cloning, nucleotide sequence and characterization of a gene encoding superoxide dismutase from *Campylobacter jejuni* and *Campylobacter coli,*'' *Microbiolog.* 140:1203–1208.

Rath, H. C., Herfath, H. H., Ikeda, J. S., Grenther, W. B., Hamm, T. E., Balish, E., Taurog, J. D., Hammer, R. E., Wilson, K. H., and Sartor, R. B. 1996. ''Normal luminal bacteria, especially *Bacteroides* species, mediate chronic colitis, gastritis and arthritis in HLA-B27/human beta2 microglobulin transgenic mice'' *J. Clin. Inv.* 98:945–953.

Rees, J. H., Soudain, S. E., Gregson, N. A., and Hughes, R. A. C. 1995a. ''*Campylobacter jejuni* infection and Guillain-Barre syndrome,'' *N. E. J. M.* 333(21):1374–1379.

Rees, J. H., Vaughan, R. W., Kondeatis, E., and Hughes, R. A. 1995b. ''HLA-class II alleles in Guillain-Barre syndrome and Miller Fisher syndrome and their association with preceding *Campylobacter jejuni* infection,'' *J. Neuroimm.* 62:53–57.

Richardson, P. T., and Park, S. F. 1995. ''Enterochelin acquisition in *Campylobacter coli:* characterization of components of a binding-protein-dependent transport system,'' *Microbiology.* 141:3181–3191.

Ritchie, A. E., Bryner, J. H., and Foley, J. W. 1983. ''Role of DNA and bacteriophage in *Campylobacter* auto-agglutination,'' *J. Med. Micr.* 16(3):333–340.

Rollins, D. M., and Colwell, R. R. 1986. ''Viable but non-culturable stage of *Campylobacter jejuni* and its role in survival in the natural aquatic environment,'' *Appl. Env. Micr.* 52(3):531–535.

Roseneau, A., Fauchere, J. L., Moyen, E. N., et al. 1987. ''Cellular factors influencing the association of *Campylobacter jejuni* with epithelial cells. Bacteriologic, cytologic and microcinematographic studies,'' in Campylobacter *IV,* ed. Kaijser, B., and Falsen, E. Sweden: University of Gothenborg. p. 207.

Ruiz, J., Goni, P., Marco, F., Gallardo, F., Mirelis, B., Jimenez De Anta, T., and Vila, J. 1998. ''Increased resistance to quinolones in *Campylobacter jejuni:* a genetic analysis of *gyrA,* gene mutations in quinolone-resistant clinical isolates,'' *Microbiol. Immunol.* 42(3):223–226.

Ruiz-Palacios, G. M., Lopez Vidal, Y., Torres, J., and Torres N. 1985. ''Serum antibodies to heat-labile enterotoxin of *Campylobacter jejuni,*'' *J. Infect. Dis.* Aug; 152(2):413–416.

Ruiz-Palacios, G. M., Torres, J., Torres, N. I., Escamilla, E., Ruiz-Palacios, B. R., and Tamayo, J. 1983. ''Cholera-like enterotoxin produced by *Campylobacter jejuni.* Characterisation and clinical significance,'' *Lancet* Jul 30; 2(8344):250–253.

Russell, R. G. 1992. ''*Campylobacter jejuni* colitis and immunity in Primates: Epidemiology of natural infection,'' in Campylobacter jejuni: *Current Status and Future Trends,* eds., Nachamkin, I., Blaser, M. J., Tompkins, L. S., Washington, D.C.: American Society for Microbiology. pp. 148–157.

Russell, R. G., and Blake, D. C. Jr. 1994. ''Cell association and invasion of Caco-2 cells by *Campylobacter jejuni,*'' *Inf. Imm.* 62(9):3773–3779.

Russell, R. G., Blaser, M. J., Sarmiento, J. I., and Fox, J. 1989. "Experimental *Campylobacter jejuni* infection in *Macacca nemestrina,*" *Inf. Imm.* 57:1438–1444.

Russell, R. G., O'Donnoghue, M., Blake, D. C. Jr., Zulty, J., and DeTolla, L. J. 1993. "Early colonic damage and invasion of *Campylobacter jejuni* in experimentally challenged infant *Macaca mulatta,*" *J. Inf. Dis.* 168(1):210–215.

Saeed, A. M., Harris, N. V., and DiGiacomo, R. F. 1993. "The role of exposure to animals in the etiology of *Campylobacter jejuni/coli* enteritis," *Am. J. Epid.* 137:108–114.

Sagara, H., Mochizuki, A., Okamura, N., and Nakaya, R. 1987. "Antimicrobial resistance of *Campylobacter jejuni* and Campylobacter coli with special reference to plasmid profiles of Japanese clinical isolates," *Antimicr. Agents Chemother.* 31:713–719.

Saha, S. K., and Sanyal, S. C. 1990. "Production and characterisation of *Campylobacter jejuni* enterotoxin in a synthetic medium and its assay in rat ileal loops," *FEMS Microbiol. Lett.* Feb; 55(3):333–338.

Saha, S. K. Singh, N. P., and Sanyal, S. C. 1988. "Enterotoxigenicity of chicken isolates of *Campylobacter jejuni* in ligated ileal loops of rats," *J. Med. Microbiol.* Jun; 26(2):87–91.

Sahay, P., West, A. P., Birkenhead, D., and Hawkey, P. M. 1995. "*Campylobacter jejuni* in the stomach," *J. Med. Micr.* 43:75–77.

Salama, S., Bolton, F. J., and Hutchinson, D. N. 1989. "Improved method for the isolation of *Campylobacter jejuni* and *Campylobacter coli* bacteriophages," *Lett. Appl. Micr.* 8(1):5–7.

Salloway, S., Mermel, L. A., Seamans, M., Aspinall, G. O., Nam-Shin, J. E., Kurjanczyk, L. A., and Penner, J. L. 1996. "Miller-Fisher syndrome associated with *Campylobacter jejuni* bearing lipopolysaccharide molecules that mimic human ganglioside GD3," *Inf. Imm.* 64(8):2945–2949.

Salyers, A. A., and Whitt, D. D., 1994. *Bacterial Pathogenesis: A Molecular Approach* Washington, DC.: ASM Press, p. 177.

Schoeni, J. L., and Doyle, M. P. 1992. "Reduction of *Campylobacter jejuni* colonization of chicks by cecum-colonizing bacteria producing anti-*C. jejuni* metabolites," *Appl. Env. Micr.* 58(2):664–670.

Skirrow, M. B. 1990. "Foodborne illness: *Campylobacter,*" *Lancet* 336(8720):921–923.

Skirrow, M. B., and Blaser, M. J. 1992. "Clinical and epidimiologic considerations," in Campylobacter jejuni: Current Status and Future Trends. Eds. Nachamkin, I. Blaser, M. J., Tompkins, L. S. Washington, D.C.: American Society for Microbiology. pp. 3–8.

Smith, J. L. 1995. "Arthritis, Guillain-Barre syndrome and other sequelae of *Campylobacter jejuni* enteritis," *J. Food Prot.* 58(10):1153–1170.

Smith, J. L., Palumbo, S. A., and Wallis, I. 1993. "Relationship between foodborne bacterial pathogens and the reactive arthritides," *J. Food Safety* 13(3):209–236.

Sorvillo, F. J., Lieb, L. E., and Waternan. S. H. 1991. "Incidence of campylobacteriosis among patients with AIDS in Los Angeles County," *J.A.I.D.S.* 4:598–602.

Spira, W. M., Sack, R. B., and Froehlich, J. L. 1981. "Simple adult rabbit model for *Vibrio cholerae* and enterotoxigenic *Escherichia coli* diarrhea," *Inf. Imm.* 32:739–747.

Steffen, C., and Matzura, H. 1989. "Nucleotide sequence analysis of a chloramphenicol-acetyltransferase gene from *Clostridium perifringens,*" *Gene* 75:349–354.

Stelma, G. N. Jr., and McCabe, L. J. 1992. "Nonpoint pollution from animal sources and shellfish sanitation," *J. Food Prot.* 55(8):649–656.

Stern, N. J. 1992. "Reservoirs for *Campylobacter jejuni* and approaches for intervention in poultry," in Campylobacter jejuni: Current Status and Future Trends, eds., Nachamkin, I, Blaser, M. J., Tompkins, L. S. Washington, D.C.: American Society for Microbiology. pp. 49–60.

Suzuki, S., Kawaguchi, M., Mizuno, K., Takama, K., and Yuki, N. 1994. "Immunological properties and ganglioside recognitions by *Campylobacter jejuni*-enterotoxin and cholera toxin,'' *FEMS Imm. Med. Micr.* 8(3):207–211.

Szymanski, C. M., and Armstrong, G. D. 1996. ''Interactions between *Campylobacter jejuni* and lipids,'' *Inf. Imm.* 64(9):3467–3474.

Szymanski, C. M., King, M., Haardt, M., and Armstrong, G. D. 1995. ''*Campylobacter jejuni* motility and invasion of Caco-2 cells,'' *Inf. Imm.* 63(11):4295–4300.

Tauxe, R. V. 1992. ''Epidemiology of Campylobacter jejuni. Infections in the United States and other industrialized nations,'' in Campylobacter jejuni: *Current Status and Future Trends,* eds., Nachamkin, I., Blaser, M. J., Tompkins, L. S. Washington, D.C.: American Society for Microbiology. pp. 9–19.

Tauxe, R. V., Pegues, D. A., and Hargrett-Bean, N. 1987. ''*Campylobacter* infections: the emerging national pattern,'' *Am. J. Pub. Health* 77(9):1219–1221.

Tauxe, R. V., Hargrett-Bean, N., Patton, C. M., and Wachsmuth, I. K. 1988. ''*Campylobacter* isolates in the United States, 1982–1986.'' CDC Surveillance Summaries, June 1988. *MMWR* 37(SS-2):1–14.

Taylor, D. E. 1992. ''Genetic analysis of *Campylobacter* spp.,'' in Campylobacter jejuni: *Current Status and Future Trends,* eds., Nachamkin, I., Blaser, M. J., Tompkins, L. S. Washington, D.C.: American Society for Microbiology. pp. 255–266.

Taylor, D. E., and Chang, N. 1991. ''In vitro susceptibilities of *Campylobacter jejuni* and *Campylobacterc coli* to azithromycin and erythromycin,'' *Antimicr. Agents Chemother.* 35:1917–1918.

Taylor, D. E. and Courvalin, P. 1988. ''Mechanisms of antibiotic resistance in *Campylobacter* species,'' *Antimicr. Agents Chemother.* 32:1107–1112.

Taylor, D. E., Garner, R. S., and Allan, B. J. 1983. ''Characterization of tetracycline resistance plasmids from *Campylobacter jejuni* and *Campylobacter coli,*'' *Antimicr. Agents Chemother.* 24(6):930–935.

Taylor, D. E., DeGrandis, S. A., Karmali, M. A., and Fleming, P. C. 1981. ''Transmissible plasmids from *Campylobacter jejuni,*'' *Antimicr. Agents Chemother.* 19:831–835.

Taylor, D. E., Eaton, M., Yan, W., and Chang, N. 1992. ''Genome maps of *Campylobacter jejuni* and *Campylobacter coli,*'' *J. Bact.* 174(7):2332–2337.

Taylor, D. N. 1992. ''*Campylobacter* infections in developing countries,'' in Campylobacter jejuni: *Current Status and Future Trends,* eds., Nachamkin, I., Blaser, M. J., Tompkins, L. S. Washington, D.C.: American Society for Microbiology. pp. 20–30.

Taylor, D. N. and Blaser, M. J. 1991. ''*Campylobacter* infections,'' in *Bacterial Infections in Humans,* eds., Evans, A. S., and Brachman, P. S. New York: Plenum Pub. Corp. pp. 151–172.

Taylor, D. E., Johnson, W. M., and Lior, H. 1987. ''Cytotoxic and enterotoxic activities of *Campylobacter jejuni* are not specified by tetracycline resistance plasmids pMAK175 and pUA466,'' *J. Clin. Microbiol.* Jan; 25(1):150–151

Tenover, F. C., Bronsdon, M. A., Gordon, K. P., and Plorde, J. J. 1983. ''Isolation of plasmids encoding tetracycline resistance from *Campylobacter jejuni* strains isolated from simians,'' *Antimicr. Agents Chemother.* 23(2):320–322.

Tenover, F. C., Williams, S., Gordon, K. P., Nolan, C., and Plorde, J. J. 1985. ''Survey of plasmids and resistance factors in *Campylobacter jejuni* and *Campylobacter coli,*'' *Antimicr. Agents Chemother.* 27(1):37–41.

Terzolo, H. R., Lawson, G. H., Angus, K. W., and Snodgrass, D. R. 1987. ''Enteric *Campylobacter* infection in gnotobiotic calves and lambs.'' *Res. Vet. Sci.* 43(1):72–77.

Thies, F. L., Hartung, H. P., and Giegerich G. 1998. ''Cloning and expression of the *Campylobacter jejuni ion* gene detected by RNA arbitrarily primed PCR," *FEMS Microbiology Letters* 165(1998):329–334.

Toivanen, A., Yli-Kerttula, T., Luukkainen, R., Merilahti-Palo, R., Granfors, K., and Seppala, J. 1993. "Effect of antimicrobial treatment on chronic reactive arthritis," *Clin. Exp. Rheum.* 11(3):301–307.

Tompkins, L. S. 1992. "Genetic and molecular approach to *Campylobacter* pathogenesis," in Campylobacter jejuni: *Current Status and Future Trends,* eds., Nachamkin, I., Blaser, M. J., Tompkins, L. S. Washington, D.C.: American Society for Microbiology. pp. 241–254.

Trieu-Cuot, P. and Courvalin, P. 1983. "Nucleotide sequence of the *Streptococcus faecalis* plasmid gene encoding the 3′5′-aminoglycoside resistance phosphotransferase type III," *Gene* 23:331–334.

Trieu-Cuot, P., Arthur, M., and Courvalin, P. 1987. "Origin, evolution and dissemination of antibiotic resistance genes," *Micr. Sci.* 4(9):263–266.

Trieu-Cuot, P., Gerbaud, G., Lambert, T., and Courvalin, P. 1985. "In vivo transfer of genetic information between gram-positive and gram-negative bacteria," *EMBO J.* 4:3583–3587.

Trust, T. J., Logan, S. M., Gustafson, C. E., Romaniuk, P. J., Kim, N. W., Chan, V. L., Ragan, M. A., and Guerry, P. 1994. "Phylogenetic and molecular characterization of a 23s rRNA gene positions the genus *Campylobacter* in the epsilon subdivision of the Proteobacteria and shows that the presence of transcribed spacers is common in *Campylobacter* spp," *J. Bact.* 176:4597–4609.

Uhl, M. A., and Miller, J. F. 1996. "Integration of multiple domains in a two-component sensor-protein: the Bordetella pertussis BvgAS phosphorelay," *EMBO J.* 15(5):1028–1036.

Urban, J. F., Fayer, R., Sullivan, C., Goldhill, J., Shea-Donohue, T., Madden, K., Morris, S. C., Katona, I., Gause, W., Ruff, M., Mansfield, L. S., and Finkelman, F. D. 1996. "Local Th1 and Th2 responses to parasitic infection in the intestine: regulation by IFN-gamma and IL-4," *Vet. Immunol. Immunopathol.* 54:337–344.

Us, D. G. Hascelik, S. Diker, S. Ustacelebi. 1995. "Cytotoxin production by *Campylobacter jejuni* isolated from Turkish patients with diarrhoeal" *J. Diarrhoeal Dis. Res.* 13(2):130–131.

Vandamme, P. E., Falsen, R., Rossau, B., Hoste, R. Tytgat, and DeLey, J. 1991. "Revision of *Campylobacter, Helicobacter,* and *Wolinella* taxonomy: emendation of generic descriptions and proposal of *Arcobacter* gen. nov." *Int. J. Syst. Bact.* 41:88–103.

van Vliet A. H. M., Woodridge, K. G., and Ketley, J. M. 1998. "Iron-responsive gene regulation in a *Campylobacter jejuni* fur mutant," *J. Bacteriol.* 180(20):5291–5298.

Velazquez, J. B., Jimenez, A., Chomon, B., Villa, T. G. 1995. "Incidence and transmission of antibiotic resistance in *Campylobacter jejuni* and *Campylobacter coli,*" *J. Antimicr. Chemother.* 35:173–178.

Vieira, J. and Messing, J. 1982. "The pUC plasmids, an M13mp7-derived system for insertion mutagenesis and sequencing with synthetic universal primers," *Gene* 19(3):259–268.

Vitovec, J., Koudela, B., Sterba, J., Tomancová, I., Matyás, Z., and Vladik, P. 1989. "The gnotobiotic piglet as a model for the pathogenesis of *Campylobacter jejuni* infection," *Zbl. Bakt.* 271:91–103.

Wadstrom, T., Baloda, S. B., Krovacek, K., Faris, A., Bengtson, S., and Walder, M. 1983. "Swedish isolates of *Campylobacter jejuni/coli* do not produce cytotonic or cytotoxic enterotoxins [letter]," *Lancet* Oct 15; 2(8355):911.

Wai, S. N., Nakayama, K., Umene, K., Moriya, T., and Amako, K. 1996. "Construction of a ferritin-deficient mutant of *Campylobacter jejuni:* contribution of ferritin to iron storage and protection against oxidative stress," *Mol. Micr.* 20:1127–1134.

Wai, S. N., Takata, T., Takade, A., Harnsaki, N., and Amako, K. 1995. "Purification and characterization of ferritin from *Campylobacter jejuni,*" *Arch. Micr.* 164:1–6.

Walker, R. I., Rollins, D. M., and Burr, D. H. 1992. "Studies of *Campylobacter* infection in the adult rabbit," in Campylobacter jejuni: *Current Status and Future Trends*, eds., Nachamkin, I., Blaser, M. J., Tompkins, L. S. Washington, D.C.: American Society for Microbiology. pp. 139–147.

Walker, R. I., Schmauder-Chock, E. A., and Parker, J. L. 1988. "Selective association and transport of *Campylobacter jejuni* through M cells of rabbit Peyer's patches," *Can. J. Microbiol.* 34:1142–1147.

Wallis, M. R. 1994. "The pathogenesis of *Campylobacter jejuni*," *Brit. J. Biomed. Sci.* 51:57–64.

Wang, Y. 1993. "Cloning and nucleotide sequence of the *Campylobacter jejuni* gyrA gene and characterization of quinolone resistance mutations," *Antimicr. Agents Chemother.* 37(3): 457–463.

Wang, Y. and Taylor, D. E. 1990. "Natural transformation of *Campylobacter* species," *J. Bacteriol.* 172:949–955.

Warner, D. P., Bryner, J. H., and Beran, G. W. 1986. "Epidemiologic study of campylobacteriosis in Iowa cattle and the possible role of unpasteurized milk as a vehicle of infection," *Am. J. Vet. Res.* 47(2):254–258.

Wassenaar, T. M., Bleumink-Pluym, N. M., and van der Zeijst, B. A. 1991. "Inactivation of the *Campylobacter jejuni* flagellin genes by homologous recombination demonstrates that flaA but not flaB is required for invasion," *EMBO J.* 10:2055–2061.

Wassenaar, T. M., Fry, B. N., and van der Zeijst, B. A. 1993. "Genetic manipulation of *Campylobacter*: evaluation of natural transformation and electro-transformation," *Gene* 132:131–135.

Wassenaar, T. M., Engelskirchen, M., Park, S., and Lastovica, A. 1997. "Differential uptake and killing potential of *Campylobacter jejuni* by human peripheral monocytes/macrophages," *Med. Microbiol. Immunol. Berl.* 186(2–3):139–144.

Whitehouse, C. A., Balbo, P. B., Pesci, E. C., Cottle, D. L., Mirabito, P. M., and Pickett, C. L. 1998. "*Campylobacter jejuni* cytolethal distending toxin causes a G_2-phase cell cycle block," *Inf. Imm.* 66(5):1934–1940.

Wick, M. J., Harding, C. V., Twesten, N. J., Normark, S. J., and Pfeifer, J. D. 1995. "The phoP locus influences processing and presentation of *Salmonella typhimurium* antigens by activated macrophages," *Mol. Micr.* 16(3):465–476.

Wirguin, I., Suturkova-Milosevic, L., Della-Latta, P., Fisher, T., Brown, R. H. Jr., and Latov, N. 1994. "Monoclonal IgM antibodies to GM1 and asialo-GM1 in chronic neuropathies cross-react with *Campylobacter jejuni* lipopolysaccharides," *Ann. Neurol.* 35(6):698–703.

Wong, P. Y., Putchucheary, S. D., and Pang, T. 1983. "Demonstration of a cytotoxin from *Campylobacter jejuni*," *J. Clin. Pathol.* Nov; 36(11):1237–1240.

Wooldridge, K. G. and Ketley, J. M. 1997. "*Campylobacter*-host cell interactions," *Trends Microbiol.* 5(3):96–102.

Wooldridge, K. G., Williams, P. H., and Ketley, J. M. 1994. "Iron-responsive genetic regulation in *Campylobacter jejuni*: cloning and characterization of a *fur* homolog," *J. Bact.* 176:5852–5856.

Wooldridge, K. G., Williams, P. H., and Ketley, J. M. 1996. "Host signal transduction and endocytosis of *Campylobacter jejuni*," *Microb. Pathog.* 21(4):299–305.

World Health Organization. 1995. "Progress towards health for all: third monitoring report," *World Health Stat. Q.* 48:190.

Wosten, M. M., Dubbink, H. J., and van der Zeijst, B. A. 1996. "The *aroA* gene of *Campylobacter jejuni*," *Gene* 181:109–112.

Wosten, M. M., Boeve, M., Koot, M. G., vanNuene, A. C., and van der Zeijst, B. A. 1998. "Identification of *Campylobacter jejuni* promoter sequences," *J. Bacteriol.* 180(3):594–599.

Wren, B. W., Henderson, J., and Ketley, J. M. 1994. ''A PCR-based strategy for the rapid construction of defined bacterial deletion mutants,'' *Biotech.* 16:994–996.

Wren, B. W., Mullany, P., Clayton, C., and Tabaqchali, S. 1989. ''Nucleotide sequence of a chloramphenicol acetyl transferase gene from *Clostridium dificile,*'' *Nucleic Acids Res.* 17(12):4877.

Yan, W. 1990. ''Characterization of erythromycin resistance in *Campylobacter* spp.'' PhD thesis. University of Alberta, Edmonton, Alberta, Canada.

Yan, W. and Taylor, D. E. 1991. ''Characterization of erythromycin resistance in *Campylobacter jejuni* and *Campylobacter coli,*'' *Antimicr. Agents Chemoth.* 35:1989–1996.

Yao, R. and Guerry, P. 1996. ''Molecular cloning and site-specific mutagenesis of a gene involved in arylsulfatase production in *Campylobacter jejuni,*'' *J. Bact.* 178:3335–3338.

Yao, R., Alm, R. A., Trust, T. J., and Guerry, P. 1993. ''Construction of new *Campylobacter* cloning vectors and a new mutational *cat* cassette,'' *Gene* 130(1):127–130.

Yao, R., Burr, D. H., Doig, P., Trust, T. J., Niu, H., and Guerry, P. 1994. ''Isolation of motile and non-motile insertional mutants of *Campylobacter jejuni:* the role of motility in adherence and invasion of eukaryotic cells,'' *Mol. Micr.* 14:883–893.

Yuki, N., Ichihashi, Y., and Taki, T. 1995. ''Subclass of IgG antibody to GM1 epitope-bearing lipopolysaccharide of *Campylobacter jejuni* in patients with Guillain-Barre syndrome,'' *J. Neuroimm.* 60(1–2):161–164.

Yuki, N., Tagawa, Y., Irie, F., Hirabayashi, Y., and Handa, S. 1997. ''Close association of Guillain-Barre syndrome with antibodies to minor monosialogangliosides GM1b and GM1 alpha,'' *J. Neuroimm.* 74:30–34.

Yuki, N., Taki, T., Takahashi, M., Saito, K., Tai, T., Miyatake, T., and Handa, S. 1994. ''Penner's serotype 4 of *Campylobacter jejuni* has a lipopolysaccharide that bears a GM1 ganglioside epitope as well as one that bears a GD1 a epitope,'' *Inf. Imm.* 62(5):2101–2103.

Zhan, Y., Liu, Z., and Cheers, C. 1996. ''Tumor necrosis factor alpha and interleukin-12 contribute to resistance to the intracellular bacterium *Brucella abortus* by different mechanisms,'' *Inf. Imm.* 64:2782–2786.

Zhuang, J., Engel, A., Pages, J. M., and Bolla, J. M. 1997. ''The *Campylobacter jejuni* porin trimers pack into different lattice types when reconstituted in the presence of lipid,'' *Eur. J. Biochem.* 244(2):575–579.

Zychlinsky, A., Fitting, C., Cavaillon, J.-M., and Sansonetti, P. J. 1994. ''Interleukin-1 is released by murine macrophages during apoptosis induced by *Shigella flexneri,*'' *J. Clin Invest.* 94:1328–1332.

PART II

GRAM POSITIVE FOODBORNE BACTERIAL PATHOGENS

CHAPTER 7

The Action, Genetics, and Synthesis of *Clostridium perfringens* Enterotoxin

BRUCE A. McCLANE

1. INTRODUCTION

THE gram-positive anaerobe *Clostridium perfringens* is ideally suited for its role as a major foodborne pathogen (see McClane, 1997, for review). The widespread natural distribution of *C. perfringens* in both soil and the gastrointestinal (GI) tract of humans and other animals provides this bacterium with ample opportunities to contaminate foods. *C. perfringens* has an excellent ability to survive in incompletely cooked foods due to the relative heat tolerance of its vegetative cells (which will grow at temperatures up to at least 50°C) and its ability to form heat-resistant endospores. The exceptionally short doubling time of *C. perfringens* (reportedly <10 minutes in some studies) makes it relatively easy for this bacterium to contaminate foods at levels ($\sim 10^6$–10^7 vegetative cells/gram of food) necessary for causing *C. perfringens* foodborne disease (see below). Finally, strains of this bacterium associated with food poisoning produce a protein toxin, named *C. perfringens* enterotoxin (CPE), that is highly active on the human Gl tract. As will be discussed in detail below, CPE is considered the virulence factor responsible for the GI symptoms of *C. perfringens* type A food poisoning.

2. EPIDEMIOLOGY OF *C. PERFRINGENS* TYPE A FOOD POISONING

Recent statistics from the Centers for Disease Control (see Table 7.1) indicate that *C. perfringens* currently ranks as the second most common cause

TABLE 7.1. Bacterial Foodborne Disease in the USA from 1989–1992.

Bacterium	Outbreaks	Cases	Mean # of Cases per Outbreak
Bacillus cereus	16	382	24
Campylobacter spp.	23	598	26
Clostridium botulinum	40	84	2
Clostridium perfringens	40	3801	95
E. coli	9	135	15
Listeria monocytogenes	1	2	2
Salmonella spp.	455	18,190	64
Shigella spp.	19	1207	64
Staphylococcus aureus	42	1433	34
Vibrio spp.	9	57	6
Group A *streptococci*	2	135	68

Compiled from Bean et al. (1996).

of foodborne disease in the U.S.A. Almost all *C. perfringens* foodborne illness in the U.S. and other industrialized countries involves *C. perfringens* type A food poisoning (McClane, 1997), which is so named because cases of this illness are nearly always caused by type A isolates of *C. perfringens* (McClane, 1997). The designation "type A" refers to a widely used classification scheme for *C. perfringens* that assigns (see Table 7.2) isolates to one of five types (A-E) depending upon their ability to express four "typing" toxins (i.e., α, β, ι and ϵ toxins). Note from Table 7.2 that CPE expression is not a component of this typing scheme; in fact, type A, C, D, and (possibly) B isolates of CPE-producing *C. perfringens* exist in nature (Skjelkvale and Duncan, 1975; Songer and Meer, 1996; Meer and Songer, 1997; Markovic et al., 1993). The overwhelming involvement of type A isolates in *C. perfringens* type A food poisoning may simply reflect the preponderance of CPE-producing *C. perfringens* type A isolates in the environment, as suggested by some recent epidemiological surveys (Songer and Meer, 1996; Meer and Songer, 1997).

TABLE 7.2. Toxin Typing Classification Scheme for *C. perfringens* Isolates.

	Toxins Produced			
Type:	Alpha	Beta	Epsilon	Iota
A	+	–	–	–
B	+	+	(+)	–
C	+	+	–	–
D	+	–	(+)	–
E	+	–	–	+

Adapted from McDonel (1986); (+) indicates that epsilon toxin is initially produced as an inactive prototoxin.

One epidemiologically interesting aspect of *C. perfringens* type A food poisoning is the unusually large size of most outbreaks, e.g., data shown in Table 7.1 indicate that the typical *C. perfringens* type A food poisoning outbreak involves ~100 cases. The fact that recognized *C. perfringens* type A food poisoning type outbreaks are typically of such large size is not surprising considering that most of these confirmed outbreaks occur in institutional settings. Institutions such as nursing homes, prisons, and hospitals represent favorable environments for *C. perfringens* type A food poisoning outbreaks since these establishments depend heavily on serving foods involving large meat items, such as roasts and turkeys, in order to feed many people at peak meal times. Large meat items are the most common food vehicles for *C. perfringens* type A food poisoning (Bean et al., 1996), at least in part, because they are fairly difficult to thoroughly cook. Incomplete cooking increases the probability that contaminating *C. perfringens* cells/spores will survive the cooking process and multiply to pathogenic levels (>10^6–10^7 cells/gram of food) before these foods are consumed. It is also relevant that institutional food is frequently prepared in advance and held for long periods before serving. If these prepared foods are improperly held, this could lead to a *C. perfringens* type A food poisoning outbreak (in fact, holding foods under improper conditions is considered the single most common contributing factor to *C. perfringens* food poisoning outbreaks (Bean et al., 1996).

The large size of most recognized *C. perfringens* type A food poisoning outbreaks is probably somewhat artificial. Because victims of *C. perfringens* food poisoning usually suffer relatively mild symptoms (see below), this illness is much more likely to receive the full attention of public health authorities when large numbers of people, in a common setting (such as an institution), simultaneously become sick, i.e., smaller outbreaks of *C. perfringens* type A food poisoning often go unrecognized. Supporting the view that the ~1000 cases/year of *C. perfringens* type A food poisoning indicated in Table 7.1 significantly underestimate the prevalence of this illness, Todd has estimated (Todd, 1989a, 1989b) that *C. perfringens* type A food poisoning actually affects >650,000 North Americans each year, resulting in ~8 deaths/year and annual costs of >$120 million.

3. THE PATHOGENESIS OF *C. PERFRINGENS* TYPE A FOOD POISONING

In contrast to many cases of foodborne botulism, *C. perfringens* type A food poisoning is rarely, or never, an intoxication resulting from consumption of foods containing preformed CPE (McClane, 1997). Instead, as shown in Figure 7.1, the pathogenesis of *C. perfringens* type A food poisoning involves the in vivo production of CPE (McClane, 1997).

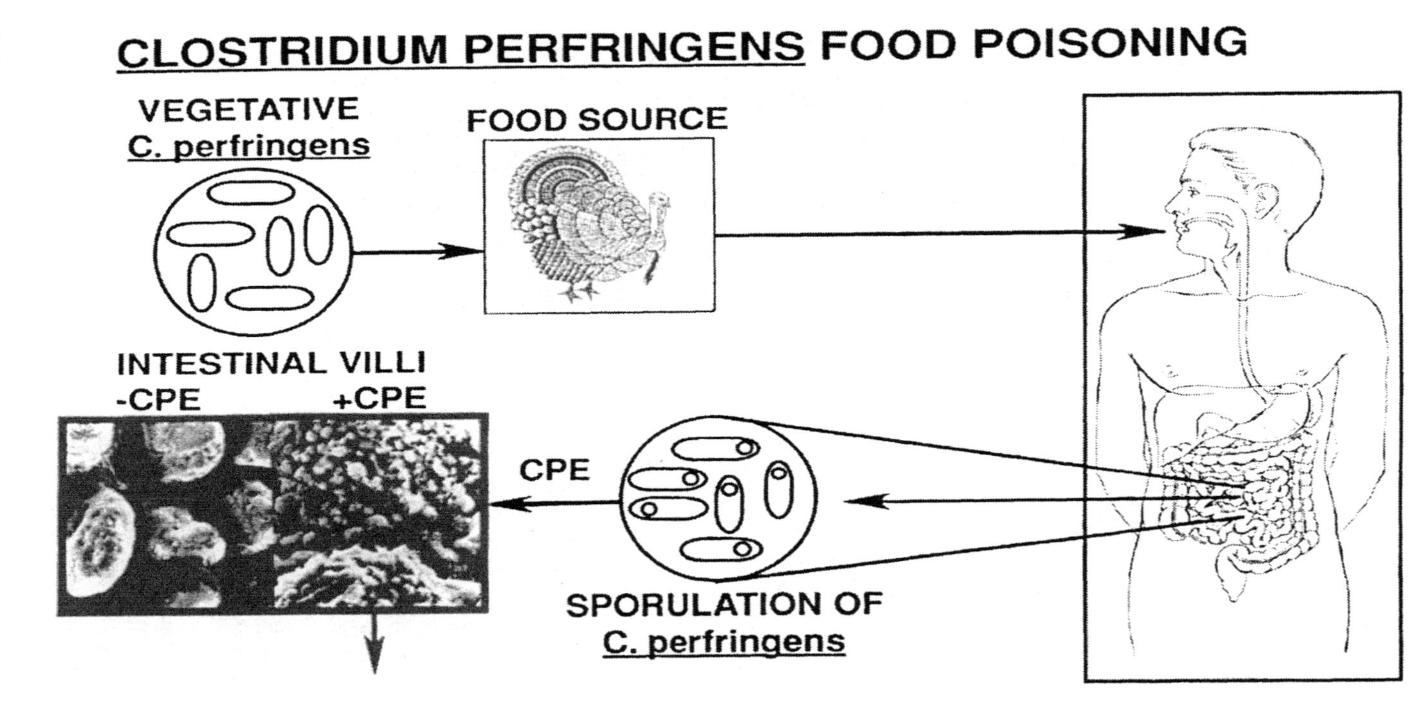

Figure 7.1 Pathogenesis of *Clostridium perfringens* type A food poisoning. Vegetative cells of a CPE-producing strain of *C. perfringens* multiply rapidly in contaminated food (typically a meat- or poultry-containing product). After ingestion, these bacteria pass into the small intestine, where they sporulate and produce CPE. CPE causes intestinal tissue damage, which culminates in the diarrheal and cramping symptoms associated with *C. perfringens* type A food poisoning. Reproduced with publishers' permission from McClane (1992).

The first step in acquiring *C. perfringens* type A food poisoning is consuming food, usually a meat or poultry product (see above), that has become contaminated with large numbers of vegetative cells of a CPE-positive *C. perfringens* isolate. Most of these ingested bacteria are killed in the stomach by gastric acid (McClane, 1997). However, if the ingested food contained sufficiently high numbers (as mentioned previously, $>10^6$–10^7 cells/gram of food) of vegetative *C. perfringens* cells, some of these bacteria may survive exposure to gastric acid long enough to escape into the small intestine.

Once present in the small intestines, surviving vegetative *C. perfringens* cells initially multiply, but later undergo sporulation. Sporulation of *C. perfringens* in the intestines may be triggered by exposure of these bacteria to the acidic conditions of the stomach (Wrigley et al., 1995) or to bile salts in the intestines (Heredia et al., 1991). Recent studies (Shih and Labbe, 1996) indicate that both CPE-positive and CPE-negative isolates of *C. perfringens* can produce a low-molecular weight, heat- and acid-resistant factor(s) that stimulates *C. perfringens* sporulation. If this sporulation-stimulating factor(s) is produced in vivo, it could contribute to the in vivo sporulation required for the development of *C. perfringens* type A food poisoning.

As discussed in more detail later in this chapter, it is during this in vivo sporulation that CPE expression occurs (McClane, 1997). The newly synthesized CPE accumulates inside the cytoplasm of the mother cell until it is released into the intestinal lumen when the mother cell lyses to free its now mature endospore (McClane, 1997). Once present in the lumen, CPE quickly binds to receptors on the intestinal epithelium and (through it molecular action discussed below) induces desquamation of the intestinal epithelium. Studies with animal models (Sherman et al., 1994; McDonel and Duncan, 1975) have demonstrated that development of this CPE-induced intestinal tissue damage strongly correlates with the onset of physiologic symptoms such as fluid/ electrolyte loss (an effect that corresponds clinically to diarrhea). However, it is possible that other CPE-induced effects, e.g., intestinal inflammation, could also contribute to the gastrointestinal symptoms of *C. perfringens* type A food poisoning (Sherman et al., 1994; Krakauer et al., 1997). There have been reports (Bowness et al., 1992; Nagata et al., 1997) that CPE possesses superantigenic activity, which could contribute to inflammation. However, this hypothesis has been called into question by another recent study (Krakauer et al., 1997) reporting that CPE lacks superantigenic activity.

Animal model studies suggest that *C. perfringens* type A food poisoning primarily involves the small intestine (McDonel and Duncan, 1977; McDonel and Demers, 1982). While all regions of the rabbit small intestine respond to CPE treatment (McDonel and Duncan, 1977), the rabbit ileum appears to be particularly CPE-sensitive. Interestingly, the rabbit colon does not significantly respond to CPE treatment (McDonel and Demers, 1982), which could indicate that *C. perfringens* type A food poisoning of humans does not involve the large intestine (however, this hypothesis still needs to be tested with human colonic tissue).

C. perfringens type A food poisoning is clinically characterized by diarrhea and abdominal cramps that develop about 8–16 hours after ingestion of contaminated foods (McClane, 1997). This incubation period stems primarily from the time required for *C. perfringens* to complete its in vivo sporulation. As mentioned, no significant release of CPE into the intestinal lumen occurs until sporulation is completed (this typically takes ~12 hours). Animal model studies indicate that, once CPE has been released into the intestinal lumen, it exerts its effects very rapidly, i.e., CPE-induced intestinal tissue damage can develop within 15–30 minutes (Sherman et al., 1994). In most affected people, symptoms of *C. perfringens* type A food poisoning continue for ~12–24 hours before self-resolving. Fatalities from this illness are relatively rare, but do occur in some elderly or debilitated individuals. Treatment for *C. perfringens* type A food poisoning is primarily symptomatic and no vaccine is currently available (further discussion later). Although *C. perfringens* type A food poisoning victims often develop substantial levels of serum IgG against CPE (Birkhead et al., 1988), there is no evidence that prior exposure to *C. perfringens* food poisoning provides any significant long-term protection against future bouts of this illness (McClane, 1997).

4. EVIDENCE OF CPE INVOLVEMENT IN *C. PERFRINGENS* TYPE A FOOD POISONING

A considerable amount of epidemiologic evidence now implicates CPE as the major (if not only) virulence factor responsible for the diarrheal and cramping symptoms of *C. perfringens* type A food poisoning. Some of this evidence includes;

(1) A strong correlation exists between illness and the presence of CPE in the feces of *C. perfringens* type A food poisoning victims (Batholomew et al., 1985; Birkhead et al., 1988).

(2) CPE is present in the feces of food poisoning victims at levels (Batholomew et al., 1985; Birkhead et al., 1988) shown to cause significant intestinal effects in animal models (McDonel and Duncan, 1975).

(3) Human volunteers who ingested highly purified CPE developed the same diarrheal and cramping symptoms that are characteristic of *C. perfringens* type A food poisoning (Skjelkvale and Uemura, 1977).

(4) CPE-positive *C. perfringens* strains are much more effective than CPE-negative strains at producing either fluid accumulation in rabbit ileal loops or diarrhea in human volunteers (McClane, 1997).

(5) CPE-specific antibodies can neutralize the intestinal effects of culture lysates from CPE-positive *C. perfringens* strains (Hauschild et al., 1971).

5. THE MECHANISM OF ACTION OF CPE

5.1. INTRODUCTION

The first major insight into the molecular mechanism by which CPE induces intestinal tissue damage was provided by electron microscopy studies of CPE-treated rabbit intestinal epithelial cells (McDonel et al., 1978). Those studies showed that CPE-treated intestinal cells rapidly develop extensive damage to their brush border membranes (BBMs). Since this damage precedes detectable damage to internal organelles, McDonel et al., (1978) suggested that CPE may kill sensitive mammalian cells by damaging their plasma membranes.

A series of studies from several laboratories (McDonel and McClane, 1979; Matsuda and Sugimoto, 1979; McClane and McDonel, 1980; McClane and McDonel, 1981; McClane, 1984; Matsuda et al., 1986; McClane et al., 1988) confirmed that CPE is a membrane-active toxin by demonstrating that this toxin induces alterations in the normal permeability properties of sensitive mammalian cells. Within 5 minutes of treatment, CPE damages plasma membranes of sensitive mammalian cells so they become highly permeable to small molecules (<200 Daltons in size); this CPE-induced membrane ''lesion'' is nonselective, as CPE-treated cells show increased permeability to cations, anions and small organic molecules such as amino acids.

These small molecule permeability changes contribute to CPE-induced cytotoxicity in at least two ways. First, CPE-induced plasma membrane permeability alterations profoundly disturb cytoplasmic pools of small molecules, which causes a rapid shutdown in vital metabolic processes such as macromolecular synthesis (Hulkower et al., 1989). Second, these permeability alterations disrupt the cellular osmotic equilibrium, which causes a significant water influx into the CPE-treated cell. This water influx can ''stretch'' the plasma membrane of CPE-sensitive cells to the point of lysis (McClane and McDonel, 1981; McClane, 1984).

Experiments conducted during the past 10 years have shed light on how CPE induces these small molecule permeability alterations. The initial event in CPE action clearly involves the binding of CPE to its receptor(s); for example, mammalian cells lacking CPE receptor(s) are totally nonresponsive to CPE (Horiguchi et al., 1985; Wieckowski et al., 1994). CPE receptor(s) are not only expressed by intestinal cells, i.e., many, but not all (see above), cell types from most, if not all, mammalian species are able to bind CPE at high levels (McDonel, 1980; Horiguchi et al., 1985; Wieckowski et al., 1994).

While the CPE receptor(s) is clearly proteinaceous (McDonel, 1980;

McClane et al., 1988), the precise number and identity of CPE receptor(s) remain unclear. Biochemical studies (Wieckowski et al., 1994) suggested that CPE associates with a mammalian membrane protein of ~40–50 kDa after binding to membranes, forming a "small complex" of ~90 kDa (see Figure 7.2). This small complex was detected following CPE-treatment of all enterotoxin-sensitive cells examined by Wieckowski et al. No small complex was detected in CPE-treated cells that do not bind this toxin (or in detergent extracts of these cells), as would be expected if the 40–50 kDa membrane protein in this small complex was a/the functional CPE receptor.

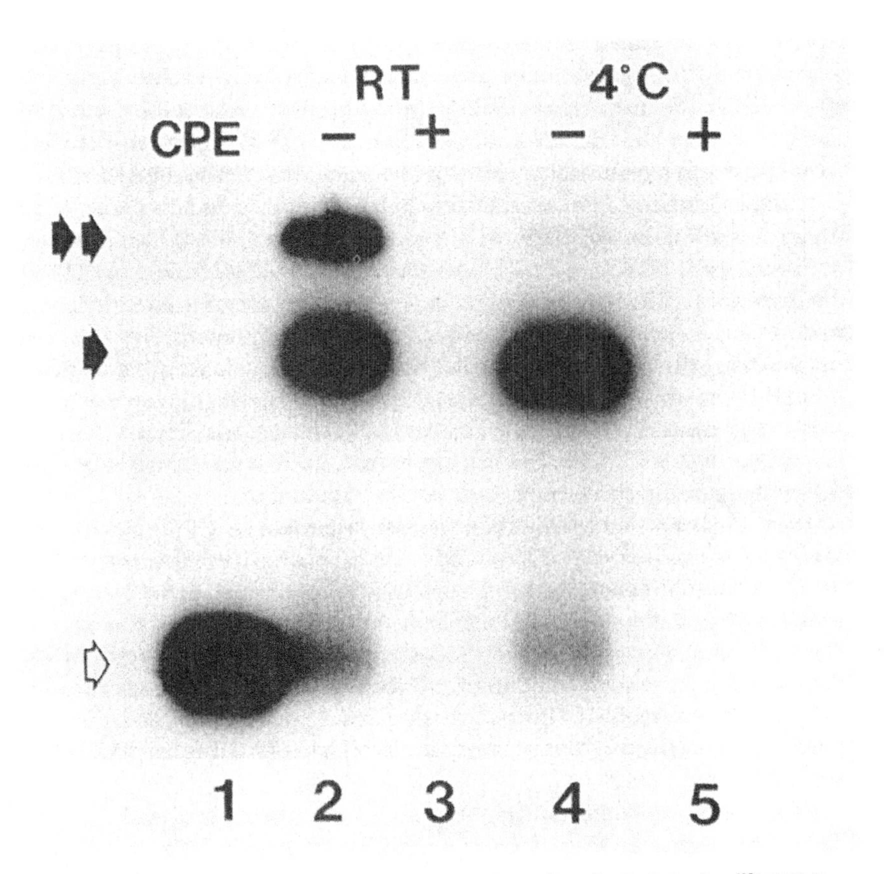

Figure 7.2 Visualization of CPE small and large complexes. Samples include: free ^{125}I-CPE (lane 1) and ^{125}I-CPE incubated with intact BBMs in the presence (+) or absence (–) of 50-fold excess unlabeled CPE at either room temperature (RT: lanes 2 and 3) or 4°C (lanes 4 and 5) prior to extraction with Triton X-100, electrophoresis under nondenaturing conditions, and autoradiography. The migration of small complex and large complex are indicated by single- and double-closed arrows, respectively. Note that similar large complex formation occurs at 37°C and 22°C (data not shown). Reproduced with the publisher's permission from Wieckowski et al. (1994).

However, recent studies (Katahira et al., 1997a,b) using expression cloning techniques now implicate members of the Claudin family (Morita et al., 1999) as functional CPE receptors. Expression of Claudin-3 or -4 in Mouse L cells (which cannot naturally bind CPE and, therefore, are naturally CPE-insensitive) produced transfectants capable of binding CPE at high levels. These transfectants were also highly sensitive to CPE, confirming that these Claudins can serve as "functional" receptors capable of mediating CPE cytotoxicity.

Although essential for CPE action, binding of CPE to its receptor(s) is not, by itself, sufficient to induce membrane permeability alterations. Several lines of experimental evidence indicate that CPE has a multistep action requiring post-binding events; (1) while rabbit colonic cells specifically bind CPE at high levels, these cells do not respond to the enterotoxin (McDonel and Demers, 1982), (2) recombinant CPE fragments have been identified that occupy the CPE receptor of sensitive cells yet are noncytotoxic (Hanna et al., 1991; Hanna and McClane, 1991; Hanna et al., 1989; Horiguchi et al., 1987; and Kokai-Kun and McClane, 1997), and (3) binding of CPE to sensitive cells at low temperatures does not induce any toxicity (McClane and Wnek, 1990).

The nature of these post-binding events in CPE action remains unclear. However, these events cannot correspond to the internalization of CPE into the cytoplasm of the mammalian cell since CPE remains plasma membrane-associated throughout its action (Tolleshaug et al., 1982). McDonel (1980) proposed that, because membrane-bound CPE becomes resistant to protease-induced release from membranes, this toxin inserts into the lipid bilayer of the plasma membrane. Other evidence has also been obtained that indirectly supports McDonel's hypothesis, including (1) bound CPE is not released from membranes by chemical treatments known to release peripherally bound proteins from membrane surfaces (McDonel, 1980; McClane et al., 1988), and (2) bound CPE does not dissociate from cells or isolated membranes, whether this binding occurs at 4°C or higher temperatures (McDonel, 1980; McClane et al., 1988).

While the evidence cited above is consistent with a post-binding step in CPE action involving the insertion of CPE into membranes, the existing data also appear fully compatible with the hypothesis that a conformational change to CPE occurs after this toxin becomes sequestered in small complex, with the end-result being that the CPE present in small complex becomes "locked" onto the membrane surface (further discussion later).

To better understand the nature of the post-binding steps in CPE action, studies have explored the membrane topology of CPE at 4°C, a temperature where CPE binding and small complex formation occur, but subsequent steps in CPE action are inhibited (McClane and Wnek, 1990, further discussion below). Kokai-Kun and McClane (1996) showed that CPE antibodies still specifically recognize membranes containing CPE allowed to complete the post-binding physical change. While this result, by itself, does not rule out the possibility that a portion(s) of CPE might be inserted into membranes when the toxin becomes localized in small complex, it does indicate that at least some region(s) of the enterotoxin remains exposed on membrane surfaces during this step in CPE action. Recent followup experiments (Wieckowski et al., 1999) demonstrated that Pronase treatment of membranes containing bound CPE results in substantial degradation of this toxin when it is present in small complex. This new result provides further evidence that most, if not all, of the CPE molecule remains surface-exposed following small complex formation and thereby raises additional doubts about the hypothesis that CPE in small complex becomes inserted into the lipid bilayer of the plasma membrane.

Perhaps the single most important insight into CPE's molecular action was the discovery (Wnek and McClane, 1989; McClane and Wnek, 1990) that, following completion of the post-binding physical change step in CPE action, CPE becomes associated with a second, larger (~160 kDa) complex in mammalian membranes (see Figure 7.3). Formation of this "large" CPE-containing complex appears to be directly responsible for CPE-induced membrane permeability changes, based upon the following observations (1) the inability of bound and "physically changed" CPE to induce membrane permeability alterations at 4°C directly correlates with the inhibition of large complex formation that also occurs at this low temperature (McClane and Wnek, 1990), (2) if cells containing CPE bound at 4°C are shifted to warmer temperatures, the subsequent onset of membrane permeability alterations closely coincides with large complex formation (McClane and Wnek, 1990), and (3) recent studies (Kokai-Kun and McClane, 1997) using recombinantly derived CPE fragments have established a strong correlation between the amount of large complex present in mammalian cells and the extent of CPE-induced membrane permeability alterations occurring in these cells (further discussion later).

Initial studies (Wnek and McClane, 1989) suggested that the large CPE-containing complex contains, at a minimum, one CPE molecule, one 50 kDa membrane protein, and one 70 kDa protein. More recent results from CPE receptor cloning studies (Katahira et al., 1997a) indicate that, at least in some CPE-sensitive mammalian cells, a Claudin may also be present in this large complex. Possible steps leading to the formation of the large CPE-containing complex will be discussed in the following section.

One interesting feature of the large CPE-containing complex is its unusual stability (Wnek and McClane, 1989; Wieckowski et al., 1994). For example, the large CPE complex is considerably more stable in SDS than is the small CPE complex (Wieckowski et al., 1994). Further, the large complex only partially dissociates upon boiling, while the small complex can be easily dissociated by boiling. There is not yet any biochemical explanation for this stability of the large CPE complex.

Finally, the same antibody and protease-challenge techniques that were used to probe the membrane topology of "physically changed" CPE have also been applied to study the membrane topology of CPE sequestered in large complex. CPE antibodies were shown (Kokai-Kun and McClane, 1996) to specifically recognize membranes containing CPE sequestered in the large complex. This implies that at least a portion of many CPE molecules remains exposed on the surface of membranes after this toxin becomes sequestered in the large complex. This conclusion is supported by recent studies indicating that the CPE molecules in the large complex become slightly smaller after Pronase treatment of large complex-containing membranes (Wieckowski et al., 1998). However, the fact that most of the CPE molecule sequestered in the large complex remains intact in these Pronase-treated membranes indicates that membranes do offer CPE sequestered in the large complex substantial protection against Pronase, which is consistent with these CPE molecules being closely associated with, or possibly even inserting into, membranes.

5.2. A WORKING MODEL FOR CPE ACTION

The recent findings regarding CPE binding and membrane topology demand some changes in our thinking about CPE action. In response, a new four-step working model for CPE action is presented in Figure 7.3 and discussed below.

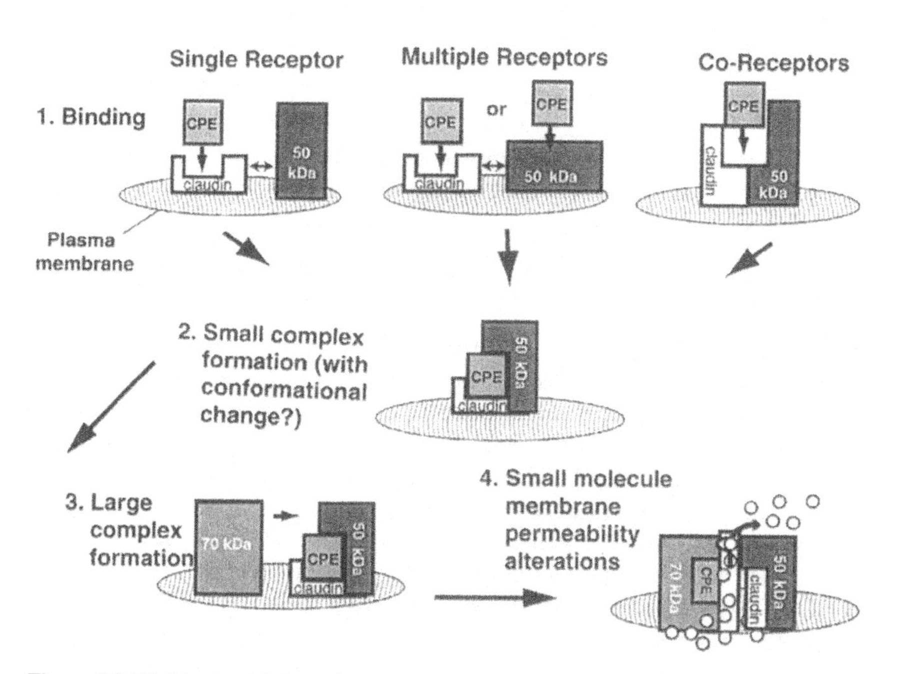

Figure 7.3 Working model for early events in CPE action. See text for a discussion of each proposed step in CPE action. Steps 1–4 in this model can occur in as little as 5 minutes at 37° and that steps 3–4 are inhibited at 4°C.

5.2.1. Binding of CPE to Its Receptor(s)

While the recent studies by Katahira et al. (1997) have clearly demonstrated that Claudins can serve as functional CPE receptors, many questions about CPE binding/receptors remain unresolved. In fact, at least three possible scenarios for CPE binding, as shown in Figure 7.3, appear equally compatible with existing information about CPE binding/receptors. The first scenario envisions Claudins as the only functional CPE receptor(s) used by all CPE-binding cells. The second scenario hypothesizes the existence of multiple types of CPE receptors. In this view, which is consistent with some kinetic studies of CPE binding suggesting that some mammalian cells express two classes of CPE receptor with very different affinities (McDonel, 1980; McDonel and McClane 1979), both Claudins and the 40–50 kDa membrane protein described by Wieckowski et al. (1994) may serve as "functional" CPE receptors, i.e., binding of CPE to either of these proteins would initiate a cytotoxic response. If there are two different types of CPE receptors, it should be resolved whether both of these CPE receptors can be expressed by a single CPE-sensitive mammalian cell. A third plausible scenario for CPE binding is that a Claudin and the 40–50 kDa protein serve as co-receptors for CPE binding, i.e., a cytotoxic response is initiated when CPE binds to both a Claudin and the 40–50 kDa membrane protein.

5.2.2. A Post-Binding Physical Change Occurs to CPE

As discussed above, results from recent antibody probe and Pronase-challenge studies cast increasing doubt on previous proposals that the second step in CPE action involves insertion of CPE into lipid bilayers. However, some sort of post-binding physical change does appear to occur immediately after binding since, as mentioned, CPE binding at 4°C (where large complex formation and membrane permeability alterations are inhibited) is irreversible (McDonel, 1980; McClane et al., 1988).

An appealing hypothesis is that the post-binding physical change step in CPE action corresponds to small complex formation, or to a conformational change immediately following small complex formation. For example, if CPE uses only a Claudin receptor, the resultant CPE:Claudin complex might subsequently interact with the 40–50 kDa membrane protein to form small complex (this would be consistent with immunoprecipitation studies indicating that a 40–50 kDa eucaryotic protein is present in the small complex (Wieckowski et al., 1994)). Formation of this small complex (perhaps coupled with a conformational change to the small complex) could effectively "lock" CPE onto the surface of the plasma membrane, explaining why CPE does not dissociate from membranes (even under conditions inhibiting large complex formation) and is not released from membranes by the addition of chemicals known to release peripherally bound membrane proteins (McClane, 1997). The possibility that small complex formation, and/or a conformational change to the newly formed small complex, is responsible for the post-binding physical change step in CPE action also appears to be similarly compatible with the multiple receptor scenario in Figure 7.3. If multiple types of CPE receptor do exist, then CPE bound to one type of receptor might subsequently interact with the second receptor to form small complex. Lastly, if CPE binds simultaneously to co-receptors, the post-binding physical change step in CPE action could be envisioned as involving a post-binding conformational change to the CPE: co-receptor complex (i.e., small complex), which effectively locks CPE onto the membrane surface.

5.2.3. Formation of CPE Large Complex

Since evidence indicates that CPE-sensitive cells form both small complex and large complex (see Figure 7.2) and that small complex formation precedes large complex formation (Wieckowski et al., 1994), it can be hypothesized that large complex formation results from an interaction between the ''physically changed'' small complex (which, at least in some cell types, apparently consists of CPE, a Claudin and/or a 40–50 kDa protein, see above) and a 70 kDa membrane protein previously linked to large complex (Wnek and McClane, 1989). The strong inhibition of large complex formation observed at low temperatures (McClane and Wnek, 1990) suggests that the interaction between physically changed small complex and the 70 kDa protein requires diffusion of membrane proteins through the lipid bilayer of membranes.

5.2.4. Onset of Small Molecule Permeability Alterations

While there is now considerable evidence implicating large complex formation in the onset of CPE-induced small molecule membrane permeability alterations, the direct mechanism by which large complex formation causes these membrane permeability effects remains unknown.

One appealing and simple mechanistic explanation would be that large complex corresponds to a pore-like structure, which allows free passage of small molecules across the plasma membranes of mammalian cells. If true, then large complex would represent an unusual pore structure that is comprised of a heterogeneous mixture of both eucaryotic and procaryotic proteins. Such a putative pore structure might result from CPE inserting into the membrane as part of the pore structure, which would be consistent with Pronase challenge studies indicating that most of the CPE molecule sequestered in large complex is inaccessible to Pronase challenge. Other evidence supporting CPE forming part or all of a pore, includes observations (Sugimoto et al., 1988) indicating that sonication of purified CPE into artificial membranes induces channel-like permeability alterations (however, note that no membrane proteins were present in this model system).

6. THE CPE PROTEIN

6.1. BIOCHEMISTRY

CPE is a 35,317 M_r, protein with an isoelectric point of 4.3 (McClane, 1997). This 319 amino acid, single polypeptide has a unique primary sequence (Czeczulin et al., 1993), except for some limited homology with the Antp70/Cl protein of *Clostridium botulinum* (Melville et al., 1997). The significance, if any, of the limited homology between CPE and Antp70/C1 is unclear. The primary sequence of CPE appears to be highly conserved among CPE-positive *C. perfringens* isolates, based upon DNA sequencing studies (Collie et al., 1998) demonstrating that identical *cpe* open reading frame (ORF) sequences are present in seven different *cpe*-positive *C. perfringens* isolates. Circular dichroism studies (Granum and Stewart, 1993) indicated that the secondary structure of CPE contains ~80% β-sheet and ~20% random coil. Due to difficulties in obtaining diffraction-grade crystals for x-ray analysis, no information is available regarding the 3-D structure of the enterotoxin.

Unlike several enterotoxins produced by other gram-positive bacteria (e.g., the staphylococcal enterotoxins), CPE has a heat-labile biologic activity. Heating CPE to 56°C for 5 min destroys its biologic activity (McClane, 1997). CPE's toxicity is also sensitive to pH extremes, i.e., biologic activity is lost when the enterotoxin is exposed to pH <5 or >10 (McClane, 1997). Interestingly, while some proteases (e.g., Pronase and subtilisin) inactivate the biologic activity of CPE (McClane, 1997), there is a 2–3 fold increase in biologic activity when CPE is digested, in vitro with intestinal proteases such as trypsin and chymotrypsin (Granum et al., 1981; Granum and Richardson, 1991). It is possible that, during food poisoning, intestinal proteases such as trypsin and chymotrypsin (Granum et al., 1981; Granum and Richardson, 1991) similarly activate CPE after this toxin has been released into the intestinal lumen upon the completion of sporulation (McClane, 1997).

6.2. CPE STRUCTURE/FUNCTION RELATIONSHIPS

Although the 3-D structure of CPE has not yet been solved, steady progress has nonetheless been achieved towards understanding how the CPE molecule exerts its action. As shown in Figure 7.4 and discussed below, CPE's binding and toxic activity domains appear to be segregated on discrete regions of the toxin.

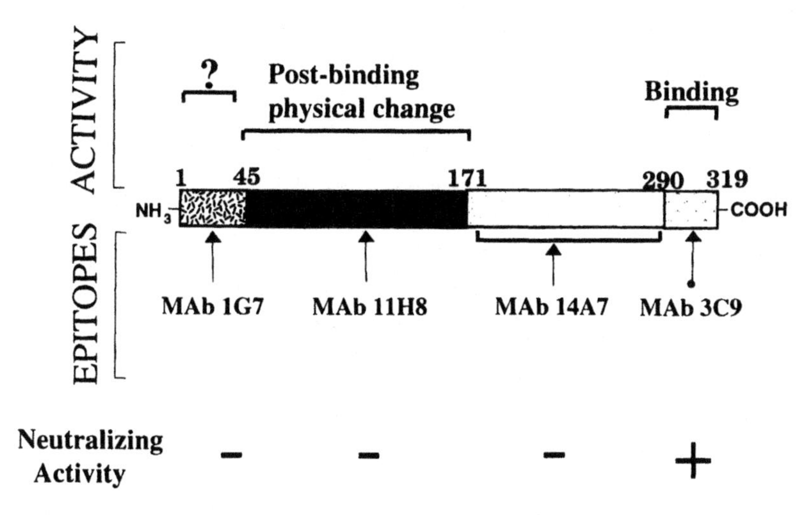

Figure 7.4 Map of the CPE structure/function relationship. CPE regions that appear to be required for the post-binding physical change, binding and various epitopes are noted. Of the four MAbs shown, only MAb 3C9 neutralizes CPE cytotoxicity.

Receptor binding activity was initially mapped to the C-terminal half of the CPE molecule in a series of studies using CPE fragments produced by either chemical cleavage or recombinant DNA approaches (Horiguchi et al., 1987; Hanna et al., 1989). CPE's receptor binding activity was then further localized to the extreme C-terminus of the toxin in studies demonstrating (1) a synthetic peptide possessing the same sequence present in the 30 C-terminal amino acids of CPE exhibits similar, if not identical, binding properties as native CPE (Hanna et al., 1991), and (2) deletion of the last five C-terminal amino acids from the CPE molecule is sufficient to abolish binding of this toxin to mammalian cells (Kokai-Kun and McClane, 1997).

CPE fragments have also proven invaluable for probing function(s) present on the N-terminal half of the enterotoxin protein. For example, it has been demonstrated (Kokai-Kun and McClane, 1997) that removing up to the first 44 N-terminal amino acids from native CPE causes a 2–3 fold increase in CPE's ability to induce membrane permeability alterations. This increased biologic activity was shown (Kokai-Kun and McClane, 1997) to result from these N-terminal CPE fragments being able to form, on a molar basis, approximately 2–3 fold more large complex than native CPE (this effect also explains the 2–3 fold increase in CPE activity induced by trypsin or chymotrypsin treatment, since these two proteases remove the first 25 and 36 amino acids, respectively, from native CPE (Granum et al., 1981; Granum and Richardson, 1991). Besides supporting the direct involvement of large complex formation in CPE-induced membrane permeability alterations, these results also indicate that the extreme N-terminal region of the CPE plays no apparent role in toxicity and serves no obvious function (recall that this N-terminal region cannot be involved in CPE secretion, since the enterotoxin retains an intracellular location until the mother cell lyses upon the completion of sporulation).

CPE completely loses biologic activity if further N-terminal amino acids beyond residue 44 are deleted (Kokai-Kun and McClane, 1997). While CPE fragments containing deletions of N-terminal amino acids beyond residue 44 were found to possess similar binding properties as native CPE (which is consistent with binding activity being localized exclusively to the C-terminus of the native CPE molecule), these fragments did not form any large complex. These results suggest that large complex formation is dependent on some amino acid residues that are present in the N-terminal half of CPE. To clarify this relationship, CPE point mutants are now being generated and characterized.

6.3. CPE VACCINE STUDIES

CPE epitope mapping studies (Hanna et al., 1992) localized the linear epitope recognized by monoclonal antibody (MAb) 3C9 to the extreme C-terminus of CPE (see Figure 7.4). Since the MAb 3C9 neutralizes the activity of native CPE by blocking the binding of this toxin to its receptor (Wnek et al., 1985), localization of the MAb 3C9 epitope to the extreme C-terminus of CPE further supports the assignment of receptor binding activity to this CPE region (as discussed above).

Localization of the MAb 3C9 epitope to the extreme C-terminus of CPE also suggested that C-terminal CPE fragments might hold promise as candidates for developing a CPE vaccine, i.e., these CPE fragments possess one or more linear neutralizing epitope(s), yet are nontoxic themselves (because they lack the N-terminal CPE sequences required for biologic activity). To test this hypothesis, Mietzner et al. (1992) prepared a 30-mer synthetic peptide with the same sequence as the C-terminal receptor-binding region of CPE. This synthetic peptide, which contains the linear neutralizing epitope recognized by MAb 3C9, was then chemically coupled to a thyroglobulin carrier protein and the resultant conjugate was administered parentally to mice. Serum from mice immunized with this conjugate contained high titers of IgG antibodies capable of neutralizing the biologic activity of native CPE.

While these initial immunization results are encouraging regarding the potential usefulness of C-terminal CPE sequences for vaccination purposes, many challenges remain with respect to developing a practical vaccine against CPE-mediated gastrointestinal diseases. Most notably, some strategy must be developed for eliciting the strong intestinal IgA responses necessary for preventing the binding of CPE to intestinal cells.

7. GENETICS AND EXPRESSION OF CPE

7.1. CPE GENETICS

The recent cloning of the intact *cpe* gene (Czeczulin et al., 1993) and technical advances in clostridial genetics (Rood, 1997) have led to considerable improvements in our understanding of *cpe* genetics. For example, it is now apparent that the *cpe* gene is present in only a small fraction (<5%) of the global *C. perfringens* population (Kokai-Kun et al., 1994; Daube et al., 1996). It has also recently been established that most, or all, food poisoning strains carry a single chromosomal copy of *cpe* (Comillot et al., 1995; Katayama et al., 1996; Collie and McClane, 1998). Further, the *cpe* gene of strain 8–6 (which is a derivative of food poisoning strain NCTC 8797) has been mapped (Canard et al., 1992) to a highly variable chromosomal region, suggesting that *cpe* is located on a mobile genetic element in some, or all, food poisoning strains. This possibility is also supported by another recent study (Brynestad et al., 1997) demonstrating that IS 1470 insertion sequences are present both upstream and downstream of the *cpe* gene in food poisoning strain NCTC 8239 (Figure 7.5). These new results suggest that, at least in some food poisoning strains, *cpe* is present on a chromosomally integrated, 6.3 kb transposon. However, mobilization of this putative *cpe*-containing transposon has not yet been demonstrated.

In contrast to the chromosomal location of *cpe* in most, if not all, food poisoning strains, *cpe* appears to be located on a plasmid in most *cpe*-positive *C. perfringens* isolates obtained from other sources. For example, Cole's group demonstrated (Comillot et al., 1995; Katayama et al., 1996) that *cpe* is present on a large plasmid in a large number of European *C. perfringens* type A isolates obtained from veterinary sources, including isolates from diseased

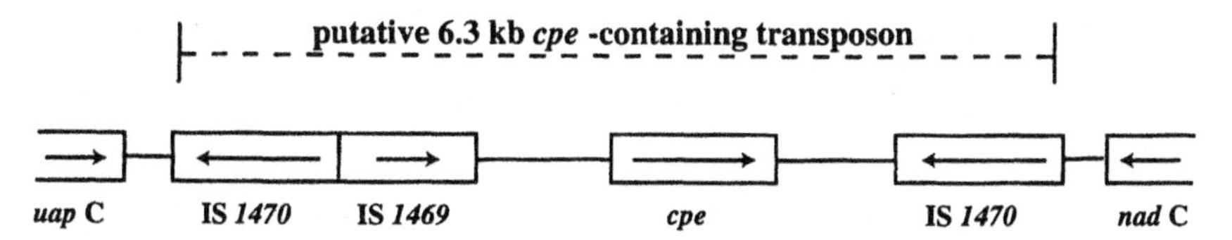

Figure 7.5 The *cpe* gene region of *C. perfringens* strain NCTC 8239. The open boxes indicate ORFs, while arrows indicate the direction of transcription of each ORF. The designation "IS" refers to insertion sequences. The *uapC* gene encodes purine permease, while *nadC* encodes quinolate phosphoribosyltransferase. The postulated 6.3 kb *cpe*-containing transposon present in this strain is also indicated on this figure. Adapted from Brynestad et al. (1997).

animals. Similarly, Collie and McClane (1998) have shown that North American veterinary disease isolates also carry a plasmid-borne *cpe* gene. Further, these researchers also demonstrated (Collie and McClane, 1998; Collie et al., 1998) that *cpe* has a plasmid location in most, if not all, type A isolates associated with non-foodborne human gastrointestinal diseases (e.g., antibiotic-associated diarrhea and sporadic diarrhea). This important new finding implies that genotypically distinct subpopulations of type A *cpe*-positive *C. perfringens* are responsible for CPE-associated foodborne vs. CPE-associated non-foodborne human gastrointestinal illnesses.

These *cpe* gene location findings raise several interesting questions that should be resolved in future research, including: What is the natural reservoir for the plasmid and chromosomal subpopulations of *cpe*-positive type A isolates? Are the chromosomal or plasmid *cpe*-positive *C. perfringens* type A isolates more common in nature? Can the recent associations between chromosomal vs. plasmid *cpe*-positive type A isolates and specific CPE-mediated diseases be used to improve diagnosis of CPE-mediated diseases?

Recently, Billington et al. (1988) made another interesting finding regarding *cpe* genetics when they screened a large collection of *C. perfringens* animal disease isolates using a multiplex PCR for detecting *C. perfringens* toxin genes. Twelve of these *C. perfringens* isolates, which were all from cases of animal enteritis, tested PCR-positive for gene sequences encoding both α toxin and ι toxin. Five of the 12 isolates were randomly selected and shown to express α and ι toxins, confirming their identity as type E isolates (see Table 7.2).

All 12 of these type E animal enteritis isolates also carried *cpe* sequences. Initially this finding suggested that CPE might play a role in the pathogenesis of type E animal enteritis infections. However, when this hypothesis was tested, the type E animal enteritis isolates carrying *cpe* sequences were found to be unable to express CPE under either sporulating or vegetative growth conditions.

Sequencing analysis of these silent type E *cpe* sequences provided several explanations for why type E animal enteritis isolates do not express CPE, namely, (1) their *cpe* sequences lack the upstream promoter sequences, ribosome binding site, and initiation codon of the functional type A cpe gene, and (2) two frame-shift mutations and multiple nonsense mutations are present in these silent type E sequences.

This sequencing analysis also revealed that the silent *cpe* sequences present in type E animal enteritis isolates are remarkably conserved. However, the type E animal enteritis isolates themselves do not appear to be clonally related. Further, these silent type E *cpe* sequences map to a plasmid, where they reside near the iota toxin genes. Collectively, these findings suggest that the highly conserved silent *cpe* sequences present in most, or all, type E animal enteritis isolates may have evolved fairly recently and then distributed on a plasmid (which also carries the iota toxin genes) to type A *C. perfringens* isolates.

7.2. CPE EXPRESSION

There are at least two interesting features to CPE expression. First, CPE expression is tightly regulated, since it is strongly associated with sporulation. For example, Western blot analyses (Czeczulin et al., 1993) have demonstrated that *C. perfringens* food poisoning strain NCTC 8239 produces >1000-fold more CPE during sporulation vs. vegetative growth. Recently, similar Western blot studies (Collie et al., 1998) demonstrated that CPE expression is also sporulation-associated for most, if not all, *cpe*-positive *C. perfringens* type A isolates, regardless of whether the isolate carries a plasmid-borne or a chromosomal *cpe* gene.

The second notable feature of CPE expression concerns the remarkably large amounts of enterotoxin that are made by sporulating cultures of many *C. perfringens* strains. CPE can account for 15% or more of the total protein present inside a sporulating *C. perfringens* cell (McClane, 1997). These high CPE expression levels do not appear to be significantly influenced by whether a *C. perfringens* isolate carries a plasmid or chromosomal *cpe* (Collie et al., 1988).

Molecular studies are now starting to address why CPE expression is so strongly sporulation-associated. For example, it has been demonstrated that CPE expression involves regulation at the transcriptional level, with several studies (Melville et al., 1994; Czeczulin et al., 1996) showing the presence of *cpe* message in sporulating, but not vegetative, cultures of CPE-positive *C. perfringens*. Northern blot analyses (Czeczulin et al., 1996) indicated that the *cpe* message is ~1.2 kb in size, consistent with the *cpe* ORF being transcribed as a monocistronic message. Two other observations offer indirect support for the transcription of *cpe* as a monocistronic message, including (1) a putative rho-independent transcriptional terminator has been identified immediately downstream of the *cpe* ORF (Czeczulin et al., 1993) and (2) at least three putative promoters have been identified immediately upstream of the *cpe* ORF (Zhao and Melville, 1998).

The identity of factor(s) specifically involved in this transcriptional regulation of CPE expression is also coming under study. In a recent study (Czeczulin et al., 1996), naturally *cpe*-negative *C. perfringens* isolates (which, as mentioned, represent >95% of the global *C. perfringens* population) were transformed with a low copy number *E. coli*-*C. perfringens* shuttle plasmid carrying the *cpe* gene. Western blots demonstrated that *C. perfringens* type A, B and C isolates transformed with this shuttle plasmid express CPE in a sporulation-associated manner. The regulated expression of CPE by these transformants suggests that most, if not all, *C. perfringens* isolates are routinely producing the factor(s) involved in regulating CPE expression. Since it is unlikely that expression of these regulatory factors would be conserved in most, or all, *C. perfringens* isolates (including the >95% of isolates that are *cpe*-negative) for no physiologic reason, some or all factors involved in regulating *cpe* transcription are probably also involved in regulating the expression of other *C. perfringens* genes.

How do these regulatory factors work? One potential insight into this question came from studies (Czeczulin et al., 1996) demonstrating that recombinant *E. coli* cells do not express CPE when transformed with the shuttle plasmid used to obtain sporulation-associated CPE expression by naturally *cpe*-negative *C. perfringens* transformants (as described above). The failure of these *E. coli* transformants to express CPE suggests that at least one factor(s) involved in regulating CPE expression is a positive regulator(s) that turns on *cpe* transcription during sporulation. Candidates for such a positive regulator(s) include the sporulation-associated alternative sigma factors; unfortunately, the limited information currently available about clostridial sigma factors makes this hypothesis relatively challenging to test.

However, even if *cpe* transcription is positively regulated during sporulation, it remains possible that CPE expression is also under negative regulation during vegetative growth. Consistent with this possibility, Hpr protein consensus binding sites have been identified both upstream and downstream of the *cpe* ORF (Brynestad et al., 1994). This finding is potentially significant for understanding the regulation of CPE expression because Hpr protein is a regulator of transition state gene expression in *Bacillus subtilus,* another gram-positive, sporulating bacterium. Therefore, the presence of putative Hpr-binding sequences adjacent to *cpe* has led to suggestions (Brynestad et al., 1994) that a *C. perfringens* Hpr-like homolog may help regulate CPE expression, presumably by repressing *cpe* transcription during vegetative growth.

The abundant expression of CPE by naturally enterotoxigenic *C. perfringens* isolates could also involve post-transcriptional regulatory effects. In this respect it is intriguing that an older study (Labbe and Duncan, 1977) reported that the *cpe* message has a half-life of ~45 min. If true, *cpe* mRNA would have a much longer half-life than most procaryotic transcripts, suggesting that *cpe* message has an unusual stability that contributes to high CPE expression levels during sporulation. A predicted stem-loop structure located downstream of the *cpe* ORF could contribute to this putative message stability (Czeczulin et al., 1993).

8. FUTURE CHALLENGES

C. perfringens type A food poisoning is neglected by some public health agencies, in favor of research on more trendy, "emerging" foodborne illnesses. Given the continuing prevalence of *C. perfringens* type A food poisoning (and the seriousness of this illness for the elderly and debilitated), this view appears extremely short-sighted, i.e., it is difficult to envision how food safety can substantially improve unless the second most common microbial cause of foodborne illness in the U.S. receives full research attention.

Notwithstanding the exciting recent progress described in this chapter, a substantial amount of additional information is still needed regarding *C. perfringens* type A food poisoning. While there are many unanswered questions concerning basic science aspects (e.g., *cpe* genetics, CPE action) of *C. perfringens* type A food poisoning, even some of the most fundamental/applied aspects of this disease remain poorly understood. For example: Why do only those *C. perfringens* isolates carrying chromosomal *cpe* genes cause *C. perfringens* type A food poisoning? How/when do these chromosomal *cpe* isolates get into our food? How can contamination of food with these food poisoning isolates be prevented?

Given sufficient public health commitment, the technology to address both the basic science and fundamental/applied questions about *C. perfringens* type A food poisoning now exists.

9. ACKNOWLEDGMENTS

Research in my laboratory on *C. perfringens* type A food poisoning has been generously supported for many years by the National Institutes of Health, grant AI 19844–15. The author thanks Eva Wieckowski and John Kokai-Kun for help with preparing figures used in this chapter.

10. REFERENCES

Batholomew, B. A., Stringer, M. F., Watson, G. N. and Gilbert, R. J. 1985. "Development and application of an enzyme-linked immunosorbent assay for *Clostridium perfringens* type A enterotoxin," *J. Clin. Pathol.* 38:222–228.

Bean, N. H., Goulding, J. S., Lao, C., and Angulo, F. J. 1996. "Surveillance for foodborne-disease outbreaks—United States, 1988–1982," *Morbidity and Mortality Weekly Reports* 45:1–54.

Billington, S. J., Wieckowski, E. U., Sarker, M. R., Bueschel, D., Songer, J. G., and McClane, B. A. 1998. "*Clostridium perfringens* type E animal enteritis isolates carry silent enterotoxin gene sequences," *Infect. Immun.* 66:4531–4536.

Birkhead, G., Vogt, R. L., Heun, E. M., Snyder, J. T., and McClane, B. A. 1988. "Characterization of an outbreak of *Clostridium perfringens* food poisoning by quantitative fecal culture and fecal enterotoxin measurement." *J. Clin. Microbiol.* 26:471–474.

Bowness, P., Moss, P. A., Tranter, H., Bell, J. T., and McMichael, A. J. 1992. "*Clostridium perfringens* enterotoxin is a superantigen reactive with human T cell receptors Vβ 6.9 and Vβ 22d," *J. Exp. Med.* 176:893–896.

Brynestad, S., Synstad, B., and Granum, P. E. 1997. "The *Clostridium perfringens* enterotoxin gene is on a transposable element in type A human food poisoning strains," *Microbiology* 143:2109–2115.

Brynestad, S., Iwanejko, L. A., Stewart, G. S. A. B. and Granum, P. E. 1994. "A complex array of Hpr consensus DNA recognition sequences proximal to the enterotoxin gene in *Clostridium perfringens* type A," *Microbiology* 140:97–104.

Canard, B., Saint-Joanis, B., and Cole, S. T. 1992. "Genomic diversity and organization of virulence genes in the pathogenic anaerobe *Clostridium perfringens,*" *Molec. Microbiol.* 6:1421–1429.

Collie, R. C., Kokai-Kun, J. F., and McClane, B. A. 1998. "Phenotypic characterization of enterotoxigenic *Clostridium perfringens* isolates from nonfoodborne human gastrointestinal diseases," *Anaerobe* 4:69–79.

Collie, R. E. and McClane, B. A. 1998. "Evidence that the enterotoxin gene can be episomal in *Clostridium perfringens* isolates associated with nonfoodborne human gastrointestinal diseases," *J. Clin. Microbiol.* 36:30–36.

Cornillot, E., Saint-Joanis, B., Daube, G., Katayama, S., Granum, P. E., Carnard, B., and Cole, S. T. 1995. "The enterotoxin gene (*cpe*) of *Clostridium perfringens* can be chromosomal or plasmid-borne," *Molec. Microbiol.* 15:639–647.

Czeczulin, J. R., Collie, R. E., and McClane, B. A. 1996. "Regulated expression of *Clostridium perfringens* enterotoxin in naturally *cpe*-negative type A, B, and C isolates of *C. perfringens,*" *Infect. Immun.* 64:3301–3309.

Czeczulin, J. R., Hanna, P. C., and McClane, B. A. 1993. "Cloning, nucleotide sequencing, and expression of the *Clostridium perfringens* enterotoxin gene in *Escherichia coli,*" *Infect Immun.* 61:3429–3439.

Daube, G., Simon, P., Limbourg, B., Manteca, C., Mainil, J., and Kaeckenbeeck A. 1996. "Hybridization of 2,659 *Clostridium perfringens* isolates with gene probes for seven toxins (α, β, ϵ, ι, τ, μ and enterotoxin) and for sialidase," *Am. J. Vet. Res.* 57:496–501.

Granum, P. E. and Richardson, M. 1991. "Chymotrypsin treatment increases the activity of *Clostridium perfringens* enterotoxin," *Toxicon.* 29:445–453.

Granum, P. E. and Stewart, G. S. A. B. 1993. "Molecular biology of *Clostridium perfringens* enterotoxin," in *Genetics and Molecular Biology of Anaerobic Bacteria,* eds., M. Sebald. New York: Springer-Verlag. pp. 235–247.

Granum, P. E., Whitaker, J. R., and Skjelkvale, R. 1981. "Trypsin activation of enterotoxin from *Clostridium perfringens* type A," *Biochim. Biophys. Acta* 668:325–332.

Hanna, P. C. and McClane, B. A. 1991. "A recombinant C-terminal toxin fragment provides evidence that membrane insertion is important for *Clostridium perfringens* enterotoxin cytotoxicity," *Molec. Microbiol.* 5:225–230.

Hanna, P. C., Wnek, A. P., and McClane, B. A. 1989. "Molecular cloning of the 3′ half of the *Clostridium perfringens* enterotoxin gene and demonstration that this region encodes receptor-binding activity," *J. Bacteriol.* 171:6815–6820.

Hanna, P. C., Mietzner, T. A., Schoolnik, G. K., and McClane, B. A. 1991. "Localization of the receptor-binding region of *Clostridium perfringens* enterotoxin utilizing cloned toxin fragments and synthetic peptides. The 30 C-terminal amino acids define a functional binding region." *J. Biol. Chem.* 266:11037–11043.

Hanna, P. C., Wieckowski, E. U., Mietzner, T. A., and McClane, B. A. 1992. "Mapping functional regions of *Clostridium perfringens* type A enterotoxin," *Infect. Immun.* 60:2110–2114.

Hauschild, A. H., Niilo, L., and Dorward, W. J. 1971. "The role of enterotoxin in *Clostridium perfringens* type A enteritis," *Can. J. Microbiol.* 17:987–991.

Heredia, N. L., Labbe, R. G., Rodriguez, M. A., and Garcia-Alvarodo, J. S. 1991. "Growth, sporulation and enterotoxin production by *C. perfringens* type A in the presence of human bile salts," *FEMS Microbiol Lett* 84:15–22.

Horiguchi, Y., Akai, T., and Sakaguchi, G. 1987. "Isolation and function of a *Clostridium perfringens* enterotoxin fragment," *Infect. Immun.* 55:2912–2915.

Horiguchi, Y., Uemura, T., Kozaki, S., and Sakaguchi, G. 1985. "The relationship between cytotoxic effects and binding to mammalian cultures cells of *Clostridium perfringens* enterotoxin,'' *FEMS Microbiol. Let.* 28:131–135.

Hulkower, K. I., Wnek, A. P., and McClane, B. A. 1989. ''Evidence that alterations in small molecule permeability are involved in the *Clostridium perfringens* type A enterotoxin-induced inhibition of macromolecular synthesis in Vero cells,'' *J. Cell. Physiol.* 140:498–504.

Katahira, J., Inoue, N., Horiguchi, Y., Matsuda, M., and Sugimoto, N. 1997a. ''Molecular cloning and functional characterization of the receptor for *Clostridium perfringens* enterotoxin,'' *J. Cell Biol.* 136:1239–1247.

Katahira, J., Sugiyama, H., Inoue, N., Horiguchi, Y., Matsuda, M., and Sugimoto, N. 1997b. ''*Clostridium perfringens* enterotoxin utilizes two structurally related membrane proteins as functional receptors in vivo,'' *J. Biol. Chem.* 272:26652–26658.

Katayama, S. I., Dupuy, B., Daube, G., China, B., and Cole, S. T. 1996. ''Genome mapping of *Clostridium perfringens* strains with I-*Ceu* 1 shows many virulence genes to be plasmid-borne,'' *Mol. Gen. Genet.* 251:720–726.

Kokai-Kun, J. F. and McClane, B. A. 1996. ''Evidence that region(s) of the *Clostridium perfringens* enterotoxin molecule remain exposed on the external surface of the mammalian plasma membrane when the toxin is sequestered in small or large complex,'' *Infect. Immun.* 64:1020–1025.

Kokai-Kun, J. F. and McClane, B. A. 1997. ''Deletion analysis of the *Clostridium perfringens* enterotoxin,'' *Infect. Immun.* 65:1014–1022.

Kokai-Kun, J. F., Songer, J. G., Czeczulin, J. R., Chen, F., and McClane, B. A. 1994. ''Comparison of Western immunoblots and gene detection assays for identification of potentially enterotoxigenic isolates of *Clostridium perfringens,*'' *J. Clin. Microbiol.* 32:2533–2539.

Krakauer, T., Fleischer, B., Stevens, D. L., McClane, B. A., and Stiles, B. G. 1997. ''*Clostridium perfringens* enterotoxin lacks superantigenic activity but induces an interleukin-6 response from human peripheral blood mononuclear cells,'' *Infect. Immun.* 65:3485–3488.

Labbe, R. G. and Duncan, C. L. 1977. ''Evidence for stable messenger ribonucleic acid during sporulation and enterotoxin synthesis by *Clostridium perfringens* type A,'' *J. Bacteriol.* 129:843–849.

Markovic, L., Asanin, R., and Dimitrijevic, B. 1993. ''In vitro effects of *Clostridium perfringens* enterotoxin on Vero cells and *in vivo* effects in animals,'' *Acta Veterinaria* (Beograd) 43:191–198.

Matsuda, M. and Sugimoto, N. 1979. ''Calcium-independent and calcium-dependent steps in action of *Clostridium perfringens* enterotoxin,'' *Biochem. Biophys. Res. Commun.* 91:629–636.

Matsuda, M., Ozutsumi, K., Iwashi, H. and Sugimoto, N. 1986. ''Primary action of *Clostridium perfringens* type A enterotoxin on HeLa and Veto cells in the absence of extracellular calcium: rapid and characteristic changes in membrane permeability,'' *Biochem. Biophys. Res. Commun.* 141:704–710.

McClane, B. A. 1984. ''Osmotic stabilizers differentially inhibit permeability alterations induced in Vero cells by *Clostridium perfringens* enterotoxin,'' *Biochim. Biophys. Acta* 777:99–106.

McClane, B. A. 1992. ''*Clostridium* perfringens enterotoxin: structure, action and detection,'' *J. Food Safety* 12:237–252.

McClane, B. A. 1997. ''*Clostridium perfringens,*'' in *Food Microbiology: Fundamentals and Frontiers,* eds., M. P. Doyle, L. R. Beuchat and T. J. Montville, Washington, D.C.: ASM Press. pp. 305–326.

McClane, B. A. and McDonel, J. L. 1980. ''Characterization of membrane permeability alterations induced in Vero cells by *Clostridium perfringens* enterotoxin,'' *Biochim, Biophys. Acta* 600:974–985.

McClane, B. A. and McDonel, J. L. 1981. ''Protective effects of osmotic stabilizers on morphological and permeability alterations induced in Vero cells by *Clostridium perfringens* enterotoxin," *Biochim. Biophys Acta* 641:401–409.

McClane, B. A. and Wnek, A. P. 1990. "Studies of *Clostridium perfringens* enterotoxin action at different temperatures demonstrate a correlation between complex formation and cytotoxicity," *Infect. Immun.* 58:3109–3115.

McClane, B. A., Wnek, A. P., Hulkower, K. I., and Hanna, P. C. 1988. "Divalent cation involvement in the action of *Clostridium perfringens* type A enterotoxin," *J. Biol. Chem.* 263:2423–2435.

McDonel, J. L. 1980. "Binding of *Clostridium perfringens* ^{125}I-enterotoxin to rabbit intestinal cells," *Biochem.* 21:4801–4807.

McDonel, J. L. and Demers, G. W. 1982. "*In vivo* effects of enterotoxin from *Clostridium perfringens* type A in rabbit colon: binding vs. biologic activity," *J. Infect. Dis.* 145:490–494.

McDonel, J. L. and Duncan, C. L. 1975. "Histopathological effect of *Clostridium perfringens* enterotoxin in the rabbit ileum," *Infect. Immun.* 12:1214–1218.

McDonel, J. L. and Duncan, C. L. 1977. "Regional localization of activity of *Clostridium perfringens* type A enterotoxin in rabbit illeum, jejunum, and duodenum," *J. Infect. Dis.* 136:661–666.

McDonel, J. L. and McClane, B. A. 1979. "Binding vs. biological activity of *Clostridium perfringens* enterotoxin in Vero cells," *Biochem. Biophys. Res. Commun.* 87:497–504.

McDonel, J. L., Chang, L. W., Pounds, J. L. and Duncan, C. L. 1978. "The effects of *Clostridium perfringens* enterotoxin on rat and rabbit ileum: an electron microscopy study," *Lab. Invest.* 39:210–218.

Meer, R. R. and Songer, J. G. 1997. "Multiplex polymerase chain reaction assay for genotyping *Clostridium perfringens*," *Am. J. Vet. Res.* 58:702–705.

Melville, S. B., Collie, R. E., and McClane, B. A. 1997. "Regulation of enterotoxin production in *Clostridium perfringens*," in *The Clostridia. Molecular Biology and Pathogenesis*, eds., J. I. Rood, McClane, B. A., Songer, J. G. and R. Titball. London: Academic Press. pp. 471–487.

Melville, S. B., Labbe, R., and Sonenshein, A. L. 1994. "Expression from the *Clostridium perfringens cpe* promoter in *C. perfringens* and *Bacillus subtilus*," *Infect. Immun.* 62:5550–5558.

Mietzner, T. A., Kokai-Kun, J. F., Hanna, P. C., and McClane, B. A. 1992. "A conjugated synthetic peptide corresponding to the C-terminal region of *Clostridium perfringens* type A enterotoxin elicits an enterotoxin neutralizing antibody response in mice," *Infect. Immun.* 60:3947–3951.

Morita, K., Furuse, M., Fujimoto, K., and Tskita, S. 1999. "Claudin multigene family, encoding four transmembrane domain protein components of tight junction strands," *Proc. Nat. Acad. Sci. USA* 96:511–516.

Nagata, K., Okamura, H., Kunitoh, D. and Uemura, T. 1997. "Mitogenic activity of *Clostridium perfringens* enterotoxin in human peripheral lymphocytes," *J. Vet. Med. Sci.* 59:U2–U3.

Rood, J. I. 1997. "Genetic analysis in *Clostridium perfringens*," in *The Clostridia. Molecular Biology and Pathogenesis*, eds., J. I. Rood, B. A. McClane, J. G. Songer and R. W. Titball. London: Academic Press. pp. 65–72.

Sherman, S., Klein, E., and McClane, B. A. 1994. "*Clostridium perfringens* type A enterotoxin induces concurrent development of tissue damage and fluid accumulation in the rabbit ileum," *J. Diarrheal Dis. Res.* 12:200–207.

Shih, N. J. and Labbe, R. G. 1996. "Sporulation-promoting ability of *Clostridium perfringens* culture fluids," *Appl. Environ. Microbiol.* 62:1441–1443.

Skjelkvale, R. and Duncan, C. L. 1975. "Enterotoxin formation by different toxigenic types of *Clostridium perfringens*," Infect. Immun. 11:563–575.

Skjelkvale, R. and Uemura, T. 1977. "Experimental diarrhea in human volunteers following oral administration of *Clostridium perfringens* enterotoxin," *J. Appl. Bacteriol.* 46:281–286.

Songer, J. G. and Meer, R. M. 1996. "Genotyping of *Clostridium perfringens* by polymerase chain reaction is a useful adjunct to diagnosis of clostridial enteric disease in animals," *Anaerobe* 2:197–203.

Sugimoto, N., Takagi, M., Ozutsumi, K., Harada, S., and Matsuda, M. 1988. "Enterotoxin of *Clostridium perfringens* type A forms ion-permeable channels in a lipid bilayer membrane," *Biochem. Biophys. Res. Commun.* 156:551–556.

Todd, E. C. D. 1989a. "Cost of acute bacterial foodborne disease in Canada and the United States," *Int. J. Food Microbiol.* 9:313–326.

Todd, E. C. D. 1989b. "Preliminary estimates of costs of foodborne disease in the United States," *J. Food Prot.* 52:595–601.

Tolleshaug, H., Skjelkvale, R., and Berg, T. 1982. "Quantitation of binding and subcellular distribution of *Clostridium perfringens* enterotoxin in rat liver cells," *Infect. Immun.* 37:486–491.

Wieckowski, E. U., Kokai-Kun, J. F., and McClane, B. A. 1998. "Characterization of membrane-associated *Clostridium perfringens* enterotoxins following Pronase treatment," *Infect. Immun.* 66:5897–5905.

Wieckowski, E. U., Wnek, A. P., and McClane, B. A. 1994. "Evidence that an ~50kDa mammalian plasma membrane protein with receptor-like properties mediates the amphiphilicity of specifically-bound *Clostridium perfringens* enterotoxin," *J. Biol. Chem.* 269:10838–10848.

Wnek, A. P. and McClane, B. A. 1989. "Preliminary evidence that *Clostridium perfringens* type A enterotoxin is present in a 160,000-M_r complex in mammalian membranes," *Infect. Immun.* 57:574–581.

Wnek, A. P., Strouse, R. J., and McClane, B. A. 1985. "Production and characterization of monoclonal antibodies against *Clostridium perfringens* type A enterotoxin," *Infect Immun.* 50:442–448.

Wrigley, D., Hanwella, H., and Thon, B. 1995. "Acid exposure enhances sporulation of certain strains of *Clostridium perfringens,*" *Anaerobe* 1:163–269.

Zhao, Y. and Melville, S. B. 1998. "Identification and characterization of sporulation-dependent promoters upstream of the enterotoxin gene (*cpe*) of *Clostridium perfringens,*" *J. Bacteriol.* 180:136–142.

CHAPTER 8

Mechanisms of Pathogenesis and Toxin Synthesis in *Clostridium botulinum*

KEIJI OGUMA
YUKAKO FUJINAGA
KAORU INOUE
KENJI YOKOTA

1. INTRODUCTION

SEVEN immunologically distinct neurotoxins (NTX), types A to G, are produced by *Clostridium botulinum* strains. The NTXs inhibit the release of acetylcholine (Ach) at the neuromuscular junctions and synapses, and cause botulism in humans and animals. Molecular mass (Mr) of all the types of NTXs is approximately 150 kDa. The NTXs associate with nontoxic components in cultures, and become large complexes that are designated progenitor toxins. Recently, the genes coding for type A to G NTXs have been cloned, and their entire nucleotide sequences determined. Furthermore, it has become clear that the NTXs are Zn^{++}-binding proteins and possess protease activity. The structure and function of the nontoxic components of the progenitor toxins have also been investigated.

In this chapter, the characteristics of organisms and the disease are first briefly summarized. The gene organization, structure and function of the NTXs and nontoxic components are then described in order to understand the mechanisms of pathogenesis and toxin synthesis.

2. CLASSIFICATION OF *C. BOTULINUM* BASED ON TOXIN ANTIGENICITY

C. botulinum is a gram-positive, obligatory anaerobic rod-producing endospores (or spores). The organisms produce poisonous NTXs, and are classified into seven types based on the antigenicity of their NTXs. These NTXs, approximately 150 kDa proteins, exist in the bacterial cultures as stable, large complexes designated as progenitor toxins, which are found in three forms: 12S toxin (Mr ca. 300 kDa), 16S toxin (Mr ca. 500 kDa), and 19S toxin (Mr ca. 900 kDa). Type A progenitor toxin consists of three forms, 19S, 16S, and 12S. Type B, C, and D consist of two forms, 16S and 12S. Type E and F consist of a single form, 12S, while type G has a single 16S form (Table 8.1). Type A, B, E, and F cause botulism in humans, whereas type C and D cause botulism in birds, cattle, and other animals.

There is some confusion as to the classification of type C and D toxins. The sera obtained from animals highly immunized by type C toxoid can cross-neutralize type D toxin with varying rates, and vice versa. It has also been reported that type C cultures produced antigenically different NTXs, C1 and C2, and the cultures were classified into two groups, Cα and Cβ (Jansen, 1971). Later it became clear by employing a number of monoclonal antibodies against these toxins that there exist many antigenic determinants common to type C1 and D toxins as well as epitopes specific only for type C1 or D toxin (Oguma et al., 1984, 1985). Since some of the common and specific antigens were involved in toxin neutralization, it was proposed that the different distribution of specific or common antigens in C1 and D toxins may cause the varying rates of (cross-) neutralization. Some type C1 and D toxins seemed to be a "mosaic" of C1 and D (Moriishi et al., 1996). The existence of common antigens was supported by the amino acid sequences of the toxins predicted from the nucleotide sequences of their genes. Oguma et al. (1986) proposed that type C and D cultures could be classified into four groups based on their biochemical properties.

As for C2 toxin, it has become clear that the antigenicity and toxin structure of C2 toxin are quite different from those of C1 and D toxins, and that C2 toxin is not a NTX but rather an entero- or cytotoxin, which has the ability to enzymatically ADP-ribosylate nonmuscle actin (Ohishi, 1983; Aktories et al., 1986). Some type C and D cultures produce only C2 toxin, while others may produce C2 toxin in addition to C1 or D. It was reported that C1 NTX, but not A and E NTXs, is cytotoxic for mice primary neuron cells (Kurokawa et al., 1987). In addition, only type C and D cultures produce C3 enzyme, which ADP-ribosylates the small-molecular weight, GTP-binding proteins such as rho. At first, it was presumed that C1 and D NTXs possess the ADP-ribosylation activity, but later it has become clear that this activity belongs to the C3 enzyme contaminating the NTX preparations (Aktories and Frevert, 1987; Moriishi et al., 1990).

The heterogeneity in antigenicity or amino acid sequence of NTX in a single type has also been reported in type A and B toxins. Type A toxin produced by a strain, Kyoto-F, isolated from an infant botulism in Japan was somewhat different from those previously reported (Tabita et al., 1991; Willems et al., 1993), and the toxins produced by proteolytic and nonproteolytic

TABLE 8.1. Classification of *C. botulinum*.

Classification Based on Toxin-Antigenicity (A ~ G)	Classification Based on Biochemical Properties		Type of Toxins		Susceptible Species
	(Protease Production)	(I ~ IV)	Neurotoxin (Progenitor Toxin)	Others	
A	+	I	A (12, 16, 19 S)		Man, Chicken
B	+	I	B (12, 16 S)		Man, Horse
	−	II			
C	−*	III	C_1 (12, 16 S)	C_2, (C_3)**	Bird, Turtle, Mink
D	−[4]	III	D (12, 16 S)	C_2, (C_3)**	Cattle, Sheep, Horse
E	−	II	E (12 S)		Man, Bird, Fish
F	+	I	F (12 S)		Man
	−	II			
G	+ (nonsaccharolytic)	IV	G (16 S)		No outbreak

* Many strains produce a protease(s) which makes a nick in the neurotoxin.
** C_3 is not a toxin but an enzyme.

type B strains were not identical (Hutson et al., 1994). On the other hand, some type A and B cultures are thought to produce two different types of toxins. Type A strain, 84, isolated from vineyard soil in Argentina, produced A and F toxins with about 100 times as much type A toxin as type F (Sugiyama et al., 1972). Type B strain, 657, isolated from an infant botulism in the U.S. produced B and A toxins in a ratio of about 10:1 (Gimenez, 1984). Therefore, these cultures are designated as Af and Ba, respectively. A strain designated, AB-I.P. 7212, isolated from a foodborne botulism in France, is also thought to produce both A and B toxin (Sakaguchi et al., 1986). Recently, we obtained data indicating that strain AB-I.P. 7212 does possess both type A and B toxin genes (Fujinage et al., 1995). Whether one strain produces one ''mosaic'' toxin or two different toxins should be analyzed genetically as well as immunologically.

3. CLASSIFICATION OF *C. BOTULINUM* BASED ON BIOCHEMICAL PROPERTIES

C. botulinum can be divided into four groups on the basis of cultural properties, cell wall composition, serological relationships, and DNA or RNA homologies (Smith, 1977). These groups are: (1) all of type A strains and the proteolytic strains of types B and F; (2) the nonproteolytic strains of types B and F and all type E strains; (3) the strains of types C and D; and (4) the proteolytic but nonsaccharolytic type G strains (type G strain is now proposed to be excluded from *C. botulinum* species, and to form a new species, *C. argentinense,* based on their biochemical properties). The property of Group I spores being highly resistant and the Group II strain's ability to grow at 4°C seem to be significant factors in causing botulism outbreaks.

4. BOTULISM

There are three different types of botulism: foodborne, infant, and wound botulism.

4.1. FOODBORNE BOTULISM

In the 19th century, it was well known that foodborne poisoning with a high incidence of fatality sometimes occurred from eating ham or sausage (sausage is coined ''botulus'' in Latin). The medical term for sausage poisoning was referred to as ''botulismus.'' Thus, botulinum became the name for the organism. In 1897, Van Ermengem isolated the cells and coined it Bacillus botulinus (this organism is now considered to be nonproteolytic type B, *C. botulinum*).

C. botulinum spores can survive in the ground or in lake or sea sediments for a long time. When foods contaminated with *C. botulinum* spores are sustained under conditions conducive to germination, they grow into active vegetative cells and toxins are produced. The toxins ingested with foods by animals, including humans, can pass through the stomach and be absorbed from the upper small intestine. Thereafter, the toxins react with the target organs (end-plates or synapses of parasympatic nerves) through the lymphatic and blood vessels, where the toxins block the release of Ach (Figure 8.1, details will be discussed later).

As described in the next section, infant botulism caused by *C. butyricum* and *C. barartii,* which produce type E and F toxins, respectively, has been reported. Recently, foodborne botulism caused by neurotoxin (type E)-producing *C. butyricum* was reported in China (1994) and India (1996, in this case, toxin type was not described). The food stuff of the former was salted and fermented paste made of soybeans and wax gourds (Meng et al., 1997), the latter was a crisp made of gram flour termed ''sevu'' (Chaudhry et al., 1998).

4.2. INFANT BOTULISM

In 1976, Pickett et al. reported two cases of botulism in infants. The clinical diagnosis was confirmed by the identification of *C. botulinum* toxin and organisms in the feces of these and other similarly afflicted infants. To date more than 1200 cases have been reported in the world (mainly in the U.S.). When infants, usually less than one year old, ingest weaning foods such as honey, or ingest house dust, soil, etc., contaminated with spores (in some cases, the source could not be identified), the spores can germinate and grow in the intestine, especially in the large intestine. Therefore, infant botulism is caused by the toxin produced in vivo. The lethal rate of infant botulism is not considered high, but it has been theorized to be involved in some cases of sudden infant death syndrome (Arnon et al., 1981).

As described in the previous section, the antigenic characteristics of toxins causing infant botulism are sometimes different from those obtained from foodborne cases. Furthermore, infant botulism caused by type F toxin-producing *C. baratii,* and type E toxin-producing *C. butyricum,* was reported in the U.S. (1977) and Italy (1984, 1985), respectively (Aureli et al., 1986, Hall et al., 1985). We also reported on infant botulism caused by a type C organism, which is very rare in humans (Oguma et al., 1990). Botulinum spores seldom germinate and grow in the gut of healthy adults. It has been proposed that an insufficient enteric flora in an infant's intestine is the main reason why the organisms are able to grow and produce toxin (Moberg and Sugiyama, 1979). The ''premature'' gastrointestinal condition of infants might also explain why unusual toxins or organisms thrive so well in the intestine, causing infant botulism.

4.3. WOUND BOTULISM

The cause of this type of botulism is similar to tetanus. The number of

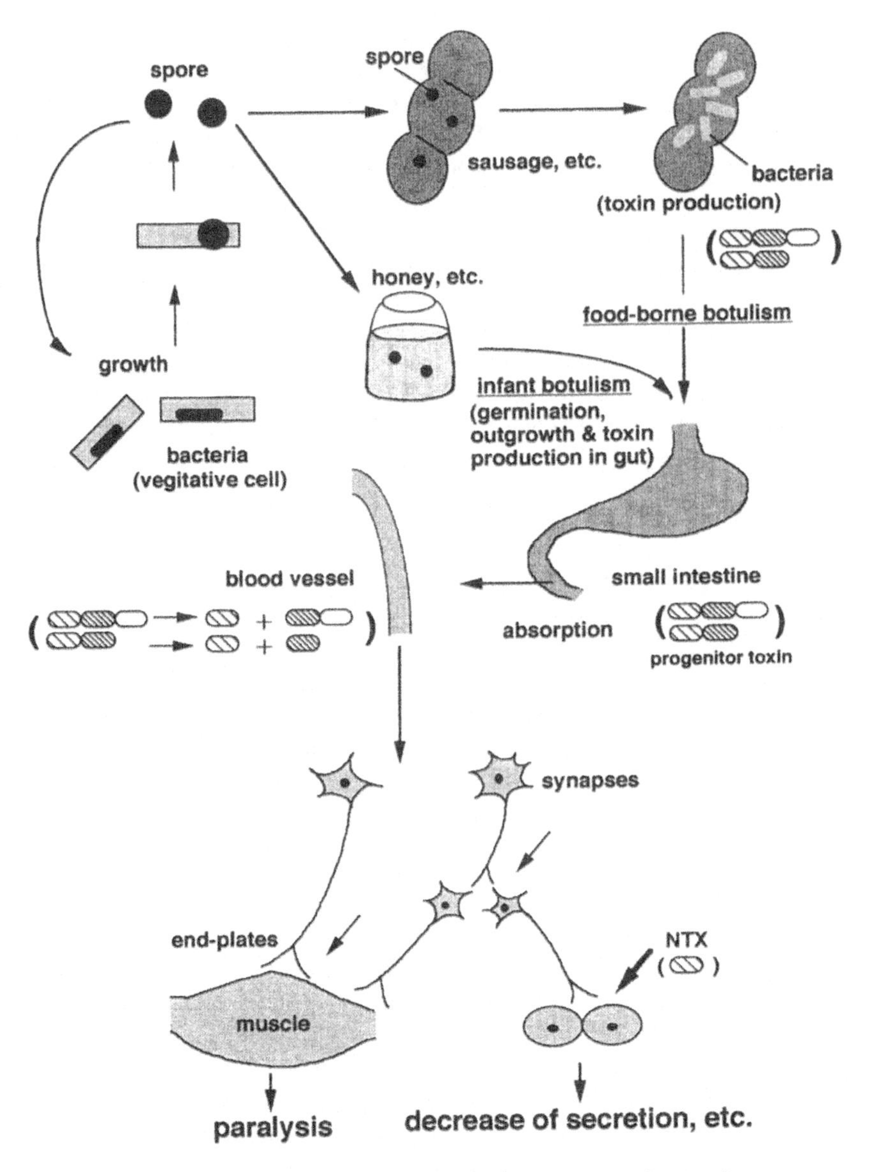

Figure 8.1 Type of botulism. The mechanisms how foodborne and infant botulism occurs are illustrated. Also, the toxin molecules existing in each environment are shown (see Figure 8.4). In addition to these two types of botulism, there exists one more type, wound botulism.
, NTX; NTNA; , HA; , 12S toxin;
, 16S, 19S toxin(s).

outbreaks are not as frequent as tetanus. However, more than 20 cases were reported among drug users since 1990 in the U.S.

5. IDENTIFICATION OF BOTULISM

"Botulism" can be diagnosed from a patient's symptoms. In the case of foodborne botulism, it is very important to determine whether the patient ingested food that was not heated properly, resulting in survival of the heat-labile botulinum toxin present in the food. Diagnosis is confirmed by isolation of the neurotoxigenic organisms and/or by detection of botulinum toxin in the suspected food and in the feces or other samples from the patient. The procedure for isolation and identification of organisms causing botulism has been established (Dowell and Hawkin, 1973). Since the entire DNA sequences of botulinum type A to G toxin genes have been determined, polymerase chain reaction (PCR) detection of toxin genes has been utilized by many researchers. The toxin gene regions can be amplified from DNA present in either vegetative cells or spore preparations. As little as 10 to 100 fg of DNA (approximately 3 to 30 cells) was needed for successful amplification of toxin genes (Szabo et al., 1993). The identity of the amplified fragments was confirmed by hybridization with oligonucleotide probes specific to each toxin gene type. We recommended amplifying a specific region of approximately 300 bp of each toxin gene. The PCR product will have only one cleavage site for a specific endonuclease, allowing easy confirmation of its identity by restriction enzyme digestion profiles (Takeshi et al., 1996).

The production of toxin is usually detected by mouse bioassay, the sensitivity of which is approximately 10 pg/mL. As an alternative to bioassay, an immunoassay technique with high sensitivity has been studied. Recently, a modified, enzyme-linked immunosorbent assay (ELISA), which relies on the detection of sandwich complexes on microtiter plates by a solid-phase coagulation assay known as enzyme-linked coagulation assay (ELICA), has been established (Doellgast et al., 1994). Using this procedure, type A, B, and E toxins have been detected at mouse bioassay levels. If appropriate antibodies are employed, the other types of toxins also may be detected at the same levels using this procedure.

Franciosa et al. (1994) reported that 43 out of 79 type A toxigenic strains contained a nonexpressed type B toxin gene in addition to an expressed type A toxin gene as determined by PCR-dot blot analysis. As described above, some strains designated as Af, Ba, and AB are thought to produce two different types of toxins in varying quantities. These facts should be taken into account when the toxin type is being determined by PCR or immunoassay.

6. STRUCTURE AND FUNCTION OF NTXs

Type A to G NTXs are synthesized as single-chain polypeptides with a Mr of approximately 150 kDa. The single-chain toxins are cleaved by a protease(s) produced by the organisms, or artificially by the addition of trypsin. Cleavage of the polypeptide occurs at about one-third of the distance from their N-terminus, resulting in a dichain form, which is joined by a single disulfide bond (Sugiyama, 1980). After reduction, these nicked toxins can be separated into light (L) chain (Mr of ca. 50~55 kDa) and heavy (H) chain (Mr of ca. 90~100 kDa) components. Since the toxicity of type E (or nonproteolytic culture) is activated by trypsin treatment, and the fully activated NTXs have a nick at this position, it was presumed that nicking is needed for activation. However, partial or full activation without nicking was also reported (Ohishi and Sakagudchi, 1977; Ksipinski and Sugiyama, 1981; DasGupta, 1981; Yokosawa et al., 1986). The mechanism of activation is still not understood.

Data concerning the molecular structure of the NTXs and their enzymatic activity have been well documented in the past decade. It was proposed that the toxin blocks Ach release by proceeding through a series of three reactions, including an extracellular binding step, an internalization step, and an intracellular lytic step (Simpson, 1980). The 50 kDa C-terminus of the H chain (binding domain in Figure 8.2) seems to be involved in neurospecific cell binding (Agui et al., 1985; Shone et al., 1985; Kozaki et al., 1986; Yokosawa et al., 1989). For type C1 toxin, it was suggested that less than 50 amino acid residues of the C-terminus are essential for its binding to the receptor (Kimura et al., 1992). Gangliosides and a glycoprotein have been proposed as the toxin receptors. Recently, it was reported that type B toxin binds to synaptotagmin (58 kDa) in the presence of gangliosides GT1b or GD1a (Nishiki et al., 1994, 1996). Although the precise structure of the complex of synaptotagmin and gangliosides remains to be established, these are the first reports identifying the receptor for the NTX. The internalization step through receptor (acceptor)-mediated endocytosis was minutely observed under electron microscopy (Black and Dolly, 1986a, b). Thereafter, the L chain is thought to penetrate (translocate) from (coated) vesicles or prelysosomes into the cytosol through the channels formed in membranes by the 50 kDa N-terminal region of the H chain (channel-forming domain) (Blaustein et al., 1987; Shone et al., 1987). It is here that the L chain (active domain) demonstrates its activity by blocking neurotransmitter release (Figure 8.3).

The genes of type A to G NTXs have been cloned, and their nucleotide sequences determined (Table 8.2). Homology at the amino acid level among these toxins is approximately 30–60%. The observation that several monoclonal antibodies react with different types of NTXs (Tsuzuki et al., 1988) can be explained by the existence of several highly conserved regions among different types of botulinum NTXs (Figure 8.2). One of the most conserved segments of the NTXs is located in the central region of the L chain, which includes the HExxH zinc binding motif of metalloendopeptidases. Both botulinum and tetanus toxins contain a zinc atom coordinated by the two histidine

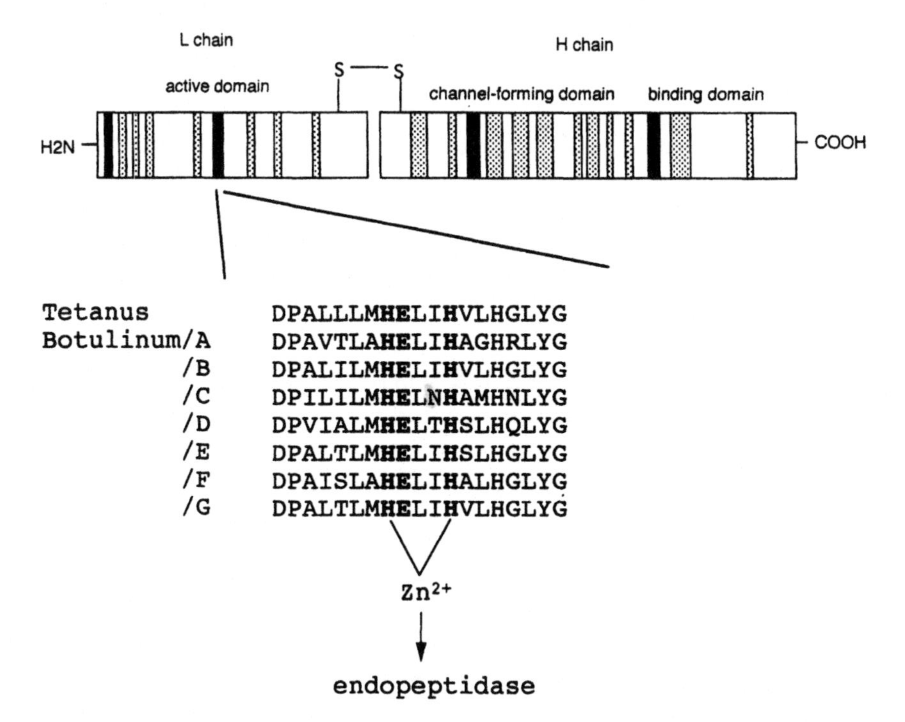

Figure 8.2 Schematic structure of NTXs. The NTXs are formed as a single-chain polypeptide. After cleavage and reduction, the NTXs separate into L and H chains. L chain, N-terminus and C-terminus of C chain are considered to be active, channel-forming, and binding domains, respectively. Highly homologous regions (■) including the HExxH motif in the L chain and relatively high homologous regions (▩) among the NTXs are shown.

residues of this motif. Both toxins possess a proteolytic activity specific for the synaptic vesicle membrane protein, VAMP (or synaptobrevin-1 and/or -2), and the presynaptic plasma membrane proteins, SNAP-25 (synaptosomal-associated protein of M_r 25 kDa) and syntaxin (Jahn and Niemann, 1994; Montecucco and Schiavo, 1994). These proteins are considered to be key players in constitutive vesicle-exocytosis in neurons. They are implicated in a docking/fusion process by interacting with a NSF (n-ethylmaleimide sensitive factor)/α- or β-SNAP (soluble NSF attachment protein)/γ-SNAP complex. In summary, botulinum and tetanus toxin inhibit neurotransmitter release by cleaving the membrane associated/transmembrane proteins involved in docking/fusion processes. The target proteins and the cleavage sites of each type of toxin are illustrated in Figure 8.3.

7. STRUCTURE AND FUNCTION OF PROGENITOR TOXINS

In 1946, Lamanna et al. succeeded in purifying a crystalline type A toxin

TABLE 8.2. **Structural Genes of *C. botulinum* Neurotoxin and Nontoxic Components.**

	Origin of the Gene	Strain	Number of Amino* Acid Residues	Molecular Mass** (kDa)
NTX (140 ~ 150 kDa)†				
Tet	Plasmid	E88 variant of Massachusetts	1,315	150.0
		CN3911	1,315	150.5
A	Chromosome	NCTC2916	1,296	149.5
		62A	1,296	149.4
B	Chromosome	Danish (proteolytic)	1,291	150.8
		Eklund 17B (non-proteolytic)	1,291	150.5
C	Phage	Stockholm	1,291	148.9
		N.D. (468)‡	1,291	148.9
D	Phage	N.D. (BVD/ – 3)‡	1,276	146.9
		CB16	1,276	146.9
E	Chromosome	NCTC11219	1,252	143.6
		Beluga	1,251	143.8
F	Chromosome	202F	1,274	146.7
G	Plasmid	NCFB3012	1,297	149.0
NTNH (130 ~ 140 kDa)†				
A	Chromosome	A-NIH	1,193	138.1
		NCTC2916	1,193	138.2
B	Chromosome	NCTC7273 (proteolytic)	1,197	138.8
		Eklund 17B (non-proteolytic)‡	1,197‡	139.0†
C	Phage	Stockholm	1,196	138.7
		468	1,196	138.7
D	Phage	CB16	1,196	138.7
E	Chromosome	Mashike	1,162	136.9
F	Chromosome	202F	1,165	136.5
G	Plasmid	ATCC 27322*	1,198*	139.1*

TABLE 8.2. (continued)

	Origin of the Gene	Strain	Number of Amino* Acid Residues	Molecular Mass** (kDa)
HA1 (33 ~ 35 kDa)†				
A	Chromosome	NCTC7272	293	33.9
		NCTC2916	291	34.0
B	Chromosome	NCTC7273 (proteolytic)	294	33.7
		Lamanna (proteolytic)	293	33.4
		Eklund 17 B (non-proteolytic)	292	33.7
C	Phage	Stockholm	286	33.8
		468	286	33.8
D	Phage	CB16	286	33.8
HA2(15 ~ 17 kDa)†				
A	Chromosome	NCTC2916	147	17.0
B	Chromosome	Lamanna (proteolytic)	146	16.9
		Eklund 17 B (non-proteolytic)*	146*	16.9*
C	Phage	Stockholm	146	16.7
		468	146	16.7
D	Phage	CB16	146	16.7
G	Plasmid	ATCC 27322*	147	17.4
HA3				
A	Chromosome	NCTC2916	626	71.2
C	Phage	Stockholm	623	70.6
		468	623	70.6
D	Phage	CB16	623	70.6

* Amino acid residues from first Met (stop codon is eliminated).
** Calculated by GENETYX-MAC (Software Development Co., Ltd. Tokyo, Japan).
† Mr. estimated from SDS-PAGE.
‡ Data obtained from Genbank (N.D., not described in the paper).

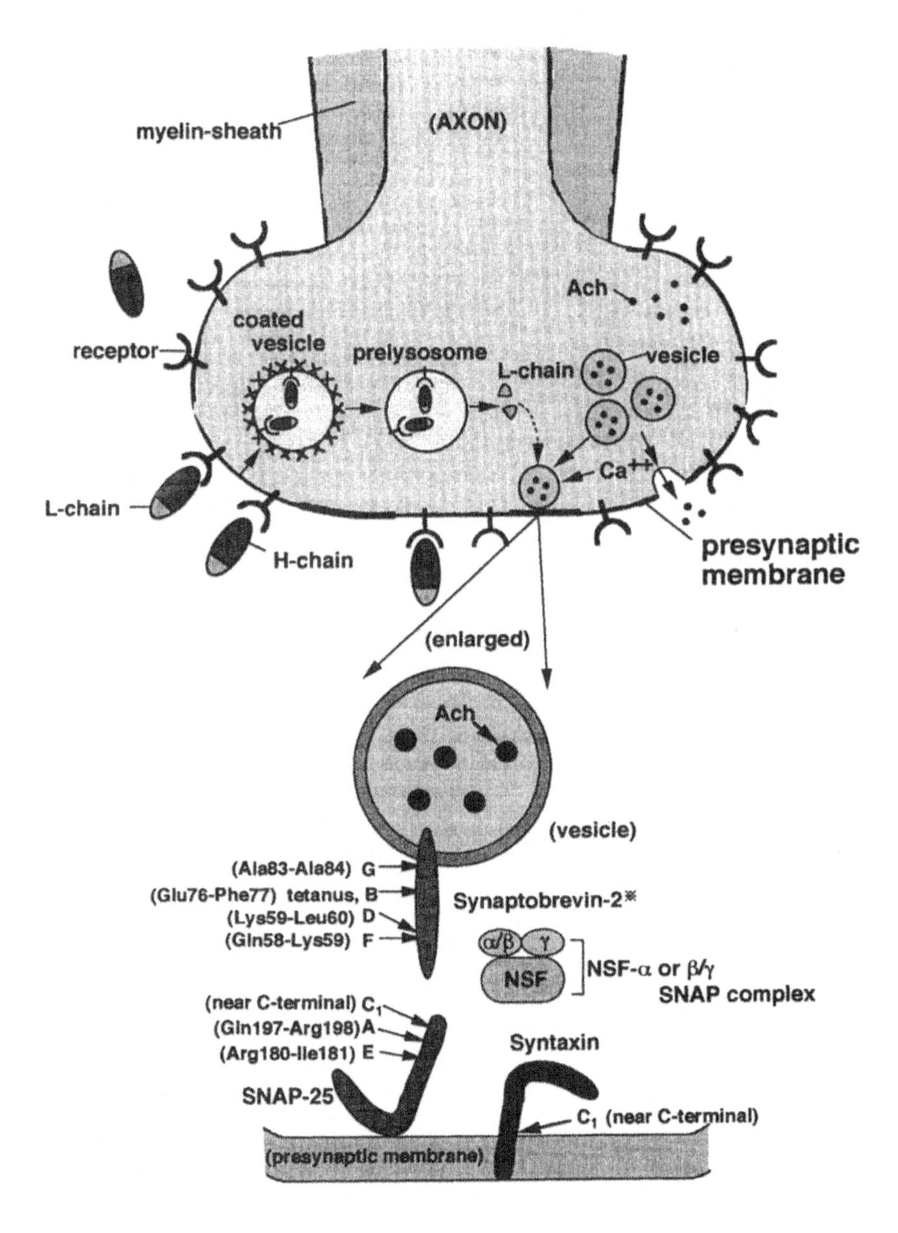

*Synaptobrevin-1 is also cleaved by types D, F, and G toxins at the same peptide sequences.

Figure 8.3 Function of the NTXs. The mechanisms of how the NTXs develop their activity are illustrated. In an enlarged figure, the target proteins of the NTXs are shown.

of 900 kDa, which had hemagglutinin (HA) activity. This HA activity could be removed by mixing the crystalline toxin with erythrocytes in an alkaline solution without reducing the toxin activity, indicating that the crystalline toxin consists of NTX and HA (Lamanna and Lowenthal, 1951). The behavior of the toxins in type A to G cultures, designated progenitor toxin, was clarified by Sakaguchi's group in Japan (Sakaguchi et al., 1981). The following conclusions were drawn: (1) type A strains produce three different-sized progenitor toxins designated LL (19S, 900 kDa), L (16S, 500 kDa), and M (12S, 300 kDa), and type B, C, and D produce two toxins, L and M. Types E and F produce only M toxin while type G produces only L toxin (Table 8.1); (2) M toxin consists of a NTX (7S, 150 kDa) and a nontoxic component showing no HA activity, which is described here as a nontoxic, non-HA (NTNH); (3) L and LL toxins are formed by the conjugation of M toxin with HA; (4) M, L, and LL toxins dissociate into NTX and nontoxic components under alkaline conditions. M toxin dissociates into a NTX and a NTNH, and L and LL toxins dissociate into a NTX and a nontoxic component, which is a complex of a NTNH and a HA (see Figures 8.1 and 8.4); (5) in the small intestine where intestinal juice exists, the progenitor toxins do not dissociate into a NTX and nontoxic components even though the pH is high. The dissociation occurs in the lymphatic or blood vessels after absorption of the progenitor toxins; and (6) nontoxic components protect the NTX when the progenitor toxin passes through the stomach.

The M_r of the NTNHs for all types of progenitor toxins was determined to be approximately 140 kDa via SDS-PAGE. The M_r of HAs, however, is not as clear because no one has yet succeeded in separating and purifying the HAs from the progenitor toxins. Using SDS-PAGE, the purified type A, B, and C 16S progenitor toxins demonstrated several bands other than the NTX and the NTNH bands, indicating that the HA consists of several subcomponents (Tsuzuki et al., 1990; Somers and DasGupta, 1991). Recently, we elucidated the structure of types A, C, D, and E progenitor toxins via genetic and protein chemical analyses (Oguma et al., 1995, 1997, 1999). The conclusions drawn were: (1) any type of HA consists of four subcomponents designated here as HA1 (M_r ca. 33~35 kDa), HA2 (M_r ca. 15~17 kDa), HA3a (M_r ca. 19~23 kDa), and HA3b (M_r ca. 52~53 kDa) (Figure 8.4); (2) the HA3a subcomponent consists of several proteins having slightly different M_rs (see next item); (3) type A 19S and 16S toxins consist of the same protein components. The molar ratios of the components of 19S and 16S toxins are similar, with the exception of the HA1 subcomponent. The ratio of HA1 of the 19S toxin is about twofold higher than that of the 16S toxin. Therefore, it was postulated that the 19S toxin is a dimer of the 16S toxin crosslinked by the HA1 (Figure 8.4). (4) The antigenicity of HA1 subcomponents of types A and C was quite different among their nontoxic components. Since the 19S toxin is produced only by type A, the HA1 seems to be a key protein in forming the 19S toxin; (5) the NTNHs of type A, C, and D 12S toxins are cleaved at their N-terminal regions,

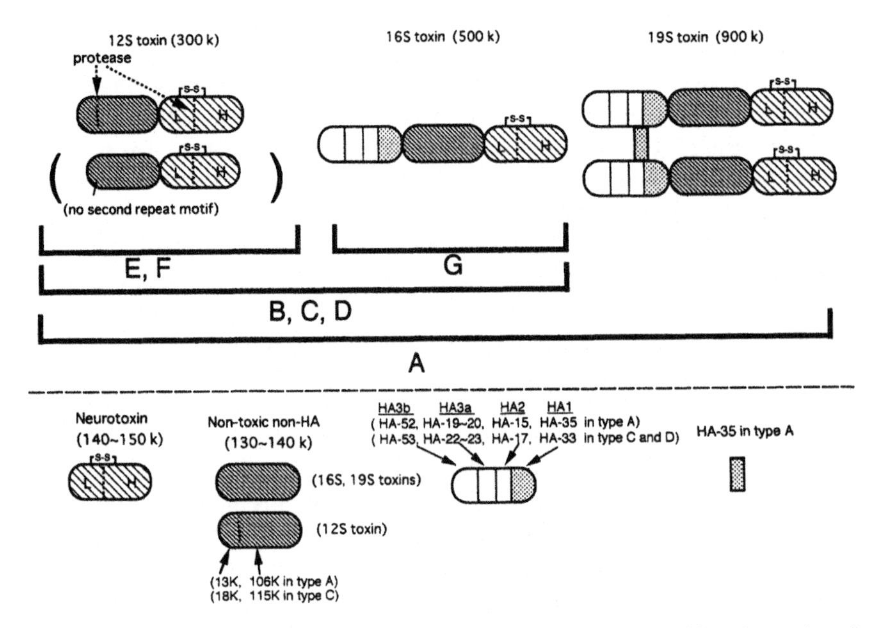

Figure 8.4 Scheme of the progenitor toxins (12S, 16S, and 19S toxins). 12S toxin consists of a NTX and a NTNH. The repeated motifs are identified in the N-terminal region of types A, B, C, and D (see Figure 8.5). The 16S and 19S toxins are formed by conjugation of HA with 12S toxin. HA consists of subcomponents designated as HA1, HA2, HA3a, and HA3b. The 19S toxin is thought to be a dimer of 16S toxin crosslinked by HA1.

giving rise to two protein fragments (for example, 13 and 106 kDa in type A, and 18 and 115 kDa in type C) on SDS-PAGE with or without 2-mercapto-ethanol (2-ME) (Figure 8.4). A multiple alignment of N-terminal regions in the NTNHs of types A to F demonstrated that cleavage occurs in a region that contains a short repeat sequence, and that types E and F, which produce only 12S toxin, lack this region. It was thought that the processing of NTNH is the reason why the 12S, 16S, and 19S toxins exist in the same culture. The processing may prevent the binding of HA to the 12S toxin or may cause the dissociation of HA from the 16S and 19S toxin.

Furthermore, we obtained data indicating that HA, especially HA1 and HA3b, plays an important role in the binding of the progenitor toxin to the epithelial cells of the small intestine in addition to its ability to protect the NTX from gastric juice. This leads to efficient absorption of the HA-positive toxins in the small intestine as compared to HA-negative toxin (Fujinaga et al., 1997). The three-dimensional structure of the progenitor toxins, as well as the mechanism/route by which such macromolecules are absorbed in the small intestine, is still unclear.

8. GENE ORGANIZATION

One of the important findings concerning the genetic control of toxigenicity in *C. botulinum* is a phage-conversion phenomenon in types C and D. The nontoxigenic variants obtained from types C and D can be converted to a toxigenic state by infection with specific phages induced from toxigenic C and D strains (Inoue and Iida, 1971; Eklund and Poysky, 1974). HA production is also governed by bacteriophages, and it is transmitted with toxin production either concomitantly or separately, indicating that genes for toxin and HA are linked (Oguma et al., 1976). Later, this was confirmed by cloning of these genes.

The efforts of numerous researchers have resulted in the cloning and characterization of almost all of the genes encoding the NTXs and nontoxic components from cell chromosomes, phages, and a plasmid. The nucleotide sequences of type A to G NTXs, type A to G NTNHs, and type A, C, and D HAs have been determined (Table 8.2; Minton, 1995; Oguma et al., 1997, 1999). Homology among the nontoxic components is higher than that of the NTXs with the nontoxic components of types C and D demonstrating almost 100% identity. The genes for NTX (*ntx*), NTNH (*ntnh*), and HA (*ha*) are organized as a cluster (Figure 8.5). The HA genes consist of three ORFs (designated ha1, ha2, and ha3), which lie just upstream of *ntnh,* while *ntx* lies just downstream of *ntnh.* It was presumed that *ntx* is transcribed alone (monocistronic transcription) or in association with *ntnh* (polycistronic transcription) (Hauser et al., 1994). The three ORFs of HA are in the opposite orientation from that of *ntnh* and *ntx,* and they are also presumed to be transcribed independently or possibly as one operon (Henderson et al., 1996). From the N-terminal amino acid sequences of each subcomponent of HA, we concluded that after translation, the gene product (70 kDa) of ha3 is split into a HA3b protein (52 kDa in type A, 53 kDa in type C) and the remaining polypeptide. Also, the remaining polypeptide is cleaved at several different sites in its N-terminal region to form proteins with slightly different M_rs (HA3a, 19~20 kDa in type A, 22~23 kDa in type C) (Figure 8.6) (Fujinaga et al., 1994; Inoue et al., 1996).

In addition to the gene cluster, one other gene, *p-21* (or *orf-22*), was identified (Hauser et al., 1994). In types A and B, *p-21* resides between genes *ha1* and *ntnh,* whereas in types C and D, this gene resides downstream of *ha3* (Figure 8.5). Recently, it became clear in type A that the product (21 kDa) of this gene activates the transcription of genes for NTX and nontoxic components. Electrophoretic mobility shift and immunoprecipitation assays suggested that the gene product binds to the promoter region of the *ntx* and *ha1* genes (Marvaud et al., 1998). As for the strains of types E and F and Kyoto-F (type A), which produce only 12S toxin, no gene for HA has been identified, but new genes, *p-47, x1* (435 bp), and *x2* (partially sequenced) were identified

(East et al., 1996; Kubota et al., 1998). The amino acid sequence of these gene products showed little homology with any of the nontoxic components, and their functions are still unclear. In type G, *ha1* has not yet been identified, and *p-21* is located between *ha2* and *ntnh.*

As described in the previous section, some strains produce two types of toxins. The genetic map of these genes and the mechanism by which the expression of these genes are regulated should be analyzed in the near future.

9. CONCLUSION

The mechanisms of pathogenesis and toxin synthesis in *C. botulinum* have been summarized. The NTXs and the progenitor toxins were first purified. Thereafter, the genes encoding them were cloned, and their nucleotide se-

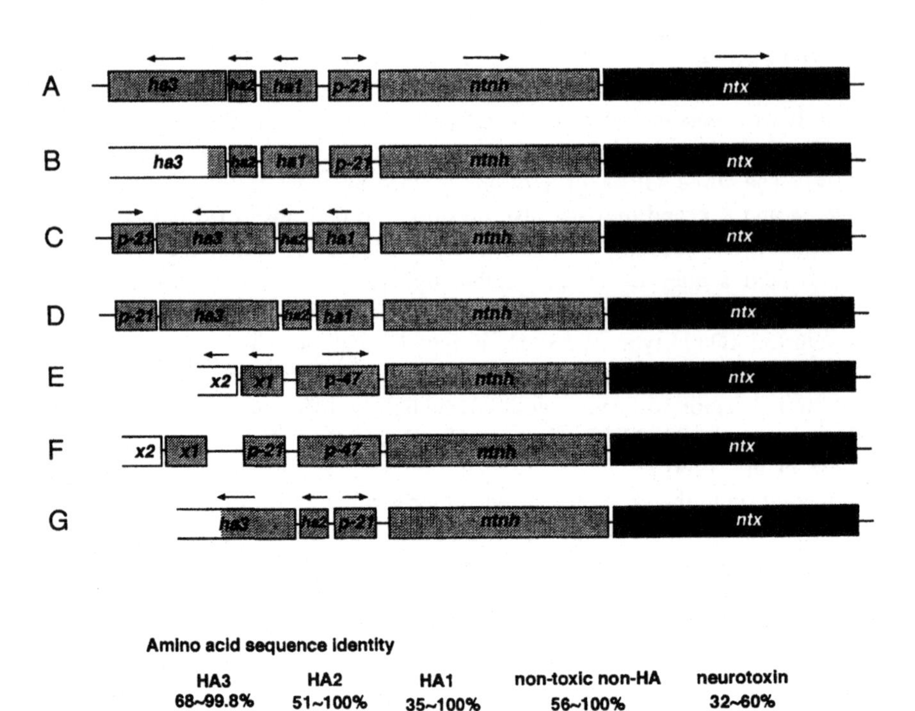

Figure 8.5 Gene organization of progenitor toxins. Organization of the gene cluster encoding NTX (*ntx*), NTNH (*ntnh*), and HA (*ha*) is represented. The genes *p-21* (or *orf-22*), *p-47*, *x1*, and *x2* are also shown. The entire nucleotide sequences of the *ha* genes in types B and G, and *x2* in type E have not yet been determined.

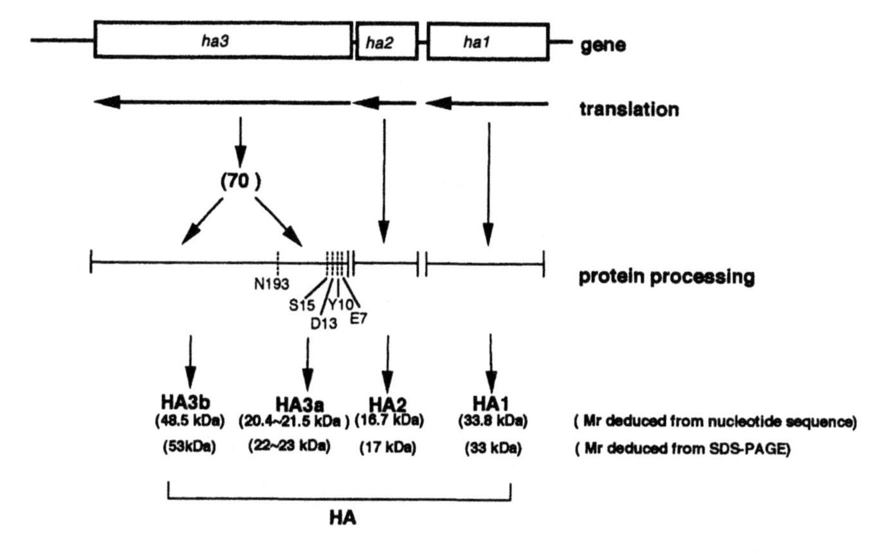

Figure 8.6 Processing of the *ha3* gene product after translation in type C. The genes for HA consist of three ORFs, *ha1, ha2,* and *ha3.* The gene product (70 kDa) of *ha3* is cleaved into a HA3a protein and a HA3b (53 kDa) protein. HA3a consists of several proteins (four in this case) having slightly different Mrs (22~23 kDa), which are formed by processing its N-terminal region. The N-terminal amino acids of HA3a and HA3b and their Mr as deduced from their nucleotide sequence are shown. Similar data were also obtained for types A, B (unpublished data), and D. Recently, we found that HA1 and HA3b play an important role in the ability of the 16S and 19S progenitor toxins to bind to epithelial cells of the small intestine and also to red blood cells (unpublished data).

quences determined. From these studies it became clear that the NTXs have endopeptidase activity similar to that of tetanus toxin. Unlike tetanus toxin, botulinum NTXs are associated with nontoxic components in cultures and, therefore, only the botulinum toxin can cause foodborne poisoning. However, the detailed mechanisms as to how the progenitor toxins are absorbed from the small intestine, and how NTXs block the release of Ach, are still not clear. The three-dimensional structure of the toxins and the detailed mechanism of regulating the transcription of toxin genes also remain to be resolved. These represent future areas of study.

10. REFERENCES

Agui, T., Syuto, B., Oguma, K., Iida, H., and Kubo, S. 1985. "The structural relation between the antigenic determinants to monoclonal antibodies and binding sites to rat brain synaptosomes and GT1b ganglioside in *Clostridium botulinum* type C neurotoxin," *J. Biochem. Tokyo.* 97(1):213–218.

Aktories, K. and Frevert, J. 1987. "ADP-ribosylation of a 21–24 kDa eukaryotic protein(s) by C3, a novel botulinum ADP-ribosyltransferase, is regulated by guanine nucleotide," *Biochem. J.* 247(2):363–368.

Aktories, K., Barmann, M., Ohishi, I., Tsuyama, S., Jakobs, K. H., and Habermann, E. 1986. "Botulinum C2 toxin ADP-ribosylates actin," *Nature* 322(6077):390–392.

Arnon, S. S., Damus, K. H., and Chin, J. 1981. "Infant botulism: epidemiology and relation to sudden infant death syndrome," *Epidemiol. Rev.* 3(45):45–66.

Aureli, P., Fenicia, L., Pasolini, B., Gianfranceschi, M., McCroskey, L. M., and Hatheway, C. L. 1986. "Two cases of type E infant botulism caused by neurotoxigenic *Clostridium butyricum* in Italy," *J. Infect. Dis.* 154(2):207–211.

Black, J. D. and Dolly, J. O. 1986a. "Interaction of ^{125}I-labeled botulinum neurotoxins with nerve terminals. I. Ultrastructural autoradiographic localization and quantitation of distinct membrane acceptors for types A and B on motor nerves," *J. Cell. Biol.* 103(2):521–534.

Black, J. D. and Dolly, J. O. 1986b. "Interaction of ^{125}I-labeled botulinum neurotoxins with nerve terminals. II. Autoradiographic evidence for its uptake into motor nerves by acceptor-mediated endocytosis," *J. Cell. Biol.* 103(2):535–544.

Blaustein, R. O., Germann, W. J., Finkelstein, A., and DasGupta, B. R. 1987. "The N-terminal half of the heavy chain of botulinum type A neurotoxin forms channels in planar phospholipid bilayers," *FEBS Lett.* 226(1):115–120.

Chaudhry, R., Dhawan, B., Kumar, D., Bhatia, R., Gandhi, J. C., Patel, R. K., and Purohit, B. C. 1998. "Outbreak of suspected *Clostridium butyricum* botulism in India," in *Emerging Infectious Diseases, CDC,* 4(3):506–507.

DasGupta, B. R. 1981. "Structure and structure function relation of botulinum neurotoxins," in *Biomedical Aspects of Botulism,* ed., Lewis, G. E. Jr., New York: Academic Press. pp. 1–19.

DasGupta, B. R. and Sugiyama, H. 1977. "Biochemistry and pharmacology of botulinum and tetanus neurotoxins," in *Perspectives in Toxinology,* ed., Bernheimer A. W. New York: John Wiley and Sons. pp. 88–119.

Doellgast, G. J., Beard, G. A., Bottoms, J. D., Cheng, T., Roh, B. H., Roman, M. G., Hall, P. A., and Triscott, M. X. 1994. "Enzyme-linked immunosorbent assay and enzyme-linked coagulation assay for detection of *Clostridium botulinum* neurotoxins A, B, and E and solution-phase complexes with dual-label antibodies," *J. Clin. Microbiol.* 32(1):105–111.

Dowell, V. R., Jr. and Hawkin, T. M. 1973. *Laboratory Methods in Anaerobic Bacteriology,* CDC laboratory manual. Atlanta, GA, CDC.

East, A. K., Bhandari, M., Stacey, J. M., Campbell, K. D., and Collins, M. D. 1996. "Organization and phylogenetic interrelationships of genes encoding components of the botulinum toxin complex in proteolytic *Clostridium botulinum* types A, B, and F: evidence of chimeric sequences in the gene encoding the nontoxic nonhemagglutinin component," *Int. J. Syst. Bacteriol.* 46(4):1105–1112.

Eklund, M. W. and Poysky, F. T. 1974. "Interconversion of type C and D strains of *Clostridium botulinum* by specific bacteriophages," *Appl. Microbiol.* 27(1):251–258.

Franciosa, G., Ferreira, J. L., and Hatheway, C. L. 1994. "Detection of type A, B, and E botulism neurotoxin genes in *Clostridium botulinum* and other *Clostridium* species by PCR: evidence of unexpressed type B toxin genes in type A toxigenic organisms," *J. Clin. Microbiol.* 32(8):1911–1917.

Fujinaga, Y., Inoue, K., Shimazaki, S., Tomochika, K., Tsuzuki, K., Fujii, N., Watanabe, T., Ohyama, T., Takeshi, K., Inoue, K., and Oguma, K. 1994. "Molecular construction of *Clostridium botulinum* type C progenitor toxin and its gene organization," *Biochem. Biophys. Res. Commun.* 205(2):1291–1298.

Fujinaga, Y., Inoue, K., Watanabe, S., Yokota, K., Hirai, Y., Nagamachi, E., and Oguma, K. 1997. "The haemagglutinin of *Clostridium botulinum* type C progenitor toxin plays an essential role in binding of toxin to the epithelial cells of guinea pig small intestine, leading to the efficient absorption of the toxin," *Microbiol.* 143(12):3841–3847.

Fujinaga, Y., Takeshi, K., Inoue, K., Fujita, R., Ohyama, T., Moriishi, K., and Oguma, K. 1995. "Type A and B neurotoxin genes in a *Clostridium botulinum* type AB strain," *Biochem. Biophys. Res. Commun.* 213(3):737–745.

Gimenez, D. F. 1984. "Clostridium botulinum subtype Ba," *Zentralbl Bakteriol Mikrobiol Hyg A,* 257(1):68–72.

Hall, J. D., McCroskey, L. M., Pincomb, B. J., and Hatheway, C. L. 1985. "Isolation of an organism resembling *Clostridium barati* which produces type F botulinal toxin from an infant with botulism," *J. Clin. Microbiol.* 21(4):654–655.

Hauser, D., Eklund, M. W., Boquet, P., and Popoff, M. R. 1994. "Organization of the botulinum neurotoxin C1 gene and its associated non-toxic protein genes in *Clostridium botulinum* C 468," *Mol. Gen. Genet.* 243(6):631–640.

Henderson, I., Whelan, S. M., Davis, T. O., and Minton, N. P. 1996. "Genetic characterisation of the botulinum toxin complex of *Clostridium botulinum* strain NCTC 2916," *FEMS Microbiol. Lett.* 140(2–3):151–158.

Hutson, R. A., Collins, M. D., East, A. K., and Tompson, D. E. 1994. "Nucleotide sequence of the gene coding for non-proteolytic *Clostridium botulinum* type B neurotoxin; Comparison with other clostridial neurotoxins," *Curr. Microbiol.* 28(2):101–110.

Inoue, K. and Iida, H. 1971. "Phage-conversion of toxigenicity in *Clostridium botulinum* types C and D," *Jpn. J. Med. Sci. Biol.* 24(1):53–6.

Inoue, K., Fujinaga, Y., Watanabe, T., Ohyama, T., Takeshi, K., Moriishi, K., Nakajima, H., Inoue, K., and Oguma, K. 1996. "Molecular composition of *Clostridium botulinum* type A progenitor toxins," *Infect. Immun.* 64(5):1589–1594.

Jahn, R. and Niemann, H. 1994. "Molecular mechanisms of clostridial neurotoxins," *Ann. N. Y. Acad. Sci.* 733(245):245–255.

Jansen, B. C. 1971. "The toxic antigenic factors produced by *Clostridium botulinum* types C and D," *Onderstepoort J. Vet. Res.* 38(2):93–98.

Kimura, K., Fujii, N., Tsuzuki, K., Yokosawa, N., and Oguma, K. 1992. "The functional domains of *Clostridium botulinum* type C neurotoxin," in *Recent Advances in Toxinology Research,* eds., Gopalakrishnakone, P. and Tan, C. K. Singapore: Venom and Toxin Research Group, National University of Singapore. pp. 375–385.

Kozaki, S., Kamata, Y., Nagai, T., Ogasawara, J., and Sakaguchi, G. 1986. "The use of monoclonal antibodies to analyze the structure of *Clostridium botulinum* type E derivative toxin." *Infect. Immun.* 52(3):786–791.

Krysinski, E. P. and Sugiyama, H. 1981. "Nature of intracellular type A botulinum neurotoxin," *Appl. Environ. Microbiol.* 41(3):675–678.

Kubota, T., Yonekura, N., Hariya, Y., Isogai, E., Isogai, H., Amano, K., and Fujii, N. 1998. "Gene structure in upstream region of *Clostridium botulinum* type E and *Clostridium butyricum* BL6340 progenitor toxin genes is different from those of other types," *FEMS Letter* 158(2):215–221.

Kurokawa, Y., Oguma, K., Yokosawa, N., Syuto, B., Fukatsu, R., and Yamashita, I. 1987. "Binding and cytotoxic effects of *Clostridium botulinum* type A, C1 and E toxins in primary neuron cultures from foetal mouse brains," *J. Gen. Microbiol.* 133(9):2647–2657.

Lamanna, C. and Lowenthal, J. P. 1951. "The lack of identity between hemagglutinin and the toxin of type A botulinal organism," *J. Bacteriol.* 61:751–752.

Marvaud, J. C., Gibert, M., Inoue, K., Fujinaga, Y., Oguma, K., and Popoff, M. R. 1998. "botR/A is a positive regulator of botulinum neurotoxin and associated non-toxin protein genes in *Clostridium botulinum* A," *Mol. Microbiol.* 29(4):1009–1018.

Meng, X., Karasawa, T., Zou, K., Kuang, X., Wang, X., Lu, C., Wang, C., Yamakawa, K., and

Nakamura, S. 1997. "Characterization of a neurotoxigenic *Clostridium butyricum* strain isolated from the food implicated in an outbreak of food-borne type E botulism," *J. Clin. Microbiol.* 35(8):2160–2162.

Minton, N. P. 1995. "Molecular genetics of clostridial neurotoxins," *Curr. Top. Microbiol. Immunol.* 195(161):161–194.

Moberg, L. J. and Sugiyama, H. 1979. "Microbial ecological basis of infant botulism as studied with germfree mice," *Infect. Immun.* 25(2):653–657.

Montecucco, C. and Schiavo, G. 1994. "Mechanism of action of tetanus and botulinum neurotoxins," *Mol. Microbiol.* 13(1):1–8.

Moriishi, K., Syuto, B., Oguma, K., and Saito, M. 1990. "Separation of toxic activity and ADP-ribosylation activity of botulinum neurotoxin D," *J. Biol. Chem.* 265(27):16614–16616.

Moriishi, K., Koura, M., Abe, N., Fujii, N., Fujinaga, Y., Inoue, K., and Ogumad, K. 1996. "Mosaic structures of neurotoxins produced from *Clostridium botulinum* types C and D organisms," *Biochim. Biophys. Acta* 1307(2):123–126.

Nishiki, T., Kamata, Y., Nemoto, Y., Omori, A., Ito, T., Takahashi, M., and Kozaki, S. 1994. "Identification of protein receptor for *Clostridium botulinum* type B neurotoxin in rat brain synaptosomes," *J. Biol. Chem.* 269(14):10498–10503.

Nishiki, T., Tokuyama, Y., Kamata, Y., Nemoto, Y., Yoshida, A., Sato, K., Sekiguchi, M., Takahashi, M., and Kozaki, S. 1996. "The high-affinity binding of *Clostridium botulinum* type B neurotoxin to synaptotagmin II associated with gangliosides GT1b/GD1a," *FEBS Lett.* 378(3):253–257.

Oguma, K., Fujinaga, Y., and Inoue, K. 1995. "Structure and function of *Clostridium botulinum* toxins," *Microbiol. Immunol.* 39(3):161–168.

Oguma, K., Fujinaga, Y., and Inoue, K. 1997. "*Clostridium botulinum* toxin," *J. Toxicol. Toxin Reviews* 16(4):253–266.

Oguma, K., Iida, H., and Shiozaki, M. 1976. "Phage conversion to hemagglutinin production in *Clostridium botulinum* types C and D," *Infect. Immun.* 14(3):597–602.

Oguma, K., Syuto, B., and Iida, H. 1985. "Analysis of antigenic structure of *Clostridium botulinum* type C1 and D toxins by monoclonal antibodies," in *Monoclonal Antibodies against Bacteria,* vol II, eds., Macario, A. J. L. and de Macario, E. C. New York: Academic Press. pp. 159–184.

Oguma, K., Murayama, S., Syuto, B., Iida, H., and Kubo, S. 1984. "Analysis of antigenicity of *Clostridium botulinum* type C1 and D toxins by polyclonal and monoclonal antibodies," *Infect. Immun.* 43(2):584–588.

Oguma, K., Yamaguchi, T., Sudou, K., Yokosawa, N., and Fujikawa, Y. 1986. "Biochemical classification of *Clostridium botulinum* type C and D strains and their nontoxigenic derivatives," *Appl. Environ. Microbiol.* 51(2):256–260.

Oguma, K., Fujinaga, Y., Inoue, K., Yokota, K., Watanabe, T., Ohyama, T., Takeshi, K., and Inoue, K. 1999. "Structure and function of *Clostridium botulinum* progenitor toxin," *J. Toxicol.-Toxin Reviews* 18(1):17–34.

Oguma, K., Yokota, K., Hayashi, S., Takeshi, K., Kumagai, M., Itoh, N., Tachi, N., and Chiba, S. 1990. "Infant botulism due to *Clostridium botulinum* type C toxin [letter]," *Lancet* 336(8728):1449–1450.

Ohishi, I. 1983. "Lethal and vascular permeability activities of botulinum C2 toxin induced by separate injections of the two toxin components," *Infect. Immun.* 40(1):336–339.

Ohishi, I. and Sakaguchi, G. 1977. "Activation of botulinum toxins in the absence of nicking," *Infect. Immun.* 17(2):402–407.

Pickett, J., Berg, B., Chaplin, E., and Brunstetter, S. M. 1976. "Syndrome of botulism in infancy: clinical and electrophysiologic study," *N. Engl. J. Med.* 295(14):770–772.

Sakaguchi, G., Ohishi, I., and Kozaki, S. 1981. "Purification and oral toxicities of *Clostridium botulinum* progenitor toxins," in *Biomedical* aspects of botulism, ed., Lewis, G. E. Jr. New York: Academic Press. pp. 21–34.

Sakaguchi, G., Sakaguchi, S., Kozaki, S., and Takahashi, M. 1986. "Purification and some properties of *Clostridium botulinum* type AB toxin," *FEMS Microbiol. Lett.* 33(1):23–29.

Shone, C. C., Hambleton, P., and Melling, J. 1985. "Inactivation of *Clostridium botulinum* type A neurotoxin by trypsin and purification of two tryptic fragments. Proteolytic action near the COOH-terminus of the heavy subunit destroys toxin-binding activity," *Eur. J. Biochem.* 151(1):75–82.

Shone, C. C., Hambleton, P., and Melling, J. 1987. "A 50-kDa fragment from the NH2-terminus of the heavy subunit of *Clostridium botulinum* type A neurotoxin forms channels in lipid vesicles," *Eur. J. Biochem.* 167(1):175–80.

Simpson, L. L. 1980. "Kinetic studies on the interaction between botulinum toxin type A and the cholinergic neuromuscular junction," *J. Pharmacol. Exp. Ther.* 212(1):16–21.

Smith, L. D. 1977. *Botulism;* the organism, *Its toxins, the Disease.* Springfield: Charles C. Thomas.

Somers, E. and DasGupta, B. R. 1991. "*Clostridium botulinum* types A, B, C1, and E produce proteins with or without hemagglutinating activity: Do they share common amino acid sequences and genes?" *J. Protein. Chem.* 10(4):415–425.

Sugiyama, H. 1980. "*Clostridium botulinum* neurotoxin," *Microbiol. Rev.* 44(3):419–448.

Sugiyama, H., Mizutani, K., and Yang, K. H. 1972. "Basis of type A and F toxicities of *Clostridium botulinum* strain 84," *Proc. Soc. Exp. Biol. Med.* 141(3):1063–1067.

Szabo, E. A., Pemberton, J. M., and Desmarchelier, P. M. 1993. "Detection of the genes encoding botulinum neurotoxin types A to E by the polymerase chain reaction," *Appl. Environ. Microbiol.* 59(9):3011–3020.

Tabita, K., Sakaguchi, S., Kozaki, S., and Sakaguchi, G. 1991. "Distinction between *Clostridium botulinum* type A strains associated with food-borne botulism and those with infant botulism in Japan in intraintestinal toxin production in infant mice and some other properties," *FEMS Microbiol. Lett.* 63(2–3):251–256.

Takeshi, K., Fujinaga, Y., Inoue, K., Nakajima, H., Oguma, K., Ueno, T., Sunagawa, H., and Ohyama, T. 1996. "Simple method for detection of *Clostridium botulinum* type A to F neurotoxin genes by ploymerase chain reaction," *Microbiol. Immunol.* 40(1):5–11.

Tsuzuki, K., Kimura, K., Fujii, N., Yokosawa, N., Indoh, T., Murakami, T., and Oguma, K. 1990. "Cloning and complete nucleotide sequence of the gene for the main component of hemagglutinin produced by *Clostridium botulinum* type C," *Infect. Immun.* 58(10):3173–3177.

Tsuzuki, K., Yokosawa, N., Syuto, B., Ohishi, I., Fujii, N., Kimura, K., and Oguma, K. 1988. "Establishment of a monoclonal antibody recognizing an antigenic site common to *Clostridium botulinum* type B, C1, D, and E toxins and tetanus toxin," *Infect. Immun.* 56(4):898–902.

van Ermengem, E. 1897. "Ueber einen neuen anaeroben Bacillus und seine Beziehungen zum Botulismus," *Z. Hyg. Infektkr.* 26:1–56.

Willems, A., East, A. K., Lawson, P. A., and Collins, M. D. 1993. "Sequence of the gene coding for the neurotoxin of *Clostridium botulinum* type A associated with infant botulism: Comparison with other clostridial neurotoxins," *Res. Microbiol.* 144(7):547–556.

Yokosawa, N., Tsuzuki, K., Syuto, B., and Oguma, K. 1986. "Activation of *Clostridium botulinum* type E toxin purified by two different procedures," *J. Gen. Microbiol.* 132(7):1981–1988.

Yokosawa, N., Kurokawa, Y., Tsuzuki, K., Syuto, B., Fujii, N., Kimura, K., and Oguma, K. 1989. "Binding of *Clostridium botulinum* type C neurotoxin to different neuroblastoma cell lines" [published erratum appears in *Infect Immun* 1989 Jul; 57(7):2265]. *Infect. Immun.* 57(1):272–277.

CHAPTER 9

Pathogenesis Determinants of *Listeria monocytogenes*

SOPHIA KATHARIOU

1. INTRODUCTION

WIDE occurrence in the environment, the ability to grow at refrigeration temperatures, and the potential to cause severe disease render *Listeria monocytogenes* an organism of significant concern to public health and the food industry. In addition to its implication in sporadic incidences of disease, the pathogen has the potential to cause foodborne epidemics associated with high disease burden. A significant portion of sporadic cases and most common-source outbreaks involve strains of serotype 4b, and one clonal serotype 4b lineage with unique molecular markers has been implicated in numerous epidemics in North America and Europe. A high-impact outbreak of foodborne listeriosis, that resulted in numerous cases of illness and several deaths, took place only recently in the United States. As in previous major outbreaks, the implicated food product was found to be contaminated with bacteria of serotype 4b (Anonymous, 1999).

Much research in the last decade has focused on the pathogenic mechanisms of *L. monocytogenes,* with an emphasis on the interactions of the bacteria with the host cell and the immune system. As a facultative intracellular pathogen, *L. monocytogenes* has evolved a number of remarkable adaptations, which have been extensively studied, mostly in nonphagocytic cell model systems. These include invasion, intracellular multiplication and direct cell-to-cell spread. Surface components influencing uptake by macrophages and the subsequent intracellular fate of the pathogen are less clearly understood. Recent research has identified additional areas that may further our understanding of pathogenesis. These include the molecular responses of the pathogen to environmental stresses, such as salinity, acidity, and extremes of temperature as well as serotype-specific genes and surface antigens.

Listeria monocytogenes is a facultative intracellular gram-positive bacterium capable of causing serious illness in humans and animals (Farber and Peterkin, 1991). Infection is commonly induced by means of contaminated food, and individuals at risk include pregnant women and newborn babies, the elderly, and the immunocompromised. The pathogen is widely distributed in the environment, where its primary habitat may be soil and decaying vegetation. It has the ability to grow at refrigeration temperatures and to tolerate a wide range of pH and osmolarity (Gray and Killinger, 1966). The psychrotrophic nature of *L. monocytogenes* has rendered it a major threat to the safety of cold-stored foods, especially dairy products, and several outbreaks of disease (listeriosis) have been traced to contaminated cold-stored products (Schuchat et al., 1991).

Several excellent reviews have addressed in detail the bacteriology and epidemiology of *L. monocytogenes* (Farber and Peterkin, 1991; Schuchat et al., 1991). The extensive literature on the immunological and physiological impact of infection on the host cell/organism has been addressed in other reviews (Kaufmann, 1993; Campbell, 1994; Theriot, 1995; Finlay and Cossart, 1997; Ireton and Cossart, 1997). Here the focus will be on the bacteriologic and molecular genetic features of *L. monocytogenes* involved in pathogenesis, with an emphasis on recent developments in the field. The discussion of known virulence determinants will be followed by a review of *Listeria*-host cell interactions in macrophages and nonphagocytic cell systems. Determinants involved in key aspects of these interactions such as invasion, intracellular growth and antigen processing will be addressed in this section. Aspects of the adaptive physiology of *L. monocytogenes,* which have recently been shown to have impact on virulence, will be then discussed. The review will conclude with a discussion of serogroup-associated genomic divisions within *L. monocytogenes,* clonal lineages of special epidemiological importance and recent developments in serotype-specific genes and surface antigens.

2. KNOWN VIRULENCE DETERMINANTS

2.1. HEMOLYSIN (LISTERIOLYSIN O, LLO): INVOLVEMENT IN VIRULENCE

Perhaps no other virulence factor of *L. monocytogenes* has been the subject of such intense and intriguing investigations as hemolysin (a pore-forming cytolysin also called listeriolysin O, or LLO). A number of early investigations had pointed to the key involvement of this factor in virulence: strains from clinical isolates were always hemolytic, and conversely, nonhemolytic isolates were not isolated from patients and were non-pathogenic in animal models (Seeliger, 1961). This was the first determinant to be shown, by a combination of genetic manipulation and animal model studies, to be essential for virulence (Gaillard et al., 1986; Kathariou et al., 1987; Portnoy et al., 1988), and the first to be molecularly characterized (Mengaud et al., 1988). LLO plays a pivotal role in pathogenesis because it allows the bacteria to escape the phagosome and enter the host cell cytoplasm (Gaillard et al., 1987). Expression of the LLO gene (*hly*) in *Bacillus subtilis* allows this nonpathogenic soil microorganism to escape from the phagocytic vacuole of macrophage-like cells (Bielecki et al., 1990).

2.2. OTHER GENES IN THE VIRULENCE REGULON

DNA sequencing of the regions adjacent to the gene encoding LLO (*hly*) revealed additional virulence-related genes, all within a ca. 10 kb region on the chromosome (reviewed in Portnoy et al., 1992; Sheehan et al., 1994). Upstream of *hly*, but divergently transcribed, are the genes *plcA*, encoding phosphoinositol-specific phospholipase C, followed by *prfA*, which encodes a positive regulator of transcription of the virulence genes in this region. Downstream of *hly*, and transcribed in the same direction, is the lecithinase operon, which includes genes encoding a metalloprotease, the actin-nucleation protein ActA, and another phospholipase, commonly termed lecithinase (Mengaud et al., 1989, 1991b). With the exception of the metalloprotease, the role of which in vivo is not yet clear, extensive studies have shown that the other gene products are involved in key steps of the cellular infection process: LLO mediates escape from the phagosome, ActA is essential for actin-based intracellular motility, whereas the two phospholipases have overlapping roles in vacuolar escape and cell-to-cell spread (Sheehan et al., 1994; Smith et al., 1995). Properly regulated transcription and coordinate expression of all the virulence genes in this region requires the regulatory protein PrfA, encoded by *prfA* (Leimeister-Wachter et al., 1990; Chakraborty et al., 1992). 14-bp consensus palindromic sequences in the promoter regions of these genes serve as binding sites for PrfA (Sheehan et al., 1995, 1996). Evidence has been provided for negative autoregulation of *prfA* (Mengaud et al., 1991a).

PrfA activates transcription of additional genes, besides those in the *hly* region. The gene encoding internalin (InlA), a surface protein essential for normal host cell invasion of certain cell lines by *L. monocytogenes*, has been shown to be under at least partial control by the *prfA* gene product and to harbor the 14-bp palindromic consensus sequences in its promoter region (Gaillard et al., 1991; Dramsi et al., 1993). Another member of the internalin family of proteins, InlC (IrpA), was identified on the basis of its transcriptional dependence on PrfA (Engelbrecht et al., 1996; Domann et al., 1997). The PrfA-dependent genes thus constitute a virulence regulon: a number of genes and operons, not necessarily linked, which are under at least partial regulatory control by the *prfA* gene product. In following discussions the term "*prfA* regulon" will be used for these genes.

2.3. MODULATION OF PrfA ACTIVITY AND VIRULENCE GENE EXPRESSION

Temperature strongly influences production of LLO, which is produced optimally at ca. 37°C but not at temperatures below 25°C: transcription of *hly* and the other virulence-related genes in this region is thermoregulated accordingly (Leimeister et al., 1992). It is possible that the mechanisms underlying this effect may be operating at the level of binding of PrfA to its recognition sites in the promoters of the virulence genes. Similar mechanisms may be operating to bring about repression of virulence gene expression in response to other environmental signals. Such repression has been shown to be affected by low temperature, as well as by presence of the plant-derived sugar cellobiose (Park and Kroll, 1993), the phenolic compound arbutin (Park, 1994) and other readily metabolizable mono- and disaccharides (Milenbachs et al., 1997). Similarities have been noted between PrfA and transcriptional regulators in the CAP-FNR family (Lampidis et al., 1994; Sheehan et al., 1994). Several studies suggest that a small co-factor may interact with PrfA to control its activity (Sheehan et al., 1996; Renzoni et al., 1997; Ripio et al., 1997a; Milenbachs et al., 1997).

2.4. VIRULENCE DETERMINANTS OUTSIDE THE prfA REGULON

To date only few virulence-related determinants have been identified outside the *prfA* regulon. These include the ClpC ATPase, a general stress protein (Rouquette et al., 1995, 1996), determinant(s) involved in acid tolerance (Marron et al., 1997) and the secreted protein p60 (Kohler et al., 1990). These determinants will be discussed later in this review.

3. *LISTERIA*-HOST CELL INTERACTION

The cellular and molecular aspects of *Listeria*-host cell interactions have been investigated extensively. Both macrophages and nonprofessional phagocytes can serve as host cells for *L. monocytogenes.* Studies with animal models suggest that in the course of an infection, the bacteria may invade enterocytes or M cells in Peyer's patches (Racz et al., 1972; McDonald and Carter, 1980), are taken up by phagocytes in the *lamina propria,* and disseminated to the primary target organs (liver and spleen). Only a small fraction of the bacteria reaching the liver avoid getting killed by resident macrophages (Mackaness,

1962; Lepay et al., 1985). Unless cleared by the host immune cells (Kaufmann, 1993), infection of hepatocytes by the small fraction of surviving Listeriae may lead to systemic infection and invasion of the secondary target organs (central nervous system, placenta and fetus).

3.1. MACROPHAGE UPTAKE AND INTRACELLULAR SURVIVAL

A hallmark of pathogenic *Listeria* is the ability to survive and multiply in macrophages (Mackaness, 1962). Replication rate of *L. monocytogenes* in murine bone marrow-derived macrophages was ca. 40 min, and intracellular growth appeared to reflect the competition between intracellular killing and expression of virulence factors (de Chastellier and Berche, 1994). Macrophages differ in terms of their listericidal potential; some permit escape of the bacterium into the cytoplasm and intracellular growth, whereas others kill the bacteria. Phagocytosis of serum-opsonized *L. monocytogenes* by listericidal macrophages was shown to be mediated by the complement receptor type 3 (CR3), whereas phagocytosis of the bacteria by nonlistericidal macrophages appeared to be CR3-independent (Drevets et al., 1991, 1993; Campbell, 1944). CR3 was found to bind to C3 deposited onto the bacterial surface (Drevets and Campbell, 1991) and *L. monocytogenes* was shown to activate the alternative pathway of human complement, suggesting that macrophages bearing C3 receptors can bind to opsonized bacteria (Croize et al., 1993). The cell wall component(s) targeted by C3 remains to be determined. The first component of complement, C1q, has also been shown to participate in phagocytosis of *L. monocytogenes* by macrophage-like cells (Alvarez-Dominguez et al., 1993).

3.2. INVASION OF NONPHAGOCYTIC CELLS

3.2.1. Internalin

The cellular and molecular aspects of the interaction of *L. monocytogenes* with cultured nonphagocytic cells have been studied extensively and have been the subject of a recent review (Ireton and Cossart, 1997). During invasion, the host membrane appears to appose the surface of the bacterial cell in a ''zipper-like'' mechanism (Mengaud et al., 1996; Finlay and Cossart, 1997). The surface proteins internalin A (InlA) and internalin B (InlB) have been implicated in invasion of Caco-2 (human intestinal epithelial cell line) and other cell lines (hepatocytes, HeLa, CHO cells), respectively (Gaillard et al., 1991, 1996; Dramsi et al., 1995). Several other members of the internalin family have been shown not to be required for invasion (Dramsi et al., 1997). The epithelial cell glycoprotein E-cadherin has been shown to serve as receptor for InlA (Mengaud et al., 1996), whereas the receptor for In1B remains unknown. InlA has several prominent protein sequence features, including repeated leucine-rich regions, a feature shared with at least seven other proteins of the "internalin family" Dramsi et al., 1997). The protein is covalently linked onto the peptidoglycan using the LPXTG motif (Schneewind et al., 1995), and surface exposure is required for bacterial contact with the host cell and invasion (Lebrun et al., 1996).

The *inlA* gene has been shown to be conserved and unique to *L. monocytogenes* (Poyart et al., 1996) but its possible role in virulence remains to be determined (Ireton and Cossart, 1997). In addition to its role in hepatocyte invasion (Dramsi et al., 1995), InlB has been shown to also be involved in intracellular growth of the bacteria in these cells (Gregory et al., 1997).

In contrast to mutants lacking LLO, which can still invade cells (Kuhn et al., 1988), a class of nonhemolytic mutants, which is now known to have lesions in *prfA*, was found to be noninvasive in several cell lines (Kathariou et al., 1990). The noninvasive phenotype of *prfA* mutants may be due to the fact that PrfA controls, at least partially, transcription of *inlA* (Dramsi et al., 1993) and may, in addition, regulate transcription of other, yet to be identified genes required for invasion.

3.2.2. Other Invasion-Associated Determinants: p60, ActA, and Teichoic Acid Decoration with Galactose

The major extracellular protein p60 (Iap) is a murein hydrolase essential for cell division and growth, which, when deficient, leads to production of cell chains (filamentous variants) and the "rough" colony phenotype (Kohler et al., 1990; Wuenscher et al., 1993). P60 appears to be required for adherence and invasion of certain cell lines (e.g., 3T6) but not others (e.g., Caco2) (Kuhn and Goebel, 1989; Gutekunst et al., 1992). Expression of secreted p60 in a vaccine (attenuated) strain of *Salmonella typhimurium* promoted the invasiveness of the vaccine strain into hepatocytes (Hess et al., 1995) and rendered it capable of conferring protection against infection by *L. monocytogenes* in mice (Hess et al., 1996).

Host heparan sulfate proteoglycans were recently shown to mediate attachment and entry of *L. monocytogenes,* using as ligand the surface protein ActA (which also mediates actin-based intracellular motility of the pathogen) (Alvarez-Dominguez et al., 1997). In addition, invasion of *L. monocytogenes* serotype 4b into several cell lines was shown to require galactosyl residues on wall teichoic acid, which, interestingly, were also required for bacteriophage adsorption (Promadej et al., unpublished results). The mechanism of involvement of teichoic acid galactosyl decoration in invasion remains to be elucidated.

Invasion of certain cell types has been shown to be independent of the known invasion determinants (Ireton and Cossart, 1997; Greiffenberg et al., 1997), suggesting the presence of additional, yet to be identified invasion-related mechanisms in *L. monocytogenes.*

3.2.3. Intracellular Motility and Cell-to-Cell Spread

The actin-based motility of *L. monocytogenes* is a key feature of cellular pathogenesis documented both in nonphagocytic cells and in macrophages (Tilney and Portnoy, 1989; Kocks et al., 1992; De Chastellier and Berche, 1994). A single *L. monocytogenes* protein with polarized surface distribution has been shown to be necessary and sufficient for actin-based motility of the pathogen. ActA mutants cannot spread from cell to cell and are avirulent in the mouse model (Kocks et al., 1992, 1993; Domann et al., 1992; Ireton and Cossart, 1997). Interestingly, the unrelated facultative intracellular pathogen *Shigella flexneri* also has actin-mediated intracellular motility, mediated by the IcsA protein (Bernardini et al., 1989). The ActA protein of *L. monocytogenes* and the *S. flexneri* IcsA, which have no apparent sequence similarity, were shown to be sufficient to confer actin-based motility on *L. innocua* and *Escherichia coli*, respectively (Kocks et al., 1995). The typical ActA-mediated accumulation of host-derived actin microfilaments around intracytoplasmic *L. monocytogenes* has been documented not only in cell cultures but also in actual clinical material from a human listeriosis patient (Kirk, 1993). The cellular aspects of ActA function have been reviewed in detail (Ireton and Cossart, 1997) and will not be discussed here.

3.2.4. T Cell Epitopes Produced by Intracellular Antigen Processing

Intracellular replication of the pathogen appears to be intricately connected with both virulence and the processing of selected antigens required for effective immune responses (Berche et al., 1987). It has been long known that heat-killed *L. monocytogenes* cannot confer protective immunity and that such immunity is cell-mediated (Mackaness, 1962). In the mouse model of infection LLO has been shown to be a major antigen recognized by specific classes of protective T cells (Berche et al., 1987; Safley et al., 1991; Archie Bouwer et al., 1992; Vijh and Pamer, 1997). LLO may owe its special role as an intracellular antigen not so much to its amounts as to its ability to be efficiently degraded and processed into epitopes recognized by MHC class I-restricted T cells (Villanueva et al., 1995). LLO has been shown to be a target not only for protective T cells but also for specific antibodies (Berche et al., 1990).

Epitopes specific for MHC class I-associated cytolytic T lymphocytes have been identified and characterized on p60 (Pamer, 1994; Bouwer and Hinrichs, 1996; Sijts et al., 1996, 1997; Vijh and Pamer, 1997), and such T cells have been shown to be protective against *L. monocytogenes* infection in mice (Harty and Pamer, 1995). T cell epitopes have also been identified on the metalloprotease, encoded by the first gene in the lecithinase operon (Busch et al., 1997), and on the amino-terminal hexapeptide of a novel protein (LemA) (Lenz et al., 1996). The extent to which the latter two proteins may also function as virulence determinants remains to be determined.

4. ADAPTIVE RESPONSES AND POSSIBLE IMPACT ON PATHOGENESIS OF *L. MONOCYTOGENES*

The relationship between physiological/metabolic parameters and pathogenesis is currently poorly understood in *L. monocytogenes.* The pathogen can utilize glucose-1-phosphate under conditions that lead to expression of PrfA-induced proteins, and *prfA* mutants were incapable of such utilization (Ripio et al., 1997b). It has been suggested that PrfA-dependent glucose-1-phosphate utilization may be related to pathogenesis, since this compound is found intracellularly as a degradation product of glycogen (Ripio et al., 1997b).

As mentioned earlier, low temperature represses transcription of the *prfA* regulon (Leimeister-Wachter et al., 1992). Furthermore, molecules such as cellobiose and the phenolic compound arbutin, present in the soil environment that the bacteria may be encountering in non-pathogenic situations, may serve as signals to induce repression of virulence-related genes (Park and Kroll, 1993; Park, 1994). These environmental signals may operate, directly or indirectly, on the ability of PrfA to recognize its target sequences on the genome (Ripio et al., 1997a; Milenbachs et al., 1997), but the signal-transducing mechanisms that may be involved remain to be elucidated. Recent data showed that cellobiose, glucose and several other readily metabolizable carbohydrates could also repress transcription of *prfA*-dependent genes, wet growth was concomitantly enhanced, suggesting that a growth rate-dependent metabolite might serve as cofactor for PrfA (Milenbachs et al., 1997). Nutrient depletion has been shown to lead to induction of several PrfA-dependent proteins, as has heat shock (Sokolovic et al., 1993).

The proven importance of certain members of the *prfA* regulon on virulence notwithstanding, it has yet to be demonstrated that the ability to mount physiological responses to such environmental stress signals (heat shock, nutrient depletion) is of importance to the pathogenesis and virulence of *L. monocytogenes.* Similarly, of the 32 oxidative stress and heat shock proteins identified by Hanawa and co-workers, none was among the proteins shown to be induced during growth in macrophage-like cells (Hanawa et al., 1995). Several features of the host environment, such as iron deprivation, may, however, serve as signals for induction of stress responses of relevance to pathogenesis. It has been shown that mutations in the ClpC ATPase of *L. monocytogenes,* a general stress response protein, rendered the bacteria unable to grow in iron-depleted media and, in addition, severely impaired virulence in the mouse and the ability to grow in bone marrow-derived macrophages (Rouquette et al., 1995, 1996). The mutants were sensitive not only to iron limitation but also to heat, salt, and oxidative stress in synthetic media (but not in complex media such as Brain Heart Infusion) (Rouquette et al., 1996). An acid-sensitive mutant of *L. monocytogenes* was also found to be reduced in virulence in the mouse model (Marron et al., 1997). It has been shown that following phagocytosis of *L. monocytogenes* in macrophages the phagosome is rapidly acidified, and that this acidification is a prerequisite for escape of the bacteria from the phagosome (de Chastellier and Berche, 1994). It is thus conceivable that acid-sensitive mutants owe their reduced virulence to reduced survival in the acidified phagosome and to the associated inability to produce LLO and escape into the cytoplasm.

5. EPIDEMIOLOGICAL ASPECTS OF PATHOGENESIS

5.1. PREVALENCE OF SPECIFIC SEROTYPES

Even though *L. monocytogenes* strains of various serotypes are found in foods and in the environment, most clinical isolates are of three serotypes: 1/2a, 1/2b and 4b. Although it is tempting to speculate that this reflects differences in virulence, actual data to this effect using animal or cell culture models are still lacking. Correlation between certain serotypes and types of clinical presentation (e.g., prevalence of serotype 4b in pregnancy-associated cases) (McLaughlin, 1990) may suggest potential differences among serotypes in human virulence/pathogenesis.

The serotypic designation of a given strain, determined according to the scheme of Seeliger and Hoehne (Seeliger and Hoehne, 1979), appears to be a stable characteristic of the organism. In this designation, the letter (a, b, c, etc.) refers to flagellar antigens, whereas the numerical portion (1/2, 3, 4, etc.) refers to somatic (mostly teichoic acid-associated) antigens. Several investigations using multilocus gel electrophoresis have suggested that strains of *L. monocytogenes* constitute two major genetic groups, one of which includes strains of the "a" and "c" class (serotypes 1/2a, 1/2c 3a, 3c) whereas the other includes strains of the "b" class (serotypes 1/2b, 3b, 4b) (Bibb et al., 1990; Piffaretti et al., 1989). Similar results were obtained with ribotyping (Graves et al., 1994: Wiedman et al., 1997) and pulse-field gel electrophoresis (Brosch et al., 1994). Sequencing and/or restriction fragment length polymorphism (RFLP) analysis of numerous *L. monocytogenes* genes (including genes encoding LLO and other virulence genes, flagellin, p60, and a genomic region essential for low temperature growth) could differentiate strains of serotypes 1/2a, 1/2c, 3a, 3c from those of serotypes 1/2b, 3b, 4b (Rasmussen et al., 1995; Vines et al., 1992; Zheng and Kathariou, 1995; Wiedman et al., 1997).

The strains that have been used in most molecular genetic and immunologic studies of virulence determinants in *L. monocytogenes* have been of either serotype 1/2a (strains 10403S, EGD, NCTC 7973, Mack) or serotype 1/2c (strain LO28), thus representing only one of the major genetic groups of the pathogen. The omission of strains representing the other major group is unfortunate and all the more noticeable in view of the apparent genetic distance between the groups and the significant clinical incidence of other serotypes, especially 4b (see below).

5.2. SEROTYPE 4B: EPIDEMIOLOGY AND MOLECULAR ASPECTS

Strains of serotype 4b account not only for a substantial fraction (ca. 40%) of sporadic infections but also for the majority of common-source outbreaks of listeriosis that have been epidemiologically monitored, including the coleslaw-associated epidemic in Nova Scotia (1981), the Jalisco cheese outbreak in Los Angeles (1985), the Swiss Vacherin Mont d'Or outbreak (1987–89) (Schuchat et al., 1991), the 1992 outbreak in France, traced to a contaminated deli meat product (Jacquet et al., 1995), and the latest multistate outbreak of listeriosis in the United States, traced to contaminated hot dogs (Anonymous, 1999). It is noteworthy that the strains implicated in several of these temporally and geographically unlinked outbreaks were not only of the same serotype (4b) but were also closely related genetically (Pifaretti et al., 1989; Jacquel et al., 1995), suggesting that they may represent an epidemic clonal lineage. Recently it was found that strains in this epidemic clonal group harbor a unique type of DNA modification (cytosine methylation at GATC sites), which renders the DNA resistant to digestion by enzymes such as *Sau*3AI. The strains may also produce a restriction endonuclease, which can serve as protective strategy against attack by bacteriophages in foods and/or the environment (Zheng and Kathariou, 1997). In addition, RFLP in a region essential for cold growth could differentiate these strains from other strains of *L. monocytogenes*, including other serotype 4b strains (Zheng and Kathariou, 1995). At this point it is not clear whether these strains owe their prevalence in epidemic listeriosis to special features of their adaptive physiology or to unique and yet-to-be identified aspects of their virulence and pathogenesis to humans. Interestingly, the strain implicated in the latest multistate outbreak of foodborne listeriosis in the United States (Anonymous, 1999), appears to have an unusual genotype based on pulsed-field gel electrophoresis, and is likely to be distinct from the previously characterized outbreak strains.

In contrast to their clinical prevalence, serotype 4b strains are not frequently recovered in routine surveys of contaminated foods (Hayes et al., 1991). Only 3% of the *L. monocytogenes* strains in a survey of cheese, meat and poultry were of serotype 4b (Schoenberg et al., 1989). There may be several reasons for this apparent paradox. For instance, strains of this serotype may contaminate food rarely, but once in the food environment they may have a higher ability to persist and multiply; alternatively, the strains may be present in foods, and capable of infection, but not readily culturable from the food environment. Lastly, the strains may have unique pathogenesis features that render them more likely to cause human infection and disease.

5.3. SEROTYPE-SPECIFIC GENES AND SURFACE ANTIGENS

Earlier studies suggested the that sugar moieties on the cell wall of *L. monocytogenes* varied among serotypes and were important antigenic determinants (Ullman and Cameron, 1969). A proven contributor to cell-surface structural diversity and antigenicity is the anionic polymer teichoic acid, consisting of a polyribitol phosphate backbone (with serotype-specific glycosylation) covalently linked to peptidoglycan. Figure 9.1 shows a diagrammatic representation of teichoic acid structures of serotypes 1/2, 3, and 4b. Serotype 4b strains have a unique teichoic acid backbone composition characterized by integral N-acetylglucosamine and a complex glycosylation involving both galactose and glucose substituents on the N-acetylglucosamine (Kamisango et al., 1983; Fiedler et al., 1984; Uchikawa et al., 1986). Recently, genes that are unique to serotype 4 or to specific serotype 4 groups (serotypes 4b, 4d and 4e) have been described (Lei et al., 1997). Mutations in these genes rendered serotype 4b bacteria negative with monoclonal antibodies that had been shown to react with serotype 4b, 4d and 4e bacteria but not with other strains of *L. monocytogenes* (Kathariou et al., 1994). Recent findings suggest that the serotype 4-unique genes are involved in glycosylation of teichoic acid in serotype 4b *L. monocytogenes* (Promadej et al., unpublished results). Two

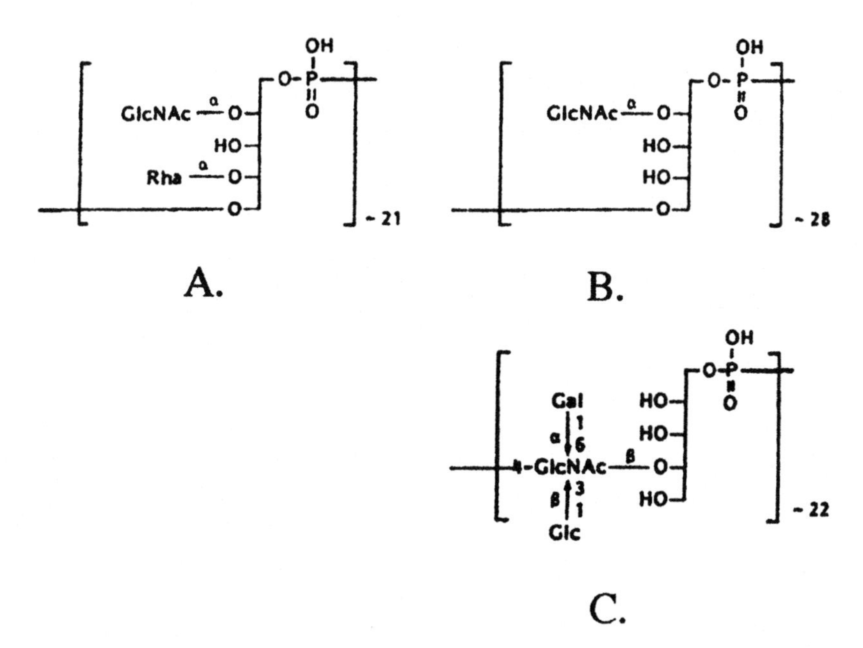

Figure 9.1 Wall teichoic acid structures of *Listeria monocytogenes* serotype 1/2a (A), serotype 3a (B), and serotype 4b (C). Modified from Uchikawa et al., 1986.

such genomic regions have been identified to date, each apparently representing a serotype-specific gene cassette that is flanked by genes conserved among different serotypes (Lei, 1998; Promadej, 1998). Serotype 4b teichoic acid glycosylation may have important roles, directly or indirectly, in pathogenesis and in the ecology of the microorganism. As mentioned earlier, data from our laboratory suggest that galactosyl residues on the teichoic acid are essential not only for normal invasion of cultured nonphagocytic cells but also for bacteriophage adsorption. Similar roles for glucosyl residues have not yet been identified. Nonphagocytic cells may represent an inappropriate model for the study of such residues. It is conceivable, for instance, that these residues may be involved in binding of complement components and macrophage recognition/uptake. As described earlier, recognition by different macrophage receptors may determine the intracellular fate of the bacteria (intracellular growth vs. killing) and is thus of importance to pathogenesis.

5.4. TEICHOIC ACID GLYCOSYLATION IN OTHER SEROTYPES

Studies with serotype 1/2 *L. monocytogenes* have shown that the rhamnose substituents on the teichoic acid were essential for phage adsorption (Wendlinger et al., 1996). It remains to be determined whether the rhamnose substituents, which are unique to serotype 1/2, are also involved in interactions with nonphagocytic cells and/or macrophages, or whether the corresponding gene(s) are unique to serotype 1/2. Rhamnose may be involved in attachment of the first component of complement, C1q, to cell walls of serotype 1/2 *L. monocytogenes* (Alvarez-Dominguez et al., 1993).

6. CONCLUSIONS

In this review we have focused on specific determinants and bacteriological/molecular features of *L. monocytogenes* involved in virulence and pathogenesis. During the past decade our understanding of the cellular and molecular aspects of the *L. monocytogenes* host-pathogen interaction has expanded remarkably. Nevertheless, several key aspects of this interaction, especially those involving invasion and determination of the intracellular fate of the pathogen, need to be further elucidated, and their relevance to human and animal infection and disease further characterized. It is clear that this pathogen represents a complex system that has evolved multiple, sophisticated mechanisms to interact with the host cell's cellular biology and physiology, while, at the same time, maintaining the ability to lead a mostly uncharacterized but seemingly well-poised lifestyle in the nonhuman environment.

The well-characterized virulence factors discussed here (e.g., LLO, PrfA, internalin, ActA etc.) appear to be shared by all strains of *L. monocytogenes,* but distinct lineages within the species may differ in their genetic and antigenic endowment and, possibly, in important properties related to pathogenesis and/or adaptive physiology. The extent to which lineage-specific features may render certain host-pathogen encounters more likely to lead to human infection and disease is one of the areas of *L. monocytogenes* pathogenesis that remains to be addressed. Future investigations may elucidate this and other, related areas of pathogenesis that are currently understudied and poorly understood. These include the evolution of pathogenicity-related genomic regions, the impact of the adaptive physiology of *L. monocytogenes* on the infectious process, and the yet-to-be defined role(s) of serotype-specific genes and surface antigens in pathogenesis.

7. ACKNOWLEDGMENTS

I apologize to colleagues who may feel that their contributions were incompletely or incorrectly reviewed here. I would like to thank F. Fiedler and the members of my laboratory, in particular, W. Zheng, X.-H. Lei, N. Promadej, T. Truong and Z. Lan, for their support, comments and discussion, and E. Lanwermeyer for critical reading of the manuscript and editing. I am grateful to my sabbatical host, J. Tiedje, for offering me an academic environment that facilitated the writing of this review. Work in my laboratory is supported by USDA grants 92–37 201–8095 and 95–37201–2031 and by a grant from the International Life Sciences Institute-North America.

This manuscript is dedicated to the memory of Professor Heinz Seeliger, who, throughout his career, provided thrust, motivation and encouragement for the study of Listeria and listeriosis to numerous scientists, including myself.

8. REFERENCES

Alvarez-Dominguez, C., Carasso-Marin, E., and Leyva-Cobian, F. 1993. "Role of complement component C1q in phagocytosis of *Listeria monocytogenes* by murine macrophage-like cell lines," *Infect. Immun.* 61:3664–3672.

Alvarez-Dominguez, C., Vazquez-Boland, J.-A., Carrasco-Marin, E., Lopez-Mato, P., and Leyva-Cobian, F. 1997. "Host cell heparan sulfate proteoglycans mediate attachment and entry of *Listeria monocytogenes*, and the listerial surface protein ActA is involved in heparan sulfate receptor recognition," *Infect. Immun.* 65:78–88.

Anonymous. 1999. "Update: multistate outbreak of listeriosis, United States, 1989–1999," *Morbid. Mortal. Weekly Rep.* 47:1117–1118.

Archie Bouwer, H. G., Nelson, C. S., Gibbins, B. L., Portnoy, D. A., and Hinrichs, D. J. 1992. "Listeriolysin O is a target of the immune response to *Listeria monocytogenes*," *J. Exp. Med.* 175:1467–1471.

Berche, P., Gaillard, J. L., and Sansonetti, P. J. 1987. "Intracellular growth of *Listeria monocytogenes* as a prerequisite for in vivo induction of T cell-mediated immunity," *J. Immunol.* 138:2266–2271.

Berche, P., Gaillard, J. L., Geoffroy, C., and Alouf, J. E. 1987. ''T cell recognition of listeriolysin O is induced during infection with *Listeria monocytogenes*,'' *J. Immunol.* 139:3813–3821.

Berche, P., Reich, K. A., Bonnichon, M., Beretti, J. L., Geoffroy, C., Raveneau, J., Cossssart, P., Gaillard, J. L., Geslin, P., Kreis, H., and Veron, M. 1990. ''Detection of anti-LLO for serodiagnosis of human listeriosis,'' *Lancet* 335:624–627.

Bibb, W. F., Gellin, B. G., Weaver, R., Schwartz, B., Plikaytis, B. D., Reeves, M. W., Pinner, R. W., and Broome, C. V. 1990. ''Analysis of clinical and food-borne isolates of *Listeria monocytogenes* in the United States by multilocus enzyme electrophoresis and application of the method to epidemiological investigations,'' *Appl. Environ. Microbiol.* 56:2133–2141.

Bielecki, J., Youngman, P., Connelly, P., and Portnoy, D. A. 1990. *Bacillus subtilis* expressing a haemolysin gene from *Listeria monocytogenes* can grow in mammalian cells,'' *Nature* 345:175–176.

Bouwer, H. G. and Hinrichs, D. J. 1996. ''Cytotoxic-T-lymphocyte responses to epitopes of listeriolysin O and p60 following infection with *Listeria monocytogenes*,'' *Infect. Immun.* 64:2515–2522.

Bouwer, H. G., Gibbins, B. L., Jones, S. and Hinrichs. D. J. 1994. ''Antilisterial immunity includes specificity to listeriolysin O (LLO) and non-LLO-derived determinants,'' *Infect. Immun.* 62:1039–1045.

Brosch, R., Chen, J., and Luchansky, J. B. 1994. ''Pulsed-field fingerprinting of listeriae: identification of genomic divisions for *Listeria monocytogenes* and their correlation with serovar,'' *Appl. Environ. Microbiol.* 60:2584–2592.

Busch, D. H., Archie Bouwer, H. G., Hinrichss, D., and Pamer, E. G. 1997. ''A nonamer peptide derived from *Listeria monocytogenes* metalloprotease is presented to cytolytic T lymphocytes,'' *Infect. Immun.* 65:5326–5329.

Campbell, P. A. 1994. ''Macrophage-Listeria interactions,'' *Immunol. Ser.* 60:313–328.

Chakraborty, T., Leimeister-Wachter, M., Domann, E., Hartl, M., Goebel, W., Nichterlein, T., and Notermans, S. 1992. ''Coordinate regulation of virulence genes in *Listeria monocytogenes* requires the product of the *prfA* gene,'' *J. Bacteriol.* 174:568–574.

Cossart, P. and Kocks, C. 1994. ''The actin-based motility of the facultative intracellular pathogen *Listeria monocytogenes*,'' *Mol. Microbiol.* 13:395–402.

Croize, J., Arvieux, J., Berche, P., and Colomb, M. G. 1993. ''Activation of the human complement alternative pathway by *Listeria monocytogenes:* evidence for direct binding and proteolysis of the C3 component on bacteria,'' *Infect Immun.* 61:5134–5139.

De Chastellier C. and Berche, P. 1994. ''Fate of *Listeria monocytogenes* in murine macrophages: Evidence for simultaneous killing and survival of intracellular bacteria,'' *Infect. Immun.* 62:543–553.

Domann, E., Zechel, S., Lingnau, A., Hain, R., Darji, A., Nichterlein, T., Wehland, J., and Chkraborty, T. 1997. ''Identification and characterization of a novel PrfA-regulated gene whose product, IrpA, is highly homologous to internalin proteins, which contain leucine-rich repeats,'' *Infect. Immun.* 65:101–109.

Domann, E., Wehland, J., Rohde, M., Pistor, S., Hartl, M., Goebel, W., Leimeister-Wachter, M., Wuenscher, M., and Chakraborty, T. 1992. ''A novel bacterial virulence gene in *Listeria monocytogenes* required for host cell microfilament interaction with homology to the proline-rich region of vinculin,'' *EMBO J.* 11:1981–1990.

Dramsi, I. Biswas, Maguin, E., Braun, L., Mastroeni, P., and Cossart, P. 1995. ''Entry of *L. monocytogenes* into hepatocytes requires expression of InlB, a surface protein of the internalin multigene family,'' *Mol. Microbiol.* 16:251–261.

Dramsi, P. Dehoux, Lebrun, M., Goossens, P., and Cossart, P. 1997. ''Identification of four new members of the internalin multigene family in strain EGD,'' *Infect. Immun.* 65:1615–1625.

Dramsi, S. C. Kochs, Forestier, C., and Cossart, P. 1993. "Internalin-mediated invasion of epithelial cells by *Listeria monocytogenes* is regulated by the bacterial growth state, temperature, and the pleiotropic activator, prfA," *Mol. Microbiol.* 9:931–941.

Drevets, D. A. and Campbell, P. A. 1991. "Roles of complement and complement receptor type 3 in phagocytosis of *Listeria monocytogenes* by inflammatory mouse peritoneal macrophages," *Infect. Immun.* 59:2645–2652.

Drevets, D. A., Canono, B. P. and Campbell, P. A. 1992. "Listericidal and non-listericidal mouse macrophages differ in complement receptor type 3-mediated phagocytosis of *L. monocytogenes* and in preventing escape of the bacteria into the cytoplasm," *J. Leucocyte Biol.* 52:70–79.

Drevets, D. A., Leenen, P. J. M. and Campbell, P. A. 1993. "Complement receptor type 3 (CD11b/CD18) involvement is essential for killing of *Listeria monocytogenes* by murine macrophages," *J. Immunol.* 151:5431–5439.

Engelbrecht, F., Chun, S.-K., Ochs, C., Hess, J., Lottspeich, F., Goebel, W., and Sokolovic, Z. 1996. "A new PrfA-regulated gene of *Listeria monocytogenes* encoding a small, secreted protein which belongs to the family of internalins," *Mol. Microbiol.* 21:823–837.

Farber, J. M. and Peterkin, P. I. 1991. "*Listeria monocytogenes,* a food-borne pathogen," *Microbiol. Rev.* 55:476–511.

Fiedler, F. Seger, J., Schrettenbrunner, A. and Seeliger, H. P. R. 1984. "The biochemistry of murein and cell wall teichoic acids in the genus of *Listeria,*" *System. Appl. Microbiol.* 5:360–376.

Finlay, B. B. and Cossart, P. 1997. "Exploitation of mammalian host cell functions by bacterial pathogens," *Science* 276:718–725.

Gaillard, J. L., Berche, P. and Sansonetti, P. 1986. "Transposon mutagenesis as a tool to study the role of hemolysin in the virulence of *Listeria monocytogenes,*" *Infect. Immun.* 52:50–55.

Gaillard, J. L., Jaubert, F. and Berche, P. 1996. "The inlAB locus mediates the entry of *Listeria monocytogenes* into hepatocytes in vivo," *J. Exp. Med.* 183:359–369.

Gaillard, J. L., Berche, P., Frehel, C., Gouin, E. and Cossart, P. 1991. "Entry of *L. monocytogenes* into cells is mediated by internalin, a repeat protein reminiscent of surface antigens from gram-positive cocci," *Cell* 65:1127–1141.

Gaillard, J. L., Berche, P., Mounier, J., Richard, S. and Sansonetti, P. 1987. "In vitro model of penetration and intracellular growth of *Listeria monocytogenes* in the human enterocyte-like cell line Caco-2," *Infect. Immun.* 55:2822–2829.

Graves, L. M., Swaminathan, B., Reeves, M. W., Hunter, S. B., Weaver, R. E., Plikaytis, B. D. and Schuchat, A. 1994. "Comparison of ribotyping and multilocus enzyme electrophoresis for subtyping of *Listeria monocytogenes* isolates," *J. Clin. Microbiol.* 32:2936–2943.

Gray, M. L. and Killinger, A. H. 1966. "*Listeria monocytogenes* and listeric infections," *Bacteriol. Rev.* 30:309–382.

Gregory, S. H., Sagnimeni, A. J. and Wing, E. J. 1997. "Internalin B promotes the replication of *Listeria monocytogenes* in mouse hepatocytes," *Infect. Immun.* 65:5137–5141.

Greiffenberg, L., Sokolovic, Z., Schnittler, H. J., Spory, A., Bockmann, R., Goebel, W. and Kuhn, M. 1997. "*Listeria monocytogenes*-infected human umbilical vein endothelial cells: internalin-independent invasion, intracellular growth, movement, and host cell responses," *FEMS Microbiol. Lett.* 157:163–170.

Gutekunst, K. A., Pine, L., White, E., Kathariou, S. and Carlone, G. M. 1992. "A filamentous-like mutant of *Listeria monocytogenes* with reduced expression of a 60-kDa extracellular protein invades and grows in 3T6 and Caco-2 cells," *Canad. J. Microbiol.* 38:843–851.

Hanawa, T., Yamamoto, T. and Kamiya, S. 1995. "*Listeria monocytogenes* can grow in macrophages without the aid of proteins induced by environmental stresses," *Infect. Immun.* 63:4595–4599.

Harty, J. T. and Pamer, E. G. 1995. "CD8 T lymphocytes specific for the secreted p60 antigen protect against *Listeria monocytogenes* infection," *J. Immunol.* 154:4642–4650.

Hayes, P. S., Graves, L. M., Ajello, G. W., Swaminathan, B., Weaver, R. E., Wenger, J. D., Schuchat, A. and Broome, C. V. 1991. "Comparison of cold-enrichment and U.S. Department of Agriculture methods for isolating *Listeria* spp. from naturally contaminated foods," *Appl. Environ. Microbiol.* 57:2109–2113.

Hess, J., Gentschev, I., Miko, D., Welzel, M., Ladel, C., Goebel, W. and Kaufmann, S. H. 1996. "Superior efficacy of secreted over somatic antigen display in recombinant *Salmonella* vaccine-induced protection against listeriosis," *Proc. Natl. Acad. Sci. USA* 93:1458–1463.

Hess, J., Gentschev, I., Szalay, G., Ladel, C., Bubert, A., Goebel, W. and Kaufmannn, S. H. 1995. "*Listeria monocytogenes* p60 supports host cell invasion by and in vivo survival of attenuated *Salmonella typhimurium*," *Infect. Immun.* 63:2047–2053.

Ireton, K. and Cossart, P. 1997. "Host-pathogen interactions during entry and actin-based movement of *Listeria monocytogenes*," *Annu. Rev. Genet.* 31:113–138.

Jacquet, C., Catimel, B., Brosch, R., Buchrieser, C., Dehaumont, P., Coulet, V., Lepoutre, A., Veit, P. and Rocourt, J. 1995. "Investigations related to the epidemic strain involved in the French listeriosis outbreak in 1992," *Appl. Environ. Microbiol.* 61:2242–2246.

Kamisango, K., Fujii, H., Okumura, H., Saiki, I., Araki, Y., Yamamura, Y. and Azuma, I. 1983. "Structural and immunological studies of teichoic acid of *Listeria monocytogenes*," *J. Biochem.* 93:1401–1409.

Kathariou, S., Metz, P., Hof, H. and Goebel, W. 1987. "Tn916-induced mutations in the hemolysin determinant affecting virulence of *Listeria monocytogenes*," *J. Bacteriol.* 169:129–137.

Kathariou, S., Mizumoto, C., Allen, R. D., Fok, A. K. and Benedict, A. A. 1994. "Monoclonal antibodies with a high degree of specificity for *Listeria monocytogenes* serotype 4b," *Appl. Environ. Microbiol.* 60:3548–3552.

Kathariou, S., Pine, L., George, V., Carlone, G. M. and Holloway, B. P. 1990. "Nonhemolytic *Listeria monocytogenes* that are also noninvasive to mammalian cells in culture: evidence for coordinate regulation of virulence," *Infect. Immun.* 58:3988–3995.

Kaufmann, S. H. E. 1993. "Immunity to intracellular bacteria," *Annu. Rev. Immunol.* 11:129–163.

Kirk, J. 1993. "Diagnostic ultrastructure of *Listeria monocytogenes* in human central nervous tissue," *Ultrustruct. Pathol.* 17:583–592.

Kocks, C., Hellio, R., Gounon, P., Ohayon, H. and Cossart, P. 1993. "Polarized distribution of *Listeria monocytogenes* surface protein ActaA at the site of directional actin assembly," *J. Cell Sci.* 105:699–710.

Kocks, C., Gouin, E., Tabouret, M., Berche, P., Ohayon, H. and Cossart, P. 1992. "*L. monocytogenes*-induced actin assembly requires the *actA* gene product, a surface protein," *Cell* 68:521–531.

Kocks, C., Marchand, J. B., Gouin, E., d'Hauteville, H., Sansonetti, P. J., Carlier, M. F. and Cossart, P. 1995. "The unrelated surface proteins ActA of *Listeria monocytogenes* and IcsA of *Shigella flexneri* are sufficient to confer actin-based motility to *L. innocua* and *E. coli*, respectively," *Mol. Microbiol.* 18:413–4423.

Kohler, S., Leimeister-Wachter, M., Chakraborty, T., Lottspeich, F. and Goebel, W. 1990. "The gene coding for protein p60 of *Listeria monocytogenes* and its use as a specific probe for *Listeria monocytogenes*," *Infect. Immun.* 58:1943–1950.

Kuhn, M. and Goebel, W. 1989. "Identification of an extracellular protein of *Listeria monocytogenes* possibly involved in intracellular uptake by mammalian cells," *Infect. Immun.* 57:55–61.

Kuhn, M., Kathariou, S. and Goebel, W. 1988. "Hemolysin supports survival but not entry of the intracellular bacterium *Listeria monocytogenes*," *Infect. Immun.* 56:79–82.

Lampidis, R., Gross, R., Sokolovic, Z., Goebel, W. and Kreft, J. 1994. ''The virulence regulator protein of *Listeria ivanovii* is highly homologous to PrfA from *Listeria monocytogenes* and both belong to the Crp-Fnr family of transcription regulators,'' *Mol. Microbiol.* 13:141–151.

Lebrun, M., Mengaud, J., Ohayon, H., Nato, F. and Cossart, P. 1996. ''Internalin must be on the bacterial surface to mediate entry of *Listeria monocytogenes* into epithelial cells,'' *Mol. Microbiol.* 21:579–592.

Lei, X.H. 1998. ''Molecular genetic studies of surface antigens specific for serotype 4b *Listeria monocytogenes.*'' Ph.D. Dissertation, University of Hawaii.

Lei, X. H., Promadej, N. and Kathariou, S. 1997. ''DNA fragments from regions involved in surface antigen expression specifically identify *Listeria monocytogenes* serovar 4 and a subset thereof: cluster IIB (serotypes 4b-4d-4e)'', *Appl. Environ. Microbiol.* 63:1077–1082.

Leimeister-Wachter, M., Domann, E. and Chakraborty, T. 1992. ''The expression of virulence genes in *Listeria monocytogenes* is thermoregulated,'' *J. Bacteriol.* 1744:947–952.

Leimeister-Wachter, M., Haffner, C., Domann, E., Goebel, W. and Chakraborty, T. 1990. ''Identification of a gene that positively regulates expression of listeriolysin, the major virulence factor of *Listeria monocytogenes,*'' *Proc. Natl. Acad. Sci. USA* 87:8336–8340.

Lenz, L. L., Dere, B. and Bevan, M. J. 1996. ''Identification of an H2-M3-restricted *Listeria* epitope: implications for antigen presentation by M3,'' *Immunity* 5:63–72.

Lepay, D. A., Steinman, R. M., Nathan, C. F., Murray, H. W. and Cohn, Z. A. 1985. ''Liver macrophages in murine infection. Cell-mediated immunity is correlated with an influx of macrophages capable of generating reactive oxygen intermediates,'' *J. Exp. Med.* 161:1503–1512.

MacDonald, T. T. and Carter, P. B. 1980. ''Cell-mediated immunity to intestinal infection,'' *Infect. Immun.* 28:516–523.

Mackaness, G. B. 1962. ''Cellular resistance to infection,'' *J. Exp. Med.* 116:381–406.

McLaughlin, J. 1990. ''Distribution of serovars of *Listeria monocytogenes* isolated from different categories of patients with listeriosis,'' *Eur. J. Clin. Microbiol. Infect. Dis.* 9:210–213.

Marron, L. N. Emerson, Gahan, C. G. M. and Hill, C. 1997. ''A mutant of *Listeria monocytogenes* LO28 unable to induce an acid tolerance response displays diminished virulence in the murine model.'' *Appl. Environ. Microbiol.* 63:4945–4947.

Mengaud, J., Geoffroy, C. and Cossart, P. 1991b. ''Identification of a novel operon involved in virulence of *Listeria monocytogenes:* Its first gene encodes a protein homologous to bacterial metalloproteases,'' *Infect. Immun.* 59:1043–1049.

Mengaud, J., Vincente, M. F. and Cossart, P. 1989. ''Transcriptional mapping and nucleotide sequence of the *Listeria monocytogenes hly* region reveal structural features that may be involved in regulation,'' *Infect. Immun.* 57:3695–3701.

Mengaud, J., Dramsi, S., Gouin, E., Milon, G. and Cossart, P. 1991a. ''Pleiotropic control of *Listeria monocytogenes* virulence factors by a gene which is autoregulated,'' *Mol. Microbiol.* 5:2273–2283.

Mengaud, J., Ohayon, H., Gounon, P., Mege, R.-M. and Cossart, P. 1996. ''E-Cadherin is the receptor for internalin, a surface protein required for entry of *L. monocytogenes* into epithelial cells,'' *Cell* 84:923–932.

Mengaud, J., Vicente, M.-F., Chenevert, J., Moniz Pereira, J., Geoffroy, C., Gicquel-Sanzey, B., Baquero, F., Perez-Diaz, J. C. and Cossart, P. 1988. ''Expression in *Escherichia coli* and sequence analysis of the listeriolysin O determinant of *Listeria monocytogenes,*'' *Infect. Immun.* 56:766–772.

Milenbachs, A. A., Brown, D. P., Moors, M. and Youngman, P. 1997. ''Carbon-source regulation of virulence gene expression in *Listeria monocytogenes,*'' *Mol. Microbiol.* 23:1075–1085.

Pamer, E. G. 1994. "Direct sequence identification and kinetic analysis of an MHC class I-restricted *Listeria monocytogenes* CTL epitope," *J. Immunol.* 152:686–694.

Park, S. F. 1994. "The repression of listeriolysin O expression in *Listeria monocytogenes* by the phenolic beta-D-glucoside, arbutin," *Lett. Appl. Microbiol.* 19:2658–2660.

Park, S. F. and Kroll, R. G. 1993. "Expression of listeriolysin and phsophatidylinositol-specific phospholipase C is repressed by the plant-derived molecule cellobiose in *Listeria monocytogenes,*" *Mol. Microbiol.* 8:653–661.

Piffaretti, J. C., Kressebuch, H., Aeschbacher, M., Bille, J., Bannerman, E., Musser, J. M., Selander, K. K. and Rocourt, J. 1989. "Genetic characterization of clones of the bacterium *L. monocytogenes* causing epidemic disease," *Proc. Natl. Acad. Sci. USA* 86:3818–3822.

Portnoy, D. A., Jacks, P. S. and Hinrichs, D. J. 1988. "Role of hemolysin for the intracellular growth of *Listeria monocytogenes,*" *J. Exp. Med.* 167:1459–1471.

Portnoy, D. A., Chakraborty, T., Goebel, W. and Cossart, P. 1992. "Molecular determinants of *Listeria monocytogenes* pathogenesis," *Infect. Immun.* 60:1263–1267.

Poyart, C., Trieu-Cuot, P. and Berche, P. 1996. "The *inl*A gene required for cell invasion is conserved and specific to *Listeria monocytogenes,*" *Microbiology* 142:173–180.

Promadej, N. 1998. "Molecular cloning, identification, and characterization of an invasion and serotype-specific surface determinant in *Listeria monocytogenes,* serotype 4b." Ph.D. Thesis, University of Hawaii.

Racz, P., Tenner, K. and Mero, E. 1972. "Experimental *Listeria enteritis.* I. An electron microscopic study of the epithelial phase in experimental *Listeria infection,*" *Lab. Invest.* 26:694–700.

Rasmussen, O. F., Skouboe, P., Dons, L., Rossen, L. and Olsen, J. E. 1995. "*Listeria monocytogenes* exists in at least three evolutionary lines: Evidence from flagellin, invasion associated protein, and listeriolysin O," *Microbiology* 141:2053–2061.

Renzoni A., Klarsfeld, A., Dramsi, S. and Cossart, P. 1997. "Evidence that PrfA, the pleiotropic activator of virulence genes in *Listeria monocytogenes,* can be present but inactive," *Infect. Immun.* 65:1515–1518.

Ripio, M.-T., Brehm, K., Lara, M., Suarez, M. and Vazquez-Boland, J.-A. 1997b. "Glucose-1-phosphate utilization by *Listeria monocytogenes* is PrfA-dependent and coordinately expressed with virulence factors." *J. Bacteriol.* 179:7174–7180.

Ripio, M.-T., Dominguez-Bernal, G., Lara, M., Suarez, M. and Vazquez-Boland, J.-A. 1997a. "A Gly 145Ser substitution in the transcriptional activator PrfA causes constitutive overexpression of virulence factors in *Listeria monocytogenes,*" *J. Bacteriol.* 179:1533–1540.

Rouquette, C., Bolla, J.-M. and Berche, P. 1995. "An iron-dependent mutant of *Listeria monocytogenes* of attenuated virulence," *FEMS Microbiol. Lett.* 133:77–83.

Rouquette, C., Ripio, M.-T., Pellegrini, E., Bolla, J.-M., Tascon, R. I., Vasquez-Boland, J.-A. and Berche, P. 1996. "Identification of a ClpC ATPase required for stress tolerance and in vivo survival of *Listeria monocytogenes.*" Mol. Microbiol. 21:977–987.

Safley, S. A., Cluff, C. W., Marshall, N. E. and Ziegler, H. K. 1991. "Role of listeriolysin-O (LLO) in the T lymphocyte response to infection with *Listeria monocytogenes.* Identification of T cell epitopes of LLO," *J. Immunol.* 146:3604–3616.

Schneewind, O., Fowler, A. and Faull, K. F. 1995. "Structure of the cell wall anchor of surface proteins in *Staphylococcus aureus.*" *Science* 168:103–106.

Schoenberg, A., Teufel, P. and Weise, E. 1989. "Serovars of *Listeria monocytogenes* and *L. innocua* from food," *Acta Microbiol. Hung.* 36:249–253.

Schuchat, A., Swaminathan, B. and Broome, C. V. 1991. "Epidemiology of human listeriosis," *Clin. Microbiol. Rev.* 4:169–183.

Seeliger, H. P. R. 1961. *Listeriosis.* Basel: Karger Verlag.

Seeliger, H. P. R. and Hoehne, K. 1979. "Serotypes of *Listeria monocytogenes* and related species," *Methods Microbiol.* 13:31–49.

Sheehan, B., Klarsfeld, A. D., Ebright, R. and Cossart, P. 1996. "A single substitution in the putative helix-turn-helix motif of the pleiotropic activator PrfA attenuates *Listeria monocytogenes* virulence," *Mol. Microbiol.* 20:785–797.

Sheehan, B., Klarsfeld, A. D., Msadek, T. and Cossart, P. 1995. "Differential activation of virulence gene expression by *prfA*, the *Listeria monocytogenes* virulence regulator," *J. Bacteriol.* 177:6469–6476.

Sheehan, B., Kocks, C., Dramsi, S., Gouin, E., Klarsfeld, A. D., Mengaud, J. and Cossart, P. 1994. "Molecular and genetic determinants of the *Listeria monocytogenes* infectious process," *Curr. Top. Microbiol. Immunol.* 192:187–216.

Sijts, A. J., Pilip, I. and Pamer, E. G. 1997. "The *Listeria monocytogenes*-secreted p60 protein is an N-end rule substrate in the cytosol of infected cells. Implications for major histocompatibility complex class I antigen processing of bacterial proteins," *J. Biol. Chem.* 272:19261–19268.

Sijts, A. J., Neisig, A., Neefjes, J. and Pamer, E. G. 1996. "Two *Listeria monocytogenes* CTL epitopes are processed from the same antigen with different efficiencies," *J. Immunol.* 156:683–692.

Smith, G., Helene, M., Jones, S., Johnston, N. C., Portnoy, D. A. and Goldfine, H. 1995. "The two distinct phospholipases of *Listeria monocytogenes* have overlapping roles in escape from a vacuole and cell-to-cell spread," *Infect. Immun.* 63:4231–4237.

Sokolovic, Z., Riedel, J., Wuenscher, M. and Goebel, W. 1993. "Surface-associated, PrfA-regulated proteins of *Listeria monocytogenes* synthesized under stress conditions," *Mol. Microbiol.* 8:219–227.

Theriot, J. A. 1995. "The cell biology of infection by intracellular bacterial pathogens," *Annu. Rev. Cell Dev. Biol.* 11:213–239.

Tilney, L. G. and Portnoy, D. A. 1989. "Actin filaments and the growth, movement and spread of the intracellular bacterial parasite, *Listeria monocytogenes*," *J. Cell Bio.* 109:1597–608.

Uchikawa, K., Sekikawa, J. and Azuma, I. 1986. "Structural study on teichoic acids in cell walls of several serotypes of *Listeria monocytogenes*," *J. Biochem.* 99:315–327.

Ullman, W. W. and Cameron, J. A. 1969. "Immunochemistry of the cell walls of *Listeria monocytogenes*," *J. Bacteriol.* 98:486–493.

Vijh, S. and Pamer, E. G. 1997. "Immunodominant and subdominant CTL responses to *Listeria monocytogenes* infection," *J. Immunol.* 158:3366–3371.

Villanueva, M. S., Sijts, A. J. and Pamer, E. G. 1995. "Listeriolysin is processed efficiently into an MHC class I-associated epitope in *Listeria monocytogenes*-infected cells," *J. Immunol.* 155:5227–5233.

Vines, A., Reeves, M. W., Hunter, S. and Swaminathan, B. 1992. "Restriction fragment length polymorphism in four virulence-associated genes of *Listeria monocytogenes*," *Res. Microbiol.* 1443:281–294.

Wendlinger, G., Loessner, M. J. and Scherer, S. 1996. "Bacteriophage receptors on *Listeria monocytogenes* cells are the N-acetyl-D-glucosamine and rhamnose substituents of teichoic acids or the peptidoglycan itself," *Microbiology* 142:985–992.

Wiedmann, M., Bruce, J. L., Keating, C., Johnson, A. E., McDonough, P. L. and Batt, C. A. 1997. "Ribotypes and virulence gene polymorphisms suggest three distinct *Listeria monocytogenes* lineages with differences in pathogenic potential," *Infect. Immun.* 65:2707–2716.

Wuenscher, M. D., Koehler, S., Bubert, A., Gerike, U. and Goebel, W. 1993. "The *tap* gene of *Listeria monocytogenes* is essential for cell viability, and its gene product, p60, has bacteriolytic activity," *J. Bacteriol.* 175:3491–3501.

Zheng, W. and Kathariou, S. 1995. ''Differentiation of epidemic-associated strains of *Listeria monocytogenes* by restriction fragment length polymorphism in a gene region essential for growth at low temperatures, (4°C),'' *Appl. Environ. Microbiol.* 61:4310–4314.

Zheng, W. and Kathariou, S. 1997. ''Host-mediated modification of *Sau*3AI restriction in *Listeria monocytogenes:* Prevalence in epidemic-associated strains,'' *Appl. Environ. Microbiol.* 63:3085–3089.

PART III

FUNGAL AND MARINE TOXINS

CHAPTER 10

Aflatoxins: Biological Significance and Regulation of Biosynthesis

JEFFREY W. CARY
DEEPAK BHATNAGAR
JOHN E. LINZ

1. INTRODUCTION

AFLATOXINS are extremely potent naturally occurring carcinogens that are found in feed for livestock as well as in food for human consumption. The two fungi, *Aspergillus flavus* and *A. parasiticus,* that most commonly produce aflatoxins do so on a number of substrates; but preharvest aflatoxin contamination is most serious on corn, peanuts, cottonseed, and tree nuts. *Aspergillus flavus* appears to be the primary aflatoxin-producing fungus on these commodities, although *A. parasiticus* also occurs frequently on peanuts (Payne, 1992). Both fungi produce a family of related aflatoxins (Bhatnagar et al., 1991a); the most prevalent aflatoxins produced by *Aspergillus flavus* are B_1 and B_2 (Figure 10.1). In addition to B_1 and B_2, *A. parasiticus* produces two additional aflatoxins, G_1 and G_2. B_1 is the most carcinogenic of the aflatoxins as well as the most abundant, and thus receives the most attention in mammalian toxicology (Bhatnagar et al., 1992).

Aflatoxins have received increased attention from the food industry and the general public for two main reasons. First, members of the aflatoxin family (aflatoxin B_1) are not only extremely toxic to animals and humans, but also are the most carcinogenic of all known natural compounds in several animal models (for reviews, see various chapters in Eaton and Groopman, 1994). In fact, aflatoxin B_1 is second in carcinogenicity only to the most carcinogenic family of chemicals known, the synthetically derived polychlorinated biphenyls (PCBs). Aflatoxin B_1 is a hepatocarcinogen in rats and trout, and can induce carcinomas when ingested at rates below 1 μg per kg body weight (Robens and Richard, 1992). Second, aflatoxin contamination has received

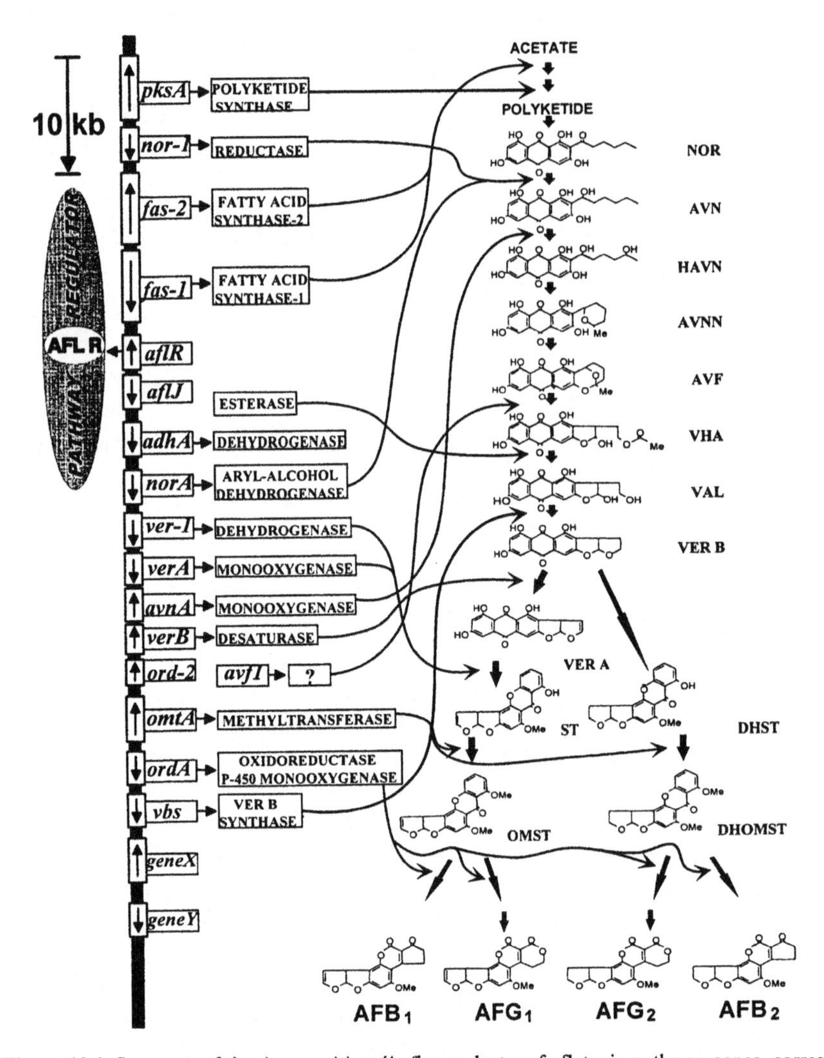

Figure 10.1 Summary of the *A. parasiticus/A. flavus* cluster of aflatoxin pathway genes, corresponding biosynthetic enzymes, and precursors involved in the synthesis of aflatoxins B_1, B_2, G_1, and G_2. The regulatory gene, *aflR* encoding a pathway regulatory protein (AFLR), controls the expression of the structural genes at the level of transcription. The vertical bar on the left represents at least a 75 kb AF pathway gene cluster with identified gene open reading frames (ORFs) shown in the open boxes. Arrows inside the open boxes denote the direction of transcription of these genes. The genotypic designation for the individual genes are shown in italics next to the open boxes. Arrows indicate the relationships from the genes to the enzymes they encode (if known); from the enzymes to the bioconversion steps they are involved in; and from the precursors to the products in the biosynthetic pathway. Abbreviations for the pathway precursors are explained in section 1.3, on chemistry. The *A. nidulans* ST pathway homologs of AF pathway genes are denoted in parentheses adjacent to the *A. parasiticus/flavus* gene designations in section 2.2, on genes involved in AF/ST biosynthesis.

significant publicity, since the incidence of these compounds in food and feed is ubiquitous, having occurred throughout the United States as well as the rest of the world (Jelinek et al., 1989; Cotty et al., 1994). Aflatoxin content in foods and feeds is, therefore, regulated in many countries (van Egmond, 1989; Stoloff et al., 1991). Of the countries that attach a numerical value to their tolerance, the difference between the limits vary significantly (van Egmond, 1989). A guideline of 20 parts aflatoxin per billion parts of food or feed substrate (ppb) is the maximum allowable limit imposed by the U.S. Food and Drug Administration for interstate shipment of foods and feeds. European countries are expected to introduce more stringent guidelines that may restrict aflatoxin levels in imported foods (3 to 5 ppb). Since approximately 25% of the world's food crops are affected by mycotoxin contamination every year (Mannon and Johnson, 1985; Charmley et al.,) 1995), serious food safety and economic implications for the entire agriculture industry are the result. As an example, a detailed analysis of the economic effects of the presence of aflatoxins in corn in Southeastern U.S. compiled by Nichols (1983) demonstrated a loss of $237 million to just the farming community in that region. A broad spectrum of financial losses of crop and animal agriculture, extending through the food chain to the consumer, have been enumerated by Shane (1994).

Domestic growers and food processors are under increased pressure from consumer groups, merchants, and regulatory agencies to eliminate these toxic compounds from food and feed. Significant emphasis has focused on preharvest control of aflatoxin contamination, because that is when the fungi first colonize host tissues (Cleveland and Bhatnagar, 1992; Cleveland et al., 1997). Several agronomic practices can reduce preharvest aflatoxin contamination of certain crops. These include the use of pesticides (fungicides and insecticides), altered cultural practices (such as irrigation), and the use of resistant varieties. However, such procedures have demonstrated only a limited potential for reducing aflatoxin levels in the field, especially in years of drought when environmental conditions favor the contamination process. Therefore, there is an increasing need to develop new technology to reduce and eventually to eliminate preharvest aflatoxin contamination (Bhatnagar et al., 1995). Broad areas are being studied for control of aflatoxin contamination. These include: (a) fundamental molecular and biological mechanisms that regulate the biosynthesis of aflatoxin by the fungi, and the ecological and biological factors that influence toxin production in the field; and (b) biochemistry of host-plant resistance to aflatoxin and/or aflatoxigenic fungi. Knowledge in these areas has already aided in significant developments of novel methods to manipulate the chain of events in aflatoxin contamination (Cleveland et al., 1997). In general, procedures for eliminating aflatoxin B_1 would be applicable to controlling B_2, G_1, and G_2 and other members of the aflatoxin family that contaminate food and feed.

1.1. AFLATOXIN TOXICITY AND CARCINOGENICITY

Biological effects of aflatoxins in animals are related to the levels of AFB_1 in the feed and to the animal's susceptibility. The biological effects may occur as: (a) an acute and clinically obvious disease; (b) a chronic, less clinically apparent, impairment of health and productivity; and (c) an impairment of resistance and immune responsiveness that is not clinically apparent as being associated with aflatoxin consumption (Roebuck and Maxuitenko, 1994). The principal target organ in aflatoxicosis is the liver (Cullen and Newberne, 1994). Acute aflatoxin poisoning has been described in poultry, swine, cattle, dogs, fish and several laboratory animals. Perhaps the single, most impressive, aflatoxin-related episode reported in the scientific literature is an acute poisoning in an area in India in 1974 involving some 400 people and resulting in 106 deaths (van Rensburg et al., 1977).

Data related to clinical aflatoxicosis in humans are limited but ample evidence exists for substantial exposure of human populations to aflatoxins in many areas of the world. Extensive studies in a large population of mainland Chinese has confirmed significant exposure to aflatoxin, and a relationship of this exposure to liver cancer seems likely (Cullen and Newberne, 1994). High incidence of hepatocellular carcinoma is recorded in these populations, but the role of aflatoxin in carcinogenesis is usually found to be concomitant with immuno-compromised livers in these populations due to causes such as hepatitis B.

Aflatoxin B_1, as part of its biotransformation in the human system, undergoes oxidative hydroxylation, *O*-demethylation and epoxidation to form aflatoxins M_1, Q_1, P_1 and the 8,9-oxide of AFB_1, respectively (reviewed in Hsieh, 1987; Hsieh and Wong, 1994; Eaton et al., 1994). The 8,9-oxide or AFB_1-epoxide is the active electrophilic form of AFB_1, which may attack nucleophilic nitrogen, oxygen and sulfur, heteroatoms in cellular constituents. The AFB_1-8,9-epoxide reacts with N^7 of guanine bases of DNA via a precovalent intercalation complex between double-stranded DNA and the highly electrophilic, unstable AFB_1-exo-8,9-epoxide isomer (reviewed in Bailey, 1994). It is this AFB_1-DNA adduction that gives aflatoxin its carcinogenic, mutagenic and teratogenic properties, particularly from aflatoxin-specific mutations in the p53 tumor-suppressor gene (reviewed in various chapters in Eaton and Groopman, 1994).

1.2. AFLATOXIN BIOSYNTHESIS

Aflatoxins are polyketide-derived secondary metabolites. Biosynthesis of secondary metabolites in fungi shares a common lipid and protein pool with primary metabolites (Luchese and Harrigan, 1993). The relationship between fatty acids and secondary biosynthesis has been reported in several species of fungi. A close association between aflatoxins and lipid biosynthesis has also been established, and it is known that protein synthesis declines during the aflatoxin-producing phase (idiophase). The mode of action, metabolism, and biosynthesis of the toxins have been extensively studied (see reviews Dutton, 1988; Bhatnagar et al., 1992; Cleveland et al., 1997; Minto and Townsend, 1997, Payne and Brown, 1998). Aflatoxin synthesis has no obvious physiological role in primary growth and metabolism of the organism and therefore is considered to be a ''secondary'' process. As yet, there is no confirmed biological role of aflatoxins in the ecological survival of the fungal organism. However, since aflatoxins are toxic to certain potential competitor microbes in the ecosystem (Detroy et al., 1971), a survival benefit to the producing fungi is implied. It should be noted, however, that aflatoxin per se is a poor antibiotic (Ciegler, 1983; Lillehoj, 1992). Theories have also been proposed about a possible biological role of aflatoxins or related compounds as deterrents to insect feeding activity on fungal overwintering bodies (sclerotia) (Wicklow and Shotwell, 1982; Dowd, 1991). Understanding the regulatory mechanism of toxin biosynthesis could provide information into why these fungi produce aflatoxins.

Elimination of preharvest aflatoxin contamination through the inhibition of toxin biosynthetic or secretory processes responsible for toxin contamination (Bhatnagar et al., 1995; Cleveland et al., 1997) could benefit significantly by additional knowledge about the fundamental molecular and biological mechanisms that regulate the biosynthesis of aflatoxin by the fungus.

1.3. CHEMISTRY

Mutants of *A. parasiticus* and *A. flavus* that are impared in aflatoxin synthesis have been useful in studying the aflatoxin biosynthesis pathway. Bioconversion experiments using blocked mutants, metabolic inhibitors, and radiolabeled precursors have produced an accepted general scheme of prescursors in aflatoxin B_1 biosynthesis: acetate → polyketide → anthraquinones → xanthones → aflatoxins. The first step in formation of most acetate-derived secondary metabolites is chain elongation by polyketide synthesis. Polyketide biosynthesis represents the initial steps in which the organism is committed to the production of secondary metabolic products. Therefore, similar to fatty acid biosynthesis, many of the enzymatic steps in polyketide biosynthesis are essentially irreversible.

In both fatty acid biosynthesis and polyketide biosynthesis, chain elongation begins with the condensation of acetyl CoA and malonyl CoA. This reaction is catalyzed by a transferase activity (the fatty acid and polyketide synthase) fixed into a multienzyme complex in the form of a flexible protein ''arm'' (Minto and Townsend, 1997). In aflatoxin synthesis, the initial decaketide condensate has the most sites available for potential modification and re-arrangement. A schematic for the biosynthesis of aflatoxin B_1 starting from the decaketide condensation product Nor A is given in Figure 10.1. The presumptive condensation product, an anthrone, is first oxidized to an anthraquinone norsolorinic acid.

The generally accepted scheme for aflatoxin biosynthesis is: polyketide precursor → norsolorinic acid (NOR) → averantin (AVN) → 5′-hydroxyaverantin (HAVN) → averufanin (AVNN) → averufin (AVF) → versiconal hemiacetal acetate (VHA) → versiconal (VAL) → versicolorin B (VERB) → versicolorin A (VERA) → demethylsterigmatocystin (DMST) → sterigmatocystin (ST) → *O*-methylsterigmatocystin (OMST) → aflatoxin B_1 (AFB_1) (Figure 10.1) (Yu et al., 1997). VERB was demonstrated to be a precursor of VERA (McGuire et al., 1989; Yabe, et al., 1991a). A branch point in the pathway has been established, following VHA production, leading to different aflatoxin structural forms B_1 and B_2 (Bhatnagar et al., 1991b; Cleveland et al., 1987a; Yabe et al., 1991b; Yabe and Hamasaki, 1988; Yu et al., 1998) according to following scheme: VERB → dihydro-DMST (DHDMST) → dihydro-ST (DHST) → dihydro-OMST (DHOMST) → aflatoxin B_2 (AFB_2).

Aflatoxin B_1 was assumed to be a precursor of AFG_1, and AFB_2 of AFG_2 based mainly on two studies. Maggon and Venkitasubramanian (1973) used [^{14}C] AFB_1 as a substrate for in vitro incubation with a cell-free homogenate of *A. parasiticus* and reported recovery of low levels of ^{14}C label in AFB_2, AFG_1, and AFG_2; Heathcote et al. (1976) reported similar results using an in vitro system with wild-type *A. flavus*. However, using mutant strains blocked in the aflatoxin pathway, Floyd et al. (1987) and Henderberg et al. (1988) did not observe any biotransformation of AFB_1 or AFB_2 into either AFG_1 or AFG_2; their results suggest independent pathways to the B and G aflatoxins. Other studies of in vivo feeding of precursors indicate that AFG_1 is synthesized from ST/OMST and AFG_2 from DHST/DHOMST, namely, the AFB_1 and AFB_2 precursors, respectively (Bhatnagar et al., 1987; Yabe and Hamasaki, 1988; Yu et al., 1998). Aflatoxin B_1 is therefore probably not an intermediate, but rather an end product of the aflatoxin biosynthetic pathway.

1.4. ENZYMOLOGY

The enzymes required for the production of aflatoxins have not been characterized in as much detail as have the enzymes involved in primary metabolism. Several efforts have, however, been made toward developing a cell-free system for identifying and purifying the enzymes involved in aflatoxin biosynthesis (Bhatnagar et al., 1989, 1992; Dutton, 1988; Hsieh et al., 1989; Yabe et al. 1989).

Several specific enzyme activities have been associated with precursor conversions in the aflatoxin pathway (Bhatnagar et al., 1989, 1992; Matsushima, et al., 1994; Cleveland et al., 1987b; Dutton, 1988; Yabe et al., 1989, 1991b) (Figure 10.1). Some of these have been partially purified (Bhatnagar et al., 1991b; Chuturgoon and Dutton, 1991; Hsieh et al., 1989; Yabe et al., 1989, 1991a, b), whereas other enzymes have been purified to homogeneity (Bhatnagar et al., 1988, 1996; Keller et al., 1993b; Lin and Anderson, 1992; McGuire et al., 1996; Kusumoto and Hsieh, 1996). It has been postulated (Bhatnagar et al., 1992; Bennett et al., 1997) that alternate pathways may exist at several steps in aflatoxin synthesis, hence, different enzymes with similar catalytic functions may be isolated from pertinent fungal cells. It has also been demonstrated that independent reactions and different chemical precursors involved in AFB_1 and AFB_2 syntheses are catalyzed by common enzyme systems (Bhatnagar et al., 1991b; Yabe et al., 1988; Keller et al., 1993b; Yu et al., 1998); AFB_1 precursors are, however, the preferred substrates for the relevant enzymes (Bhatnagar, et al., 1991b).

1.5. DETECTION AND ANALYTICAL METHODS OF AFLATOXIN ANALYSES

Development of the detection and quantitation methods for mycotoxins is dependent on their chemical nature, molecular weight and the functional groups in its chemical makeup. Aflatoxins were one of the first mycotoxins to be regulated and a variety of analytical methods for its detection and analyses have been developed. Several of these methods have been reviewed extensively in the last few years (Trucksess and Wood, 1994; Wilson et al., 1998; Holcomb et al.,1992). Use of thin layer chromatography, high performance liquid chromatography, gas chromatography, mass spectrometry, and immunoassays are the methods of choice for visualization and analyses of aflatoxins.

Aflatoxins are easily extracted into water-saturated chloroform, aqueous methanol, acetonitrite, acetone, and other polar solvents. The choice of the solvent depends on the assay used for analysis (Wilson et al., 1998). A comprehensive accumulation of chemical and physical properties of aflatoxins is provided by Cole and Cox (1981). Briefly, these toxins are highly oxygenated, heterocyclic compounds characterized by dihydrodifurano or tetrahydrodifurano moieties fused to a substituted coumarin moiety. Aflatoxins are highly fluorescent under UV light and, therefore, can be easily visualized.

Thin layer chromatography (TLC), also known as flat bed chromatography or planer chromatography, is one of the most widely used separation techniques in aflatoxin analysis (reviewed in Trucksess and Wood, 1994). Many TLC methods for aflatoxins in foods such as corn, peanut butter, cottonseed, milk, meat, and eggs are included in the compendium *Official Methods of Analysis* (AOAC, 1990). Aflatoxin analysis using liquid chromatography (HPLC) with fluorescence detection is similar to TLC in many respects. All of the common aflatoxins can be separated and quantified using HPLC (reviewed in Wilson et al., 1998). TLC and HPLC methods for determining aflatoxins in food are laborious and time consuming. Often these techniques require knowledge and experience of chromatographic techniques to solve separation and interference problems. With advances in hybridoma technology and progress in the medical diagnostic area, rapid progress in the development of immunoassays for aflatoxins has been made in the last 10 years (Wilson et al., 1998; Chu, 1994). These tests are based on the affinities of the monoclonal or polyclonal antibodies for aflatoxins. In the past few years, several immunoassay kits (ELISA kits) for aflatoxins have been marketed under various trade names; these tests can identify and measure aflatoxins in food in less than 10 minutes.

2. MOLECULAR BIOLOGY OF AFLATOXIN BIOSYNTHESIS

2.1. BACKGROUND

Over the last seven years, isolation of aflatoxin (AF) biosynthetic genes, elucidation of their function, and the molecular mechanisms regulating their expression were made possible, in large part, through the employment of a variety of molecular biological techniques for gene cloning and analysis. However, many of the successes reported to date on the use of gene cloning can be directly traced to the availability of a large body of fundamental genetic, chemical and biochemical observations. Bioconversion studies using blocked mutants, metabolic inhibitors, and radiolabeled aflatoxin precursors led to a better understanding of the order and enzymatic steps involved in the synthesis of aflatoxin and its precursors (reviewed in Bhatnagar et al., 1992). Perhaps the most critical research that opened the door for cloning of genes involved in aflatoxin biosynthesis has been the isolation and characterization of several mutants of *A. flavus* and *A. parasiticus* blocked in aflatoxin biosynthesis (reviewed in Bennett and Papa, 1988). This coupled with the development of transformation systems for the introduction of DNA into *A. flavus* (Woloshuk et al., 1989) and *A. parasiticus* (Horng et al., 1990; Skory et al., 1990) enabled the use of genetic complementation of *A. parasiticus* mutants to identify the first genes involved in aflatoxin biosynthesis, the *nor-1* gene (Chang et al., 1992) and the *ver-1* gene (Skory et al., 1992). Shortly thereafter, a gene, *omtA*, was cloned by reverse genetics (Yu et al., 1993). These discoveries also led to the finding that the genes involved in AF biosynthesis in *A. parasiticus* and *A. flavus* (Trail et al., 1995b; Yu et al., 1995a) were clustered within an approximate 14 kb region of their respective genomes greatly accelerated our understanding of this biosynthetic pathway. Hybridization studies using radioactive probes for the *A. flavus nor-1* and *aflR* genes with electrophoretically separated *A. flavus* 656–2 chromosomes showed that these two genes and therefore the AF gene cluster reside on a 4.9 Mb chromosome band (Foutz et al., 1995). Electrophoretic karyotyping has also been used in an attempt to identify the chromosome on which the *A. parasiticus* AF gene cluster resides

(Wright et al., unpublished data). However, due to resolution difficulties (Keller et al., 1992) the pathway cluster could only be assigned to a chromosomal band(s) of greater than 6 Mb.

In addition, many advances in our understanding of the molecular biology of AF biosynthesis can be traced to information gained from the study of sterigmatocystin (ST) biosynthesis in *Aspergillus nidulans,* a distant relative of *A. flavus* and *A. parasiticus. A. nidulans* as well as other deuteromycetes and ascomycetes produce ST, which is the penultimate precursor to AFB1. The enzymatic steps leading to ST in the AF pathway are identical to those involved in ST production in *A. nidulans* (Figure 10.1). The genes involved in ST biosynthesis were also shown to be clustered within an 60 kb region of the *A. nidulans* genome (Brown et al., 1996). A total of 25 co-regulated transcripts were identified in the ST gene cluster and many of the genes have been cloned and their function determined. Transcript mapping, DNA sequence analysis, and functional analysis of genes within the AF and ST clusters have shown that with the exception of the *A. nidulans stcI* (encodes a putative esterase), all the genes necessary for ST production have a homolog present in the AF pathway (Brown et al., 1996; Yu et al., 1995a; Trail et al., 1995a). DNA sequence analysis has shown greater than 95% identity at the nucleotide level and greater than 97% identity at the amino acid level between aflatoxin pathway genes in *A. parasiticus* and *A. flavus* (Chang et al., 1993; Yu et al., 1995b), conservation is somewhat lower (80% or less) between gene homologs of *A. nidulans* and *A. flavus/parasiticus* (Yu et al., 1996). In addition, the organization of the ST genes within the cluster differs somewhat from that of the AF biosynthetic gene cluster. Both *A. flavus* and *A. parasiticus* lack a known sexual stage. This has limited their use in genetic analyses of AF production. *A. nidulans,* however, has a sexual stage, which has already lent itself to detailed genetic analyses of regulatory mechanisms controlling ST/AF production as well as the relationship between toxin production and fungal development (see section on regulation of AF/ST production).

Information presented here on the genes involved in AF/ST biosynthesis will be condensed and emphasis will be on novel aspects of the pathway. Particular emphasis has been placed on the regulation of AF/ST biosynthesis (following section), which we feel has the most potential as a target for studies on the control of aflatoxin contamination of food and feed. More detailed information on the genes involved in AF/ST biosynthesis can be obtained from a number of excellent review articles (Trail et al., 1995a; Woloshuk and Prieto, 1998; Minto and Townsend, 1997, Payne and Brown, 1998).

2.2. GENES INVOLVED IN AF/ST BIOSYNTHESIS

2.2.1. Polyketide to Averantin

The C20 polyketide backbone of AF/ST is derived from the head-to-tail condensation of one acetyl-CoA and two malonyl-CoA molecules via a fatty acid synthetase (FAS), which produces a 6-carbon hexanoate starter unit that is subsequently extended by a polyketide synthase (PKS) to produce a noranthrone (Minton and Townsend, 1997). The noranthrone then undergoes oxidative conversion by a hypothetical oxidase to form NA, the first stable intermediate of AF/ST biosynthesis. Two genes, *fas-1A* (*stcJ*) and *fas-2A* (*stcK*), encoding fatty acid synthetases were identified by nucleotide sequence and gene disruption analyses (Mahanti et al., 1996). The sequence analyses have shown that the FAS-1 and FAS-2 proteins have a high degree of identity to the yeast FAS α and β subunits, respectively, and in combination have all of the catalytic domains required for fatty acid synthesis (Mahanti et al., 1996, unpublished results). The products of these two genes in concert with the polyketide synthase (PKS) encoded by the *pksA* (*stcA*) gene (Chang et al., 1995b; Trail et al., 1995a) have been shown to be required for the synthesis of the noranthrone precursor of norsolorinic acid (Watanabe et al., 1996). Feng and Leonard (1995) also identified the *A. parasiticus pksL1* gene (the equivalent of *pksA*) by PCR of DNAs taken from a pool of 19 clones that were differentially expressed under conditions conducive to aflatoxin biosynthesis (Feng et al., 1992) using degenerate oligonucleotide primers based on conserved sequences of known PKS genes. Sequence analysis of the PKS-A revealed that it contains four recognizable domains common to other known PKS proteins: an acylcarrier protein, β-ketoacyl synthase, acyltransferase, and thioesterase. However, the reduction/dehydration domains characteristic of FAS proteins were not present. A large body of genetic and biochemical evidence indicates the FAS-1 and FAS-2 enzymes synthesize a 6-carbon hexanoate primer molecule to which the PKS-A subsequently condenses malonyl units to initiate formation of NA (Brobst and Townsend, 1994; Watanabe et al., 1996). Unlike typical type 1 PKS enzymes, the efficient synthesis of the noranthrone appears to require a close physical association of both FASs and the PKS (Watanabe et al., 1996).

The *nor-1* (*stcE*) gene (Chang et al., 1992) was cloned by complementation of an *A. parasiticus* "color" mutant, which accumulated the readily identifiable, pigmented aflatoxin pathway intermediate NOR (brick-red). The predicted amino acid sequence of 29 kDa *nor-1* gene product demonstrated significant identity (23%) with several NAD(P)-binding dehydrogenase/reductase enzymes (Trail et al., 1994). This is consistent with the proposed role of the *nor-1* gene product as a ketoreductase involved in the conversion of NOR to AVN. Inactivation of the *nor-1* gene resulted in accumulation of NA: however, small quantities of aflatoxin were still being produced (Trail et al., 1994). Bhatnagar et al. (1992) previously reported that the NOR accumulating mutant of *A. parasiticus* accumulated significant levels of aflatoxins. This supported the theory that there may be an alternate route(s) or enzyme(s) that can be used to synthesize AVF from NOR (Yabe et al., 1993; Cary et al.,

1996). The *norA* (*stcV*) gene was isolated after screening an *A. parasiticus* cDNA library with monoclonal antibody raised to a purified 43 kDa enzyme (Lee et al., 1995) that demonstrated NOR reductase activity (Cary et al., 1996; Bhatnagar et al., 1996). Though proposed to play a role in the conversion of NOR to AVN, which would explain the observed ''leakiness'' of *nor-1* mutants, the exact function of the *norA* gene product in aflatoxin biosynthesis has yet to be determined. The deduced 43 kDa NORA protein demonstrated 43% identity at the amino acid level with an aryl-alcohol dehydrogenase gene from *Phanerochaete chrysosporium.*

2.2.2. Averantin to Versicolorin A

Transcript mapping of the *A. parasiticus* gene cluster identified a gene transcript localized between the *ver-1* and *omtA* genes. DNA sequencing of this region identified a gene, designated *avnA* (stcF), capable of encoding a 56.3 kDa protein that when disrupted in wild-type *A. parasiticus* resulted in the accumulation of AVN (Yu et al., 1997). This indicated that *avnA* is involved in an early step in aflatoxin biosynthesis, namely the conversion of AVN to HAVN. A search of nucleotide sequence databases indicated that *avnA* shared a high degree of similarity to cytochrome P-450-type monooxygenase genes. Transformation of a mutant of *A. flavus* lacking the entire aflatoxin gene cluster, with two overlapping *A. flavus* cosmids (5E6 and 8B9 DNA spanned from the *pksA* to *omtA* gene), resulted in transformants that accumulated AVF (Prieto et al., 1996). Transformation of the AVF-accumulating transformant with an additional cosmid whose DNA spanned from *ver-1* to well past the *omtA* gene resulted in restoration of AFB1 biosynthesis. Deletion analysis of cosmid 13B9 localized the gene involved in the conversion of AVF to VHA, designated *avf1,* to a 7 kb region of the fungal DNA. DNA sequence analysis of this region of cosmid 13B9 has identified at least three potential coding regions, two of which show similarity to the *A. nidulans stcB* and *stcW* genes (Prieto and Woloshuk, unpublished data) and geneX and geneY of *A. parasiticus* (Yu et al., unpublished data). The function of these two genes in ST/AF biosynthesis has yet to be determined.

The *A. parasiticus* versicolorin B synthase (*vbs*) gene (*stcN*) was cloned (Silva et al., 1996) using reverse transcriptase-mediated PCR amplification of 48- and 60-h mRNA and degenerate oligonucleotide probes designed from amino acid sequences of peptide fragments from the purified 78 kDa enzyme (Silva and Townsend, 1996). The *vbs* gene product catalyzes the dehydrative, side-chain cyclization of VHA to the bisfuran ring system of VB, a critical step in the pathway as the metabolically activated epoxide form of the bisfuran ring is responsible for the mutagenic nature of aflatoxin. The deduced amino acid sequence of the 70.3 kDa *vbs* gene product showed significant identity to flavin-dependent oxidoreductases. The *vbs* coding sequence lacked the region presumed to be responsible for FAD binding and no bound flavin chromophore was detected in the native protein suggesting its evolution from a flavin-binding ancestor. Following production of VB, studies have shown that the AF pathway branches, with VB being converted to either dihydrosterigmatocystin (DHST) or desaturated to produce VA, which is subsequently converted to ST. The *A. nidulans stcL* gene has been shown to encode a predicted cytochrome P-450 monoxygenase that is responsible for the desaturation of VB to VA (Kelkar et al., 1997). The *A. parasiticus* homolog of *stcL*, designated *verB*, has been sequenced and found to have 80% amino acid identity with *stcL* (Bhatnagar et al., unpublished).

2.2.3. Versicolorin A to AFB1

The *ver-1* (*stcU*) gene was also cloned by complementation of an *A. parasiticus* "color" mutant that accumulated the readily identifiable, pigmented aflatoxin pathway intermediate VA (yellow) (Skory et al., 1992). Southern hybridization analysis showed that there were two copies of the *ver-1* gene (designated *ver-1*A and *ver-1*B) in *A. parasiticus* SU-1 as well as several other strains tested (Liang et al., 1996). Further investigation identified a 12 kb duplication of the AF gene cluster that spanned from *aflR* to *ver-1*. To date, no duplication of the AF cluster has been noted in *A. flavus*. This duplication of AF pathway genes may explain why nearly all isolates of *A. parasiticus* produce AF while many strains of *A. flavus* (no duplication) do not produce AF or lose the ability to produce AF during culture. It has been hypothesized that this redundancy in the pathway, especially with respect to the *aflR* regulatory gene, may allow the fungus to overcome the effects of mutations that would normally eliminate AF production. DNA sequencing determined that *ver-1*B had 95% identity with *ver1*A but it contained a translational stop codon resulting in a truncated, nonfunctional gene product (Liang et al., 1996). The deduced amino acid sequence of the *ver-1*A gene displayed 66% identity with a polyhydroxynaphthalene reductase involved in melanin biosynthesis in *Magnaportha grisea*. This finding supported the theory that the *ver-1* gene product is responsible for the deoxygenation of VA to 6-deoxy VA. Inactivation of the *ver-1*A gene in a toxigenic *A. parasiticus* strain resulted in accumulation of VA (Liang et al., 1996).

The first genetic evidence that more than one type of enzyme activity is needed for the conversion of VA to ST was demonstrated by Keller et al. (1995) when they identified the *A. nidulans* gene, *stcS* (*verA* in *A. flavus/parasiticus*), that, when disrupted, caused the fungus to accumulate VA. Analysis of the gene showed that it encoded a cytochrome P-450 monooxygenase capable of catalyzing one of the proposed oxidative steps required for the conversion of VA to ST. The *A. nidulans stcP* gene is predicted to encode an O-methyl transferase involved in the conversion of DMST to ST based upon studies in which *stcP* disruptants accumulated DMST (Kelkar et al., 1996). A 40 kDa O-methyltransferase (OMT) enzyme from *A. parasiticus,* which functions in the conversion of ST to OMST and DHST to DHOMST, was purified (Keller et al., 1993b). Polyclonal antibody was raised to the OMT enzyme and it was subsequently used to screen an *A. parasiticus* cDNA library (Yu et al., 1993). A full-length cDNA clone, *omt-1* (redesignated *omtA*), whose expressed protein reacted with the antibody, was isolated and nucleotide and deduced amino acid sequence data revealed a proposed S-adenosylmethionine (SAM)-binding site motif found in other SAM-dependent methyltransferases. The *A. flavus omtA* homolog was sequenced and shown to be 97% identical to the *A. parasiticus* gene while no *omtA* homolog appears to be present in *A. nidulans* (Yu et al., 1995). Other methyltransferases have been purified in *A. parasiticus* and have been shown to function in the conversion of DMST to ST and DHDMST to DHST, though the gene(s) responsible for encoding this enzyme(s) has yet to be identified (Bhatnagar et al., 1988; Yabe et al., 1998). However, a gene homologous to *stcP* is located on the aflatoxin biosynthetic gene cluster in *A. parasiticus* (Yu et al., unpublished). Metabolite feeding studies were carried out on mycelia of the *A. flavus* mutant which had been transformed with overlapping cosmids containing various regions of the aflatoxin gene cluster (see above). Only transformants carrying a cosmid (8B9) that contained the *aflR* regulatory gene (discussed later) and downstream DNA regions were able to convert exogenously supplied OMST to aflatoxin. Further analysis of the cosmid DNA identified a gene, *ord1,* within a 3.3 kb region of the DNA that demonstrated significant identity to P450-type monooxygenases and has been assigned to a new cytochrome P450 family, CYP64. Confirmation of the function of the *ord1* gene was obtained by heterologous expression of the *ord1* cDNA under control of the *S. cerevisiae gal1* promoter. This resulted in the ability of the yeast to convert exogenously supplied OMST to aflatoxin B1 (Prieto and Woloshuk, 1997). The *A. parasiticus ord1* homolog, *ordA,* demonstrated 97% identity at the amino acid level to *ord1,* with both genes located adjacent to the *omtA* gene (Yu et al., 1998). As in *A. flavus,* galactose-induced expression of the *ordA* gene in yeast allowed for the conversion of exogenously supplied OMST to AFB1 (Yu et al., 1998). Additionally, comparison of the *ordA* gene from an OMST-accumulating mutant of *A. parasiticus* with wild-type *ordA* noted nucleotide differences that resulted in three amino acid changes within the coding region. Feeding studies were conducted in yeast, which were transformed with a wild-type *ordA* gene that had been mutagenized to change His-400 to Leu-400. This mutation eliminated the monooxygenase activity completely while an Ala-143 to Ser-143 change reduced but did not eliminate activity. Complementation of the OMST-accumulating *A. parasiticus* strain by transformation with the either the wild-type *A. parasiticus ordA* gene or *A. flavus ord1* gene restored the mutants' ability to convert OMST and DHOMST to AFB1/AFB2 and

AFG1/AFG2. The fact that the *A. flavus ord1* gene product is able to restore production of AFG1/AFG2 in *A. parasiticus* but *A. flavus* strains have only been found to produce AFB1/AFB2 indicates that an additional enzyme(s) is required for production of AFG1/AFG2.

2.2.4. Regulatory Genes

The *aflR* gene from *A. flavus* and *A. parasiticus* was found to be a positive regulator of transcription of aflatoxin biosynthetic genes (see Regulation of AF/ST Biosynthesis). The *A. flavus aflR* gene was identified by complementation of a mutant that did not produce AF or convert exogenously supplied pathway precursors to AF (Payne et al., 1993; Woloshuk et al., 1994). The *A. parasiticus aflR* gene was identified from observations that pathway intermediates were being overproduced following transformation of an *O*-methylsterigmatocystin (OMST)-accumulating *A. parasiticus* strain with a cosmid DNA that harbored the *aflR* gene region (Chang et al., 1993; Chang et al., 1995c). The *A. nidulans aflR* homolog was isolated by utilizing a single degenerate oligonucleotide based on *A. flavus aflR* and yeast GAL4 DNA-binding domain sequence data as a sequencing primer used with plasmid templates harboring ST cluster DNA (Yu et al., 1996). While the predicted *A. flavus* and *A. parasiticus* AFLR amino acid sequences showed greater than 95% identity, they shared only 31% identity to the *A. nidulans aflR* (Yu et al., 1996). Disruption of the *A. nidulans aflR* gene resulted in loss of expression of ST pathway genes and therefore ST production.

The *aflJ* gene resides in the cluster adjacent to the pathway regulatory gene, *aflR*, and the two genes are divergently transcribed (Meyers et al., 1998). Disruption of *aflJ* in *A. flavus* results in failure to produce aflatoxins and to convert exogenously added pathway intermediates norsolorinic acid, sterigmatocystin, and *O*-methylsterigmatocystin to aflatoxin even though *pksA, nor-1, ver-1,* and *omtA* transcripts were detected in the disrupted strain. Therefore, the disruption of *aflJ* does not affect transcription of these genes and does not appear to have regulatory function similar to that of *aflR*. Sequence analysis of *aflJ* and its putative polypeptide, AflJ, did not reveal any enzymatic motifs or significant similarities to proteins of known function. The putative peptide does contain three regions predicted to be membrane-spanning domains and a Microbodies C-terminal targeting signal (CMTS).

2.3. IDENTIFICATION OF AF/ST GENES IN OTHER ASPERGILLI

The three fungal species known to produce AF, *A. flavus, A. parasiticus,* and *A. nomius,* have similar morphological characteristics and are members of *Aspergillus* section *Flavi.* However, within this section there are a number of non-aflatoxigenic species that are also morphologically similar to the AF-producing members. Two of the most economically important species of the Flavi section are *A. oryzae* and *A. sojae,* both or which are ''koji molds'' used in Asian food and beverage fermentations. The taxanomic status of these four species has been the subject of much debate, especially with regard to regulatory concerns with respect to food production strains (Kurtzman et al., 1986; Barbesgaard et al., 1992; Chang et al., 1995a). A comprehensive screen of all known industrial isolates used in the production of foods and beverages for the presence of AF genes would allow for identification of strains with definitive lesions that would eliminate the potential for AF production. Woloshuk et al. (1994) and Chang et al. (1995a) identified non-aflatoxigenic isolates of *A. sojae* and *A. oryzae* whose DNA hybridized with *A. flavus/parasiticus aflR* gene probes. Klich et al. (1995) identified non-aflatoxigenic isolates of *A. sojae* and *A. oryzae* whose DNA hybridized with an *A. parasiticus aflR* gene probe and an *omtA* gene probe. Three strains of non-aflatoxigenic *A. tamarii* (also a Flavi section ''koji mold'') did not show hybridization to either of the probes. However, Goto et al. (1996) did report on the isolation of a strain of *A. tamarii* that produced both AF and cyclopiazonic acid. Further Northern blot analyses of RNA from three *A. sojae* strains found that only two *A. sojae* strains demonstrated hybridization with the *aflR* and *fas-1A* gene probes but not with five other AF pathway gene probes (Klich et al., 1997). Geiser et al. (1998) studied phylogenetic relationships between 31 Australian isolates of *A. flavus* and 5 strains of *A. oryzae* utilizing PAUP analysis software of single restriction site polymorphisms present in 11 protein encoding genes. Results showed that the *A. flavus* isolates could be placed into two distinct clades (groups I and II), which indicated a history of recombination in *A. flavus.* In addition, the *A. oryzae* were very similar at the gene level to group I isolates even though there are obvious morphological differences and lack of AF production. They concluded that *A. oryzae* is a species that evolved by domestication from group I. Further experimentation is needed to determine if atoxigenic strains exhibiting homology with AF gene probes do not produce AF due to lack of other genes required for AF biosynthesis (that were not examined) or possibly lack of transcription of these genes. Recently, Klich et al. (unpublished data) showed that DNA of an unusual, aflatoxigenic strain of *A. tamarii* hybridized with an *A. parasiticus aflR* probe but not with an *A. nidulans aflR* probe. In addition, DNA from a AF-producing strain of *A. ochraceoroseus* hybridized with the *A. nidulans aflR* probe but not the *A. parasiticus aflR* gene probe. These results indicate that through evolution there has been some degree of dissemination and divergence from the original AF/ST progenitor *Aspergillus* spp. Further study of the molecular biology of the AF/ST pathway genes in these strains may yield some insight as to methods for more accurate phylogenetic classification of Aspergilli as well as possible molecular biological approaches to better control AF contamination of foods and feeds.

3. REGULATION OF AF/ST SYNTHESIS IN *ASPERGILLUS*

3.1. BACKGROUND

The polyketides are a large and diverse family of secondary metabolites that are produced primarily by actinomycetes, fungi and higher plants but are also synthesized in other organisms including animals (reviewed in Hopwood and Khosia, 1992). Regulation of the synthesis of these secondary metabolites is distinct from regulation of primary metabolism although synthesis of AF and ST, like other secondary metabolites, relies on primary metabolism to provide energy, enzyme cofactors (NADPH), and building blocks (acetyl CoA, malonyl-CoA, S-adenosyl methionine; SAM). The effect of primary metabolism on the biosynthesis of AF has been reviewed by Luchese and Harrigan (1993).

In liquid culture media that induce AF synthesis, *A. parasiticus* and *A. flavus* produce AF at maximum rates during a transition from exponential growth to stationary phase when growth has slowed or ceased (reviewed in Trail et al., 1995a). Using transcription and translation inhibitors, Buchanan (Abdollahi and Buchanan, 1981a, b; Buchanan et al., 1987) demonstrated that *de novo* RNA and protein synthesis is required for AF production. Other studies demonstrated that the activities of four of the enzymes involved in the AF biosynthetic pathway were not detected until just prior to appearance of AF (Anderson and Green, 1994; Chuturgoon et al., 1990; Cleveland and Bhatnagar, 1990). Polyclonal antibodies specific to two enzymes involved in AF synthesis, *Nor-1* and *Ver-1*, confirmed that these proteins begin to accumulate late in active growth and reach high levels during the time when maximum rates of AF synthesis are observed (Liang et al., 1997; Zhou, 1997). In agreement with these data, the *ver-1, nor-1,* and *omtA* RNA transcripts also accumulated most rapidly during the transition between active growth and stationary phase (Skory et al., 1993; Yu et al., 1993). RNA transcripts from the *aflR* gene, proposed to encode a key regulatory protein, and transcripts derived from other genes involved in AF synthesis (*pks*A, *fas*-1A) accumulated in a similar pattern (Payne et al., 1993; Trail et al., 1995b). Together, the data suggest that AF genes are not expressed at measurable levels during exponential growth but are induced in the transition between active growth and stationary phase, just before toxin synthesis occurs. It also appears that most if not all AF genes are coordinately regulated, possibly by a common set of regulatory factors.

3.2. FACTORS THAT AFFECT AF/ST SYNTHESIS

A variety of factors influence the level of AF and ST produced by *Aspergillus* in laboratory culture and in the field including nutrients, environmental factors, and the interaction between the fungus and the host plant (reviewed in Cotty et al., 1994; Payne and Brown, 1998; Davis and Diener, 1986; Gourama and Bullerman, 1995; Ellis et al., 1991; Dutton, 1988; Zaika and Buchanan, 1987). This section will review the regulation of AF and ST synthesis with emphasis on those factors that appear to directly or indirectly influence the expression of the genes involved in toxin synthesis. It is important to emphasize that the regulatory influences summarized below were analyzed by different laboratories using different strains of fungus under different growth conditions. Therefore the interpretations of the data provided must be viewed with this in mind.

3.2.1. Nutritional Factors

3.2.1.1. Influence of Growth Media on AF and ST Synthesis

The following simple experiment is useful to illustrate the effect of media composition on levels of AF produced in culture. *A. parasiticus* was grown in batch fermentation (29°C, shaking at 150 RPM, 100 mL media in a 250 mL Erlenmeyer flask, 2×10^4 conidiospores/mL, in the dark, 5 glass beads—3–4 mm dia) to measure levels of AF (ELISA), *nor-1* and *ver-1* transcripts (Northern analysis) and proteins (Western analysis), and growth (dry weight) at 24, 48, and 72 h after inoculation of the growth medium (Miller and Rarick, unpublished data). Under these growth conditions, AF synthesis normally initiates after 24 h and reaches maximum rates between 36 and 60 h during a transition from active growth to stationary phase (Skory et al., 1993; Trail et al., 1995a). YES, a rich media containing yeast extract and sucrose, supported high levels of AF synthesis (approximately 6,000 ng/ml/g) measured in the growth medium. GMS (glucose minimal salts), a defined media containing glucose as the carbon source, supported medium levels (600 ng/ml/g). Both media resulted in accumulation of *nor-1* and *ver-1* transcripts and proteins although the levels of each of these were several fold higher in YES consistent with differences in toxin levels. However, dry weight was approximately 3–4 fold greater in YES than in GMS, which suggested that growth rate may influence toxin synthesis. PMS (peptone minimal salts), which contains peptone as the carbon source, supported only basal levels of toxin synthesis (approximately 50 ng/ml/g) while AF gene transcripts, and protein accumulation were not detectable even though growth (as measured by dry weight) was approximately the same in GMS and PMS. The data suggested that components of the growth media and growth rate can influence levels of toxin synthesis.

3.2.1.2. Glucose Induction (the Glucose Effect)

Buchanan developed the nutritional shift assay (sequential culture replacement technique; Abdollahi and Buchanan, 1981a, b; Buchanan and Lewis,

1984; Buchanan, et al., 1987) to minimize the potential influence of growth rate on AF synthesis. In one version of the nutritional shift assay, the fungus is grown to stationary phase (72 h) in YES, the mycelium or homogenized with a blender and transferred to an AF noninducing medium for 24 h (PMS; basal levels of AF). The resulting mycelia were washed and then shifted to fresh PMS (control) or to a defined media containing an alternative carbon source. In this assay procedure, a variety of carbohydrates including glucose, sucrose, fructose, and sorbose stimulated AF synthesis while other carbon sources such as lactose, galactose, pyruvate, lactate, and acetate did not support toxin synthesis. GMS, for example, stimulated AF more than 100 fold within 24 to 40 h after nutritional shift. Glucose added to PMS after the shift also stimulated AF synthesis to the same extent, which suggested that PMS was not inhibitory. In related experiments, it was determined that glucose induced certain enzymes in the glycolytic pathway and repressed enzymes in the hexose monophosphate shunt (HMS) and the tricarboxylic acid cycle (TCA) (Buchanan and Lewis, 1984). Growth on glucose reportedly resulted in inactivation of mitochondria, potentially reducing the ability of cells to oxidize acetate (Buchanan et al., 1987). It was proposed that the net ''glucose effect'' was to shunt acetate toward synthesis of polyketides and away from oxidation via the TCA cycle or use in fatty acid synthesis. In support of this notion, cerulenin, a fatty acid synthesis inhibitor, stimulated AF production (Fanelli et al., 1983a). One simple interpretation of these data is that cerulenin prevented fatty acid synthesis, shunting acetate toward polyketide synthesis. However, since the fatty acid content and composition was apparently unchanged in cells treated with cerulenin as compared to control, the authors proposed that the epoxide structure of this lipid molecule stimulated AF synthesis by an alternative mechanism (see ''the lipoxygenase theory''). The situation seems more complex than the simple availability of acetate because addition of acetate to the medium after nutritional shift failed to support significant levels of AF synthesis (Buchanan and Stahl, 1984).

Northern hybridization analysis (*nor-1* and *ver-1* probes) of RNA samples isolated from *A. parasiticus* grown in a modified nutritional shift assay demonstrated that the net effect of ''glucose induction'' is to stimulate AF gene expression at the level of transcript accumulation (Skory et al., 1993; Liang et al., 1997; Zhou, 1997). To determine if AF gene expression was regulated at the transcriptional level, GUS (*uidA*; encodes β-glucuronidase; Jefferson, 1987) reporter constructs were developed by fusing the *ver-1* A promoter to the GUS gene (Liang et al., 1997; Flaherty et al., 1995). Using these reporter constructs, *ver-1* A promoter activity can be measured indirectly by measuring the activity of the GUS enzyme in cell extracts. Northern hybridization analysis and a GUS activity assay were used to analyze transformants grown in batch fermentation in rich media (YES) or in a nutritional shift assay (PMS to GMS shift) (Liang et al., 1997). The timing of appearance and pattern of accumulation of GUS transcript and GUS activity in transformants was consistent with the timing of appearance and pattern of accumulation of *ver-1* transcript and *Ver-1* protein. These data validated the use of the GUS reporter system for analysis of *ver-1* promoter function and confirmed that transcriptional regulation plays an important role in *ver-1*A expression under conditions of glucose induction. A *nor-1*/GUS reporter construct also was generated (Wilson et al., 1996) and transformed into *A. parasiticus.* The data from experiments conducted under the same growth conditions were very similar to those observed for the *ver-1* A promoter, suggesting that both genes are coordinately regulated at the level of transcription.

3.2.1.3. Nitrate Repression

Addition of nitrate to an AF-inducing culture medium (containing ammonium ion as the nitrogen source) results in a dose-dependent repression of AF synthesis (Kacholz and Demain, 1983; Niehaus and Jiang, 1989). Sodium nitrate added as the sole nitrogen source completely inhibits AF synthesis, but ammonium nitrate supports toxin synthesis (Kacholz and Demain, 1983). This suggest that the presence of ammonium ion can partially overcome the repression by nitrate. Using buffered media, Kacholz and Demain (1983) reported that the nitrate effect was not due to differences in pH resulting from metabolism of nitrate versus ammonium ion. One possible mechanism to explain the influence of nitrate of AF synthesis is that nitrate represses AF synthesis directly or indirectly via the fungal regulatory network for nitrogen assimilation/metabolism utilized in primary metabolism and that ammonium ion, which is the preferred nitrogen source of many organisms for growth, relieves the repression of AF due to nitrate. An alternative explanation proposed by Niehaus is that nitrate stimulates glucose-6-phosphate dehydrogenase and mannitol dehydrogenase activities resulting in increased reducing power in the cell (high NADPH/NADP ratio). This in turn would result in increased fatty acid synthesis and decreased polyketide synthesis (Niehaus and Jiang, 1989). More recently, it was observed that overexpression of *aflR* (one protein that is important in regulation of AF synthesis; see below) relieves nitrate repression (Chang et al., 1995c). These data suggest that nitrate may have a direct or indirect influence on AF gene expression at the level of transcription.

3.2.1.4. Regulation by Trace Metals

Previous research has demonstrated a strong positive correlation between the level of zinc and AF synthesis in culture (Bennett et al., 1979; Coupland and Niehaus, 1987). The absence of added zinc in the growth media completely eliminated AF synthesis (Bennett et al., 1979) even though growth (as measured by dry weight) decreased by only a factor of two compared to complete media (with added zinc). Since zinc is an essential trace element, it appears clear that the base media used in this study contained sufficient zinc to support growth but not to support AF synthesis. The media in this study contained 5% sucrose-as the carbon source. In a separate study, the concentration of zinc was varied between 0.3 μM to 10 μM in a chemically defined media containing glucose (3.6%) plus peptone (0.2%) as the carbon sources. The quantity of toxin (versicolorin A was analyzed in an aflatoxin-blocked mutant strain) increased by 60 fold, going from the lowest to highest concentration of zinc (Coupland and Niehaus, 1987). Zinc induction also occurs at higher levels (50–400 μM; Tiwari et al., 1986), though the net effect is less pronounced than at lower zinc concentrations. No differences were observed in the level of TCA enzymes in cultures with low or high zinc (Gupta et al., 1977a, b): however, zinc did stimulate several enzymes in the glycolytic pathway. Niehaus and colleagues proposed that zinc influences AF production via metabolism by lowering the reducing power in the cell (NADPH/NADP ratio) based on the observation that zinc inhibited mannitol dehydrogenase and glucose-6-phosphate dehydrogenase activity in *A. parasiticus* (Niehaus and Dilts 1982, 1984). This presumably would shunt acetate into polyketide biosynthesis instead of fatty acid synthesis where more reducing power is required. In apparent agreement, in a related study, the ratio of NADPH/NADP was lowest in media that supported highest levels of AF synthesis—the ratio was high in media that did not support AF synthesis (Bhatnagar et al., 1986). In addition, the level of AMP and ADP was high during stationary phase and low during exponential growth. Davis and Diener (1986) proposed that high levels of ADP and AMP could result in shunting of acetate toward polyketide synthesis. More recently, preliminary data obtained using *A. parasiticus* transformed with the *nor-1*/GUS reporter construct suggest that zinc can influence AF synthesis by directly or indirectly altering the level of AF gene expression at the level of transcription (Miller and Rarick, unpublished data). The presence of zinc on the growth medium was also shown to affect the level of *aflR* expression (Liu and Chu, 1998).

Manganese has been observed to stimulate AF synthesis while molybdenum and vanadium inhibited toxin synthesis (Rabie et al., 1981). Therefore, several trace metals can have strong positive or negative influences on AF synthesis. Manganese has also been demonstrated to regulate the expression of a gene encoding a key enzyme in patulin (a mycotoxin and secondary metabolite) biosynthesis in the filamentous fungus *Penicillium urticae,* although the mechanism of regulation was not determined (Wang et al., 1991). This may suggest that regulation of polyketide biosynthesis in filamentous fungi by metals is conserved during evolution and likely is relevant to regulation during growth in the soil or on plants.

In support of this notion, experimental data suggest that zinc induction may be an important influence on AF production on certain plants. For example, there was a strong positive correlation between AF accumulation and zinc concentration in corn kernels and peanuts, two major crops affected by aflatoxin contamination (Failla et al., 1986; Reding and Harrison, 1994). When peanuts were cultivated in a field with gypsum supplementation (calcium sulfate), significantly reduced levels of zinc and AF (40 fold reduction) were observed on treated plants when compared with untreated plants (Reding and Harrison, 1994). However, on other plant species, including soy, sunflower, and cottonseed, the positive correlation did not hold up (Chulze et al., 1987; Stossel, 1986). The simplest explanation is that zinc levels are not relevant to levels of toxin synthesis on plants. An alternative interpretation of these data is that an interaction exists between zinc and other factors, which together determine the level of AF production. The chemical composition of seeds from these different plant species varies in the level and nature of the carbon and nitrogen sources as well as other nutrients and potential cofactors. It is important, therefore, to study the relative influence of these factors and how they interact during regulation in culture to begin to understand how they determine AF levels on plants.

3.2.1.5. Other Nutritional Factors

Several other nutrients influence the level of AF synthesis including certain amino acids (Payne and Hagler, 1983), phosphorus (Niehaus and Coupland, 1987; Dutton and Anderson, 1980), and nucleotides (Tice and Buchanan, 1981; Bhatnagar et al., 1986) (reviewed in Zaika and Buchanan, 1987; Davis and Diener, 1986; Ellis et al., 1991). These other factors appear to have more moderate influences on AF synthesis and relatively little information is available with respect to the mechanism by which they influence AF synthesis.

3.2.2. Environmental Factors

3.2.2.1. Temperature

A. parasiticus synthesizes AF at maximum levels in the range between 25 to 30°C on laboratory growth media, although this range is influenced by media composition (reviewed in Gourama and Bullerman, 1995; Davis and Diener, 1986). Above and below this range, toxin synthesis drops off sharply. For example, in one study, growth of an *A. parasiticus* aflatoxin-blocked mutant at 37°C (which accumulates versicolorin A, VA) completely inhibited toxin synthesis (Skory et al., 1992). Especially interesting is that AF synthesis and development of survival structures called sclerotia (see below) were both inhibited, which suggests that the regulation of toxin synthesis and fungal development might be linked.

3.2.2.2. Other Environmental Factors

Other environmental factors influence AF synthesis including pH (Cotty, 1988), atmospheric gases (Clevstrom et al., 1983) water activity, and light (Bennett et al., 1981). The effects of these factors on AF synthesis have been reviewed (Ghourama and Bullerman, 1995; Davis and Diener, 1986). Little is known about the mechanism(s) the fungus utilizes to alter AF synthesis in response to changes in these factors. This clearly represents an interesting and potentially fruitful avenue of study. Recently, Keller et al., (1997) have initiated studies to help define the role of pH in regulation of AF/ST gene expression. They demonstrated that acidic pH can result in higher levels of AF/ST transcript accumulation and toxin synthesis. They proposed that pH directly regulates gene expression at the transcriptional level and that this regulation may be mediated through *AflR* or a transcription factor, PacC, which has been demonstrated to regulate penicillin gene expression in *A. nidulans* (Tilburn et al., 1995).

3.3. PLANT/PATHOGEN INTERACTIONS

Since AF is a potential health problem when it is associated with contamination of food and feed crops, several researchers are beginning to use newly developed technology to study fungal growth, toxin synthesis, and AF gene expression in association with plant tissue. For example, it was determined that glucose induction may influence toxin production by the fungus growing on the host plant. Woloshuk (1995) demonstrated that extracts of corn kernels colonized with a toxigenic strain of *A. flavus* could induce toxin synthesis and GUS expression in *A. flavus* containing a GUS reporter construct fused to the *ver-1* promoter. They proposed that fungal amylase, released upon infection of the kernel, hydrolyses starch-releasing glucose, maltose and maltotriose, which in turn induce AF synthesis (Woloshuk et al., 1997).

GUS reporter constructs have been useful in screening natural plant compounds for effective inhibitors of AF gene expression and fungal infection. For example, Payne and colleagues developed the original GUS reporter strains in *A. flavus,* which they utilized to identify compounds in extracts from an aflatoxin-resistant cultivar of corn (Tex6), which inhibits AF synthesis and/or fungal growth (Payne, 1996; Huang et al., 1997). Extract of black pepper completely inhibited expression at the level of transcription in an *A. parasiticus* strain transformed with a *nor-1*/GUS construct (Trail and Linz, 1995). The effect was apparently specific because the pepper extract did not affect the expression of the β-tubulin (a ''housekeeping'' gene) promoter fused to the GUS reporter gene. Brown and his colleagues have also used GUS reporter strains (β-tubulin/GUS) to study mechanisms of resistance in corn and identified that a waxy coating on kernels of a resistant variety of corn may provide physical and chemical barriers to fungal infection (Brown et al., 1996). Keller utilized a blocked mutant of *A. parasiticus,* which accumulates the bright red pathway intermediate norsolorinic acid (NA), to study the interaction of the fungus and the host plant. Using this strain in the laboratory, Keller identified that AF synthesis appears to occur at highest levels in specific regions of the corn kernel that are lipid rich (embryo and aleurone; Keller et al., 1993a). In agreement with these data, a chemical and microscopic study performed on cotton seed also determined that toxin synthesis does not occur in all regions of the seed where fungal growth occurs (Goynes and Lee, 1989). Trail has initiated similar studies on peanut, which investigate plant/pathogen interaction and regulation of toxin synthesis (Trail and Linz, 1996). These studies are extremely important because they utilize knowledge and tools gained in the research laboratory and apply them to a real-life environment.

A wide variety of plant or plant-related products have been shown to have positive or negative influences on fungal growth and toxin synthesis (Hitokoto et al., 1978; Chatterjee, 1990; Zeringue, 1991; Buchanan et al., 1993; Mellon and Cotty, 1996, Greene-McDowelle et al., 1999; Zeringue et al., 1996; reviewed in Zaika and Buchanan, 1987). Using the reporter strains that have been developed, it is possible to begin to identify the mechanism by which these compounds affect AF synthesis. Once this information is known, the next step is to utilize this knowledge to develop strategies to inhibit toxin synthesis by the fungus associated with plants using biological control, classical plant breeding or genetic engineering of these crops.

3.4. TRANSCRIPTIONAL REGULATION

There is clear precedent for transcriptional regulation of fungal primary and secondary metabolism via nutritional and environmental factors. In *A. nidulans,* a "close" relative of *A. parasiticus,* certain of these effects are mediated by specific transacting transcription factors (TAF), which recognize specific binding sites in promoters. For example, penicillin synthesis (a secondary metabolite) is repressed at the transcriptional level by glucose (Perez-Estaban et al., 1993). This negative effect is mediated by a carbon catabolite repressor, which binds cis-acting sites in the promoters of the three penicillin genes. In filamentous fungi, glucose exerts carbon catabolite repression via *cre*A (and related TAF), which encodes a binuclear zinc cluster TAF (Brakhage et al., 1992; Stossel, 1986). In bacteria, carbon catabolite repression is mediated via signal transduction resulting in production of the second messenger, cAMP. cAMP then interacts with specific DNA sequences in promoters via the cAMP receptor protein (CRP; reviewed in Kolb et al., 1993). What is unique about AF synthesis is that glucose induces the expression of genes involved in a secondary metabolic pathway, which in many well-studied cases are subject to carbon catabolite repression.

AREA is responsible for mediating nitrogen metabolite repression (Peters and Caddick, 1994) in *A. nidulans* by binding to a core sequence GATA in gene promoters. Ammonium ion inhibits AREA binding to genes involved in utilization of alternative nitrogen sources (global regulation). Growth in an alternative nitrogen source, such as nitrate, relieves nitrogen repression by allowing AREA to bind to the promoters of the structural genes *nii*A and *nia*D (encoding nitrite and nitrate reductase). This results in binding of a pathway-specific positive regulator, which in the example of nitrate utilization is called *nir*A. The catalytic and regulatory mechanisms of nitrate assimilation are well conserved in *Aspergillus* (Chang et al., 1996).

Vertebrate and fungal cells synthesize metallothionein proteins in response to influx of heavy metals. In vertebrates, the metallothionein promoters contain metal responsive elements, which mediate the effect of heavy metals (Otsuka et al., 1994). Is there any evidence that *Aspergillus* may use related mechanisms to mediate the response to glucose, nitrate and metal ions? A survey of the *ver-1* and *nor-1* promoter regions for similar TAF binding sites failed to find exact matches to those reported in the literature.

3.5. ROLE OF *aflR* IN GENE REGULATION

The *aflR* gene has been demonstrated to encode a protein that functions as a key control point in regulation of AF gene expression. Mutations in *aflR* eliminate AF or ST synthesis and block the accumulation of transcripts and enzymatic activities encoded by toxin genes (Payne et al., 1993; Woloshuk et al., 1994). Complementation of the *aflR* mutation simultaneously restores function of the pathway (Payne et al., 1993). *aflR* from *A. flavus* was shown to function in *A. nidulans* and stimulated ST biosynthesis (Yu et al., 1996) at a time when synthesis normally does not occur. Induced expression of *aflR* before the normal onset of ST synthesis stimulated the early onset of ST gene expression and ST biosynthesis. These data together suggest that *aflR* encodes a protein that is a necessary positive regulator of AF and ST biosynthesis and directly or indirectly influences transcription of several genes. The data also suggest that regulation via *aflR* is highly conserved among toxin-producing species.

A cysteine-rich zinc cluster motif (Cys-Xaa2-Cys-Xaa6-Cys-Xaa6-Cys-Xaa2-Cys-Xaa6-Cys) and an adjacent, highly acidic domain were observed in the predicted amino acid sequences of *apa*-2 and *afl*-2 (Chang et al., 1993; Woloshuk et al., 1994). These are characteristic signatures of fungal and yeast transacting regulatory factors (TAF) that bind DNA and regulate gene expression. The GAL4 protein in the yeast *Saccharomyces* is a well-known example of this class of TAF (Johnston, 1987). GAL4 regulates galactose utilization via interaction with a specific palindromic motif in the promoters of the GAL genes.

Recently, two lines of evidence suggested that AFLR may bind to the palindromic sequence, TCG{n}CGA, in the promoters of many structural genes in the ST cluster in *A. nidulans* (Fernandes et al., 1998). An AFLR protein generated in *E. coli* bound to an oligonucleotide and a subfragment of the native *stc*U (a structural gene involved in ST synthesis) promoter region in vitro. The bound region was confirmed by eliminating binding by mutagenesis of the entire binding site or of specific bases and by a methylation interference footprint assay. Finally, mutagenesis of this site in a *stc*U/GUS reporter reduced GUS expression in *A. nidulans* grown under ST-inducing conditions, which demonstrates the functionality of this site in vivo. Ehrlich et al. (1998) performed EMSA assays on the promoters of 11 AF biosynthetic genes and found all of them demonstrated bonding of *AflR* to the conserved palindromic sequence.

It is clear that AFLR protein is a key feature of the switch that regulates AF and ST synthesis. Based on the data presented, it is reasonable to hypothesize that AFLR plays one of three alternative roles. (1) AFLR is the only TAF that interacts with promoters of the structural genes—nutritional and environmental factors influence pathway function by determining the level of synthesis and/or activity of AFLR. (2) AFLR interacts with other TAF (whose synthesis/activity is induced by nutritional and/or environmental factors) to induce transcription. (3) AFLR does not induce gene expression directly—AFLR induces the synthesis or activity of one (or more) TAF, which acts independently or together with other TAF to induce transcription. For example, AFLR may make pathway gene promoters available for binding by the general transcription machinery (RNA Pol II and associated general transcription factors) by directly or indirectly influencing chromatin structure (i.e., unwinding chromatin in the AFB_1 cluster). However, the data presented above seem to demonstrate direct interaction of AFLR with its own promoter in *A. parasiticus* and to a site common in many structural genes in *A. nidulans* and argue that models 1 or 2 are operative in the fungal cells.

3.6. OTHER REGULATORY INFLUENCES ON AF SYNTHESIS AT THE TRANSCRIPTIONAL LEVEL

3.6.1. *afl*-1

A second putative regulatory locus (besides *aflR*), called *afl*-1, was identified by Leaich and Papa (1974) using UV mutagenesis and was later determined to be linked to *nor-1* by parasexual analysis (reviewed by Bennett and Papa, 1987). *Afl*-1 mutants are functionally dominant in diploids resulting in loss of aflatoxin production. More recent studies (Woloshuk et al., 1995) determined that the *afl*-1 mutation results in suppression of transcription of the three structural genes tested (*nor-1, ver-1,* and *omtA*). Transcription of *aflR* was reported to be normal in these strains, which might suggest that *aflR* is necessary but not sufficient for AF gene expression under the growth conditions tested.

3.6.2. *afl*-4

One additional Papa mutant was demonstrated to be of importance to the AF regulatory scheme (Brown and Payne, 1996). The *afl*-4 mutation was not genetically linked to the AF gene cluster and resulted in lack of AF synthesis and lack of the *aflR* transcript under growth conditions that induce AF synthesis. An *aflR*/GUS reporter construct was not active in the *afl*-4 genetic background. Expression of *aflR* using a heterologous promoter restored AF synthesis. These data suggest that multiple factors may regulate expression of *aflR*. Probable candidates are reported to include two positive regulators, AFLR and *afl*-4 protein, and one negative regulator.

3.6.3. Structure/Function of *nor-1* and *ver-1* Promoters

Based on the complete nucleotide sequences of *nor-1* and *ver-1* (Trail et al., 1994; Skory et al., 1992) the polymerase chain reaction (PCR) was used to generate three overlapping DNA fragments, right (R), middle (M), left (L), and the full-length fragment for each promoter (T) so that TAF binding sites located at the junctions of adjacent fragments would not be missed. To identify proteins that bind to these promoter fragments and presumably act as trans-acting regulatory factors (TAF), a gel-mobility shift assay (GMSA) was performed (Miller et al., 1996). Fungal cells were grown in PMS (control; basal levels of AF synthesis), YES and GMS (high and medium levels of AF synthesis, respectively) in batch fermentation. Nuclear extracts, from cells grown in PMS did not result in a mobility shift of R, L, M, or T promoter fragments, whereas nuclear extracts from cells grown in GMS showed multiple shifted complexes for the *ver-1* and *nor-1* promoter fragments. The YES cell and nuclear extracts also induced multiple shifts in the *nor-1* promoter fragments. The data suggested that TAF involved in AF synthesis were not present in the PMS extract and that AF gene expression is induced by positively acting TAF. The presence of multiple shifted complexes may also suggest that regulation is more complex than simple binding of AFLR to the promoters of structural genes although it is also possible that AFLR has multiple binding sites within promoters.

3.7. REGULATORY AND PHYSICAL ASSOCIATION BETWEEN AF/ST GENE EXPRESSION AND FUNGAL DEVELOPMENT

In many organisms including actinomycetes and filamentous fungi, it is clear that production of secondary metabolites is directly related to the regulation of cellular development (reviewed in Beppu, 1992). These chemicals often act as hormones in yeast and fungi as mating signals and for differentiation of structures involved in sexual reproduction. *Aspergillus parasiticus* and *A. flavus* are imperfect fungi with no identified sexual stage. Large numbers of asexual spores (conidiospores) for reproduction and spread of the organism are generated on specialized aerial hyphae called conidiophores. In addition, the vegetative mycelium differentiates to generate condensed masses of cells, which become dehydrated and pigmented. These "sclerotia" allow survival of the organism during times of stress. Evidence is emerging, which links morphological differentiation (conidial and sclerotial development) to chemical differentiation. This evidence is primarily provided by genetic and biochemical studies. A brief review of this evidence is presented followed by a discussion of the significance of this potential relationship and a hypothesis about the biological relevance to the fungus.

3.7.1. Genetic Evidence

3.7.1.1. Conidiospore Development

In *Aspergillus,* Bennett observed a close association between AF synthesis and conidiospore development when she isolated variants of *A. parasiticus* called "fan" and "fluff". These strains did not produce AF and were deficient or totally lacking in conidiospore production (Bennett et al., 1981; reviewed in Bennett and Papa, 1987). For example, when Bennett serially transferred an *A. parasiticus* mutant, which accumulates norsolorinic acid (NA), accumulation of NA and conidiospore development disappeared simultaneously in the isolates analyzed (Bennett et al., 1981), which suggests that these processes are genetically linked or depend on a common regulator. More recently, Kale et al. (1994) reported the isolation of secondary metabolite (sec) variants with similar characteristics to fan and fluff. Kale et al. (1996) also demonstrated, using the variant strains, that the regulation of aflatoxin synthesis and conidiogensis may be interlinked. Subsequently, Zhou (1997) isolated a mutant, called *flu*P, with an interesting phenotype.

*Flu*P grows 3 to 5 times more slowly than wild-type cells, does not produce conidiospores on solid media, but does produce abundant aerial hyphae giving the colony a fluffy appearance. Based on nucleotide sequence data, the *flu*P gene is predicted to encode a large polyfunctional enzyme with similarity to the polyketide synthetase (methyl salicylic acid synthetase; MSAS) involved in patulin synthesis in *Penicillium patulum* (Beck et al., 1990). AF production is reduced 3 to 5 fold in comparison to wild-type strains. The fluffy colony morphology in the *flu*P mutant was similar to that reported for the "fluffy" mutants of *A. nidulans* first isolated by Adams and colleagues. The genes in the fluffy class activate *brl*A, which initiates a cascade of regulatory factors that results in conidiospore development (Lee and Adams, 1994a, b; Lee and Adams, 1995). Among these genes, identified initially by the phenotype of the corresponding mutant, are two which are of particular interest in this discussion. *Flu*G encodes a protein that is involved in the synthesis of a soluble extracellular signal which is somehow linked to *brl*A activation (Lee and Adams, 1994a). The second gene, called *flb*A, is reported to interact with *flu*G and a signal transduction pathway involving a G protein, one subunit of which (Gα) is encoded by *fad*A (Yu et al., 1996). It was proposed that *flb*A antagonizes the G protein pathway resulting in growth inhibition and stimulation of conidiospore development. Data from a recent study strongly suggest that this is true (Hicks et al., 1997). Recently, it was observed that *flb*A mutants, in addition to the lack of conidiospore development, also are unable to synthesize ST (Keller and Adams, personal communication). Complementation of the *flb*A mutation simultaneously restored ST and conidiospore production. Overexpression of *flb*A initiates *brl*A activation, conidiospore development and ST synthesis (Lee and Adams, 1995). Taken together, these data implicate *flb*A as a global regulator of growth, conidiospore development, and ST synthesis.

In related studies, it was determined that the activity of adenylate cyclase and cAMP phosphodiesterase, enzymes involved in the synthesis and degradation of cAMP, respectively, varied in phase with the synthesis of AF in *A. parasiticus,* which suggested that the cyclic nucleotide may be related to levels of AF synthesis (Khan and Venkitasubramanian, 1987). Addition of cAMP to the fungus grown in liquid culture resulted in a dose-related increase (up to 18 fold) in AF synthesis (Tice and Buchanan, 1981). These data suggest that a signaling pathway involving cAMP may help regulate AF synthesis. Is this putative pathway in some way connected to the G protein signaling pathway which interacts with *flb*A?

3.7.1.2. Sclerotial Development

Cotty conducted experiments which investigated the potential association between AF synthesis and sclerotial development (Cotty, 1988). The fungus was grown on solid media which varied in nitrogen content (ammonium ion vs. nitrate) and pH (buffered vs. not buffered). Sclerotial production was observed to increase with increased pH on buffered ammonium medium but AF production decreased. On nitrate media, growth resulted in an increase in pH and production of a large quantity of sclerotia, which contained very little AF. Sclerotia produced on nitrate media contained 20 to 30 fold less AF than sclerotia produced on ammonium media. These results suggest that pH and nitrogen source can regulate sclerotial number and AF content. In related research, it was observed that genetic blocks in the AF pathway, which result in accumulation of AF pathway intermediates, have a significant influence on the number of sclerotia produced under similar growth conditions on solid culture media. Accumulation of versicolorin A or averufin, intermediates near the middle of the AF pathway, nearly eliminated sclerotial production, whereas mutations that resulted in total loss of synthesis of AF or pathway intermediates, enhanced sclerotial production (Mahanti et al., 1996; Trail et al., 1995a). Other studies report that there is not an association between AF synthesis and sclerotial production (Bennett and Horowitz, 1979). Recent studies have shown a relationship between regulatory elements of aflatoxin synthesis and sclerotia shape, size and number (Chang et al., unpublished). Some strains that produce high levels of AF do not make sclerotia and some strains that produce little or no AF produce abundant sclerotia. However, these experiments do not directly address the issue of regulation of these processes. Further experiments are needed to look at this issue in more detail.

3.7.2. Biochemical Evidence

3.7.2.1. Physical Association

One of the most convincing pieces of evidence that AF synthesis and development are related is the observation that large quantities of AF (3–132 ppm; 1000 fold more than the allowable level of AF in food) are found in conidiospores and sclerotia. The presence of AF in these structures has been proposed to play a protective role against competitors in the soil and on plants (Cotty, 1988; Wicklow, 1988).

3.7.2.2. Spatial Regulation of AF Enzyme Accumulation (Ver-1 and Nor-1 Proteins)

Other biochemical evidence shows that AF gene expression and accumulation of AF enzymes is physically associated with the structures that generate conidiospores. Most studies on the regulation of AF gene expression have been conducted in liquid culture. Fungal development normally occurs on solid surfaces such as those in the soil and on plant material. To more accurately model the regulation of gene expression that occurs in the field, it was important to initiate studies of AF gene expression on solid media (Liang, 1996; Zhou, 1997).

Inoculation of *A. parasiticus* in the center of an agar plate under "standard growth conditions" (29°C in the dark) results in radial growth outward from the site of inoculation. Approximately six days (144 h) are required for growth to reach the outer margin of the petri dish on rich media such as YES. Three concentric zones that differ in their morphology were observed in colonies grown for 72 to 144 h on YES (rich) agar medium. The central zone contained abundant pigmented conidia. The middle zone (M) contained more aerial hyphae but with only scattered conidia. The peripheral zone consisted of an outer margin (3–5 mm) with a vegetative hyphal network and an adjacent inner region where the formation of aerial hyphae and immature conidiophores was initiated. The width of the C, M, and P zones increased as growth proceeded outward but the zones were still easily distinguishable. Colonies grown for 72 to 144 h were cut in concentric zones based on morphology (C, M, and P) and the quantity of Ver-1 and Nor-1 protein was measured over time by Western blot analysis of proteins extracted from each zone. The highest quantity of these proteins was detected in zone P where conidiophore development was initiated. In zone C, where mature conidiophores are abundant, little Ver-1 and Nor-1 protein were detected. Highest levels of Ver-1 and Nor-1 protein accumulation were detected at similar times and in similar locations as conidiophore development. One reasonable interpretation of these data is that a physical and regulatory association exists between toxin synthesis and conidiospore development.

3.8 PHYSICAL LOCALIZATION OF AF ENZYMES

Disruption of protoplasts with a homogenizer was employed to prepare a cell lysate, which was then fractionated using differential centrifugation. Using this protoplast disruption method, 14% of the Ver-1 protein co-fractionated with mitochondria, which suggested that the protein associates with an "organelle". The relatively low quantity of protein associated with these organelles may suggest a loose association or that these "organelles" are fragile (single-membrane structure?) and break during the homogenization step. The Nor-1 protein co-fractionated with ribosomes, which suggests that this protein is associated with smaller particles distributed primarily in the cytoplasm although loose association with larger organelles appears to occur at low frequency (Liang, 1996; Zhou, 1997).

3.9. THE "LIPOXYGENASE THEORY"

One intriguing theory recently proposed by Keller and her colleagues (Burow et al., 1997) appears to provide a framework for integrating a large body of data related to AF/ST synthesis, conidiospore development, as well as the propensity for *Aspergillus* to produce AF/ST on oilseed crops. The theory is based on the fact that several fungi contain lipoxygenase (LOX) activities that are involved in the synthesis of pheromones, which regulate development. For example, *A. nidulans* contains LOX activity involved in synthesis of psi factors (precocious sexual inducer) (Champe et al., 1987; Champe and El-Zayat, 1989; Mazur et al., 1990), which regulate sexual and asexual sporulation in this organism. Because of the close regulatory link between conidiospore development and ST synthesis, it was not unreasonable to propose that psi factors help the fungus achieve ''competence'' to produce ST. What is the connection to the host plant?

Early observations showed that the level of AF production by *Aspergillus* is higher on oilseeds than on starch seeds and that there is a preference for synthesis in oil-rich tissues. Mixtures of lipids and specific lipids or their oxidized derivatives also were found to affect the levels of ST/AF synthesis in culture and on plant tissue. For example, Fanelli and Fabbri and their colleagues reported that a variety of lipid compounds and hydroperoxy derivatives of lipids had significant effects (up to 200 fold increase in certain cases) on AF synthesis (Fabbri et al., 1983; Fanelli et al., 1983b; Passi et al., 1985; reviewed in Fanelli and Fabbri, 1989). They proposed that unsaturated lipids in plant oils could undergo peroxidation during aging/storage and induce AF production. Further studies by Keller and colleagues were based on key observations: (a) the polyunsaturated fatty acids linoleic and linolenic acid are major constituents of lipid storage bodies in seeds (Bewley and Black, 1985); (b) metabolites from plant lipoxygenases seemed to have a large stimulatory affect (up to 200 fold) on toxin synthesis (DeLuca et al., 1995); (c) lipid storage bodies are metabolized by *A. flavus* on corn before the starch granules (Smart et al., 1990). They, therefore, studied the effects of lipoperoxide derivatives of linolenic and linoleic acids on ST and AF synthesis in *A. parasiticus* and *A. nidulans* and showed that 13S hydroperoxy derivatives repressed toxin synthesis while the 9S derivatives appeared to have little effect (Burow et al., 1997; Gardner et al., 1998). These derivatives were speculated to affect toxin synthesis because these lipids may mimic the structure of psi factors and interfere with the ability of the fungus to gain competence. They propose that the stereochemistry of the lipoxygenase products have differential effects on AF/ST synthesis and that the effects of mixtures of lipoxygenase products (as in the Fanelli/Fabri studies) may be difficult to predict.

One reasonable scenario to explain this observed effect is as follows. Initially, production of hydroperoxy derivatives of linoleoic acid and linolenic acid is induced to occur (by the fungus?) in the host plant via host lipoxygenases. These then mimic psi factor and alter regulation of development and ST synthesis. This presumably occurs via the global regulator *flb*A discussed above.

In related work, Wicklow and colleagues proposed that plant lipoxygenase activity may result in resistance of certain crops like soybean to *Aspergillus* and AF contamination (Doehlert et al., 1993). Keller and colleagues report preliminary data that seem to support this proposal. Expression of soybean LOX-1 in *A. nidulans* interferes with sexual development presumably by interfering with psi factor activity (Keller et al., unpublished). Workers at USDA in New Orleans have also noted strong regulatory effects of plant volatiles, including alkenal and alkanal compounds, on AF synthesis (Greene-McDowelle et al., 1996a; Zeringue et al., 1996). Although lacking formal proof at this stage, the lipoxygenase theory puts a novel perspective on secondary metabolism and the regulation of this process.

4. FUTURE FOCUS

Since specific nutrients, environmental factors, and developmental factors influence the level of AF produced in culture and/or on plants, it is reasonable to propose that these regulatory schemes arose as a result of selective pressure in a soil environment or in association with the host plant. It is also reasonable to propose that alteration of plant biochemistry at the site of plant infection (seed or nut) will significantly alter the level of AF produced by reducing AF gene expression. One major focus is to utilize knowledge gained through studies on gene regulation and apply it directly to prevent AF or ST contamination of food or feed crops.

One promising approach to interfering with AF gene expression is to utilize natural plant products. Several natural plant products have been identified that influence the level of AF produced by the fungus including caffeine (Zaika and Buchanan, 1987), black pepper extract, and neem extracts (Bhatnagar and McCornick, 1988; Zeringue and Bhatnagar, 1994), which inhibit AF synthesis, and plant lipids/oils and their peroxidation products (Fabbri et al., 1983; Fanelli et al., 1983b), which stimulate AF synthesis (reviewed in Zaika and Buchanan, 1987). An alternative approach that is being tested is to transform crops with lipoxygenase that produces 13-HDPODE. The products of this enzyme are inhibitory to ST synthesis in *A. nidulans*. Other natural plant products have been identified that directly affect fungal growth including oils of cassia, clove, and cinnamon (Chaterjee, 1990; Valcarel et al., 1986). If these natural plant products could be synthesized directly by the host plant or be introduced into the growth environment (spray or fertilizer), they could significantly reduce aflatoxin levels on plants. Progress has been reported in the development of transformation systems for cotton and peanuts, allowing the potential for genetic engineering of these host plants to introduce genes that encode or synthesize factors for inhibition of AF biosynthesis (like natural plant products) or resistance to fungal infection (Cary et al., 1993; Chlan et al., 1993; Cleveland et al., 1992; Moyne et al., 1993).

5. REFERENCES

Abdollahi, A. and Buchanan, R. L. 1981a. "Regulation of aflatoxin biosynthesis: Characterization of glucose as an apparent inducer of aflatoxin production," *J. Food Sci.* 46:143–146.

Abdollahi, A. and Buchanan, R. L. 1981b. "Regulation of aflatoxin biosynthesis: Induction of aflatoxin production by various carbohydrates," *J. Food Sci.* 46:633–635.

Anderson, J. A. and Green, L. D. 1994. "Timing of appearance of versiconal herniacetal acetate esterase and versiconal cyclase activity in cultures of *Aspergillus parasiticus,*" *Mycopathologia* 126:169–172.

Association of Official Analytical Chemists. 1990. *Official Methods of Analysis,* 15th Ed. Arlington, Virginia: AOAG.

Bailey, G. S. 1994. "Role of aflatoxin—DNA adducts in the cancer process," in The Toxicology of Aflatoxins, ed., Eaton, D. L. and Groopman, J. D. eds. Academic Press. New York: pp. 137–148.

Barbesgaard, P., Heldt-Hansen, H. P., and Diderichsen, B. 1992. "On the safety of *Aspergillus oryzae:* a review," *Appl. Microbiol. Biotechnol.* 36:569–572.

Beck, J., Ripka, S., Siegner, A., Schilz, E., and Schweizer, E. 1990. "The multifunctional 6-methylsalicylic acid synthase gene of *Penicillium patulum,*" *Eur. J. Biochem.* 192:487–498.

Bennett, J. W., Chang, P.-K., Bhatnagar, D. 1997. "One gene to whole pathway: the role of norsolorinic acid in aflatoxin research," *Adv. Appl. Microbiol.* 45:1–15.

Bennett, J. W., Dunn, J. J., and Goldsman, C. I. 1981. "Influence of white light on production of aflatoxins and anthraquinones in *Aspergillus parasiticus,*" *Appl. Environ. Microbiol.* 41:488–491.

Bennett, J. W. and Horowitz, P. C. 1979. "Production of sclerotia by aflatoxigenic and nonaflatoxigenic strains of and *A. parasiticus,*" LXXI:415–422.

Bennett, J. W. and Papa, K. E. 1987. "Genetics of aflatoxigenic *Aspergillus* species," *Adv. in Plant Pathol.* 6:263–280.

Bennett, J. W. and Papa, K. E. 1988. "The aflatoxigenic *Aspergillus,*" in *Genetics of Plant Pathogenic Fungi,* ed., G. S. Sidhu. London: Academic Press. pp. 263–278.

Bennett, J. W., Rubin, P. L., Lee, L. S., and Chen, P. N. 1979. "Influence of trace elements and nitrogen sources on versicolorin production by a mutant strain of *Aspergillus parasiticus,*" *Mycopathologia* 69:161–166.

Beppu, T. 1992. "Secondary metabolites as chemical signals for cellular differentiation," *Gene* 115:159–165.

Bewley, J. D. and Black, M. 1985. *Seeds: Physiology of Development and Germination.* New York: Plenum Press.

Bhatnagar, D. and Cleveland, T. E. 1990. "Purification and characterization of a reductase from *Aspergillus parasiticus* SRRC 2043 involved in aflatoxin biosynthesis," *FASEB J.* 4:2727.

Bhatnagar, D. and McCormick, S. P. 1988. "The inhibitory effect of neem (*Azadirachta indica*) leaf extracts on aflatoxin in *Aspergillus parasiticus,*" *JAOCS* 65:1166–1168.

Bhatnagar, D., Cleveland, T. E., and Kingston, D. G. I. 1991b. "Enzymological evidence for separate pathways for aflatoxin B_1 and B_2 biosynthesis," *Biochemistry* 30:4343–4350.

Bhatnagar, D., Cleveland, T. E., and Lillehoj, E. B. 1989. "Enzymes in aflatoxin B_1 biosynthesis—strategies for identifying pertinent genes," *Mycopathologia* 107:75–83.

Bhatnagar, D., Ehrlich, K. C., and Cleveland, T. E. 1992. "Oxidation-reduction reactions in biosynthesis of secondary metabolites," in *Mycotoxins in Ecological Systems,* eds., D. Bhatnagar, E. B. Lillehoj, and D. K. Arora, New York: Marcel Dekker, Inc. pp. 255–286.

Bhatnagar, D., Lillehoj, E. B., and Bennett, J. W. 1991a. "Biological detoxification of mycotoxins," in *Mycotoxins and Animal Foods,* eds., J. E. Smith and R. S. Henderson. Boca Raton, FL: CRC Press Inc. pp. 815–826.

Bhatnagar, D., Ullah, A. H. J., and Cleveland, T. E. 1988. "Purification and characterization of a methyltransferase from *Aspergillus parasiticus* SRRC 163 involved in aflatoxin biosynthetic pathway," *Prep. Biochem.* 18:321–369.

Bhatnagar, R. K., Ahmad, S., Mukerji, K. G., and Venkitsubramanian, T. A. 1986. "Pyridine nucleotides and redox state regulation of aflatoxin biosynthesis in *Aspergillus parasiticus* NRRL 3240," *J. Appl. Bacteriol.* 60:135–141.

Bhatnagar, D., McCormick, S. P., Lee, L. S., and Hill, R. A. "Identification of *O*-methylsterigmatocystin as an aflatoxin B_1 and G_1 percursor in *Aspergillus parasiticus*," *Appl. Environ. Microbiol* 53:1028–1033, 1987.

Bhatnagar, D., Payne, G. A., Linz, J. E., and Cleveland, T. E. 1995. "Molecular biology to eliminate aflatoxins," *Inform* 6:262–271.

Bhatnagar, D., Lax, A. R., Prima, B. P., Cary, J. W., and Cleveland, T. E. 1996. "Purification of a 43 kDa enzyme that catalyzes the reduction of norsolorinic acid to averantin in aflatoxin biosynthesis," *FASEB J.* 10:A1522.

Bhatnagar, D., Cary, J. W., Ehrlich, K. C., Cleveland, T. E., and Payne, G. A. 1998. "Molecular characterization of an aflatoxin B_2 producing mutant strain of *Aspergillus flavus*," *FASEB J.* 12:A1472.

Brakhage, A., Browne, P., and Turner, G. 1992. "Regulation of *Aspergillus nidulans* penicillin biosynthesis and penicillin biosynthesis genes *acv*A and *ipn*A by glucose," *J. Bacteriol.* 174:3789–3799.

Brobst, S. W. and Townsend, C. A. 1994. "The potential role of fatty acid initiation in the biosynthesis of the fungal aromatic polyketide aflatoxin B1," *Can. J. Chem.* 72:200–207.

Brown, D. W., Adams, T. H., and Keller, N. P. 1996. "A fatty acid synthase required for secondary metabolism," *Proc. Natl. Acad. Sci. USA* 93:14873–14877.

Brown, D. W., Yu, J.-H., Kelkar, H. S., Fernandes, M., Nesbitt, T. C., Keller, N. P., Adams, T. H., and Leonard, T. J. 1996. "Twenty-five coregulated transcripts define a sterigmatocystin gene cluster in *Aspergillus nidulans*," *Proc. Natl. Acad. Sci.*, USA, 93:1418–1422.

Brown, M. P. and Payne, G. A. 1996. "Molecular and genetic analysis of a putative aflatoxin biosynthesis regulatory mutant form," Proceedings: USDA-ARS Aflatoxin Elimination *Workshop*, October 28–29, 1996, Fresno, CA. p. 80.

Brown, R. L., Cleveland, T. E., Guo, G. Z., Lax, A. R., Russin, J. S., Williams, W. P., Windham, G. L., Payne, G. A., Flaherty, J. E., Woloshuk, C. P., and Widstrom, N. W. 1996. "Advances in identifying and characterizing maize kernel biochemical resistance traits for incorporation into a commercial breeding program," *Proceedings: USDA-ARS Aflatoxin Elimination Workshop*, October 28–29, 1996, Fresno, CA. p. 17.

Buchanan, R. L. and Lewis, D. F. 1984. "Regulation of aflatoxin biosynthesis: effects of glucose on activities of various glycolytic enzymes," *Appl. Environ. Microbiol.* 48:306–310.

Buchanan, R. L. and Stahl, H. G. 1984. "Ability of different carbon sources to induce and support aflatoxin synthesis by *Aspergillus parasiticus*," *J. Food Safety* 6:271–279.

Buchanan, R. L., Hoover, D. G., and Jones, S. B. 1993. "Caffeine-inhibition of aflatoxin production: Mode of action," *Appl. Environ. Microbiol.* 46:1193–1200.

Buchanan, R. L., Jones, S. B., Gerasimowicz, W. V., Zaika, L. L., Stabl, H. G., and Ocker, L. A. 1987. "Regulation of aflatoxin biosynthesis: assessment of the role of cellular energy status as a regulator of the induction of aflatoxin production" *Appl. Environ. Microbiol.* 53:1224–1231.

Burow, G. B., Nesbitt, T. C., Dunlap, J., and Keller, N. P. 1997. "Seed lipoxygenase products modulate *Aspergillus* mycotoxin biosynthesis," *Mol. Plant-Microbe Int.* 10:380–387.

Cary, J. W., Wright, M., Bhatnagar, D., Lee, R., and Chu, F. S. 1996. "Molecular characterization of an *Aspergillus parasiticus* dehydrogenase gene, *norA*, located on the aflatoxin biosynthesis gene cluster," *Appl. Environ. Microbiol.* 62:360–366.

Cary, J. W., Cleveland, T. E., Jacks, T. J., Chian, C. A., Kunning, S., and Jaynes, J. 1993. "Construction of plant transformation vectors for expression of GUS and antifungal traits," *Proceedings: USDA-ARS Aflatoxin Elimination Workshop*, Oct. 25–26, Little Rock, AR. p. 83.

Champe, S. P. and El-Zayat, A. A. E. 1989. "Isolation of a sexual sporulation hormone from *Aspergillus nidulans*," *J. Bacteriol.* 171:3982–3988.

Champe, S. P., Rao, P., and Chang, A. 1987. "An endogenous inducer of sexual development in *Aspergillus nidulans*," *J. Gen. Microbiol.* 133:1383–1387.

Chang, P.-K., Skory, C. D., and Linz, J. E. 1992. "Cloning of a gene associated with aflatoxin B1 biosynthesis in *Aspergillus parasiticus*," *Curr. Genet.* 21:231–233.

Chang, P.-K., Bhatnagar, D., Cleveland, T. E., and Bennett, J. W. 1995a. "Sequence variability in homologs of the aflatoxin pathway gene *aflR* distinguishes species in *Aspergillus* section *flavi*," *Appl. Environ. Microbiol.* 61:40–43.

Chang, P.-K., Cary, J. W., Yu, J.-J., Bhatnagar, D., and Cleveland, T. E. 1995b. "*Aspergillus parasiticus pks*A, a homolog of *Aspergillus nidulans w*A, is required for aflatoxin B_1 biosynthesis," *Mol. Gen. Genet.* 248:270–277.

Chang, P.-K., Ehrlich, K. E., Yu, J.-J., Bhatnagar, D., and Cleveland, T. E. 1995c. "Increased expression of *aflR*, encoding a sequence specific DNA binding protein, relieves nitrate inhibition of aflatoxin biosynthesis," *Appl. Environ. Microbiol.* 61:2372–2377.

Chang, P.-K., Ehrlich, K. E., Linz, J. E., Bhatnagar, D., Cleveland, T. E., and Bennett, J. W. 1996. "Characterization of the *Aspergillus parasiticus niaD* and *niiA* gene clusters," *Curr. Genet.* 30:68–75.

Chang, P.-K., Cary, J. W., Bhatnagar, D., Cleveland, T. E., Bennett, J. W. Linz, J. E., Woloshuk, C. P., and Payne, G. A. 1993. "Cloning of the *Aspergillus parasiticus apa*-2 gene associated with the regulation of aflatoxin biosynthesis," *Appl. Environ. Microbiol.* 59:3273–3279.

Charmley, L. L., Trenholm, H. L., Preluskey, D. B., and Rosenberg, A. 1995. "Economic losses and decontamination," *Natural Toxins* 3:199–203.

Chatterjee, D. 1990. "Inhibition of fungal growth and infection in maize grains by spice oils," *Lett. Appl. Microbiol.* 11:148–151.

Chlan, C. A., LaPorte, R. O., Willis, L., Lin, J., Cleveland, T. E., and Cary, J. W. 1993. "Progress toward generation of transgenic cotton plants resistant to *A. flavus*," *Proceedings: USDA-ARS Aflatoxin Elimination Workshop*, Oct. 25–26, Little Rock, AR. p. 94.

Chu, F. S. 1994. "Development of antibodies against aflatoxins," in *The Toxicology of Aflatoxins: Human Health, Veterinary and Agricultural Significance*, eds., E. L. Eaton and J. D. Groopman. San Diego, CA: Academic Press. pp. 451–490.

Chulze, S., Fusero, S., Dalcero, A., Etcheverry, M., and Varsavsky, E. 1987. "Aflatoxins in sunflower seed: The effects of zinc in aflatoxin production by two strains of *Aspergillus parasiticus*," *Mycopathologia* 99:91–94.

Chuturgoon, A. A. and Dutton, M. F. 1991. "The affinity purification and characterization of a dehydrogenase from *Aspergillus parasiticus* involved in aflatoxin B1 biosynthesis," *Mycopathologia* 113:41–44.

Chuturgoon, A. A., Dutton, M. F., and Berry, R. K. 1990. "The preparation of an enzyme associated with aflatoxin biosynthesis by affinity chromatography," *Biochem. Biophys. Res. Commun.* 166:38–42.

Ciegler, A. 1983. "Evolution, ecology and mycotoxins: somes mussings," in *Secondary Metabolism and Differentiation in Fungi*, eds., J. W. Bennett and A. Ciegler. New York: Marcel Dekker, pp. 429–440.

Cleveland, T. E. and Bhatnagar, D. 1990. "Evidence for the *de novo* synthesis of an aflatoxin pathway methyltransferase near the cessation of active growth and the onset of aflatoxin biosynthesis in *Aspergillus parasiticus* mycelia," *Can. J. Microbiol.* 36:1–5.

Cleveland, T. E. and Bhatnagar, D. 1991. "Aflatoxins: elimination through biotechnology," in *Encyclopedia of Food Science and Technology*, ed., Hui, Y. H. New York: John Wiley and Sons, Inc. pp. 6–11.

Cleveland, T. E. and Bhatnagar, D. 1992. ''Molecular strategies for reducing aflatoxin levels in crops before harvest,'' in *Molecular Approaches to Improving Food Quality and Safety,* eds., Bhatnagar, D. and Cleveland, T. E. New York: Van Nostrand Reinhold. pp. 205–228.

Cleveland, T. E., Bhatnagar, D., Foell, C. J., and McCormick, S. P. 1987a. ''Conversion of a new metabolite to aflatoxin B2 by *Aspergillus parasiticus,*'' *Appl. Environ. Microbiol.* 53:2804–2807.

Cleveland, T. E., Lax, A. R., Lee, L. S., and Bhatnagar, D. 1987b. ''Appearance of enzyme activities catalyzing conversion of sterigmatocystin to aflatoxin B_1 in late growth-phase *Aspergillus parasiticus cultures,*'' *Appl. Environ. Microbiol.* 53:1711–1713.

Cleveland, T. E., Cary, J. W., Bhatnagar, D., Yu, J.-J., Chang, P.-K., Chlan, C. A., and Rajasekaran, K. 1997. ''Use of biotechnology to eliminate aflatoxin in preharvest crops,'' *Bull. Inst. Compr. Agr. Sci.* 5:75–90.

Cleveland, T. E., Brown, R. L., Cary, J. W., Bhatnagar, D., Cotty, P. J., Mellon, J. E., Jacks, T. J., Neucere, J. N., Guy, P. A., Tuzun, S., Chlan, C. A., Ozias-Akins, P., and Weissinger, A. 1992. ''Induction of growth and aflatoxin biosynthesis inhibitors: sources of resistance genes for genetic engineering of crops,'' *Proceedings: USDA-ARS Aflatoxin Elimination Workshop,* October 28–29, Fresno, CA, p. 18.

Clevström, G., Ljunggren, H., Tegelström, S., and Tideman, K. 1983. ''Production of aflatoxin by an isolate cultured under a limited oxygen supply,'' *Appl. Environ. Microbiol.* 46:400–405.

Cole, R. J. and Cox, R. H. 1981. *Handbook of Toxic Fungal Metabolites.* New York: Academic Press. pp. 1–66.

Cotty, P. J. 1988. ''Aflatoxin and sclerotia production by influence of pH,'' *Phytopathology* 78:1250–1253.

Cotty, P. J., Bayman, P., Egel, D. S., and Elias, K. S. 1994. ''Agriculture, aflatoxins, and *Aspergillus,*'' in *The Genus Aspergillus,* ed., Powell, K. New York: Plenum Press. pp. 1–27.

Coupland, K. and Niehaus, W. G. 1987. ''Effect of nitrogen supply, Zn^{2+}, and salt concentration on kojic acid and versicolorin biosynthesis by *Aspergillus parasiticus,*'' *Exptl. Mycol.* 11:206–213.

Cullen, J. M. and Newberne, P. M. 1994. ''Acute hepatotoxicity of aflatoxins, ''in *The Toxicology of Aflatoxins,* eds., Eaton, D. L. and Groopman, J. D. New York: Academic Press. pp. 3–26.

Davis, N. D. and Diener, U. L. 1986. ''Biology of and *A. parasiticus*'' in *Aflatoxin in maize, Proceedings of the Workshops Sponsored by CIMMYT, UNDP, and USAIS.* pp. 33–40.

DeLuca, C., Passi, S., Fabbri, A. A., and Fanelli, C. 1995. ''Ergosterol oxidation may be considered a signal for fungal growth and aflatoxin production in *Aspergillus parasiticus,*'' *Food Addit. Contam.* 12:445–450.

Detroy, R. W., Lillehoj, E. B., and Ciegler, A. 1971. ''Aflatoxin and related compounds,'' in *Microbial Toxins,* Vol. 6, ed., A. Ciegler, S. Kadis, and S. J. Ajl. (New York: Academic Press,) pp. 3–178.

Doehlert, D. C., Wicklow, D. T., and Gardner, H. W. 1993. ''Evidence implicating the lipoxygenase pathway for providing resistance to soybeans against,'' *Proceedings: USDA-ARS Aflatoxin Elimination Workshop,* Oct. 25–26, Little Rock, AR.

Dowd, P. F. 1991. ''Insect interactions with mycotoxin producing fungi and their hosts,'' in *Mycotoxins in Ecological Systems* New York: eds., D. Bhatnagar, E. B. Lillehoj, and D. K. Arora. Marcel Dekker. pp. 137–155.

Dutton, M. F. 1988. ''Enzymes and aflatoxin biosynthesis,'' *Microbiol. Rev.* 52:274–295.

Dutton, M. F. and Anderson, M. S. 1980. ''Inhibition of aflatoxin biosynthesis by organophosphorus compounds,'' *J. Food Protec.* 43:381–384.

Eaton, D. L. and Groopman, J. D. 1994. *The Toxicology of Aflatoxins: Human Health, Veterinary and Agricultural Significance.* London: Academic Press. p. 544.

Eaton, D. L., Ramsdell, H. S., and Neal, G. E. 1994. ''Biotransformation of aflatoxins,'' in *The Toxicology of Aflatoxins,* eds., D. L. Eaton and J. D. Groopman. New York: Academic Press. pp. 45–72.

Ehrlich, K. C., Montalbano, B. G., and Cary, J. W. 1999. ''Bonding of the C6-zinc cluster protein, AFLR, to the promoters of aflatoxin pathway biosynthesis genes in *Aspergillus parasiticus,*'' *Gene* 230:249–257.

Ellis, W. O., Smith, J. P., and Simpson, B. K. 1991. ''Aflatoxin in food: Occurrence, biosynthesis, effects on organisms, detection, and methods for control,'' *Critical Reviews in Food Science and Human Nutrition* 30:403–439.

Fabbri, A. A., Fanelli, C., Panfili, G., Passi, S., and Fasella, P. 1983. ''Lipoperoxidation and aflatoxin biosynthesis by *Aspergillus flavus* and *Aspergillus parasiticus,*'' *J. Gen. Microbiol.* 129:3447–3452.

Fanelli, C. and Fabbri, A. A. 1989. ''Relationship between lipids and aflatoxin biosynthesis,'' *Mycopathologia* 107:115–120.

Fanelli, C., Fabbri, A. A., Finotti, E., and Panfili, G. 1983a. ''Cerulenin and tetrahydrocerulenin: Stimulating factors of aflatoxin biosynthesis,'' *Trans. Br. Mycol. Soc.* 81:201–204.

Fanelli, C., Fabbri, A. A., Finotti, E., and Passi, S. 1983b. ''Stimulation of aflatoxin biosynthesis by lipophilic epoxides,'' *J. Gen. Microbiol.* 129:1721–1723.

Failla, L. J., Lynn, D., and Niehaus, W. G. 1986. ''Correlation of Zn^{2+} content with aflatoxin content of corn,'' *Appl. Environ. Microbiol.* 52:73–74.

Feng, G. H. and Leonard, T. J. 1995. ''Characterization of the polyketide synthase gene (*pksL1*) required for aflatoxin biosynthesis in *Aspergillus parasiticus,*'' *J. Bacteriol.* 177:6246–6254.

Feng, G. S., Chu, F. S., and Leonard, T. J. 1992. ''Molecular cloning of genes related to aflatoxin biosynthesis by differential screening,'' *Appl. Environ. Microbiol.* 58:445–460.

Fernandes, M., Keller, N. P., and Adams, T. H. 1998. ''Sequence specific binding by *Aspergillus nidulans AflR,* a C_6 zinc cluster protein regulating mycotixin biosynthesis,'' *Mol. Microbiol.* 28:1355–1365.

Flaherty, J. E., Weaver, M. A., Payne, G. A., and Woloshuk, C. P. 1995. ''A β-glucuronidase reporter gene construct for monitoring aflatoxin biosynthesis in *Aspergillus flavus,*'' *Appl. Environ. Microbiol.* 61:2482–2486.

Floyd, J. C., Mills, J., and Bennett, J. W. 1987. ''Biotransformation of sterigmatocystin and absence of aflatoxin biotransformation by blocked mutants of *Aspergillus parasiticus,*'' *Exp. Mycol.* 11:109–114.

Foutz, K. R., Woloshuk, C. P., and Payne, G. A. 1995. ''Cloning and assignment of linkage group loci to a karyotype map of the filamentous fungus *Aspergillus flavus,*'' *Mycopathologia* 87:787–794.

Gardner, H. W., Grove, M. J., and Keller, N. P. 1998. ''Soybean lipoxygenase is active on nonaqueous media at low moisture: A constraint to xerophilic fungi and aflatoxins,'' *JAOCS* 75:1801–1808.

Geiser, D. M., Pitt, J. I., and Taylor, J. W. 1998. ''Cryptic speciation and recombination in the aflatoxin-producing fungus *Aspergillus flavus,*'' *Proc. Natl. Acad. Sci.* USA, 95:388–393.

Goto, T., Wicklow, D. T., and Ito, Y. 1996. ''Aflatoxin and cyclopiazonic acid production by a sclerotium-producing *Aspergillus tamarii,* strain,'' *Appl. Environ. Microbiol.* 62:4036–4038.

Gourama, H. and Bullerman, L. B. 1995. ''*Aspergillus flavus* and *Aspergillus parasiticus:* aflatoxigenic fungi of concern in food and feeds: a review,'' *J. Food Protection,* 58:1395–1404.

Goynes, W. R. and Lee, L. S. 1989. ''*Aspergillus flavus* infection of developing cottonseed: Microscopical determination of mycelial progression and associated aflatoxin formation,'' *Arch. Environ. Contam. Toxicol.* 18:421–429.

Greene-McDowelle, D. M., Ingber, B., Wright, M. S., Zeringue, Jr., H. J. Bhatnagar, D., and Cleveland, T. E. 1999. The Effects of selected cotton-leaf volatiles on growth, development and aflatoxin production of *Aspergillus parasiticus*," *Toxicon* 37:883–893.

Gupta, S. K., Maggon, K. K., and Venkitsubramanian, T. A. 1977a. "Regulation of aflatoxin biosynthesis: 1. Comparative study of mycelial composition and glycolysis in aflatoxigenic and non-aflatoxigenic strains," *Microbiol.* 18:27–33.

Gupta, S. K., Maggon, K. K., and Venkitsubramanian, T. A. 1977b. "Regulation of aflatoxin biosynthesis. 2. Comparative study of tricarboxylic acid cycle in aflatoxigenic and non-aflatoxigenic strains of *Aspergillus flavus*," *Microbiol.* 19:7–15.

Harris, J. L. 1986. "Modified method for fungal slide culture," *J. Clin. Microbiol.* 24:460–461.

Heathcote, J. G., Dutton, M. F., and Hibbert, J. R. 1976. "Biosynthesis of aflatoxins. Part II," *Chem. Ind.* 1976:270–273.

Henderberg, A., Bennett, J. W., and Lee, L. S. 1988. "Biosynthetic origin of aflatoxin G_1: confirmation of sterigmatocystin and lack of confirmation of aflatoxin B_1 as precursors," *J. Gen. Microbiol.* 134:661–667.

Hicks, J. K., Yu, J.-H., Keller, N. P., and Adams, T. H. 1997. "*Aspergillus* sporulation and mycotoxin production both require inactivation of the FadA G protein-dependent signaling pathway," *EMBO J.* 16:4916–4923.

Hitokoto, H., Miorozumi, S., Wauke, T., Sakai, S., and Ueno, I. 1978. "Inhibitory effects of condiments and herbal drugs on the growth and toxin production of toxigenic fungi," *Mycopathologia* 66:161–167.

Holcomb, M., Wilson, D. M., Trucksess, M. W., and Thompson, H. C. 1992. "Determination of aflatoxins in food products by chromatography," *J. Chromatog.* 624:341–352.

Hopwood, D. A. and Khosla, C. 1992. "Genes for polyketide secondary metabolic pathways in microorganisms and plants," in *Secondary Metabolites: Their Function and Evolution, Ciba Foundation Symposium,* vol. 171, eds., Chadwick, D. J., Whelan, J., Chichester, J. New York: Wiley & Sons. pp. 88–112.

Horng, J. S., Chang, P.-K., Pestka, J. J., and Linz, J. E. 1990. "Development of a homologous transformation system for *Aspergillus parasiticus* with the gene encoding nitrate reductase," *Mol. Gen. Genet.* 224:294–296.

Hsieh, D. P. H. 1987. "Mode of action of mycotoxins," in *Mycotoxins in Foods,* ed., P. Krogh. New York: Academic Press. pp. 149–176.

Hsieh, D. P. H. and Wong, J. J. 1994. "Pharmacokinetics and excretion of aflatoxins," in *The Toxicology of Aflatoxins,* ed., Eaton, D. L. and Groopman, J. D. New York: Academic Press. pp. 73–88.

Hsieh, D. P. H., Wan, C. C., and Billington, J. A. 1989. "A versiconal hemiacetal acetate converting enzyme in aflatoxin biosynthesis," *Mycopathologia* 107:121–126.

Huang, Z., White, D. G., and Payne, G. A. 1997. "Corn seed proteins inhibitory to and aflatoxin biosynthesis," *Phytopathology* 87:622–627.

Jefferson, R. A. 1987. *GUS Gene Fusion System: User's Manual.* Palo Alto, CA: CLONTECH. Laboratories.

Jelinek, C. F., Pohland, A. E., and Wood, G. E. 1989. "Worldwide occurrence of mycotoxins in foods and feeds—an update," *J. Assoc. Off. Anal. Chem.* 72:223–230.

Johnston, M. 1987. "A model fungal gene regulatory mechanism: the GAL genes of *Saccharomyces cerevisiae,*" *Microbiol. Rev.* 51:458–476.

Kachholz, T. and Demain, A. L. 1983. "Nitrate repression of averufin and aflatoxin biosynthesis." *J. Natural Products* 46:499–506.

Kale, S. P., Bhatnagar, D., and Bennett, J. W. 1994. "Isolation and characterization of morphological variants of *Aspergillus parasiticus* deficient in secondary metabolite production,'' *Mycol. Res.* 98:645–652.

Kale, S. P., Cary, J. W., Bhatnagar, D., and Bennett, J. W. 1996. ''Characterization of experimentally induced, nonaflatoxigenic variant strains of *Aspergillus parasiticus,*'' *Appl. Environ. Microbiol.* 62:3399–3404.

Kelkar, H. S., Keller, N. P., and Adams, T. H. 1996. ''*Aspergillus nidulans stcP* encodes an O-methyltransferase that is required for sterigmatocystin biosynthesis,'' *Appl. Environ. Microbiol.* 62:4296–4298.

Kelkar, H. S., Skloss, T. W., Haw, J. F., Keller, N. P., and Adams, T. H. 1997. ''*Aspergillus nidulans stcL* encodes a putative cytochrome P-450 monooxygenase required for bisfuran desaturation during aflatoxin/sterigmatocystin biosynthesis,'' *J. Biol. Chem.* 272:1589–1594.

Keller, N. P. 1996. ''Products of the lipoxygenase pathway in plants that regulate aflatoxin biosynthesis,'' *Proceedings: USDA-ARS Aflatoxin Elimination Workshop,* October 28–29, 1996, Fresno, CA. p. 64.

Keller, N. P., Cleveland, T. E., and Bhatnagar, D. 1992. ''Variable electrophoretic karyotypes of members of *Aspergillus* section *flavi,*'' *Curr. Genet.* 21:371–375.

Keller, N. P., Butchko, R. A., Sarr, B., and Phillips, T. D. 1993a. ''Visualization of mycotoxin production in maize tissues,'' *Proceedings: USDA-ARS Aflatoxin Elimination Workshop,* Oct. 25–26, Little Rock, AR. p. 78.

Keller, N. P., Segner, S., Bhatnagar, D., and Adams, T. H. 1995. ''*stcS,* a putative P-450 monooxygense, is required for the conversion of versicolorin to sterigmatocystin in *Aspergillus nidulans,*'' *Appl. Environ. Microbiol.* 61:3628–3632.

Keller, N. P., Dischinger, C. J., Bhatnagar, D., Cleveland, T. E., and Ullah, A. H. J. 1993b. ''Purification of a 40-kilodalton methyltransferase active in the aflatoxin biosynthetic pathway,'' *Appl. Environ. Microbiol.* 59:479–484.

Keller, N. P., Nesbitt, C., Sarr, B., Phillips, T. D., and Burrow, G. D. 1997. ''pH regulation of sterigmatocystin and aflatoxin biosynthesis in *Aspergillus* spp,'' *Phytopathology* 87:643–648.

Khan, S. N. and Venkitasubramanian, T. A. 1987. ''Cyclic AMP pool and aflatoxin production in *Aspergillus parasiticus* NRRL 3240 and NRRL 3537,'' *Indian J. of Biochem. Biophys.* 24:308–313.

Klich, M. A., Montalbano, B., and Ehrlich, K. 1997. ''Northern analysis of aflatoxin biosynthesis genes in *Aspergillus parasiticus* and *Aspergillus sojae,*'' *Appl. Microbiol. Biotechnol.* 47:246–249.

Klich, M. A., Yu, J., Chang, P. K., Mullaney, E. J., Bhatnagar, D., and Cleveland, T. E. 1995. ''Hybridization of genes involved in aflatoxin biosynthesis to DNA of aflatoxigenic and non-aflatoxigenic aspergilli,'' *Appl. Microbiol. Biotechnol.* 44:439–443.

Kolb, A., Busby, S., Buc, H., Garges, S., and Adhya, S. 1993. ''Transcriptional regulation by cAMP and its receptor protein,'' *Ann. Rev. Biochem.* 62:749–795.

Kurtzman, C. P., Smiley, M. J., Robnett, C. J., and Wicklow, D. T. 1986. ''DNA relatedness among wild and domesticated species in the *Aspergillus flavus* group,'' *Mycologia* 78:955–959.

Kusumoto, K. and Hsieh, D. P. H. 1996. ''Purification and characterization of the esterases involved in aflatoxin biosynthesis in *Aspergillus parasiticus,*'' *Can. J. Microbiol.* 42:804–810.

Leaich, L. L. and Papa, K. E. 1974. ''Aflatoxins in mutants of *Aspergillus flavus,*'' *Mycopathol. Mycol. Appl.* 52:223–229.

Lee, B. N. and Adams, T. H. 1994a. ''Overexpression of *flbA,* an early regulator of *Aspergillus* asexual sporulation, leads to activation of *brlA* and premature initiation of development,'' *Mol. Microbiol.* 14:323–334.

Lee, B. N. and Adams, T. H. 1994b. ''The *Aspergillus nidulans flu*G gene is required for production of an extracellular developmental signal and is related to prokaryotic glutamine synthetase I," *Genes Devel.* 8:641–651.

Lee, B. N. and Adams, T. H. 1995. "*flu*G and *fib*A function interdependently to initiate conidiophore development in *Aspergillus nidulans* through *brlA* activation," *EMBO J.* 15:299–309.

Lee, R. C., Cary, J. W., Bhatnagar, D., and Chu, F. S. 1995. "Production and characterization of polyclonal antibodies against norsolorinic acid reductase involved in aflatoxin biosynthesis," *Food. Agri. Immunol.* 7:21–32.

Liang, S.-H. 1996. The function and expression of the *ver-1* gene and localization of the *Ver-1* protein involved in aflatoxin biosynthesis in *Aspergillus parasiticus.* PhD Dissertation, Michigan State University.

Liang, S.-H., Skory, C. D., and Linz, J. E. 1996. "Characterization of the function of the *ver-1*A and *ver-1*B genes involved in aflatoxin biosynthesis in *Aspergillus parasiticus,*" *Appl. Environ. Microbiol.* 62:4568–4575.

Liang, S.-H., Wu, T.-S., Lee, R., Chu, F. S., and Linz, J. E. 1997. "Analysis of the mechanisms regulating the expression of the *ver-1* gene involved in aflatoxin biosynthesis," *Appl. Environ. Microbiol.* 63:1058–1065.

Lillehoj, E. B. 1992. "Aflatoxins: genetic mobilization agent," in *Handbook of Applied Mycology, Vol. 5. Mycotoxins in Ecological Systems,* eds., D. Bhatnagar, E. B. Lillehoj and D. K. Anora. New York: Marcel Decker, pp. 1–22.

Lin, B.-K. and Anderson, J. A. 1992. "Purification and properties of versiconal cyclase from *Aspergillus parasiticus,*" *Arch. Biochem. Biophys.* 293:67–70.

Liu, B.-H. and Chu, F. S. 1998. "Regulation of *aflR* and its product, AflR, associated with aflatoxin biosynthesis," *Appl. Environ. Microbiol.* 64:3718–3727.

Luchese, R. H. and Harrigan, W. F. 1993. "Biosynthesis of aflatoxin—the role of nutritional factors," *J. Appl. Bacteriol.* 74:5–14.

Mahanti, N., Bhatnagar, D., Cary, J. W., Joubran, J., and Linz, J. E. 1996. "Structure and function of *fas*-1A, a putative fatty acid synthetase directly involved in aflatoxin biosynthesis in *Aspergillus parasiticus,*" *Appl. Environ. Microbiol.* 62:191–195.

Maggon, K. K. and Venkitasubramanian, T. A. 1973. "Metabolism of aflatoxin B_1 and G_1 by *Aspergillus parasiticus,*" *Experientia* 29:1211–1222.

Mannon, J. and Johnson, E. 1985. "Fungi down on the farm," *New Scientist* 105:12–16.

Matsushima, K.-I., Ando, Y., Hamasaki, T., and Yabe, K. 1994. "Purification and characterization of two versiconal hemiacetal acetate reductases involved in aflatoxin biosynthesis," *Appl. Environ. Microbiol.* 60:2561–2567.

Mazur, P., Meyers, H. V., and Nakanishi, K. 1990. "Structural elucidation of sporogenic fatty acid metabolites from *Aspergillus nidulans,*" *Tetrahedron Letts.* 31:3837–3840.

McGuire, S. M., Silva, J. C., Casillas, E. G., and Townsend, C. A. 1996. "Purification and characterization of versicolorin B synthase from *Aspergillus parasiticus:* Catalysis of the stereodifferentiating cyclization in aflatoxin biosynthesis essential to DNA interaction," *Biochemistry* 35:11470–11486.

McGuire, S. M., Brobst, S. W., Graybill, T. L., Pal, K., and Townsend, C. A. 1989. "Partitioning of tetrahydro- and dihydrobisfuran formation in aflatoxin biosynthesis defined by cell-free and direct incorporation experiments," *J. Am. Chem. Soc.* 111:8308–8309.

Mellon, J. E. and Cotty, P. J. 1996. "Aflatoxin inhibitors and stimulators from oilseeds," *Proceedings: USDA-ARS Aflatoxin Elimination Workshop,* October 28–29, Fresno, CA. p. 90.

Meyers, D. M., O'Brian, G., Du, W. L., Bhatnagar, D., and Payne, G. A. 1998. "Characterization of *aflJ,* a gene required for the conversion of pathway intermediates to aflatoxin," *Appl. Environ. Microbiol.* 64:3713–3717.

Miller, M. J., Rarick, M. D., and Linz, J. E. 1996. ''Analysis of the *nor-1* and *ver-1* promoters in the filamentous fungus *Aspergillus parasiticus,*'' Institute for Environmental Toxicology Forum, Lansing Center, Lansing MI.

Minto, R. E. and Townsend, C. A. 1997. ''Enzymology and molecular biology of aflatoxin biosynthesis,'' *Chem. Rev.* 97:2537–2555.

Moyne, A. L., Karyalas, G., Cleveland, T. E., Cary, J. W., and Tuzun, S. 1993. ''Bacterial chitinases for control of aflatoxin contamination,'' *Proceedings: USDA-ARS Aflatoxin Elimination Workshop,* Oct. 25–26, Little Rock, AR. p. 72.

Nichols, Jr., T. E. 1983. ''Economic impact of aflatoxin in corn,'' in *Aflatoxin and Aspergillus flavus in Corn,* eds. Diener, U. L. et al., Auburn University. pp. 67–71.

Niehaus, W. G., Jr. and Coupland, K. 1987. ''Effect of orthophosphate on versicolorin synthesis by a mutant strain of *Aspergillus parasiticus,*'' *Exper. Mycol.* 11:154–158.

Niehaus, W. G. and Dilts, R. P. 1982. ''Biosynthesis of the aflatoxin precursor sterigmatocystin by *Aspergillus parasiticus,*'' *J. Bacteriol.* 151:243–250.

Niehaus, W. G. and Dilts, R. P. 1984. ''Purification and characterization of glucose-6-phosphate dehydrogenase from *Aspergillus parasiticus,*'' *Arch. Biochem. Biophys.* 228:113–119.

Niehaus, W. G. and Jiang, W. 1989. ''Nitrate induces enzymes of the mannitol cycle and suppresses versicolorin synthesis in *Aspergillus parasiticus,*'' *Mycopathologia* 107:131–137.

Otsuka, F., Iwamatsu, A., Suzuki, K., Uhsawa, M., Hamer, D. H., and Koizumi, S. 1994. ''Purification and characterization of a protein that binds to metal responsive elements on the human metallothionein IIA gene,'' *J. Biol. Chem.* 269:23700–23707.

Passi, S., Fanelli, C., Fabbri, A. A., Finotti, E., Panfili, G., and Nazzaro-Porro, M. 1985. ''Effect of halomethanes on aflatoxin induction in cultures of *Aspergillus parasiticus,*'' *J. Gen. Microbiol.* 131:687–691.

Payne, G. A. 1992. ''Aflatoxin in maize,'' *Critical Rev. Plant Sci.* 10:423–440.

Payne, G. A. 1996. ''A corn metabolite that inhibits aflatoxin synthesis,'' *Proceedings: USDA-ARS Aflatoxin Elimination Workshop.* October 28–29, 1996, Fresno, CA. p. 86.

Payne, G. A. and Brown, M. P. 1998. ''Genetics and physiology of aflatoxin biosynthesis,'' *Annu. Rev. Phytopathol.* 36:329–362.

Payne, G. A. and Hagler, Jr., W. M. 1983. ''Effect of specific amino acids on growth and aflatoxin production by *Aspergillus parasiticus* and in defined media,'' *Appl. Environ. Microbiol.* 46:805–812.

Payne, G. A., Nystrom, G. J., Bhatnagar, D., Cleveland, T. E., and Woloshuk, C. P. 1993. ''Cloning of the *afl*-2 gene involved in aflatoxin biosynthesis from *Aspergillus flavus,*'' *Appl. Environ. Microbiol.* 59:156–162.

Perez-Estaban, B., Orejas, M., Gomez-Pardo, E., and Penalva, M. 1993. ''Molecular characterization of a secondary metabolism promoter. Transcription of the *Aspergillus nidulans* isopenicillin N synthetase is modulated by upstream negative elements,'' *Molec. Microbiol* 9:881–895.

Peters, D. G. and Caddick, M. X. 1994. ''Direct analysis of native and chimeric GATA specific DNA binding proteins from *Aspergillus nidulans,*'' *Nucleic Acids Res.* 22:51674–51672.

Prieto, R. and Woloshuk, C. P. 1997. ''*ord1,* an oxidoreductase gene responsible for conversion of *O*-methylsterigmatocystin to aflatoxin in *Aspergillus flavus,*'' *Appl. Environ. Microbiol.* 63:1661–1666.

Prieto, R., Yousibova, G. L., and Woloshuk, C. P. 1996. ''Identification of aflatoxin biosynthetic genes by genetic complementation in a mutant of *Aspergillus flavus* lacking the aflatoxin gene cluster,'' *Appl. Environ. Microbiol.* 62:3567–3571.

Rabie, C. J., Meyer, C. J., van Heerden, L., and Lubben, A. 1981. ''Inhibitory effect of molybdenum and vanadium salts on aflatoxin B_1 synthesis by *Aspergillus flavus*," *Can. J. Microbiol.* 27:962–967.

Reding, C. L. and Harrison, M. 1994. "Possible relationship of succinate dehydrogenase activity and fatty acid synthetase activities to *Aspergillus parasiticus* (NRRL 5139) growth and aflatoxin production," *Mycopathologia* 127:175–181.

Robens, J. F. and Richards, J. L. 1992. "Aflatoxins in animal and human health," *Rev. Environ. Contam. Tox.* 127:69–94.

Roebuck, B. D. and Maxuitenko, Y. Y. 1999. "Biochemical mechanisms and biological implications of the toxicity of aflatoxins as related to aflatoxin carcinogenesis," in *The Toxicology of Aflatoxins,* eds., D. L. Eaton and J. D. Groopman. New York: Academic Press. pp. 27–43.

Shane, S. M. 1994. "Economic issues associated with aflatoxins," in *The Toxicology of Aflatoxins,* eds., D. L. Eaton and J. D. Groopman. London: Academic Press, Inc. pp. 513–527.

Silva, J. C. and Townsend, C. A. 1996. "Heterologous expression, isolation, and characterization of versicolorin B synthase from *Aspergillus parasiticus,*" *J. Biol. Chem.* 272:804–813.

Silva, J. C., Minto, R. E., Barry, C. E., III, Holland, K. A., and Townsend, C. A. 1996. "Isolation and characterization of the versicolorin B synthase gene from *Aspergillus parasiticus*; expansion of the gene cluster," *J. Biol. Chem.* 271:13600–13608.

Skory, C. D., Chang, P.-K., and Linz, J. E. 1993. "Regulated expression of the *nor-1* and *ver-1* genes associated with aflatoxin biosynthesis," *Appl. Environ. Microbiol.* 59:1642–1646.

Skory, C. D., Chang, P.-K., Cary, J. W., and Linz, J. E. 1992. "Isolation and characterization of a gene from *Aspergillus parasiticus* associated with conversion of versicolorin A to sterigamatocystin in aflatoxin biosynthesis," *Appl. Environ. Microbiol.* 58:3527–3537.

Skory, C. D., Horng, J. S., Pestka, J. J., and Linz, J. E. 1990. "A transformation system for *Aspergillus parasiticus* based on a homologous gene involved in pyrimidine biosynthesis (pyrG)," *Appl. Environ. Microbiol.* 56:3315–3320.

Smart, M. G., Wicklow, D. T., and Caldwell, R. W. 1990. "Pathogenesis in *Aspergillus* ear rot of maize: Light microscopy of fungal spread from wounds," *Phytopathology* 80:1287–1294.

Stoloff, L., van Egmond, H. P., and Park, D. L. 1991. "Rationales for the establishment of limits and regulations for mycotoxins," *Food Add. Contam.* 8:213–222.

Stossel, P. 1986. "Aflatoxin contamination in soybeans: role of proteinase inhibitors, zinc availability, and seed coat integrity," *Appl. Environ. Microbiol.* 52:68–72.

Tice, G. and Buchanan, R. L. 1981. "Regulation of aflatoxin biosynthesis: Effect of exogenously supplied cyclic nucleotides," *J. Food Sci.* 47:153–157.

Tilburn, J., Sarkar, S., Widdick, D. A., Espeso, E. A., Orejas, M., Mungroo, J., Penalva, M. A., and Arst, H. N. 1995. "The *Aspergillus* PacC zinc finger mediates regulation of both acid- and alkaline-expressed genes by ambient pH," *EMBO J.* 14:779–790.

Tiwari, R. P., Mittal, V., Bhalla, T. C., Saini, S. S., Singh, G., and Vadhera, D. V. 1986. "Effect of metal ions on aflatoxin production by *Aspergillus parasiticus,*" *Folia Microbiol.* 124:124–128.

Trail, F. and Linz, J. E. 1995. Regulation of the *nor-1* gene involved in aflatoxin biosynthesis in *Aspergillus parasiticus.* 18th Fungal Genetics Conference, Asilomar, CA.

Trail, F. and Linz, J. E. 1996. "Characterizing the natural resistance of peanut to *Aspergillus parasiticus* and aflatoxin," *Proceedings: USDA-ARS Aflatoxin Elimination Workshop,* October 28–29, 1996, Fresno, CA. p. 42.

Trail, F., Mahanti, N., and Linz, J. E. 1995a. "Review—Molecular biology of aflatoxin biosynthesis," *Microbiol.* 141:755–765.

Trail, F., Chang, P.-K., Cary, J. W., and Linz, J. E. 1994. "Structural and functional analysis of the *nor-1* gene involved in the biosynthesis of aflatoxins by *Aspergillus parasiticus,*" *Appl. Environ. Microbiol.* 60:4078–4085.

Trail, F., Mahanti, N., Rarick, M., Mehigh, R., Liang, S. H., Zhou, R., and Linz, J. E. 1995b. "A physical and transcriptional map of an aflatoxin gene cluster in *Aspergillus parasiticus* and the functional disruption of a gene involved early in the aflatoxin pathway," *Appl. Environ. Microbiol.* 61:2665–2673.

Trucksess, M. W. and Wood, G. E. 1994. "Recent methods of analysis for aflatoxins in foods and feeds," in *The Toxicology of Aflatoxins: Human Health Veterinary and Agricultural Significance,* eds., E. L. Eaton and J. D. Groopman. San Diego, CA: Academic Press. pp. 409–431.

Valcarcel, R., Bennett, J. W., and Vitanza, J. 1986. "Effect of selected inhibitors on growth, pigmentation, and aflatoxin production by *Aspergillus parasiticus,*" *Mycopathologia* 94:7–10.

van Egmond, H. P. 1989. "Current situation on regulations of mycotoxins. Overview of tolerance and status of standard methods of sampling and analysis," *Food Addit. Contam.* 6:139–188.

van Rensburg, S. J. (1977). "Role of epidemiology in the elucidation of mycotoxin health risks," in *Mycotoxins in Human and Animal Health,* eds., J. V. Rodricks, C. W. Hesseltine, and M. A. Mehlman. Forest Park South, Ill: Pathotox. pp. 699–711.

Wang, I. K., Reeves, C., and Gaucher, M. 1991. "Isolation and sequencing of a genomic DNA clone containing the 3′ terminus of the 6-methylsalicylic acid polyketide synthase gene of *Penicillium urticae,*" *Can. J. Microbiol.* 37:86–95.

Watanabe, C. M. H., Wilson, D., Linz, J. E., and Townsend, C. A. 1996. "Demonstration of the catalytic roles and evidence for the physical association of type I fatty acid synthases and a polyketide synthase in the biosynthesis of aflatoxin B1," *Chem. Biol.* 3:463–469.

Wicklow, D. T. 1988. "Metabolites in the coevolution of fungal defense systems" in *Coevolution of fungi with plants and animals,* eds. K. A. Pirozynski and D. L. Hawksworth, London: Academic Press. pp. 173–201.

Wicklow, D. T. and Shotwell, O. L. 1982. Intrafungal distribution of aflatoxins among conidia and sclerotia of *Aspergillus flavus* and *Aspergillus parasiticus,*" *Can. J. Microbiol.* 29:1–5.

Wilson, D. L., Wu, T.-S., Trail, F., and Linz, J. E. 1996. "The expression of β-glucuronidase reporter constructs of aflatoxin biosynthesis genes is position dependent," Institute for Environmental Toxicology Forum, Lansing Center, Lansing MI.

Wilson, D. M., Sydenham, E. W., Lombaert, G. A., Trucksess, M. W., Abramson, D., and Bennett, G. A. 1998. *Mycotoxins in Agriculture and Food Safety.* New York: Marcel Dekker, Inc. pp. 135–182.

Woloshuk, C. 1995. "Controlling aflatoxin production in maize through the identification of inducing metabolites," *Proceedings: USDA-ARS Aflatoxin Elimination Workshop,* October 23–24, 1995, Atlanta, GA. p. 39.

Woloshuk, C. P. and Prieto, R. 1998. "Genetic organization and function of the aflatoxin B1 biosynthetic genes," *FEMS Microbiol. Lett.* 160:169–176.

Woloshuk, C. P., Cavaletto, J. R., and Cleveland, T. E. 1997. "Inducers of aflatoxin biosynthesis from colonized maize kernels are generated by an amylase activity from *Aspergillus flavus,*" *Phytopathology* 87:164–169.

Woloshuk, C. P., Seip, E. R., Payne, G. A., and Adkins, C. R. 1989. "Genetic transformation system for the aflatoxin-producing fungus *Aspergillus flavus,*" *Appl. Environ. Microbiol.* 55:86–90.

Woloshuk, C. P., Yousibova, G. L., Rollins, J. A., Bhatnagar, D., and Payne, G. A. 1995. "Molecular characterization of the *afl-l* locus in *Aspergillus flavus,*" *Appl. Environ. Microbiol.* 61:3019–3023.

Woloshuk, C. P., Foutz, K. R., Brewer, J. F., Bhatnagar, D., Cleveland, T. E., and Payne, G. A. 1994. "Molecular characterization of *aflR*, a regulatory locus for aflatoxin biosynthesis," *Appl. Environ. Microbiol.* 60:2408–2414.

Yabe, K., Ando, Y., and Hamasaki, T. 1988. "Biosynthetic relationship among aflatoxins B_1, B_2, G1, G_2," *Appl. Environ. Microbiol.* 54:2101–2106.

Yabe, K., Ando, Y., and Hamasaki, T. 1991a. "Desaturase activity in the branching step between aflatoxins B_1 and G_1 aflatoxins B_2 and G_2," *Agric. Biol. Chem.* 55:1907–1911.

Yabe, K., Ando, Y., and Hamasaki, T. 1991b. "A metabolic grid among versiconal hemiacetal acetate, versiconol acetate, versiconol and versiconal during aflatoxin biosynthesis," *J. Gen. Microbiol.* 137:2469–2475.

Yabe, K., Matsushima, K.-I., Koyama, T., and Hamasaki, T. 1998. "Purification and characterization of *O*-methyltransferase I involved in conversion of demethylsterigmatocystin to sterigmatocystin and of dihydrodemethylsterigmatocystin to dihydrosterigmatocystin during aflatoxin biosynthesis," *Appl. Environ. Microbiol.* 64:166–171.

Yabe, K., Matsuyama, Y., Ando, Y., Nakajima, H., and Hamasaki, T. 1993. "Steriochemistry during aflatoxin biosynthesis: conversion of norsolorinic acid to averufin," *Appl. Environ. Microbiol.* 59:2486–2492.

Yu, J., Chang, P.-K., Cary, J. W., Bhatnagar, D., and Cleveland, T. E. 1997. "*avnA*, a gene encoding a cytochrome P450 monooxygenase is involved in the conversion of averantin to averufin in aflatoxin biosynthesis in *Aspergillus parasiticus*," *Appl. Environ. Microbiol.* 63:1349–1356.

Yu, J., Cary, J. W., Bhatnagar, D., Cleveland, T. E., Keller, N. P., and Chu, F. S. 1993. "Cloning and characterization of a cDNA from *Aspergillus parasiticus* encoding *O*-methyltransferase involved in aflatoxin biosynthesis," *Appl. Environ. Microbiol.* 59:3564–3571.

Yu, J., Chang, P.-K., Payne, G. A., Cary, J. W., Bhatnagar, D., and Cleveland, T. E. 1995b. "Comparison of the *omtA* genes encoding *O*-methyltransferases involved in aflatoxin biosynthesis from *Aspergillus parasiticus* and *A. flavus*," *Gene*, 163:121–125.

Yu, J., Chang, P.-K., Cary, J. W., Wright, M., Bhatnagar, D., Cleveland, T. E., Payne, G. A., and Linz, J. E. 1995a. "Comparative mapping of aflatoxin pathway gene clusters in *Aspergillus parasiticus* and *Aspergillus flavus*," *Appl. Environ. Microbiol*, 61:2365–2371.

Yu, J., Chang, P.-K., Ehrlich, E. C., Cary, J. W., Montalbano, B., Dyer, J. M., Bhatnagar, D., and Cleveland, T. E. 1998. "Characterization of the critical amino acids of an *Aspergillus parasiticus* cytochrome P-450 mono-oxygenase encoded by *ord*A involved in the biosynthesis of aflatoxin B_1, G_1 and B_2 and G_2," *Appl. Environ. Microbiol.* 64:4834–4841.

Yu, J.-H. and Leonard, T. J. 1995. "Sterigmatocystin biosynthesis in *Aspergillus nidulans* requires novel type I polyketide synthase," *J. Bact.* 177:4792–4800.

Yu, J.-H., Wieser, J., and Adams, T. H. 1996. "The *Aspergillus flb*A RGS domain protein antagonizes G protein signalling to block proliferation and allow development," *EMBO J.* 15:5184–5190.

Yu, J.-H., Butchko, R. E., Fernandes, M., Keller, N. P., Leonard, T. J., and Adams, T. H. 1996. "Conservation of structure and function of the aflatoxin regulatory gene *aflR* from *Aspergillus nidulans* and *A. flavus*," *Curr. Genet.* 29:549–555.

Zaika, L. L. and Buchanan, R. L. 1987. "Review of compounds affecting the biosynthesis of aflatoxins," *J. Food Protec.* 50:691–708.

Zeringue, H. J., Jr. 1991. "Effect of C6 to C9 alkenals on aflatoxin production in corn, cottonseed, and peanuts," *Appl. Environ. Microbiol.* 2433–2434.

Zeringue, H. J., Jr. and Bhatnagar, D. 1994. "Effects of neem leaf volatiles on submerged culture of aflatoxigenic *Aspergillus parasiticus*," *Appl. Environ. Microbiol.* 60:3543–3547.

Zeringue, H. J., Jr. Brown, R. L., Neucere, J. N., and Cleveland, T. E. 1996. "Relationships between C6-C16 alkenals and alkenal volatile contents and resistance of maize genotypes to *Aspergillus flavus* and aflatoxin production," *Proceedings: USDA-ARS Aflatoxin Elimination Workshop,* October 28–29, 1996, Fresno, CA. p. 103.

Zhou, R. 1997. The function, expression, and localization of the Nor-1 protein involved in aflatoxin B_1 biosynthesis; the function of *fluP,* a gene associated with sporulation in *Aspergillus parasiticus.* PhD Dissertation, Michigan State University.

CHAPTER 11

Fusarium Toxins: Trichothecenes and Fumonisins

ROBERT H. PROCTOR

1. INTRODUCTION

FUSARIUM is a cosmopolitan genus of filamentous fungi that includes plant pathogens, saprophytes, and opportunistic human and animal pathogens. Some of the plant pathogenic species of *Fusarium* are problematic to food and feed safety because of the mycotoxins they produce in infected crop plants. These *Fusarium* toxins are low-molecular weight metabolites but vary markedly in structure and the apparent mechanisms by which they induce toxicoses. Some of the more commonly reported toxins are beauvericin, fumonisins, fusaric acid, fusarins, moniliformin, trichothecenes, and zearelanone.

Although a role for *Fusarium* toxins in human diseases has not been demonstrated, epidemiological data from several areas of the world have linked outbreaks of some diseases to the consumption of food contaminated with certain toxins or toxin-producing strains of *Fusarium.* In addition, the harmful effects of some toxins, such as fumonisins and trichothecenes, to human cells has been demonstrated in vitro. In animals, these toxins induce a variety of health problems from inappetence to death. Therefore, ample evidence exists to warrant serious consideration of *Fusarium* toxins as a threat to human health.

The goal of this chapter is to provide an overview of the toxicity and biosynthesis of trichothecenes and fumonisins. Among *Fusarium* toxins, these two groups are currently considered the most significant threat to food and feed safety because of their widespread occurrence in grains and their potent toxicity. For more thorough treatments of various aspects of these toxins, readers are encouraged to refer to several recent reviews (Desjardins et al.,

1993; Desjardins et al., 1996b; Munkvold and Desjardins, 1997; Riley et al., 1996; Rotter et al., 1996).

2. TRICHOTHECENES

2.1. STRUCTURE AND OCCURRENCE

Trichothecenes consist of a family of over 60 sesquiterpenoid compounds. All trichothecenes share the core tricyclic trichothecene structure and typically have a epoxide residue at carbons 12 (C-12) and 13 (C-13) and a double bond between C-9 and C-10 (Figure 11.1). Individual trichothecenes vary by the presence/absence, position, and number of hydroxyl, carbonyl, and ester groups on the core structure. The number of *Fusarium* species that produce these toxins is unclear because of state of flux of *Fusarium* taxonomy and uncertainties in the identification of this fungus. However, some of important trichothecene-producing species include *F. culmorum, F. graminearum, F. poae, F. sambucinum,* and *F. sporotrichioides.* Species within several other fungal genera (e.g., *Myrothecium, Stachybotrys, Trichothecium*) and the plant genus *Baccharis* also produce trichothecenes. However, the trichothecenes produced by *Baccharis* and most of the other fungi are structurally more complex than those produced by *Fusarium* (Sharma and Kim, 1991).

Collectively, trichothecene-producing species of *Fusarium* cause disease

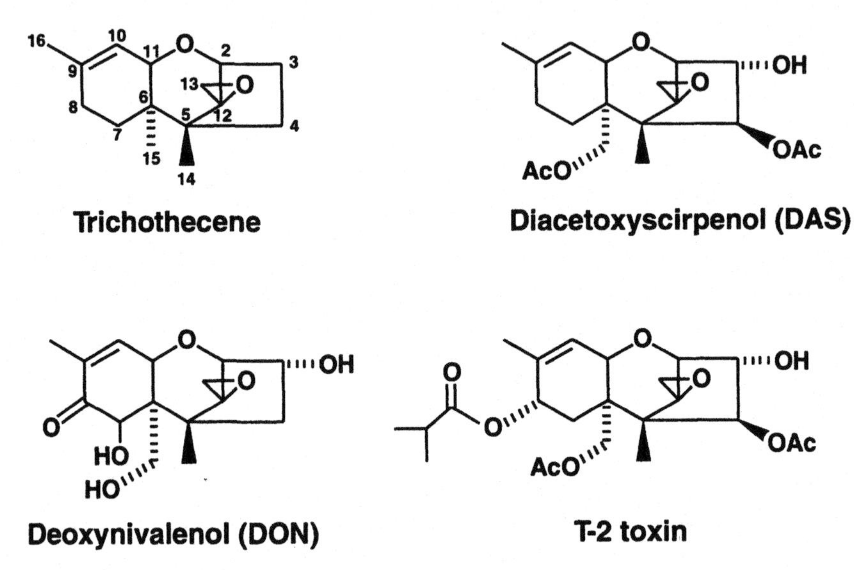

Figure 11.1 The structure of the trichothecenes, trichothecene, deoxynivalenol, diacetoxyscirpenol, and T-2 toxin.

on a multitude of plant species. Although trichothecenes have been detected sporadically in a variety of crops, it is trichothecene contamination of grains, such as maize, wheat, barley, and rye, that poses the most significant threat to food and feed safety. For example, *F. graminearum* incites a disease known as head scab (blight) of wheat and barley and produces high levels of deoxynivalenol (DON, or vomitoxin) and closely related trichothecenes in the infected grain. In North America during the 1990s, severe epidemics of this disease and the resulting contamination of grain have resulted in losses to the agricultural industry in the billions of dollars.

2.2. TOXICOLOGY

Trichothecenes exhibit both dermal and oral toxicity and are associated with several human and animal diseases. Perhaps the most infamous of these is alimentary toxic aleukia, a disease that was widespread in the former Soviet Union during World War II and reported to have been responsible for the deaths of thousands of people (Joffe, 1986). Other human diseases associated with trichothecenes have also been reported in China, India, Japan, and Korea (Beardall and Miller, 1997). Although there is no direct evidence demonstrating that trichothecenes caused these diseases, the toxins are thought to be the involved because the diseases occurred following the consumption of grain severely contaminated with trichothecenes or infected trichothecene-producing species of *Fusarium* and the symptoms are the same as some of those induced by trichothecenes in animals. In addition, when the trichothecene diacetoxyscirpenol was administered to cancer patients as a potential chemotheraputic agent, it caused side effects that were the same as some of the symptoms of diseases attributed to trichothecenes (Committee on Protection Against Mycotoxins, 1983). The symptoms attributed to acute trichothecene toxicoses in several animal species and/or humans include erythema, fever, chills, nausea, vomiting, diarrhea, stomatitis, angina, hemorrhaging of internal organs and gastrointestinal mucosa, generalized burning sensation, central nervous system dysfunction, necrosis of lymphoid tissue and bone marrow and, in some cases, death (Sharma and Kim, 1991).

The effects of trichothecenes on the human immune system are unclear. However, extensive studies with animals have demonstrated that trichothecenes can both suppress and stimulate immunoglobulin production, host resistance to microbes, and tissues associated with the immune response (Pestka and Bondy, 1994). Acute doses of T-2 toxin and DON inhibited the mitogen-induced proliferation of lymphocytes and severely damaged bone marrow, thymus and spleen tissue (Sharma and Kim, 1991). Even though DON inhibited lymphocyte proliferation in vitro, it stimulated the secretion of the cytokines interleukin (IL)-2, IL-5, and IL-6 from these cells (Azcona-Olivera et al., 1995). In mice and swine, DON and T-2 toxin reduced antibody titers to sheep red blood cells while exposure to DON led to an increase in serum IgA levels in mice (Rosenstein et al., 1979; Rotter et al., 1996). The increased production of IgA following exposure to DON may be the result of increased secretion of cytokines, which are known to enhance differentiation of IgA-producing cells (Rotter et al., 1996).

Trichothecenes have a multitude of cytotoxic effects, including the inhibition of respiration and synthesis of DNA, RNA, and protein, at concentrations of less than 1 μM. The toxins also induce cell death over a range of concentrations (Holt and DeLoach, 1988; Koshinsky et al., 1988; Sharma and Kim, 1991). For example, T-2 toxin induced apoptosis, or programmed cell death, in human cell line HL-60 at concentrations of 0.02–0.2 μM (Yoshino et al., 1996).

2.3. MECHANISM OF ACTION

Although a number of biochemical processes are altered by trichothecenes, which alteration(s) is responsible for the toxicological effects induced by these toxins has not been demonstrated unequivocally. However, inhibition of protein synthesis is often considered the most likely mechanisms by which these toxins induce disease symptoms.

Trichothecenes fall into two distinct groups based on whether they disrupt translation initiation or the elongation/termination steps of translation (Cundliffe et al., 1974; Wei et al., 1974). The current model for inhibition of protein synthesis is that trichothecenes bind at or near the peptidyl transferase center and as a result inhibit this enzyme. This model is supported by the fact that trichothecenes inhibited peptidyl transferase, regardless of whether they disrupted translation initiation or elongation/termination (Wei and McLaughlin, 1974). In addition, all trichothecenes that inhibited protein synthesis bound to eukaryotic ribosomes and competed with one another for the same, or perhaps overlapping, binding sites (Gilly et al., 1985; Middlebrook and Leatherman, 1989). In yeast mutants and in some species of filamentous fungi, trichothecene resistance was localized to the 60S ribosomal subunit, the subunit in which peptidyl transferase is located, and the mutant ribosomes bound trichothecenes less efficiently than wild-type ribosomes (Berry et al., 1978; Iglesias and Ballesta, 1994; Schindler et al., 1974).

The inhibition of protein synthesis as the primary mechanism of action of trichothecenes is an attractive hypothesis, because it would account for many of the effects induced by these toxins. Pestka and co-workers have also proposed how this model could account for the stimulation of some immune responses. They noted that the rise in serum IgA levels after exposure of mice to trichothecenes may result from increased production of cytokines. Ouyang et al. (1996) hypothesized that increased cytokine production induced by trichothecenes results from inhibition of the synthesis of IκB proteins. These proteins are inactivators of the transcription factor NF-κB/Rel, which is a positive regulator of cytokine gene expression. Experimental evidence in support of this hypothesis includes the observations that exposure of CD4+ cells to DON results in reduced cellular concentrations of the inhibitor IκBα, increased NF-κB/Rel activity, and increased levels of cytokine mRNAs (Azcona-Olivera et al., 1995; Ouyang et al., 1996). Why synthesis of IκB proteins but not cytokines, which are also proteins, would be inhibited by DON is unclear but may involve temporal fluctuations of DON within cells due to its metabolism (Rotter et al., 1996). Although the inhibition of protein synthesis is an attractive model to explain trichothecene toxicity, other mechanisms such as interactions with cell membranes and inhibition of respiration have not been ruled out.

2.4. BIOSYNTHESIS

Trichothecenes are sesquiterpenoids and, therefore, products of isoprenoid metabolism. The cyclization of the isoprenoid intermediate farnesyl diphosphate to trichodiene marks the point at which trichothecene biosynthesis diverges from general isoprenoid metabolism. The subsequent conversion of trichodiene to complex trichothecenes occurs via a linear series of reactions involving a number of intermediate compounds (Figure 11.2). Overall, these reactions consist of the oxygenation of various carbon positions to form hydroxyl groups and the subsequent modification of these group via isomerizations, cyclizations and esterifications (Desjardins et al., 1993). During the formation of T-2 toxin, for example, six carbons of the original trichodiene become oxygenated. Of these, the oxygen at C-2 forms the pyran ring, the C-13 oxygen becomes part of the epoxide group, and the oxygens at positions C-4, C-8, and C-15 are esterified to acetyl or isovaleryl, while the C-3 oxygen persists as part of an hydroxyl group (Figure 11.2) (Desjardins et al., 1993). Biochemical analyses of *F. culmorum* and *F. sporotrichioides* indicate that these species share most of the initial oxygenation and cyclization steps in trichothecene biosynthesis (Figure 11.2) (Corley et al., 1987; Desjardins et al., 1993; Yang et al., 1996). However, the pathway in these two species appears to diverge at the intermediate 15-decalonectrin (Desjardins et al., 1993). The pathway in *F. culmorum* leads to the formation of type B trichothecenes (e.g., deoxynivalenol), which have a keto group at C-8 while the *F. sporotrichioides* pathway leads to the formation of type A trichothecenes (e.g., diacetoxyscirpenol and T-2 toxin) in which the C-8 position is either saturated with hydrogens or hydroxylated with or without esterification.

To my knowledge, the only enzyme involved in trichothecene biosynthesis that has been purified and characterized is trichodiene synthase, which catalyzes the cyclization of farnesyl diphosphate to trichodiene, the first committed step in the trichothecene biosynthetic pathway (Figure 11.2) (Hohn and Van-Middlesworth, 1986). The purification of trichodiene synthase from *F. sporotrichioides* and subsequent cloning of the gene, *TRI5*, that encodes it facilitated

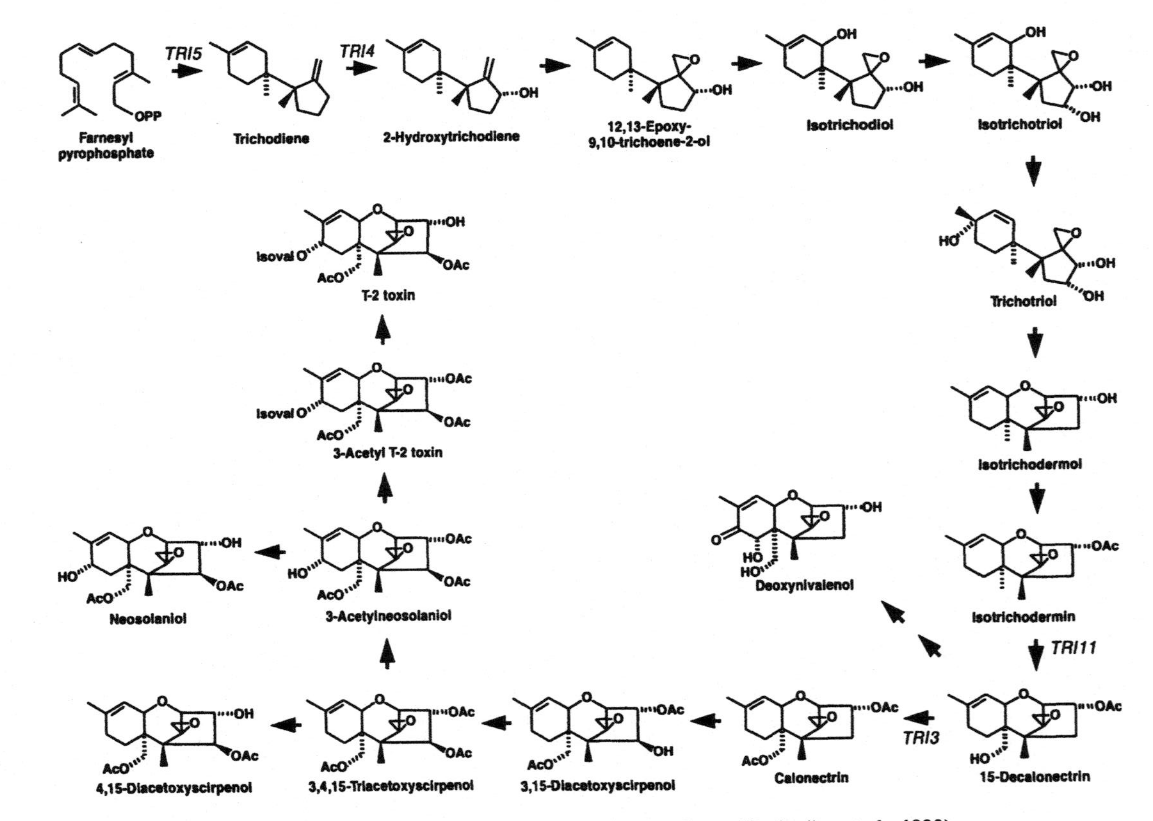

Figure 11.2 Proposed trichothecene biosynthetic pathway (Desjardins et al., 1993).

rapid progress in studies of the molecular genetics, biochemistry, and regulation of trichothecene biosynthesis because other genes involved in this process are tightly linked to *TRI5*. This linkage was first established when two overlapping cosmid clones carrying the *F. sporotrichioides TRI5* gene were identified and introduced into mutants of the fungus that were blocked at various steps of T-2 toxin production (Hohn et al., 1993). When either cosmid clone was introduced into mutants defective at the *TRI3* or *TRI4* locus, toxin production was restored. This indicated that wild-type alleles of the mutated genes were present on the cosmid clones and, therefore, tightly linked to *TRI5*. Thus far, analysis of a 25 kb region of DNA spanning the two cosmids has revealed a cluster of 10 genes that appear to be involved in trichothecene biosynthesis (Figure 11.3) (Beremand and Hohn, 1995; Hohn, 1997). The genes other than *TRI5* for which functions have now been assigned are *TRI3, TRI4, TRI6, TRI11,* and *TRI12.* The predicted amino acid sequences for *TRI4* and *TRI11* are similar to those of cytochrome P450 monooxygenases (Alexander et al., 1998; Hohn et al., 1995). The trichothecene production phenotype of strains of *F. sporotrichioides* in which *TRI4* has been disrupted indicated that this gene catalyzes the oxygenation of trichodiene at the C-2 position while the phenotype of *TRI11*-disrupted strains indicated that it encodes a monooxygenase that catalyzes the C-15 oxygenation of isotrichodermin to yield 15-decalonectrin (Figure 11.2) (Alexander et al., 1998; McCormick and Hohn, 1997). Although the predicted amino acid sequence for *TRI3* is unrelated to other known protein sequences, disruption of this gene revealed that it encodes an *O*-acetyltransferase that catalyzes the conversion of 15-decalonectrin to calonectrin via acetylation of the C-15 position (Figure 11.2) (McCormick et al., 1996). The predicted amino acid sequence for *TRI6* contains three sequence motifs that resemble the Cys_2His_2 DNA-binding domains found in many eukaryotic transcription factors (Proctor et al., 1995b). Disruption of *TRI6* in *F. sporotrichioides* blocked T-2 toxin production and markedly reduced expression of other trichothecene biosynthetic genes. Together these results indicated that *TRI6* is a transcription factor responsible for the positive regulation of trichothecene production. Finally, the predicted amino acid sequence of *TRI12* is similar to members of the major facilitator superfamily of transport proteins and suggests that this gene is a transport proteins possibly involved in movement of trichothecenes from inside the fungal cell to the surrounding environ-

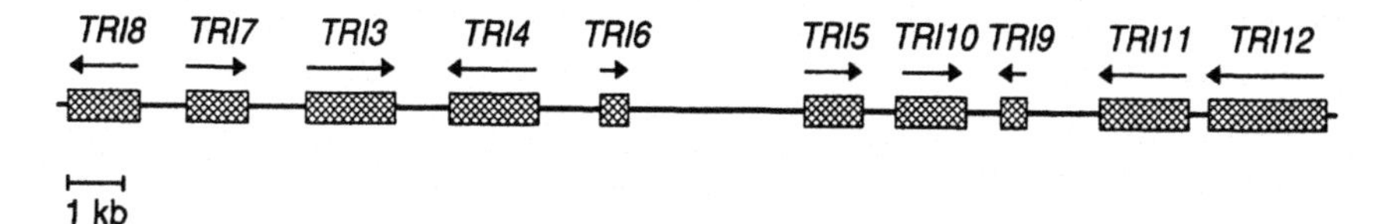

Figure 11.3 Trichothecene biosynthetic pathway gene cluster (Hohn, 1997). Arrows indicate direction of transcription.

ment (Alexander et al., 1999). Another gene, *TRI101* lies outside the 25 kb of the *TRI* gene cluster that has already been characterized, and may be located completely outside the cluster (Kimura et al. 1998). The predicted *TRI101* product does not share amino acid sequence homology with other known gene products. However, functional analysis revealed that the *TRI101* product catalyzes the acetylation of the C-3 position of trichothecenes, and thereby markedly reduces their toxicity (Kimura et al. 1998). The activities of the *TRI12* and *TRI101* products suggest they may both protect Fusarium from the trichothecenes that it produces.

Homologues of *TRI5* from *F. graminearum, F. poae,* and *F. sambucinum* have been characterized and are 91–96% identical to the *F. sporotrichioides TRI5* and each other at the amino acid level (Hohn and Desjardins, 1992; Hornok et al., 1996; Proctor et al., 1995a). Although the existence of the trichothecene biosynthetic gene cluster has not been demonstrated in species of *Fusarium* other than *F. sporotrichioides,* homologues of *TRI4, TRI5,* and *TRI6* are clustered in the macrocyclic trichothecene-producing fungus *Myrothecium* (Trapp et al., 1997). The clustering of genes for trichothecenes biosynthesis represents what appears to be a common feature of fungal genes involved in the biosynthesis of toxins and other secondary metabolites (Keller and Hohn, 1997). The biological function of these gene clusters is unclear; however, several hypotheses have been proposed. For example, the clusters may function in gene regulation or they may facilitate the horizontal transfer of the pathways between organisms (Keller and Hohn, 1997).

3. FUMONISINS

3.1. STRUCTURE AND OCCURRENCE

Fumonisins are a recently identified group of mycotoxins that were first reported by South African researchers investigating the association between human esophageal cancer and the consumption of maize infected with *F. moniliforme* (synonym *F. verticillioides*) (Bezuidenhout et al., 1988; Gelderblom et al., 1988). These toxins are aminopolyols with a core structure consisting of a 20-carbon backbone (eicosane) with an amino group at C-2. There are also one to three hydroxyl groups at C-3, C-5, and/or C-10, methyl groups at C-12 and C-16 and propane-1,2,3-tricarballylic acid moieties esterified to C-14 and C-15 (Figure 11.4). Fumonisin B_1 (FB_1) is the most abundant fumonisin in maize, while fumonisins B_2 (FB_2) and B_3 (FB_3) typically occur but at lower levels. Fumonisin B_4 also occurs in maize but at still lower levels than FB_2 and FB_3. Four other series of fumonisins (A, AK, C, and P), which are structurally closely related to the B series, also occur at low levels in cultures of *F. moniliforme.* In the A series, the amino group on the fumonisin backbone

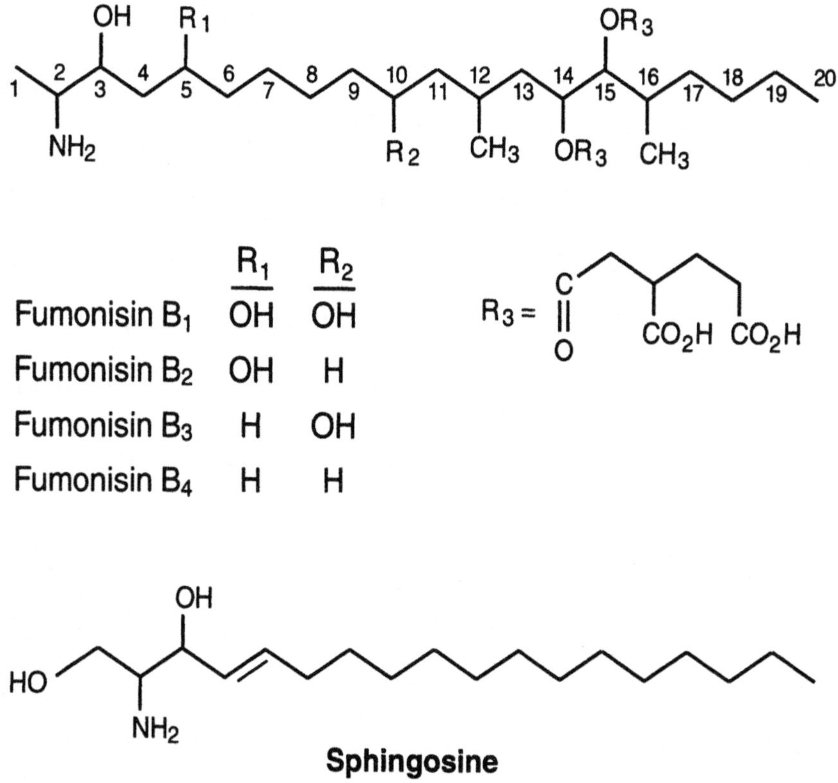

	R_1	R_2
Fumonisin B_1	OH	OH
Fumonisin B_2	OH	H
Fumonisin B_3	H	OH
Fumonisin B_4	H	H

Sphingosine

Figure 11.4 Structure of the B series of fumonisins and sphingosine.

is acetylated (Bezuidenhout et al., 1988). In the AK series, the amino group is also acylated and the C-15 tricarballylic acid moiety is replaced with a keto group (Musser and Plattner, 1997). In the C series, the amino-terminal methyl group (C-1) is absent and in the P series, the amino group is replaced with a 3-hydroxypyridinium moiety (Musser and Plattner, 1997).

Since fumonisins were originally identified in cultures of *F. moniliforme,* these toxins have been reported in cultures of *F. anthophilum, F. dlamini, F. napiforme, F. nygamai, F. oxysporum,* and *F. proliferatum* (Munkvold and Desjardins, 1997; Musser and Plattner, 1997; Seo et al., 1996). Fumonisin production by species in other fungal genera has not been reported; however, the tomato pathogen *Alternaria alternata* produces a structurally closely related compound AAL toxin, another aminopolyol with a linear 17-carbon backbone and a single tricarballylic acid moiety (Wang et al., 1996). The production of fumonisins by *F. moniliforme* and *F. proliferatum* poses the greatest threat to food and feed safety because of the worldwide prevalence of these species on maize. *F. moniliforme* is one of the most common pathogens of maize and is associated with diseases of the roots, stalk, and ears of this crop. In addition, the fungus is frequently isolated from apparently healthy maize tissue. The association of *F. proliferatum* and maize is less well understood, but it appears to be similar to the *F. moniliforme*-maize association (Munkvold and Desjardins, 1997).

Surveys of fumonisins in maize and maize-based foods and feed indicate that fumonisins can occur in maize throughout the world where this crop is grown and that good as well as poor quality maize can be contaminated. However, when FB_1 was detected in good quality maize and refined corn products, it was generally present at 1–3 μg/g sample or less (Munkvold and Desjardins, 1997). On the other hand, mean levels of FB_1 in poor quality grain associated with human or animal diseases were 13–63 μg/g. These surveys also revealed that FB_1 levels in maize-based foods are generally low (<1 μg/g) but can occasionally be high enough (5–10 μg/g) to be of concern.

3.2. TOXICOLOGY

Fumonisins have been associated with a high incidence of human esophageal cancer in Linxian County in Henan, China, and the Transkei region of South Africa, where large quantities of maize contaminated with fumonisins are consumed (Marasas, 1996). The toxins also cause liver cancer in rat, pulmonary edema in swine, and leukoencephalomalacia in equines, a disease that results in necrosis and liquefaction of white matter of the brain (Marasas, 1996). Kidney and liver damage has also been observed in rats and sheep exposed to fumonisins or fumonisin-containing cultures of *F. moniliforme,* but across species, the liver appears to be the most sensitive organ (Bondy et al., 1996; Edrington et al., 1995; Tolleson et al., 1996a). Kidney and liver damage in rat resulted, at least in part, from apoptosis of cells within these organs (Tolleson et al., 1996a). Fumonisin-induced apoptosis has also been observed in vitro in cultures of turkey lymphocytes, monkey kidney cells, and human keratinocytes, esophageal epithelial cells, and hepatoma cells (Dombrink-Kurtzman et al., 1994; Tolleson et al., 1996b; Wang et al., 1996). FB_1 also stimulated DNA synthesis, inhibited proliferation and/or altered the morphology of a number of cell types in culture, but these effects varied depending on the toxin concentration and cell type (Meivar-Levy et al., 1997; Schroeder et al., 1994; Tolleson et al., 1996b).

3.3. MECHANISM OF ACTION

Fumonisins are structurally very similar to sphingosine (Figure 11.4), the long-chain backbone of sphingolipids, and there is a growing body of evidence indicating that these toxins induce disease via disruption of sphingolipid metabolism (Merrill et al., 1997; Riley et al., 1996). Sphingolipids are a diverse group of membrane lipids that are an important structural component of membranes and are involved in the regulation of cell growth, differentiation, and apoptosis. Sphingolipid biosynthesis proceeds via the condensation of palmitoyl-CoA and serine to form 3-ketosphinganine followed by ketoreduction to yield sphinganine (Figure 11.5) (Merrill et al., 1997). Sphinganine is combined with a long-chain fatty acid-CoA to yield dihydroceramide, which is converted to complex sphingolipids via ceramide. Turnover of sphingolipids results from the hydrolysis of complex sphingolipids to ceramide and subsequent deacylation to sphingosine. Sphingosine is thought to be reincorporated into complex sphingolipids via ceramide or shunted into phosphatidylethanolamine metabolism. Fumonisins inhibit the enzyme sphinganine (sphingosine) *N*-acyltransferase, which catalyzes the conversion of sphinganine to dihydroceramide and of sphingosine to ceramide (Figure 11.5) (Wang et al., 1991). The outcome of this inhibition includes the accumulation of sphinganine and reduction in complex sphingolipid biosynthesis in cell cultures and in animals exposed to fumonisins (Riley et al., 1996).

Given the functions of sphingolipids in cell growth, differentiation, and death, it seems likely that disruption of sphingolipid metabolism could cause the cytotoxic effects and diseases induced by fumonisins. Evidence for the link between sphingolipid disruption and disease include the following: (1) sphinganine is cytotoxic to many cell types under at least some conditions

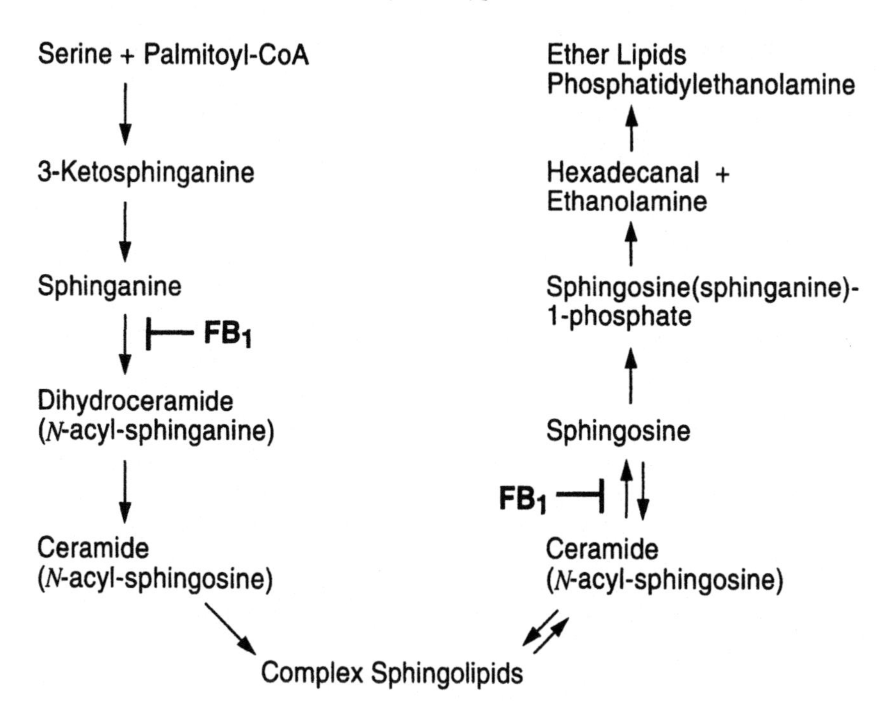

Figure 11.5 Sphingolipid metabolism (Merrill et al., 1997).

and complex sphingolipids appear to be essential for cell growth, survival, and adhesion; (2) the rise in sphinganine in animals following exposure to fumonisins is rapid and occurs before or at the same time as disease symptoms are observed; (3) in pigs and rats there is a dose response association between the amount of fumonisin in the diet and the rise in sphinganine or decline in complex sphingolipids; and (4) in cell culture, some of the toxic effects induced by FB_1 can be eliminated or reduced by addition of β-haloamines, which prevent sphinganine accumulation by inhibiting the enzyme serine palmitoyltransferase that catalyzes the formation of the sphinganine precursor 3-ketosphinganine (Figure 11.5) (Riley et al., 1996; Schroeder et al., 1994).

When fed to rats at high doses, FB_1 acted as a complete liver carcinogen (Gelderblom et al., 1991). However, FB_1 acted primarily as a tumor promoter when fed to rats at lower doses and was neither mutagenic to *Salmonella* nor genotoxic (Gelderblom et al., 1996; Riley et al., 1996). The mechanism by which fumonisins promote tumor development is not clear but is currently an area of intense study. Due to the importance of sphingolipids in cell growth and differentiation, disruption of sphingolipid metabolism has been proposed to alter the controls governing cell proliferation (Riley et al., 1996). Fumonisin-induced apoptosis has also been proposed to play a role in tumor promotion via an increased proliferation of living cells adjacent to dead cells (Riley et al., 1996; Tolleson et al., 1996b). Finally, the tumor-promoting activity of fumonisins has also been proposed to result from the stimulation or suppression of key signal transduction enzymes, such as mitogen-activated protein kinase and protein kinase C (Huang et al., 1995; Wattenberg et al., 1996). However, the possibility that changes in the activities of these enzymes were mediated via disruption of sphingolipid metabolism has not been ruled out.

3.4. BIOSYNTHESIS

The 20-carbon chain that forms the backbone of fumonisins structurally resembles fatty acids and linear polyketides and therefore its biosynthesis may be analogous to fatty acid/polyketide biosynthesis (Blackwell et al., 1994; Desjardins et al., 1996b). Overall, fatty acid biosynthesis involves the successive condensation of acetyl units (CH_3CO–) to yield the long carbon chain characteristic of these acids (Hopwood and Sherman, 1990). The β-carbonyl groups, introduced on the growing carbon chain with the addition of each acetyl unit, are reduced to hydroxyls, then to enoyls (i.e., C=C), and finally to fully saturated carbons before the next condensation. Polyketide biosynthesis is similar to fatty acid biosynthesis but can differ in several respects. For example, in polyketide formation the reduction of the β-carbonyl groups does not always occur or proceed to a fully saturated carbon. This can result in the presence of carbonyl, hydroxyl and/or enoyl functions at various positions along the nascent polyketide chain. Studies in which ^{13}C and ^{14}C labeled acetate indicated that carbon atoms from C-3 through C-20 of the fumonisin backbone originate from acetate and that odd-numbered carbons are derived from the carbonyl group of acetate while even-numbered carbons originate from the methyl group of acetate (Blackwell et al., 1994). These data are consistent with the hypothesis that the fumonisin backbone is synthesized in a manner similar to fatty acids and polyketides. Fungal polyketide synthases consist of one large multifunctional polypeptide while fungal fatty acid synthases consist of two polypeptides with the multiple functional domains divided between them (Hopwood and Sherman, 1990; Yang et al., 1996). Recently Proctor et al. (1999) cloned and characterized a gene from *F. moniliforme* that encodes a polyketide synthase and that is required by the fungus for fumonisin production. The identification of such a gene indicates that the fumonisin backbone is a polyketide rather than a fatty acid.

Experiments in which ^{13}C and ^{2}H labeled alanine were added to *F. moniliforme* cultures indicated that amino nitrogen at C-2 as well as the carbon atoms at C-1 and C-2 are derived directly from alanine (Branham and Plattner, 1993). Thus, fumonisin biosynthesis may include a step analogous to the condensation of serine and palmitoyl CoA in sphingolipid biosynthesis (Figure 11.5), except with fumonisins the condensation would involve alanine and presumably a linear 18-carbon polyketide (Desjardins et al., 1996b). Additional experiments with isotope-labeled compounds indicate that the methyl groups at C-12 and C-16 are derived from the S-methyl of methionine (Plattner and Branham, 1994).

The structure of fumonisins as well as the isotope incorporation experiments mentioned above indicate that multiple enzymatic reactions are involved in the biosynthesis of these toxins after the backbone is formed. Although none of the genes coding for these enzymes has been isolated and characterized, four loci (designated *fum1, fum2, fum3,* and *fum4*) involved in fumonisin biosynthesis have been identified (Figure 11.6) (Desjardins et al., 1995; Desjardins et al., 1996a; Plattner et al., 1996). The identification of these loci was accomplished via classical genetic analyses utilizing *Gibberella fujikuroi,* the

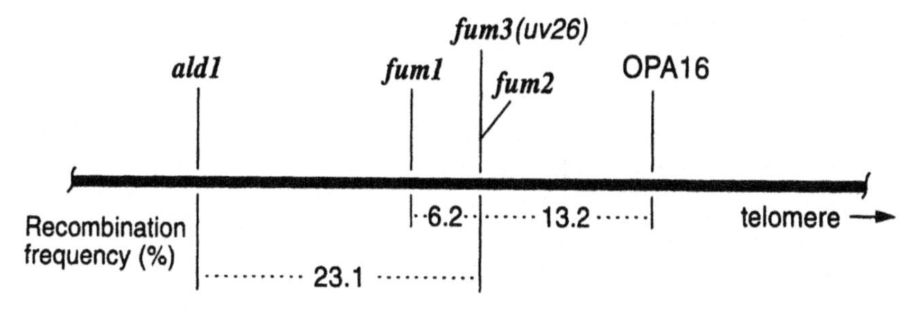

Figure 11.6 Genetic map of *fum* loci and RAPD markers *aldl* and OPA16.

sexual stage of *F. moniliforme,* and rare naturally occurring mutants of the fungus with unusual fumonisin production phenotypes. Strains of the fungus carrying the naturally occurring, mutated *fum1* allele (*fum1*) did not produce detectable levels of fumonisins, suggesting that the wildtype allele of *fum1* is involved in regulation of fumonisin production or encodes an enzyme that catalyzes an early reaction in fumonisin biosynthesis. Fumonisin production in strains of *F. moniliforme* carrying the mutated *fum4* allele was less than 4% of wild-type production (Plattner et al., 1996). This phenotype suggests that *fum4* may also be involved in regulation or encode an enzyme for an early step in fumonisin biosynthesis. Strains carrying the mutant allele of *fum2* (*fum2*) only produced fumonisins that lack a hydroxyl at C-10 (Desjardins et al., 1996a). This suggests that *fum2* is involved in hydroxylation of fumonisins at C-10. Similarly, strains carrying the mutant allele of *fum3* produced only fumonisins that lack the C-5 hydroxyl, indicating that *fum3* confers the ability to hydroxylate the C-5 position (Desjardins et al., 1996a). An additional mutant, generated in the lab by UV mutagenesis, exhibited the same fumonisin production phenotype as strains carrying the *fum3* allele. Genetic analyses suggest that the phenotype of the UV mutant is conferred by a mutation at a single locus, designated *uv26,* and that it is very tightly linked, perhaps allelic, to *fum3* (Proctor et al., 1997).

Genetic mapping by Xu and Leslie (1996) localized *fum1* to chromosome 1 of the *F. moniliforme* genome. Subsequently, the other *fum* loci were found to be linked to *fum1* and, therefore, also located on chromosome 1. Additional mapping studies that followed the recombination frequencies between the various *fum* loci, *uv26,* and two RAPD markers, *ald1* (OPH3) and OPA16, facilitated the ordering of these loci along chromosome 1 and provided an estimate of the genetic distance between them (Figure 11.6) (Desjardins et al., 1996b; Desjardins et al., 1996a; Proctor et al., 1997). The genetic linkage of the *fum* loci suggests that they may be part of a cluster of genes involved in fumonisin biosynthesis, perhaps in an analogous manner to the clustering of trichothecene and aflatoxin biosynthetic genes.

4. FUTURE AREAS OF RESEARCH

Future research on trichothecenes and fumonisins will undoubtedly include further studies on the molecular mechanisms by which these toxins induce disease in humans and animals. In fact, efforts are currently underway to elucidate the molecular events leading to fumonisin-induced apoptosis and the relationship between apoptosis and tumor promotion (Tolleson et al., 1996a). In addition, long-term feeding studies are underway to verify the carcinogenicity of fumonisins in experimental animals (Tolleson et al., 1996a). Such studies should help to define safe levels of fumonisins in feed and food.

Further research on the DON-induced activation of transcription factor NF-κB/Rel should help to resolve the mechanism by which trichothecenes stimulate the immune system and perhaps induce other toxic effects (Rotter et al., 1996).

Further molecular characterization of trichothecene and fumonisin biosynthetic genes will also continue to be an important area of research. Such studies should improve our fundamental understanding of the biosynthetic enzymes involved and the regulation of toxin production in *Fusarium*. Future research on the trichothecene regulatory gene *TRI6* should help to elucidate the regions of the *TRI6* protein involved in DNA binding and transcriptional activation as well as elements of *TRI* gene promoters that facilitate their activation by the *TRI6* product (Proctor et al., 1995b).

One of the major goals of fumonisin and trichothecene research is to reduce the occurrence of these toxins in grain crops. Perhaps the most effective way to achieve this goal is to improve the resistance of these crops to diseases caused by *Fusarium*. Thus, considerable resources are being and will continue to be expended to improve the resistance of grain crops to *Fusarium* diseases via classical plant breeding and molecular genetic approaches. In addition, the observation that trichothecene production in *F. graminearum* is required by the fungus to cause high levels of head scab on wheat suggests that trichothecene resistance in wheat may also confer resistance to this diseases (Desjardins et al., 1996c). This disease resistance should, in turn, prevent the accumulation of trichothecenes in wheat.

5. REFERENCES

Alexander, N. J., McCormick, S. P., and Hohn, T. M. 1999. "*TRI12*, a trichothecene efflux pump from *Fusarium sporotrichioides:* gene isolation and expression in yeast," *Mol. Gen. Genet.* (in press).

Alexander, N. J., Hohn, T. M., and McCormick, S. P. 1998b. "The *TRI11* gene of *Fusarium sporotrichioides* encodes a cytochrome P450 monooxygenase required for C-15 hydroxylation in trichothecene biosynthesis," *Appl. Environ. Microbiol.* 64:221–225.

Azcona-Olivera, J. I., Ouyang, Y. L., Warner, R. L., Linz, J. E., and Pestka, J. J. 1995. "Effects of vomitoxin (deoxynivalenol) and cycloheximide on IL-2, 4, 5 and 6 secretion and mRNA levels in murine CD4+ cells," *Food Chem. Toxicol.* 33:433–441.

Beardall, J. M. and Miller, J. D. 1997. "Diseases in humans with mycotoxins as possible causes," in *Mycotoxins in Grain: Compounds Other than Aflatoxin*, eds., J. D. Miller and H. L. Trenholm. St. Paul: Eagan Press. pp. 487–539.

Beremand, M. N. and Hohn, T. M. 1995. "*Tri10:* a new gene in the trichothecene gene cluster in *Fusarium sporotrichioides,*" *Fungal Genet. Newsl.* 42A:100.

Berry, C. H. J., Ibrahim, M. A. K., and Coddington, A. 1978. "Characterisation of ribosomes from drug resistant strains of *Schizosaccharomyces pombe* in a poly U directed cell free protein synthesising system," *Mol. Gen. Genet.* 167:217–225.

Bezuidenhout, C. S., Gelderblom, W. C. A., Gorst-Allman, C. P., Horak, R. M., Marasas, W. F. O., Spiteller, G., and Bleggaar, R. 1988. "Structure elucidation of fumonisins, mycotoxins from *Fusarium moniliforme,*" *J. Chem. Soc. Chem. Commun.* 1988:743–745.

Blackwell, B. A., Miller, J. D., and Savard, M. E. 1994. "Production of carbon 14-labeled fumonisin in liquid culture," *J. AOAC Int.* 77:506–511.

Bondy, G., Barker, M., Mueller, R., Fernie, S., Miller, J. D., Armstrong, C., Hierlihy, S. L., Rowsell, P., and Suzuki, C. 1996. "Fumonisin B_1 toxicity in male Sprague-Dawley rats," in *Fumonisins in Food,* eds., L. S. Jackson, J. W. DeVries, and L. B. Bullerman. New York: Plenum Press. pp. 251–264.

Branham, B. E. and Plattner, R. D. 1993. "Alanine is a precursor in the biosynthesis of fumonisin B1 by *Fusarium moniliforme,*" *Mycopathologia* 124:99–104.

Committee on Protection Against Mycotoxins. 1983. *Protection Against Trichothecene Mycotoxins.* Washington, D.C.: National Academy Press. pp. 227.

Corley, D. G., Rottinghaus, G. E., and Tempesta, M. S. 1987. "Toxic trichothecenes from *Fusarium sporotrichioides* (MC-72083)," *J. Org. Chem.* 52:4405–4408.

Cundliffe, E., Cannon, M., and Davies, J. 1974. "Mechanism of inhibition of eukaryotic protein synthesis by trichothecene fungal toxins," *Proc. Natl. Acad. Sci. USA* 71:30–34.

Desjardins, A. E., Hohn, T. M., and McCormick, S. P. 1993. "Trichothecene biosynthesis in *Fusarium* species: chemistry, genetics, and significance," *Microbiol. Rev.* 57:595–604.

Desjardins, A. E., Plattner, R. D., and Proctor, R. H. 1996a. "Linkage among genes responsible for fumonisin biosynthesis in *Gibberella fujikuroi* mating population A," *Appl. Environ. Microbiol.* 62:2571–2576.

Desjardins, A. E., Plattner, R. D., and Proctor, R. H. 1996b. "Genetic and biochemical aspects of fumonisin production," in *Fumonisins in Food,* eds., L. S. Jackson, J. W. DeVries and L. B. Bullerman. New York: Plenum Press. pp. 165–173.

Desjardins, A. E., Plattner, R. D., Nelsen, T. C., and Leslie, J. F. 1995. "Genetic analysis of fumonisin production and virulence of *Gibberella fujikuroi* mating population A (*Fusarium moniliforme*) on maize (*Zea mays*) seedlings," *Appl. Environ. Microbiol.* 61:79–86.

Desjardins, A. E., Proctor, R. H., Bai, G., McCormick, S. P., Shaner, G., Buechley, G., and Hohn, T. M. 1996c. "Reduced virulence of trichothecene antibiotic-nonproducing mutants of *Gibberella zeae* in wheat field tests," *Mol. Plant-Microbe Interact.* 9:775–781.

Dombrink-Kurtzman, M. A., Bennett, G. A., and Richard, J. L. 1994. "Induction of apoptosis in turkey lymphocytes by fumonisin," *FASEB J.* 8:488.

Edrington, T. S., Kamps-Holtzapple, C. A., Harvey, R. B., Kubena, L. F., Elissalde, M. H., and Rottinghaus, G. E. 1995. "Acute hepatic and renal toxicity in lambs dosed with fumonisin-containing culture material," *J. Anim. Sci.* 73:508–515.

Gelderblom, W. C. A., Kriek, N. P. J., Marasas, W. F. O., and Thiel, P. G. 1991. "Toxicity and carcinogenicity of the *Fusarium moniliforme* metabolite, fumonisin B1, in rats," *Carcinogenisia* 12:1247–1251.

Gelderblom, W. C. A., Snyman, S. D., Lebepe-Mazur, S., van der Westhuizen, L., Kriek, N. P. J., and Marasas, W. F. O. 1996. "The cancer-promoting potential of fumonisin B1 in rat liver using diethylnitrosamine as a cancer initiator," *Cancer Lett.* 109:101–108.

Gelderblom, W. C. A., Jaskiewicz, K., Marasas, W. F. O., Thiel, P. G., Horak, R. M., Vleggaar, R., and Kriek, N. P. J. 1988. "Fumonisins—Novel mycotoxins with cancer-promoting activity produced by *Fusarium moniliforme,*" *Appl. Environ. Microbiol.* 54:1806–1811.

Gilly, M., Benson, N. R., and Pellegrini, M. 1985. "Affinity labeling the ribosome with eukaryotic-specific antibiotics: (bromoacetyl)trichodermin," *Biochemistry* 24:5787–5792.

Hohn, T. M. 1999. "Cloning and expression of terpene synthase genes," in *Comprehensive Natural Products Chemistry, Vol. 2, Isoprenoids Including Carotenoids and Steroids,* ed., D. E. Cane. Elsevier Science: Oxford. pp. 201–215.

Hohn, T. M. and Desjardins, A. E. 1992. "Isolation and gene disruption of the *Tox5* gene encoding trichodiene synthase in *Gibberella pulicaris,*" *Mol. Plant-Microbe Interact.* 5:249–256.

Hohn, T. M. and VanMiddlesworth, F. 1986. "Purification and characterization of the sesquiterpene cyclase trichodiene synthetase from *Fusarium sporotrichioides,*" *Arch. Biochem. Biophys.* 251:756–761.

Hohn, T. M., McCormick, S. P., and Desjardins, A. E. 1993. "Evidence for a gene cluster involving trichothecene-pathway biosynthetic genes in *Fusarium sporotrichiodes,*" *Curr. Genet.* 24:291–295.

Hohn, T. M., Desjardins, A. E., and McCormick, S. P. 1995. "The *Tri4* gene of *Fusarium sporotrichioides* encodes a cytochrome P450 involved in trichothecene biosynthesis," *Mol. Gen. Genet.* 248:95–102.

Holt, P. S. and DeLoach, J. R. 1988. "Cellular effects of T-2 myctoxin on two different cell lines," *Biochim. Biophys. Acta.* 971:1–8.

Hopwood, D. A. and Sherman, D. H. 1990. "Molecular genetics of polyketides and its comparison to fatty acid biosynthesis," *Annu. Rev. Genet.* 24:37–66.

Hornok, L., Fekete, C., and Giczey, G. 1996. "Molecular characterization of *Fusarium poae,*" *Sidowia* 48:23–31.

Huang, C., Dickman, M., Henderson, G., and Jones, C. 1995. "Repression of protein kinase C and stimulation of cyclic AMP response elements by fumonisin, a fungal encoded toxin which is a carcinogen," *Cancer Res.* 55:1655–1659.

Iglesias, M. and Ballesta, J. P. G. 1994. "Mechanism of resistance to the antibiotic trichothecin in the producing fungi," *Eur. J. Biochem.* 223:447–453.

Joffe, A. Z. 1986. *Fusarium Species: Their Biology and Toxicology.* New York: John Wiley & Sons. pp. 588.

Keller, N. P. and Hohn, T. M. 1997. "Metabolic pathway gene clusters in filamentous fungi," *Fungal Genet. Biol.* 21:17–29.

Kimura, M., Kaneko, I., Komiyama, M., Takatsuki, A., Koshino, H., Yoneyama, K., and Yamaguchi, I. 1998. "Trichothecene 3-*O*-acetyltransferase protects both the producing organism and transformed yeast from related mycotoxins, cloning and characterization of *TRI101,*" *J. Biol. Chem.* 273:1654–1661.

Koshinsky, H., Honour, S., and Khachatourians, G. 1988. "T-2 toxin inhibits mitochondrial function in yeast," *Biochem. Biophys. Res. Commun.* 151:809–814.

Marasas, W. F. O. 1996. "Fumonisins: history, world-wide occurrence and impact," in *Fumonisins in Food,* eds., L. S. Jackson, J. W. DeVries and L. B. Bullerman. New York: Plenum Press. pp. 1–17.

McCormick, S. P. and Hohn, T. M. 1997. "Accumulation of trichothecenes in liquid cultures of a *Fusarium sporotrichioides* mutant lacking a functional trichothecene C-15 hydroxylase," *Appl. Environ. Microbiol.* 63:1685–1688.

McCormick, S. P., Hohn, T. M., and Desjardins, A. E. 1996. "Isolation and characterization of *Tri3,* a gene encoding 15-*O*-acetyltransferase from *Fusarium sporotrichioides,*" *Appl. Environ. Microbiol.* 62:353–359.

Meivar-Levy, I., Sabanay, H., Bershadsky, A. D., and Futerman, A. H. 1997. "The role of sphingolipids in the maintenance of fibroblast morphology," *J. Biol. Chem.* 272:1558–1564.

Merrill, A. H., Schmelz, E., Dillehay, D. L., Spiegel, S., Shayman, J. A., Schroeder, J. J., Riley, R. T., Voss, K. A., and Wang, E. 1997. "Sphingolipids—the enigmatic lipid class: biochemistry, physiology, and pathophysiology," *Toxicol. Appl. Pharmacol.* 142:208–225.

Middlebrook, J. L. and Leatherman, D. L. 1989. "Binding of T-2 toxin to eukaryotic cell ribosomes," *Biochem. Pharmacol.* 38:3103–3110.

Munkvold, G. P. and Desjardins, A. E. 1997. "Fumonisins in maize: can we reduce their occurrence?" *Plant Dis.* 81:556–565.

Musser, S. M. and Plattner, R. D. 1997. "Fumonisin composition in cultures of *Fusarium moniliforme, Fusarium proliferatum,* and *Fusarium nygami,*" *J. Agric. Food. Chem.* 45:1169–1173.

Ouyang, Y. L., Li, S., and Pestka, J. J. 1996. "Effects of vomitoxin (deoxynivalenol) on transcription factor NF-κB/Rel binding activity in murine EL-4 thymoma and primary CD4+ T cells," *Toxicol. Appl. Pharmacol.* 140:328–336.

Pestka, J. J. and Bondy, G. S. 1994. "Immunotoxic effects of mycotoxins," in *Mycotoxins in Grain: Compounds Other than Aflatoxin,* eds., J. D. Miller and H. L. Trenholm. St. Paul: Eagan Press. pp. 339–358.

Plattner, R. D. and Branham, B. E. 1994. "Labeled fumonisins: production use of fumonisin B1," *J. AOAC Int.* 77:525–532.

Plattner, R. D., Desjardins, A. E., Leslie, J. F., and Nelson, P. E. 1996. "Identification and characterization of strains of *Gibberella fujikuroi* mating population A with rare fumonisin production phenotypes," *Mycologia* 88:416–424.

Proctor, R. H., Desjardins, A. E., and Plattner, R. D. 1997. "Analysis of a *Gibberella fujikuroi* mutant deficient in a hydroxylation step of fumonisin biosynthesis," *Proceedings of the 19th Fungal Genetics Conference.* Pacific Grove, CA. [Abstract] p. 119.

Proctor, R. H., Hohn, T. M., and McCormick, S. P. 1995a. "Reduced virulence of *Gibberella zeae* caused by disruption of a trichothecene toxin biosynthectic gene," *Mol. Plant-Microbe Interact.* 8:593–601.

Proctor, R. H., Desjardins, A. E., Plattner, R. D., and Hohn, T. M. 1999. A polyketide synthase gene required for biosynthesis of fumonisin mycotoxins in *Gibberella fujikuroi* mating population A, *Fungal Genet. Biol.* (in press).

Proctor, R. H., Hohn, T. M., McCormick, S. P., and Desjardins, A. E. 1995b. "*Tri6* encodes an unusual zinc finger protein involved in regulation of trichothecene biosynthesis in *Fusarium sporotrichioides,*" *Appl. Environ. Microbiol.* 61:1923–1930.

Riley, R. T., Wang, E., Schroeder, J. J., Smith, E. R., Plattner, R. D., Abbas, H., Yoo, H., and Merrill, A. H. 1996. "Evidence for disruption of sphingolipid metabolism as a contributing factor in the toxicity and carcinogenicity of fumonisins," *Nat. Toxins,* 4:3–15.

Rosenstein, Y., LaFarge-Frayssinet, C., Lespinats, G., Loisillier, F., FaFont, P., and Frayssinet, C. 1979. "Immunosupprissive activity of *Fusarium* toxins: effects on antibody synthesis and skin grafts of crude extracts, T-2 toxin and diacetoxyscirpenol," *Immunology* 39:111–117.

Rotter, B. A., Prelusky, D. B., and Pestka, J. J. 1996. "Toxicology of deoxynivalenol (vomitoxin)," *J. Toxicol. Environ. Health* 48:1–34.

Schindler, D., Grant, P., and Davies, J. 1974. "Trichodermin resistance—mutation affecting eukaryotic ribosomes," *Nature* 248:535–536.

Schroeder, J. J., Crane, H. M., Xia, J., Liotta, D. C., and Merrill, A. H. 1994. "Disruption of sphingolipid metabolism and stimulation of DNA synthesis by fumonisin B1," *J. Biol. Chem.* 269:3475–3481.

Seo, J. A., Kim, J.-C., and Lee, Y.-W. 1996. "Isolation and characterization of two new type C fumonisins produced by *Fusarium oxysporum,*" *J. Nat. Prod.* 59:1003–1005.

Sharma, R. P. and Kim, Y. W. 1991. "Trichothecenes," in *Mycotoxins and Phytoalexins,* eds., R. P. Sharma and D. K. Salunkhe. Boca Raton, FL: CRC Press. pp. 339–359.

Tolleson, W. H., Dooley, K. L., Sheldon, W. G., Thurman, J. D., Bucci, T. J., and Howard, P. C. 1996a. "The mycotoxin fumonisin induces apoptosis in cultured human cells and in livers and kidneys of rats," in *Fumonisins in Food,* eds., L. S. Jackson, J. W. DeVries and L. B. Bullerman. New York: Plenum Press. pp. 237–250.

Tolleson, W. H., Melchior, W. B., Morris, S. M., McGarrity, L. J., Domon, O. E., Muskhelishvili,

L., James, S. J., and Howard, P. C. 1996b. "Apoptotic and anti-proliferative effects of fumonisin B1 in human keratinocytes, fibroblasts, esophageal epithelial cells and hepatoma cells," *Carcinogenisis* 17:239–249.

Trapp, S. C., Hohn, T. M., McCormick, S. P., and Jarvis, B. B. 1998. "Characterization of the macrocyclic trichothecene gene cluster in *Myrothecium roridum,*" *Mol. Gen. Genet.* 257:421–432.

Wang, E., Norred, W. P., Bacon, C. W., Riley, R. T., and Merrill, A. H. 1991. "Inhibition of sphingolipid biosynthesis by fumonisins: implications for diseases associated with *Fusarium moniliforme,*" *J. Biol. Chem.* 266:14486–14490.

Wang, H., Jones, C., Ciacci-Zanella, J., Holt, T., Gilchrist, D. G., and Dickman, M. B. 1996. "Fumonisins and *Alternaria alternata lycopersici* toxins: sphinganine analog mycotoxins induce apoptosis in monkey kidney cells," *Proc. Natl. Acad. Sci. USA* 93:3461–3465.

Wattenberg, E. V., Badria, F. A., and Shier, W. T. 1996. "Activation of mitogen-activated protein kinase by the carcinogenic mycotoxin fumonisin B1," *Biochem. Biophys. Res. Commun.* 227:622–627.

Wei, C., Hansen, B. S., Vaughan, M. H., and McLaughlin, C. S. 1974. "Mechanism of action of the mycotoxin trichodermin, a 12,13-epoxytricothecene," *Proc. Natl. Acad. Sci. USA* 71:713–717.

Wei, C.-M. and McLaughlin, C. S. 1974. "Structure-function relationship in the 12,13-epoxytrichothecene: novel inhibitors of protein synthesis," *Biochem. Biophys. Res. Commun.* 57:838–844.

Xu, J.-R. and Leslie, J. F. 1996. "A genetic map of *Gibberella fujikuroi* mating population A (*Fusarium moniliforme*)," *Genetics,* 143:175–189.

Yang, G., Rose, M. S., Turgeon, B. G., and Yoder, O. C. 1996. "A polyketide synthase is required for fungal virulence and production of the polyketide T-toxin," *Plant Cell* 8:2139–2150.

Yoshino, N., Takizawa, M., Akiba, H., Okumura, H., Tashiro, F., Honda, M., and Ueno, Y. 1996. "Transient elevation of intracellular calcium ion levels as an early event in T-2 toxin-induced apoptosis in human promyelotic cell line HL-60," *Nat. Toxins* 4:234–241.

CHAPTER 12

PSP Toxins: Their Biosynthesis by Marine Dinoflagellates and Molecular Identification

YUZABURO ISHIDA
YOSHIHIKO SAKO

1. INTRODUCTION

Harmful algal blooms including toxic algae have a severe economic impact on aquaculture as well as environmental and human health implications. In recent years there has been a global increase in the number of toxic events: for example, paralytic shellfish poisoning (PSP) is an accumulation of toxins in shellfish and can be fatal to humans upon ingestion. The distribution of PSP appears to be expanding all over the world (Hallegraeff, 1993).

PSP in molluscan bivalves was the first health hazard to be traced to dinoflagellates containing PSP toxin. Clinical symptoms of PSP include muscular paralysis, pronounced respiratory difficulty, choking sensation, and respiratory paralysis which can lead to death within 2–24 h of ingestion. The dinoflagellate, *Alexandrium catenella* (formally *Gonyaulax catenella*) was first recognized as the causative organism by Sommer and Meyer (1937). Saxitoxin (STX), a major PSP toxin, was isolated in a large quantity from Alaska butter clam, *Saxidomus giganteus,* in 1957 and is the sole known toxin responsible for the poisoning (Schantz et al., 1957; Schantz et al., 1975). Later, gonyautoxins designated GTX1, 2, 3, and 4 were isolated from the soft-shell clam *Mya arenaria* exposed to a massive red tide caused by *Alexandrium tamarense* (formally *Gonyaulax tamarensis*) in Massachusetts in 1975 (Burkley et al., 1976; Shimizu et al., 1975), and to date more than 20 analogues (neoSTX, GTX1, 2, 3, 4, 5, 6, 7(C1), 8(C2), etc.) of STX have been reported (Oshima, 1995). Analogues have been detected not only from several dinoflagellates (*A. tamarense, A. catenella, A. cohorticula, A. fundyense, A. fraterculus, A. minutum, A. lusitanicum, Gymnodinium catenatum,* and *Pyrodinium baha-*

mense var *compressum*), but also cyanobacteria (*Aphanizomenon flos-aquae, Anabaena circinalis* and *Lyngbya wollei*). However, it has been recognized that dinoflagellate toxicity varies between isolates of a species from different geographic areas and for individual isolates under varying growth conditions, although the relative toxin profile (STXs and GTXs) does not change (Shimizu, 1979; Boyer et al., 1987; Ogata et al., 1987). Several hypotheses about the biological production of PSP have been proposed.

That toxic and nontoxic dinoflagellates are considered to represent a single species led Silva (1982) to first suspect bacterial involvement in PSP toxicity. She suggested that dinoflagellate toxin production was due to bacteria living in association with *A. lusitanicum.* Kodama (Kodama and Ogata, 1988; Kodama, 1990) also pointed out the possible association of intracellular bacteria with toxin production by *A. tamarense,* and suggested that toxin production is not a hereditary characteristic. However, this hypothesis had little if any supporting evidence, given that (1) the algal cultures were not axenic; (2) electron microscopy revealed only 2~5 bacteria-like particles lacking membrane-bound nuclei in only a small percentage of the algal cells present. Also, the algal cells were completely deficient in permanently condensed, banded chromosomes (normally 100~150); and (3) a so-called marine PSP-producing bacterium, *Moraxella* sp., which was isolated from an antibiotic-treated, non-axenic culture of *A. tamarense,* may have been a contaminant in the algal culture (Sako et al., 1992; Ishida et al., 1993). Recently Sato and Shimizu (1997) have shown that the neosaxitoxin (neoSTX)-like compound produced by the *Moraxella* sp. has no resemblance to neoSTX in 1H and 13C NMR spectra, and it also gave a negative result in the neuroblastoma assay. This strongly suggests that HPLC-fluorometric analysis and neuroblast cell culture assays alone are not sufficient for identification of PSP components and that axenic cultures of toxic dinoflagellates should be used in biological analyses on toxin production.

In this chapter, we will discuss PSP production throughout the cell cycle and under various environmental conditions, the biosynthesis of the STX skeleton, and the biochemical and molecular analysis and molecular identification of enzymes involved in PSP toxin production in *Alexandrium* spp. and *G. catenatum.*

2. ENVIRONMENTAL CONDITIONS AFFECT PSP PRODUCTION

Investigators have analyzed cellular toxin content of cultures grown under a number of environmental variables and have recognized that toxicity is highly variable both between isolates of the same species and in a single isolate grown under different conditions (Shimizu, 1979; Boyer et al., 1987; Ogata et al. 1987). They have confirmed a decline in the total molar toxin content per cell during the transition from exponential to stationary growth in non-axenic cultures of *A. catenella* and *A. tamarense.* However, they demonstrated that the toxin profile of a particular isolate does not change significantly with culture age. It was concluded that toxin profile is a relatively stable or conserved property in *Alexandrium* species (Figure 12.1). However, using an axenic batch culture of *A. tamarense* that produced only neoSTX and STX, Boczar et al. (1988) observed variability in toxin profile under different growth conditions; that is, the mole percent of neoSTX increased substantially from 8 to 44% as total toxin levels per cell decreased, and a concomitant decrease in the mole percent of STX during the stationary phase was noted, although the toxin profile remained relatively constant during exponential growth. In batch cultures, even with axenic cultures, the physiological conditions during stationary growth may change. A decrease in the levels of carbon, nitrogen, phosphorus and other minerals, as well as the pH of the medium, brings about a reduction in light levels and increases the accumulation substances excreted from the cells. These factors may impact directly on toxin production.

In nitrogen- and phosphorus-limited semi-continuous cultures, the toxin profile varied systematically with growth rate (Anderson, 1990). Nitrogen limitation favored the production of toxins C1, C2 and GTX1 and 4, whereas phosphorus limitation produced cells with a high relative abundance of GTX2

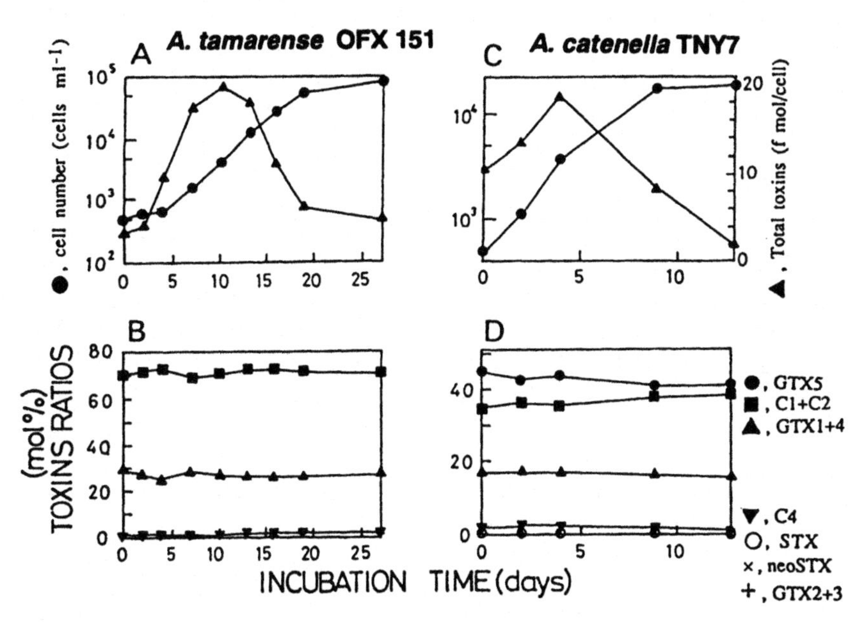

Figure 12.1 Stability of PSP toxin profile during growth of *A. tamarense* OFX 151 and *A. catenella* TNY7 (Kim et al., 1993a). A and C, growth curves and total toxins. B and D, toxin profiles.

and 3. In all cases, STX reached its highest relative abundance when growth was most rapid. Anderson et al. (1990) concluded, upon comparison of toxin profiles between cultures grown under identical conditions and harvested at the same stage of exponential growth, that the observed levels of different toxins had a genetic basis, thus confirming the validity of the clustering analyses and other comparisons.

A question was raised as to during which stage, G1, S, G2 or M phase of the cell cycle, toxin synthesis occurs. Anderson's group (Anderson, 1990; Taroncher-Oldenburg and Anderson, 1997) revealed that synchronized cells of *A. fundyense* synthesize toxin for only 8–10 h during the G1 phase of the cell cycle. To analyze the actual mechanisms of toxin biosynthesis and their regulation at the molecular level, it is important to identify PSP toxin-associated genes (Taroncher-Oldenburg and Anderson, 1997).

3. BIOSYNTHESIS OF THE SAXITOXIN (STX) SKELETON

Biosynthesis of the deoxydecarbamoyl STX skeleton of PSP toxins from precursors such as the purine derivatives, guanine and adenine, was ruled out by Shimizu et al. (1984a), who confirmed the involvement of arginine or its precursor, α-ketoglutarate, in its formation (Shimizu et al., 1984b; Shimizu, 1996). Subsequently it was established that ornithine, the direct precursor of arginine, can be incorporated into the toxin with the loss of the carboxyl group (Shimizu et al., 1984a). The PSP-producing cyanobacterium, *A. flos-aquae,* was ideal for this experiment as it was able to take up various organic compounds into the cell. In comparison, the photoautotroph *A. tamarense* does not utilize exogenous amino acids. Feeding of [2–13C, 2–15N]-double-labeled ornithine to *A. flos-aquae* resulted in the intact incorporation of the 13C-15N combination into C-4 and N-9 of neoSTX as shown by a 15N-13C spin-spin coupling in the NMR spectrum (Shimizu et al., 1984a; Shimizu, 1996). [13C2] acetate was incorporated into C-5 and C-6 of neoSTX, and [1, 2–13C] glycine and [methyl-13C] methionine were incorporated into C-13 of the toxin (Shimizu et al., 1986). However, very little is known about the enzymes involved in STX biosynthesis, nor of the mechanisms regulating toxin production.

4. BIOCHEMICAL AND MOLECULAR ANALYSIS OF PSP TOXIN PRODUCTION

To obtain direct evidence that the genes encoding PSP toxin-synthesizing enzymes, including sulfotransferases, are present in the chromosomal DNA of *Alexandrium* and other toxic dinoflagellates (Sako et al., 1995; Yoshida et al., 1996, 1998; Ishida et al., 1997), it is necessary to isolate and purify the enzymes and determine their DNA sequence based on N-terminal amino acid sequence information. However, there is little information available on the enzymes involved in STX biosynthesis, or on which enzymes are associated with N-1 hydroxylation, C-13 carbamoylation, C-11 sulfation, and N-21 sulfation of the STX skeleton and its analogues.

Sako et al. (1995) focused on sulfation according to the putative pathway (Figure 12.2) of PSP biosynthesis in *G. catenatum.* Axenic cultures of this organism were used because sulfated toxins predominate its toxin profile. Moreover, in mammals, sulfation is known to be an important step in the metabolism of bioactive compounds, i.e., hormones and drugs. In mammals, sulfation is catalyzed by a sulfotransferase (ST), which transfers a sulfate group of 3′-phosphoadenosine 5′-phosphosulfate (PAPS) to target compounds (Mulder and Jakoby, 1990; Saidha and Schiff, 1994; Homma et al., 1992; Barnes et al., 1989; Hondoh et al., 1993).

A PSP toxin sulfotransferase has been detected in *G. catenatum* GC21V

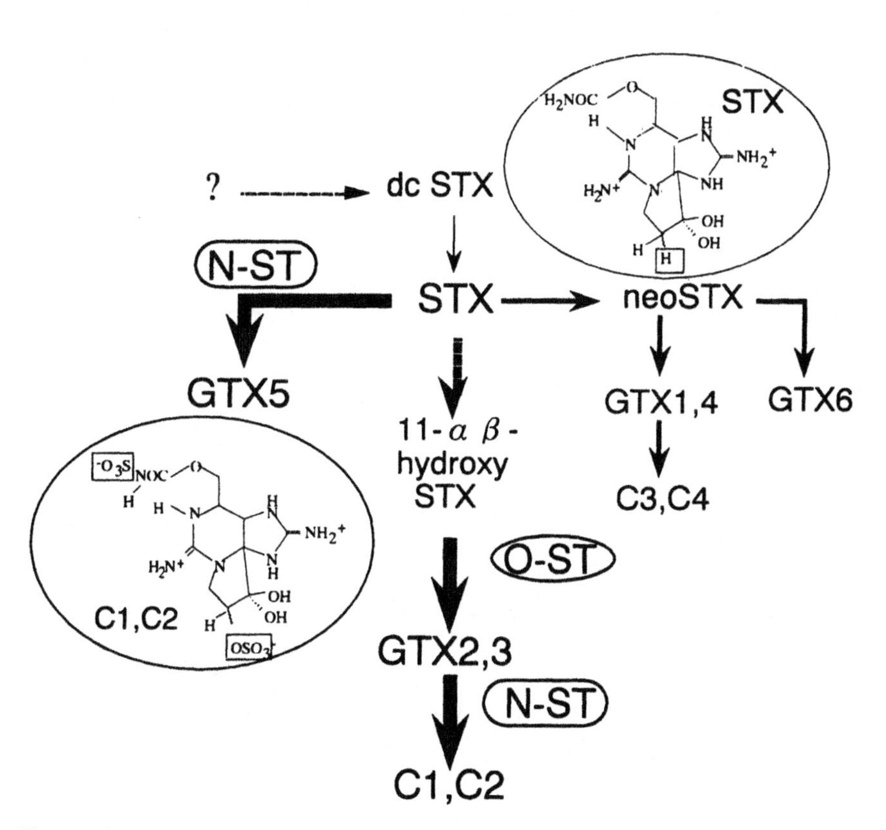

Figure 12.2 Putative pathway of STX analogue biosynthesis from STX in *Gymnodinium catenatum.*

from Ria de Vigo, Spain (Sako et al., 1995; Yoshida et al., 1996, & 1998). This enzyme transfers the sulfate group of PAPS to N-21 of STX and GTX2+3 producing GTX5 and C1+2, respectively (Figure 12.3). A N-sulfotransferase (N-ST) of *G. catenatum* GC21V was purified by $(NH_4)_2SO_4$ precipitation and DEAE-cellulose, Blue Toyopearl, Mono Q and Superose 6 column chromatography. A single band of about 59 kDa was observed for the purified enzyme upon SDS-PAGE, which was in agreement with results of Superose 6 gel-filtration chromatography. Thus this enzyme seemed to be a monomer with a larger molecular mass than those reported for other mammalian sulfotransferases (M_r of 30–35 kDa) (Homma et al., 1992; Barnes et al., 1989; Hondoh et al., 1993) and phenol sulfotransferase of *Euglena* (M_r of 26 kDa) (Saidha and Schiff, 1994). This enzyme was active toward STX and GTX2+3, but not the N-1 hydroxy toxins, neoSTX and GTX1+4. The activity for STX or GTX2+3 was not inhibited by addition of neoSTX, GTX1+4 or other analogues of PSP toxins. This enzyme required PAPS as a sulfate donor, but not adenosine 5′-phosphosulfate (APS) and $MgSO_4$. The activity under standard conditions with PAPS was inhibited by addition of the PAPS analogue, 3′-phosphoadenosine 5′-phosphate (PAP), which is known to be a competitive inhibitor of sulfotransferase. N-ST activity was stimulated by Mg^{2+} or Co^{2+}, but was inhibited by Ca^{2+}, Cu^{2+}, Fe^{2+}, Mn^{2+}, Ni^{2+}, or Zn^{2+}. In the pH range of 4 to 10, the highest activity was observed at pH 6. This enzyme has an optimum temperature of 25°C (Table 12.1).

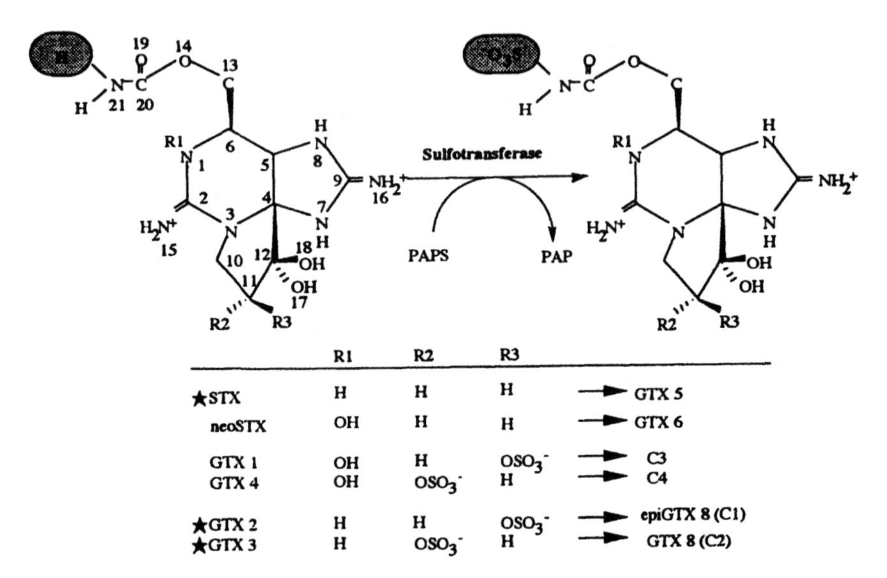

	R1	R2	R3	
★STX	H	H	H	→ GTX 5
neoSTX	OH	H	H	→ GTX 6
GTX 1	OH	H	OSO_3^-	→ C3
GTX 4	OH	OSO_3^-	H	→ C4
★GTX 2	H	H	OSO_3^-	→ epiGTX 8 (C1)
★GTX 3	H	OSO_3^-	H	→ GTX 8 (C2)

Figure 12.3 Possible scheme for the enzymatic step catalyzed by the PSP N-sulfo transferase of *G. catenatum* (Yoshida et al., 1996).

TABLE 12.1. Comparison between Properties of PSP N-sulfotransferase (N-ST) and O-sulfotransferase (O-ST) in *G. catenatum.*

	N-ST		O-ST
Substrate	STX	GTX2+3	11-α, β-Hydroxy STX
	↓	↓	↓
Products	GTX5	C1+C2	GTX2+3
SO_4 Donor	PAPS		PAPS
Opt pH	6		6
Opt Temp	25°C		35°C
Metal	Mg^{2+}, Co^{2+}		NO

The sulfotransferases (N-ST) of *G. catenatum* MZ5 and MZ12, isolated from Miyazu Bay, and Kyoto, Japan, respectively, were partially purified. Characteristics of both enzymes were similar to those of the enzyme from strain GC21V. Furthermore, *A. catenella* Acko5 from Uranouchi Bay, Japan, which produced mainly C1+2, also had a similar N-ST. The N-ST from *A. catenella* produced C1+2 and GTX5 from GTX2+3 and STX, respectively. It was optimally active at 15°C. The activity was not stimulated by any metals, and was inhibited not only by Ca^{2+}, Cu^{2+}, Fe^{2+}, Mn^{2+}, Ni^{2+} or Zn^{2+}, but also Co^{2+} and Mg^{2+}. These results suggest that the properties of N-ST from *G. catenatum* are somewhat different from those of *A. catenella* (Yoshida et al., 1998).

The other sulfotransferase (O-ST) activity of *G. catenatum* associated with conversion of 11-α, β-hydroxySTX into GTX2+3 was detected in a fraction bound to the DEAE-cellulose column. When this fraction was applied to a Blue-Toyopearl column, an active fraction of O-ST was eluted without binding and separated from N-ST. Recently the O-ST was purified and characterized from *G. catenatum* (Sako et al., unpublished). This enzyme had an optimum pH of 6 and an optimum temperature of 35°C, and did not require divalent cation for activity (Table 12.1). These results suggest that the O-ST specific to C-11 of 11-α, β-hydroxy STX was different from N-ST (Yoshida et al., 1996). Oshima et al. (1993) showed that crude extracts of *A. tamarense* transformed GTX2+3 to GTX1+4 while those of *G. catenatum* JP02 transformed GTX2+3 to C1+C2, and the extracts in both toxic (DE05) and nontoxic (DE09) isolates from Tasmania, Australia, produced the same reaction.

Cloning and sequencing of the N-ST and O-ST genes of *G. catenatum* will likely reveal a family of genes encoding PSP toxin-associated enzymes in *G. catenatum* and *A. catenella* that are located on the chromosomal DNA. Biochemical and molecular experiments focusing on PSP biosynthesis will be more readily performed with PSP-producing cyanobacteria such as *A. circinalis* (Onodera et al., 1996) and *L. wollei* (Onodera et al., 1997), which have a much higher growth rate and yield than dinoflagellates, and also a smaller genome.

5. GENETIC ANALYSIS OF PSP TOXIN PRODUCTION BY MATING REACTIONS

It is not yet clear whether the enzymes for PSP toxin biosynthesis and for transforming STX to its analogues are encoded by chromosomal, chloroplast, or mitochondrial DNA, or possibly other DNA sources such as bacteria, viruses or plasmids. One approach to elucidation of the mechanisms of PSP toxin production is through studies of the genetics of toxin production and its regulation.

According to the hypothesis of Gillham (1978), there are two inheritance patterns in heterothallic *Chlamydomonas reinhardtii,* in which the vegetative cells are haploid, as is also found in *A. catenella.* In crosses, chromosomal genes are inherited in a 1:1 Mendelian pattern, whereas chloroplast genes and mitochondrial genes are inherited uniparentally. Therefore, it is important to establish whether the production of PSP toxins is genetically regulated by the dinoflagellate's own chromosomal DNA.

Hypotheses about the mechanisms of PSP toxin production in *A. catenella* and other dinoflagellates are controversial (Kodama and Ogata, 1988; Silva, 1982; Franca et al., 1995; Sako et al., 1992; Ishida et al., 1993, 1998; Ishida, 1993; Kim et al., 1993a; Sako et al., 1995). Sako et al. (1992) tried to determine whether PSP toxins in toxic dinoflagellates are encoded by genes present on chromosomal DNA, chloroplast DNA, mitochondrial DNA, or contaminating bacteria, viruses or plasmids. The inheritance of the toxin profile in F1 progeny from parents with different toxin profiles was studied by means of heterothallic sexual reproduction of *A. tamarense, A. catenella* and *G. catenatum.* The proportion of the PSP toxin component was used as a phenotypic marker, because the toxin profile is stable in *Alexandrium* spp. (Cembella et al., 1987; Ogata et al., 1987; Kim et al., 1993b), as mentioned before (Figure 12.1)

Crosses of parental mating types (mt+ and mt–) of *A. tamarense* and *A. catenella* with different PSP toxin profiles were made and the toxin profiles of the F1 cells were compared with those of the parents, as shown in Table 12.2 (Sako et al., 1995). When a characteristic is encoded by a gene residing on chromosomal DNA, it is inherited in a 1:1 Mendelian pattern, whereas if the gene is located on the chloroplast or mitochondrial DNA, it is inherited uniparentally from the mt+ or mt– parent. Other symbiotic or contaminant factors such as bacteria, viruses and plasmids are also inherited uniparentally or at random.

Table 12.2 shows that a cross of *A. tamarense* OFX181 (mt+) and OFY184 (mt–), both having similar toxin profiles produced four F1 progeny of both

TABLE 12.2. Relationship between Toxin Type and Mating Type of Parent and F1 Strains in *A. catenella* and *A. tamarense* (Sako et al., 1995).

Parentage	Mating Type	Toxin Type	No. of Cysts	No. of F_1 Progeny	Mating Type	Toxin Type
A. catenella						
TNX 22	mt+	A[a]	2	1	mt−	A
				1	mt+	A
				1	mt−	B
OFY 101	mt−	B[b]		1	mt+	B
TNY 7	mt−	A	17	9	mt+	A
				8	mt−	A
				8	mt+	C
				9	mt−	C
OFX 072	mt+	C[c]		2	mt+	A′
				2	mt−	C′
A. tamarense						
OFX 181	mt+	D[d]	4	4	mt+	D
				4	mt−	D
				4	mt+	D
OFY 184	mt−	D		4	mt−	D
OFY 152	mt−	E[e]	4	2	mt+	E
				2	mt−	E
				2	mt+	F
OFX 191	mt+	F[f]		1	mt−	F

[a] GTX 4 + 5 and C2.
[b] C2 and neo STX.
[c] C2, neoSTX and STX.
[d] GTX4 and C2.
[e] GTX1 + 4.
[f] GTX4, C2 and neoSTX.

mt+ and mt– mating types, and having the same toxin profile as their parents. However, when the *A. tamarense* parent strain OFY152 (mt–), which has GTX1+4 as its major toxin component, and OFX191 (mt+), which has C1+C2, neoSTX and GTX1+4, were crossed, two of the F1 progeny showed the same toxin profile as OFY152, and the other two the same profile as OFX191 (Sako et al., 1992; Ishida et al., 1993).

In pairing *A. catenella* TNX22 (mt+) with OFY101 (mt–), both having different toxin profiles, F1 progeny of mt– and mt+ showed the same toxin profile as parents TNX22 and OFY101, respectively. Furthermore, the toxin profile of the F1 and F2 progeny was the same. The toxins of all F1 progeny were inherited biparentally. From these inheritance patterns, we concluded that a family of PSP toxin genes in *Alexandrium* was inherited in a 1:1 Mendelian pattern, and not in a uniparental or random pattern (Sako et al., 1992).

Differing from the normal Mendelian pattern of biparent type mentioned above, crosses with F1 and F2 cells of TNY7(mt–) and OFX072(mt+) produced, at high frequency, toxin profiles different from both parents. In the case of parent strains OFX072 (mt+) and TNY7 (mt–), the former produced neoSTX and STX and the latter, GTX4 and C4. Among 38 F1 progeny, 9 mt+ clones and 8 mt– clones showed the same toxin profile (A) as TNY7 (mt–), and 8 mt+ clones and 9 mt– clones showed the same toxin profile (C) as OFX072 (mt+). The four remaining F1 progeny had toxin profiles (A′ and C′), which represented a modification of the parents toxin profiles (Table 12.2).

Next, backcrosses of parent TNY7 (mt–) and F1 progeny 707-4 (mt+, generated from TNY7 and OFX072 crosses and with the same profile as OFX072) produced two B1 cells (B1 means first backcross generation; TN74–1 and TN74–2) having the same profile as F1 progeny 707-4, and two B1 cells (TN74–3 and TN74–4) having the same profile as TNY7. But when OFX072 (mt+) and F1 progeny 707-6 (mt–, generated from TNY7 and OFX072 and with the same profile as TNY7) were backcrossed, two (OF76–1 and OF76–2) of the four B1 cells produced had the same profile as OFX072 and F1 707–6, respectively, while the two other B1 cells (OF76–3 and OF76–4) did not inherit either the parental or F1 toxin profiles (Figure 12.4).

These results support the hypothesis that the genes encoding enzymes required for biosynthesis of PSP toxin analogues must reside on the chromosomal DNA of *Alexandrium* and other toxic dinoflagellates (Sako et al., 1992; Ishida et al., 1993, 1998). The mating types (mt+ and mt–) were not necessarily associated with toxin inheritance, as shown in Table 12.2. Genes involved in

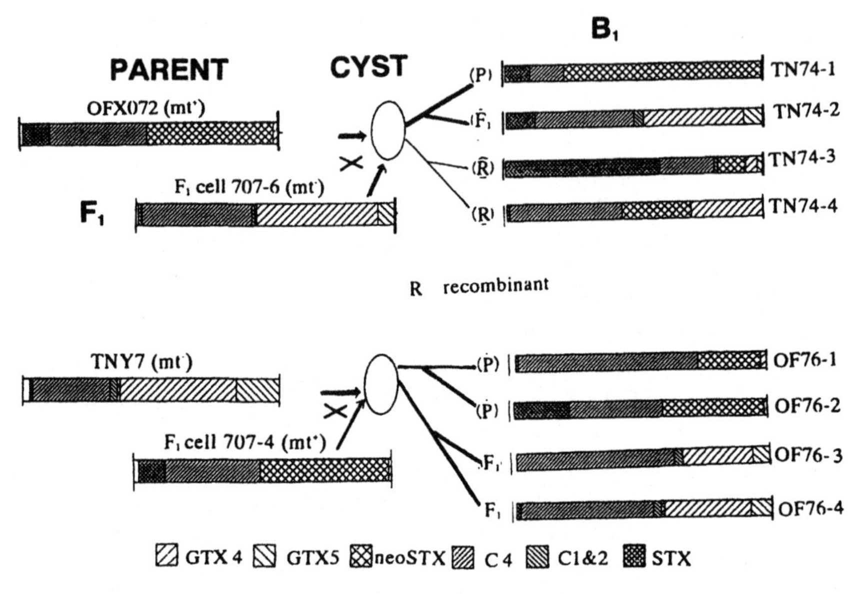

Figure 12.4 PSP toxin profile in B1 cells produced by backcrosses of parent and F1 (Kim et al., unpublished data).

the synthesis of PSP toxins are not linked with the mating type gene (Sako et al., 1992; Ishida et al., 1993).

6. MOLECULAR IDENTIFICATION OF *ALEXANDRIUM* SPP. AND RELATED DINOFLAGELLATES

As mentioned, the PSP toxin profile of F1 progeny in *A. tamarense* and *A. catenella* is inherited in a 1:1 Mendelian pattern (Sako et al., 1992). Then, PSP production must be indigenous to the microalgae. This indicates that molecular taxonomy based on genotypic characteristics is useful for identification and monitoring of toxic dinoflagellates.

To date, identification of dinoflagellates has been based on fine-scale morphological features such as general cell form, shape of apical pore plates, presence or absence of a ventral pore on the 1′ apical plate, and position of a posterior attachment pore (Taylor, 1984; Fukuyo, 1985). However, the occurrence of morphological "intermediates" between *A. tamarense* and *A. catenella* from the Northeast Pacific coast region (Taylor, 1984; Cembella and Taylor, 1986) calls into question the validity of this approach. Also of concern are reports that environmental conditions or growth stage alter the morphological features used to define these two species (Taylor, 1984; Cembella and Taylor, 1986). The morphological criteria have been inconclusive and remain controversial. Thus a more accurate discrimination at the intraspecies and interspecies levels of these toxic dinoflagellates is necessary for predicting the occurrence of toxic algal blooms.

Several methods have been used to differentiate species boundaries in order to resolve taxonomic ambiguities. Sexual compatibility is useful for discrimination of biological species (Mayr, 1982), but is not applicable to *Alexandrium* spp. because the sexual reproduction rate between mt+ and mt− mating types of the same species is remarkably low (Kim et al., 1993b). Protein-based comparisons at the species level generally are limited to enzyme electrophoresis and immunological approaches. The former has been used to solve systematic problems in *A. catenella, A. tamarense* and *A. fundyense* (Taylor, 1984; Cembella and Taylor, 1986; Hayhome et al., 1989; Sako et al., 1990). Sako et al. (1990) found a positive correlation between isozyme banding patterns and morphospecies designations among Japanese isolates of *A. catenella* and *A. tamarense.* The latter provides an estimate of similarity based on antigen-antibody cross-reactivity that can be used to delineate taxonomic boundaries (Hiroishi et al., 1988; Nagasaki et al., 1991). Sako et al. (1993) and Adachi et al. (1993a) developed interspecific identification between Japanese *A. catenella* and *A. tamarense* as well as intraspecific discrimination between Japanese strains and Thai strains of *A. tamarense* (Adachi et al.,

1993b; 1993c). However, the antigens recognized by these antibodies change with differing environmental conditions.

To resolve the taxonomic debate and define genetic markers useful for classifying these organisms, Lenaers et al. (1991) and Scholin and Anderson (1993) used small subunit (SS) and large subunit (LS) rDNA sequences as phylogenetic and taxonomic indicators for several species of the genus *Alexandrium,* and clarified their phylogenetic relationships. For discrimination at the inter- and intraspecies level of *Alexandrium,* Adachi et al. (1994 & 1995) targeted the regions containing the 5.8S rDNA and internal transcribed spacers (ITSs). The 5.8S rRNA gene and the ITS1 and ITS2 regions among eucaryotes are located between the 18S and 28S rRNA genes. The nucleotide sequence of the ITS regions is less tightly conserved than that in the 18S and 28S rDNA (Appels and Honeycutt, 1986). This diversity in the ITS1 and ITS2 is thus likely to be a valuable taxonomic character among intraspecies.

Table 12.3 shows the analysis of 23 clonal axenic strains of seven species of *Alexandrium* for restriction fragment length polymorphisms (RFLP) of 5.8S rDNA and ITS1 and 2 regions. Morphospecies identifications followed the morphological classification according to the criteria of Fukuyo (1985).

PCR amplification of the ITSs and 5.8S rDNA region from all *Alexandrium* species resulted in a single product of approximately 610 bp. RFLP analysis using four restriction enzymes revealed six distinct ITS types of rDNA. The restriction patterns of *A. catenella* (Catenella type) were uniform at the intraspecific level and clearly distinguishable from those of *A. tamarense.* The patterns associated with *A. tamarense* (Tamarense type) were also uniform except for strain CU-1 and a strain WKS-1 (Adachi et al., 1994). *A. tamarense* CU-1 from Thailand was different from the Tamarense type and showed the same RFLP pattern as *A. affine,* and the species name was changed to *A. affine* CU-1. *A. affine* (Affine type), *A. insuetum* (Insuetum type), *A. lusitanicum* (Lusitanicum type) and *A. pseudogonyaulax* (Pseudogonyaulax type) carry unique ITS types.

The sum of the lengths of digestion products from the Tamarense type were greater than 610 bp, indicating heterogeneity within the multicopy rDNA cistrons. Scholin and Anderson (1993) reported the existence of two distinct SS rDNA. All isolates of *A. catenella* and *A. tamarense* with the exception of the WKS-1 strain are toxic. By contrast, the isolates of WKS-1, *A. affine, A. insuetum* and *A. pseudogonyaulax,* are nontoxic. RFLP analysis of the 5.8S rDNA and flanking ITS regions from *Alexandrium* species revealed that they are useful for discrimination at the species and/or population levels.

To further assess the degree of variability in the 5.8S rDNA and ITS regions, the rDNA amplification products from 23 strains of five species of *Alexandrium* were cloned and sequenced (Adachi et al., 1995, 1996c). The precise lengths of ITS regions containing the 5.8S rDNA ranged from 518–535 bp in all cases. Alignments of the 5.8S rDNA, ITS1 and ITS2 sequences of these strains

TABLE 12.3. List of *Alexandrium* Strains.

Species	Strains	Abbreviation	Toxicity[a]
Alexandrium catenella	TN11 (Tanabe Bay, Japan)	TNY11	+
A. catenella	TN12 (Tanabe Bay, Japan)	TNX12	+
A. catenella	TN22 (Tanabe Bay, Japan)	TNX22	+
A. catenella	OFX072 (Ofunato Bay, Japan)	OFX072	+
A. catenella	OFY101 (Ofunato Bay, Japan)	OFY101	+
A. catenella	OFX102 (Ofunato Bay, Japan)	OFX102	+
A. catenella	M17 (Harima-Nada, Japan)	M17	+
A. catenella	ko-3 (Uranouchi Bay, Japan)	ko-3	+
A. catenella	Y-2 (Yamakawa Bay, Japan)	Y-2	+
A. tamarense	OFX151 (Ofunato Bay, Japan)	OFX151	+
A. tamarense	OFX181 (Ofunato Bay, Japan)	OFX181	+
A. tamarense	OFX191 (Ofunato Bay, Japan)	OFX191	+
A. tamarense	OK875-1 (Okirai Bay, Japan)	OK875-1	+
A. tamarense	OK875-6 (Okirai Bay, Japan)	OK875-6	+
A. tamarense	At304A (Mikawa Bay, Japan)	At304A	+
A. tamarense	At503A-A (Mikawa Bay, Japan)	At503A-A	+
A. tamarense	FK-788 (Funka Bay, Japan)	FK-788	+
A. tamarense	HIAI (Hiroshima Bay, Japan)	HIAI	+
A. tamarense	HI38 (Hiroshima Bay, Japan)	HI38	+
A. tamarense	AT4 (Harima-Nada, Japan)	AT4	+
A. tamarense	WKS-1 (Kushimoto, Japan)	WKS-1	–
A. tamarense	PE1V (Vigo, Spain)	PE1V	–
A. tamarense[b]	CU-1 (Gulf of Thailand, Thailand)	CU-1	–
A. tamarense	CU-15 (Gulf of Thailand, Thailand)	CU-15	–
A. tamarense	PW06 (Araska, U.S.A.)	PW06	+
A. fundyense	GtCA29 (Gulf of Maine, U.S.A.)	GtCA29	+
A. affine	T-6 (Tachibana Bay, Japan)	T-6	–
A. affine	(Harima Nada, Japan)		–
A. lusitanicum	(Vigo, Spain)		
A. insuetum	(Shoudoshima, Japan)		–
A. pseudogonyaulax	(Harima Nada, Japan)		–

[a] The toxicities of these strains were determined by HPLC fluorescence method (Oshima et al., 1989).
[b] The present species name has been changed to *A. affine* CU-1.

of *Alexandrium* spp. were compared. The phylogenetic tree of these species (*A. catenella, A. tamarense, A. fundyense* and *A. affine*) showed that there are six divergent ITS-types among the four species (Figure 12.5). Nine strains of Japanese *A. tamarense* from six different locations were identical at the intraspecies level. *A. fundyense* GtCA29 from the U.S. belonged to the same branch as nine Japanese strains of *A. tamarense* (Tamarense type) except for strain WKS-1. Strain WKS-1 from Kushimoto (Japan) and two strains CU-1 and CU-15 from Thailand belonged to branches different from the Tamarense type. Strain CU-1 from Thailand belonged to the same branch as *A. affine* from Japan. The ITS types of the other strains corresponded with the morpho-

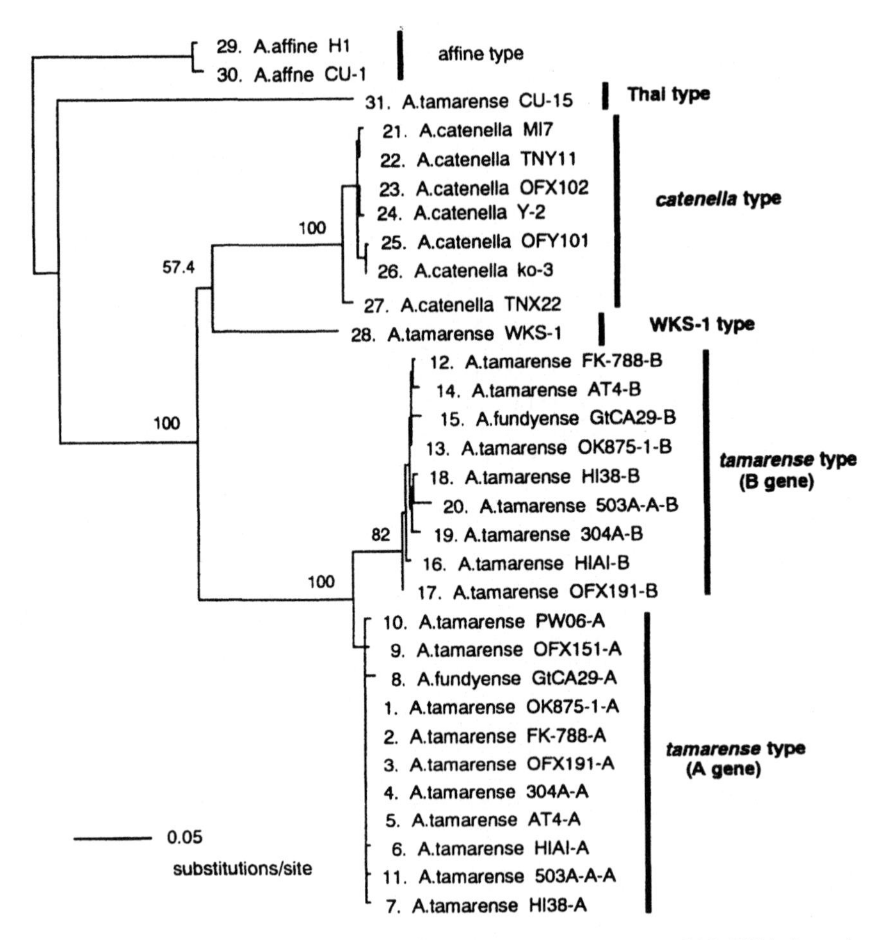

Figure 12.5 Molecular phylogenetic tree inferred from ITS 1, ITS 2 and 5.8S rDNA (base: 1-557) in *Alexandrium* species by means of the NJ method (Adachi et al., 1996c).

species. In relation to this grouping, the interspecies differences between the Japanese strains of *A. catenella* and *A. tamarense* as well as intraspecies discrimination between Japanese strains of *A. tamarense* and those from Thailand were demonstrated using several different monoclonal antibodies (Sako et al., 1993; Adachi et al., 1993). Results of studies using monoclonal antibody analysis also indicated that *A. fundyense* GtCA29 and *A. tamarense* PW06 from the U.S. belonged, immunologically, to the same ITS type as the Japanese *A. tamarense* (Adachi et al., 1993).

The existence of at least two major genes within the rRNA gene family in the Tamarense type was suggested based on the increased size of fragments generated from restriction endonuclease digestion of PCR products and from the two distinct classes of ITS genes in the sequence (Adachi et al., 1995). These two genes (A and B) may correspond to the two distinct SS rDNAs in *A. tamarense* and *A. fundyense* from North America, although B is a pseudogene (Scholin and Anderson, 1993). The ITS region in the four other groups exhibited a homogeneity within the gene family.

Sequence analysis of the 5.8S rDNA and ITS regions of genus *Alexandrium* revealed that variations in the 3′ end of the ITS 1 region (bases 150–177) was very low within the species but extremely high at the inter-species level. Two types (cTAM-F1 and cCAT-F1) of fluorescein-conjugated DNA probes based on this hypervariable region of ITS 1 were prepared, and the reactivity of these probes to targeted chromosomal DNA in various species of *Alexandrium* was analyzed by a fluorescent in situ hybridization method (Adachi et al., 1996a, b). cTAM-F1 and cCAT-F1 hybridized with chromosomal DNA in the cells of *A. tamarense* and *A. catenella*, respectively, but did not react with the DNA of other species. These hypervariable regions might be useful as specific probes for molecular identification of an individual cell in toxic species. Orias et al. (1991) reported on the PCR amplification of *Tetrahymena* rDNA segments starting with individual cells. To apply both methods to field samples of *Alexandrium* from natural seawaters would be very useful for molecular identification and early monitoring of toxic *Alexandrium* species.

Implementation of the methods discussed above will contribute to forecasting. The findings will contribute to forecasting the occurrence of PSP toxin-producing dinoflagellates, thus reducing the incidence of paralytic shellfish poisoning.

7. ACKNOWLEDGMENTS

We are indebted to Prof. A. Uchida, Dr. I. Yoshinaga and Dr. T. Yoshida, Graduate School of Agriculture, Kyoto University: Dr. M. Adachi, Faculty of Agriculture, Kochi University: and Dr. C-H. Kim, Pukyong National University, Korea; for valuable discussions and useful comments. This work was partly supported by a Grant-in-Aid for Scientific Research from the Ministry of Education, Science and Culture of Japan (04404014, 07456096), and by grants (1990–1997) from the Ministry of Agriculture, Forestry and Fisheries, Japan.

8. REFERENCES

Adachi, M., Sako, Y., and Ishida, Y. 1993a. "The identification of nonspecific dinoflagellates *Alexandrium tamarense* from Japan and Thailand by monoclonal antibodies," *Nippon Suisan Gakkaishi* 59:327–332.

Adachi, M., Sako, Y., and Ishida, Y. 1993b. "Application of monoclonal antibodies to field samples of *Alexandrium* species," *Nippon Suisan Gakkaishi* 59:1171–1175.

Adachi, M., Sako, Y., and Ishida, Y. 1994. "Restriction fragment length polymorphism of ribosomal DNA internal transcribed spacer and 5.8S regions in Japanese *Alexandrium* species (Dinophyceae)," *J. Phycol.* 30:857–863.

Adachi, M., Sako, Y., and Ishida, Y. 1995. "Ribosomal DNA internal transcribed spacer regions (ITS) define species of the genus *Alexandrium,*" in *Harmful Marine Algal Blooms,* eds., P. Lassus, F. Arzul, E. Erard-Le Denn, P. Gentien, C. Marcaillou-Le Baut. France: Lavoisier Publishing Inc. pp. 15–20.

Adachi, M., Sako, Y., and Ishida, Y. 1996a. "Cross-reactivity of fluorescent DNA probes to isolates of the genus *Alexandrium* by in situ hybridization," in *Harmful and Toxic Algal Blooms,* eds., T. Yasumoto, Y. Oshima and Y. Fukuyo. Intergov. Sendai: Oceanogr. Comm of UNESCO. pp. 455–458.

Adachi, M., Sako, Y., and Ishida, Y. 1996b. "Identification of the toxic dinoflagellates *Alexandrium catenella* and *A. tamarense* (Dinophyceae) using DNA probes and whole-cell hybridization," *J. Phycol.* 32:1049–1052.

Adachi, M., Sako, Y., and Ishida, Y. 1996c. "Analysis of *Alexandrium* (Dinophyceae) species using sequences of the 5.8S ribosomal DNA and internal transcribed spacer regions," *J. Phycol.* 32:424–432.

Adachi, M., Sako, Y., Ishida, Y., Anderson, D. M., and Reguera, B. 1993c. "Cross-reactivity of five monoclonal antibodies to various isolates of *Alexandrium* as determined by an indirect immunofluorescence method," *Nippon Suisan Gakkaishi* 59:807.

Anderson, D. M. 1990. "Toxin variability in *Alexandrium* species," in *Toxic Marine Phytoplankton,* eds., E. Graneli, B. Sundstrom, L. Edler, and D. M. Anderson. Elsevier Sci. Publ. Co., Inc. Amsterdam pp. 41–51.

Anderson, D. M., Kulis D. M., Sullivan, J. J. and Hall, S. 1990. "Toxin composition variations in one isolate of the dinoflagellate *Alexandrium fundyense,*" *Toxicon,* 28:885–893.

Appels, R. and Honeycutt, R. L. 1986. "rDNA: evolution over a billion years," in *DNA systematics, vol. II, Plants.* Boca Raton, FL: CRC Press., pp. 81–135.

Barnes, S., Buchina, E. S., King, R. J., McBurnett, T., and Taylor, K. B. 1989. "Bileacid sulfotransferase 1 from rat river sulfates bile acids and 3-hydroxy steroids: purification, N-terminal amino acid sequence, and kinetic properties," *J. Lipid. Res.* 30:529–540.

Boczar, B. A., Beitler, M. K., Liston, J., Sulivan, J. J., and Cattolico, R. A. 1988. "Paralytic shellfish toxins in *Protogonyaulax tamarensis* and *Protogonyaulax catenella* in axenic culture," *Plant Physiol.* 88:1285–1290.

Boyer, G. L., Sullivan, J. J., Andersen, R. J., Harrison, P. J., and Taylor, F. J. R. 1987. "Effects of nutrient limitation on toxin production and composition in the marine dinoflagellate *Protogonyaulax tamarensis,*" *Mar. Biol.* 96:123–128.

Burkley, L. J., Ikawa, M., and Sasnen, J. J. 1976. "Isolation of *G. tamarensis* toxins from soft shell clams (*Mya arenaria*) and a thin-layer chromatographic-fluorometric method for their detection," *J. Agri. Food Chem.* 24:107–111.

Cembella, A. D. and Taylor, F. J. R. 1986. "Electrophoretic variability within the *Protogonyaulax tamarensis/catenella* species complex: pyridine linked dehydrogenases," *Bioch. Syst. Ecol.* 14:311–321.

Cembella, A. E., Sullivan, J. J., Boyer, G. L., Taylor, F. J. R., and Andersen, R. J. 1987. "Variation in paralytic shellfish toxin composition within the *Protogonyaula tamarensis/catenella* species complex: Red tide dinoflagellates," *Biochem. Syst. Ecol.* 15:171–186.

Franca, S., Viega, S. S., Mascarenhas, V., Pinto, L. and Doucette, G. J. 1995. "Prokaryotes in association with a toxic *Alexandrium lusitanicum* in culture," in *Harmful Marine Algal Blooms,* eds., Lassus, P., Arzul, C., Erard-Le Denn, E., Gentien, P. and Marcaillou-Le Baut, C. Lavoisier Publ. Inc. Paris: pp. 45–51.

Fukuyo, Y. 1985. "Morphology of *Protogonyaulax tamarensis* (Lebour) Taylor and *Protogonyaulax catenella* (Whedon and Kofoid) Taylor from Japanese coastal waters," *Bull. Mar. Sci.* 37:529–537.

Gillham, N. W. 1978. *Organella Heredity,* New York: Raven Press.

Hallegraeff, G. M. 1993. "A review of harmful algal blooms and their apparent global increase," *Phycologia* 32:79–99.

Hayhome, B., Anderson, D. M., Kulis, D. M., and Whitten, D. J. 1989. "Variation among congeneric dinoflagellates from the north eastern United States and Canada," *Mar. Biol.* 101:427–435.

Herzog, M. and Maroteaux, L. 1986. "Dinoflagellate 17S rRNA sequence inferred from the gene sequence: Evolutionary implications," *Proc. Nat. Acad. Sci. USA* 83:8644–8648.

Hiroishi, S., Uchida, A., Nagasaki, K., and Ishida, Y. 1988. "A new method for identification of inter- and intra-species of red tide algae *Chattonella antiqua* and *Chattonella marina* (Raphidophyceae) by means of monoclonal antibodies," *J. Phycol.* 24:442–444.

Homma, H., Nakagome, I., Kamakura, M., and Matsui, M. 1992. "Immunochemical characterization of developmental changes in rat hepatic hydroxysteroid sulfotransferase," *Biochem. Biophys. Act.* 1121:1169–1174.

Hondoh, T., Suzuki, T., Hirato, K., Saitoh, H., Kadofuku, T., Sato, T., Yanaihara, T. 1993. "Purification and properties of estrogen sulfotransferase of human fetal liver," *Biomedical Res.* 14:129–136.

Ishida, Y. 1993. "Who produces PSP?," *Harmful Algal News,* 7:1–2.

Ishida, Y., Uchida, A., and Sako, Y. 1998. "Genetic and biochemical approaches to PSP toxin production of toxic dinoflagellates," in *The Physiological Ecology of Harmful Algal Blooms,* eds., D. M. Anderson, A. Cembella and G. Hallegraeff. NATO ASI Series. Berlin: Springer-Verlag. pp. 49–58.

Ishida, Y., Kim, C.-H., Sako, Y., Hirooka, N., and Uchida, A. 1993. "PSP toxin production is chromosome dependent in *Alexandrium* spp.," in *Toxic Phytoplankton Blooms in the Sea,* eds., T. J. Smayda and Y. Shimizu. Amsterdam: Elsevier Sci. Publ. pp. 881–887.

Kim, C.-H., Sako, Y., and Ishida, Y. 1993a. "Variation of toxin production and composition in axenic cultures of *Alexandrium catenella* and *A. tamarense,*" *Nippon Suisan Gakkaishi* 59:633–639.

Kim, C.-H., Sako, Y., and Ishida, Y. 1993b. "Comparison of toxin composition between populations of *Alexandrium* spp. from geographically distant areas," *Nippon Suisan Gakkaishi* 59:641–646.

Kodama, M. 1990. "Possible links between bacteria and toxin production in algal blooms," in *Toxic Marine Phytoplankton,* eds., E. Graneli, B. Sundstrom, L. Edler, and D. M. Anderson. Amsterdam: *Elsevier Sci. Publ. Co., Inc.* pp. 52–61.

Kodama, M. and Ogata, T. 1988. "New insights into shellfish toxins," *Mar. Pollut. Bull.* 19:559–564.

Lenaers, G., Scholin, C. A., Bhaud, Y., Saint-Hilaire, D., and Herzog, M. 1991. "A molecular phylogeny of dinoflagellate protists (Pyrrhophyta) inferred from the sequence of the 24S rRNA divergent domains D1 and D8," *J. Mol. Evol.* 32:53–63.

Mayr, E. 1982. *The Growth of Biological Thought: Diversity, Evolution and Inheritance.* Cambridge, MA: Harvard Univ. Press. 273 pp.

Mulder, G. J. and Jakoby, W. B. 1990. "Sulfation," in *Conjugation Reactions in Drug Metabolism,* ed., Mulder, G. J. London: Taylor & Francis. pp. 107–161.

Nagasaki, K., Uchida, A. and Ishida, Y. 1991. "A monoclonal antibody which recognized the cell surface of red tide alga *Gymnodinium nakasakiennse,*" *Nippon Suisan Gakkaishi* 57:1211–1214.

Ogata, Y., Ishimaru, T., and Kodama, M. 1987. "Effect of water temperature and light intensity on growth rate and toxicity change in *Protogonyaulax tamarensis,*" *Mar. Biol.* 95:217–220.

Onodera, H., Satake, M., Oshima, Y., Yasumoto, T., and Carmichael, W. W. 1997. "Detection of PSP toxins and six new saxitoxin analogs in the freshwater filamentous cyanobacterium *Lyngbya wollei,*" *Abstracts of VIII Internl Conf. Harmful Algae,* June 25–29, 1997, Vigo, Spain. p. 153.

Onodera, H., Oshima, Y., Watanabe, M. F., Watanabe, M., Bolch, C. J., Blackburn, S., and Yasumoto, T. 1996. "Screening of paralytic shellfish toxins in freshwater cyanobacteria and chemical confirmation of the toxins in cultured *Anabaena circinalis* from Australia," in *Harmful and Toxic Algal Blooms,* eds., T. Yasumoto, Y. Oshima and Y. Fukuyo. Intergovern. Sendai: Oceanogr. Comm. UNESCO. pp. 563–566.

Oshima, Y. 1995. "Post-column derivatization HPLC methods for paralytic shellfish poisons," in *Manual on Harmful Marine Microalgae,* eds., G. M. Hallegraeff, D. M. Anderson, A. D. Cembella. Intergovern. Paris: Oceanogr. Comm. UNESCO. pp. 81–94.

Oshima, Y., Sugino, K., and Yasumoto, T. 1989. "Latest advances in HPLC analysis of paralytic shellfish toxins," in *Mycotoxins and Phycotoxins '88,* eds., S. Natori, K. Hashimoto and Y. Ueno. Amsterdam: Elsevier Sci. Publ. pp. 319–326.

Oshima, Y., Itakura, H., Lee, K. C., Yasumoto, T., Blackburn, S., and Hallegraeff, G. 1993. "Toxin production by the dinoflagellate *Gymnodinium catenatum,*" in *Toxic Phytoplankton Blooms in the Sea,* eds., T. Smayda and Y. Shimizu. Amsterdam: Elsevier Sci. Publ. pp. 907–912.

Orias, E., Hashimoto, N., Chau, M.-F., and Higashinakagawa, T. 1991. "PCR amplification of *Tetrahymena* rDNA segments starting with individual cells," *J. Protozool* 38:306–311.

Saidha, T. and Schiff, J. A. 1994. "Purification and properties of a sulfotransferase from *Euglena* using L-tyrosine as substrate," *Biochem. J.* 298:45–50.

Sako, Y., Adachi, M., and Ishida, Y. 1993. "Preparation and characterization of monoclonal antibodies to *Alexandrium* species," in *Toxic Phytoplankton Blooms in the Sea,* eds., T. J. Smayda and Y. Shimizu, Amsterdam: Elsevier Sci. Publ. B.V. p. 87–93.

Sako, Y., Kim, C.-H., and Ishida, Y. 1992. "Mendelian inheritance of paralytic shellfish poisoning toxin in the marine dinoflagellate *Alexandrium catenella,*" *Biosci. Biotech. Biochem.* 56:692–694.

Sako, Y., Kim, C.-H., Ninomiya, H., Adachi, M., and Ishida, Y. 1990. "Isozyme and cross analysis of mating populations on the *Alexandrium catenella/tamarense* species complex," in *Toxic Marine Phytoplankton,* eds., E. Graneli, B. Sundstrom, L. Edler, and D. M. Anderson. Amsterdam: Elsevier Sci. Publ. Co., Inc. pp. 320–323.

Sako, Y., Naya, N., Yoshida, T., Kim, C.-H., Uchida, A., and Ishida, Y. 1995. "Studies on stability and heredity of PSP toxin composition in the toxic dinoflagellate *Alexandrium,*" in *Harmful Marine Algal Blooms,* eds., Lassus, P., Arzul, F., Denn, E. E. L., Gentien, P. and Marcaillou-Le Baut, C. Paris: Lavoisier Publ. Inc. pp. 345–350.

Sato, S. and Shimizu, Y. 1997. "Purification of a fluorescent product of the bacterium, *Moraxella* sp.: a neosaxitoxin impostor in HPLC," *Abstracts of VIII Internatl. Conf. Harmful Algae,* June 25–29, 1997, Vigo, Spain. p. 180.

Schantz, E. J., Mold, J. D., Stanger, D. W., Riel, F. J., Bowden, J. P., Lynch, J. M., Wyle, R. S., Riegel, B., and Sommer, H. 1957. "Paralytic shellfish poison, VI, A procedure for the isolation and purification of the poison from toxic clams and mussel tissues," *J. Am. Chem. Soc.* 79, 5230–5235.

Schantz, E. J., Ghazarossian, V. E., Schnoes, H. K., Strong, F. M., Springer, J. P., Pezzanite, J. O., and Clardy, J. 1975. "The structure of saxitoxin," *J. Am. Chem. Soc.* 97:1238–1239.

Scholin, C. A. and Anderson, D. M. 1993. ''Population analysis of toxic and nontoxic *Alexandrium* species using ribosomal RNA signature sequences,'' in *Toxic Phytoplankton Blooms in the Sea,* eds., T. J. Smayda and Y. Shimizu. Amsterdam: Elsevier Sci. Publ. pp. 95–102.

Shimizu, Y. 1979. ''Developments in the study of paralytic shellfish toxins,'' in *Toxic Dinoflagellate Blooms,* eds., D. L. Taylor and H. H. Seliger. New York: Elsevier Sci. Publ. pp. 321–326.

Shimizu, Y. 1986. ''Chemistry and biochemistry of saxitoxin analogues and tetrodotoxin,'' in *Tetrodotoxin, Saxitoxin and the molecular Biology of the Sodium Channel,* eds., C. Y. Kao and S. R. Levinson. *Ann. NY Acad. Sci.* 479:24–31.

Shimizu, Y. 1996. ''Microalgal metabolites: A new perspective,'' *Annu. Rev. Microbiol.* 50:431–465.

Shimizu, Y., Alam, M., Oshima, Y., and Fallon, W. E. 1975. ''Presence of four toxins in red tide infested clams and cultured *Gonyaulax tamarensis* cells,'' *Biochem. Biophys. Res. Commun.* 66:731–737.

Shimizu, Y., Kobayashi, M., Genenah, A., and Ichihara, I. 1984b. ''Biosynthesis of paralytic shellfish toxins,'' in *Seafood Toxins,* ed., E. P. Ragelis, ACS Symposium Series 262. Washington D.C.: *Amer. Chem. Soc.* pp. 151–160.

Shimizu, Y., Norte, M., Hori, A., Genenah, A., and Kobayashi, M. 1984a. ''Biosynthesis of saxitoxin analogues: the unexpected pathway,'' *J. Am. Chem. Soc.* 106:6433–6434.

Silva, E. S. 1982. ''Relationship between dinoflagellates and intracellular bacteria,'' in *Marine Algae in Pharmaceutical Science 2,* ed., L. Hoppe. Berlin: Walter de Gruyter & Co. pp. 269–288.

Sommer, H. and Meyer, K. F. 1937. ''Paralytic shellfish poisoning,'' *Arch. Path.* 24:560–598.

Taroncher-Oldenburg, G. and Anderson, D. M. 1997. ''Identification and characterization of genes related to saxitoxin biosynthesis in the red tide dinoflagellate *Alexandrium fundyense,*'' *Abstracts of VIII Internatl. Conf. Harmful Algae,* June 25–29, 1997, Vigo, Spain. p. 195.

Taylor, F. J. R. 1984. ''Toxic dinoflagellates: taxonomic and biogeographic aspects with emphasis on *Protogonyaulax,*'' in *Seafood Toxins,* ed., E. P. Ragelis, ACS Symposium Series 262. Washington, D.C.: *Amer. Chem. Soc.* pp. 77–97.

Yoshida, T., Sako, Y., Uchida, A., Ishida, Y., Arakawa, O., and Noguchi, T. 1996. ''Purification and properties of paralytic shellfish poisoning toxins sulfotransferase from toxic dunoflagellate *Gymnodinium catenatum,*'' in *Harmful and Toxic Algal Blooms,* eds., T. Yasumoto, Y. Oshima and Y. Fukuyo. Intergovern. Oceanogr. Comm. UNESCO. p. 499–502.

Yoshida, T., Sako, Y., Fujii, A., Uchida, A., Ishida, Y., Arakawa, O., and Noguchi, T. 1998. ''Comparative study on two sulfotransferases involved with sulfation to N-21 of PSP toxins from *Gymnodinium catenatuma* and *Alexandrium catenella,*'' *Abstracts of VIII Internatl. Conf. Harmful Algae,* June 25–29, 1997, Vigo, Spain. p. 218.

Yoshida, T., Sako, Y., Kakutani, T., Fujii, A., Uchida, A., Ishida, Y., Arakawa, O., and Noguchi, T. 1998. ''Comparative study of two sulfotransferases for sulfation to N-21 of *Gymnodinium catenatum* and *Alexandrium catenella* toxins,'' in *Harmful Algae,* eds., B. Reguera et al. Vigo: Xunta de Galicia & TOC of UNESCO. pp. 366–369.

PART IV

PARASITIC PROTOZOA

CHAPTER 13

Toxoplasma gondii Strain Variation and Pathogenicity

JUDITH E. SMITH
NAOMI REBUCK

1. INTRODUCTION

TOXOPLASMA gondii is one of the most common of all parasites. Its distribution is global, ranging from Alaska to Australia, and it is estimated to infect about one third of the human population (Jackson and Hutchinson, 1989). The widespread distribution of the parasite may in part be due to its dual mechanisms of transmission, as infection may be either foodborne, due to ingestion of tissue cysts in uncooked meat, or "environmental" due to ingestion of oocysts excreted by cats. The spectrum of disease caused by the parasite is broad, but toxoplasmosis is mainly known as a cause of congenital disease and abortion both in humans and in livestock (Remington and Desmonts, 1990; Dubey and Beattie, 1988) and as a potentially lethal infection of AIDS patients (Luft and Remington, 1992). The cellular and molecular organization of *Toxoplasma* and host immune response to the parasite are well understood but parasite population biology has only recently been investigated (Sibley and Howe, 1996). The advent of molecular markers to analyze strain variation allows a reappraisal of several unresolved issues. Firstly, it is important to evaluate the extent to which parasite genotype influences the pathogenesis of infection and, related to this, whether the molecular mechanism of virulence can be determined. Secondly, the markers enable closer investigation of parasite epidemiology and in particular, allow us to assess the relative importance of the two transmission routes.

In this chapter we review current understanding of disease pathogenesis, giving a general overview of parasite biology, life cycle, epidemiology and immunoregulation before considering in detail the impact of molecular markers on analysis of population structure and disease epidemiology.

2. *TOXOPLASMA:* AN UNUSUAL APICOMPLEXAN PARASITE

Toxoplasma gondii is a protozoan parasite belonging to the Phylum Apicomplexa, which contains many human and veterinary disease agents. In the human population the most important of these are *Plasmodium falciparum* and *Plasmodium vivax,* the causative agents of human malaria, while other species cause severe disease in domestic animals such as cattle (*Babesia, Theileria*), sheep (*Sarcocystis*), and birds (*Eimeria*). Apicomplexan parasites have three common characteristics: they are obligate intracellular parasites, they have complex life cycles involving both sexual and asexual reproduction and they are highly host specific. *Toxoplasma* is an unusual member of the group in that it can theoretically sustain itself purely by asexual reproduction and infects an extremely broad range of intermediate hosts.

2.1. CELLULAR STRUCTURE

In terms of its cellular structure *Toxoplasma* is a typical apicomplexan. It has a complex life cycle during which three distinct invasive stages (sporozoites, tachyzoites and bradyzoites) mediate host cell entry prior to intracellular multiplication. These individual parasites possess the classic ''zoite'' structure which shows many adaptations for host cell invasion and intracellular survival (Figure 13.1). The polarity of the cell is determined by the underlying microtubular cytoskeleton which forms a basket-like structure, the conoid, at the anterior of the zoite, from which radiate the subpellicular microtubules (Nichols and Chiappino, 1987). Associated with these anterior structures are two sets of secretory organelles, rhoptries and micronemes, while a third type, known as dense granules, are distributed throughout the cytoplasm (Sheffield and Melton, 1968). Each of these organelles contains complex families of proteins which play specific roles in establishing the parasite within the host cell.

The process of invasion and intracellular growth has been most widely studied in vitro with the disease-associated, tachyzoite stage of the parasite. Early attachment to the host cell appears to be mainly mediated by the major surface protein SAG 1 (Grimwood and Smith, 1992; Mineo et al., 1993) although several other surface molecules such as SAG 2 (Grimwood and Smith, 1996) and SAG 3 (Tomavo, 1996) may play supplementary roles. Following this initial interaction, the parasite becomes apically oriented and forms a tight junction with the host membrane (Figure 13.1c). The most likely candidates for the formation of this junction are the microneme molecules (Wan et al., 1997; Fourmaux et al., 1996). These are members of the TRAP (Thrombospondin-related anonymous protein) family of proteins, which have been widely implicated in adhesion of apicomplexan parasites (Robson et al., 1988, Tomley et al., 1991). One of these molecules (MIC2) has been localized in the tight junction during invasion (Carruthers and Sibley, 1997) while a

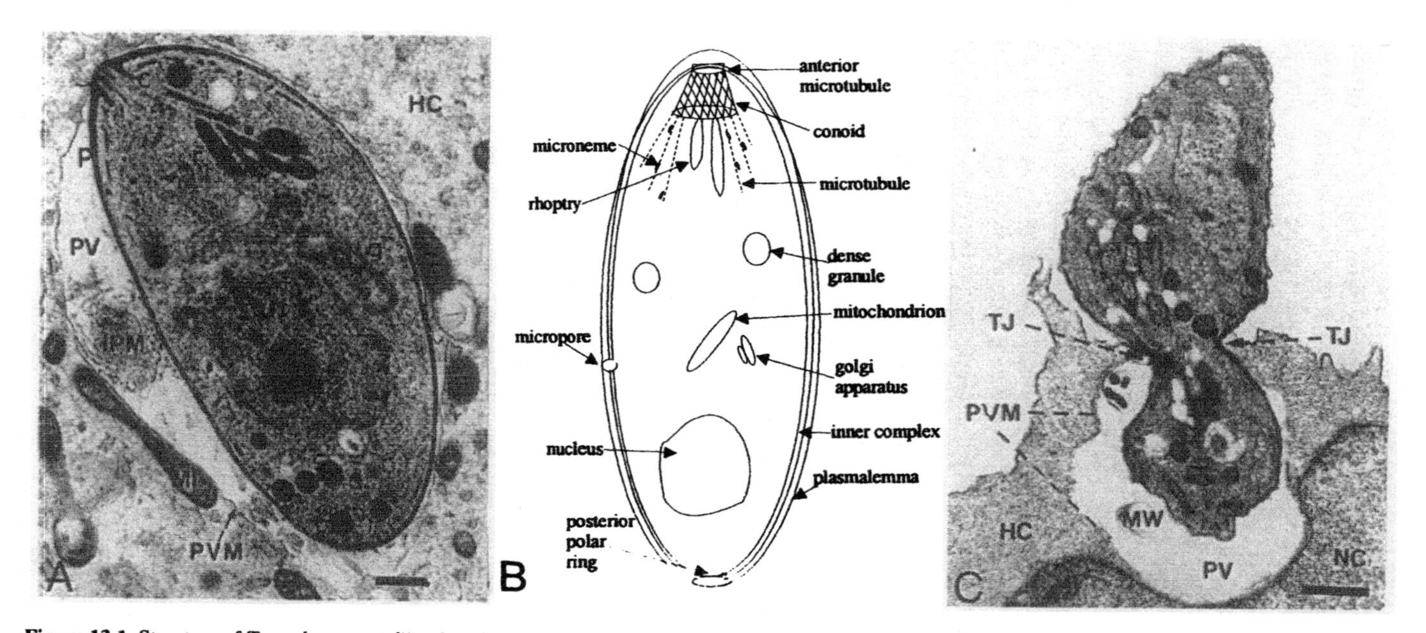

Figure 13.1 Structure of *Toxoplasma gondii* tachyzoite. A). Transmission electron micrograph of intracellular tachyzoite 30 mins post invasion showing parasite plasmalemma (p), golgi (g), nucleus (n), conoid (c), rhoptries (r), dense granules (d). The parasite is within the parasitophorous vacuole (PV) surrounded by the PV membrane (PVM) which is closely associated with host cell mitochondria (MI), note the network of tubules (IPM) secreted by the parasite into the vacuole, bar = 200nm. B) Schematic diagram of tachyzoite. C) Tachyzoite entering host cell (HC) note tight junction (TJ) and newly formed parasitophorous vacuole (PV), containing membranous whorls released by the parasite (MW) bar = 500nm.

second molecule (MIC1) has been shown to bind to host cells (Fourmaux et al., 1996). Once formed, this junction is translocated backwards along the parasite, shedding membrane and propelling the parasite into the cell (Dubremetz et al., 1985; Grimwood and Smith, 1995). The resulting vacuole is derived from the host cell plasma membrane but modified by the parasite. This process of entry is linked with sequential exocytosis, first from the rhoptries, which immediately secrete their contents into the parasitophorous vacuole (Nichols et al., 1983; Beckers et al., 1994), then from the dense granules (Dubremetz et al., 1993). The contents of the dense granules are clearly important in modifying and stabilizing the vacuole as they are differentially integrated either into the parasitophorous vacuole membrane (GRA 3, 5), or into the membranous network which extends between the parasite and the vacuole membrane (GRA 1, 2, 4, 6), or released as soluble proteins (NTPase) (Achbarou et al., 1991; Lecordier et al., 1993; Mercier et al., 1993; Sibley et al., 1994). The resulting vacuole forms a close relationship with host cell endoplasmic reticulum and mitochondria (de Melo et al., 1992; Sinai et al., 1997) and allows the traffic of small nutrient molecules (up to M_r 1900) from the cytoplasm (Schwab et al., 1994). Within the vacuole the parasite divides by repeated endodyogeny (Sheffield and Melton, 1968) generating 16 to 32 parasites over 48 hours. It was originally suggested that the parasite was released following lysis of the host cell but recent evidence suggests that exit from the host cell may be parasite-directed (Stommel et al., 1997; Silverman et al., 1998).

Less is known about invasion of the bradyzoite and sporozoite stages. Both are structurally similar to the tachyzoite, although the density and abundance of organelles differ (Dubey, 1998) and there are significant molecular differences (Kasper, 1989, Woodison and Smith, 1990; Manger et al., 1998b). The interaction of these stages with the host cell is, however, very different. The entry of bradyzoites into MDBK cells in vitro is reminiscent of induced phagocytosis and results in the formation of a vacuole that does not initially associate with host mitochondria (Sasono and Smith, 1998). The inner vacuole membrane becomes associated with protein to form the cyst wall, and bradyzoites within the cyst divide slowly and become embedded in a dense proteinaceous matrix (Ferguson and Hutchinson 1987; Lane et al., 1996). Sporozoite invasion has been monitored in vitro by video microscopy and appears similar to tachyzoite invasion. However, it results in the formation of a vacuole that appears larger than normal and devoid of an IPM network. The parasite does not divide in this vacuole but undergoes stage conversion to become a tachyzoite. At around 20 hrs post-infection the parasite escapes to form and divide within a typical tachyzoite vacuole (Speer et al., 1995; Tilley et al., 1997).

2.2. LIFE CYCLE

Parasitic protozoa typically have complex life cycles, which involve a highly coordinated series of developmental changes which produce phenotypically distinct stages. In apicomplexan parasites there is an obligate cycle with alternating phases of sexual and asexual reproduction. The simplest form of the cycle is seen in directly transmitted parasites such as *Eimeria* where several rounds of asexual amplification are followed by the generation of gametes that fuse to form the transmission stage, the oocyst (Figure 13.2a). More commonly, parasites alternate between two hosts: for example, malaria parasites have a sexual cycle in the mosquito and two distinct cycles of asexual division (exoerythrocytic and erythrocytic) in the mammalian host (Figure 13.2b). *Toxoplasma* follows the general life cycle pattern, with a sexual cycle in the cat and an asexual cycle in a variety of intermediate hosts (Figure 13.2c), but there are two important exceptions. Firstly, the parasite does not take a one-way route around the life cycle as interconversion can occur between two stages, the tachyzoite and the bradyzoite. Secondly, the parasite can be transmitted directly via the asexual bradyzoite stage and is not obliged to use the sexual cycle.

The asexual cycle begins with ingestion of oocysts or tissue cysts, which release infective sporozoites and bradyzoites, respectively. Sporozoites penetrate gut enterocytes and pass the lamina propria prior to undergoing endodyogeny and transforming into tachyzoites (Dubey et al., 1997b). The tachyzoite stage, which divides rapidly and infects a wide range of cells, is responsible for dissemination of the infection to host tissues (Dubey et al., 1997b). This acute stage of the infection is associated with pathology, but is normally transient due to strong immune responses targeted against the tachyzoite. Tachyzoites transform to bradyzoites, which are predominantly found in brain and muscle tissue and which divide slowly, forming cysts. Tissue cysts turn over very slowly and are responsible for maintaining chronic infection in intermediate hosts (Ferguson et al., 1989).

The sexual, or enteroepithelial, cycle occurs only in felines and is initiated when the cat ingests either tissue cysts or bradyzoites. Several rounds of division occur in gut enterocytes before the production of micro- and macrogametes (Hutchinson et al., 1971). These stages are poorly characterized, but five morphologically distinct stages, named A-E forms, have been described (Dubey and Frenkel 1972; Ferguson et al., 1974). Fertilization occurs in the gut leading to the formation of an oocyst. On contact with air the oocyst differentiates to produce two sporocysts, each of which contains four sporozoites. The entire cycle is very rapid, 3–10 days from ingestion of bradyzoites, 18 days from ingestion of oocysts, but very prolific as a single infected cat can produce millions of oocysts (Dubey and Frenkel 1972; Dubey 1996).

2.3. A ZOONOTIC DISEASE

The second feature that separates *Toxoplasma* from the majority of the apicomplexa is its host range. Most apicomplexan genera consist of numerous species each of which interacts with a specific host. The genus *Toxoplasma*

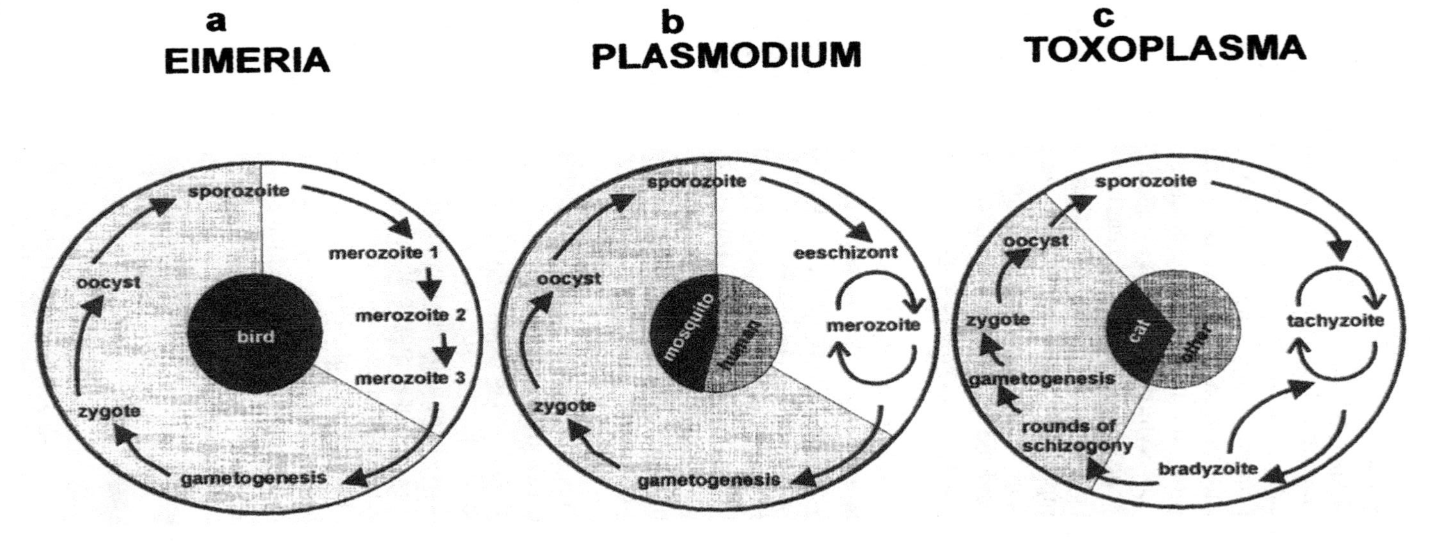

Figure 13.2 Life cycles of apicomplexan parasites. Diagrammatic representation of the life cycles of apicomplexan parasites. Sexual stages and asexual stages are represented by the outer wheel, host species in which these life cycle stages occur by the inner wheel.

contains only one species, *Toxoplasma gondii,* which is known to infect an extremely broad range of intermediate hosts. The parasite was originally described as an infection in an African rodent, the gondii (*Ctenodactylus gundi:* Nicolle and Manceaux, 1908), but has subsequently been reported in humans, in domestic and wild mammals and in birds (Table 13.1). The parasite has successfully colonized a diverse range of host species from very different environment. These include pigs (Dubey et al., 1991), deer (Vanek et al., 1996), bears (Dubey et al., 1995), marsupials (Cranfield et al., 1990) and even dolphins (Inskeep et al., 1990). Although the parasite is found in many animal species, the prevalence varies among species and between populations. It is, however, usually highest among carnivores such as red fox (90%) and mink (66%) (Smith and Frenkel, 1995).

3. *TOXOPLASMA:* A HUMAN PATHOGEN

3.1. DISEASE PATHOLOGY

Toxoplasma gondii is common in human populations. Seroprevalence studies show rates of infection that vary from 12% in Japan (Ko et al., 1980) to 21–36% in the U.K. and U.S. (Jackson et al., 1987; Feldman and Miller, 1956) and 84–90% in France and El Salvador (Desmonts and Couvreur, 1974; Remington et al., 1970). Within populations there is a clear relationship between age and infection (Remington et al., 1970). Variation in infection rates between populations is generally attributed to differences in cooking and eating habits, but climatic factors can influence transmission (Feldman and Miller, 1956; Yamaoka and Konishi, 1993). The majority of infected individuals

TABLE 13.1. Prevalence of *Toxoplasma* among Animal Species.

Animal	Country	N =	Prevalence	Test	Reference
Pigs 'finishers'	Illinois, U.S.	4552	2.3%	sera	Dubey et al., 1995b
Sheep	Mexico	495	30%	sera	Garciavasquez et al., 1990
Goats	Mexico	211	44%	sera	Garciavasquez et al., 1990
Raccoons	Illinois, U.S.	188	67%	sera	Dubey et al., 1995b
House mice	Iowa, U.S.	588	0.3%	sera	Smith et al., 1992
Otter	North Carolina, U.S.	103	47%	sera	Tocidlowski et al., 1997
Birds	Czech Republic	5880	8%	para	Literak et al., 1992
Bears	Pennsylvania, U.S.	665	80%	sera	Briscoe et al., 1993
Deer	Minnesota, U.S.	1367	30%	sera	Vanek et al., 1996

suffer from asymptomatic or mild, transient, flu-like infection in which the most common symptoms are pyrexia, lymphadenopathy and myalgia (Ho-Yen and Joss, 1992; McCabe et al., 1987). More rarely, serious, or even life threatening, symptoms such as pneumonia (Pomeroy and Filice, 1992), myocarditis (Montoya et al., 1997), hepatitis (Venthanyagam and Bryceson, 1976) or encephalitis (Grant and Klein, 1987) occur. The chronic phase of infection is rarely associated with clinical disease, although occasionally symptoms such as retinochoroiditis have been reported (Couvreur and Thuillez, 1996). It is somewhat surprising, given the high prevalence of toxoplasmosis and its location in brain and muscle, that subclinical effects of the parasite on the host have rarely been studied. There is no cause for complacency on this issue as clear behavioral differences have been observed between infected and uninfected mice and rats (Hutchinson et al., 1980; Webster et al., 1994), which may find parallels in the human population (Flegr et al., 1996). The long-term influence of the parasite on heart, gut, lung and other tissues has never been specifically addressed.

There is a clear relationship between disease symptoms and the dissemination of the parasite during the acute phase of disease. Most pathology is related to the rapid division of the tachyzoite stage of the parasite which infects many cell types causing localized tissue damage. It is not possible to follow the course of parasitaemia in humans; however, this is likely to follow the pattern seen in other animals (Dubey and Beattie, 1988). The lymphatic system is believed to mediate early dissemination of the parasite from the gut as parasites are seen first in the gut mesentery and Peyer's patches, then in lung and liver, then in brain (Sumyuen et al., 1995; Dubey et al., 1997b). In mouse models mortality and morbidity peak at around 10–15 days post-infection at which point parasite-induced damage and inflammation are seen in lung, liver, gut, muscle and brain (Suzuki et al., 1989b; Dubey et al., 1997b). By 21 days post-infection the inflammation has largely subsided and from this point parasites are primarily seen as tissue cysts (Ferguson et al., 1991). Bradyzoites can first be recognized 5–6 days post-infection and form cysts which are found in many tissues but are more abundant in muscle and brain (Dubey et al., 1997b; Odaert et al., 1996). Cyst numbers peak at around 2–12 weeks post infection and thereafter there appears to be a slow turnover with periodic cyst rupture (Burke et al., 1994; Ferguson et al., 1989). Bradyzoites released from the cyst may either enter surrounding cells to form new cysts, which are frequently clustered around the ruptured cyst (Conley and Jenkins, 1981), or they may convert into tachyzoites. Tachyzoite-bradyzoite interconversion is a common feature of chronic infection and is of great importance as it can lead to disease reactivation. Evidence suggests that control of interconversion is critical and that chronic phase pathology is inevitably linked to increased cyst rupture and localized increases in tachyzoites (Figure 13.3) (Odaert et al., 1996; Suzuki et al., 1989i).

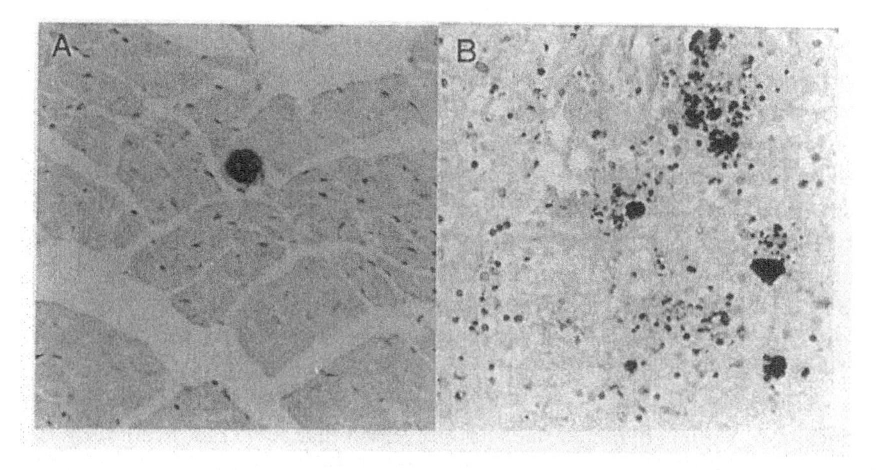

Figure 13.3 Toxoplasmic encephalitis in an AIDS patient. A) Section of mouse brain from animal with chronic toxoplasmosis, stained with anti-toxoplasma IgG Immunoperoxidase. Note the absence of inflammation around the cyst. B) Section of human brain, from an AIDS patient with fatal toxoplasmic encephalitis, stained with anti-toxoplasma IgG Immunoperoxidase. Note presence of small cysts and free parasites indicative of cyst rupture and reactivation.

3.2. IMMUNOREGULATION

The immune system is most important in control of toxoplasmosis, and in most cases primary infection elicits protective immunity. Characterization of parasite antigens has revealed that the tachyzoite and bradyzoite stages are antigenically distinct (Kasper, 1989; Woodison and Smith, 1990) and that there is virtually no overlap in serological recognition of the two forms (Zhang and Smith, 1995a; Smith et al., 1996). In particular, the surface molecules of the two stages are radically different (Couvreur et al., 1988; Tomavo et al., 1991) but other bradyzoite-specific molecules have been identified in the cyst wall (Weiss et al., 1992), matrix (Parmley et al., 1994; Zhang and Smith, 1995b), and cytoplasm (Bonhe et al., 1995). The existence of stage-specific antigens is a typical feature of parasitic protozoa and, as with many other examples, evidence suggests that regulation of the parasite is stage-specific. A strong contrast may be drawn between immunoregulation of the tachyzoite and the bradyzoite. The tachyzoite elicits strong humoral and cellular responses, which have been well characterized (Handman and Remington, 1980; Canessa et al., 1988), while the bradyzoite appears to escape recognition (Zhang and Smith, 1995a), and there is little evidence of inflammation associated with intact cysts (Ferguson et al., 1989). While humoral mechanisms are clearly important in the destruction of extracellular tachyzoites (Schreiber and Feldman, 1980; Pavia, 1986), there is broad consensus that cell-mediated responses are paramount in disease control. During the acute phase of infection tumor necrosis factor alpha (TNF-α) and interleukin 12 (IL-12) are produced by macrophages and these cytokines induce the production of interferon gamma (IFN-γ) by natural killer (NK) cells (Gazzinelli et al., 1993). IFN-γ is a key cytokine in control of infection (Suzuki et al., 1989a; Suzuki and Remington, 1990), inducing macrophage activation and acting with IL-12 to stimulate Th1 CD4+ lymphocyte expansion and stimulation of cytotoxic CD8+ lymphocytes (Denkers et al. 1993), which are important in protective immunity (Brown and McLeod, 1990). Evidence from in vivo and in vitro studies suggests that IFN-γ induced upregulation of nitric oxide stimulates tachyzoite:bradyzoite conversion (Suzuki and Remington, 1990; Bohne et al., 1994).

Host genes are important in determining resistance and susceptibility to infection. Studies in inbred mice have revealed that acute phase mortality and cyst burden are separately regulated. Acute phase survival is regulated by at least five genes and is associated with the H-2a haplotype (McLeod et al., 1989) while control of cyst burden is mainly regulated by the Class I L^d gene (Brown et al., 1995). In the human population severe pathology in congenitally infected infants and AIDS patients has been linked with the *HLA DQ3* allele (Smith et al., 1997; Suzuki et al., 1996).

3.3. AT-RISK GROUPS

The importance of the immune system in control of disease is highlighted by enumeration of "at-risk" groups, all of which can be regarded as immunocompromised. The most common and significant of these groups is congenital infection, which affects between 1:2000 (U.K.; Williams et al., 1981) and 1:100 (France; Desmonts and Couvreur, 1974) births. In mothers who seroconvert during pregnancy, the risk of transmission to the fetus increases with gestational age but the severity of symptoms declines (Remington and Desmonts, 1990). In the first trimester the parasite may induce abortion or cause severe symptoms such as hydrocephalus while in the last trimester retinochoroiditis is the most common presentation (Cook, 1990). The parasite also causes serious disease in immunosupressed adults. This was first noted in patients on immunosupressive therapy (Derouin et al., 1992; Wreghitt et al., 1989) but more recently has come to be associated with AIDS (Luft and Remington, 1992). It has been estimated that without prophylaxis 50% of *Toxoplasma* seropositive AIDS patients will suffer reactivation and serious or fatal encephalitis (Zangerle et al., 1991).

In summary, the clinical spectrum of toxoplasmosis ranges from asymptomatic to severe or fatal disease. Host genes, particularly those that direct the immune response, are important in disease regulation. The question arises as to whether parasite genes also influence the course of infection.

4. *TOXOPLASMA:* STRAIN VARIATION AND VIRULENCE

4.1. THE VIRULENCE PHENOTYPE

It has long been known that isolates of *Toxoplasma* vary in their pathogenicity (Jacobs, 1974). Virulent strains are defined as having an LD50 of a single viable parasite (Howe et al., 1996). Mortality occurs during the acute phase of infection and is associated with high parasitaemia (Eyles and Coleman, 1956). Virulent strains are not lethal to all animals but are frequently associated with severe symptoms; for example, the mouse virulent strain RH was isolated from a lethal human infection (Sabin, 1941). There are signs that these strains have altered or incomplete life cycles in that tissue cysts are rarely seen (Lecompte et al., 1992) and in some cases the ability to produce oocysts has been lost (Darde et al. 1992). By contrast avirulent isolates rarely cause mortality, and have an LD 50 of 10^2 parasites (Howe et al., 1996). Variation in pathogenicity is also evident among avirulent strains, and this can be seen both in the severity of the acute phase and in the risk of encephalitis (Suzuki et al., 1989b).

4.2. MOLECULAR MARKERS AND VIRULENCE

With the advent of molecular markers it has become evident that there is a genetic basis to observed patterns of virulence. Early indications of strain variation came from comparison of parasite antigen profiles, but this was not related to pathogenicity (Ware and Kasper, 1987). The first studies to describe virulence markers used isoelectric focusing to analyze isoenzymes in tachyzoites purified from 35 *Toxoplasma* isolates (Darde et al., 1992). One surprising observation was that despite the wide geographical and host range of these isolates, there was very little variation. Initially 15 enzymes were studied but only six were found to be polymorphic and each of these had two or three isoforms. There was little suggestion that the isoenzymes altered with passage in culture and only one new isoform (lactate dehydrogenase, LDH) was seen, and this may have been due to the induction of bradyzoites. On the basis of these six enzymes, the 35 isolates could be subdivided into four zymodemes Z1–4. In this initial analysis all mouse virulent strains fell into zymodeme 1, while Z2–4 were avirulent strains with Zymodeme 2 being overrepresented. Finally the patterns of isoenzymes implied co-inheritance and very little genetic exchange. This study was later expanded to include 61 isolates (Darde, 1996). In this new analysis the number of zymodemes has increased to 11 but the majority of strains remain in Zymodemes 1–4; Z1 contains only virulent strains and Z2 is overrepresented (29/61 strains). Zymodemes 5–11 are all represented by a single strain and vary in virulence.

Simultaneous to this study a variety of other probes were being developed for direct analysis of parasite DNA. The high level of conservation first noted between *Toxoplasma* isolates in the isoenzyme studies was confirmed in these studies. Techniques such as riboprinting (Brindley et al., 1993) and comparative sequence analysis of the small subunit ribosomal RNA gene (Luton et al., 1995) have proved useful for comparison between *Toxoplasma* and other apicomplexan parasites, but have shown insufficient variation for strain analysis. Analysis of polymorphism within single-copy loci has proven to be the most valuable approach to analysis of the parasite population structure (Sibley and Boothroyd, 1992; Howe and Sibley, 1995) while repetitive DNA probes may be useful in ''fingerprinting'' isolates (Cristina et al., 1991; Sibley and Boothroyd, 1992). Sibley and Boothroyd (1992) analyzed 28 parasite isolates using a variety of single-locus and repetitive DNA probes selected randomly from a larger set devised for genome mapping. Probes were used to amplify segments of genomic DNA and the amplified products were analyzed by RFLP and Southern blotting. The most striking finding was that the *SAG1* locus had two alleles one of which was specific to virulent strains, the other to avirulent strains. On the basis of available *SAG1* sequence data, it was argued that the polymorphism was not within the coding region of the gene and did not imply any structural change in the major tachyzoite surface protein. However, more recently PCR amplification and sequencing of the 3′ region of the *SAG1* gene has revealed three polymorphisms within the coding region that do correlate with virulence (Rinder et al., 1995). Further comparison, using the wider range of probes (Sibley and Boothroyd, 1992) revealed that there was virtually no polymorphism among nine virulent strains, implying that they constituted a clonal lineage. In a later more comprehensive study, Howe and Sibley (1995) analyzed the population structure of 106 isolates by mapping polymorphism in six independent single-locus genes. This larger-scale analysis confirmed that there was very little diversity among these strains, with only 15 different genotypes represented from a possible total of over 1700 different combinations of alleles. The level of recombination was very low and only four isolates had mixed genotypes. The population structure of the parasite was clonal and could be subdivided into three lineages designated type I, II and III. Type I strains corresponded to the virulent phenotype, type II and III strains were avirulent. Highly conserved repeated DNA sequences from *Toxoplasma* have also been used in RFLP analysis of genomic DNA. In initial studies these proved to be useful indicators of strain diversity but showed no relationship with virulence (Cristina et al., 1991), but later numerical taxonomic analysis of RFLP data from these probes and comparative isoenzyme data resulted in similar clustering of strains into three groups (Cristina et al., 1995). Other techniques also have been successfully used to generate virulence markers. Binas and Johnson (1998) found virulence associated polymorphisms in an intron of the DNA polymerase gene. Guo and

Johnson (1995) used the random amplified polymorphic DNA (RAPD) technique, and with seven arbitrary 10mer primers were able to separate virulent and avirulent strains. Finally Costa et al. (1997) have described a microsatellite in an intron of the β-tubulin gene that can be amplified directly from amniotic fluid and can be used to distinguish virulent and avirulent lines.

The level of agreement between these studies is very high (Table 13.2). The clustering of virulent isolates into a single group is the most common observation but features such as the low level of polymorphism and lack of recombination have also been frequently reported. In some cases specific comparisons have been made by the authors to illustrate overlap in results (Cristina et al., 1995; Darde, 1996). The overwhelming conclusion is that the parasite is subdivided into three clonal lineages. This factor together with lack of recombination implies that the asexual transmission cycle is of overriding importance and the sexual cycle rare.

4.3. ASSOCIATION BETWEEN STRAINS AND DISEASE PATHOLOGY

One major question that arises from these studies is the extent to which an association can be made between strains and clinical disease (Table 13.3). The obvious relationship lies with the type I mouse virulent strains, which have been frequently isolated from severe or fatal infection. Beyond this

TABLE 13.2. **Molecular Markers of Virulence.**

Study	Method	Strain Classification			
Darde, 1996	Isoenzymes	Z1 *RH, ENT*	Z2 BOU, BEV	Z4 DEG, CEP	Z3 C56
Sibley and Boothroyd, 1992	PCR:RFLP	Type 1	Type 2		Type 3
Howe and Sibley, 1995	sc loci	*RH, ENT, BK*	BOU, BEV, DEG, CEP		C56
Cristina et al., 1995	Isoenzyme PCR:RFLP rep probe	Group 1 *RH, ENT, BK*	Group 2 CEP		Group 3 C56
Rinder et al., 1995	PCR:seq	virulent *RH, BK*	non virulent C56		
	PCR:seq	virulent *RH*	non virulent ME49		
Costa et al., 1997	microsatellite	virulent *RH, ENT*	non-virulent BOU, BEV, CEP, DEG, C56		
Guo and Johnson, 1995	RAPD:PCR	virulent *RH, ENT*	non-virulent CEP, ME49		

Strains in italics have been demonstrated to be mouse virulent.

TABLE 13.3. Strain Variation and Disease Pathology.

Study Strain	Percentage of Cases			No. Isolates
	Type 1	Type 2	Type 3	Total
Howe and Sibley, 1995*				
Congenital	28	60	12	41
AIDS	20	70	10	27
Animal	9	44	47	34
Howe et al., 1997				
Congenital	0	100	0	13
AIDS	13	76	11	45
Mondragon et al., 1998				
Pigs	0	84	16	43

* All figures calculated from histogram.

indirect association, few authors have related strains to symptoms. In their analysis of 106 isolates Howe and Sibley (1995) found a bias in strain distribution that related to disease and transmission. Type I strains were more commonly found in congenital disease, type II in AIDS-associated infection, and type III in strains derived from animal hosts. This exciting preliminary suggestion was conditional on the isolates being a representative collection. There were, however, two sources of bias, a disproportionate number of isolates were associated with clinical disease and the majority were laboratory-maintained strains selected for ease of passage. It was clear that the representation of strains in nature might vary from this sample and this question was in part addressed by two further studies. Howe et al. (1997) analyzed a group of 68 clinical isolates from immunosuppressed and congenitally infected individuals. Any bias due to laboratory culture was removed by extracting and amplifying DNA from microscopic preparations made at the time of diagnosis. A sensitive PCR-based screen was developed to amplify the *SAG2* locus and strains were typed on the basis of RFLP analysis. The results of this study suggested that all forms of clinical disease were predominantly caused by type II strains (81%). Type I strains were not isolated from congenital infection in this study, in keeping with the findings of Costa et al. (1997), who found no virulent isolates in 37 cases of congenital infection. A further study analyzed isolates obtained from pigs on the basis of polymorphism at the *SAG1* and *SAG2* loci (Mondragon et al., 1998). Again type II isolates were the most common (83%) and no type I isolates were seen. These latter studies imply that the distribution of strains is the same in humans and animals and it is not possible to link clinical presentation with parasite strain at this level. It is, however, clear that a greater range of samples should be studied before we can draw any firm conclusions as to whether pathology or host specificity might be lineage or strain dependent.

4.4. CELLULAR ASPECTS OF VIRULENCE

Given the genetic homogeneity between strains, one might suggest that virulence is determined by a small number of genes or even by a single genetic modification. One approach to understanding the basis of virulence has been to study the phenotype of strains in vitro. This has revealed differences that correspond to observed patterns of virulence in vivo. In many systems the growth rate of a pathogenic microorganism is an indicator of virulence. Early studies revealed differences in the growth of parasites in vitro, but direct comparison of growth rates was rare (Kaufman and Maloney, 1962). In recent studies we have demonstrated differences in the growth of three type I strains cultured under identical conditions (Appleford and Smith, 1997). In a follow-up study the average doubling time of the type I ENT strain was found to be 6.8 hours as opposed to 15.2 hours with the type II RRA strain (Figure 13.4). Division of type I strains also exhibited a higher degree of synchrony, suggesting that the cell cycle is operating at maximum efficiency.

As part of the same study rates of tachyzoite: bradyzoite conversion were monitored and, under identical culture conditions, bradyzoites were found to be much more abundant in type II than in virulent type I strains. Tachyzoite:bradyzoite interconversion has been widely studied in vitro. Under normal culture conditions where host cells and nutrients are abundant, tachyzoites prevail and bradyzoites are present in low numbers (Lane et al., 1996). However,

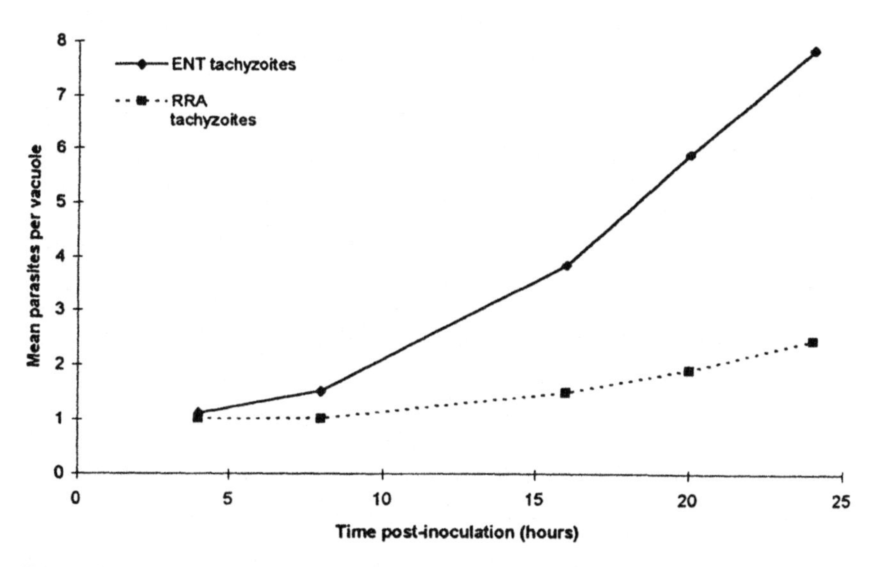

Figure 13.4 Growth of virulent and avirulent strains. Graph showing the growth rate of virulent (ENT) and avirulent (RRA) strain parasites in MBDK cells. Growth is expressed as the mean number of parasites per vacuole over 24 hrs in culture.

conversion to bradyzoites can be induced in culture by a variety of stress agents (nitric oxide, pH shift) and, although direct comparisons have not been made, this induced stage conversion appears to be more efficient among avirulent strains (Bonhe et al., 1993; Soete et al., 1994). There appears to be a direct link between parasite growth and stage conversion as the induction of bradyzoites in vitro is always preceded by a reduction in tachyzoite replication (Bonhe et al., 1994). In summary, differences in the growth and interconversion rates of acute virulent and avirulent strains can be measured in vitro. These studies provide strong evidence that parasite virulence is genetically determined and is not simply a function of the host immune response, although the growth characteristics of the parasite will clearly be further modified by environmental factors in vivo.

4.5. MOLECULAR DETERMINANTS OF VIRULENCE

Although genetic markers were primarily developed for analysis of parasite strain variation, they might also prove useful in determining the molecular basis of virulence. The most striking observation from these studies was perhaps the association between the presence of allele 1 at the *SAG1* locus (SAG1–1) and mouse virulence (Sibley and Boothroyd, 1992; Howe and Sibley, 1995) and the additional, virulence-associated, sequence changes at the 3′ end of the *SAG1* gene (Rinder et al. 1995). To further probe the importance of this locus in determining virulence, Howe et al. (1996) studied recombinant strains which carried SAG1–1 on a type III background genotype. Possession of the SAG1–1 allele was not sufficient to determine virulence in these strains but wider probing of the genotype revealed a disproportionate number of virulence-associated polymorphisms on chromosome VIII. In particular, a novel restriction site approximately 1 kb upstream of the *SAG1* initiator was shown to associate with virulence in both recombinant strains and in 18 of 19 type reference strains.

The question arises as to whether modification of the *SAG1* locus might be directly responsible for determining acute virulence. The SAG1 protein has important roles in invasion and immunity, but it is unlikely that the minor modifications noted in the coding region of the gene (Rinder et al., 1995) would influence either the structure or function of the molecule. The changes in the upstream flanking region of *SAG1,* however, may well influence its expression. In support of this, Windeck and Gross (1996) found the level of mRNA transcripts for *SAG1* higher in the type I RH strain than in the avirulent NTE. It may also be relevant that a second gene, *SRS1*, is located immediately (approximately 1.2 kb) upstream of SAG1 and is controlled by the same bidirectional promoter (Hehl et al. 1997), and it is possible that the polymorphic site might influence this molecule rather than SAG1. Recent screening of *Toxoplasma*-expressed sequence tags (ESTs) revealed that SAG1, SAG3 and

SRS1 are members of a family of glycophosphotidylinositol (GPI)-linked surface proteins with conserved cysteine residues (Manger et al., 1998a). Preliminary observations suggest that expression of these molecules may be strain-specific. It is likely that there is redundancy in attachment molecules and that changes in the relative abundance of the SAG1 family of surface proteins might alter the efficiency and specificity of binding to host cells and hence influence pathology. Finally it should be noted that SAG1 is found only on the surface of tachyzoites, and it is possible that virulence is linked to disruption of inducible genomic elements which have been shown to regulate stage conversion (Bonhe and Roos, 1997).

Other molecular markers have been shown to vary between virulent and avirulent strains but the strength of association is not as great as seen between the SAG1 locus and acute virulence. Changes in molecules that modify the parasitophorous vacuole, supply nutrients, coordinate the cell cycle, induce stage conversion and mediate escape from the vacuole could all influence parasite growth and pathogenicity. Several examples of this have been found. Dense granule molecules GRA5 and GRA6 have altered posttranslational modification in virulent strains, which may relate to adaptation of vacuole (Smith et al., 1997). Higher levels of the heat shock protein HSP70 have been found in virulent strains in response to stress, while under similar conditions avirulent strains produce no HSP70 but induce conversion to bradyzoites (Lyons and Johnson, 1995, 1998). Finally virulent strains express two rather that one form of the NTPase enzyme which may be linked to vacuolar escape (Asai et al., 1995; Silverman et al., 1998).

In summary, molecular markers have proven very useful in mapping virulence-associated loci. In the case of acute virulence (type I strains), there is evidence that this is linked to a single mutation close to the SAG1 locus on chromosome VIII. Other molecular markers have been associated with virulence, but further work is needed in order to understand their relevance to disease pathogenesis. Molecular genetic techniques are well established for *Toxoplasma* and it is likely that the role of candidate virulence genes will be clarified over the next few years using a reverse genetic approach (Kim and Boothroyd, 1995; Roos et al., 1997).

5. A FOODBORNE PATHOGEN

One of the major remaining questions about toxoplasmosis relates to the source of human infection. Transmission may be due to ingestion either of undercooked infected meat or oocysts. Most of the evidence on disease transmission relates to assessment of the relative risk of consuming infected meat. It has been demonstrated that levels of infection are high in pigs, sheep and poultry and low in cattle (Dubey and Beattie, 1988; Jackson and Hutchinson,

1989). The risk of transmission is lowered by food processing as cysts are susceptible to freezing, irradiation and cooking (Dubey et al., 1986; Dubey, 1988; Dubey et al., 1990). Although these studies point to likely sources of infection, they cannot be taken as direct evidence that transmission occurs and evaluation of the risk of contracting the disease from oocysts is even more problematic. There are sporadic accounts of both foodborne (Choi et al., 1997) and oocyst-related (Benenson et al., 1982) outbreaks of disease where the source of infection has been documented. More commonly, indirect evidence from epidemiological studies is cited. For example, a large-scale study of the disease on pig farms in Illinois suggested that both transmission routes were important (Dubey et al., 1995b). A later study (Dubey et al., 1997a) found a reduced prevalence of infection among feral pigs on a remote island lacking cats (0.9%) compared to their counterparts on the mainland (18.2%). Only one attempt has been made to discriminate between infection source in humans. Kasper and Ware (1985) reported serological recognition of oocyst-specific antigens among patients from an oocyst-related outbreak, but this method was not developed for further use in epidemiological studies. Appropriate use of molecular markers should allow us to reexamine and trace strains more effectively.

The recent evidence from studies of parasite population biology demonstrates the existence of three clonal lineages and of many isolates with essentially identical genotypes (Howe and Sibley, 1995). This structure implies a high degree of inbreeding and the low frequency of recombinant strains appears to confirm this suggestion. It is possible that parasite transmission is mainly mediated via the asexual cycle (carnivory) and that the sexual cycle is rarely used or in some strains even redundant. However, it is also likely that self-fertilization is common during the sexual cycle and this may also contribute to the clonal population structure. The true frequency of recombination in the cat cannot be easily assessed from current studies due to the small sample size and bias in the collection of strains. It would be of interest to obtain a direct estimate of recombination frequency in the field by analyzing the parasite genotype directly from collections of oocysts. A further application would be to use repetitive DNA probes to fingerprint the strains in sympatric groups of animals and thus identify transmission cycles directly.

One obvious area of interest relates to the transmission of type I virulent strains. These strains are rarely isolated in epidemiological surveys (Mondragon et al., 1998; Howe et al., 1997) but are overrepresented in laboratory-maintained lines (Darde et al., 1992). Since type I strains have been isolated from both animals and man and do not generally form oocysts, they are likely to be passed on by carnivory. However, there is also some doubt about this transmission route as virulent strains are reported to be either fatal or nonpersistent (Waldeland et al., 1983; Howe et al., 1996). In some cases cyst-like structures have been reported in animals surviving RH infection (Lecompte et al., 1992; Appleford, unpublished data) but these have never conclusively been shown to contain bradyzoites. Recent information, however, suggests that tachyzoites may be able to mediate oral infection (Dubey, 1998) and may be responsible for transmission. It is important to understand the transmission cycles that sustain these parasites, particularly as "nonpersistent" virulent strains are deemed suitable anti-toxoplasma vaccines for animals (Buxton et al., 1991) and may thus be inadvertently introduced into the human food chain.

6. ACKNOWLEDGMENTS

We are indebted to P. M. D. Sasono and P. J. Appleford for providing illustrations and to A. M. Dunn, D. Tanneyhill and M. F. Hemingway for comments on the manuscript. Naomi Rebuck is supported by a grant from The Wellcome Trust.

7. REFERENCES

Achbarou, A., Mercereau-Puijalon, O., Sadak, A., Fortier, B., Leriche, M. A., Camus, D., and Dubremetz, J. F. 1991. "Differential targeting of dense granule proteins in the parasitophorous vacuole of *Toxoplasma gondii*," *Parasitology* 103:321–329.

Appleford, P. J. and Smith, J. E. 1997. "*Toxoplasma gondii:* the growth characteristics of three virulent strains," *Acta Tropica* 65:97–104.

Asai, T., Miura S., Sibley, L. D., Okabayashi, H., and Takeuchi T. 1995. "Biochemical and molecular characterization of nucleoside triphosphate hydrolase isozymes from the parasitic protozoan *Toxoplasma gondii*," *Journal of Biological Chemistry* 270:11391–11397.

Beckers, C. J. M., Dubremetz, J. F., Mercereau-Puijalon, O., and Joiner K. A. 1994. "The *Toxoplasma gondii* rhoptry protein ROP 2 is inserted into the parasitophorous vacuole membrane, surrounding the intracellular parasite, and is exposed to the host cell cytoplasm," *Journal of Cell Biology*, 127:947–961.

Benenson, M. W., Takafuji, E. T., Lemon, S. M., Greenup, R. L. and Sulzer, A. J. 1982. "Oocyst transmitted toxoplasmosis associated with ingestion of contaminated water," *New England Journal of Medicine*, 307:666–669.

Binas, M. and Johnson, A. M. 1998. "A polymorphism in a DNA polymerase α gene intron differentiates between murine virulent and avirulent strains of *Toxoplasma gondii*," *International Journal for Parasitology*, 28:1033–1040.

Bonhe, W. and Roos, D. S. 1997. "Stage specific expression of a selectable marker in *Toxoplasma gondii* permits selective inhibition of either tachyzoites or bradyzoites," *Molecular and Biochemical Parasitology* 88:115–126.

Bonhe, W., Heeseman, J., and Gross, U. 1993. "Induction of bradyzoite specific *Toxoplasma gondii* antigens in gamma interferon-treated mouse macrophage," *Infection and Immunity*, 61:1141–1145.

Bonhe, W., Heesemann J. and Gross, U. 1994. "Reduced replication of *Toxoplasma gondii* is necessary for induction of bradyzoite specific antigens; a possible role for nitric oxide in triggering stage conversion," *Infection and Immunity*, 62:1761–1767.

Bonhe, W., Gross, U., Ferguson, D. J. P., and Heeseman J. 1995. "Cloning and characterization of a bradyzoite-specifically expressed gene (hsp30/bag1) of *Toxoplasma gondii*, related to genes encoding small heat shock proteins of plant," *Molecular Microbiology*, 16:1221–1230.

Brindley, P. J., Gazzinelli, R. T., Denkers, E. Y., Davis, S. W., Dubey J. P., Belfort R., Martins M. C., Silveira, C., Jamra, L., Waters, A. P., and Sher, A. 1993. "Differentiation of *Toxoplasma gondii* from closely related coccidia by riboprint analysis and a surface antigen gene polymerase chain reaction," *American Journal of Tropical Medicine and Hygiene* 48:447–456.

Briscoe, N., Humphreys, J. G., and Dubey, J. P. 1993. "Prevalence of *Toxoplasma gondii* infections in Pennsylvania black bears, *Ursus americanus*," *Journal of Wildlife Diseases* 29:599–601.

Brown, C. R., and McLeod R. 1990. "Class 1 MHC genes and CD8+ T cells determine cyst number in Toxoplasma gondii infection," *Journal of Immunology* 145:3435–3441.

Brown, C. R., Hunter, C. A., Estes, R. G., Beckmann, E., Forman, J., David, C., Remington, J. S., and McLeod, R. 1995. "Definitive identification of a gene that confers resistance against *Toxoplasma* cyst burden and encephalitis," *Immunology*, 85:419–428.

Burke, J. M., Roberts, C. W., Hunter, C. A., Murray, M., and Alexander, J. 1994 "Temporal differences in the expression of mRNA for IL-10 and IFN-μ in the brains and spleens of C57BL/10 mice infected with *Toxoplasma gondii*," *Parasite Immunology*, 16:305–314.

Buxton, D., Thomson, K., Maley, S., Wright, S., and Bos, H. J. 1991. "Vaccination of sheep with a live incomplete strain (S48) of *Toxoplasma gondii* and their immunity to challenge when pregnant," *The Veterinary Record* Aug 3rd:89–93.

Canessa, A., Pistoia, V., Poncella, S., Merli, A., Melioli, G., Terragna, A., and Ferrarini, M. 1988. "An in vitro model for *Toxoplasma* infection in man. Interaction between CD4+ monoclonal T cells and macrophages results in killing of trophozoites," *Journal of Immunology* 140:3580–3588.

Carruthers, V. B. and Sibley, L. D. 1997. "Sequential protein secretion from three distinct organelles of *Toxoplasma gondii* accompanies invasion of human fibroblasts," *European Journal of Cell Biology* 73:114–123.

Choi, W. Y., Nam, H. W., Kwak, N. H., Huh, W., Kim, Y. R., Kang, M. W., Cho, S. Y., and Dubey, J. P. 1997. "Foodborne outbreaks of human toxoplasmosis," *Journal of Infectious Diseases* 175:1280–1282.

Conley, F. K. and Jenkins, K. A. 1981. "Immunohistological study of the anatomic relationship of *Toxoplasma* antigens to the inflammatory response in the brains of mice chronically infected with *Toxoplasma gondii*," *Infection and Immunity* 31:1184–1192.

Cook, G. C. 1990. "*Toxoplasma gondii* infection: a potential danger to the unborn fetus and AIDS sufferer," *Quarterly Journal of Medicine* 74:3–19.

Costa, J. M., Darde, M. L., Assouline, B., Vidaud, M., and Bretagne, S. 1997. "Microsatellite in the Beta-tubulin gene of *Toxoplasma gondii* as a new genetic marker for use in direct screening of amniotic fluids," *Journal of Clinical Microbiology* 35:2542–2545.

Couvreur, J., and Thuillez, P. 1996. "Toxoplasmose acquise a localisation oculaire ou neurologique," *La Nouvelle Presse Medicale* 25:438–442.

Couvreur, G., Sadak, A., Fortier, B., and Dubremetz, J. F. 1988. "Surface antigens of *Toxoplasma gondii*," *Parasitology* 97:1–10.

Cranfield, P. J., Hartley, W. J., and Dubey, J. P. 1990. "Lesions of toxoplasmosis in Australian marsupials," *Journal of Comparative Pathology* 103:159–167.

Cristina, N., Liaud, M. F., Santoro, F., Oury, B., and Ambroise-Thomas, P. 1991. "A family of repeated DNA sequences in *Toxoplasma gondii*: cloning, sequence analysis, and use in strain characterisation," *Experimental Parasitology* 73:73–81.

Cristina, N., Darde, M. L., Boudin, C., Tavernier, G., Pestre-Alexandre, M., and Ambroise-

Thomas, P. 1995. "A DNA fingerprinting method for individual characterisation of *Toxoplasma gondii* strains: combination with isoenzymatic characters for determination of linkage groups," *Parasitology Research* 81:32–37.

de Melo, E. J., de Cavalho, T. U. and de Souza W. 1992. "Penetration of *Toxoplasma gondii* into host cells induces changes in the distribution of the mitochondria and the endoplasmic reticulum," *Cell Structure and Function* 17:311–317.

Darde, M. L. 1996. "Biodiversity in *Toxoplasma gondii*," *Current Topics in Microbiology and Immunology* 219:27–41.

Darde, M. L., Bouteille, B., and Pestre-Alexandre, M. 1992. "Isoenzyme analysis of 35 *Toxoplasma gondii* isolates and the biological and epidemiological implications," *Journal of Parasitology* 78:786–794.

Denkers, E. Y., Sher, A., and Gazinelli, R. T. 1993. "T cell interactions with *Toxoplasma gondii:* implications for processing of antigen for class I-restricted recognition," *Research in Immunology* 144:51–57.

Derouin, F., Devergie, A., Auber, P., Gluckman, E., Beauvais, B., Garin, Y. J. F., and Lariviere, M. 1992. "Toxoplasmosis in bone marrow-transplant recipients: report of seven cases and review," *Clinical Infectious Diseases* 15:267–270.

Desmonts, G. and Couvreur, J. 1974. "Congenital toxoplasmosis. A prospective study of 378 pregnancies," *The New England Journal of Medicine* 290:1110–1116.

Dubey, J. P. 1988. "Long term persistence of *Toxoplasma gondii* in tissues of pigs inoculated with *T. gondii* oocysts and effect of freezing on viability of tissue cysts in pork," *American Journal of Veterinary Research* 49:901–903.

Dubey, J. P. 1996. "Infectivity and pathogenicity of *Toxoplasma gondii* oocysts for cats," *Journal of Parasitology* 82:957–961.

Dubey, J. P. 1998. "Re-examination of resistance of *Toxoplasma gondii* tachyzoites and bradyzoites to pepsin and trypsin digestion," *Parasitology* 116:43–50.

Dubey, J. P. and Beattie, C. P. 1988. Toxoplasmosis of animals and man. Boca Raton, FL: CRC Press.

Dubey, J. P. and Frenkel, J. K. 1972. "Cyst induced toxoplasmosis in cats," *Journal of Protozoology* 19:155–177.

Dubey, J. P., Humphreys, J. G., and Thuillez, P. 1995a. "Prevalence of viable *Toxoplasma gondii* tissue cysts and antibodies to *T. gondii* by various serologic tests in black bears (*Ursus americanus*) from Pennsylvania," *Journal of Parasitology,* 81:109–112.

Dubey, J. P., Lindsay, D. S., and Speer, C. A. 1998. "Structures of *Toxoplasma gondii* tachyzoites, bradyzoites and sporozoites and biology and development of tissue cysts," *Clinical Microbiology* Reviews 11:267–299.

Dubey, J. P., Brake, R. J., Murrel, K. D. and Fayer, R. 1986. "Effect of irradiation on the viability of *Toxoplasma gondii* cysts in the tissues of mice and pigs," *American Journal of Veterinary Research* 47:518–522.

Dubey, J. P., Kotula, A. W., Sharar, A., Andrews, C. D., and Lindsaay, D. S. 1990. "Effect of high temperature on infectivity of *Toxoplasma gondii* tissue cysts in pork," *Journal of Parasitology* 76:201–204.

Dubey, J. P., Rollor, E. A., Smith, K., Kwok, O. C. H., and Thuillez, P. 1997a. "Low seroprevalence of *Toxoplasma gondii* in feral pigs from a remote island lacking cats," *Journal of Parasitology* 83:839–841.

Dubey, J. P., Speer, C. A., Shen, S. K., Kwok, O. C. H., and Blixt, J. A. 1997b. "Oocyst induced murine toxoplasmosis: life cycle, pathogenicity, and stage conversion in mice fed *Toxoplasma gondii* oocysts," *Journal of Parasitology* 83:870–882.

Dubey, J. P., Leighty, J. C., Beal, V. C., Anderson, W. R., Andrews, C. D., and Thuillez, P. 1991. "National seroprevalence of *Toxoplasma gondii* in pigs," *Journal of Parasitology* 81:48–53.

Dubey, J. P., Weigel, R. M., Siegal, A. M., Thuillez, P., Kitron, U. D., Mitchell, M. A., Mannelli, A., Mateus-Pinilla, N. E., Shen, S. K., Kwok, O. H. C., and Todd, K. S. 1995b. "Sources and reservoirs of *Toxoplasma gondii* on 47 swine farms in Illinois," *Journal of Parasitology* 81:723–729.

Dubremetz, J. F., Rodriguez, C., and Ferreira, E. 1985. "*Toxoplasma gondii:* redistribution of monoclonal antibodies on tachyzoites during host cell invasion," *Experimental Parasitology* 59:24–32.

Dubremetz, J. F., Achbarou, A., Bermudes, D., and Joiner, K. A. 1993. "Kinetics and pattern of organelle exocytosis during *Toxoplasma gondii* host cell interaction," *Parasitology Research* 79:402–408.

Eyles, D. E. and Coleman, N. 1956. "Relationship of size of inoculum to time of death in mice infected with *Toxoplasma gondii,*" *Journal of Parasitology* 42:272–275.

Feldman, H. A. and Miller, L. T. 1956. "Serological study of toxoplasmosis prevalence," *American Journal of Hygiene* 64:320–335.

Ferguson, D. J. P., and Hutchinson, W. M. 1987. "An ultrastructural study of the early development and tissue cyst formation of *Toxoplasma gondii* in the brains of mice," *Parasitology Research,* 73:483–491.

Ferguson, D. J. P., Dunachie, W. M., and Siim, J. C. 1974. "Ultrastructural study of early stages of asexual multiplication and microgametogony of *Toxoplasma gondii* in the small intestine of the cat," *Acta Pathologica Microbiolgica Scanda,* Sect B, 82:167–181.

Ferguson, D. J. P., Graham, D. I., and Hutchinson, W. M. 1991. "Pathological changes in the brains of mice infected with *Toxoplasma gondii:* a histological, immunocytochemical and ultrastructural study," *International Journal of Experimental Pathology,* 72:463–474.

Ferguson, D. J. P., Hutchinson, W. M., and Pettersen, E. 1989. "Tissue cyst rupture in mice chronically infected with *Toxoplasma gondii,*" *Parasitology Research* 75:599–603.

Flegr, J., Zitkova, S., Kodym, P., and Frynta, D. 1996. "Induction of changes in human behavior by the parasitic protozoan *Toxoplasma gondii,*" *Parasitology,* 113:49–54.

Fourmaux, M. N., Achbarou, A., Mercereau-Puijalon, O., Biderre, C., Briche, I., Loyens, A., Odberg-Ferragut, C., Camus, D., Dubremetz, J. F. 1996. "The MIC1 microneme protein of *Toxoplasma gondii* contains a duplicated receptor-like domain and binds to host cell surface," *Molecular and Biochemical Parasitology* 83:201–210.

Garciavasquez, Z., Rosariocruz, R. and Solorzanosalgado, M. 1990. "Prevalence of antibodies against *Toxoplasma gondii* in sheep and goats in three states in Mexico," *Preventive Veterinary Medicine* 10:25–29.

Gazzinelli, R. T., Hieny, S., Wynn, T., Wolf, S. and Sher A. 1993. "Interleukin-12 is required for the T-cell independent induction of interferon-γ by an intracellular parasite and induces resistance in T lymphocyte deficient hosts," *Proceedings of the National Academy of Sciences USA,* 90:6115–6119.

Grant, S. C. and Klein, C. 1987. "*Toxoplasma gondii* encephalitis in an immunocompetent adult. A case report," *South African Medical Journal,* 71:585–587.

Grimwood, J. and Smith, J. E. 1992. "*Toxoplasma gondii:* the role of a 30kDa surface protein in host cell invasion," *International Journal for Parasitology* 74:106–111.

Grimwood, J. and Smith, J. E. 1995. "*Toxoplasma gondii:* redistribution of tachyzoite surface protein during invasion and intracellular development," *Parasitology Research* 81:657–661.

Grimwood, J. and Smith, J. E. 1996. "*Toxoplasma gondii:* the role of parasite surface and secreted proteins in host cell invasion," *International Journal for Parasitology* 26:169–173.

Guo, Z. G. and Johnson, A. M. 1995. "Genetic characterization of *Toxoplasma gondii* strains by random amplified polymorphic DNA polymerase chain reaction," *Parasitology* 111:127–132.

Handman, E. and Remington, J. S. 1980. "Antibody responses to *Toxoplasma* antigens in mice infected with strains of different virulence," *Infection and Immunity,* 29:215–220.

Hehl, A., Krieger, T. and Boothroyd, J. C. 1997. "Identification and characterization of SRS1, a *Toxoplasma gondii* surface antigen upstream of and related to SAG1," *Molecular and Biochemical Parasitology,* 89:271–282.

Ho-Yen, D. O. and Joss, A. W. L. 1992. *Human Toxoplasmosis.* Oxford, UK: Oxford University Press.

Howe, D. K. and Sibley, L. D. 1995. "*Toxoplasma gondii* comprises three clonal lineages: correlation of parasite genotype with human disease," *Journal of Infectious Diseases* 172:1561–1566.

Howe, D. K., Summers, B. C., and Sibley, L. D. 1996. "Acute virulence in mice is associated with markers on chromosome VIII in *Toxoplasma gondii,*" *Infection and Immunity* 64:5193–5198.

Howe, D. K., Honore, S., Derouin, F. and Sibley, L. D. 1997. "Determination of genotypes of *Toxoplasma gondii* strains isolated from patients with toxoplasmosis," *Journal of Clinical Microbiology* 35:1411–1414.

Hutchinson, W. M., Aitken, P. P., and Wells, B. P. W. 1980. "A chronic *Toxoplasma* infection and familiarity-novelty discrimination in the mouse," *Annals of Tropical Medicine and Parasitology* 74:337–345.

Hutchinson, W. M., Dunachie, J. E., Work K., and Siim, J. C. 1971. "The life cycle of the coccidian parasite *Toxoplasma gondii* in the domestic cat," *Transactions of the Royal Society of Tropical Medicine and Hygiene* 65:380–399.

Inskeep, W., Gardiner, C. H., Harris, R. K., Dubey, J. P. and Goldston, R. T. 1990. "Toxoplasmosis in Atlantic bottle nosed dolphins (*Tursiops truncatus*)," *Journal of Wildlife Diseases* 26:377–382.

Jackson, M. H. and Hutchinson, W. M. 1989. "The prevalence and source of Toxoplasma infection in the environment," *Advances in Parasitology* 28:55–105.

Jackson, M. H., Hutchinson, W. M., and Siim, J. C. 1987. "A seroepidemiological survey of toxoplasmosis in Scotland and England," *Annals of Tropical Medicine and Parasitology* 81:359–365.

Jacobs, J. 1974. "*Toxoplasma gondii:* parasitology and transmission," *Bulletin of the New York Academy of Medicine* 50:128–145.

Kasper, L. H. 1989. "Identification of stage specific antigens of *Toxoplasma gondii,*" *Infection and Immunity* 57:668–672.

Kasper, L. H. and Ware, P. L. 1985. "Recognition and characterization of stage-specific oocyst/sporozoite antigens of *Toxoplasma gondii* by human antisera," *Journal of Clinical Investigation* 75:1570–1577.

Kaufman, H. E. and Maloney, E. D. 1962. "Multiplication of three strains of *Toxoplasma gondii* in tissue culture," *Journal of Parasitology* 48:358–361.

Kim, K. and Boothroyd, J. C. 1995. "*Toxoplasma gondii:* stable complementation of a sag1 (p30) mutants using SAG1 transfection and fluorescence activated cell sorting," *Experimental Parasitology* 80:46–53.

Ko, R. C., Wong, F. W. T., Todd, D. and Lam, K. C. 1980. "Prevalence of *Toxoplasma gondii* antibodies in the Chinese population of Hong Kong," *Transactions of the Royal Society of Tropical Medicine and Hygiene* 74:351–354.

Lane, A., Soete, M., Dubremetz, J. F., and Smith, J. E. 1996. "*Toxoplasma gondii:* appearance of specific markers during the development of tissue cysts *in vitro,*'' *Parasitology Research,* 82:340–346.

Lecompte, V., Chumpitazi B. F. F., Pasquier B., Ambroise-Thomas, P., and Santoro, F. C. 1992. ''Brain-tissue cysts in rats infected with the RH strain of *Toxoplasma gondii,*'' *Parasitology Research* 78:267–269.

Lecordier, L., Mercier, C., Torpier, G., Tourvielle, B., Darcy F., Lui, J. L., Maes, P., Tartar, A., Capron, A., and Cesbron-Delauw, M. F. 1993. ''Molecular structure of *Toxoplasma gondii* dense granule antigen (GRA 5) associated with the parasitophorous vacuole membrane,'' *Molecular and Biochemical Parasitology* 59:143–153.

Literak, I., Hejlicek, K., Nezval, J., and Folk C. 1992. ''Incidence of *Toxoplasma gondii* in populations of wild birds in the Czech republic,'' *Avian Pathology,* 21:659–665.

Luton, K., Gleeson, M., and Johnson, A. M. 1995. ''rRNA gene sequence heterogeneity among *Toxoplasma gondii* strains,'' *Parasitology Research* 81:310–315.

Luft, B. J., and Remington, J. S. 1992. ''Toxoplasmic encephalitis in AIDS,'' *Clinical Infectious Diseases* 15:211–222.

Lyons, R. E. and Johnson, A. M. 1995. ''Heat shock proteins of *Toxoplasma gondii,*'' *Parasite Immunology* 17:353–359.

Lyons, R. E. and Johnson, A. M. 1998. ''Gene sequence and transcription differences in 70kDa heat shock protein correlate with murine virulence of *Toxoplasma gondii,*'' *International Journal for Parasitology* 28:1041–1051.

Manger, I. D., Hehl, A. B., and Boothroyd, J. C. 1998a. ''The surface of toxoplasma tachyzoites is dominated by a family of glycophosphatidylinositol-anchored antigens related to SAG1,'' *Infection and Immunity* 66:2237–2244.

Manger, I. D., Hehl, A., Parmley, S., Sibley, L. D., Marra, M., Hillier, L., Waterston, R., and Boothroyd, J. C. 1998b. ''Expressed sequence tag analysis of the bradyzoites stage of *Toxoplasma gondii:* Identification of developmentally regulated genes,'' *Infection and Immunity* 66:1632–1637.

McCabe, R. E., Brooks, R. G., Dorfman, R. F., and Remington, J. S. 1987. ''Clinical spectrum in 107 cases of toxoplasmic lymphadenopathy,'' *Reviews of Infectious Diseases* 9:754–774.

McLeod, R., Skamene, E., Brown, C. R., Eisenhauer, P. B., and Mack, D. G. 1989. ''Genetic regulation of early survival and cyst number after oral *Toxoplasma gondii* infection of AxB/BxA recombinant inbred and B10 congenic mice,'' *Journal of Immunology* 143:3031–3034.

Mercier C., Lecordier, L., Darcy, F., Deslee D., Murray, A., Tourvieille, B., Maes P., Capron, A., and Cesbron-Delauw, M. F. 1993. ''Molecular characterization of a dense granule antigen (GRA 2) associated with the network of the parasitophorous vacuole in *Toxoplasma gondii,*'' *Molecular and Biochemical Parasitology* 58:71–82.

Mineo, J. R., Mcleod, R., Mack, D., Khan I. A., Ely K. H., Smith, J. E., and Kasper, L. H. 1993. ''Antibodies to *Toxoplasma gondii* major surface protein (SAG-1, P30) inhibit infection of host cells and are produced in murine intestine after peroral infection,'' *Journal of Immunology* 150:3951–3964.

Mondragon, R., Howe, D. K., Dubey, J. P., and Sibley, L. D. 1998. ''Genotypic analysis of *Taxoplasma gondii* isolates from pigs,'' *Journal of Parasitology* 84:639–641.

Montoya, J. G., Jordan, R., Lingamneni, S., Berry, G. J., and Remington, J. S. 1997. ''Toxoplasmic myocarditis and polymyositis in patients with acute acquired toxoplasmosis diagnosed during life,'' *Clinical Infectious Diseases* 24:676–683.

Nichols, B. A., and Chiappino, M. L. 1987. ''Cytoskeleton of *Toxoplasma gondii,*'' *Journal of Protozoology* 34:217–226.

Nichols, B. A., Chiappino, M. L., and O'Connor, G. R. 1983. ''Secretion from the rhoptries of *Toxoplasma gondii* during host-cell invasion,'' *Journal of Ultrastructure Research* 83:85–98.

Nicolle, C. and Manceaux, L. 1908. "Sur un infection a corps de Leishman (ou organisms voisins) du gondii," *Comptes Reundus du Academie des Sciences* 147:763.

Odaert, H., Soete, M., Fortier, B., Camus, D., and Dubremetz, J. F. 1996. "Stage conversion of *Toxoplasma gondii* in mouse brain during infection and immunodepression," *Parasitology Research* 82:28–31.

Parmley, S. F., Yang, S., Harth, G., Sibley, L. D., Sucharczuk, A., and Remington, J. S. 1994. "Molecular characterization of a 65kilodalton *Toxoplasma gondii* antigen expressed abundantly in the matrix of tissue cysts," *Molecular and Biochemical Parasitology* 66:283–296.

Pavia, C. S. 1986. "Protection against experimental toxoplasmosis by adoptive immunotherapy," *Journal of Immunology* 9:2985–2990.

Pomeroy, C. and Filice, G. A. 1992. "Pulmonary toxoplasmosis: a review," *Clinical Infectious Diseases* 14:863–870.

Remington, J. S. and Desmonts, G. 1990. "Toxoplasmosis," In *Infectious Diseases of the Fetus and Newborn Infant,* ed., Remington, J. S., and Klein, J. O. Philadelphia: WB Saunders pp. 89–198.

Remington, J. S., Efron, B., Cavanaugh, E., Simon, H. J., and Trejos, A. 1970. "Studies on toxoplasmosis in El Salvador: prevalence and incidence of toxoplasmosis as measured by the Sabin-Feldman dye test," *Transactions of the Royal Society for Tropical Medicine and Hygiene* 64:252–267.

Rinder, H., Thomschke, A., Darde, M. L., and Loscher, T. 1995. "Specific DNA polymorphisms discriminate between virulence and non-virulence to mice in nine *Taxoplasma gondii* strains," *Molecular and Biochemical Parasitology* 69:123–126.

Robson, K. J. H., Hall, J. R. S., Jennings, M. W., Harris, T. J. R., Marsh, K., Newbold, C. I., Tate, V. E., and Wetherall, D. J. 1988. "A highly conserved amino acid sequence in thrombospondin, properdin and in proteins from sporozoites and blood stages of a human malarial parasite," *Nature* 335:79–82.

Reos, D. S., Sullivan, W. J., Striepen, B., Bonhe, W., and Donald, R. G. K. 1997. "Tagging genes and trapping promoters in *Toxoplasma gondii* by insertional mutagenesis," *Methods* 13:112–122.

Sabin, A. B. 1941. "Toxoplasmic encephalitis in children," *Journal of the American Medical Association* 116:800–807.

Sasono, P. M. D. and Smith, J. E. 1998. "*Toxoplasma gondii:* an ultrastructural study of host cell invasion by the bradyzoite stage," *Parasitology Research* 84:640–645.

Schreiber, R. D. and Feldman, H. A. 1980. "Identification of the activator system for antibody to *Toxoplasma* as the classical complement pathway," *Journal of Infectious Diseases* 141:366–369.

Schwab, J. C., Beckers, C. J. M., and Joiner, K. A. 1994. "The parasitophorous vacuole membrane surrounding intracellular *Toxoplasma gondii* functions as a molecular sieve," *Proceedings of the National Academy of Sciences, USA* 91:509–513.

Sheffield, H. G. and Melton, M. L. 1968. "The fine structure and reproduction of Toxoplasma gondii," *Journal of Parasitology* 54:209–226.

Sibley, L. D., and Boothroyd, J. C. 1992. "Virulent strains of *Toxoplasma gondii* comprise a single clonal lineage," *Nature* 359:82–85.

Sibley, L. D., and Howe, D. K. 1996. "Genetic basis of pathogenicity in toxoplasmosis," *Current Topics in Microbiology and Immunology* 219:3–15.

Sibley, L. D., Niesman, I. R., Asai, T., and Takeuchi T. 1994. "*Toxoplasma gondii:* secretion of a potent nucleoside triphosphate hydrolase into the parasitophorous vacuole," *Experimental Parasitology* 79:301–311.

Silverman, J. A., Huilin, Q., Riehl, A., Beckers, C., Nakaar, V., and Joiner K. 1998. "Induced activation of the *Toxoplasma gondii* nucleoside triphosphate hydrolase leads to depletion of host cell ATP levels and rapid exit of intracellular parasites from infected cells," *Journal of Biological Chemistry* 273:12352–12359.

Sinai, A. P., Webster, P., and Joiner, K. A. 1997. "Association of host endoplasmic reticulum and mitochondria with the *Toxoplasma gondii* parasitophorous vacuole membrane: a high affinity interaction," *Journal of Cell Science* 110:2117–2128.

Smith, D. D., and Frenkel, J. K. 1995. "Prevalence of antibodies to *Toxoplasma gondii* in wild mammals of Missouri and East Central Kansas—Biologic and ecologic considerations of transmission," *Journal of Wildlife Diseases* 31:15–21.

Smith, J. E., Boothroyd, J. C., Hunter, C. J., and Peterson, E. 1997. "Progress in toxoplasmosis research," *Parasitology Today* 13:245–246.

Smith, K. E., Zimmerman, J. J., Patton, S., Beran, G. W., and Hill, H. T. 1992. "The epidemiology of toxoplasmosis on Iowa swine farms with an emphasis on the roles of free-living mammals," *Veterinary Parasitology* 42:199–211.

Smith, J. E., McNeil, G., Zhang, Y. W., Dutton, S., Biswas-Hughes G., and Appleford P. 1996. "Serological recognition of *Toxoplasma gondii* cyst antigens," *Current Topics in Microbiology and Immunology* 219:67–73.

Soete, M., Camus, D., and Dubremetz, J. F. 1994. "Experimental induction of bradyzoite-specific antigen expression and cyst formation by the RH strain of *Toxoplasma gondii* in vitro," *Experimental Parasitology,* 78:361–370.

Speer, C. A., Tilley, M., Temple, M. E., Blixt, J. A., Dubey J. P. and White, M. A. 1995. "Sporozoites of *Toxoplasma gondii* lack dense granule protein GRA3 and form a unique parasitophorous vacuole," *Molecular and Biochemical Parasitology* 75:75–86.

Stommel, E. W., Ely, K. H., Schwartzman, J. D., and Kasper L. H. 1997. "*Toxoplasma gondii:* Dithiol induced Ca^{2+} flux causes egress of parasites from the parasitophorous vacuole," *Experimental Parasitology* 87:88–97.

Sumyuen, M. H., Garin, Y. J. F., and Derouin F. 1995. "Early kinetics of *Toxoplasma gondii* infection in mice infected orally with cysts of an avirulent strain," *Journal of Parasitology* 81:327–329.

Suzuki, Y. and Remington, J. S. 1990. "Effect of anti-IFN-gamma antibody on the protective effect of Lyt-2+ immune T cells against toxoplasmosis in mice," *Journal of Immunology* 144:1954–1956.

Suzuki, Y., Conley, F. K., and Remington, I. S. 1989a. "Importance of endogenous IFN-γ for prevention of toxoplasmic encephalitis in mice," *Journal of Immunology* 143:2045–2050.

Suzuki, Y., Conley, F. K., and Remington, J. S. 1989b. "Differences in virulence and development of encephalitis during chronic infection vary with the strain of *Toxoplasma gondii,*" *Journal of Infectious Diseases* 159:790–794.

Suzuki, Y., Wong, S. Y., Grumet, F. C., Fessel, J., Montoya, J. G., Zolopa, A. R., Portmore, A., Schumacher-Perdreau, F., Schrappe, M., Koppen, S., Ruf B., Brown, B. W., and Remington, J. S. 1996. "Evidence for genetic regulation of susceptibility to toxoplasmic encephalitis in AIDS patients," *Journal of Infectious Diseases* 173:265–268.

Tilley, M., Fichera, M. E., Jerome, M. E., Roos D. S. and White M. W. 1997. "*Toxoplasma gondii* sporozoites form a transient parasitophorous vacuole that is impermeable and contains only a subset of dense-granule proteins," *Infection and Immunity* 65:4598–4605.

Tocidlowski, M. E., Lapin, M. R., Summer, P. W., and Stoskopf, M. K. 1997. "Serological survey for toxoplasmosis in river otters," *Journal of Wildlife Diseases* 33:649–652.

Tomavo, S. 1996. "The major surface proteins of *Toxoplasma gondii:* structures and functions," *Current Topics in Microbiology and Immunology* 219:45–54.

Tomavo, S., Fortier, B., Soete, M., Ansel, C., Camus, D., and Dubremetz, J. F. 1991 "Characterization of bradyzoite specific antigens of *Toxoplasma gondii,*" *Infection and Immunity* 59:3750–3753.

Tomley, F. M., Clarke, L. E., Kawazoe, U., Dijkema, R., and Kok J. J. 1991. "Sequence of the gene encoding an immunodominant microneme protein of *Eimeria tenella,*" *Molecular and Biochemical Parasitology* 49:277–288.

Vanek, J. A., Dubey, J. P., Thuillez, P., Riggs M. R. and Stromberg B. E. 1996. "Prevelance of *Toxoplasma gondii* antibodies in hunter-killed white-tailed deer (*Odocoileus virginianus*) in four regions of Minnesota," *Journal of Protozoology* 82:41–44.

Venthanyagam, A. and Bryceson, A. D. M. 1976. "Aquired toxoplasmosis presenting as hepatitis," Transactions of the Royal Society of Tropical Medicine and Hygiene 70:524–527.

Waldeland, H., Pfefferkorn, E. R., and Frenkel, J. K. 1983. "Temperature-sensitive mutants of *Toxoplasma gondii* pathogenicity and persistence in mice," *Journal of Parasitology* 69:171–175.

Wan, K. L., Carruthers, V. B., Sibley, L. D., and Ajioka J. W. 1997. "Molecular characterization of an expressed sequence tag locus of *Toxoplasma gondii* encoding the micronemal protein MIC 2," *Molecular and Biochemical Parasitology,* 84:203–214.

Ware, P. L. and Kasper, L. H. 1987. "Strain specific antigens of *Toxoplasma gondii,*" *Infection and Immunity* 55:778–783.

Webster, J. P., Brunton, C. F. A., and Macdonald, D. W. 1994. "Effect of *Toxoplasma gondii* upon neophobic behavior in wild brown rats, *Rattus norvegicus,*" *Parasitology* 109:37–43.

Weiss, L. M., LaPlace, D., Tanowitz, H. B., and Wittner M. 1992. "Identification of *Toxoplasma gondii* bradyzoite specific monoclonal antibodies," Journal of Infectious Diseases 166:213–215.

Williams, K. A., Scott, I. M., Macfarlane, D. E., Williamson, J. M., Elias-Jones, T. F., and Williams, H. 1981. "Congenital toxoplasmosis: a prospective study in the West of Scotland," *Journal of Infection* 3:219–229.

Windeck, T. and Gross, U. 1996. "*Toxoplasma gondii* strain specific transcript levels of SAG1 and their association with virulence," *Parasitology Research* 82:715–719.

Woodison, G. and Smith, J. E. 1990. "Identification of the dominant cyst antigens of *Toxoplasma gondii,*" *Parasitology* 100:389–342.

Wreghitt, T., Hakim, M., Gray, J. J., Balfour, A. H., Stovin, P. G. I., Stewart, S., Scott, J., English, T. A. H., and Wallwork, J. 1989. "Toxoplasmosis in heart and heart and lung transplant recipients," *Journal of Clinical Pathology* 42:194–199.

Yamaoka, M. and Konishi, E. 1993. "Prevalence of antibody to *Toxoplasma gondii* among inhabitants under different geographical and climatic conditions in Hyogo prefecture, Japan," *Japanese Journal of Medical Science and Biology* 46:121–129.

Zangerle, R., Allerberger F., Pohl, P., Fritsch, P., and Dierich, M. P. 1991. "High risk of developing toxoplasmic encephilitis in AIDS patients seropositive to *Toxoplasma gondii,*" *Medical Microbiology and Immunology,* 180:59–66.

Zhang, Y. W. and Smith, J. E. 1995a. "*Toxoplasma gondii:* reactivity of murine sera against tachyzoite and cyst antigens via FAST-ELISA," *International Journal for Parasitology* 25:637–640.

Zhang, Y. W. and Smith, J. E. 1995b. "*Toxoplasma gondii:* identification and characterization of a cyst molecule," *Experimental Parasitology* 80:228–233.

CHAPTER 14

Entamoeba histolytica and *Cryptosporidium parvum*

UPINDER SINGH
THEODORE S. STEINER
BARBARA J. MANN

1. INTRODUCTION

ENTAMOEBA *histolytica* and *Cryptosporidium parvum* are pathogenic protozoa that are transmitted by the fecal-oral route. Infection occurs by ingestion of environmentally resistant cyst forms of the parasites in contaminated food or water. Neither parasite can grow or multiply in food or water and requires a host to complete its life cycle. *E. histolytica* infection, or amebiasis, is now an uncommon cause of disease in developed countries but is still a threat in many developing nations that have areas with poor sanitation and unclean water. Cryptosporidiosis is an example of an emerging disease. The parasite has long been known to cause disease in farm animals but has only been recognized as a human pathogen since the early 1980s. Cryptosporidiosis is also common in developing nations but is now recognized as one of the leading causes of waterborne disease outbreaks in the United States.

The ability to apply molecular techniques to the study of *E. histolytica* has opened the doors for scientists in the field to begin to study the molecular basis of pathogenesis. Since cryptosporidiosis is a relatively newly recognized human pathogen, there are many avenues of research to be pursued. Molecular research on *Cryptosporidium* is hampered by the inability to grow infectious forms of the parasite in culture. However DNA libraries are now available and data are beginning to emerge.

In this chapter life cycles, disease, epidemiology, detection, vaccine development and a detailed description of advancements in understanding of pathogenesis for *E. histolytica* and *C. parvum* are described.

2. *ENTAMOEBA HISTOLYTICA*

E. histolytica is a pseudopod-forming protozoan parasite and is the only medically important species in the genus *Entamoeba.* This genus includes *E. dispar, E. hartmanni, E. polecki, E. coli* and *E. gingivalis.* For many years *Entamoeba histolytica* was classified as having "pathogenic" or "nonpathogenic" zymodemes based on the migration patterns of isoenzymes in starch gels (Sargeaunt et al., 1978; Sargeaunt et al., 1982; Sargeaunt, 1987). While strains bearing the pathogenic zymodemes were consistently isolated from patients with disease, strains carrying the nonpathogenic zymodeme pattern were never associated with disease (Sargeaunt et al., 1982; Sargeaunt, 1987). As early as 1925 Emile Brumpt suggested that *E. histolytica* consisted of two different, but morphologically similar species (Brumpt, 1925). More recently an increasing number of biochemical, immunological, and genetic differences between *E. histolytica* and *E. dispar* have been reported (Garfinkel et al., 1988; Tannich et al., 1989; Edman et al., 1990; Petri, et al., 1990; Tachibana et al., 1990; Tannich and Burchard, 1991). In 1993 a formal redescription of *E. histolytica* and *E. dispar* as two separate species was published (Diamond and Clark, 1993). A joint report of a consultation of experts on amebiasis from the World Health Organization, the Panamerican Health Organization and UNESCO was issued in January 1997, which recommended that only *E. histolytica* infection need be treated and that treatment of *E. dispar* is unnecessary (WHO/PAHO/UNESCO, 1997).

The World Health Organization reports that *E. histolytica* infects 50 million people worldwide and results in 70,000 deaths annually. Amebiasis is surpassed only by malaria and schistosomiasis as the leading parasitic cause of death (WHO, 1995). Although the organism has a global distribution, *E. histolytica* is endemic in many tropical and subtropical countries where poor sanitation is prevalent. Severe disease is most likely to occur in the very young, the elderly, the malnourished and pregnant women (Armon, 1978; Walsh, 1986). In developed nations the majority of disease occurs in immigrants from or travelers to endemic regions (Pehrson, 1983). Populations at increased risk in developed nations include residents of institutions for the mentally retarded (Krogstad et al., 1978; Petri and Ravdin, 1988), members of communal groups (Hart et al., 1984), and sexually active homosexual males (Schmerin et al., 1977; Quinn et al., 1983; Druckman and Quinn, 1988; Takeuchi et al., 1989; Ohnishi et al., 1994).

In the United States the Center for Disease Control reported that cases of amebiasis have averaged around 3500 per year from 1945–1995 (CDC, 1995b). However, since 1995 it is no longer a reportable disease. Perhaps the most significant reported outbreak of amebiasis in the United States occurred at the Chicago World's Fair in 1933 (Markell, 1986). During the six months of this epidemic 1409 people were affected and 98 deaths occurred. The majority of the cases were guests or employees of two hotels. An investigation revealed that the source of infection was cross-connecting sewage and water pipes and leakage from an overhead sewer pipe into the drinking water tank that was shared by the two hotels. The investigation also concluded that food directly contaminated by food handlers played little if any part in the spread of infection.

Most individuals infected with *E. histolytica* are asymptomatic cyst passers. Only about 10% of infected individuals develop symptoms which can, however, be severe. The primary clinical manifestations of amebiasis include colitis and dysentery. Occasionally *E. histolytica* will spread hematogeously and form abscesses usually in the liver and rarely at other sites such as the brain, heart, lung and spleen.

2.1. LIFE CYCLE

The *E. histolytica* life cycle is relatively simple, consisting of infective cysts and invasive trophozoite forms (Figure 14.1) (Brown, 1969). Cysts are oval or spherical in shape, have up to four nuclei and range from 9–25 μm in diameter (Mirelman and Avron, 1988). The cyst form is resistant to chlorination and gastric acidity and can survive in a moist environment for several weeks. Infection occurs when cysts are ingested from fecally contaminated food or water (Figure 14.2). As few as 100–200 cysts are needed to establish an infection (Guerrant, 1994). Once the ingested cyst reaches the small bowel it excysts, undergoes nuclear division followed by cytoplasmic division to release eight trophozoites. Trophozoites are ameboid in shape and range from 12–60 μm in diameter. The single nucleus characteristically has peripheral chromatin and a prominent central karyosome (Figure 14.1). Ingested red blood cells are sometimes visible in the cytoplasm of clinical

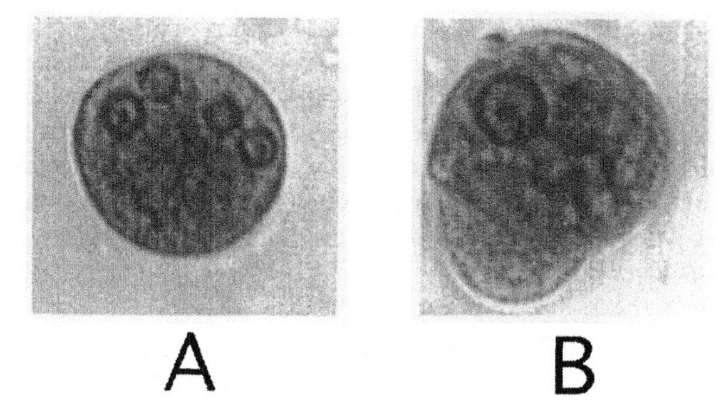

Figure 14.1 (A) A quadrinucleate cyst of *E. histolytica.* (B) A *E. histolytica* trophozoite extending a pseudopod. In both pictures, note the evenly distributed dark-staining peripheral chromatin and central karyosome in the nucleus. (Reprinted with permission from *Procedure Manual for the Diagnosis of Intestinal Parasites,* D. L. Price, Editor. Copyright ©-CRC Press.)

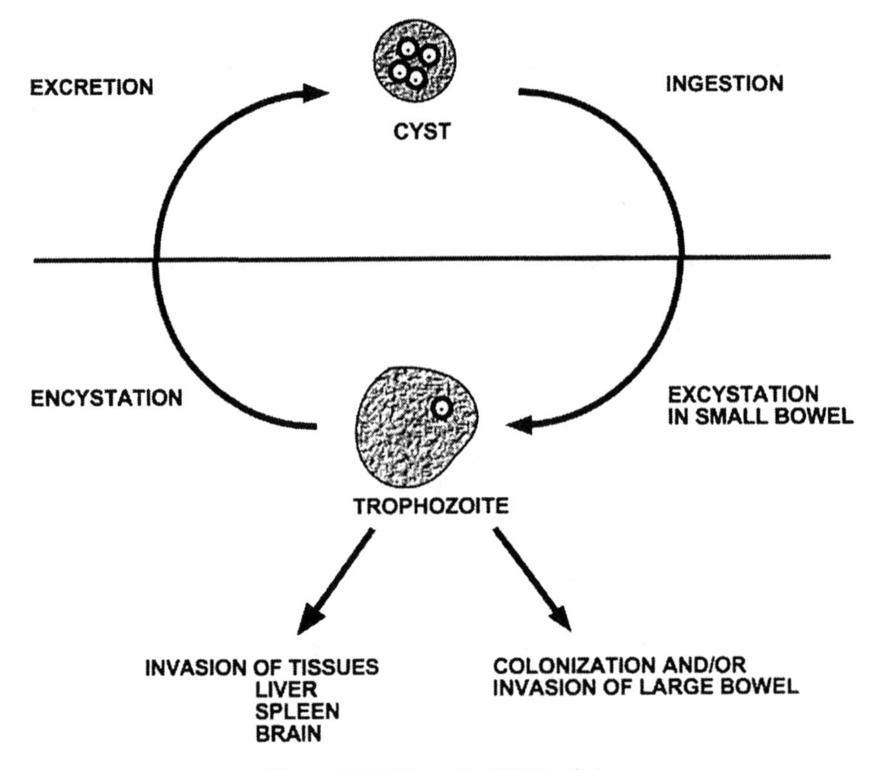

Figure 14.2 Life cycle of *E. histolytica*

isolates. Trophozoites are the invasive form of the parasite and have the ability to colonize, and/or invade the large bowel. Trophozoites can also encyst and be excreted to start a new round of infection. Invasion of the intestinal mucosal barrier by the trophozoite leads to the formation of flask-shaped colonic ulcers, and subsequent migration from colonic lesions to the liver is thought to occur via the portal vein. Factors influencing the disease progression, cyst formation and occurrence of colonization versus invasion are poorly understood and may be influenced by the *E. histolytica* strain and its interaction with bacterial flora, host genetic susceptibility, and factors such as malnutrition, sex, age, and immunocompetence.

2.2. DIAGNOSIS AND DETECTION

Reports of detection of *E. histolytica* cysts in food and water are rare (Walsh, 1988). The majority of outbreaks or infections have been characterized by identifying cysts or trophozoites in human materials. Traditionally, microscopy has been used for identification of the parasite in stool, liver abscess pus, or colonic biopsies. Unfortunately this method is insensitive and nonspecific. Microscopy has been reported to detect only 33–50% of cases (Petri et al., 1993; Haque et al., 1995; Haque et al., 1997) and it is not capable of distinguishing between pathogenic *E. histolytica* and nonpathogenic *E. dispar.* It has been suggested that erythrophagocytic amebae are more likely to be *E. histolytica* than *E. dispar* (Gonzalez-Ruiz et al., 1994), but *E. dispar* trophozoites have also been found to contain ingested red blood cells (Haque et al., 1995). The insensitivity of microscopy is illustrated by a pseudo-outbreak of amebiasis in California (CDC, 1992). In 1983 a clinical laboratory reported 38 cases of *E. histolytica* to the Los Angeles County Health Department over a three-month period. Since this represented an usually large increase in incidence and the cases did not seem to be related the slides from these 38 patients were reviewed. Only two were found to contain *E. histolytica* (possibly *E. dispar*) and the remaining 36 slides contained polymorphonuclear neutrophils and/or macrophages, and two contained nonpathogenic protozoa.

The morphologically identical *E. histolytica* and *E. dispar* can be distinguished on the basis of isoenzyme analysis, typing by monoclonal antibodies to surface antigens, and restriction fragment length polymorphisms (Sargeaunt et al., 1978; Garfinkel et al., 1988; Edman et al., 1990; Clark and Diamond, 1991; Tannich and Burchard, 1991; Diamond and Clark, 1993; Petrie et al., 1990b). A stool antigen detection test that is specific for *E. histolytica* is now commercially available for clinical use (Haque et al., 1993; Haque et al., 1995; Haque et al., 1997). The *E. histolytica* antigen test, which is based on detection of an amebic galactose/N-acetyl D-galactosamine inhibitable (GalNAc) lectin in stool by monoclonal antibody, is rapid and has a sensitivity and specificity of 86% and 98% respectively (Haque et al., 1997). This is slightly less sensitive than the "gold standard" of culture/isoenzyme analysis. Other ELISA-based kits are available for detecting *Entamoeba* but they do not appear to be specific for *E. histolytica* (Mirelman et al., 1997).

Polymerase Chain Reaction (PCR)-based detection is not yet available for *E. histolytica.* Preliminary results utilizing PCR detection of *E. histolytica* DNA in stool and liver abscess pus appear promising and its use to distinguish amongst isolates of *E. histolytica* should prove useful for epidemiological purposes (Garfinkel et al., 1988; Tannich and Burchard, 1991; Tachibana et al., 1992; Acuna-Soto, Samuelson et al., 1993). No testing of food or water sources by PCR techniques has been reported.

2.3. PATHOGENESIS

One of the central features of the pathogenicity of *E. histolytica* is its ability to kill target cells. Amebic-mediated cell killing relies on the organism's ability to adhere to target cells. Other activities that are thought to be important for pathogenesis are phagocytosis of target cells and the ability to evade lysis by complement. The molecular mechanisms of these activities and the virulence factors involved in pathogenesis are described in detail below.

2.3.1. Adherence to Target Cell

Galactose-inhibitable amebic adherence to Chinese hamster ovary (CHO) cells was first described by Ravdin and Guerrant (1981). Pretreatment of amebae with α- or β-galactosyl containing oligosaccharides results in the complete inhibition of amebic adherence and cytolysis of CHO cells (Ravdin and Guerrant, 1981; Saffer and Petri, 1991). Other carbohydrates such as N-acetyl-D-glucosamine, mannose and neuraminic acid have no effect. Galactose-inhibitable amebic adherence has been shown for a variety of targets including human colonic mucin glycoproteins (Chadee et al., 1987), human colonic epithelium (Ravdin, John et al., 1985), human neutrophils and erythrocytes (Guerrant et al., 1981; Ravdin and Guerrant, 1981; Burchard and Bike, 1992; Burchard et al., 1992), certain bacteria (Bracha and Mirelman, 1983) as well as a range of cell culture lines (Li et al., 1988; Li et al., 1989; Burchard et al., 1992).

Adherence of trophozoites to target cells is mediated by the Galactose and N-Acetyl-D-galactosamine inhibitable (GalNAc) lectin. The GalNAc lectin is the only adhesin participating in the adherence event as roles for fibronectin, vitronectin, CD11/CD18 integrins, complement, or mannose binding proteins in the adherence mechanism have been excluded (Burchard and Bilke, 1992). Monoclonal antibodies directed against the GalNAc lectin affect adherence, cytotoxicity and serum resistance indicating that the lectin is a multi-functional protein and plays a central role in pathogenesis (Petri et al., 1990b; Mann et al., 1993).

2.3.2. GalNAc Lectin

The amebic lectin is a 260 kDa heterodimer of heavy (170 kDa) and light (31–35 kDa) subunits linked in a 1:1 stoichiometric ratio by disulfide bonds (Petri et al., 1989). It was originally identified and purified by Petri using carbohydrate affinity chromatography and adherence-inhibitory monoclonal antibodies (Petri et al., 1987; Petri et al., 1989). The genes encoding the heavy and light subunits are members of multigene families consisting of 5–7 loci, depending on the strain (Ramakrishnan et al., 1996). The sequence identity between products of the different gene family members ranges from 79–94% (Mann et al., 1991; Tannich et al., 1991; Tannich et al., 1992; McCoy et al., 1993a; Purdy et al., 1993). Most of the heavy and light gene family members are simultaneously transcribed and probably expressed within a given strain (Ramakrishnan et al., 1996). There is no known apparent spatial or temporal difference in the expression of the different isoforms. Whether the differences in the heavy and light subunit isoforms translate into functionally distinct proteins is an open question.

The hydropathy profile of amino acid sequence of the heavy subunit suggests that it is a transmembrane protein with a single membrane spanning region, a large cysteine-rich extracellular domain and a short cytoplasmic tail (Figure 14.3) (Mann et al., 1991; Tannich et al., 1991). The extracellular domain can be further divided into subdomains consisting of a cysteine-tryptophan-rich region, a region devoid of cysteine, and cysteine-rich region that consists of over 10% cysteine residues. The membrane orientation of the subunit was verified by ability of heavy-subunit-specific monoclonal antibodies, which recognize epitopes in the cysteine-rich region of the protein, to bind to the surface of intact trophozoites (Mann et al., 1993).

The ability of anti-heavy subunit-specific monoclonal antibodies to dramatically enhance or inhibit amebic adherence to target cells suggests that the carbohydrate binding domain of the lectin resides within the heavy subunit (Petri et al., 1990b). A carbohydrate recognition domain (CRD) has been identified within the lectin heavy subunit cysteine-rich domain (Dodson et al., 1999). This domain does not share any sequence similarity to classic lectin-carbohydrate binding domains of the mammalian C type lectins, mammalian galectins, legume lectins, wheat germ agglutinin, ricin, *E. coli* heat-labile enterotoxin, cholera toxin or the influenza virus hemagglutinin (Barondes et al., 1994; Rini, 1995). The lectin CRD does share some sequence similarity to the receptor binding domain of the hepatocyte growth factor (HGF). The lectin CRD also competes with HCF for binding to the c-Met HGF receptor and may explain the parasite's tropism for the liver (Dodson et al., 1999).

The light subunit can be resolved into 31 and 35 kDa isoforms on SDS-polyacrylamide gel electrophoresis (McCoy et al., 1993b). The 31 kDa isoform has a glycerol phosphatidyl inositol (GPI) anchor (McCoy et al., 1993b). Thus the GalNAc lectin is a rather unique heterodimeric molecule consisting of a heavy subunit, which spans the membrane, and a light subunit with a GPI anchor. Monoclonal antibodies specific for the light subunit do not affect amebic adherence, cytolysis or serum resistance. The role of the light subunit in lectin-mediated functions is unclear at present, and may be related to cell-

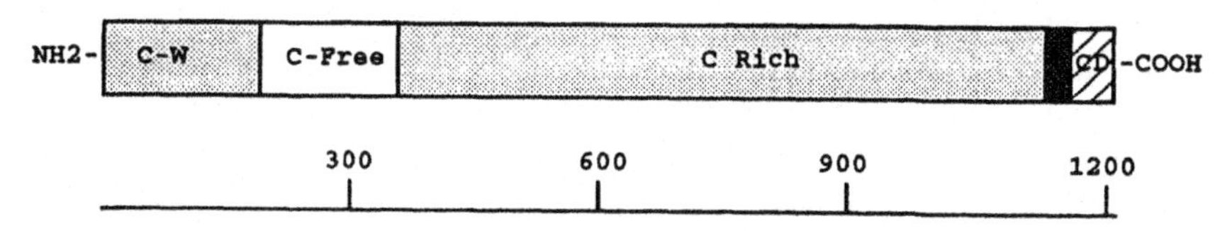

Figure 14.3 Diagram of the putative domain structure of the heavy subunit of the GalNAc lectin. Regions are designated as CW—cysteine-tryptophan rich, C-Free—cysteine-free, C-Rich—cysteine-rich, CD—cytoplasmic domain. The scale underneath indicates amino acid residue numbers.

signaling events via the GPI anchor. A cooperative role between the heavy and light subunits is implicated by the fact that neither subunit is found as a separate entity in vivo outside of the heterodimeric structure of the lectin (McCoy et al., 1993b).

2.3.3. Host Cell Receptors for the GalNAc Lectin

The human colonic mucin layer of the large intestine is the first receptor encountered by the lectin (Chadee and Meerovitch, 1985) and binds the lectin with high affinity (dissociation constant of 8.2×10^{-11} M^{-1}) (Chadee et al., 1987; Chadee et al., 1988) and with substrate specificity (i.e., Gal/GalNAc inhibitable). The trophozoites and colonic mucins interact in a dynamic manner with trophozoites, both inducing the secretion of and degrading colonic mucins (Chadee et al., 1987). Thus the mucin layer functions in a dual manner by providing a site for attachment of the parasite and in addition has a role in providing a host-protective mechanism by preventing contact dependent cytotoxicity from the lectin.

The specific host-cell receptors for the GalNac lectin have not been identified. However, Li and Ravdin have independently characterized the host receptors for the GalNAc lectin using CHO cell glycosylation mutants (Li et al., 1988; Li et al., 1989; Ravdin et al., 1989). Mutants that lack the N-linked carbohydrates on CHO cells have 80% decreased adherence and cells that lack the Gal and GalNAc residues on both the N- and O-linked carbohydrates have decreased adherence by 95%. In contrast, adherence is enhanced by 50% to CHO cell mutants that have an increased number of terminal Gal residues. These results suggest that both N- and O-linked carbohydrates may serve as receptors for the GalNAc lectin.

2.3.4. Cytolysis

E. histolytica is one of the most effective killer cells that has been described. It is capable of killing a target cell within minutes and it can effectively destroy macrophages, T lymphocytes and neutrophils (Salata et al., 1985; Burchard and Bilke, 1992), as well as a wide variety of tissue culture cell lines. The killing has been shown to be extracellular and contact dependent and is mediated in part via the GalNAc lectin (Saffer and Petri, 1991). Some monoclonal antibodies directed against the heavy subunit of the lectin block cytotoxicity but not adherence (Saffer and Petri, 1991). The ability of an anti-lectin monoclonal antibody to block cytolysis without blocking adherence suggests that either the lectin is a cytotoxin or that the lectin is involved in signaling the initiation of the cytolytic event.

A complex series of events occurs following the contact of the trophozoite with its target cell. Intracellular calcium in the target cell rises 20-fold and is associated with membrane blebbing, resulting in cell death in 5–15 minutes (Ravdin et al., 1988). Addition of extracellular EDTA and calcium channel blockers significantly reduces amebic killing (Ravdin et al., 1982). Murine myeloid cells killed by *E. histolytica* undergo a process of death that morphologically resembles the programmed cell death seen with growth factor deprivation, and is associated with a nucleosomal pattern of DNA fragmentation (Ragland, et al., 1994). This apoptotic pattern of host cell death induced by the ameba is blocked with Gal/GalNAc but not by overexpression of Bcl-2, a protein that confers resistance to apoptotic death from some stimuli (Ragland et al., 1994). This suggests that amebic apoptotic killing proceeds from a point distal to Bcl-2's inhibition of apoptosis in growth factor-deprived cells. The human leukemic cell lines HL-60 and jurkat incubated with amebic trophozoites appear to die by a necrotic mechanism based on morphological data observed by electron transmission microscopy. Necrotic features such as target cell swelling, rupture of the plasma membrane and release of cell contents were observed (Berninghausen and Leippe, 1997). In this system nucleosomal fragmentation of DNA did not occur. Studies of amebic-mediated cell death have all been conducted in vitro, thus it appears that amebae kill cells by different mechanisms depending on the type of target cell and the specific set of conditions that are used. This may also be true in vivo but the answer awaits additional investigations.

The biochemical events that occur within the amebic trophozoite during cytolysis are not well understood. Cytolysis is inhibited by cytochalsins but not colchicine, indicating a role for the actin cytoskeleton and an independence for microtubule function (Ravdin and Guerrant, 1981). In contrast to target cells, no detectable increases in regional or total free [Ca^{2+}] are observed in amebae upon contact with a target cell (Ravdin et al., 1988). The cytolytic activity of *E. histolytica* has been shown to increase in the presence of phorbol esters and is inhibited by sphingosine, suggesting a role for protein kinase C in the signaling of cytolysis (Weikel et al., 1988). Two potential types of cytolytic effector molecules have been identified, pore-forming proteins called amebapores and a family of cysteine proteases.

2.3.5. Amoebapore

The concept of pore-forming proteins as tools for cell killing has been described in effector cytotoxic lymphocytes and natural killer cells (Lowin et al., 1995). Pore-forming proteins, such as perforin, are expressed by effector cells and bind to and create aqueous channels in the target cell's plasma membrane, thus leading to osmotic lysis and cell death. A family of three peptides, called amoebapores A, B, and C, with intrinsic pore-forming properties have been isolated from the cytoplasmic granules of *E. histolytica* (Lynch et al., 1982; Young et al., 1982; Leippe et al., 1992; Leippe et al., 1994b;

Leippe and Muller-Eberhard, 1994). All three amebapores exhibit pore-forming activity toward lipid vesicles although they differ in their kinetics of activity. Although these three proteins differ fairly significantly in their primary structure (35–57% amino acid sequence identity), they all contain six conserved cysteine residues and are predicted to have four similar amphipathic α-helical domains stabilized by disulfide bonds (Leippe, et al., 1996) (Figure 14.4). The first and third α-helical domains are long enough to span a membrane and synthetic peptides corresponding to these domains exhibit pore-forming activity (Leippe et al., 1994a). Chemical modification of the positively charged amino acids results in a complete loss of pore-forming activity (Andra and Leippe, 1994). Amoebapore activity is also found in the nonpathogenic *E. dispar* although its pore-forming activity is 80% less than that of *E. histolytica* (Keller et al., 1988; Leippe et al., 1993). A structural comparison of the amoebapores from the two species reveals a proline substitution for a glutamate in the N-terminal α-helix of the protein from *E. dispar,* which shortens the

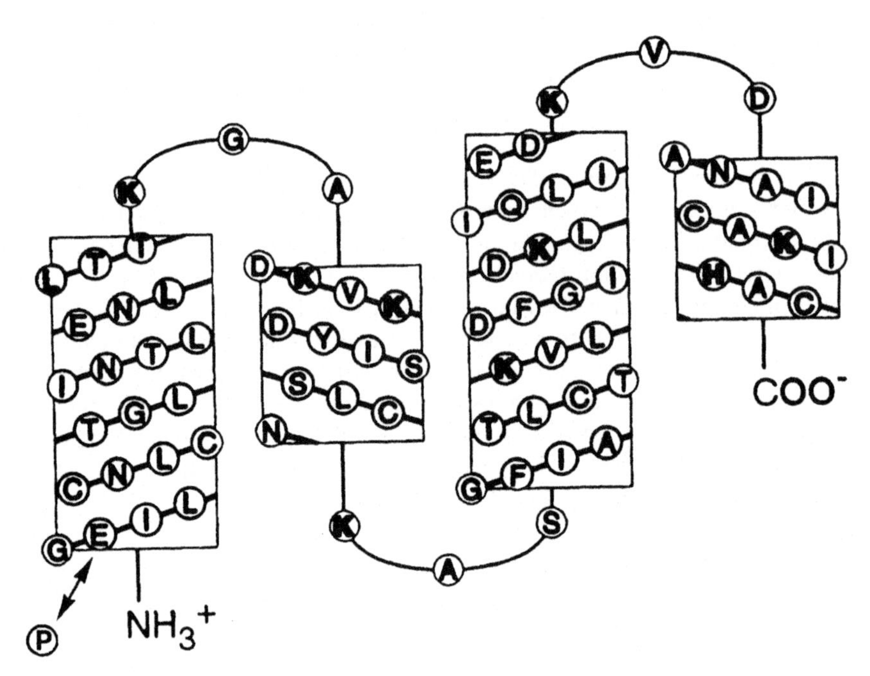

Figure 14.4 Diagram of the putative α-helical structure of the *E. histolytica* amebapore. The four predicted α-helical regions are represented as opened and flattened cylinders viewed from the side. Residues are displayed to give approximately 3.6 residues per turn of the helix. Amino acid residues are indicated by their single amino acid abbreviations. Positively charged residues are highlighted in bold. The proline (P) for glutamate (E) substitution found in *E. dispar* amebapore is indicated by an arrow in the first α-helix. [Reprinted with permission from Andra and Leippe (1994).]

first α-helix and thus is likely to be responsible for the reduction in the membranolytic activity of the protein (Leippe et al., 1993) (Figure 14.4).

Amebapores are concentrated in cytoplasmic granules in the ameba (Leippe et al., 1994b) and are not released into the culture medium from intact amebae (Leippe et al., 1995). One model for explaining the mode of amebapore action is that the amebapores are released from the granules upon a stimulus such as cell-cell contact, possibly via the Ga1NAc lectin, into the host-cell membrane. Amebapores are also able to disrupt the integrity of bacterial cytoplasmic membranes and may have a role as antimicrobial agents against engulfed bacteria (Leippe et al., 1994b). Amebapores have 45–50% similarity to NK lysin, an effector molecule of porcine cytotoxic lymphocytes that also has antibacterial activity (Leippe, 1995).

2.3.6. Cysteine Proteinases

The enzymatic destruction of extracellular matrix proteins (Bracha and Mirelman, 1984; Luaces and Barrett, 1988; Schulte and Scholze, 1989; Li et al., 1995), cytolytic/cytopathic effects on target cells (Lushbaugh et al., 1984; Keene et al., 1990), and activation of the host complement system by *E. histolytica* (Schulte and Scholze, 1989) is accomplished at least in part by cysteine proteinases which are secreted by the organism into its environment and culture supernatants. The secretion of cysteine proteinases appears to be constitutive, which is in sharp contrast to the release of amoebapores, which are secreted only upon contact with a target cell (Leippe et al., 1995). A critical role in pathogenesis for cysteine proteinases is suggested by studies that have shown that treatment of trophozoites with laminin or a specific cysteine proteinase inhibitor, E-64, blocks or greatly reduces amebic liver abscess formation in a severe combined immunodeficient (SCID) mouse model (Li et al., 1995; Stanley et al., 1995).

Amebic cysteine proteinases have a molecular size of 27 kDa and belong to the papain superfamily (Tannich et al., 1991; Reed et al., 1993). Six different genes encoding cysteine proteinases have been identified in *E. histolytica* (*ehcpl–6*), which share 40–85% DNA sequence identity (Bruchhaus et al., 1996). Essentially all of the cysteine proteinase activity in lysates from culture-grown amebae can be attributed to three of these gene products, EhCP1, EhCP2 and EhCP5. By Southern blot analysis *E. dispar* contains homologues for four out of the six *ehcp* genes. EhCP1 and EhCP5 have no homologues in *E. dispar* (Bruchhaus et al., 1996). In *E. histolytica ehcpl, ehcp2* and *ehcp5* produce the most abundant transcripts while in *E. dispar,* the *ehcp3* homologue has the highest level of steady-state mRNA (Bruchhaus et al., 1996). In general, the level of expression cysteine proteinases in *E. dispar* is markedly decreased when compared to *E. histolytica* (Reed et al., 1989; Tannich et al., 1991; Reed et al., 1993). Whether the overall level or the presence of two novel cysteine proteinase genes in *E. histolytica* is significant in contributing to the difference in the pathogenic potential between *E. histolytica* and *E. dispar* remains to be dissected by genetic analysis.

2.3.7. Resistance to Lysis by Complement

During invasion of the colon and portal spread to the liver, trophozoites are exposed to the human complement system but are successfully able to evade this component of the immune system. Amebic trophozoites isolated from patients with invasive disease are able to activate the alternative complement pathway but are resistant to C5b-9 complexes deposited on the membrane surface (Reed et al., 1986; Reed and Gigli, 1991). A monoclonal antibody specific for the 170 kDa heavy subunit of the GalNAc lectin was identified that was capable of neutralizing resistance to complement lysis by human sera and purified human complement components C5b-9, and thus revealed a role for the lectin in serum resistance (Braga et al., 1992). The heavy subunit has limited identity with CD59, a human inhibitor of the C5b-9 assembly, and purified lectin is also recognized by anti-CD59 antibodies. The lectin blocks assembly of the membrane attack complex at the level of C8 and C9 insertion. Reconstitution of the lectin from serum-resistant into serum-sensitive amebae confers resistance to the membrane attack complex, a direct demonstration of its C5b-9 inhibitory activity. The lectin, therefore, appears to function not only in adherence and host cell killing, but also in evasion of the complement system of defense via a remarkable mimicry of human CD59 (Braga et al., 1992).

2.4. HOST IMMUNE RESPONSES

Evidence for acquired natural immunity to *E. histolytica* reinfection is somewhat anecdotal. For example, one study in Mexico followed 1021 patients over a five-year period. During this time only three patients (0.29%) experienced a second episode of amebic liver abscess (DeLeon, 1970). Unfortunately the study lacked case controls. Humoral and cell-mediated immune responses have been detected in patients recovered from amebic disease but the contributions of the different arms of the immune response to protective immunity have not been clearly established. In serological studies 80–100% of patients with invasive disease have anti-*E. histolytica* antibodies (Petri et al., 1987; Ravdin et al., 1990). However, there is little evidence in humans to correlate antibody levels with protection or clinical outcome.

Stronger support for a protective role of antibodies comes from studies using a SCID mouse model of amebic liver abscess. Passive transfer of anti-rabbit or anti-human *E. histolytica* antibodies has been shown to protect against liver abscess formation in the SCID mouse (Cleslak et al., 1992; Zhang and

Stanley, 1994). *E. histolytica*-specific secretory IgA (sIgA) has also been detected in the serum and saliva of patients with invasive disease (Kelsall et al., 1994; Abou-El-Magd et al., 1996). The sIgA recognizes the Gal NAc lectin and is capable of inhibiting amebic adherence to target cells (Carrero et al., 1994).

Animal studies involving interventions that produce a depression of CMI such as silica gel, neonatal thymectomy, splenectomy, steroid treatment and anti-macrophage or anti-lymphocyte antibody result in exacerbated amebic liver disease (Ghadirian and Meerovitch, 1981a; Ghadirian and Meerovitch, 1981b; Ghadirian et al., 1983). Depression of cell-mediated immune responses may be important for the natural development of disease (Salata et al., 1990). In humans depression of cell-mediated immune responses appears to occur during acute disease as evidenced by skin test anergy to *E. histolytica* antigen, a decreased number of T cells and a decreased T cell proliferative response to amebic antigen (Ortiz-Ortiz et al., 1975; Salata et al., 1986). Abscess-derived gerbil macrophages are deficient in several activities including the ability to develop a respiratory burst and kill amebic trophozoites (Denis and Chadee, 1988). This is in contrast to macrophages derived from the spleen and peritoneal cavities from the same infected animals, which are not significantly downregulated in these same activities.

Lymphocytes from patients recovered from invasive disease develop CMI responses to total amebic antigen and the Gal NAc lectin. These responses include gamma interferon (γ-IFN), interleukin-2 (IL-2), T cell proliferation, and amebicidal activity (Salata et al., 1987; Schain et al., 1992). γ-IFN may be particularly important as γ-IFN-activated human macrophages and neutrophils have enhanced amebicidal activity, while in the absence of γ-IFN these effector cells are killed by amebae (Salata et al., 1987; Denis and Chadee, 1988). The major amebicidal molecule released by activated macrophages is nitric oxide (Lin and Chadee, 1992; Lin et al., 1995). The Gal NAc lectin has been shown to stimulate γ-IFN-primed murine bone marrow-derived macrophages to produce both TNF-α and nitric oxide (Seguin et al., 1996).

2.5. VACCINES

Until the developing world has access to clean water and adequate sanitation systems, amebiasis will continue to remain a significant health problem. Since this accomplishment is still in the future, considerable effort has been made towards the development of a protective vaccine against amebiasis. There are no insect or other animal-host intermediaries in the life cycle and humans are the only epidemiologically significant reservoir of *E. histolytica.* Therefore, an effective protective vaccine has the potential to eliminate amebiasis. Although there are still questions regarding the nature of natural immunity in humans, animal studies have demonstrated that complete or partial immunity to *E. histolytica* infection is possible (Petri and Ravdin, 1991). Two proteins, GalNAc lectin and the serine-rich protein (SREHP), have emerged as the most promising subunit vaccine candidates. SREHP is a surface protein of *E. histolytica* that was first described by Stanley and co-workers (Stanley et al., 1990). This protein is recognized by human immune sera but its function is unknown.

Antibodies specific to the GalNAc lectin 170 kDa subunit (see above discussion) completely block amebic adherence to and cytolysis of target cells (Petri et al., 1987; Petri et al., 1990). Its potential as a protective vaccine was first demonstrated by Petri and Ravdin, who showed that immunization with the native heterodimeric lectin results in 43–68% protection against liver abscess formation in gerbils (Petri and Ravdin, 1991).

Efforts to identify the protective epitopes of the GalNAc lectin have concentrated on the 170 kDa subunit. Several laboratories have expressed and purified fragments of 170 kDa subunit from *E. coli* for vaccine studies (Zhang and Stanley, 1994; Soong et al., 1995; Lotter et al., 1997; Mann et al., 1997). Fragments of the cysteine-rich region of the 170 kDa produce protection with efficacies from 71–81% (see Figure 14.1) (Zhang and Stanley, 1994; Soong et al., 1995). Oral immunization with an attenuated strain of *Salmonella* expressing a portion of the cysteine-rich region also results in some protection against liver abscess formation (Mann et al., 1997). Epitopes found on the 35 kDa subunit and created by any post-translational modifications of the heavy subunit apparently do not contribute significantly to protective immunity. The major protective epitopes are found between amino acids 939–1053 (Soong et al., 1995; Lotter et al., 1997). Vaccination with these amino acids protects 62.5% of the animals (Lotter et al., 1997). In addition, passive transfer of rabbit antisera raised against this fragment also protects 40–60% of SCID mice (Lotter et al., 1997). This passive protection correlates with an antibody response to a 25-amino acid peptide of the cysteine-rich region (Lotter et al., 1997).

Stanley and co-workers have established that recombinant SREHP is also an effective protective antigen in animal models. SREHP is a 25 kDa protein, which consists of about 22% serine residues and multiple octapeptide and dodecapetide tandem repeats (Stanley et al., 1990). The protein is located on the surface of the parasite but its exact function is unknown. Two initial immunization trials by intraperitoneal injection with a maltose-binding protein fusion protein of SREHP resulted in 64% and 100% protection against intrahepatic challenge with amebic trophozoites in the gerbil liver abscess model of amebiasis (Zhang and Stanley, 1994). Passive transfer of anti-SREHP antibodies has been shown to completely block amebic liver abscess formation in a SCID mouse model (Zhang et al., 1994). Recombinant SREHP has also been shown to be an effective oral vaccine against liver abscess formation. Oral vaccination with an attenuated strain of *Salmonella typhimurium*, expressing

SREHP, produced both specific mucosal and serum antibody responses and protected against or reduced the size of amebic liver abscesses in gerbils (Zhang and Stanley, 1996).

Since both the GalNAc lectin and the SREHP show potential as protective vaccines, it is possible that a combined vaccine may prove even more efficacious. However, in a single trial where animals were immunized with both the SREHP and the cysteine-rich region of the 170 kDa, the level of protection was not significantly different than in animals immunized with either antigen alone (Zhang and Stanley, 1994).

While significant progress has been made towards developing a vaccine against systemic disease, the ability of these antigens to protect against intestinal disease is not known. Currently an animal model of intestinal amebiasis does not exist. Since the majority of disease caused by amebiasis is colitis and dysentery, testing the ability of a vaccine to protect against mucosal disease is of primary importance. Oral immunization with the cysteine-rich region of the GalNAc lectin or SREHP in conjunction with cholera toxin produces secretory IgA, which inhibits amebic adherence to target cells, suggesting a role for secretory IgA in preventing colonization (Beving et al., 1996; Zhang and Stanley, 1996). However, true testing will require either an intestinal animal model or direct testing in human volunteers.

2.6. EMERGING RESEARCH

Control of gene expression is a critical factor in the ability of the organism to transform from a cyst to a trophozoite and adapt from an anaerobic intestinal lumen to an aerobic intestinal epithelium. Understanding regulation of gene expression is, therefore, a vital component of dissecting the molecular mechanisms of pathogenesis. Genetic manipulation of the organism has only recently become possible with the development of transient and stable transfection systems (Nickel and Tannich, 1994; Purdy et al., 1994; Vines et al., 1995). Simultaneous development of a transient transfection system via electrophoration was developed by Petri and Tannich utilizing the 5′ and 3′ flanking regions of amebic *hgl* and actin genes to drive luciferase and chloramphenicol acetyltransferase (CAT) reporter genes, respectively (Nickel and Tannich, 1994; Purdy et al., 1994). These data demonstrated that both the 5′ and 3′ regions of amebic promoters are necessary to drive gene expression in *E. histolytica.* It appears that amebic promoter regulatory regions and transcription factors are unique from those of higher eukaryotes as demonstrated by the inability of amebic promoters to function in epithelial cells and the inability of viral promoters to function in the parasite (Purdy et al., 1994). The development of a stable transfection system utilizing G418 drug selection was subsequently achieved, allowing further genetic manipulation of the organism (Hamann et al., 1995; Vines et al., 1995). Recently, the techniques of an inducible promoter system (Hamann et al., 1997; Ramakrishnan et al., 1997) and antisense expression (Alon et al., 1997) have been developed and should provide further ability to genetically manipulate the organism. In addition, characterization of promoter regions of various amebic genes is underway and may identify novel transcription factors (Bruchhaus et al., 1993; Purdy et al., 1996; Singh et al., 1997). Techniques such as antisense and gene knockout can be used to specifically address the relationship between pathogenesis and the various virulence factors and can also be used to characterize the vast differences in virulence between *E. histolytica* and *E. dispar.*

3. *CRYPTOSPORIDIUM PARVUM*

While *Cryptosporidium parvum* was first identified early in this century in animals, the first human cases were not reported until 1976 (Meisel et al., 1976; Nime et al., 1976). Only a handful more human cases were recognized until the beginning of the AIDS epidemic in the early 1980s, when *C. parvum* emerged as an important and devastating diarrheal pathogen in immunocompromised individuals. Since that time, *Cryptosporidium* has been found much more often in immunocompetent people, and is a leading cause of endemic childhood diarrhea in developing areas and waterborne diarrheal outbreaks in developed countries.

C. parvum is a coccidian protozoan in the phylum Apicomplexa of the protozoal kingdom, which contains many human pathogens, including *Plasmodium, Babesia,* and *Toxoplasma.* These protozoa are all characterized by an anterior (or apical) polar complex which allows penetration into host cells. *C. parvum* and the related diarrheal coccidia *Isospora belli* and *Cyclospora cayetanensis* share common features in their life cycle, including fecal excretion of oocysts.

The life cycle of *C. parvum* is shown in Figure 14.5. The infectious form of *Cryptosporidium* is a 4–6 μm, thick-walled oocyst, which is ingested by the host and excysts within the lumen of the small intestine to release four sporozoites. These penetrate the microvillus border and develop into trophozoites and subsequently type 1 meronts (schizonts). These reproduce asexually with three nuclear divisions to release eight merozoites, which invade nearby cells and develop into either type 2 meronts or trophozoites. Type 2 meronts in turn undergo two nuclear divisions and release four type 2 merozoites, which develop into male (microgamont) or female (macrogamete) forms. The microgamont releases microgametes which penetrate the macrogamete to form a zygote, which can then develop into a thin-walled autoinfectious oocyst or a thick-walled oocyst which is shed in the stool. The average incubation period (time from ingestion to disease manifestation) is approximately one week.

Among the many species of *Cryptosporidium* known to infect vertebrate

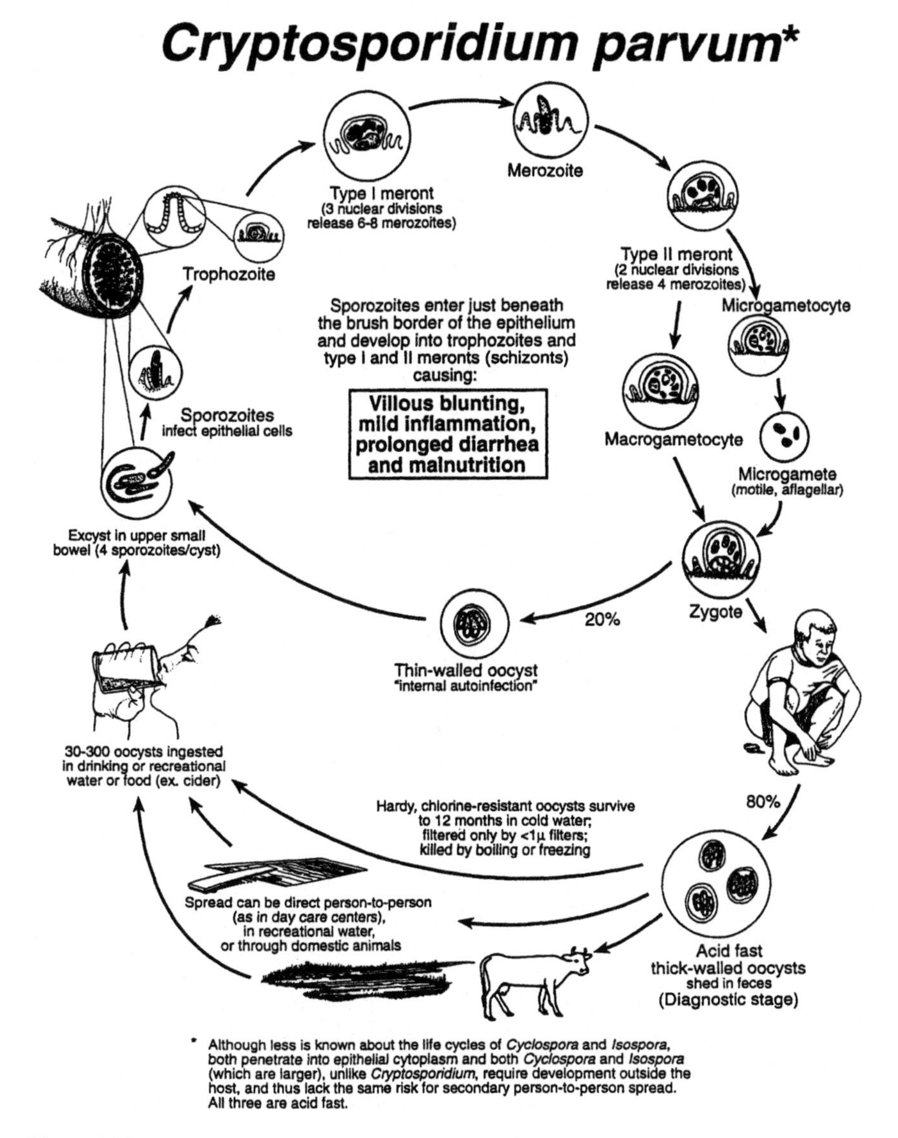

Figure 14.5 Life cycle of *Cryptosporidium parvum*. [Reprinted with permission from Steiner and Guerrant (1998).]

hosts, only *C. parvum* is clearly pathogenic for humans. It also infects a wide variety of mammals, including many domesticated species such as cows, goats, sheep, cats, and dogs. These animals may serve as a substantial reservoir for *C. parvum* and help explain its nearly universal presence in surface water in the United States and Canada (LeChevallier et al., 1991a; LeChevallier et al.,

1991b). Recent data suggest variability in the pathogenesis of different strains of *C. parvum* as well (Okhuysen et al., 1997).

3.1. EPIDEMIOLOGY

Cryptosporidial diarrhea is seen primarily in four settings: endemic childhood diarrhea in developing areas, protracted diarrhea in immunocompromised patients worldwide, traveler's diarrhea in visitors to developing areas, and water- and foodborne disease outbreaks in developed countries. Seroprevalence studies reflect these settings, with seropositivity ranging from 58% in adolescents in Oklahoma (Kuhls et al., 1994) to 95% by age 2 in northeastern Brazil (Zu et al., 1994).

C. parvum and *Giardia lamblia* have accounted for almost half of all reported waterborne disease outbreaks in the United States in the 1990s (Steiner et al., 1997). A list of the major waterborne outbreaks is provided in Table 14.1. By far the largest of these was the 1993 outbreak in Milwaukee, Wisconsin, in which an estimated 403,000 people developed cryptosporidial diarrhea due to a combination of increased source water contamination and transient reduction in the efficiency of prefiltration processing, despite water quality which met all U.S. federal drinking water standards of the time (MacKenzie et al., 1994). *C. parvum* is also a leading cause of recreational water outbreaks (Steiner et al., 1997). Finally, *C. parvum* was identified in three recent foodborne diarrheal outbreaks, two due to unpasteurized apple cider and one likely due to contaminated chicken salad (CDC, 1996; CDC, 1997).

Cryptosporidium possesses several unusual characteristics that make the organism such a threat. First, oocysts are highly infectious: The median infectious dose in healthy adult volunteers is 132 oocysts (DuPont et al., 1995), although a mathematical model based on the Milwaukee outbreak estimated that some people could have been infected after ingestion of only one oocyst (Haas and Rose, 1994). This infectivity probably accounts for the high rate of person-to-person transmission, ranging from 5.4% of household contacts of adults infected in the Milwaukee outbreak developing symptomatic disease to 19% of family members of infected children in Fortaleza, Brazil, developing disease or seroconversion (Newman et al., 1994; MacKenzie et al., 1995).

In addition, the oocysts are highly resistant to water treatment protocols. The oocysts measure 4–6 μm in diameter, which enables them to pass through most conventional filter systems. Only water filters rated at "absolute" 1 μm are reasonably effective at removing *Cryptosporidium,* but even those may fail (CDC, 1995a; Addiss, et al., 1996). Morever, *Cryptosporidium,* oocysts are highly resistant to chlorination, remaining infectious even after exposure for 30 minutes to 80 ppm chlorine (100 times the acceptable dose for elimination of coliform bacteria) (Korich et al., 1990). The cysts can be inactivated, however,

TABLE 14.1. Community-Based Outbreaks of *Cryptosporidium* Due to Treated Drinking Water.

Location	Dates	No. Affected	Filtered	Presumptive Cause
Braun Station, TX	May–July, 1984	346	No	Sewage contamination of community well
Sheffield, Eng.	Apr–Oct, 1986	104	Yes	Contamination of reservoir after heavy rain
Carroll County, GA	Jan–Feb, 1987	13,000	Yes	Increased turbidity due to replacement of mechanical agitators in flocculation basins
Ayrshire, U.K.	Apr, 1988	27	Yes	Leakage into storage tank
Wiltshire and Oxfordshire, Eng.	Dec, 1988–Apr, 1989	516	Yes	Major contamination of source water after heavy rainfall and warm winter
Oregon	Feb, 1992	3000	No	Contamination of spring with surface water
Oregon	May, 1992		Yes	Inadequate filtration due to low stream flow
Milwaukee, WI	Mar, 1993	403,000	Yes	Increased turbidity due to decreased effectiveness of coagulation/filtration and source contamination
Las Vegas, NV	Jan–Apr, 1994	≥103	Yes	None identified
Washington, D.C.	Aug, 1994	134	No	Sewage contamination of well

by ozone treatment, prolonged freezing, or heating/pasteurization (to 72°C for 1 min or 45°C for 10–20 min) (Anderson, 1985; Korich et al., 1990; Harp et al., 1996).

To further complicate matters, there is no practical "gold-standard" method for detecting viable *C. parvum* oocysts in treated water. Current techniques rely on large-volume concentration followed by light and/or immunofluorescence microscopy to identify structures likely to be oocysts. However, it is difficult to establish whether these structures are actually infectious oocysts. Water surveys after a large cryptosporidiosis outbreak in Las Vegas, New York, in 1994 found only "presumptive" oocysts, which had the characteristic size, shape, and immunofluorescence of oocysts but lacked internal structures (Goldstein et al., 1996). The tests for oocyst viability most often used are excystation (by microscopy), vital staining, animal infectivity, or cell culture infectivity, but the results of these can be discordant (Black et al., 1996). Newer techniques for oocyst detection using the polymerase chain reaction appear promising but are still under development (Laberge et al., 1996).

3.2. CLINICAL FEATURES AND PATHOPHYSIOLOGY

Cryptosporidiosis in immunocompetent individuals is characterized by diarrhea, which is usually described as watery, voluminous, and occasionally explosive and foul-smelling. Abdominal cramps, fatigue, and anorexia occur in more than 80% of affected patients (MacKenzie et al., 1994; MacKenzie et al., 1995). In addition, more than half of patients may experience weight loss, nausea, low-grade fever, chills, sweats, myalgias, and headache. In immunocompetent patients the illness is self-limited but may be prolonged, lasting 10–14 days, with an average of 12 stools per day during the peak of the illness and often significant weight loss (median 4.5 kg in the Milwaukee outbreak). While the disease is self-limited in immunocompetent people, relapses can occur; 39% of the patients infected in the Milwaukee outbreak had a brief recurrence of diarrhea after a period of normal stools.

In immunocompromised people the illness may be unrelenting, fulminant, and ultimately fatal. Moreover, these patients may develop infection in extraintestinal sites, especially the biliary tree. The lack of any curative treatment makes cryptosporidiosis one of the most feared opportunistic pathogens in these patients.

While the clinical manifestations of cryptosporidiosis are profound, the pathologic changes seen in the intestine may be relatively mild. Infection produces mild to severe villus blunting, submucosal edema, and inflammatory infiltrates in the lamina propria, but not the marked epithelial necrosis observed with more invasive pathogens like *E. histolytica*. These changes are associated with evidence of malabsorption, such as abnormal D-xylose tests (Brandborg et al., 1970; Modigliani et al., 1985; Connor et al., 1993; Adams et al., 1994; Goodgame et al., 1995; Lima et al., 1996). Various life cycle forms of cryptosporidia reside in an intracellular but extracytoplasmic location on the apical membrane; ultrastructurally, they reside in vacuoles with specialized basal membrane folds called "feeder organelles" (Adams et al., 1994). These are believed to provide a large surface area for exchange of nutrients.

A great mystery in the study of cryptosporidiosis is how it produces profound inflammatory and destructive pathophysiology without tissue invasion. A simple explanation would be the elaboration of an enterotoxin or cytotoxin, although thorough investigation has failed to discover such a toxin (Guarino et al., 1994; Sears and Guerrant, 1994; Guarino et al., 1995). Most research has focused on the direct epithelial effects of colonization with *C. parvum.* Several observations on experimental porcine infection suggest that part of the symptomatology of cryptosporidiosis may be due to malabsorption by inflamed villi coupled with intact fluid secretion from the crypts. Prostaglandins appear to play a role in this imbalance, but the details of this interaction are still unclear (Argenzio et al., 1994; Kandil et al., 1994). Two studies of intestinal epithelial cells in tissue culture have found destructive effects of *Cryptosporidium* infection, including loss of barrier integrity and moderate cell injury (Adams et al., 1994; Griffiths et al., 1994), effects that are confirmed in vivo by increased lactulose: mannitol excretion ratios in patients with cryptosporidiosis (Lima et al., 1996).

3.3. MOLECULAR PATHOGENESIS

Very little is known about the molecular determinants of pathogenicity in *C. parvum.* This is partly due to the difficulty in obtaining infective organisms, which must be purified from the stool of infected animals. While a genomic library of the organism has been produced (Spano et al., 1997), no genes encoding virulence factors have been identified. Several genes encoding metabolic and structural proteins have been cloned recently, including the oocyst wall protein (Spano et al., 1997), dihydrofolate reductase/thymidylate synthase (Vasquez et al., 1996), a putative acetyl-CoA synthetase (Khramtsov et al., 1996), elongation factor-1α (Bonafonte et al., 1997), and disulfide isomerase (Blunt et al., 1996). A recent report suggests that the genome is contained on eight chromosomes and one low-molecular-weight DNA molecule (Blunt et al., 1997).

While genetic determinants of pathogenicity are still unknown, several potentially important surface molecules have been identified. Riggs et al. (1997) identified a circumsporozoite-like glycoprotein antigen, which binds a protective monoclonal antibody. Joe et al. found that *C. parvum* contains a Gal/GalNAc lectin, which is important in attachment to MDCK cells in culture (Joe et al., 1994). Gut et al. identified an antigen shed from the posterior membranes of *C. parvum* during locomotion, which may be important in cell entry (Gut and Nelson, 1994). These molecules are being explored as targets for immunization or passive immunotherapy.

3.4. HOST RESISTANCE

The severe nature of cryptosporidiosis in patients with AIDS has led to the hypothesis that cell-mediated immunity is more important than humoral immunity in the clearance of these organisms. This is supported by several observations. First, humoral immune responses to *C. parvum* in patients with AIDS are as strong as or stronger than those of immunocompetent people (Benhamou et al., 1995). Second, humoral responses are not associated with clearance of infection in mice (Taghi-Kilani et al., 1990). Third, mitogen-responsive peripheral blood mononuclear cells from patients with AIDS do not proliferate normally to cryptosporidial antigens (Gomez Morales et al., 1995). Finally, experimental mice become susceptible to cryptosporidial infection after depletion of functional T-helper lymphocytes or treatment with dexamethasone, an inhibitor of cell-mediated immune function (Chen et al., 1993a; Rehg, 1996). On the other hand, a role for humoral immunity in the resistance to *Cryptosporidium* is suggested by the severe disease seen in patients with congenital hypo- or agammaglobulinemias but normal T cell function (Lasser et al., 1979; Current et al., 1983).

Recent evidence suggests that the cell-mediated immune response to *C. parvum* depends in part on appropriate secretion of the TH1-derived cytokine interferon-γ (IFN-γ). First, severe combined immunodeficient (SCID) mice, which are normally relatively resistant to infection with *C. parvum,* develop fulminant disease if they are treated with a single dose of antibody to IFN-γ prior to infection (Chen, et al., 1993a; Chen et al, 1993b). These infections could be aborted after reconstitution with normal mouse splenocytes, but not if the mice were subsequently treated with IFN-γ. Second, peripheral blood mononuclear cells from previously infected mice (Harp et al., 1994) and humans (Gomez Morales et al., 1995) release high concentrations of IFN-γ after exposure to cryptosporidial antigens; the same response was not seen in naive donors. Third, IL-12 can protect mice from infection via induction of IFN-γ (Urban et al., 1996). Finally, in a recent case report of an HIV-negative child with chronic and ultimately fatal cryptosporidiosis, a thorough immunologic evaluation revealed no abnormalities other than the lack of IFN-γ production by lymphocytes in response to cryptosporidial antigens (Gomez Morales et al., 1996).

The role in IFN-γ in resistance to cryptosporidiosis is especially interesting given the unique effects of this cytokine on intestinal epithelial function. Brief exposure of T84 intestinal epithelial cells (a colon carcinoma cell line that differentiates in culture to resemble small intestinal epithelium) to IFN-γ causes loss of barrier integrity, reduced expression of ion channels, and loss of responsiveness to secretory stimuli (Adams et al., 1993; Colgan et al., 1994; Planchon et al., 1994). Caco-2 cells (a related intestinal epithelial cell line) lose some of their glutamine-coupled sodium transport after treatment with

IFN-γ (Souba and Copeland, 1992). These findings raise the possibility that some of the pathophysiology of cryptosporidiosis may be due to the epithelial effects of IFN-γ and other immune moderators produced in response to the infection. Further studies will be needed to explore this hypothesis.

4. REFERENCES

Abou-El-Magd, I., Soong, C. J. D., El-Hawey, A. M. and Ravdin, J. I. 1996. "Humoral and mucosal IgA antibody response to a recombinant 52-kDa cysteine-rich portion of the *Entamoeba histolytica* galactose-inhibitable lectin correlates with detection of native 170 kDa lectin antigen in serum of patients with amebic colitis," *J. Infect. Dis.* 174(1):157–162.

Acuna-Soto, R., Samuelson, J., De Girolami, P., Zarate, L., Millan-Velasco, F., Schoolnick, G. and Wirth, D. 1993. "Application of the polymerase chain reaction to the epidemiology of pathogenic and nonpathogenic *Entamoeba histolytica,*" *Am. J. Trop. Med. Hyg.* 48(1):58–70.

Adams, R. B., Guerrant, R. L., Zu, S., Fang, G., and Roche, J. K. 1994. "*Cryptosporidium parvum* infection of intestinal epithelium: morphologic and functional studies in an in vitro model," *J. Infect. Dis.* 169(1):170–177.

Adams, R. B., Planchon, S. M. and Roche, J. K. 1993. "IFN-gamma modulation of epithelial barrier function. Time course, reversibility, and site of cytokine binding," *J. Immunol.* 150(6):2356–2363.

Addiss, D., Pond, R., Remshak, M., Juranek, D., Stokes, S. and Davis, J. 1996. "Reduction of risk of watery diarrhea with point-of-use water filters during a massive outbreak of waterborne *Cryptosporidium* infection in Milwaukee, Wisconsin, 1993," *Am. J. Trop. Med. Hyg.* 54(6):549–553.

Alon, R. N., Bracha, R. and Mirelman, D. 1997. "Transfection of *Entamoeba dispar* inhibition of expression of the lysine-rich 30 kDa antigen by the transcription of its antisense RNA," *Arch. Med. Res.* 28:S52–55.

Anderson, B. 1985. "Moist heat inactivation of *Cryptosporidium* sp," *Am. J. Pub. Health* 75(12):1433–1434.

Andra, J. and Leippe, M. 1994. "Pore-forming peptide of *Entamoeba histolytica.* Significance of positively charged amino acid residues for its mode of action," *FEBS Lett.* 354(1):97–102.

Argenzio, R. A., Rhoads, J. M., Armstrong, M. and Gomez, G. 1994. "Glutamine stimulates prostaglandin-sensitive Na^+-H^+ exchange in experimental porcine cryptosporidiosis," *Gastroenterology* 106(6):1418–1428.

Armon, P. J. 1978. "Amoebiasis in pregnancy and the puerperium," *Brit. J. Ob. Gyn.* 85(4):264–269.

Barondes, S. H., Cooper, D. N. W., Gitt, M. A. and Leffler, H. 1994. "Galectins: structure and function of a large family of animal lectins," *J. Biol. Chem.* 269(33):20807–20810.

Benhamou, Y., Kapel, N., Hoang, C., Matta, H., Meillet, D., Magne, D., Raphael, M., Gentilini, M., Opolon, P. and Gobert, J. G. 1995. "Inefficacy of intestinal secretory immune response to *Cryptosporidium* in acquired immunodeficiency syndrome," *Gastroenterology* 108(3): 627–635.

Berninghausen, O. and Leippe, M. 1997. "Necrosis versus apoptosis as the mechanism of target cell death induced by *Entamoeba histolytica,*" *Infect. Immun.* 65(9):3615–3621.

Beving, D. E., Soong, C.-J. G. and Ravdin, J. I. 1996. "Oral immunization with a recombinant cysteine-rich section of the *Entamoeba histolytica* galactose-inhibitable lectin elicits an intestinal secretory immunoglobulin A response that has in vitro adherence inhibition activity," *Infect. Immun.* 64(4):1473–1476.

Black, E. K., Finch, G. R., Taghi-Kilani, R. and Belosevic, M. 1996. ''Comparison of assays for *Cryptosporidum parvum* oocysts viability after chemical disinfection,'' *FEMS Microbiol. Lett.* 135(2–3):187–189.

Blunt, D. S., Khramtsov, N. V., Upton, S. J. and Montelone, B. A. 1997. ''Molecular karyotype analysis of *Cryptosporidium parvum:* evidence for eight chromosomes and a low-molecular-size molecule,'' *Clin. Diagn. Lab. Immunol.* 4(1):11–13.

Blunt, D. S., Montelone, B. A., Upton, S. J. and Khramtsov, N. V. 1996. ''Sequence of the parasitic protozoan, *Cryptosporidium parvum,* putative protein disulfide isomerase-encoding DNA,'' *Gene* 181(1–2):221–223.

Bonafonte, M. T., Priest, J. W., Garmon, D., Arrowood, M. J. and Mead, J. R. 1997. ''Isolation of the gene coding for elongation factor-1alpha in *Cryptosporidium parvum,*'' *Biochim. Biophys. Acta* 1351(3):256–260.

Bracha, R. and Mirelman, D. 1983. ''Adherence and ingestion of *Escherichia coli* serotype 055 by trophozoites of *Entamoeba histolytica,*'' *Infect. Immun.* 40(3):882–887.

Bracha, R. and Mirelman, D. 1984. ''Virulence of *Entamoeba histolytica* trophozoites. Effects of bacteria, microaerobic conditions and metronidizole,'' *J. Exp. Med.* 160(2):353–386.

Braga, L. L., Ninomiya, H., McCoy, J. J., Eacker, S., Wiedmer, T., Pham, C., Wood, S., Sims, P. J. and Petri, W. A. Jr. 1992. ''Inhibition of the complement membrane attack complex by the galactose-specific adhesin of *Entamoeba histolytica,*'' *J. Clin. Invest.* 90(3):1131–1137.

Brandborg, L. Goldberg, S. and Briedenback, W. 1970. ''Human coccidiosis—a possible cause of malabsorption,'' *N. Engl. J. Med.* 283(24):1306–1313.

Brown, H. W. (1969). *Basic Clinical Parasitology.* New York: Appleton-Century-Crofts.

Bruchhaus, I., Jacobs, T., Leippe, M. and Tannich, E. 1996. ''*Entamoeba histolytica* and *Entamoeba dispar:* differences in numbers and expression of cysteine proteinase genes,'' *Mol. Microbiol.* 22(2):255–263.

Bruchhaus, I., Leippe, M., Lioutas, C. and Tannich, E. 1993. ''Unusual gene organization in the protozoan parasite *Entamoeba histolytica,*'' *DNA Cell Biol.* 12(2):925–933.

Brumpt, E. 1925. ''Etude sommaire de l'*Entamoeba dispar* n. sp. Amibe a kystes quadrinuclees, parasite de l'homme,'' *Bull. Acad. Med.* (Paris) 94:943–952.

Burchard, G. D. and Bilke, R. 1992. ''Adherence of pathogenic and non-pathogenic *Entamoeba histolytica* strains to neutrophils,'' *Parasitol. Res.* 78(4):146–153.

Burchard, G. D., Moslein, C. and Brattig, N. W. 1992. ''Adherence between *Entamoeba histolytica* trophozoites and undifferentiated or DMSO-induced HL-60 cells,'' *Parasitol. Res.* 78(2): 336–340.

Carrero, J. C., Diaz, M. Y., Viveros, M., Espinoza, B., Acosta, E. and Ortiz-Ortiz, L. 1994. ''Human secretory immunoglobulin A anti-*Entamoeba histolytica* antibodies inhibit adherence of amebae to MDCK cells,'' *Infect. Immun.* 62(2):764–767.

CDC. 1992. ''Epidemiological notes and reports pseudo-outbreak of intestinal amebiasis—California,'' *MMWR* 34(9):125–126.

CDC. 1995a. ''Assessing the public health threat associated with waterborne cryptosporidiosis: report of a workshop,'' *MMWR* 44(RR-6):1–19.

CDC. 1995b. ''Summary of notifiable disease, United States 1994,'' *MMWR* 43(53):1.

CDC. 1996. ''Foodborne outbreak of diarrheal illness associated with *Cryptosporidium parvum*—Minnesota, 1995,'' *MMWR* 45(36):783–784.

CDC. 1997. ''Outbreaks of *Escherichia coli* O157:H7 infection and cryptosporidiosis associated with drinking unpasteurized apple cider—Connecticut and New York, October 1996,'' *MMWR* 46(1):4–8.

Chadee, K., Johnson, M. L., Orozco, E., Petri, W. A., Jr. and Ravdin, J. I. 1988. ''Binding and internalization of rat colonic mucins by the galactose/N-acetyl-D-galactosamine adherence lectin of *Entamoeba histolytica*," *J. Infect. Dis.* 158(2):398–406.

Chadee, K. and Meerovitch, E. 1985. "*Entamoeba histolytica:* early progressive pathology in the cecum of the gerbil (*Meriones unguiculatus*)," *Am. J. Trop. Med. Hyg.* 34(2):283–291.

Chadee, K., Petri, W. A., Jr., Innes, D. J. and Ravdin, J. I. 1987. "Rat and human colonic mucins bind to and inhibit the adherence lectin of *Entamoeba histolytica*," *J. Clin. Invest.* 80(5):1245–1254.

Chen, W., Harp, J. A. and Harmsen, A. G. 1993a. "Requirements for CD4+ cells and gamma interferon in resolution of established *Cryptosporidium parvum* infection in mice," *Infect. Immun.* 61(9):3928–3932.

Chen, W., Harp, J. A., Harmsen, A. G., and Havell, E. A. 1993b. "Gamma interferon functions in resistance to *Cryptosporidium parvum* infection in severe combined immunodeficient mice," *Infect. Immun.* 61(8):3548–3551.

Cieslak, P. R., Virgin, H. W. I. and Stanley, S. L. Jr. 1992. "A severe combined immunodeficient (SCID) mouse model for infection with *Entamoeba histolytica*," *J. Exp. Med.* 176(6):1605–1609.

Clark, C. G., and Diamond, L. S. 1991. "Ribosomal RNA genes of 'pathogenic' and 'nonpathogenic' *Entamoeba histolytica* are distinct," *Mol. Biochem. Parasitol.* 49(2):97–302.

Colgan, S. P., Parkos, C. A., Matthews, J. B., D'Andrea, L., Awtrey, C. S., Lichtman, A. H., Delp-Archer, C. and Madara, J. L. 1994. "Interferon-gamma induces a cell surface phenotype switch on T84 intestinal epithelial cells," *Am. J. Physiol.* 267(2 Pt 1):C402–410.

Connor, B., Shlim, D., Scholes, J., Rayburn, J., Reidy, J. and Rajah, R. 1993. "Pathologic changes in the small bowel in nine patients with diarrhea associated with a coccidia-like body," *Ann. Intern. Med.* 119:377–382.

Current, W. L., Reese, N. C., Ernst, J. V., Bailey, W. S., Heyman, M. B. and Weinstein, W. M. 1983. "Human cryptosporidiosis in immunocompetent and immunodeficient persons. Studies of an outbreak and experimental transmission," *N. Engl. J. Med.* 308(21):1252–1257.

DeLeon, A. 1970. "Prognostico tardio en el absceso hepatica amebiano," *Arch. Invest. Med.* (Mex.) Suppl. 1:S205.

Denis, M. and Chadee, K. 1988. "In vivo and in vitro studies of macrophage functions in amebiasis," *Infect. Immun.* 56(6):3126–3131.

Diamond, L. S. and Clark, C. G. 1993. "A redescription of *Entamoeba histolytica* Schaudinn, 1903 (Emended Walker, 1911) separating it from *Entamoeba dispar* Brumpt, 1925," *J. Euk. Microbiol.* 40(3):340–344.

Dodson, J. M., Lenkowski, P. W. Jr., Eubanks, A. C., Jackson, T. F. G. H., Napodano J., Lyerly, D. M., Lockhart, L. A., Mann, B. J. and Petri, W. A. Jr. 1999. "Infection and immunity mediated by the carbohydrate recognition domain of the *Entamoeba histolytica* Gal/GalNAc lectin," *J. Infect. Dis.* 179(2):460–466.

Druckman, D. A. and Quinn, T. C. 1988. "*Entamoeba histolytica* infections in homosexual men," in *Amebiasis: Human Infection by* Entamoeba histolytica, ed., J. I. Ravdin. New York: John Wiley and Sons, pp. 563–575.

DuPont, H. L., Chappell, C. L., Sterling, C. R., Okhuysen, P. C., Rose, J. B. and Jakubowski, W. 1995. "The infectivity of *Cryptosporidium parvum* in healthy volunteers," *N. Eng. J. Med.* 332(13):855–859.

Edman, U., Meraz, M. A., Rausser, S., Agabian, N. and Meza, I. 1990. "Characterization of an immuno-dominant variable surface antigen from pathogenic and nonpathogenic *Entamoeba histolytica*," *J. Exp. Med.* 172(3):879–888.

Garfinkel, L. I., Giladi, M., Huber, M., Gitler, C., Mirelman, D., Revel, M. and Rozenblatt, S. 1988.

"DNA probes specific for *Entamoeba histolytica* possessing pathogenic and nonpathogenic zymodemes." *Infect. Immun.* 57(3):926–931.

Ghadirian, E. and Meerovitch, E. 1981a. "Effect of immunosuppression on the size and metastasis of amoebic liver abscesses in hamsters," *Parasite Immunol.* 3(4):329–338.

Ghadirian, E. and Meerovitch, E. 1981b. "Effect of splenectomy on the size of amoebic liver abscesses and metastatic foci in hamsters," *Infect. Immun.* 31(2):571–573.

Ghadirian, E., Meerovitch, E. and Kongshavn, P. 1983. "Role of macrophages in host defense against hepatic amoebiasis hamsters," *Infect. Immun.* 42(3):1017–1019.

Goldstein, S. T., Jurnaek, D. D., Ravenholt, O., Hightower, A. W., Martin, D. G., Mesnik, J. L., Giffiths, S. D., Bryant, A. J., Riech, R. R. and Herwalt, B. L. 1996. "Cryptosporidiosis: an outbreak associated with drinking water despite state-of-the-art water treatment [published erratum appear in *Ann. Intern. Med.* 125(2):1581]," *Ann. Intern. Med.* 124(5):459–468.

Gomez Morales, M., Ausiello, C. M., Guarino, A., Urbani F., Spagnuolo, M. I., Pignata, C. and Pozio, E. 1996. "Severe, protracted cryptosporidiosis associated with interferon-gamma deficient; pediatric case report," *Clin. Infect. Dis.* 22(5):848.

Gomez Morales, M. A., Ausiello, C. M., Urbani, F. and Pozio, E. 1995. "Crude extract and recombinant protein of *Cryptosporidium parvum* oocysts induce proliferation of human peripheral blood mononuclear cells in vitro," *J. Infect. Dis.* 172(1):211–216.

Gonzalez-Ruis, A., Haque, R., Castanon, G., Hall, A., Guhl, F., Ruiz-Palacios, G., Miles, M. and Warhurst, D. 1994. "Value of microscopy in the diagnosis of dysentery associated with invasive *Entamoeba histolytica,*" *J. Clin. Pathol.* 47(3):236–239.

Goodgame, R. W., Kimball, K., Ou, C. N., White, A. Jr., Genta, R. M., Lifschitz, C. H. and Chappell, C. L. 1995. "Intestinal function and injury in acquired immunodeficiency syndrome-related cryptosprodiosis," *Gastroenterology* 108(4):1075–1082.

Griffiths, J. K., Moore, R., Dooley, S., Keusch, G. T. and Tzipori, S. 1994. "*Cryptosporidium parvum* infection of caco-2 cell monolayers induces an apical monolayer defect, selectively increases transmonolayer permeability, and causes epithelial cell death," *Infect. Iummun.* 62(10):4506–4514.

Guarino, A., Canani, R., Pozio, E., Terracciano, L., Albano, F. and Mazzeo, M. 1994. "Enterotoxic effect of stool supernatant of *Cryptosporidium parvum*-infected calves on human jejunum," *Gastroenterology* 106(1):28–34.

Guarino, A., Canani, R. B., Casola, A., Pozio, E., Russo, R., Bruzzese, E., Fontana, M. and Rubino, A. 1995. "Human intestinal cryptosporidiosis: secretory diarrhea and enterotoxic activity in Caco-2 cells," *J. Infect. Dis.* 171(4):976–983.

Guerrant, R. L. (1994). "Principles and syndromes of enteric infection," *Mandell, Douglas, and Bennett's Principles and Practice of Infectious Diseases,* eds., G. L. Mandell, J. E. Bennett and R. Dolin. New York: Churchill Livingstone Inc. pp. 945–962.

Guerrant, R. L., Brush, J., Ravdin, J. I., Sullivan, J. A. and Mandell, G. L. 1981. "Interaction between *Entamoeba histolytica* and leukocytes," *J. Infect. Dis.* 143(1):83–93.

Gut, J. and Nelson, R. D. 1994. "*Cryptosporidium parvum* sporozoites deposit trails of 11A5 antigen during gliding locomotion and shed 11A5 antigen during invasion of MDCK cells in vitro," *J. Euk. Microbiol.* 41(5):42S.

Haas, C. and Rose, J. (1994). "Reconciliation of microbial risk models and outbreak epidemiology: the case of the Milwaukee outbreak," *Proceedings of the American Water Works Association, 1994.* Denver: American Water Works Association, pp. 517–523.

Hamann, L., Buß, H. and Tannich, E. 1997. "Tetracycline-controlled gene expression in *Entamoeba histolytica,*" *Mol. Biochem. Parasitol.* 84(1):83–91.

Hamann, L., Nickel, R. and Tannich, E. 1995. "Transfection and continuous expression of heterologous genes in the protozoan parasite *Entamoeba histolytica,*'' *Proc. Natl. Acad. Sci. USA* 92(19):8975–8979.

Haque, R., Faruque, A. S. G., Hahn, P., Lyerly, D. M. and Petri, W. A. Jr. 1997. ''*Entamoeba histolytica* and *Entamoeba dispar* infection in children in Bangladesh,'' *J. Infect. Dis.* 175(3):734–736.

Haque, R., Kress, K., Wood, S., Jackson, T. H. F. G., Lyerly, D., Wilkins, T. and Petri, W. A. Jr. 1993. ''Diagnosis of pathogenic *Entamoeba histolytica* infection using a stool ELISA based on monoclonal antibodies to the galactose-specific adhesin,'' *J. Infect. Dis.* 167(1):247–249.

Haque, R., Neville, L. M., Hahn, P. and Petri, W. A. Jr. 1995. ''Rapid diagnosis of *Entamoeba* infection using the *Entamoeba* and *Entamoeba histolytica* stool antigen detection kits,'' *J. Clin. Microbiol.* 33(10):2258–2261.

Harp, J. A., Fayer, R., Pesch, B. A. and Jackson, G. J. 1996. ''Effect of pasteurization on infectivity of *Cryptosporidium parvum* oocysts in water and milk,'' *Appl. Environ. Microbiol.* 62(8):2866–2868.

Harp, J. A., Whitmire, W. M. and Sacco, R. 1994. ''In vitro proliferation and production of gamma interferon by murine CD4+ cells in response to *Cryptosporidium parvum* antigen,'' *J. Parasitol.* 80(1):67–72.

Hart, J., Spirma, U. and Shattach, J. 1984. ''An outbreak of amebic infection in a kibbutz population,'' *Tran. Roy. Soc. Trop. Med.* 78(3):346–348.

Joe, A., Hamer, D. H., Kelley, M. A., Pereira, M. E., Keusch, G. T., Tzipori, S. and Ward, H. D. 1994. ''Role of a Gal/GalNAc-specific sporozoite surface lectin in *Cryptosporidium parvum*-host cell interaction,'' *J. Euk. Microbiol.* 41(5):44S.

Kandil, H. M., Berschneider, H. M. and Argenzio, R. A. 1994. ''Tumour necrosis factor alpha changes porcine intestinal ion transport through a paracrine mechanism involving prostaglandins,'' *Gut* 35(7):934–940.

Keene, W. E., Hidalgo, M. E., Orozco, E. and McKerrow, J. H. 1990. ''*Entamoeba histolytica:* correlation of the cytopathic effect of virulent trophozoites with secretion of cysteine protease,'' *Exp. Parasitol.* 71(2):199–206.

Keller, F., Walter, C., Lohden, U., Hanke, W., Bakker-Grunwald, T. and Trissl, D. 1988. ''Pathogenic and non-pathogenic *Entamoeba:* pore formation and hemolytic activity,'' *J. Protozol.* 35(3):359–365.

Kelsall, B. L., Jackson, T. G. F. H., Gathiram, V., Salig, S. B., Vaithilingum, M., Pearson, R. D. and Ravdin, J. I. 1994. ''Secretory immunoglobulin A antibodies to the galactose-inhibitable adherence protein in the saliva of patients with amebic liver disease,'' *Am. J. Trop. Med. Hyg.* 51(4):454–459.

Khramtsov, N. V., Blunt, D. S., Montelone, B. A. and Upton, S. J. 1996. ''The putative acetyl-CoA synthetase gene of *Crytptosporidium parvum* and a new conserved protein motif in acetyl-CoA synthetases,'' *J. Parasitol.* 82(3):423–427.

Korich, D., Mead, J., Madore, M., Sinclair, N. and Sterling, C. 1990. ''Effects of ozone, chlorine dioxide, chlorine, and monochloramine on *Cryptosporidium parvum* oocyst viability,'' *Appl. Environ. Microbiol.* 56(5):1432–1428.

Krogstad, D. J., Spencer, H. C., Healy, G. R., Gleason, N. N., Sexton, D. J. and Herron, C. A. 1978. ''Amebiasis: epidemiologic studies in the United States 1971–1974,'' *Ann. Intern. Med.* 88(1):89–97.

Kuhls, T. L., Mosier, D. A., Crawford, D. L. and Griffis, J. 1994. ''Seroprevalence of cryptosporidial antibodies during infancy, childhood, and adolescence,'' *Clin. Infect. Dis.* 18(5):731–735.

Laberge, I., Giffiths, M. W. 1996. ''Prevalence, detection and control of *Cryptosporidium parvum* in food,'' *Int. J. Food Microbiol.* 32(1–2):1–26.

Lasser, K. H., Lewin, K. J. and Ryning, F. W. 1979. "Cryptosporidial enteritis in a patient with congenital hypogammaglobulinemia," *Human Pathology* 10(2):234–240.

LeChevallier, M. W., Norton, W. D. and Lee, R. G. 1991a. "*Giardia* and *Cryptosporidium* spp. in filtered drinking water supplies," *Appl. Environ. Microbiol.* 57(9):2617–2621.

LeChevallier, M. W., Norton, W. D. and Lee, R. G. 1991b. "Occurrence of *Giardia* and *Cryptosporidium* spp. in surface water supplies," *Appl. Environ. Microbiol.* 57(9):2610–2616.

Leippe, M. 1995. "Ancient weapons: NK-lysin, is a mammalian homolog to pore-forming peptides of a protozoan parasite," *Cell* 83(1):17–18.

Leippe, M., Andra, J. and Muller-Eberhard, H. J. 1994a. "Cytolytic and antibacterial activity of synthetic peptides derived from amebapore, the pore-forming peptide of *Entamoeba histolytica*," *Proc. Natl. Acad. Sci. USA* 91(7):2602–2606.

Leippe, M., Andra, J., Nickel, R., Tannich, E. and Muller-Eberhard, H. J. 1994b. "Amoebapores, a family of membranolytic peptides from cytoplasmic granules of *Entamoeba histolytica:* isolation, primary structure, and pore formation in bacterial cytoplasmic membranes," *Mol. Microbiol.* 14(5):895–904.

Leippe, M., Bahr, E., Tannich, E. and Horstmann, R. D. 1993. "Comparison of pore-forming peptides from pathogenic and nonpathogenic *Entamoeba histolytica*," *Mol. Biochem. Parasitol.* 59(1):101–110.

Leippe, M. and Muller-Eberhard, H. J. 1994. "The pore-forming peptide of *Entamoeba histolytica*," *Toxicology* 87(1–3):5–18.

Leippe, M., Sieversten, E., Tannich, E., Muller-Eberhard, H. J. and Horstmann, R. D. 1995. "Spontaneous release of cysteine proteinases but not of pore-forming peptides by viable *Entamoeba histolytica*," *Parasitology* 111(part 5):569–574.

Leippe, M., Tannich, E., Nickel, R., van der Goot, G., Pattus, F., Horstmann, R. D. and Muller-Eberhard, H. J. 1992. "Primary and secondary structure of the pore-forming peptide of pathogenic *Entamoeba histolytica*," *EMBO J.* 11(10):3501–3506.

Li, E., Becker, A. and Stanley, S. L., Jr. 1988. "Use of Chinese hamster ovary cells with altered glycosylation patterns to define the carbohydrate specificity of *Entamoeba histolytica* adhesion," *J. Exp. Med.* 167(5):1725–1730.

Li, E., Becker, A. and Stanley, S. L Jr 1989. "Chinese hamster ovary cells deficient in N-acetelyglucosaminyltransferase I activity are resistant to *Entamoeba histolytica*-mediated cytotoxicity," *Infect. Immun.* 57(1):8–12.

Li, E., Yang, W.-G., Zhang, T. and Stanley, S. L. Jr. 1995. "Interaction of laminin with *Entamoeba histolytica* cysteine proteinases and its effect on amebic pathogenesis," *Infect. Immun.* 63(10):4150–4153.

Lima, A., Silva, T., Brage, L., Fang, G., Agnew, D., Newman, R., Wuhib, T., Sears, C., Nataro, J., Barrett, L., Mann, B., Petri, W. and Guerrant, R. 1996. *Etiologies, Pathogenesis and Impact of Persistent Diarrhea in Fortaleza.* Fifth Annual Meeting of International Centers for Tropical Disease Research. Bethesda, MD: National Institutes of Allergy and Infectious Diseases.

Lin, J.-Y. and Chadee, K. 1992. "Macrophage cytotoxicity against *Entamoeba histolytica* trophozoites is mediated by nitric oxide from L-arginine," *J. Immunol.* 148(12):3999–4005.

Lin, J.-Y., Keller, K. and Chadee, K. 1995. "Transforming growth factor-β1 primes macrophages for enhanced expression of the nitric oxide synthase gene for nitric oxide-dependent cytotoxicity against *Entamoeba histolytica*," *Immunology* 85(3):400–407.

Lotter, H., Zhang, T., Seydel, K. B., Stanley, S. L. Jr. and Tannich, E. 1997. "Identification of an epitope on the *Entamoeba histolytica* 170-kDa lectin conferring antibody-mediated protection against invasive amebiasis," *J. Exp. Med.* 185(10):1793–1801.

Lowin, B., Peitsch, M. C. and Tschopp, J. 1995. "Perforin and granzymes: crucial effector molecules in cytolytic T lymphocyte and natural killer cell-mediated cytotoxicity," *Curr. Topics. Microbial Immunol.* 198:1–20.

Luaces, A. L. and Barrett, A. J. 1988. "Affinity purification and biochemical characterization of histolysin, the major cysteine proteinase of *Entamoeba histolytica,*" *Biochem J.* 250(3):903–909.

Lushbaugh, W. B., Hofbauer, A. F. and Pittman, F. E. 1984. "Proteinase activities of *Entamoeba histolytica* cytotoxin," *Gastroenterology* 87(1):7–27.

Lynch, E. C., Rosenberg, I. M. and Gitler, C. 1982. "An ion-channel forming protein produced by *Entamoeba histolytica,*" *EMBO J.* 1(7):801–804.

MacKenzie, W. R., Hoxie, N. J., Proctor, M. E., Gradus, M. S., Blair, K. A., Peterson, D. E., Kazmierczak, J. J., Addiss, D. G., Fox, K. R., Rose, J. B. and Davis, J. P. 1994. "A massive outbreak in Milwaukee of cryptosporidium infection transmitted through the public water supply," *N. Engl. J. Med.* 331(3):161–167.

MacKenzie, W. R., Schell, W. L., Blair, K. A., Addiss, D. G., Peterson, D. E., Hoxie, N. J., Kazmierczak, J. J. and Davis, J. P. 1995. "Massive outbreak of waterborne cryptosporidium infection in Milwaukee, Wisconsin: recurrence of illness and risk of secondary transmission," *Clin. Infect. Dis.* 21(1):57–62.

Mann, B. J., Burkholder, B. V. and L. A., L. 1997. "Protection in a gerbil model of amebiasis by oral immunization with *Salmonella* expressing the galactose-N-acetyl D-galactosamine inhibitable lectin of *Entamoeba histolytica,*" *Vaccine* 15(6–7):659–663.

Mann, B. J., Chung, C. Y., Braga, L. L., M., D. J., Ashley, L. A. and Snodgrass, T. L. 1993. "Neutralizing monoclonal antibody epitopes of the *Entamoeba histolytica* galactose adhesin map to the cysteine-rich extracellular domain of the 170 kDa subunit," *Infect. Immun.* 61(5):1772–1778.

Mann, B. J., Torian, B. E., Vedvick, T. S. and Petri, W. A. Jr. 1991. "Sequence of the cysteine-rich heavy subunit of the galactose lectin of *Entamoeba histolytica,*" *Proc. Natl. Acad. Sci. USA* 88(8):3248–3252.

Markell, E. K. 1986. "The 1933 Chicago outbreak of amebiasis," *West. J. Med.* 144:750.

McCoy, J. J., Mann, B. J., Vedvick, T. S. and Petri, W. A., Jr. 1993a. "Sequence analysis of genes encoding the light subunit of the *Entamoeba histolytica* galactose-specific adhesin," *Mol. Biochem. Parasitol.* 61(2):325–328.

McCoy, J. J., Mann, B. J., Vedvick, T. S. and Petri, W. A. Jr. 1993b. "Structural analysis of the light subunit of the *Entamoeba histolytica* galactose-specific adherence lectin," *J. Biol. Chem.* 268(32):24223–24231.

Meisel, J., Perea, D., Melifro, C. and Rubin, C. E. 1976. "Overwhelming watery diarrhea associated with *Cryptosporidium* in an immunocompromised patient," *Gastroenterology* 70(6):1156–1160.

Mirelman, D. and Avron, B. 1988. "Cyst formation in *Entamoeba,*" in *Amebiasis: Human Infection by Entamoeba histolytica,* ed., J. I. Ravdin. New York City: John Wiley and Sons, pp. 768–781.

Mirelman, D., Nuchamowitz, Y. and Stolarsky, T. 1997. "Comparison of use of enzyme-linked immunosorbent assay-based kits and PCR amplification of rRNA genes for simultaneous detection of *Entamoeba histolytica* and *E. dispar,*" *J. Clin. Microbiol.* 35(9):2405–2407.

Modigliani, R., Bories, C., Le Charpentier, Y., Salmerson, M., Messing, B., Galian, A., Rambaud, J. C., Lavergne, A., Cochand-Priollet, B. and Desportes, I. 1985. "Diarrhoea and malabsorption in acquired immune deficiency syndrome: a study of four cases with special emphasis on opportunistic protozoan infestations," *Gut* 26(2):179–187.

Newman, R. D., Zu, S. X., Wuhib, T., Lima, A. A., Guerrant, R. L. and Sears, C. L. 1994.

"Household epidemiology of *Cryptosporidium parvum* infection in an urban community in northeast Brazil," *Ann. Int. Med.* 120(6):500–505.

Nickel, R. and Tannich, E. 1994. "Transfection and transient expression of chloramphenicol acetyltransferase gene in the protozoan parasite *Entamoeba histolytica,*" *Proc. Natl. Acad. Sci. USA* 91(15):7095–7098.

Nime, F. A., Burek, J. D., Page, D. L., Holscher, M. A. and Yardley, J. H. 1976. "Acute enterocolitis in a human being infected with the protozoan *Cryptosporidium,*" *Gastroenterology* 70(4):592–598.

Ohnishi, K., Murata, M. and Okuzawa, E. 1994. "Symptomatic amebic colitis in a Japanese homosexual AIDS patient," *Intern. Med.* 33(2):120–122.

Okhuysen, P. C., Chappell, C. L., Sterling, C. R., and Dupont, H. L. 1997. "Virulence of three distinct *C. parvum* isolates in healthy adults," *Abstracts of the IDSA 35th Annual Meeting,* San Francisco. Abstr. 396.

Ortiz-Ortiz, L., Zamacona, G., Sepulveda, B. and Capin, N. R. 1975. "Cell-mediated immunity in patients with amebic abscess of the liver," *Clin. Immunol. Immunopathol.* 4(1):127–134.

Pehrson, P. O. 1983. "Amoebiasis in a non-endemic country. Epidemiology, presenting symptoms and diagnostic methods," *Scand. J. Infect. Dis.* 15(2):207–214.

Petri, W. A. Jr., Chapman, M. D., Snodgrass, T. L., Mann, B. J., Broman, J. and Ravdin, J. I. 1989. "Subunit structure of the galactose and N-acetyl-D-galactosamine-inhibitable adherence lectin of *Entamoeba histolytica,*" *J. Biol. Chem.* 264(5):3007–3012.

Petri, W. A. Jr., Jackson, T. F. H. G., Gathiram, V., Kress, K., Saffer, L. D., Snodgrass, T. L., Chapman, M. D., Keren, Z. and Mirelman, D. 1990a. "Pathogenic and nonpathogenic strains of *Entamoeba histolytica* can be differentiated by monoclonal antibodies to the galactose-specific adherence lectin," *Infect. Immun.* 58(6):1802–1806.

Petri, W. A., Jr., Joyce, M. P., Broman, J., Smith, R. D., Murphy, C. F. and Ravdin, J. I. 1987. "Recognition of the galactose or N-acetylgalactosamine-binding lectin of *Entamoeba histolytica* by human immune sera," *Infect. Immun.* 55(10):2327–2331.

Petri, W. A., Jr. and Ravdin, J. I., eds. 1988. "Amebiasis in institutionalized populations," In *Amebiasis: Human Infection by* Entamoeba histolytica. New York: John Wiley and Sons, pp. 576–581.

Petri, W. A. Jr. and Ravdin, J. I. 1991. "Protection of gerbils from amebic liver abscess by immunization with the galactose-specific adherence lectin of *Entamoeba histolytica,*" *Infect. Immun.* 59(1):97–101.

Petri, W. A. Jr., Smith, R. D., Schlesinger, P. H., Murphy, C. F. and Ravdin, J. I. 1987. "Isolation of the galactose binding lectin of *Entamoeba histolytica,*" *J. Clin. Invest.* 80(5):1238–1244.

Petri, W. A. Jr., Snodgrass, T. L., Jackson, T. F. H. G., Gathiram, V., Simjee, A. E., Chadee, K. and Chapman, M. D. 1990b. "Monoclonal antibodies directed against the galactose-binding lectin of *Entamoeba histolytica* enhance adherence," *J. Immunol.* 144(12):4803–4809.

Petri, W. A., Jr., Haque, R., Kress, K., Wood, S. and Jackson, T. 1993. "Predictive value of tests for amebiasis," *J. Infect. Dis.* 168(2):514–515.

Planchon, S. M., Martins, C. A., Guerrant, R. L. and Roche, J. K. 1994. "Regulation of intestinal epithelial barrier function by TGF-beta 1. Evidence for its role in abrogating the effect of a T cell cytokine," *J. Immunol.* 153(12):5730–5739.

Purdy, J. E., Mann, B. J., Shugart, E. C. and Petri, W. A. Jr. 1993. "Analysis of the gene family encoding the *Entamoeba histolytica* galactose-specific adhesin 170 kDa subunit." *Mol. Biochem. Parasitol.* 62(1):53–60.

Purdy, J. E., Mann, B. J., Pho, L. T. and Petri, W. A. Jr. 1994. "Transient transfection of the enteric parasite *Entamoeba histolytica,*" *Proc. Natl. Acad. Sci. USA* 91(15):7099–7103.

Purdy, J. E., Pho, L. T., Mann, B. J. and Petri, W. A. Jr. 1996. "Upstream regulatory elements controlling expression of the *Entamoeba histolytica* lectin," *Mol. Biochem. Parasitol.* 78(1–2):91–103.

Quinn, T. C., Stamm, W. E., Goodell, S. E., Kritichian, E. M., Benedetti, J., Corey, L., Schuffler, M. and Holmes, K. K. 1983. "The polymicrobial origin of intestinal infections in homosexual men," *N. Eng. J. Med.* 309(10):576–582.

Ragland, B. D., Ashley, L. S., Vaux, D. L. and Petri, W. A. Jr. 1994. "*Entamoeba histolytica:* Target cells killed by trophozoites undergo apoptosis which is not blocked by bcl-2," *Exp. Parasitol.* 79(3):460–467.

Ramakrishnan, G., Ragland, B. D., Purdy, J. E. and Mann, B. J. 1996. "Physical mapping and expression of gene families encoding the N-acetyl D-galactosamine adherence lectin of *Entamoeba histolytica,*" *Mol. Microbiol.* 19(1):91–100.

Ramakrishnan, G., Vines, R. R., Mann, B. J. and Petri, W. A. Jr. 1997. "A tetracycline-inducible gene expression system in *Entamoeba histolytica,*" *Mol. Biochem. Parasitol.* 84(1):93–100.

Ravdin, J. I. and Guerrant, R. L. 1981. "Role of adherence in cytopathic mechanisms of *Entamoeba histolytica.* Study with mammalian tissue culture cells and human erythrocytes," *J. Clin. Invest.* 68(5):1305–1313.

Ravdin, J. I., Jackson, T. F. H. G., Petri, W. A. Jr., Murphy, C. F., Ungar, B. L. P., Gathiram, V., Skilogiannis, J. and Simjee, A. E. 1990. "Association of serum antibodies to adherence lectin with invasive amebiasis and asymptomatic infection with pathogenic *Entamoeba histolytica,*" *J. Infect. Dis.* 162(3):768–772.

Ravdin, J. I., John, J. E., Johnston, L. I., Innes, D. J. and Guerrant, R. L. 1985. "Adherence of *Entamoeba histolytica* trophozoites to rat and human colonic mucosa," *Infect. Immun.* 48(2):292–297.

Ravdin, J. I., Moreau, F., Sullivan, J. A., Petri, W. A. Jr., P. W. and Mandell, G. L. 1988. "Relationship of free calcium to the cytolytic activity of *Entamoeba histolytica.*" *Infect. Immun.* 56(6):1505–1512.

Ravdin, J. I., Sperelakis, N. and Guerrant, R. L. 1982. "Effect of ion channel inhibitors on the cytopathogenicity of *Entamoeba histolytica,*" *J. Infect. Dis.* 154(3):27–32.

Ravdin, J. I., Stanley, P., Murphy, C. F. and Petri, W. A. Jr. 1989. "Characterization of cell surface carbohydrate receptors for *Entamoeba histolytica* adherence lectin," *Infect. Immun.* 57(7):2179–2186.

Reed, S. L., Bouvier, J., Pollack, A. S., Engel, J. C., Brown, M., Hirata, K., Que, X., Eakin, A., Hagbloom, O., Gillin, F. and McKerrow, J. H. 1993. "Cloning of a virulence factor of *Entamoeba histolytica,*" *J. Clin. Invest.* 91(4):1532–1540.

Reed, S. L., Curd, J. G., Gigli, I., Gillin, F. D. and Braude, A. I. 1986. "Activation of complement by pathogenic and nonpathogenic strains of *Entamoeba histolytica,*" *J. Immunol.* 136(6):2265–2269.

Reed, S. L. and Gigli, I. 1991. "Lysis of complement-sensitive *Entamoeba histolytica* by activated terminal complement components. Initiation of complement activation by an extracellular neutral cysteine proteinase," *J. Clin. Invest.* 86(8):1815–1822.

Reed, S. L., Keene, W. E. and McKerrow, J. H. 1989. "Thiol proteinase expression and pathogenicity of *Entamoeba histolytica,*" *J. Clin. Microbiol.* 27(12):2772–2777.

Rehg, J. 1996. "Effect of interferon-γ in experimental *Cryptosporidium parvum* infection," *J. Infect. Dis.* 174(1):229–232.

Riggs, M. W., Stone, A. L., Yount, P. A., Langer, R. C., Arrowood, M. J. and Bentley, D. L. 1997. "Protective monoclonal antibody defines a circumsporozoite-like glycoprotein exoantigen of *Cryptosporidium parvum* sporozoites and merozoites," *J. Immunol.* 158(4):1787–1795.

Rini, J. M. 1995. ''Lectin structure,'' *Annu. Rev. Biophys. Biomol. Struct.* 24:551–577.

Saffer, L. D. and Petri, W. A. Jr. 1991. ''Role of the galactose lectin of *Entamoeba histolytica* in adherence-dependent killing of mammalian cells,'' *Infect. Immun.* 59(12):4681–4683.

Salata, R. A., Martinez-Palomo, A., Murray, H. W., Canales, L., Trevino, N., Segonia, E., Murphy, C. F. and Ravdin, J. I. 1986. ''Patients treated for amebic liver abscess develop cell-mediated immune responses effective in vitro against *Entamoeba histolytica,*'' *J. Immunol.* 136(7):2633–2639.

Salata, R. A., Martinez-Palomo, A., Canales, L., Murray, H. W., Trevino, N. and Ravdin, J. I. 1990. ''Suppression of T-lymphocyte responses to *Entamoeba histolytica* antigen by immune sera,'' *Infect. Immun.* 58(12):3941–3946.

Salata, R. A., Murray, H. W., Rubin, B. Y. and Ravdin, J. I. 1987. ''The role of gamma interferon in the generation of human macrophages cytotoxic for *Entamoeba histolytica* trophozoites,'' *Am. J. Trop. Med. Hyg.* 37(1):72–78.

Salata, R. A., Pearson, R. P., Murphy, C. F. and Ravdin, J. I. 1985. ''Interaction of human leukocytes with *Entamoeba histolytica:* killing of virulent amebae by the activated macrophage,'' *J. Clin. Invest.* 76(2):491–499.

Sargeaunt, P. G. 1987. ''The reliability of *Entamoeba histolytica* zymodemes in clinical diagnosis,'' *Parasitology Today* 3:40–43.

Sargeaunt, P. G., Jackson, T. F. H. G. and Simjee, A. 1982. ''Biochemical homogeneity of *Entamoeba histolytica* isolates, especially those from liver abscess,'' *Lancet* 1(8286):1386–1390.

Sargeaunt, P. G., Williams, J. E. and Grene, J. D. 1978. ''The differentiation of invasive and non-invasive *Entamoeba histolytica* by isoenzyme electrophoresis,'' *Trans. R. Soc. Trop. Med. Hyg.* 72(4):519–521.

Schain, D. C., Salata, R. A. and Ravdin, J. I. 1992. ''Human T-lymphocyte proliferation, lymphokine production, and amebicidal activity elicited by the galactose-inhibitable adherence protein of *Entamoeba histolytica,*'' *Infect. Immun.* 60(5):2143–2146.

Schmerin, M. J., Gelston, A. and Jones, T. C. 1977. ''Amebiasis, an increasing problem among homosexuals in New York,'' *JAMA* 238:1386–1387.

Schulte, E. and Scholze, J. 1989. ''Action of the major protease from *Entamoeba histolytica* on proteins of the extracellular matrix,'' *J. Protozool.* 36(6):538–543.

Sears, C. L. and Guerrant, R. L. 1994. ''Cryptosporidiosis: the complexity of intestinal pathophysiology,'' *Gastroenterology* 106(1):252–254.

Seguin, R., Mann, B. J., Keller, K. and Chadee, K. 1996. ''The tumor necrosis factor alpha-stimulating region of galactose-inhibitable lectin of *Entamoeba histolytica* activates gamma interferon-primed macrophages for amebicidal activity mediated by nitric oxide,'' *Infect. Immun.* 65(7):2522–2527.

Singh, U., Rogers, J. B., Mann, B. J. and Petri, W. A. Jr. 1997. ''Transcription initiation is controlled by three core promoter elements in the *hgl5* gene of the protozoan parasite *Entamoeba histolytica,*'' *Proc. Natl. Acad. Sci.* 94(16):8812–8817.

Soong, C.-J., G., Kain, K. C., Abd-Alla, M., Jackson, T. F. H. G. and Ravdin, J. I. 1995. ''A recombinant cysteine-rich section of the *Entamoeba histolytica* galactose-inhibitable lectin is efficacious as a subunit vaccine in the gerbil model of amebic liver abscess,'' *J. Infect. Dis.* 171(3):645–651.

Souba, W. W. and Copeland, E. M. 1992. ''Cytokine modulation of Na(+)-dependent glutamine transport across the brush border membrane of monolayers of human intestinal Caco-2 cells,'' *Ann. Surg.* 215(5):536–544.

Spano, F., Puri, C., Ranucci, L., Putignani, L. and Crisanti, A. 1997. ''Cloning of the entire

COWP gene of *Cryptosporidium parvum* and ultrastructural localization of the protein during sexual parasite development,'' *Parasitology* 114(Pt 5):427–437.

Stanley, S. L. Jr., Becker, A., Kunz-Jenkins, C., Foster, L. and Li, E. 1990. ''Cloning and expression of a membrane antigen of *Entamoeba histolytica* possessing multiple tandem repeats,'' *Proc. Natl. Acad. Sci. USA* 87(13):4976–4980.

Stanley, S. L. Jr., Zhang, T., Rubin, D. and Li, E. 1995. ''Role of the *Entamoeba histolytica* cysteine proteinase in amebic liver abscess formation in severe combined immunodeficient mice,'' *Infect. Immun.* 63(4):1587–1590.

Steiner, T. S. and Guerrant, R. L. 1998. ''Cryptosporidiosis,'' in *Tropical Infectious Diseases,* eds., R. L. Guerrant, D. H. Walker and P. F. Weller. New York: Churchhill Livingstone.

Steiner, T. S., Thielman, N. M. and Guerrant, R. L. 1997. ''Protozoal agents: what are the dangers for the public water supply?'' *Ann. Rev. Med.* 48:329–340.

Tachibana, H., Kobayashi, S., Kato, Y., Nagakura, K., Kaneda, Y. and Takeuchi, T. 1990. ''Identification of a pathogenic isolate-specific 30,000 M_r antigen of *Entamoeba histolytica* by using a monoclonal antibody,'' *Infect. Immun.* 58(4):955–960.

Tachibana, H., Kobayashi, S., Okuzawa, E. and Masuda, G. 1992. ''Detection of pathogenic *Entamoeba histolytica* DNA in liver abscess fluid by polymerase chain reaction,'' *Int. J. Parasitol.* 22(8):1193–1196.

Taghi-Kilani, R., Sekla, L. and Hayglass, K. T. 1990. ''The role of humoral immunity in *Cryptosporidium* spp. infection. Studies with B cell-depleted mice,'' *J. Immunol.* 145(5):1571–1576.

Takeuchi, T., Okuzawa, E. and Nozaki, T. 1989. ''High seropositivity of Japanese homosexual men for amebic infection,'' *J. Infect. Dis.* 159(4):808.

Tannich, E. and Burchard, G. D. 1991. ''Differentiation of pathogenic from nonpathogenic *Entamoeba histolytica* by restriction fragment analysis of a single gene amplified in vitro,'' *J. Clin. Microbiol.* 29(2):250–255.

Tannich, E., Ebert, F. and Horstmann, R. D. 1991. ''Primary structure of the 170-kDa surface lectin of pathogenic *Entamoeba histolytica,*'' *Proc. Natl. Acad. Sci. USA* 88(5):1849–1853.

Tannich, E., Ebert, F. and Horstmann, R. D. 1992. ''Molecular cloning of cDNA and genomic sequences coding for the 35-kildalton subunit of the galactose-inhibitable lectin of pathogenic *Entamoeba histolytica,*'' *Mol. Biochem. Parasitol.* 55(1–2):225–228.

Tannich, E., Horstmann, R. D., Knobloch, J. and Arnold, H. H. 1989. ''Genomic DNA differences between pathogenic and nonpathogenic *Entamoeba histolytica,*'' *Proc. Natl. Acad. Sci. USA* 86(13):5118–5122.

Tannich, E., Scholze, H., Nickel, R. and Horstmann, R. D. 1991. ''Homologous cysteine proteinase of pathogenic and nonpathogenic *Entamoeba histolytica:* differences in structure and expression,'' *J. Biol. Chem.* 266(8):4798–4803.

Urban, J. F. Jr., Fayer, R., Chen, S. J., Gause, W. C., Gately, M. K. and Finkelman, F. D. 1996. ''Il-12 protects neonatal mice against infection with *Cryptosporidium parvum,*'' *J. Immunol.* 156(1):263–268.

Vasquez, J. R., Gooze, L., Kim, K., Gut, J., Petersen, C. and Nelson, R. G. 1996. ''Potential antifolate resistance determinants and genotypic variation in the bifunctional dihydrofolate reductase-thymidylate synthase gene from human and bovine isolates of *Cryptosporidium parvum,*'' *Mol. Biochem. Parasitol.* 79(2):153–165.

Vines, R. R., Purdy, J. E., Ragland, B. D., Samuelson, J., Mann, B. J. and Petri, W. A. Jr. 1995. ''Stable transfection of *Entamoeba histolytica,*'' *Mol. Biochem. Parasitol.* 71(2):265–267.

Walsh, J. A. 1986. ''Problems in recognition and diagnosis of amebiasis: estimation of the global magnitude of morbidity and mortality,'' *Rev. Infect. Dis.* 8:228–238.

Walsh, J. A. 1988. ''Transmission of *Entamoeba histolytica,*'' in *Amebiasis: Human Infection by* Entamoeba histolytical, ed., J. I. Ravdin. New York: John Wiley and Sons. 106–119.

Weikel, C. S., Murphy, C. F., Orozco, E. and Ravdin, J. I. 1988. "Phorbol esters specifically enhance the cytolytic activity of *Entamoeba histolytica,*" *Infect. Immun.* 56(6):1485–1491.

WHO. 1995. *The World Health Report 1995: Bridging the Gaps.* Report of the Director-General., World Health Organizations, Geneva, Switzerland.

WHO/PAHO/UNESCO. 1997. *Report of a Consultation of Experts on Amoebiasis.* Mexico City, Mexico.

Young, J. D.-E., Young, T. M., Lu, L. P., Unkeless, J. C. and Cohn, Z. A. 1982. "Characterization of a membrane pore-forming protein from *Entamoeba histolytica,*" *J. Exp. Med.* 156(6):1677–1690.

Zhang, T., Cieslak, P. R., Foster, L., Kunz-Jenkins, C. and Stanley, S. L. Jr. 1994. "Antibodies to the serine rich *Entamoeba histolytica* protein (SREHP) prevent amoebic liver abscess in severe combined immunodeficient (SCID) mice," *Parasite Immunol.* 16:225–230.

Zhang, T. and Stanley, S. L. Jr. 1994. "Protection of gerbils from amebic liver abscess by immunization with a recombinant protein derived from the 170-kilodalton surface adhesin of *Entamoeba histolytica,*" *Infect. Immun.* 62(5):2605–2608.

Zhang, T. and Stanley, S. L. Jr. 1996. "Oral immunization with an attenuated vaccine strain of *Salmonella typhimurium* expressing the serine-rich *Entamoeba histolytica* protein induces an antiamebic immune response and protects gerbils from amebic liver abscess," *Infect. Immun.* 64(4):1526–1531.

Zu, S. X., Li, J. F., Barrett, L. J., Fayer, R., Shu, S. Y., McAuliffe, J. F., Roche, J. K. and Guerrant, R. L. 1994. "Seroepidemiologic study of *Cryptosporidium* infection in children from rural communities of Anhui, China and Fortaleza, Brazil," *Am. J. Trop. Med. Hyg.* 51(1):1–10.

PART V

VIRUSES AND RELATED INFECTIOUS AGENTS

CHAPTER 15

Norwalk and Other Human Caliciviruses: Molecular Characterization, Epidemiology, and Pathogenesis

KELLOGG J. SCHWAB
MARY K. ESTES
ROBERT L. ATMAR

1. IMPACT OF FOODBORNE DISEASE CAUSED BY NORWALK VIRUS AND HUMAN CALICIVIRUSES

NORWALK virus (NV), the prototype human calicivirus (HuCV), and related HuCVs are the most widely recognized agents of outbreaks of foodborne and waterborne viral gastroenteritis. Early studies estimated that at least 42% of outbreaks of nonbacterial gastroenteritis in the U.S. were caused by such infections (Kaplan et al., 1982a). Currently the CDC estimates that greater than 96% of reported outbreaks of nonbacterial gastroenteritis characterized by nausea, vomiting, diarrhea, and an illness lasting 1–3 days are caused by HuCVs (Fankhauser et al., 1998). Similarly, in the Netherlands, a group of HuCVs, previously called small round structured viruses (SRSVs), were detected in 87% of all outbreaks of gastroenteritis reported to national epidemiologists in 1996, confirming the etiologic significance of SRSVs in outbreaks of gastroenteritis (Vinje et al., 1997). For this review, we will refer to NV and related viruses as human caliciviruses (HuCVs). This terminology will not distinguish HuCVs that can be separated into different genetic groups (see below).

1.1. OUTBREAKS ASSOCIATED WITH FOOD AND WATER

With an ever-increasing population placing greater demands on limited resources, there has been a marked increase in exposure to HuCVs in food and water, resulting in outbreaks of gastroenteritis. Outbreaks have occurred in recreational camps, communities, hospitals, schools, nursing homes, and families, as well as on naval vessels and cruise ships. Outbreaks occur year-round and affect all age groups. More recent analyses have suggested outbreak prevalence may be more common in winter months (Vinje et al., 1997). Outbreaks of foodborne disease have been associated with consumption of uncooked and cooked shellfish, ice, bakery products (frosting), various types of salads (potato, chicken, fruit, tossed), cold foods (celery, melon, vermicelli consommé, sandwiches, and cold cooked ham), and hot foods such as hamburgers and french fries (Metcalf et al., 1995; Hedberg and Osterholm, 1993; Wanke and Guerrant, 1987) (Table 15.1). Outbreaks of waterborne viral gastroenteritis have been associated with ice, water supplies on cruise ships, private wells, small water systems, community water systems, and groundwater contamination (Kaplan et al., 1982b; Lawson et al., 1991; Taylor et al., 1981; Wilson et al., 1982; Beller et al., 1997; Payment et al., 1994; Khan et al., 1994; McAnulty et al., 1993; Cannon et al., 1991; Levine et al., 1990; Ramia, 1985). In addition, there have been several outbreaks associated with recreational waters (Gray et al., 1997; Baron et al., 1982; Kappus et al., 1982; Koopman et al., 1982).

Consumption of raw or cooked shellfish has resulted in numerous documented outbreaks of HuCV-associated disease (McDonnell et al., 1997; Kirkland et al., 1996; Le Guyader et al., 1996b; Sugieda et al., 1996; Lees et al., 1995; Ando et al., 1995a; Chalmers and McMillan, 1995; Kohn et al., 1995; Parker and Cubitt, 1994; Truman et al., 1987; Morse et al., 1986; Dowell et al., 1995; Murphy et al., 1979). Some reports have concluded that cooking (steaming, grilling, stewing, frying or baking) bivalve mollusks before eating them did not prevent gastroenteritis illness (McDonnell et al., 1997; Kirkland et al., 1996; Truman et al., 1987). Resistance to heat inactivation of HuCVs during documented outbreaks strengthens the experimental data suggesting that NV is very resistant to environmental degradation.

HuCVs also have been implicated in outbreaks of gastroenteritis caused by ill or asymptomatic infected food handlers who contaminate food while preparing salads, cold food items, and frosted confectionery items (Patterson et al., 1997; Kilgore et al., 1996; Lo et al., 1994; Heun et al., 1987; White et al., 1986; Lieb et al., 1985; Griffin et al., 1982; Kuritsky et al., 1984; Sekla et al., 1989; Gross et al., 1989; Fleissner et al., 1989; Reid et al., 1988; Curry et al., 1987; Iversen et al., 1987; Gordon et al., 1990; Brondum et al., 1985; Herwaldt et al., 1994; Patterson et al., 1993; Daniels et al., 1998). One outbreak in a school cafeteria was suspected to be caused by hot foods (french fries and hamburgers) served by infected food handlers (Guest et al., 1987).

HuCV outbreaks have also occurred when food (celery) was washed with contaminated water (Warner et al., 1991), or prepared in a sink contaminated one day earlier with vomit from an ill kitchen assistant (Patterson et al., 1997). The celery outbreak involved over 1500 cadets and staff at the U.S. Air Force Academy. No ill food handler was identified but celery prepared for chicken

TABLE 15.1. Representative Subset of Water or Foodborne Outbreaks Caused by Human Caliciviruses.

Vehicle of Transmission	# of People Affected	Attack Rate	Associated Cause of Outbreak and Other Comments	Reference
Water				
Drinking water	305	87%	Leaking sewage pipe contaminated drinking water	McAnulty et al., 1993
Drinking water	1500	—	Cross-connection contamination of water supply	Kaplan et al., 1982b
Ice	202	30%	Ice on a cruise ship	Khan et al., 1994
Ice	191	31%	Commercial ice distributed to many locations	Cannon et al., 1991
Shellfish				
Oysters	129	58%	67% of infected ate only cooked oysters, outbreak caused by overboard disposal of feces into the oyster bed	McDonnell et al., 1997
Oysters	27	56%	Steaming oysters did not prevent illness	Kirkland et al., 1996
Oysters	70	83%	Outbreak caused by overboard disposal of feces into the oyster bed	Kohn et al., 1995
Clams and oysters	1017	—	103 outbreaks in NY in 8 months	Morse et al., 1986
Oysters	2000	—	Large outbreak in Australia	Murphy et al., 1979
Salads				
Potato salad	55	50%	Vomit in sink subsequently cleaned with bleach, contaminated sink still source of outbreak	Patterson et al., 1997
Turkey salad	195	—	Pre-symptomatic food handler	Lo et al., 1994
Green salad	188	—	University cafeteria outbreak	Kilgore et al., 1996
Other foods				
Corgnation chicken	67	—	Post-symptomatic food handler	Patterson et al., 1993
Hamburgers and french fries	129	34%	Hot food implicated as potential vectors	Guest et al., 1987
Cake frosting	129 (3000 estimated)	52%	Large dispersal of NV from one food handler	Kuritsky et al., 1984

salad had been soaked for one hour in water obtained from a hose used earlier in the day to clear clogged drains in the kitchen after sewage had backed up. It was postulated that viruses were rinsed from the contaminated hose into a container containing soaking celery and subsequently adsorbed by the celery. The outbreak originating from the contaminated sink is of interest because the sink was cleaned of vomit with a chlorine-based disinfectant and used the next morning to prepare a potato salad, subsequently identified as the vehicle of infection. This outbreak supports the importance of vomiting in the transmission of HuCV infection and highlights HuCVs' relative resistance to disinfection and decontamination. Other reports have linked vomiting with airborne or fomite transmission of viruses to humans in outbreaks of gastroenteritis caused by HuCVs (McEvoy et al., 1996; Ho et al., 1989; Sawyer et al., 1988; Caul, 1994; Cheesbrough et al., 1997).

1.2. ECONOMIC IMPACT OF INFECTIONS WITH HuCVS

It is difficult to assess the true economic impact of HuCV infections due to the mild and self-limiting nature of the disease and the continued lack of complete epidemiological studies. Gastroenteritis is not a reportable disease and few individual illnesses are reported to health officials. Usually only epidemic outbreaks, and subsequent smaller outbreaks due to increased public awareness and media attention, are documented. However, the loss of productivity (employee absenteeism and care of dependents) can have a major economic impact based on family studies that indicate, on average, each family member experiences more than one enteric illness of unknown etiology per year (Kapikian et al., 1996). The shellfish industry has been greatly affected in recent years by decreased public confidence in the microbiological quality of bivalve mollusks. Frequent HuCV outbreaks caused by consumption of raw as well as cooked shellfish have impacted profits in this multimillion dollar per year industry. For example, an economic analysis of less than half the outbreaks that occurred in New York in 1982 estimated that the market loss to the seafood industry was $1,840,000 over a five-month period and a conservative estimate of the costs of investigations and medical care was $630,000 (Brown and Folsom, 1983).

1.3. WIDESPREAD OCCURRENCE OF INFECTIONS WITH HuCVs

Increasing numbers of studies document the common occurrence of infections with HuCVs. NV antibody prevalence was initially analyzed in relatively large studies using reagents from volunteers and a RIA or immune adherence hemagglutionation assay (Kapikian et al., 1978; Blacklow et al., 1979; Greenberg et al., 1979). These studies showed antibody to NV is acquired gradually, being present with low frequency during childhood and with increasing frequency in adult years, so that >50% of adults possess antibody to NV by age 50 (Kapikian et al., 1978). More recent studies in the United Kingdom, Japan, and Sweden using recombinant NV (rNV) capsid antigens have reported antibodies to the NV antigen in most adults (89–98%) (Gray et al., 1993; Numata et al., 1994; Hinkula et al., 1995). The higher detection rate for antibody probably reflects the greater sensitivity of newer antibody assays (Green et al., 1993). In developing countries, antibody prevalence has been high even in infants and young children, suggesting infections at an early age (Dimitrov et al., 1997; Smit et al., 1997).

2. NORWALK VIRUS AND OTHER HUMAN CALICIVIRUSES

Norwalk virus (NV) was first discovered 30 years ago following an outbreak of epidemic gastroenteritis in a Norwalk, Ohio, elementary school in 1968. Well water at the school was suspected of being the vehicle of infection and there was a high secondary attack rate (Adler and Zickl, 1969). The small (27 nm), icosahedral NV particles isolated from a stool sample were first observed by immune electron microscopy (IEM) in 1972 (Kapikian et al., 1972). Between 1968 and 1990, many similar viruses were described. However, because none of these agents could be cultivated, it was difficult to compare them directly. During this period, agents were named by the location of the outbreak. As these agents were compared using the electron microscope, they also were divided into two morphological groups: small round structured viruses and viruses with classical calicivirus morphology (Figure 15.1).

A new era began in 1990 when the genome of NV was cloned, and was

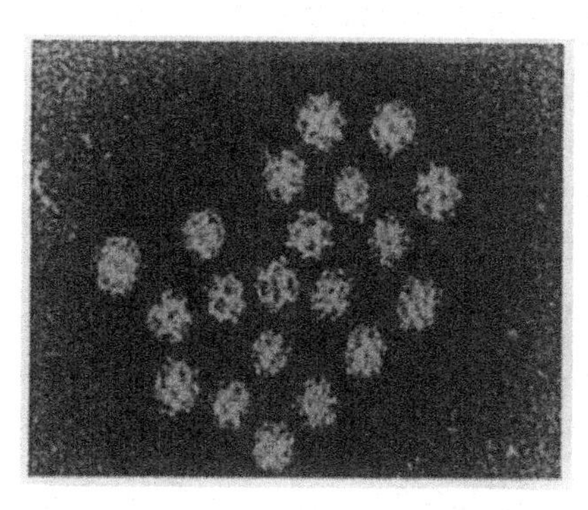

A

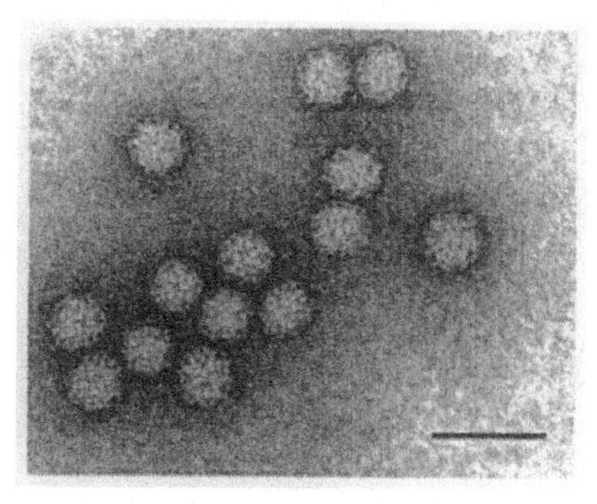

B

Figure 15.1 Electron micrographs of human caliciviruses. These micrographs illustrate the morphology of viruses with typical calicivirus structure. (A:HuCV Sapporo) and viruses with less defined structure (B: Norwalk virus). Bar in photo is 50 nm. Norwalk virus also has been called a small round structured virus. This chapter uses the term human calciviruses to refer to these viruses. Either type of virus can cause foodborne illness, although most studies have detected Norwalk-like HuCVs.

classified as an RNA virus belonging to the family *Caliciviridae* (Jiang et al., 1993; Jiang et al., 1990; Matsui et al., 1991). Knowledge of genomic information of NV was critical because this and related viruses remain noncultivable in cell cultures or in any animal model. A number of recent reviews describe the impact of and new information about the molecular characteristics of these viruses (Estes and Hardy, 1995; Estes et al., 1997; Clarke and Lambden, 1997; Caul, 1996a,b). Currently, over 100 HuCVs including the prototype viruses [NV, Snow Mountain agent (SMA), Hawaii virus (HV), and human calicivirus Sapporo (SV)], some virus strains previously characterized antigenically using a solid-phase immunosorbent method as different types of SRSVs (denoted UK types 1–4) (Norcott et al., 1994; Ando et al., 1994; Lewis, 1990) and viruses obtained from a variety of outbreaks have been at least partially sequenced. Two genogroups (I, II) and two genera, the Norwalk-like viruses (NLVs) and Sapporo-like viruses (SLVs), of HuCVs are now recognized based on the analysis of nucleotide or predicted amino acid sequences from part of the polymerase region, the entire capsid region, or open reading frame 3 (ORF3), of the genomes of these viruses (Figure 15.2) (Green et al., 1999). Within a genogroup, the amino acid identity between two different strains is >82% for the polymerase region and >65% for the capsid region. Sequence

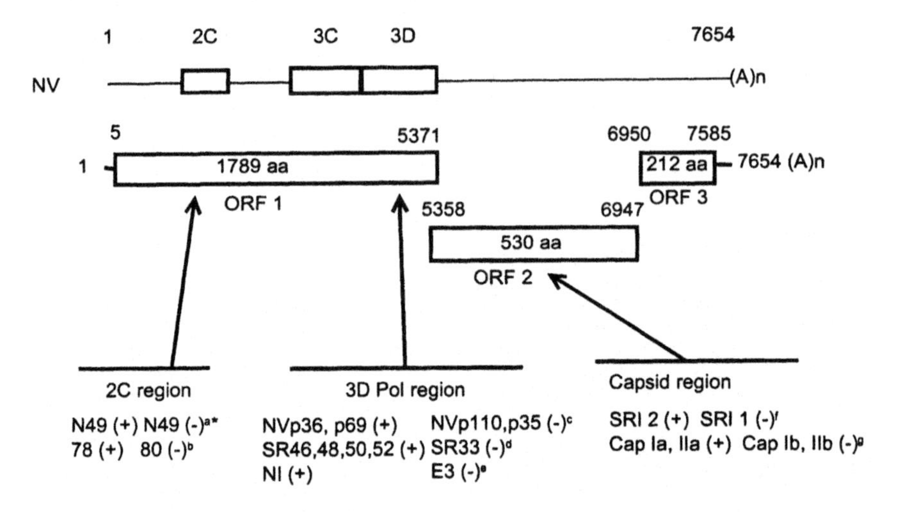

Figure 15.2 Schematic of NV genome and locations of RT-PCR primers. NV genome has 3 ORFs: ORF 1 contains regions with homology to picornavirus 2C helicase, 3C protease, and 3D RNA dependent, RNA polymerase; ORF2 encodes a single capsid protein; ORF3 function is unknown. RT-PCR primers have amplified regions of the 3D polymerase most frequently, although the predicted 2C helicase and capsid regions also have been targeted. Primer name, (polarity), references; [a]Matsui et al., 1991; [b]Wang et al., 1994; [c]Le Guyader et al., 1996a; [d]Ando et al., 1995b; [e]Green et al., 1995; [f]Hafliger et al., 1997; and [g]Green et al., 1997.

analyses of viruses from multiple outbreaks and from different geographic locations have confirmed that all HuCVs fall into these three genogroups. A striking result from the analysis of multiple viruses is that even within the RNA polymerase region, there is a high degree of genomic variation, and genogrouping has not always agreed with previous solid-phase IEM antigenic typing.

There has been considerable variation in the recent literature about how to name the different HuCVs. Recently, viruses in genogroup I and II have been put into a single genus and viruses previously called genogroup III viruses have been put into a separate genus within the *Caliciviridae* family; such taxonomic decisions were made by the International Committee on the Taxonomy of Viruses. Although the existence of the SLVs is evident based on studies with the prototype HuCV Sapporo and Manchester SLVs, some other caliciviruses with classical morphology have been shown to be in genogroup I and genogroup II (Cubitt et al., 1994; Kjeldsberg, 1977; Wang et al., 1994). Caliciviruses with typical morphology have been identified in both genera highlighting the fact that morphology should no longer be used for classification of these viruses.

2.1. VIRION MORPHOLOGY AND BIOLOGICAL PROPERTIES

Original descriptions of NV morphology were based on electron microscopic analysis of antibody-aggregated particles obtained from stools (Kapikian et al., 1972). NV particles had some surface structure and in some micrographs they looked like classical caliciviruses (Figure 15.1). In initial studies, NLVs were generally observed in outbreaks of disease affecting adults while the viruses with classical calicivirus structure were associated with disease in young children. Recently, cloned DNA (cDNA) encoding the capsid protein of NV has been inserted into a baculovirus vector. Following infection of insect cells with this recombinant baculovirus, recombinant Norwalk virus (rNV) particles are produced (Jiang et al., 1992b) These virus-like particles (VLPs) are identical to native NV in both structure and antigenicity, the only difference being that VLPs lack nucleic acid (Green et al., 1993; Hardy et al., 1996). The structure of NV has been discovered by studying the easily produced rNV VLPs. By negative stain electron microscopy (EM), rNV exhibit T = 3 icosahedral symmetry (Prasad et al., 1994). The single-capsid protein folds into 90 dimers that form a shell domain from which arch-like capsomers protrude. These arches give NV cup-like structures similar to typical caliciviruses. A number of HuCV VLPs have been produced recently, including rMexico virus (rMxV) (Jiang et al., 1995a), rHV (Green et al., 1997), rSMA (Hardy et al., 1997), rSV (Numata et al., 1997), rDesert Shield virus (rDSV) (Lew et al., 1994), rLordsdale virus (rLV) (Dingle et al., 1995), and grimsby virus (rGV) (Hale et al., 1999). These VLPs are proving useful for serodiagnosis and as candidate vaccines (Ball et al., 1998; Noel et al., 1997). A single type of VLP can detect seroresponses in adults who have had infections with a different virus type, but homologous or closely related VLP antigens have been required to achieve the highest test sensitivity in the measurement of seroresponses (Treanor et al., 1993; Noel et al., 1997; Jiang et al., 1995a).

2.2. STABILITY OF HuCVs

The lack of a cell culture or animal infectivity model has limited studies on the stability of HuCVs to environmental degradation. Human volunteer studies have shown that NV is resistant to inactivation following treatment with chlorine levels frequently found in drinking water, and NV is more resistant to inactivation by chlorine than poliovirus 1, human rotavirus (Wa), simian rotavirus (SA11), or f2 bacteriophage (Keswick et al., 1985). NV retained infectivity for volunteers following (1) exposure to pH 2.7 for 3 hours at room temperature, (2) treatment with 20% ether at 4°C for 18 hours, or (3) incubation at 60°C for 30 minutes (Dolin et al., 1972). Recently, researchers have attempted to correlate NV nucleic acid detection with infectivity following disinfection with chlorine dioxide (ClO_2) by incorporating antibody capture of NV particles prior to RT-PCR detection (Shin and Sobsey, 1997). These studies indicate that NV is as resistant as poliovirus type 3 and bacteriophage MS2 to inactivation by ClO_2.

2.3. GENOME ORGANIZATION

The NV genome is a positive-sense polyadenylated single-stranded RNA of 7654 nucleotides, excluding the 3′ polyadenylated tail; this genome is predicted to encode three ORFs (Figure 15.2). ORF1, 1789 amino acids in length, encodes a polyprotein precursor to nonstructural proteins based on identification of sequences similar to the picornavirus 2C helicase, 3C protease, and 3D RNA-dependent RNA polymerase. The 26 bases at the 5′ end of the NV genome are 88.4% identical to sequences located at the start of the capsid coding sequence in ORF2, supporting the presence of a promoter sequence for synthesis of a subgenomic RNA encoding the capsid protein of NV. Proteins expressed from ORF1 are immunoreactive (Jiang et al., 1993; Matsui et al., 1991), indicating that infected individuals make antibodies to proteins other than the capsid protein. ORF2 encodes a single capsid protein of 530 amino acids with a calculated molecular weight of 57,000 (57K). This protein contains a conserved amino acid motif of PPG, which is also found in the picornavirus capsid protein VP3 (Palmenberg, 1989). ORF3, 212 amino acids in length, at the 3′ end of the genome is predicted to encode a small protein with an unknown function, but it may be involved in nucleic acid binding due to having a very basic charge predicted (isoelectric point of 10.99).

Knowledge of the sequence of a number of the HuCVs has been important in permitting the development of sensitive methods to detect the viral genome based on using RT-PCR (Jiang et al., 1992a; Ando et al., 1995b; Le Guyader et al., 1996a; Green et al., 1995; Wang et al., 1994) (see below).

3. HOW CAN THE HuCVs BE DETECTED?

3.1. IMMUNE ELECTRON MICROSCOPY

Immune electron microscopy (IEM) is the ''gold standard'' that has been used to detect HuCVs in stools for the past 25 years (Kapikian et al., 1972). However, the concentration of HuCVs present in stool samples and foodstuffs is low, and the presence of other small, round, nonviral objects in these samples limits the diagnostic value of electron microscopy. For example, detection of viruses by IEM requires approximately 10^5 virus particles per mL, which probably explains the failure of this method to detect viruses in more than 50% of stool samples collected within the first 72 hours of Norwalk virus-induced diarrhea illness (Thornhill et al., 1975). The need for suitable reference antisera (needed to aggregate viruses by IEM) also limits the usefulness of IEM. In addition, IEM requires expensive equipment and maintenance as well as highly trained electron microscopists.

3.2. ANTIGEN DETECTION

The initial antigen detection methods were radioimmunoassays (RIAs), and later, enzyme immunoassays (EIAs). These initial assays utilized acute and convalescent sera from infected individuals as capture and detector antibodies (Greenberg et al., 1978; Gary, Jr. et al., 1985; Herrmann et al., 1985). These methods had improved sensitivity compared to IEM (Greenberg et al., 1978), probably due to the detection of both particulate and soluble antigens present in stool samples. The need for reference reagents limited the use of these tests to research laboratories.

As noted above, the cloning of Norwalk virus and expression of its capsid protein yielded VLPs that morphologically and antigenically are indistinguishable from native particles (Jiang et al., 1992b; Green et al., 1993). Hyperimmune sera to Norwalk VLPs were produced and used in EIAs. Subsequently, the VLPs of several other HuCV strains (Mexico, Snow Mountain Agent, Sapporo, Hawaii, Desert Shield, and Lordsdale) were expressed and used to produce hyperimmune sera (Jiang et al., 1995a; Green et al., 1997; Hardy et al., 1997; Numata et al., 1997; Lew et al., 1994; Dingle et al., 1995). EIAs using these new reagents have significantly improved the sensitivity of detecting these viruses over older methods using human reagents (Graham et al., 1994). However, the EIAs using hyperimmune animal serum are relatively type-specific, detecting only those strains most closely related to the strain used to produce the serum (Graham et al., 1994; Jiang et al., 1995b; Jiang et al., 1992b; Hale et al., 1996). The type specificity of the hyperimmune antisera contrasts with the broader reactivity seen using human convalescent sera. The presence of cross-reactive antibodies in human sera is thought to result from past infections with several distinct HuCV strains. The specificity of the EIAs that use animal hyperimmune sera limits the usefulness of individual assays for the detection of most HuCV infections. The production of polyclonal and monoclonal antibodies that react with HuCVs in both genera should lead to the development of more broadly reactive EIAs in the near future (Hardy et al., 1996; Herrmann et al., 1995). However, it remains unclear if the sensitivity of even broadly reactive EIAs will be sufficient for virus detection in food and water samples. More likely, such immunologic reagents will be most useful if applied to concentrate and purify low levels of virus away from inhibitors or other food products and then low amounts of the viral genome amplified by RT-PCR.

3.3. RT-PCR DETECTION

RT-PCR assays are the most sensitive diagnostic methods currently available for the detection of HuCVs. Primer pairs to amplify a number of different regions in the viral genome have been used diagnostically, although most primer pairs amplify a region of the putative RNA-dependent, RNA polymerase (Figure 15.2) (Le Guyader et al., 1996a; Ando et al., 1995b; Green et al., 1995). Unfortunately, the genetic diversity of the HuCVs has prevented the development of a universal primer pair that can amplify all such viruses. Approaches to overcome the diversity of HuCVs have included the use of a degenerate oligonucleotide, inosine containing oligonucleotides, and multiplex (use of multiple primers in the reaction mix) PCR (Le Guyader et al., 1996a; Green et al., 1995; Ando et al., 1995b). Because nonspecific amplicons that are the same approximate size as virus-specific amplicons may be generated from clinical and environmental samples, a confirmatory test should be performed following the RT-PCR reaction. Hybridization assays (dot/slot or Southern blot hybridization) are the most common confirmatory assays and have the additional advantage of increasing the sensitivity of virus detection by 10–100 fold over that obtained with agarose gel electrophoresis alone. However, the genetic diversity of the HuCVs limits the usefulness of hybridization assays. The use of as many as 10 different oligoprobes was not sufficient to confirm all PCR results obtained using a panel of different HuCV strains (Le Guyader et al., 1996a). Direct sequencing of PCR products is an alternate, although more labor-intensive, approach for confirmation of PCR results.

Stool samples and foodstuffs contain a variety of substances that may inhibit reverse transcription or PCR reactions. When working with stool and food samples, the use of appropriate controls is essential to detect any inhibition. The addition of low concentrations of an internal standard RNA control allows the detection of inhibitory substances without compromising the sensitivity of the RT-PCR assay (Atmar et al., 1995; Schwab et al., 1997). False positive results due to carryover contamination may occur due to the exquisite sensitivity of RT-PCR assays; procedures such as physical separation of pre- and post-PCR working areas, dedicated reagents and equipment for pre-PCR sample processing, and appropriate controls for each PCR reaction to prevent such occurrences should be followed (Kwok and Higuchi, 1989).

3.4. SEROLOGY

Measurement of serum antibody to HuCVs is another method that can be used to detect HuCV infection. For most assays, the antibody levels in acute and convalescent sera are compared, and a fourfold or greater increase in antibody level is used to define recent infection. Many of the early assays used to detect viral antigen were modified to detect serum antibody, including the ability of sera to immunoprecipitate virus as measured by IEM or to block reactivity of antigen in RIAs and EIAs (Kapikian et al., 1972; Greenberg et al., 1978; Gary, Jr. et al., 1985; Herrmann et al., 1985; Nakata et al., 1988). More recently, HuCV VLPs produced in baculovirus expression systems have been used to coat the wells of ELISA plates in assays that measure serum antibody. The antibody detection EIAs are more cross-reactive than the antigen detection assays, although the assays are most sensitive when the VLPs coating the ELISA plates are closely related to the infecting virus strain. Assays to measure IgM, IgA, and IgG subclass antibody are being developed. A recent study that developed a new assay to detect IgM using a single early serum sample indicates that this will be useful to assess recent infections (Brinker et al., 1998).

3.5. APPLICATION OF METHODS

A general schematic that provides an overview of methods to determine if HuCVs cause an outbreak is depicted in Figure 15.3. Outbreaks potentially caused by viruses (18–48 hr incubation, rapid onset of diarrhea and/or vomiting, fever, 2–3 days of clinical illness) can be examined for HuCVs as well as other pathogens (bacteria, protozoa, and other viruses). Samples of stool and serum from individuals and implicated water or food should be collected as soon as possible and processed for HuCV detection. Guidelines for data and sample collection have been published (CDC, 1995). Serum samples can be analyzed by EIA using reagents developed using VLPs, and it is hoped such tests will become commercially available. If acute and convalescent pairs

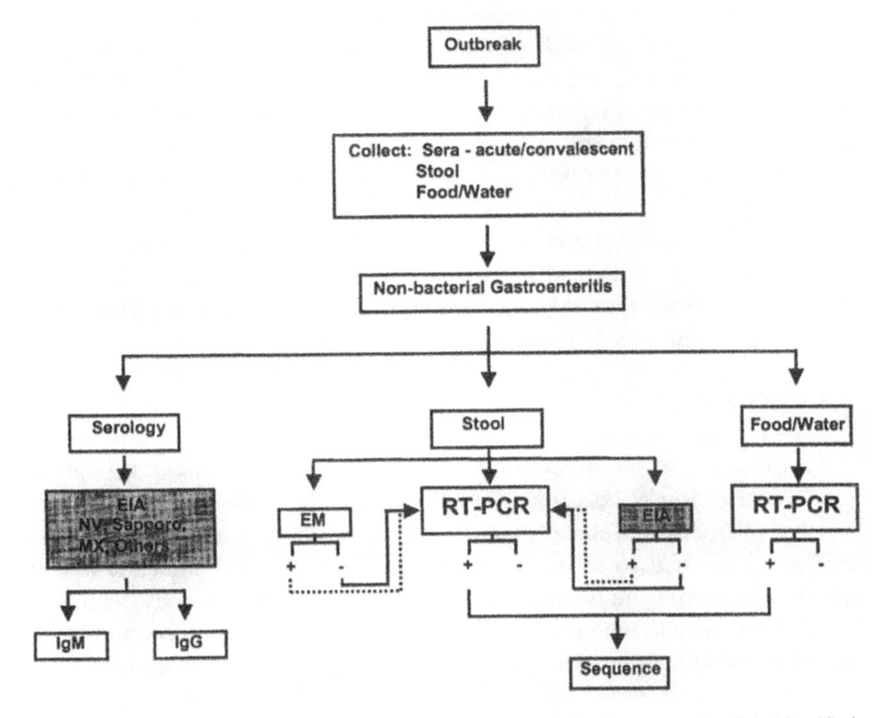

Figure 15.3 Evaluation of a gastroenteritis outbreak for HuCVs. Once an outbreak is identified, acute and convalescent (at least two weeks after acute) sera, stool samples and implicated food/water samples should be collected. The stool can be examined for virus by any of 3 methods: EM, EIA (shaded because available only in research labs), and RT-PCR. If EM or EIA are positive, no further evaluation is needed, but the virus can be further characterized by RT-PCR (dashed line). If EM or EIA are negative, RT-PCR should be performed. The food/water source also can be examined by RT-PCR. If RT-PCR is positive in both stool and food/water, the amplicons can be sequenced and compared to document the similarity between the samples. Serology can be performed by EIA (shaded because only available in research labs) using a number of different antigens. IgM tests can be performed on single sera, but paired sera are needed to evaluate changes in IgG levels.

of serum samples can be obtained, with the initial acute sample taken within 3 to 4 days post-infection, IgG assays using different VLPs can be used. Alternatively, IgM assays may be useful to detect that an outbreak is caused by a recent infection with HuCVs. IgM has the advantage of only requiring one serum sample but further research on preexisting IgM levels and cross-reactivity needs to be done. The EIA assays shown in Figure 15.3 are shaded to indicate that this detection technique is still being developed. The main diagnostic focus with HuCV outbreak investigations has been the analysis of stool samples. As with sera analysis, testing of stool by EIA has not been completely evaluated (see shaded box, Figure 15.3). Samples found positive by EIA may need no further examination. A few specialized laboratories have successfully used EM on stool samples to characterize and classify SRSV outbreaks. A positive IEM sample may need no further examination. However, negative EM and IEM results, due to the lack of sensitivity of EM analysis and the narrow specificity of current antigen EIAs, may represent false negative results. Therefore, EM or EIA negative samples should probably be examined by RT-PCR. IEM or EIA positive samples also may be characterized further using RT-PCR (dashed lines in Figure 15.3). RT-PCR is the current detection method of choice for detection of HuCVs. RT-PCR is more sensitive than EM or EIA, and most labs are able to perform RT-PCR analysis, enabling many outbreaks to be investigated. Food or water implicated in an outbreak also can be examined by RT-PCR following sample purification and concentration. Any positive RT-PCR samples from stool or food/water can be sequenced to further characterize what strains of HuCV are causing outbreaks (Figure 15.3). Sequence information may be important to establish a direct link between a suspected vehicle of transmission and the outbreak. Sequence data may also clearly establish routes of transmission.

The methods described above have been used to detect HuCVs from shellfish (Lees et al., 1995; Atmar et al., 1995), food (Gouvea et al., 1994; Schwab et al., 1999), stool (Jiang et al., 1992a; Schwab et al., 1997; Ando et al., 1995b), and the environment (Green et al., 1998; Cheesbrough et al., 1997). More recently, HuCVs have been detected in other foods associated with outbreak's of gastroenteritis (Daniels et al., 1998; Gaulin et al., 1999).

4. MECHANISMS OF PATHOGENESIS

4.1. HUMAN CHALLENGE MODEL

Information on the mechanisms of pathogenesis for HuCVs is limited because of the lack of cell culture or animal infectivity models. What information is known has come from volunteer studies in which adults were challenged with NV, Montgomery County agent, SMA, and HV. In some studies, proximal intestinal biopsies were taken (Wyatt et al., 1974; Dolin et al., 1982; Agus et al., 1973; Schreiber et al., 1974; Dolin et al., 1975; Widerlite et al., 1975). Histologic changes were seen in jejunal biopsies from ill volunteers. Symptomatic illness was correlated with a broadening and blunting of the intestinal villi, crypt cell hyperplasia, cytoplasmic vacuolization, and infiltration of polymorphonuclear and mononuclear cells into the lamina propria (Widerlite et al., 1975). The extent of small intestinal involvement remains unknown because studies have only examined the proximal small intestine, and the site of virus replication remains unknown.

Initial studies in volunteers on the duration of NV shedding using IEM detection concluded that shedding occurred in about 50% of ill volunteers, lasted approximately 72 hours and was undetectable after 100 h post-challenge (Thomhill et al., 1975). Subsequent volunteer studies using the more sensitive methods of EIA and RT-PCR revealed that while 50% of volunteers became ill, another 30% had asymptomatic infections (Graham et al., 1994). Shedding of viral antigen occurred in >90% of ill volunteers, in most asymptomatic individuals, and was detectable up to two weeks post-challenge (Okhuysen et al., 1995; Graham et al., 1994). The exact length of shedding remains unknown because stools collected later than two weeks after infection were not available. However, these data lend support to reports of outbreaks caused by post-symptomatic food handlers and of a food handler who reportedly excreted virus for over three weeks (Patterson et al., 1993; White et al., 1986; Iversen et al., 1987).

Studies of viral immunity have been limited, and have primarily monitored whether resistance to clinical infection or illness can be correlated with preinfection antibody status of volunteers administered NV, SMA, or HV, and of individuals involved in outbreaks (Parrino et al., 1977; Wyatt et al., 1974; Graham et al., 1994). A recent study of volunteers that used the new sensitive assays observed that about 50% of inoculated volunteers became ill, but a larger number (82%) of volunteers were infected, with many of these infections being asymptomatic (Graham et al., 1994). The proportion of subjects infected was similar for those with and without preexisting antibody, although uninfected individuals were more likely to have lower preexisting antibody titers. This unusual result remains unexplained but it has been hypothesized that some individuals may be genetically resistant to infection, possibly because they lack a receptor or other cellular factor needed for infection with these viruses. Individuals lacking antibody are not all resistant to infection or illness, with a 60% seroconversion rate seen in volunteers lacking preexisting serum antibody. Therefore, a lack of detectable serum antibody does not correlate absolutely with protection from infection, probably reflecting type specificity of protective antibody or absence of specific T cell responses.

Volunteer studies have shown that short-term homologous immunity develops, since volunteers who became ill following an initial NV challenge failed to become ill on rechallenge 6 to 14 weeks later with the same agent (Parrino et al., 1977). Johnson and co-workers (Johnson et al., 1990) confirmed that preexisting serum antibody to NV is not associated with protective immunity, but antibody levels become associated with protection after repeated exposure to a single antigenic type of HuCV. Short-term resistance can last greater than or equal to 6 months after challenge and a small percentage of resistant individuals maintain low antibody titers even after multiple challenges. It must be emphasized that these studies have all measured ELISA binding and not neutralizing antibodies. Correlates of protection may become clear once neutralizing antibody can be monitored. In children, serum antibody titers measured by a blocking EIA have been correlated with protection from reinfection (Nakata et al. 1985).

4.2. HOST RESPONSE

The hallmark of infection with NV and related HuCVs is the acute onset of vomiting or diarrhea, or both. No prodrome is seen, and the spectrum of illness may vary widely in individual patients. The most common symptoms with clinical illness are nausea, malaise, and abdominal cramps. Diarrhea, which is usually watery, occurs in many patients, and vomiting is seen in most (Dolin et al., 1976). The illness is generally mild and self-limited, with symptoms lasting 12 to 48 hours, and illnesses caused by the different HuCVs are clinically indistinguishable. Seroconversion occurs following infection, with IgM being detectable as early as 4–7 days post-challenge and IgG seroconversion appearing by day 9.

4.3 CULTIVATION OF NV IN TISSUE CULTURE

The HuCVs remain refractory to cultivation in tissue culture and in animal models. A variety of cell lines (including both human and animal cells) can specifically bind rNV particles, suggesting that host-range specificity may not occur at the level of cell binding (White et al., 1996). The human intestinal cell line Caco-2 has shown the highest binding efficiency, but internalization of particles into these cells is low. More work is needed on trying to understand what regulates NV replication strategy.

5. SUMMARY OF NEW AREAS OF RESEARCH

The advent of new diagnostic tools and illness criteria for detecting HuCV outbreaks has enabled epidemiologists to clearly demonstrate that HuCVs are a major cause of foodborne diseases. Further optimization of new diagnostic methods is needed to allow the development of broadly reactive, rapid, automated methods for detection of viruses in water and stool as well as rapid detection of infections with virus based on early antibody assays. Once rapid, sensitive, automated tests are commercially available, other issues will appear. A better understanding of modes of transmission will be needed to better direct public health policy. Outbreaks have been documented to originate from both presymptomatic (Lo et al., 1994) and postsymptomatic (Patterson et al., 1993; White et al., 1986) food handlers. Large, multistate outbreaks have occurred when HuCV-infected shellfish harvesters have dumped human waste overboard into shellfish harvesting waters with subsequent contamination of the shellfish (McDonnell et al., 1997; Kohn et al., 1995). These reports clearly point to the need for education of food handlers and shellfish harvesters with respect to proper hygiene as well as for better enforcement of proper sanitation handling on boats. Efforts should also focus on easier rapid detection of HuCVs. This effort could be enhanced by developing cross-reactive antibodies and improved RT-PCR primers to aid in HuCV detection. A major breakthrough will be achieved if conditions to cultivate these viruses can be identified. This will permit needed studies on how to inactivate these viruses to prevent transmission and to facilitate effective disinfection of contaminated areas. If the virus remains noncultivable, high-resolution structures of these viruses and screening of compounds to enhance virus diassembly or block virus assembly may provide new methods of disinfection. Finally, further studies on the mechanisms of pathogenesis and correlates of immunity may lead to new methods for disease treatment. Such information may come from the testing of available candidate VLP vaccines that are highly immunogenic in animals and have been safe in preliminary phase 1 trials (Ball et al., 1997, 1999). Finally, as the population continues to grow, new methods to ensure the safety of drinking water may be needed. VLPs are also showing promise as a tool to assess filtration properties of viruses through soils and natural groundwater disinfection (Redman et al., 1997).

6. ACKNOWLEDGMENTS

The authors gratefully acknowledge support for research on the HuCVs from the Environmental Protection Agency, the National Atmospheric and Oceanic Organization, and the National Institutes of Health. Kellogg J. Schwab is supported by training grant T32 AI07471 from the National Institutes of Health.

7. REFERENCES

Adler, J. L. and Zickl, R. 1969. ''Winter vomiting disease,'' *J. Infect Dis.* 119:668–673.

Agus, S. G., Dolin, R., Wyatt, R. G., Tousimis, A. J., and Northrup, R. S. 1973. ''Acute infectious nonbacterial gastroenteritis: intestinal histopathology. Histologic and enzymatic alterations during illness produced by the Norwalk agent in man,'' *Ann. Intern. Med.* 79:18–25.

Ando, T., Monroe, S. S., Gentsch, J. R., Jin, Q., Lewis, D. C., and Glass, R. I. 1995b. ''Detection and differentiation of antigenically distinct small round-structured viruses (Norwalk-like viruses) by reverse transcription-PCR and Southern hybridization,'' *J. Clin. Microbiol,* 33:64–71.

Ando, T., Mulders, M. N., Lewis, D. C., Estes, M. K., Monroe, S. S., and Glass, R. I. 1994. ''Comparison of the polymerase region of small round structured virus strains previously characterized in three serotypes by solid-phase immune electron microscopy,'' *Arch. Virol.* 135:217–226.

Ando, T., Jin, Q., Gentsch, J. R., Monroe, S. S., Noel, J. S., Dowell, S. F., Cicirello, H. G., Kohn, M. A., and Glass, R. I. 1995a. ''Epidemiologic applications of novel molecular methods to detect and differentiate small round structured viruses (Norwalk-like viruses),'' *J. Med. Virol.* 47:145–152.

Atmar, R. L., Neill, F. H., Romalde, J. L., Le Guyader, F., Woodley, C. M., Metcalf, T. G., and

Estes, M. K. 1995. ''Detection of Norwalk virus and hepatitis A virus in shellfish tissues with the PCR,'' *Appl. Environ. Microbiol.* 61:3014–3018.

Ball, J. M., Hardy, M. E., Atmar, R. L., Conner, M. E., and Estes, M. K. 1998. ''Oral immunization with recombinant Norwalk virus-like particles induces a systemic and mucosal immune response in mice,'' *J. Virol.* 72:1345–1353.

Ball, J. M., Graham, D. Y., Opekun, A. R., Gilger, M. A., Guerrero, R. A., and Estes, M. K. 1999. ''Recombinant Norwalk virus-like particles given orally to volunteers: phase 1 study,'' Gastroenterology 116:1–9.

Baron, R. C., Murphy, F. D., Greenberg, H. B., Davis, C. E., Bregman, D. J. Gary, G. W., Hughes, J. M., and Schonberger, L. B. 1982. ''Norwalk gastrointestinal illness: an outbreak associated with swimming in a recreational lake and secondary person-to-person transmission,'' *Am. J. Epidemiol.* 115:163–172.

Beller, M., Ellis, A., Lee, S. H., Drebot, M. A., Jenkerson, S. A., Funk, E. Sobsey, M. D., Simmons, O. D. III, Monroe, S. S., Ando, T., Noel, J., Petric, M., Middaugh, J. P., and Spika, J. S. 1997, ''Outbreak of viral gastroenteritis due to a contaminated well. International consequences,'' *JAMA* 278:563–568.

Blacklow, N. R., Cukor, G., Bedigian, M. K., Echeverria, P., Greenberg, H. B., Schreiber, D. S., and Trier, J. S. 1979. ''Immune response and prevalence of antibody to Norwalk enteritis virus as determined by radioimmunoassay,'' *J. Clin. Microbiol.* 10:903–909.

Brinker, J. P., Blacklow, N. R. Estes, M. K. Moe, C. L. Schwab, K. J. and Herrmann, J. E. 1998. ''Detection of Norwalk virus infections and other genogroup 1 human calciviruses by a monoclonal antibody, recombinant-antigen-based immunoglobulin M capture enzyme immunoassay.'' *J. Clin. Microbiol.* 36:1064–1069.

Brondum, J., Spitalny, K. C., Vogt, R. L., Godlewski, K., Madore, H. P., and Dolin, R. 1985. ''Snow Mountain agent associated with an outbreak of gastroenteritis in Vermont,'' *J. Infect. Dis.* 152:834–837.

Brown, J. W. and Folsom, W. D. 1983. ''Economic impact of hard clam associated outbreaks of gastroenteritis in New York State,'' *NMFS-SEFC* 121:1–44.

Cannon, R. O., Poliner, J. R., Hirschhorn, R. B., Rodeheaver, D. C., Silverman, P. R., Brown, E. A., Talbot, G. H., Stine, S. E., Monroe, S. S., Dennis, D. T. et al. 1991. ''A multistate outbreak of Norwalk virus gastroenteritis associated with consumption of commercial ice [published erratum appears in *J. Infect Dis* 1992 Sep;166(3):698],'' *J. Infect. Dis.* 164:860–863.

Caul, E. O. 1994. ''Small round structured viruses: airborne transmission and hospital control,'' *Lancet* 343:1240–1241.

Caul, E. O. 1996a. ''Viral gastroenteritis: small round structured viruses, caliciviruses and astroviruses. Part I. The clinical and diagnostic perspective,'' *J. Clin. Pathol.* 49:874–880.

Caul, E. O. 1996b. ''Viral gastroenteritis small round structured viruses, calciviruses and astroviruses. Part II. The epidemiological perspective,'' *J. Clin. Pathol.* 49:959–964.

CDC. 1995. ''Surveillance for foodborne-disease outbreaks—United States, 1988–1992,'' *MMWR Surveillance Summaries* 45:1–73.

Chalmers, J. W. and McMillan, J. H. 1995. ''An outbreak of viral gastroenteritis associated with adequately prepared oysters,'' *Epidemiol. Infect.* 115:163–167.

Cheesbrough, J. S., Barkess-Jones, L., and Brown, D. W., 1997. ''Possible prolonged environmental survival of small round structured viruses,'' *J. Hosp. Infect.* 35:325–326.

Clarke, I. N. and Lambden, P. R. 1997. ''The molecular biology of caliciviruses,'' *J. Gen. Virol.* 78:291–301.

Cubitt, W. D., Jiang, X. J., Wang, J., and Estes, M. K. 1994. ''Sequence similarity of human caliciviruses and small round structured viruses,'' *J. Med. Virol.* 43:252–258.

Curry, A., Riordan, T., Craske, J., and Caul, E. O. 1987. ''Small round structured viruses and persistence of infectivity in food handlers [letter],'' *Lancet* 2:864–865.

Daniels, N. A., Reddy, S., Rowe, S., Bergmire-Sweat, D., Hendricks, K., Fankhauser, R., Monroe, S., Atmar, R., Schwab, K., Glass, R., and Mead, P. 1998. ''An outbreak of viral gastroenteritis associated with consumption of deli sandwiches: application of reverse transcriptase-polymerase chain reaction assay,'' *Clin. Infect. Dis.* 27:1024.

Dimitrov, D. H., Dashti, S. A. H., Ball, J. M., Bishbishi, E., Alsaeid, K., Jiang, X., and Estes, M. K. 1997. ''Prevalence of antibodies to human caliciviruses (HuCVs) in Kuwait established by ELISA using baculovirus-expressed capsid antigens representing two genogroups of HuCVs,'' *J. Med. Virol* 51:115–118.

Dingle, K. E., Lambden, P. R., Caul, E. O., and Clarke, I. N. 1995. ''Human enteric *Caliciviridae:* the complete genome sequence and expression of virus-like particles from a genetic group II small round structured virus,'' *J. Gen. Virol.* 76:2349–2355.

Dolin, R., Reichman, R. C., and Fauci, A. S. 1976. ''Lymphocyte populations in acute viral gastroenteritis,'' *Infect. Immun.* 14:422–428.

Dolin, R., Levy, A. G., Wyatt, R. G., Thornhill, T. S., and Gardner, J. D. 1975. ''Viral gastroenteritis induced by the Hawaii agent. Jejunal histopathology and serologic response,'' *Am. J. Med.* 59:761–768.

Dolin, R., Reichman, R. C., Roessner, K. D., Tralka, T. S., Schooley, R. T., Gary, W., and Morens, D. 1982. ''Detection by immune electron microscopy of the Snow Mountain agent of acute viral gastroenteritis,'' *J. Infect. Dis.* 146:184–189.

Dolin, R., Blacklow, N. R., DuPont, H., Buscho, R. F., Wyatt, R. G., Kasel, J. A., Hornick, R., and Chanock, R. M. 1972. ''Biological properties of Norwalk agent of acute infectious nonbacterial gastroenteritis,'' *Proc. Soc. Exp. Biol. Med.* 140:578–583.

Dowell, S. F., Groves, C., Kirkland, K. B., Cicirello, H. G., Ando, T., Jin, Q., Gentsch, J. R., Monroe, S. S., Humphrey, C. D., Slemp, C., Dwyer, D. M., Meriwether, R. A., and Glass, R. I. 1995. ''A multistate outbreak of oyster-associated gastroenteritis: implications for interstate tracing of contaminated seafood,'' *J. Infect. Dis.* 171:1497–1503.

Estes, M. K., and Hardy, M. E. 1995. ''Norwalk virus and other enteric caliciviruses,'' in *Infections of the Gastrointestinal Tract,* ed. Blaser, M., P. Smith, J. Ravdin, H. B. Greenberg, and R. Guerrant. New York: Raven Press. p. 1009.

Estes, M. K., Atmar, R. L., and Hardy, M. E. 1997. ''Norwalk and related diarrhea viruses,'' in *Clinical Virology,* eds., Richman, D. D., R. J. Whitley, and F. G. Hayden. New York: Churchill Livingstone. p. 1073.

Fankhauser, R. L., Noel, J. S., Monroe, S. S., Ando, T., and Glass, R. I. 1998. ''Molecular epidemiology of 'Norwalk-like viruses' in outbreaks of gastroenteritis in the United States,'' *J. Infect. Dis.* 178:1571–1578.

Fleissner, M. L., Herrmann, J. E., Booth, J. W., Blacklow, N. R., and Nowak, N. A. 1989. ''Role of Norwalk virus in two foodborne outbreaks of gastroenteritis: definitive virus association,'' *Am. J. Epidemiol.* 129:165–172.

Gary, G. W., Jr., Kaplan, J. E., Stine, S. E., and Anderson, L. J. 1985. ''Detection of Norwalk virus antibodies and antigen with a biotin-avidin immunoassay,'' *J. Clin. Microbiol.* 22:274–278.

Gaulin, C., Ramsay, D., Cardinal, P., d'Halewyn, M. and Carpenter, M. 1999. ''Outbreaks of gastroenteritis associated with imported raspberries in Quebec (July 1997),'' *International Workshop on Human Caliciviruses,* CDC, Atlanta, GA. S4–5.

Gordon, S. M., Oshiro, L. S., Jarvis, W. R., Donenfeld, D., Ho, M. S., Taylor, F., Greenberg, H. B., Glass, R., Madore, H. P., Dolin, R., et al. 1990. ''Foodborne Snow Mountain agent gastroenteritis with secondary person-to-person spread in a retirement community.'' *Am. J. Epidemiol.* 131:702–710.

Gouvea, V., Santos, N., Timenetsky, M. C., and Estes, M. K. 1994. "Identification of Norwalk virus in artificially seeded shellfish and selected foods," *J. Virol. Methods* 48:177–187.

Graham, D. Y., Jiang, X., Tanaka, T., Opekun, A. R., Madore, H. P., and Estes, M. K. 1994. "Norwalk virus infection of volunteers: New insights based on improved assays," *J. Infect. Dis.* 170:34–43.

Gray, J. J., Jiang, X., Morgan Capner, P., Desselberger, U., and Estes, M. K. 1993. "Prevalence of antibodies to Norwalk virus in England: detection by enzyme-linked immunosorbent assay using baculovirus-expressed Norwalk virus capsid antigen," *J. Clin. Microbiol.* 31:1022–1025.

Gray, J. J., Green, J., Cunliffe, C., Gallimore, C. I., Lee, J. V., Neal, K., and Brown, D. W. 1997. "Mixed genogroup SRSV infections among a party of canoeists exposed to contaminated recreational water," *J. Med. Virol.* 52:425–429.

Green, J., Gallimore, C. I., Norcott, J. P., Lewis, D., and Brown, D. W. G. 1995. "Broadly reactive reverse transcriptase polymerase chain reaction in the diagnosis of SRSV-associated gastroenteritis," *J. Med. Virol.* 47:392–398.

Green, J. Wright, P. A., Galtimore, C. L., Mitchell, O., Morgan-Capner, P., and Brown, D. W. 1998. "The role of environmental contamination with small round structured viruses in a hospital outbreak investigated by reverse-transcriptase polymerase chain reaction assay," *J. Hosp. Infect.* 39:39–45.

Green, K. Y., Lew, J. F., Jiang, X., Kapikian, A. Z. and Estes, M. K. 1993. "Comparison of the reactivities of baculovirus-expressed recombinant Norwalk virus capsid antigen with those of the native Norwalk virus antigen in serologic assays and some epidemiologic observations," *J. Clin. Microbiol.* 31:2185–2191.

Green, K. Y., Kapikian, A. Z., Valdesuso, J., Sosnovtsev, S., Treanor, J. J., and Lew, J. F. 1997. "Expression and self-assembly of recombinant capsid protein from the antigenically distinct Hawaii human calicivirus," *J. Clin. Microbiol.* 35:1909–1914.

Green, K. Y., Ando, T., Balayan, M. S., Clarke, I. N., Estes, M. K., Matson, D. O., Nakata, S., Neill, J. D., Studdert, M. J., and Thiel, H.-J. 1999. "*Caliciviridae,*" in: M. vanRegenmortel, C. M. Fauquet, D. H. L. Bishop, E. Carsten, M. K. Estes, S. M. Lemon, J. Maniloff, M. Mayo, D. McGeoch, C. R. Pringle, and R. Wickner, eds., *Virus Taxonomy: 7th Report of the International Committee on Taxonomy of Viruses.* Orlando, FL: Academic Press.

Green, S. M., Lambden, P. R., Caul, E. O., and Clarke, I. N. 1997. "Capsid sequence diversity in small round structured viruses from recent UK outbreaks of gastroenteritis," *J. Med. Virol.* 52:14–19.

Green, S. M., Lambden, P. R., Deng, Y., Lowes, J. A., Lineham, S., Bushell, J., Rogers, J., Caul, E. O., Ashley, C. R., and Clarke, I. N. 1995. "Polymerase chain reaction detection of small round-structured viruses from two related hospital outbreaks of gastroenteritis using inosine-containing primers," *J. Med. Virol.* 45:197–202.

Greenberg, H. B., Wyatt, R. G., Valdesuso, J., Kalica, A. R., London, W. T., Chanock, R. M., and Kapikian, A. Z. 1978. "Solid-phase microtiter radioimmunoassay for detection of the Norwalk strain of acute nonbacterial, epidemic gastroenteritis virus and its antibodies," *J. Med. Virol.* 2:97–108.

Greenberg, H. B., Valdesuso, J., Kapikian, A. Z., Chanock, R. M., Wyatt, R. G., Szmuness, W., Larrick, J., Kaplan, J., Gilman, R. H., and Sack, D. A. 1979. "Prevalence of antibody to the Norwalk virus in various countries," *Infect. Immun.* 26:270–273.

Griffin, M. R., Surowiec, J. J., McCloskey, D. I., Capuano, B., Pierzynski, B., Quinn, M., Wojnarski, R., Parkin, W. E., Greenberg, H., and Gary, G. W. 1982. "Foodborne Norwalk virus," *Am. Epidemiol.* 115:178–184.

Gross, T. P., Conde, J. G., Gary, G. W., Harting, D., Goeller, D., and Israel, E. 1989. "An outbreak of acute infectious nonbacterial gastroenteritis in a high school in Maryland,'' *Public Health Rep.* 104:164–169.

Guest, C., Spitalny, K. C., Madore, H. P., Pray, K., Dolin, R. Herrmann, J. E., and Blacklow, N. R. 1987. ''Foodborne Snow Mountain agent gastroenteritis in a school cafeteria,'' *Pediatrics* 79:559–563.

Hafliger, D., Gilgen, M., Luthy, J., and Hubner, P. 1997. ''Seminested RT-PCR systems for small round structured viruses and detection of enteric viruses in seafood,'' *Inter. J. Food Microbiol.* 37:27–36.

Hale, A. D., Lewis, D., Green, J., Jiang, X., and Brown, D. W. G. 1996. ''Evaluation of an antigen capture ELISA based on recombinant Mexico virus capsid protein.'' *Clin. Diagn. Virol.* 5:27–35.

Hale, A. D., Crawford, S. E., Ciarlet, M., Green, J., Gallimore, C. I., Brown, D. W. G., Jiang, X., and Estes, M. K. 1999. ''Expression and self-assembly of grimsby virus: antigenic distinction from Norwalk and Mexico viruses,'' *Clin. Diagn. Lab. Immunol.* 6:142–146.

Hardy, M. E., Kramer, S. F., Treanor, J. J., and Estes, M. K. 1997. ''Human calicivirus genogroup II capsid sequence diversity revealed by analyses of the prototype Snow Mountain agent,'' *Arch. Virol.* 142:1469–1479.

Hardy, M. E., Tanaka, T. N., Kitamoto, N., White, L. J., Ball, J. M., Jiang, X., and Estes, M. K. 1996. ''Antigenic mapping of the recombinant Norwalk virus capsid protein using monoclonal antibodies,'' *Virology* 217:252–261.

Hedberg, C. W. and Osterholm, M. T. 1993. ''Outbreaks of food-borne and waterborne viral gastroenteritis,'' *Clin. Microbiol. Rev.* 6:199–210.

Herrmann, J. E., Nowak, N. A., and Blacklow, N. R. 1985. ''Detection of Norwalk virus in stools by enzyme immunoassay,'' *J. Med. Virol.* 17:127–133.

Herrmann, J. E., Blacklow, N. R., Matsui, S. M., Lewis, T. L., Estes, M. K., Ball, J. M., and Brinker, J. P. 1995. ''Monoclonal antibodies for detection of Norwalk virus antigen in stools,'' *J. Clin. Microbiol.* 33:2511–2513.

Herwaldt, B. L., Lew, J. F., Moe, C. L., Lewis, D. C., Humphrey, C. D., Monroe, S. S., Pon, E. W., and Glass, R. I. 1994. ''Characterization of a variant strain of Norwalk virus from a food-borne outbreak of gastroenteritis on a cruise ship in Hawaii,'' *J. Clin. Microbiol.* 32:861–866.

Heun, E. M., Vogt, R. L., Hudson, P. J., Parren, S., and Gary, G. W. 1987. ''Risk factors for secondary transmission in households after a common-source outbreak of Norwalk gastroenteritis,'' *Am. J. Epidemiol.* 126:1181–1186.

Hinkula, J., Ball, J. M., Lofgren, S., Estes, M. K., and Svensson, L. 1995. ''Antibody prevalence and immunoglobulin IgG subclass pattern to Norwalk virus in Sweden,'' *J. Med. Virol.* 47:52–57.

Ho, M. S., Glass, R. I., Monroe, S. S., Madore, H. P., Stine, S., Pinsky, P. F., Cubitt, D., Ashley, C., and Caul, E. O. 1989. ''Viral gastroenteritis aboard a cruise ship,'' *Lancet* 2:961–965.

Iversen, A. M., Gill, M., Bartlett, C. L., Cubitt, W. D., and McSwiggan, D. A. 1987. ''Two outbreaks of foodborne gastroenteritis caused by a small round structured virus: evidence of prolonged infectivity in a food handler,'' *Lancet* 2:556–558.

Jiang, X., Wang, J., and Estes, M. K. 1995b. ''Characterization of SRSVs using RT-PCR and a new antigen ELISA,'' *Arch. Virol.* 140:363–374.

Jiang, X., Graham, D. Y., Wang, K. N., and Estes, M. K. 1990. ''Norwalk virus genome cloning and characterization,'' *Science* 250:1580–1583.

Jiang, X., Wang, J., Graham, D. Y., and Estes, M. K. 1992a. ''Detection of Norwalk virus in stool by polymerase chain reaction,'' *J. Clin. Microbiol.* 30:2529–2534.

Jiang, X., Wang, M., Graham, D. Y., and Estes, M. K. 1992b. ''Expression, self-assembly, and antigenicity of the Norwalk virus capsid protein,'' *J. Virol.* 66:6527–6532.

Jiang, X., Wang, M., Wang, K., and Estes, M. K. 1993. ''Sequence and genomic organization of Norwalk virus,'' *Virology* 195:51–61.

Jiang, X., Matson, D. O., Ruiz-Palacios, G. M., Hu, J., Treanor, J., and Pickering, L. K. 1995a. ''Expression, self-assembly, and antigenicity of a Snow Mountain Agent-like calicivirus capsid protein,'' *J. Clin. Microbiol.* 33:1452–1455.

Johnson, P. C., Mathewson, J. J., DuPont, H. L., and Greenberg, H. B. 1990. ''Multiple-challenge study of host susceptibility to Norwalk gastroenteritis in US adults,'' *J. Infect. Dis.* 161:18–21.

Kapikian, A. Z., Estes, M. K., and Chanock, R. M. 1996. ''Norwalk group of viruses,'' in *Fields Virology,* eds., Fields, B. N., D. M. Knipe, and P. M. Howley. New York: Raven Press. p. 783–810.

Kapikian, A. Z., Wyatt, R. G., Dolin, R., Thornhill, T. S., Kalica, A. R., and Chanock, R. M. 1972. ''Visualization by immune electron microscopy of a 27 nm particle associated with acute infectious nonbacterial gastroenteritis,'' *J. Virol.* 10:1075–1081.

Kapikian, A. Z., Greenberg, H. B., Cline, W. L., Kalica, A. R., Wyatt, R. G., James, H. D. Jr., Lloyd, N. L., Chanock, R. M., Ryder, R. W., and Kim, H. W. 1978. ''Prevalence of antibody to the Norwalk agent by a newly developed immune adherence hemagglutination assay,'' *J. Med. Virol.* 2:281–294.

Kaplan, J. E., Goodman, R. A., Schonberger, L. B., Lippy, E. C., and Gary, G. W. 1982b. ''Gastroenteritis due to Norwalk virus: an outbreak associated with a municipal water system,'' *J. Infect. Dis.* 146:190–197.

Kaplan, J. E., Gary, G. W., Baron, R. C., Singh, N., Schonberger, L. B., Feldman, R., and Greenberg, H. B. 1982a. ''Epidemiology of Norwalk gastroenteritis and the role of Norwalk virus in outbreaks of acute nonbacterial gastroenteritis,'' *Ann. Intern. Med.* 96:756–761.

Kappus, K. D., Marks, J. S., Holman, R. C., Bryant, J. K., Baker, C., Gary, G. W., and Greenberg, H. B. 1982. ''An outbreak of Norwalk gastroenteritis associated with swimming in a pool and secondary person-to-person transmission,'' *Am. J. Epidemiol.* 116:834–839.

Keswick, B. H., Satterwhite, T. K., Johnson, P. C., DuPont, H. L., Secor, S. L., Bitsura, J. A., Gary, G. W., and Hoff, J. C. 1985. ''Inactivation of Norwalk virus in drinking water by chlorine.'' *Appl. Environ. Microbiol.* 50:261–264.

Khan, A. S., Moe, C. L., Glass, R. I., Monroe, S. S., Estes, M. K., Chapman, L. E., Jiang, X., Humphrey, C., Pon, E., Islander, J. K., and Schonberger, L. B. 1994. ''Norwalk virus-associated gastroenteritis traced to ice consumption aboard a cruise ship in Hawaii: Application of molecular-based assays,'' *J. Clin. Microbiol.* 32:318–322.

Kilgore, P. E., Belay, E. D., Hamlin, D. M., Noel, J. S., Humphrey, C. D., Gary, H. E. J., Ando, T., Monroe, S. S., Kludt, P. E., Rosenthal, D. S., Freeman, J., and Glass, R. I. 1996. ''A university outbreak of gastroenteritis due to a small round-structured virus. Application of molecular diagnostics to identify the etiologic agent and patterns of transmission,'' *J. Infect. Dis.* 173:787–793.

Kirkland, K. B., Meriwether, R. A., Leiss, J. K., and Mac Kenzie, W. R. 1996. ''Steaming oysters does not prevent Norwalk-like gastroenteritis,'' *Public Health Rep.* 111:527–530.

Kjeldsberg, E. 1977. ''Small spherical viruses in faeces from gastroenteritis patients,'' *Acta Pathol. Microbiol. Scand.* 85B:351–354.

Kohn, M. A., Farley, T. A., Ando, T., Curtis, M., Wilson, S. A., Jin, Q., Monroe, S. S., Baron, R. C., McFarland, L. M., and Glass, R. I. 1995. ''An outbreak of Norwalk virus gastroenteritis associated with eating raw oysters. Implications for maintaining safe oyster beds,'' *JAMA* 273:466–471.

Koopman, J. S., Eckert, E. A., Greenberg, H. B., Strohm, B. C., Isaacson, R. E., and Monto, A. S. 1982. ''Norwalk virus enteric illness acquired by swimming exposure,'' *Am. J. Epidemiol.* 115:173–177.

Kuritsky, J. N., Osterholm, M. T., Greenberg, H. B., Korlath, J. A., Godes, J. R., Hedberg, C. W., Forfang, J. C., Kapikian, A. Z., McCullough, J. C., and White, K. E. 1984. "Norwalk gastroenteritis: a community outbreak associated with bakery product consumption," *Ann. Intern. Med.* 100:519–521.

Kwok, S. and Higuchi, R. 1989. "Avoiding false positives with PCR," *Nature* 339:237–238.

Lawson, H. W., Braun, M. M., Glass, R. I., Stine, S. E., Monroe, S. S., Atrash, H. K., Lee, L. E., and Englender, S. J. 1991. "Waterborne outbreak of Norwalk virus gastroenteritis at a southwest US resort: role of geological formations in contamination of well water," *Lancet* 337:1200–1204.

Le Guyader, F., Neill, F. H., Estes, M. K., Monroe, S. S., Ando, T., and Atmar, R. L. 1996b. "Detection and analysis of a small round-structured virus strain in oysters implicated in an outbreak of acute gastroenteritis," *Appl. Environ. Microbiol.* 62:4268–4272.

Le Guyader, F., Estes, M. K., Hardy, M. E., Neill, F. H., Green, J., Brown, D., and Atmar, R. L. 1996a. "Evaluation of a degenerate primer for the PCR detection of human caliciviruses," *Arch. Virol.* 141:2225–2235.

Lees, D. N., Henshilwood, K., Green, J., Gallimore, C. I., and Brown, D. W. G. 1995. "Detection of small round structured viruses in shellfish by reverse transcription-PCR," *Appl. Environ. Microbiol.* 61:4418–4424.

Levine, W. C., Stephenson, W. T., and Craun, G. F. 1990. "Waterborne disease outbreaks, 1986–1988," *MMWR CDC Surveill. Summ.* 39:1–13.

Lew, J. F., Kapikian, A. Z., Jiang, X., Estes, M. K., and Green, K. Y. 1994. "Molecular characterization and expression of the capsid protein of a Norwalk-like virus recovered from a Desert Shield troop with gastroenteritis," *Virology* 200:319–325.

Lewis, D. C. 1990. "Three serotypes of Norwalk-like virus demonstrated by solid-phase immune electron microscopy," *J. Med. Virol.* 30:77–81.

Lieb, S., Gunn, R. A., Medina, R., Singh, N., May, R. D., Janowski, H. T., and Woodward, W. E., 1985. "Norwalk virus gastroenteritis. An outbreak associated with a cafeteria at a college," *Am. J. Epidemiol.* 121:259–268.

Lo, S. V., Connolly, A. M., Palmer, S. R., Wright, D., Thomas, P. D., and Joynson, D. 1994. "The role of the pre-symptomatic food handler in a common source outbreak of food-borne SRSV gastroenteritis in a group of hospitals," *Epidemiol. Infect.* 113:513–521.

Matsui, S. M., Kim, J. P., Greenberg, H. B., Su, W., Sun, Q., Johnson, P. C., DuPont, H. L., Oshiro, L. S., and Reyes, G. R. 1991. "The isolation and characterization of a Norwalk virus-specific cDNA," *J. Clin. Invest.* 87:1456–1461.

McAnulty, J. M., Rubin, G. L., Carvan, C. T., Huntley, E. J., Grohmann, G., and Hunter, R. 1993. "An outbreak of Norwalk-like gastroenteritis associated with contaminated drinking water at a caravan park," *Aust. J. Public Health* 17:36–41.

McDonnell, S., Kirkland, K. B., Hlady, W. G., Aristeguieta, C., Hopkins, R. S., Monroe, S. S., and Glass, R. I. 1997. "Failure of cooking to prevent shellfish-associated viral gastroenteritis," *Arch. Intern. Med.* 157:111–116.

McEvoy, M., Blake, W., Brown, D., Green, J., and Cartwright, R. 1996. "An outbreak of viral gastroenteritis on a cruise ship," *Commun. Dis. Rep.* CDR Rev. 6:R188–R192.

Metcalf, T. G., Melnick, J. L., and Estes, M. K. 1995. "Environmental virology: From detection of virus in sewage and water by isolation to identification by molecular biology-a trip of over 50 years," *Annu. Rev. Microbiol.* 49:461–487.

Morse, D. L., Guzewich, J. J., Hanrahan, J. P., Stricof, R., Shayegani, M., Deibel, R., Grabau, J. C., Nowak, N. A., Herrmann, J. E., Cukor, G., et al. 1986. "Widespread outbreaks of clam- and oyster-associated gastroenteritis. Role of Norwalk virus," *N. Engl. J. Med.* 314:678–681.

Murphy, A. M., Grohmann, G. S., Christopher, P. J., Lopez, W. A., Davey, G. R., and Millsom, R. H. 1979. "An Australia-wide outbreak of gastroenteritis from oysters caused by Norwalk virus," *Med. J. Aust.* 2:329–333.

Nakata, S., Estes, M. K., and Chiba, S. 1988. "Detection of human calicivirus antigen and antibody by enzyme-linked immunosorbent assays," *J. Clin. Microbiol.* 26:2001–2005.

Nakata, S., Chiba, S., Terashima, H., Yokoyama, T., and Nakao, T. 1985. "Humoral immunity in infants with gastroenteritis caused by human calicivirus," *J. Infect. Dis.* 152:274–279.

Noel, J. S., Ando, T., Leite, J. P. G., Green, K. Y., Dingle, K. E., Estes, M. K., Seto, Y., Monroe, S. S., and Glass, R. I. 1997. "Correlation of patient immune responses with genetically characterized small round-structured viruses involved in outbreaks of nonbacterial acute gastroenteritis in the United States, 1990 to 1995," *Med. Virol.* 53:372–383.

Norcott, J. P., Green, J., Lewis, D., Estes, M. K., Barlow, K. L., and Brown, D. W. 1994. "Genomic diversity of small round structured viruses in the UK," *J. Med. Virol.* 44:280–286.

Numata, K., Nakata, S., Jiang, X., Estes, M. K., and Chiba, S. 1994. "Epidemiological study of Norwalk virus infections in Japan and Southeast Asia by enzyme-linked immunosorbent assays with Norwalk virus capsid protein produced by the baculovirus expression system," *J. Clin. Microbiol.* 32:121–126.

Numata, K., Hardy, M. E., Nakata, S., Chiba, S., and Estes, M. K., 1997. "Molecular characterization of human calicivirus Sapporo," *Arch. Virol.* 142:1537–1552.

Okhuysen, P. C., Jiang, X., Yei, L., Johnson, P. C., and Estes, M. K. 1995. "Viral shedding and fecal IgA response after Norwalk virus infection," *J. Infect. Dis.* 171:566–569.

Palmenberg, A. C. 1989. "Sequence alignments of picornaviral capsid proteins, " in *Molecular Aspects of Picornavirus Infection and Detection,* eds., Semler, B. L. and E. Ehrenfeld. Washington, D.C.: American Society for Microbiology. p. 211.

Parashar, U. D., Dow, L., Fankhauser, R. L., Humphrey, C. D., Miller, J., Ando, T., Williams, K. S., Eddy, C. R., Noel, J. S., Ingram, T., Bresee, J. S., Monroe, S. S., and Glass, R. L. 1998. "An outbreak of viral gastroenteritis associated with consumption of sandwiches: implications for the control of transmission by food handlers," *Epidemiol. Infect.* 121:615–621.

Parker, S. P. and Cubitt, W. D. 1994. "Measurement of IgA responses following Norwalk virus infection and other human caliciviruses using a recombinant Norwalk virus protein EIA," *Epidemiol. Infect.* 113:143–151.

Parrino, T. A., Schreiber, D. S., Trier, J. S., Kapikian, A. Z., and Blacklow, N. R. 1977. "Clinical immunity in acute gastroenteritis caused by Norwalk agent," *N. Engl. J. Med.* 297:86–89.

Patterson, T., Hutchings, P., and Palmer, S. 1993. "Outbreak of SRSV gastroenteritis at an international conference traced to food handled by a post-symptomatic caterer," *Epidemiol. Infect.* 111:157–162.

Patterson, W., Haswell, P., Fryers, P. T., and Green, J. 1997. "Outbreak of small round structured virus gastroenteritis arose after kitchen assistant vomited," *Commun. Dis. Rep. CDR Rev.* 7:R101–R103.

Payment, P., Franco, E., and Fout, G. S. 1994. "Incidence of Norwalk virus infections during a prospective epidemiological study of drinking water-related gastrointestinal illness," *Can. J. Microbiol.* 40:805–809.

Prasad, B. V. V., Rothnagel, R., Jiang, X., and Estes, M. K. 1994. "Three-dimensional structure of baculovirus-expressed Norwalk virus capsids," *J. Virol.* 68:5117–5125.

Ramia, S. 1985. "Transmission of viral infections by the water route: implications for developing countries," *Rev. Infect. Dis.* 7:180–188.

Redman, J. A., Grant, S. B., Olson, T. M., Hardy, M. E., and Estes, M. K. 1997. "The filtration of recombinant Norwalk virus particles and bacteriophage MS2 in quartz sand: Importance of electrostatic interactions," *Envir. Sci. Tech.* 31:3378–3383.

Reid, J. A., Caul, E. O., White, D. G., and Palmer, S. R. 1988. "Role of infected food handler in hotel outbreak of Norwalk-like viral gastroenteritis: implications for control," *Lancet* 2:321–323.

Sawyer, L. A., Murphy, J. J., Kaplan, J. E., Pinsky, P. F., Chacon, D., Walmsley, S., Schonberger, L. B., Phillips, A., Forward, K., Goldman, C., et al. 1988. "25- to 30-nm virus particle associated with a hospital outbreak of acute gastroenteritis with evidence for airborne transmission," *Am. J. Epidemiol.* 127:1261–1271.

Schreiber, D. S., Blacklow, N. R., and Trier, J. S. 1974. "The small intestinal lesion induced by Hawaii agent acute infectious nonbacterial gastroenteritis," *J. Infect. Dis.* 129:705–708.

Schwab, K. J., Estes, M. K., Neill, F. H., and Atmar, R. L. 1997. "Use of heat release and an internal RNA standard control in reverse transcription-PCR detection of Norwalk virus from stool samples," *J. Clin. Microbiol.* 35:511–514.

Schwab, K. J., Neill, F. H., Estes, M. K., and Atmar, R. L. 1999. "Detection of HuCVs in food," *International Workshop on Human Caliciviruses,* CDC, Atlanta, GA. S3–9.

Sekla, L., Stackiw, W., Dzogan, S., and Sargeant, D. 1989. "Foodborne gastroenteritis due to Norwalk virus in a Winnipeg hotel [see comments]," *Can. Med. Assoc. J.* 140:1461–1464.

Shin, G. and Sobley, M. D. 1997. "Norwalk virus reduction by Chlorine Dioxide as determined by alternative PCR assays," *ASM General Meeting Proceedings,* May 4–8, 1997, Miami, FL. p. 457.

Smit, T. K., Steele, A. D., Peenze, I., Jiang, K., and Estes, M. K. 1997. "Study of Norwalk virus and Mexico virus infections at Ga-Rankuwa Hospital, Gs-Rankuwa, South Africa," *J. Clin. Microbiol.* 35:2381–2385.

Sugieda, M., Nakajima, K., and Nakajima, S. 1996. "Outbreaks of Norwalk-like virus-associated gastroenteritis traced to shellfish: coexistence of two genotypes in one specimen," *Epidemiol. Infect.* 116:339–346.

Taylor, J. W., Gary, G. W., Jr., and Greenberg, H. B. 1981. "Norwalk-related viral gastroenteritis due to contaminated drinking water," *Am. J. Epidemiol.* 114:584–592.

Thornhill, T. S., Kalica, A. R., Wyatt, R. G., Kapikian, A. Z., and Chanock, R. M. 1975. "Pattern of shedding of the Norwalk particle in stools during experimentally induced gastroenteritis in volunteers as determined by immune electron microscopy," *J. Infect. Dis.* 132:28–34.

Treanor, J. J., Jiang, X., Madore, H. P., and Estes, M. K. 1993. "Subclass-specific serum antibody responses to recombinant Norwalk virus capsid antigen (rNV) in adults infected with Norwalk, Snow Mountain, or Hawaii virus," *J. Clin. Microbiol.* 31:1630–1634.

Truman, B. I., Madore, H. P., Menegus, M. A., Nitzkin, J. L., and Dolin, R. 1987. "Snow Mountain agent gastroenteritis from clams," *Am. J. Epidemiol.* 126:516–525.

Vinje, J., Altena, S. A., and Koopmans, M. 1997. "The incidence and genetic variability of small round-structured viruses in outbreaks of gastroenteritis in The Netherlands," *J. Infect. Dis.* 176:1374–1378.

Wang, J., Jiang, X., Madore, H. P., Gray, J., Desselberger, U., Ando, T., Seto, Y., Oishi, I., Lew, J. F., Green, K. Y., and Estes, M. K. 1994. "Sequence diversity of small round structured viruses," *J. Virol.* 68:5982–5990.

Wanke, C. A. and Guerrant, R. L. 1987. "Viral hepatitis and gastroenteritis transmitted by shellfish and water," *Infect. Dis. Clin. North Am.* 1:649–664.

Warner, R. D., Carr, R. W., McCleskey, F. K., Johnson, P. C., Elmer, L. M., and Davison, V. E. 1991. "A large nontypical outbreak of Norwalk virus. Gastroenteritis associated with exposing celery to nonpotable water and with *Citrobacter freundii,*" *Arch. Intern. Med.* 151:2419–2424.

White, K. E., Osterholm, M. T., Mariotti, J. A., Korlath, J. A., Lawrence, D. H., Ristinen, T. L., and Greenberg, H. B. 1986. ''A foodborne outbreak of Norwalk virus gastroenteritis. Evidence for postrecovery transmission,'' *Am. J. Epidemiol.* 124:120–126.

White, L. J., Ball, J. M., Hardy, M. E., Tanaka, T. N., Kitamoto, N., and Estes, M. K. 1996. ''Attachment and entry of recombinant Norwalk virus capsids to cultured human and animal cell lines,'' *J. Virol.* 70:6589–6597.

Widerlite, L., Trier, J. S., Blacklow, N. R., and Schreiber, D. S. 1975. ''Structure of the gastric mucosa in acute infectious bacterial gastroenteritis,'' *Gastroenterology* 68:425–430.

Wilson, R., Anderson, L. J., Holman, R. C., Gary, G. W., and Greenberg, H. B. 1982. ''Waterborne gastroenteritis due to the Norwalk agent: clinical and epidemiologic investigation,'' *Am. J. Public Health* 72:72–74.

Wyatt, R. G., Dolin, R., Blacklow, N. R., DuPont, H. L., Buscho, R. F., Thornhill, T. S., Kapikian, A. Z., and Chanock, R. M. 1974. ''Comparison of three agents of acute infectious nonbacterial gastroenteritis by cross-challenge in volunteers,'' *J. Infect. Dis.* 127:709–714.

CHAPTER 16

Molecular Biology of Prion Diseases

CORINNE IDA LASMÉZAS
STEFAN WEISS

1. INTRODUCTION

1.1. IS BSE TRANSMISSIBLE TO HUMANS?

TRANSMISSIBLE spongiform encephalopathies are neurodegenerative diseases, which include Scrapie in sheep, bovine spongiform encephalopathy (BSE) in cattle, Creutzfeldt-Jakob disease (CJD), fatal familial insomnia (FFI), Gerstmann-Sträussler-Scheinker syndrome (GSS) and Kuru in humans. Today these diseases have raised great concern, especially in Europe, since 46 cases of a new variant form of CJD (nvCJD) have appeared in the United Kingdom as well as one case in France (status September 1999). nvCJD differs from "classical" CJD in many aspects, among them the fact that patients are very young (the mean age is 29 years in nvCJD versus about 60 years in sporadic CJD) and exhibit in thin sections of the brain typical amyloid plaques dubbed "florid plaques" because of their daisy-like appearance. People worldwide are worried about the risks of BSE transmission to humans and new findings suggest that BSE is in fact transmissible to humans: (1) macaques inoculated with prions originating from BSE-infected cattle developed the same florid plaques as nvCJD patients: (2) experiments employing transgenic animals carrying the *Prn*-p gene coding for human PrP developed the disease after inoculation with BSE prions and nvCJD showed the same glycosylation pattern as BSE in FVB mice: (3) wild-type mice inoculated with prions from humans suffering from nvCJD developed the disease after the same period of time as animals inoculated with BSE prions and showed the same lesion profiles: and (4) human PrP can be converted by bovine PrP^{BSE} into the proteinase K-resistant state.

1.2. TRANSMISSIBLE SPONGIFORM ENCEPHALOPATHIES AND PRIONS

The agent causing transmissible spongiform encephalopathies (TSE) is widely named *prion,* which stands for proteinaceous infectious particle. Prions define a new class of infectious agents, constituted partly or entirely (in the strict acceptance of the "protein-only" hypothesis) of an abnormal isoform of the host-encoded protein PrP. No nucleic acid specifically associated with infectivity has been detected so far. Thus TSEs seem to be different from classical infectious diseases such as AIDS caused by RNA containing HIV particles, tuberculosis caused by the DNA containing *Mycobacteria tuberculosis* or influenza caused by the RNA containing influenza virus. However, hypotheses such as the viral or the "virino" hypothesis, stating that the agent is constituted of a small nucleic acid protected by the prion protein, cannot be dismissed as they are supported by mainly pathophysiological data.

1.3. PRION PROPAGATION IN THE INFECTED ORGANISM

After prions have entered the organism, infectivity first appears in organs of the lympho-reticular system such as tonsils, thymus, lymph nodes and spleen. The kinetics of infectivity in these organs indicates that the infectious agent travels by lymphatic vessels and the bloodstream. After this first peripheral replication phase, the pathway of infection most probably involves the peripheral nervous system. Finally, infectivity and PrP become detectable in the central nervous system usually a long time after infection has been established in peripheral organs (more than half the incubation time of the disease). (Kimberlin and Walker, 1988).

The physical transfer of the infectious agent from a lymphoid cell to a neuron might require a ligand molecule as mentioned by Brown 1997. Membrane-bound molecules, which have been suggested to act as potential prion receptors by Brentani (Martins et al., 1997) and us (Rieger et al., 1997) could play an important role in this transmission process.

1.4. THE SCRAPIE AGENT CAN PERSIST IN HEALTHY ORGANISMS

A recent report completed older studies showing that the scrapie agent could survive in an organism without causing any signs of a scrapie infection. (Race and Chesebro, 1998). Mice have been inoculated with the scrapie agent from hamsters. Since mice are not susceptible to hamster prions, they did not show any symptoms as expected. Homogenates from the brain and spleen of these mice, however, when inoculated in control hamsters, again caused scrapie in these animals. Therefore, the scrapie agent is able to persist in organisms. Products of several organisms such as pigs and chicken, which lack any signs of a TSE, could be more dangerous for humans than previously thought.

1.5. PROPERTIES OF PrP

PrP^c, the cellular form of the prion protein, and PrP^{Sc}, the scrapie-inducing prion isoform, are thought to become internalized in neuronal cells via clathrin-coated pits (Shyng et al., 1994) or caveolae like domains (Vey et al., 1996). The conversion of PrP^c to PrP^{Sc} is thought to take place in endolysosomes or lysosomes. Molecular chaperones which interact with PrP^c (Edenhofer et al., 1996) are thought to be involved in this process. Newly generated PrP^{Sc} molecules convert more and more PrP^c molecules in a chain reaction into PrP^{Sc} which accumulate and destroy neuronal cells; this results in vacuoles of the brain and amyloidosis with an invariable fatal issue.

The physiological role of PrP^c, which is anchored on the cell surface by a glycosyl phosphatidyl inositol (GPI) (Stahl et al., 1993), is still unclear. Employing a series of different PrP knock-out mice, some researchers did not observe any phenotype at all, whereas others described a loss in purkinje neurons, electrophysiological abnormalities and sleep alterations (for a review, see Weissmann, 1996). These differences might be due to different deletion variants of PrP. The only proven phenotype of PrP^c, however, is the susceptibility of the organism towards a prion infection resulting in the development of TSEs.

1.6. TSE DIAGNOSTICS AND THERAPEUTICS

Although since 1997 antibodies and RNA aptamers have been available, which recognize PrP^{Sc} (Korth et al., 1997) and PrP^c (Weiss et al., 1997) respectively, a powerful diagnostic assay for the detection of TSE in living organisms has not been developed until now. The final diagnosis of a scrapie infection is performed post-mortem by an analysis of the brain of the infected individual. However, it is now possible to detect PrP^{Sc} in the tonsils of nvCJD patients (Hill et al., 1999). A powerful therapeutic tool for TSEs is also missing although some success has been obtained with MS 8209, an amphotericin B derivative, which has been proven to double the time from the onset of the disease to death (Adjou et al., 1995); and with pentosane sulfate which is able to prevent the disease caused by certain mouse strains (Farquhar et al., 1999).

2. HISTORY AND EPIDEMIOLOGY OF TSE

In 1732 scrapie appeared in the U.K. for the first time. In the early stage of the disease, the affected animals ''scrape'' on hedges, resulting in the term ''scrapie.'' Further on during the disease the animals stagger, collapse and finally die. The autopsy reveals spongiosis of the brain associated with amyloid like plaques, astrocytosis and gliosis.

Carlton Gajdusek, the later Nobelist, identified in the 1950s aborigines in New Guinea suffering from a human TSE later termed Kuru (Gajdusek and Zigas, 1957). The reason for the appearance of Kuru was tribal cannibalism. Typical of this disease are ''kuru-type'' plaques in the brain of the affected persons. Although incidence rates dramatically decreased after the prohibition of cannibalism by law, one case of Kuru, a man with an incubation time of over 43 years, is still alive in Papua, New Guinea (John Collinge, personal communication).

The German physicians Creutzfeldt and Jakob investigated in the 1920s, people with brain lesions similar to these observed later on in the aborigines suffering from Kuru: spongiosis, astrocytosis and gliosis. This disease was termed Creutzfeldt-Jakob disease (CJD) (Creutzfeldt, 1920). Of the CJD cases, 15% are due to mutations within the *Prn*-p gene, thus clustered into families, whereas 85% of the cases are of unknown origin and harbor a sporadic pattern of occurrence. Gerstmann-Sträussler-Scheinker-Syndrome (GSS) (Gerstmann, 1928) and fatal familial insomnia (FFI) (Tateishi et al., 1995), however, are strictly linked to mutations within the *Prn*-p locus.

In November 1987 the first case of bovine spongiform encephalopathy (BSE) was histologically confirmed. In the following years the BSE epidemic raised its maximum with 3500 new infections per month in 1992/1993 (Figure 16.1). What happened? According to the sheep origin hypothesis, the causative agent of the disease was transmitted from sheep to cattle by feeding of meat and bone meal prepared from sheep infected with the scrapie agent. Simultaneously, the sterilization temperature was decreased from 130°C to about 110°C. Other steps such as solvent extractions of fat have been suppressed so that the agent was not inactivated and was able to cross the species barrier from sheep to cattle (for reference see Edenhofer et al., 1997). In contrast, the bovine origin hypothesis states that the agent has infected cows before the feeding of meat and bone meal to cows. Indeed, an incidence rate for a bovine TSE (before the BSE epidemic) approximately identical to that of sporadic CJD, i.e., one case in 10^6 individuals, would have been undetectable. After introducing the ''feed ban'' on July 18, 1988, prohibiting the feeding of meat and bone meal to ruminants in the U.K. followed by the specified bovine offal ban (SBO) on November 13, 1989, the numbers of BSE-infected cattle dropped dramatically, resulting in about 150 confirmed cases of newly infected cows per month in England in September 1998 (Figure 16.1). Vertical transmissions of the agent from the mother cow to the calf seem to have occured at a low level and would be responsible for part of the cases ''born after the ban.'' BSE cases have been confirmed in 15 countries worldwide. Among them are six countries where all of the cases have been cattle imported from the U.K. Table 16.1 summarizes the number of BSE cases in these countries.

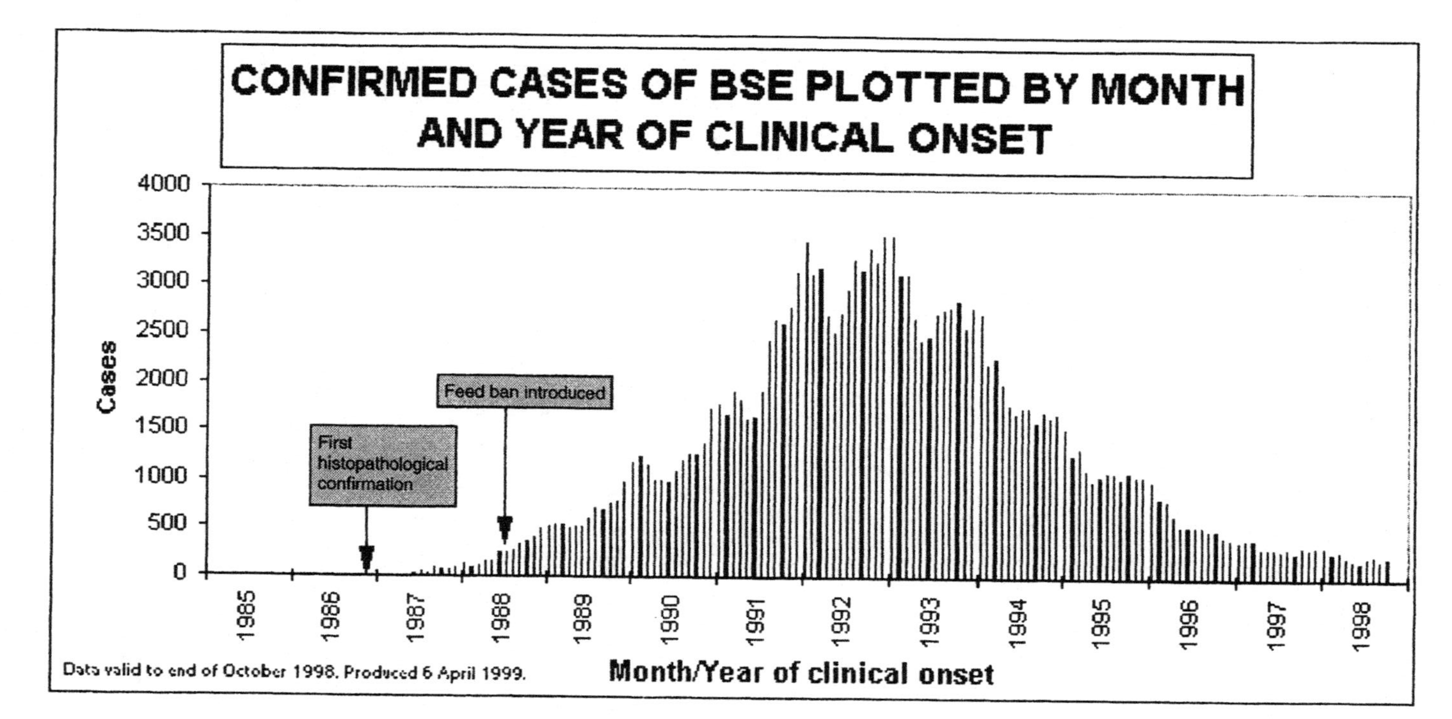

Figure 16.1 Confirmed cases of BSE in the UK plotted by month and year of clinical onset. Data have been kindly provided by the Ministery of Agriculture, Fisheries and Food (MAFF, U.K., status, April 1999).

TABLE 16.1. Numbers of Confirmed BSE Cases Worldwide (MAFF, U.K., April 1999).

Country	Number of BSE Cases
U.K. (including Great Britain, Northern Ireland, Isle of Man, Jersey and Guernsey)	177,359
Ireland	365
Switzerland	289
Portugal	244
France	57
Belgium	7
Germany	6[a]
Netherlands	4
Oman	2[a]
Italy	2[a]
Liechtenstein	2
Canada	1[a]
Falkland Islands	1[a]
Denmark	1[a]
Luxembourg	1
Africa	0
Australia	0
New Zealand	0
South America	0
United States	0

[a] These cases of BSE have been confirmed in cattle imported from the U.K.

3. HISTOPATHOLOGICAL FEATURES OF TSE

The typical histopathological features of TSEs include (1) vacuolisation of the neuropil and neuronal perikarya (Figure 16.2), (2) astrogliosis, which can be revealed by GFAP (glial fibrillary acidic protein) immunostaining and (3) amyloid plaques (Figure 16.3 shows a so-called Kuru type plaque). Figure 16.4 shows a florid plaque in the brain of a macaque inoculated with BSE (a) and of a French patient suffering from new variant (nv) CJD (b). This florid plaque is constituted of a dense amyloid core (corresponding to PrP deposition) surrounded by vacuoles, providing the typical daisy-like appearance; these plaques are specific to and constitute a hallmark of nvCJD. Figure 16.5 represents an electron microscopic image of a vacuole containing the typical pseudomembrane structures.

4. THE PRION

Prions (proteinaceous infectious particles) are the causative agents of transmissible spongiform encephalopathies (Prusiner, 1982) consisting entirely or at least in part of the misfolded scrapie prion termed PrP^{Sc} or its N-terminally

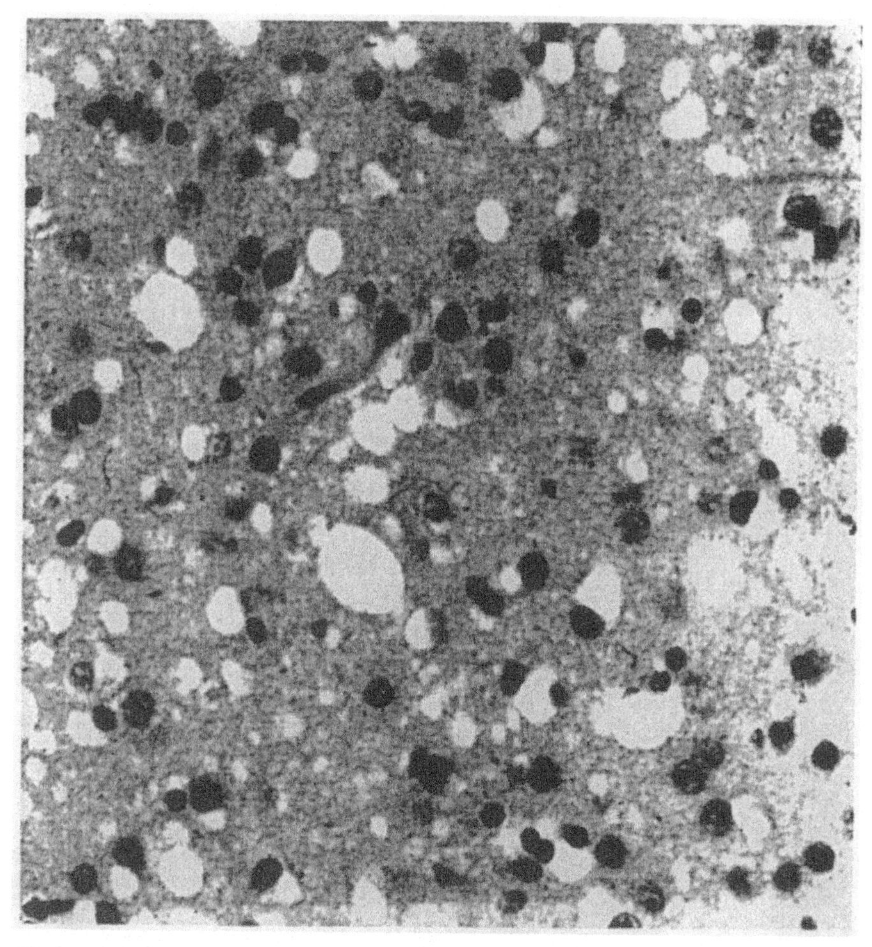

Figure 16.2 Vacuolization in the grey matter of the brain of a *Cynomolgus macaque* infected with BSE (optical microscopy, hematoxylin-eosin staining).

truncated version PrP27–30 generated by proteolysis of PrP^{Sc} (Prusiner et al., 1981; Prusiner et al., 1983; Prusiner et al., 1984). PrP^{Sc} originates most likely by a conformational change from PrP^{c}. α-Helical or unstructured regions are most likely converted to β-sheet since the α-helix content is 42% in PrP^{c} but only 30% and 21% in PrP^{Sc} and PrP27–30, respectively (Caughey et al., 1989; Pan et al., 1993). β-Sheet contents are 3% for PrP^{c} and 43% and 54% for PrP^{Sc} or PrP27–30, respectively (Table 16.2).

4.1. PHYSIOLOGICAL ROLE OF PrP^{c}

The physiological role of PrP^{c} is poorly understood. Some reports employing *Prn*-P knock-out mice ($PrP^{0/0}$) describe a role of PrP^{c} in synaptic processes

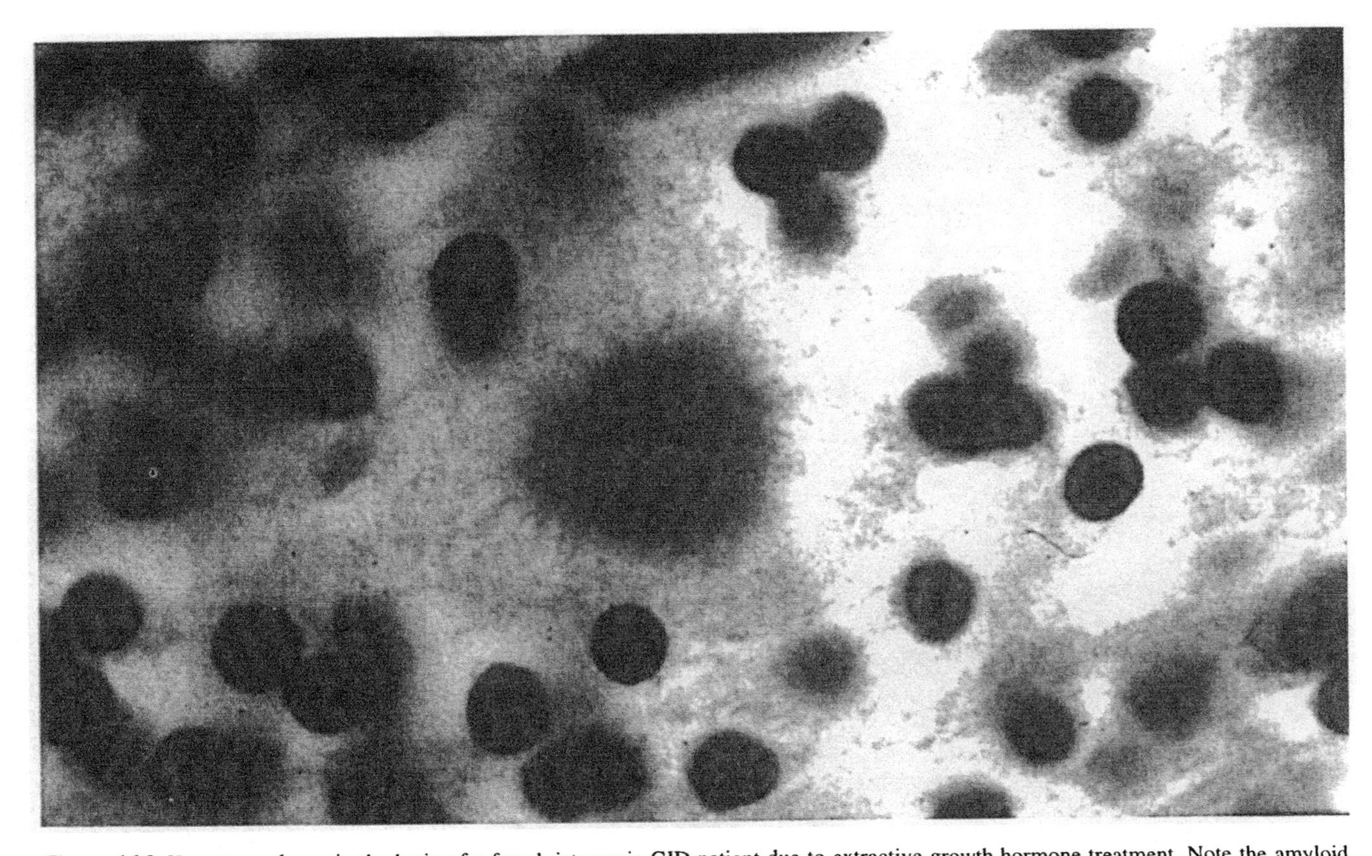

Figure 16.3 Kuru-type plaque in the brain of a french iatrogenic CJD patient due to extractive growth hormone treatment. Note the amyloid surrounded with a pale halo (optical microscopy, PAS-staining). Courtesy of Dr. O. Robain, Paris, France.

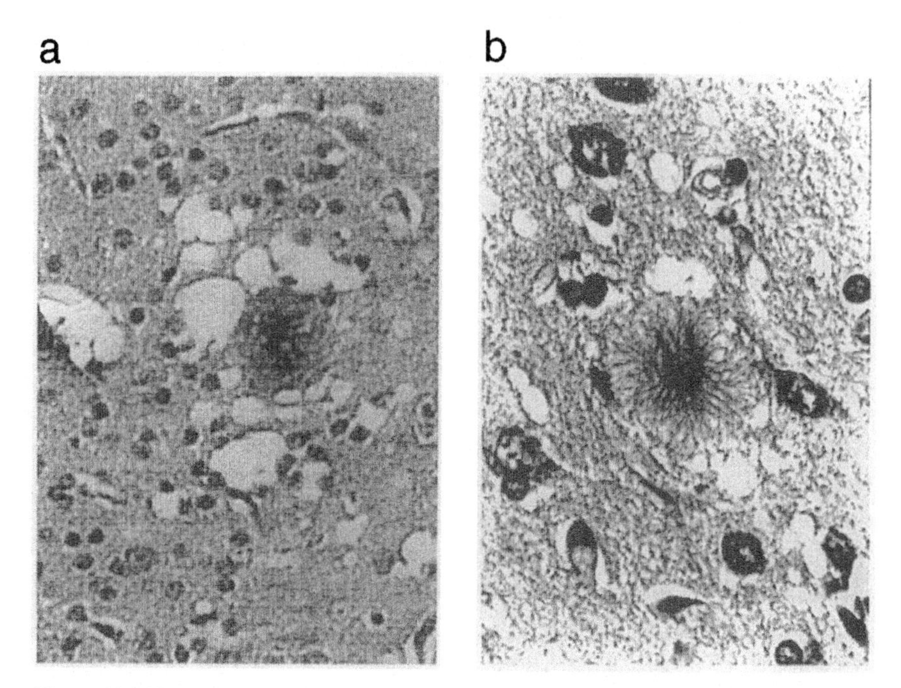

Figure 16.4 Florid plaque in the cerebral cortex of (a) a cynomolgus macaque inoculated with the BSE agent and (b) a french new variant CJD patient (optical microscopy, PAS-staining). Courtesy of Dr. Kopp, France.

(Collinge et al., 1994), others observed altered circadian activities and sleep patterns (Tobler et al., 1996) or effects on purkinje cells (Sakaguchi et al., 1996). On the contrary, some researchers did not observe any phenotype at all (Bueler et al., 1992; Manson et al., 1994; Lledo et al., 1996). The differences in these observations might be due to the different regions of the *Prn*-p gene that have been knocked out (for review, see Weissmann, 1996) and to the age of the transgenic animals. The only proven phenotype of PrP^c, however, is the necessity of the protein for the organism to be fully susceptible for PrP^{Sc}. Therefore, individuals lacking the PrP^c protein are not susceptible to infection with mouse-adapted strains (Bueler et al., 1993). Recently, Adriano Agguzzi and Charles Weissmann engineered transgenic mice that expressed amino-terminal truncated versions of the prion protein (Shmerling et al., 1998). A PrP protein lacking amino acid residues 32 to 134 caused severe ataxia and neuronal death limited to the granular layer of the cerebellum. The defect was completely abolished by introducing one copy of the wild-type gene. The authors speculated that these truncated PrPs may be nonfunctional and compete with PrP, or some other molecule with a PrP-like function, for a common ligand, which could be a receptor (Martins et al., 1997; Rieger et al., 1997). Very re-

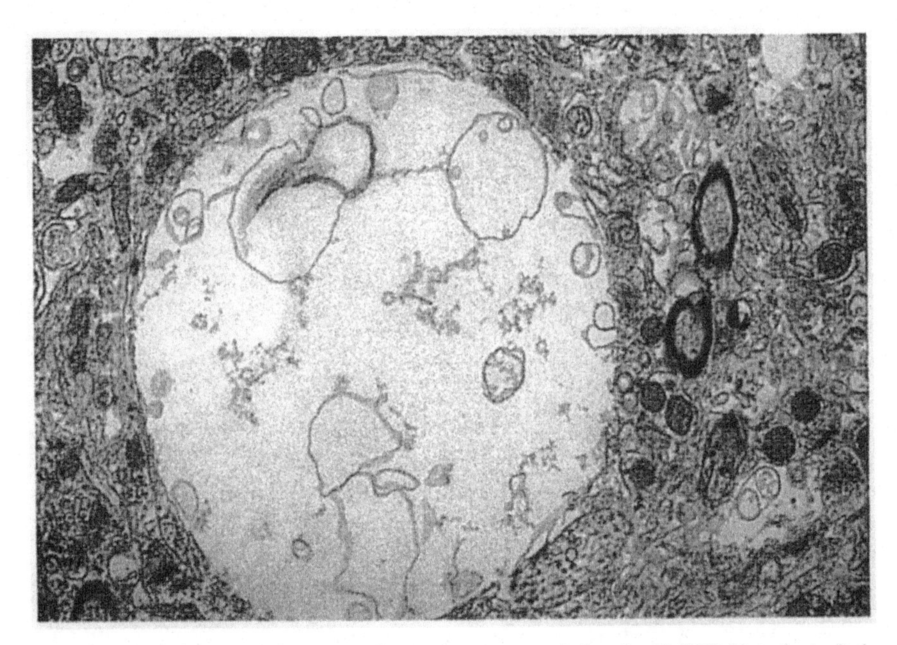

Figure 16.5 Vacuole in the brain of a *Cynomolgus macaque* infected with BSE. Note the typical pseudomembrane structures in the vacuole (electron microscopy). Courtesy of Dr. O. Robain, Paris, France.

cently, a transgenic mouse was constructed which expressed a ''miniprion'' lacking amino acids 23 to 88 and 141 to 176. This mouse [Tg (PrP106) *Prnp*$^{0/0}$] was infectable with RML prions and, subsequently with the ''miniprions'' produced in the brains of these transgenic mice (Supattapone et al., 1999).

4.2. BIOCHEMISTRY OF PrP^c

PrP is synthesized as a precursor containing a signal peptide (SP) at the N-terminus and a signal sequence (SS) at the carboxy terminus (Figure 16.6). The protein is then further processed to PrP^c by cleaving off the signal sequences from both ends of the prion protein and by further addition of a

TABLE 16.2. Analysis of the Secondary Structure of PrP^c, PrP^{Sc} and PrP27–30.

	α-Helix	β-Sheet
PrP^c	42%	3%
PrP^{Sc}	30%	43%
PrP27–30	21%	54%

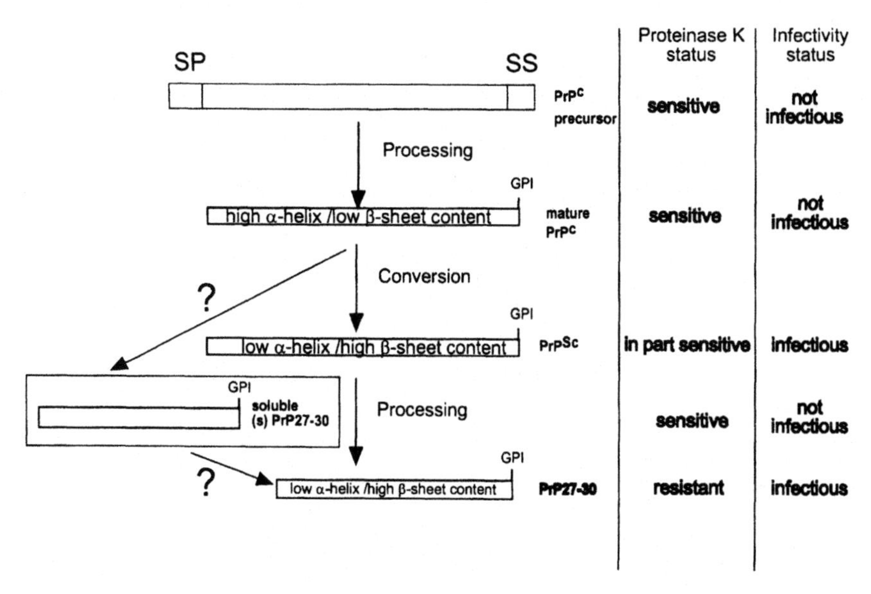

Figure 16.6 Maturation of the prion protein PrP and its individual proteinase K and infectivity status. SP = signal peptide; SS = signal sequence; GPI = glycosyl phosphatidyl inositol anchor.

glycophosphoinositol (GPI) anchor at its carboxy terminus. Moreover, the protein becomes glycosylated at two amino acid positions. The protein passes through the secretory pathway and remains GPI anchored on the cell surface. The conversion of PrP^c to PrP^{Sc} is accompanied by an increase of the β-sheet content and the aquisition of proteinase K resistance. Since PrP^c is completely sensitive to proteinase K, PrP^{Sc} is degradable to PrP27–30, which itself is very resistant to proteinase K. PrP27–30 lacks about 60 amino acids at the N-terminus compared to PrP^{Sc}. Which form of PrP is the dominant one in an affected individual is thought to depend on the presence and activity of cellular proteinases in the individual tissue or organ.

The resistant PrP27–30 molecule is most likely responsible for the extraordinary resistance of the infectious agent to reagents such as proteases, formamide or UV-radiation. Only phenol, NaOH or heat under pressure (138°C at 3 bar) inactivates the agent (Meyer et al., 1986).

4.3. SCRAPIE-ASSOCIATED FIBRILS (SAF)

In the presence of detergent and proteinase K, PrP^{Sc} can be purified in the form of scrapie-associated fibrils (SAFs) (Prusiner et al., 1983). They appear as rods of about 200 nm in length and about 20 nm in diameter. SAFs also exist in vivo as demonstrated by Fournier. Figure 16.7 (a kind gift of Dr. J. G. Fournier and Dr. N. Kopp, France) demonstrates amyloid plaques from a

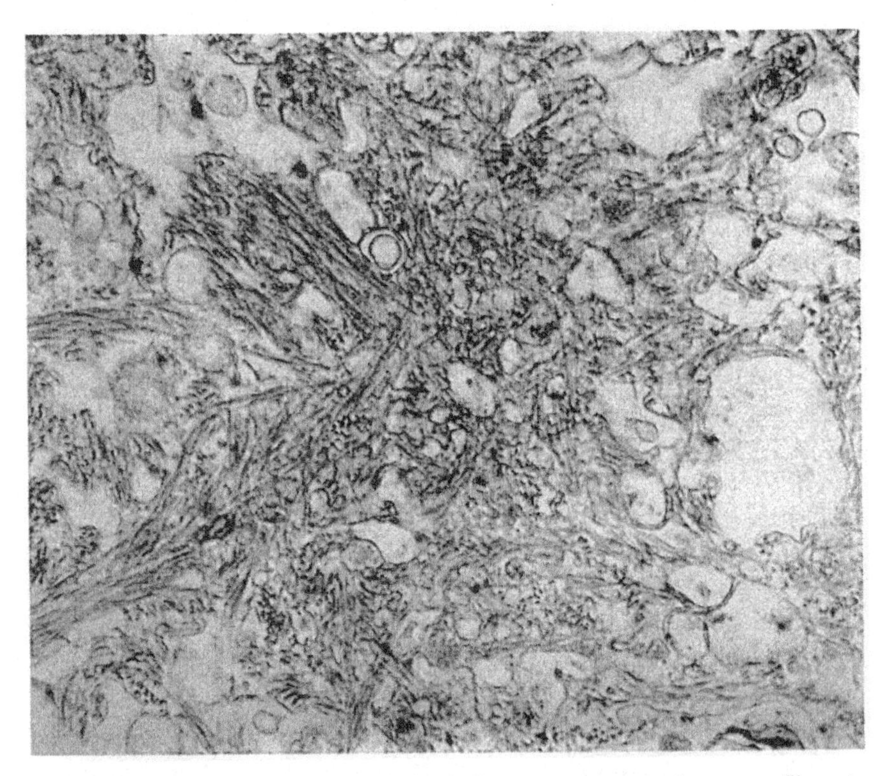

Figure 16.7 Scrapie associated fibrils (SAFs). SAFs represent numerous interwoven filaments of 12–15 nm in diameter, here mixed with cellular components (×30000). They originate from the brain of the french new variant CJD patient. Courtesy of Drs. J. G. Fournier and N. Kopp, France.

brain biopsy of nvCJD. Numerous interwoven filaments (equivalent to scrapie-associated fibrils) of 12–15 nm in diameter are mixed with cellular components.

4.4. THE NEW FORMS OF PRION PROTEINS ^{Ctm}PrP AND ^{Ntm}PrP COMPARED TO PrP^{Sec}

Recently, Lingappa and Prusiner identified two new forms of the prion protein in the ER of the scrapie-infected cell (Hegde et al., 1998). The Gerstmann-Sträussler-Scheinker (GSS) Syndrome mutation 117 Alanine to Valine resulted in the formation of ^{Ctm}PrP, the carboxy-terminal transmembrane form of the prion protein. Interestingly, the occurrence of ^{Ctm}PrP, which becomes degraded in the absence of the mutation, leads to vacuolation, astrogliosis and gliosis. Although ^{Ctm}PrP seems to adhere in a first step to the membrane of the ER with its carboxy terminus to the lumen of the ER, the molecule seems to be exported in a latter step to post-ER compartments. The orientation of ^{Ntm}PrP (amino-terminal transmembrane form of the prion protein) is just the opposite that of ^{Ctm}PrP. The role of ^{Ntm}PrP is unclear. PrP^{c}, which is transported through the secretory pathway to the cell surface where its release to the extracellular space is prevented by the GPI anchor, has been termed ^{Sec}PrP for the ''secreted'' form of PrP. We want to note that the expression, ^{Sec}PrP, is misleading, since no PrP is actually ''secreted'' to the extracellular space, due to the GPI anchor.

5. REPLICATION MECHANISMS OF PRIONS

Although the protein-only hypotheses (heterodimerization or nucleation-dependent polymerization) are favored by the majority of the scientists in the field, there are other hypotheses that cannot be discarded, such as the virino and the viral ones, stipulating the existence of a so far undiscovered nucleic acid or virus. In favor of the presence of such a nucleic acid are mainly the data of prion pathogenesis, which show a virus-like behavior of the agent in the organism with a particular tropism for the immune system and for the central nervous system, and the existence of various strains of agent that are able: (1) for about 20 of them, to infect the same host harboring the same PrP (mice), (2) for the strains known as being ''stable'' (like the BSE strain), to retain their properties upon transmission to several different hosts. Arguing against the presence of a nucleic acid as a component of the infectious agent is mainly the fact that no scrapie-specific nucleic acid has been identified to date. Riesner excluded that a RNA larger than 80 nucleotides in size could be hidden in the infectious particle (Kellings et al., 1992).

5.1. THE VIRUS

The viral hypotheses encompass the retroviral one supported by Manuelidis stipulating that possibly defective capsid proteins could be responsible for the absence of viral particles easily identifiable by electron microscopy (Manuelidis et al., 1988). The hypothesis of a small amyloidogenic virus has been documented by the observation of Diringer of small particles about 10 nm in diameter in purified hamster prions and prion preparations of sporadic and familial CJD cases (Özel and Diringer, 1994; Özel et al., 1994). There is no proof, however, that these particles are infectious.

5.2. THE VIRINO

This term was defined for the first time by Dickinson and Outram in 1979 to depict the agents responsible for TSEs: ''an appropriate name for this class of agent would be 'virino,' which (by analogy with neutrinos) are small, immunologically neutral particles with high penetration properties but needing special criteria to detect their presence'' (Dickinson and Outram, 1979, p. 20). In fact, virinos would be composed of a small noncoding nucleic acid protected by host proteins (Dickinson and Outram, 1988; Dickinson et al., 1989). This type of informational ''hybrid'' between host and agent allows one to reconcile modern data showing that the pathological form of PrP is a constituent of the infectious agent and thus suggesting the presence of an independent genome.

In 1997 it was demonstrated that interspecies transmission of BSE to the mouse could lead to a disease characterized in the brain by neuronal death in the absence of vacuolization and PrP^{Sc}. After the second and third passage in mice of this unusually silent agent, PrP^{Sc} appeared in the brains of all animals. We concluded that PrP^{Sc} is necessary for strain adaptation but that another unknown component of the agent is able, under certain conditions, to cause the disease (Lasmézas et al., 1997).

5.3. THE PROTEIN-ONLY HYPOTHESIS (HETERODIMER HYPOTHESIS)

In the mid-sixties Alper (Alper et al., 1966; Alper et al., 1967) and Griffith (1967) suspected that the scrapie agent could replicate in the absence of a specific nucleic acid. Stanley Prusiner, the later Nobel Prize winner, named the causative agent ''the prion,'' which stands for ''proteinaceous infectious particle,'' and mentioned in his ''protein-only'' hypothesis that the prion is an infectious protein that causes the disease (Prusiner, 1982). Prusiner introduced a completely new class of infectious agents to the scientific community. His hypothesis turned molecular biology upside down. Whereas the classical agents of infectious diseases such as viruses, bacteria and fungi require nucleic acids for replication, prions were stated to replicate in the absence of any nucleic acid.

First it was difficult for Prusiner to explain the replication mode of prions. He thought of reverse translation or protein-dependent protein synthesis (Prusiner, 1982). The scheme of how prions could replicate is quite simple: an infectious PrP^{Sc} molecule binds to the endogeneous PrP^{c} protein most likely in the endocytotic pathway of the scrapie-infected cell forming a heterodimer responsible for naming this hypothesis the ''heterodimer'' model (Figure 16.8). The direct interaction of PrP^{Sc} with PrP^{c} forces PrP^{c} to undergo a structural change: α-helices become converted to β-sheets. The newly generated PrP^{Sc} converts in a chain reaction more and more PrP^{c} molecules into PrP^{Sc}, resulting in astrogliosis, gliosis, vacuolization and apoptosis. Although this hypothesis has not been totally confirmed until now, Prusiner received the Nobel Prize for Medicine in 1997 for this striking and completely new principle of how infectious agents would be able to replicate. It is very conceivable that this new replication mechanism based on infectious proteins will apply to many diseases that are not yet discovered. The final proof of the protein-only hypoth-

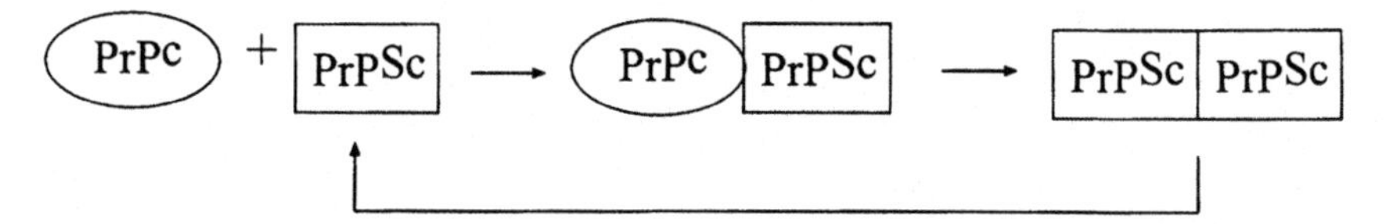

Figure 16.8 Protein-only hypothesis (heterodimer hypothesis) according to Prof. S. B. Prusiner, San Francisco. A PrP^{Sc} monomer forms together with a PrP^{c} monomer producing a heterodimer. PrP^{Sc} forces PrP^{c} to undergo a structural change resulting in the transition to PrP^{Sc}. The newly generated PrP^{Sc} molecules transform in a chain reaction more and more PrP^{c} monomers to PrP^{Sc}, a process which finally kills the entire organism.

esis is still missing, one of which could be that recombinant prion proteins (Weiss et al., 1995; Weiss et al., 1996) harbor endogenous infectivity after conversion.

5.4. IN VITRO CONVERSION OF PROTEINASE K-SENSITIVE INTO PROTEINASE K-RESISTANT PrP

The conversion of a proteinase K sensitive PrP into a proteinase K resistant PrP has been demonstrated in cell-free systems (Kocisko et al., 1994; Bessen et al., 1995; Kocisko et al., 1995; Raymond et al., 1997). The generation of infectious PrP, however, has failed so far. Generation of PrP^{Sc} by post-translational conversion of PrP^{c} (Caughey et al., 1989; Borchelt et al., 1990) in chronically infected neuroblastoma cells may be due to the presence of co-factors such as molecular chaperones (Edenhofer et al., 1996).

5.5. PERMANENT WAVES AND PRION REPLICATION

In a very simplified view the replication mechanism of prions can be compared to the conversion of plain hair to permanent-waved hair. Prion replication involves the structural conversion of PrP^{c} into PrP^{Sc} concomitant with the conversion of α-helices into β-sheets. Permanent waves, however, are generated by the breakage and reformation of disulfide bridges. *Permanent wave formation is permanent, prion conversion is thought to be irreversible.*

5.6. WHAT DO PRIONS AND VIRUSES HAVE IN COMMON?

Both prions and viruses need a host for replication. Viruses containing a nucleic acid (which in the case of HIV-1 RNA is surrounded by a nucleocapsid and an envelope, Figure 16.9) require the protein synthesis apparatus of a host cell for replication. They are inactive without host cells. Prions, however, contain a host protein but possibly lack any kind of nucleic acids. They seem to require a PrP^{c} containing cell for replication. Knock-out mice lacking the

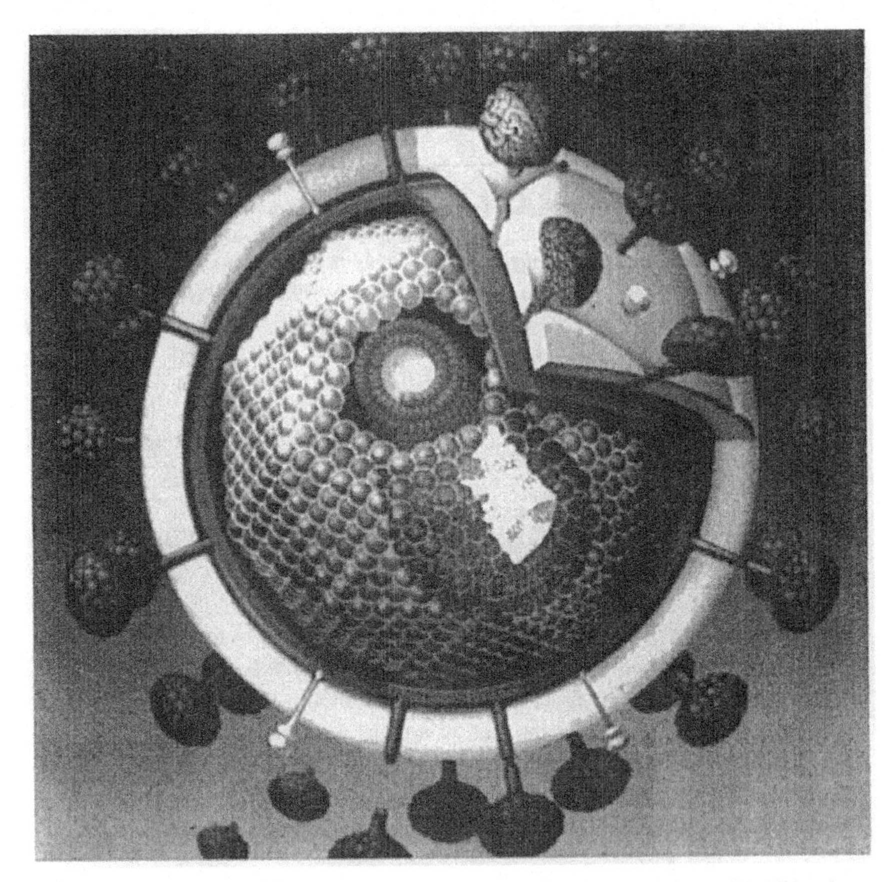

Figure 16.9 The HIV-1 Virus. Two RNA strands represent the genetic information of the virus. One strand is shown as a string in the interior of the icosaedric nucleocapsid. An envelope surrounds the nucleocapsid and is responsible for the globular shape of the virus. Prions differ from viruses by their molecular nature and their physicochemical properties. However, viruses and prions do both need an intact cell for replication.

Prn-p gene are not susceptible to a scrapie infection (Bueler et al., 1993). Since the phenotype of PrP^c is speculative (loss in purkinje neurons, synaptic dysfunction, altered circadian activity and sleep, no phenotype), is the only physiological function of PrP^c to make the individual susceptible to a TSE infection? Can an organism be cured or protected by knocking out its *Prn*-p gene? Does nature leave a set of genes functional to support infectious lethal diseases? Transgenic mice, however, lacking their own *Prn*-p gene but harboring the hamster *Prn*-p gene ($moPrP^{o/o}/haPrP^{+/+}$) (Bueler et al., 1993) or a *Prn*-p gene, which lacks 49 amino acids from its amino terminus ($moPrP^{o/o}/haPrP\Delta49$), are again susceptible to the infectious agent (Fischer et al., 1996). After grafting of a PrP-expressing tissue into the brain of a PrP-knock-out mouse and inoculation with scrapie, spongiosis and gliosis were seen only in the graft although PrP^{Sc} could be found in the PrP-negative brain outside the graft, indicating that cellular PrP expression is required for TSE-induced pathology (Brandner et al., 1996).

5.7. THE NUCLEATION-DEPENDENT POLYMERIZATION MODEL/ CRYSTAL SEED HYPOTHESIS

The crystal seed model claims the existence of a nucleus consisting of three (Byron Caughey, personal communication) to a few PrP^{Sc} molecules (Lansbury and Caughey, 1995), whereas the heterodimer model focuses on a PrP^{c}/PrP^{Sc} heterodimer (Figure 16.8). Different crystals or nuclei originate from different species or strains. Most important for the replication of prions are the similarities between PrP^{c} (which equilibrates with its unfolded form termed PrP^{u}) and one monomer of the PrP^{Sc} nucleus. The greater the similarities between PrP^{c} (PrP^{u}) and a PrP^{Sc} monomer in the seed, the better the polymerization proceeds. The nucleation-dependent polymerization model offers an explanation for the existence of different scrapie strains (within one species) and the existence of species barriers.

The model states that PrP^{c} of one species can be incorporated into PrP^{Sc} nuclei of different species. However, species barriers do appear in case of different secondary/tertiary structures of endogenous PrP^{c} and incoming PrP^{Sc}. Since the amino acid sequence of a protein determines its structure, species barriers would be determined by amino acid sequence homologies of different species. In fact, the species barrier phenomenon is more complex as it also depends on the infecting strain of the agent. The model would explain the appearance of different strains within one species with differently packed nuclei consisting of different PrP monomers with individual secondary/tertiary structures.

5.8. THE CO-PRION HYPOTHESIS—NUCLEIC ACIDS AS CO-FACTORS

Co-factors could also explain the appearance of different scrapie strains within one species which are characterized by different incubation periods and different lesion profiles in the brain. Charles Weissmann (1991) postulated that nucleic acids may serve as co-factors for determination of the strain-specific properties of prions. He named these co-factors "co-prions," the prion without the nucleic acid, the "apoprion," and the co-prion plus the apoprion, the "holoprion." This hypothesis unified the protein-only hypothesis and the viral hypothesis and was therefore named the "unified theory." Since nucleic acids have not been identified in prion preparations (Kellings et al., 1992), nucleic acids have lost their importance in this model in contrast to a nucleic acid-free co-factor.

5.9. MOLECULAR CHAPERONES AS CO-FACTORS FOR PRIONS

Studies employing transgenic mice resulted in the suggestion of the existence of a factor X, which was thought to be important for the development of TSEs (Telling et al., 1995). Promising candidates for such a factor are molecular chaperones, which usually prevent proteins from misfolding. Three different roles of molecular chaperones in the life cycle of prions are conceivable: (1) Prions themselves represent misfolded molecular chaperones which could be responsible for their own assembly (Liautard, 1991). (2) Molecular chaperones catalyze the conversion from PrP^{c} to PrP^{Sc} as observed in case of the prion-like factor sup35 in yeast, which can be reversibly converted to sup35* (the prion-like form of sup35) in the presence of the molecular yeast chaperone Hsp104 (Chernoff et al., 1995). Specific interactions of the molecular chaperones Hsp60 and its bacterial counterpart GroEL with mammalian prion proteins have been described (Edenhofer et al., 1996). The interaction site between PrP^{c} and Hsp60/GroEL has been mapped between aa 180 and 210 of the prion protein, a region encompassing helices 2 and 3 (Riek et al., 1996) (Figure 16.10) or helices B and C (Donne et al., 1997). Recently, Caughey demonstrated the in vitro conversion of recombinant proteinase K-sensitive PrP into its proteinase K resistant form (Raymond et al., 1997), suggesting that molecular chaperones could also catalyze the conversion of PrP^{c} into PrP^{Sc}. (3) Molecular chaperones prevent PrP^{c} from being converted to PrP^{Sc} as observed in the case of chemical chaperones such as DMSO, TMAO or glycerol and PrP^{c} in cell culture (Tatzelt et al., 1996).

In summary, we want to point out that none of the above mentioned theories have been demonstrated and that either theory has still to be experimentally confirmed.

6. PERIPHERAL PATHOGENESIS OF PRIONS

6.1. OCCURRENCE OF INFECTIVITY

In natural scrapie in sheep and experimental models in rodents, infectivity has been evidenced in several tissues, organs and body fluids outside the brain, which have been classified by the World Health Organization in four categories according to their infectivity level (Table 16.3).

In TSEs, the only means to measure infectivity is the bioassay using intracerebral inoculation into mice. These experiments have been repeated for BSE showing that the infectious level in peripheral compartments, with regard to the brain, is relatively lower than in sheep scrapie, and thus is undetectable with the mouse bioassay. Infectivity was found in brain, spinal cord, eye, but in cattle experimentally challenged by the oral route, also in the ileum,

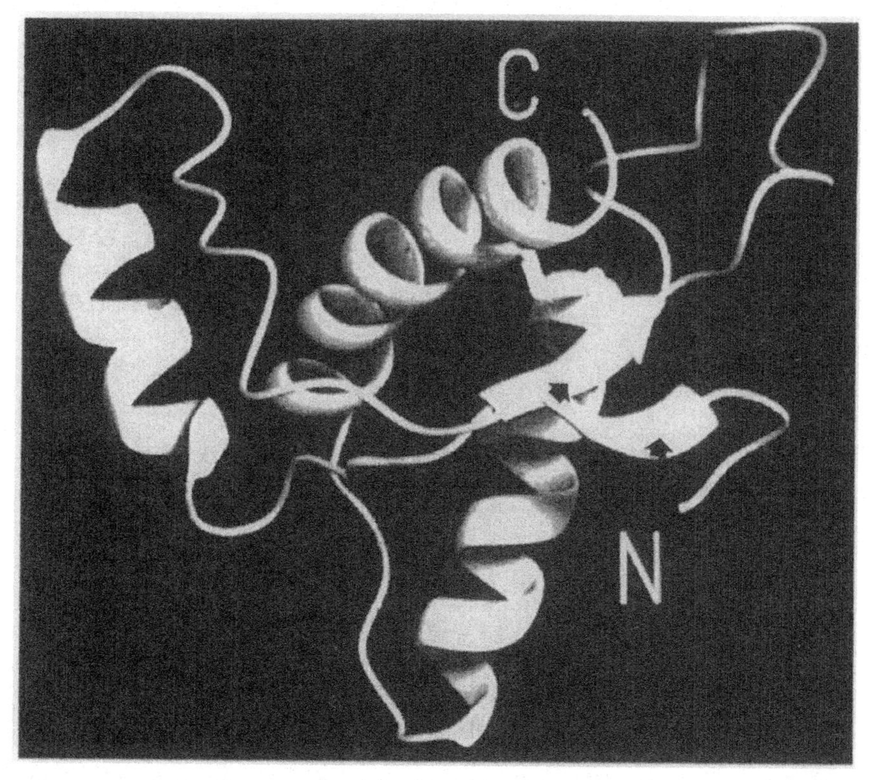

Figure 16.10 NMR-structure of the prion protein. The curled structures represent α-(or alpha) helices 1, 2 and 3 (Riek et al., 1996), and correspond in principle to helices A, B and C, (Donne et al., 1997). The two blue arrows describe two β-(beta) sheet structures. Courtesy of Dr. Wüthrich, Switzerland.

(containing the lympoid tissue Peyer's Patches), the dorsal root ganglia and the bone marrow. On the contrary, no infectivity could be detected in blood, bone marrow, tallow, gastrointestinal tract, heart, kidney, liver, lung, lymph nodes, muscle, peripheral nerves, pancreas, gonads, skin, spleen, trachea or tonsils of field cases of BSE (H. Fraser, 1994, Brüssel, 14–15, September 1993). No infectivity was detected in milk given orally to mice (Middleton and Barlow, 1993).

6.2. PATHWAY OF INFECTION

In most experimental models of sheep scrapie, infectivity can be first evidenced in organs of the lympho-reticular system (LRS) such as tonsils, thymus, lymph nodes and spleen. One might expect that a similar distribution of low-level infectivity would be found in BSE cattle with a more sensitive assay

TABLE 16.3. **WHO Classification of Tissues with Regard to Their Infectivity (Data from Sheep Scrapie, Bioassay in Mice).**

Category I: High Infectivity
Brain, spinal cord
Category II: Medium Infectivity
Spleen, lymph nodes, tonsils, ileum, proximal colon
Category III
IIIa: Low Infectivity
Sciatic nerve, distal colon, adrenal, nasal mucosa, hypophyse
IIIb: Minimal Infectivity
Thymus, bone marrow, liver, lung, pancreas, cerebrospinal fluid
Category IV: No Detectable Infectivity
Blood clot, serum, milk, colostrum, mammary gland, squelettal muscle, heart, kidney, thyroid, salivary gland, saliva, ovary, uterus, testis, seminal gland, feces.

such as homologous inoculation to cattle. The kinetics of infectivity in the LRS and other separate body organs indicates that the infectious agent travels by the lymphatic vessels and the bloodstream. Isolation of infectivity from the blood during the the clinical and preclinical stages of disease has been proven by several researchers (Kozak et al., 1996). The recent observation that B-lymphocytes participate in neuroinvasiveness could account for the hypothesis that B-lymphocytes work as a blood-borne carrier of the infectious agent (Klein et al., 1997). Alternatively, B-lymphocytes could act in the spleen and they might contact terminal nerve endings or have an indirect role by ensuring the maturation of other replication sustaining cells (Klein et al., 1998). Indeed, whole-body irradiation has shown that cells sustaining replication within the LRS are mitotically quiescent (Fraser and Farquhar, 1987), and immunomorphological analyses of cells exhibiting increased PrP labeling after scrapie infection point toward follicular dendritic cells (Kitamoto et al., 1991; McBride et al., 1992; Muramoto et al., 1992). Thus, the exact implication of either cell population for peripheral pathogenesis is still unclear. After infection has been established in peripheral organs, infectivity and PrP^{Sc} become detectable in the central nervous system; first in the thoracic region of the spinal cord, then in the lumbar and cervical cord segments and finally in the brain (Kimberlin and Walker, 1980). This targeting of the infectious agent strongly suggests that the pathway of neuroinvasion from the spleen to the central nervous system involves visceral autonomic (probably sympathetic) fibers. Studies using various types of immunodeficient mice have shown the importance of a primary replication of the agent in the LRS (which requires the integrity of at least B cell function) but also show that there probably exist(s) alternative(s), less efficient direct route(s) of infection through direct neural spread (Lasmézas et al., 1996a).

7. THE LIFE CYCLE OF PRIONS INCLUDING THE ROLE OF POSSIBLE PRION RECEPTORS

The prion protein PrP is expressed at the rough endoplasmatic reticulum (rER). Recently, Prusiner and Lingappa localized three different forms of PrP at this cellular compartment (see above): the carboxy terminal transmembrane protein ^{Ctm}PrP, the amino terminal transmembrane protein ^{Ntm}PrP and the secretory form of the prion protein ^{Sec}PrP (Hegde et al., 1998).

The third and classical isoform of the prion proteins, ^{Sec}PrP is secreted via the golgi and secretory granules to the cell surface (Figure 16.11). Preliminary data suggest that PrP^{c} could form dimers presumably already in the secretory pathway. PrP dimerization has been reported for wild-type hamster prion proteins (Priola et al., 1995) and very recently for hamster prion proteins with additional octarepeats (Priola and Chesebro, 1998). PrP^{c} appears then on the surface of the neuronal (or other scrapie-infectable) cell anchored by glycosylphosphatidyl inositol (GPI) (Rogers et al., 1991). Although treatment of the cells with phospholipase c (PIPL-c) removes PrP^{c} from the cell surface by cleaving off its GPI anchor (Caughey and Raymond, 1991), some researchers have evidence that a transmembrane region (TM) of PrP^{c} exists. (Harris and Lehmann, 1996), which is responsible for fixing a mutant PrP on the cell surface.

7.1. IS THE 37 kDa LAMININ RECEPTOR PRECURSOR (LRP) THE PRION RECEPTOR?

Recently, we identified the 37 kDa laminin receptor precursor (LRP) as a receptor interacting with PrP^{c} (Figure 16.11) (Rieger et al., 1997). We therefore propose that LRP acts as a receptor or co-receptor for PrP^{c} and/or PrP^{Sc} (alone or together with another component of the agent). LRP is present in higher amounts in several organs and tissues of scrapie-infected mice and hamsters such as brain, spleen and pancreas compared to uninfected control animals (Rieger et al., 1997). LRP and PrP^{c} are both present on the surface of neuronal cells (Rieger et al., 1997). We strongly suggest that LRP is necessary for the uptake of PrP^{c} and possibly PrP^{Sc} into the cell. In the life cycle of prions, PrP^{c} becomes internalized by caveolae-like domains (Vey et al., 1996) or clathrin-coated pits (Shyng et al., 1994), a process that is thought to require LRP (Rieger et al., 1997). PrP^{Sc} enters the nerve cell either LRP-mediated or directly, and comes in contact with PrP^{c} during the endolytic pathway (Figure 16.11) involving endolysosomes and lysosomes. The conversion of PrP^{c} into PrP^{Sc}

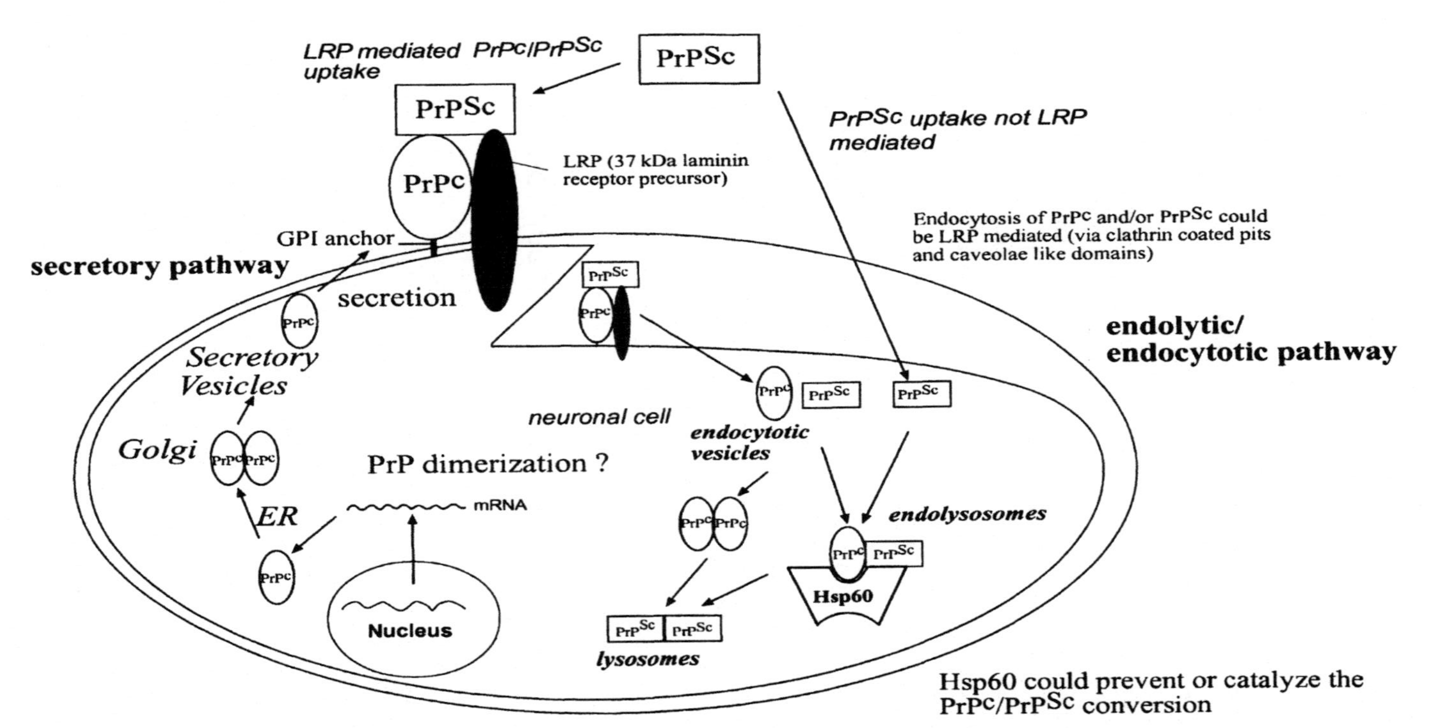

Figure 16.11 The life cycle of prions. Endocytosis of PrP^c and eventually PrP^{Sc} via caveolae-like domains or clathrin coated pits could be laminin receptor precursor (LRP) mediated. Conversion of PrP^c to PrP^{Sc} is thought to take place in the endolysosomes or lysosomes of the cell. Molecular chaperones are thought to be involved in that process. PrP dimerization could play an essential role in the life cycle of prions. PrP replication can occur in brain cells such as neuroblastoma cells but also in other cells of the organism located for example in the spleen or the pancreas.

is thought to take place in any of these compartments either independent of any additional co-factor or with the help of co-factors such as molecular chaperones (Tatzelt et al., 1995; Edenhofer et al., 1996; Tatzelt et al., 1996). There is evidence that some early steps in the generation of PrP^{Sc} from mutant PrPs might take place in compartments of the secretory pathway such as the ER, the golgi or secretory granules (Lehmann and Harris, 1996). PrP^{Sc} accumulation eventually kills the neuronal cell resulting in vacuolization, which finally destroys the brain and consequently kills the organism.

8. PRION-LIKE ELEMENTS IN THE YEAST (Sup35p, Ure2p)

The prion concept states that a normal noninfectious cellular protein converts to an infectious prion form that reveals the same primary structure as the cellular form but acquires additional features such as the autocatalytic activity involved in the conversion of the normal protein to the infectious form, an altered behavior towards proteinases and the formation of insoluble aggregates.

The non-Mendelian genetic elements in *Saccharomyces cerevisiae* [PSI^+] and [URE3] are equivalent to mammalian prions such as PrP^{Sc}. Sup35p* and Ure2p*, respectively, represent the protein determinants of these genetic elements (Table 16.4). The corresponding non-prion-associated genetic elements are [psi^-] and [*ure*3], the protein determinants Sup35p and Ure2p, respectively. The normal functions of Sup35p and Ure2p are the promotion of the termination of translation and blockage of ureidosuccinate uptake, respectively. The conversion of both protein factors to the prion-like form leads either to the production of nonfunctional proteins or the uptake of ureidosuccinate. Both events are harmful to the yeast, corresponding to the TSEs of mammals leading to the death of the infected individual.

The knock-out of the entire gene encoding Ure2p or deletion of the N-terminus of Ure2p prevents synthesis of Ure2p* (for review, see Wickner, 1995). Overproduction of Ure2p or its N-terminal truncated version, however, leads to the *de novo* synthesis of Ure2p* (Masison and Wickner, 1995).

The conversion of Sup35p to Sup35p* is triggered by the molecular chaperone Hsp104 (Chernoff et al., 1995). Dependent on the concentration of Hsp104, the chaperone can either catalyze the formation of Sup35p* and/or the refolding of Sup35p* to Sup35p. Hsp104 interacts directly with Sup35p and with mammalian prions (Patino et al., 1996) (for review, see Lindquist, 1996). Protease resistant Sup35p* forms, analogous to mammalian prion scrapie-associated fibrils (SAFs) (Paushkin et al., 1996), can refold in the presence of guanidinium hydrochloride, analogous to ure2p* (Wickner et al., 1995), into Sup35 (Tuite et al., 1981).

TABLE 16.4. **Comparison of Mammalian Prions and Prion-Like Elements in Yeast and Fungi.**

	Mammalian Prions	*S. cerevisiae* Prion-Like Element	*S. cerevisiae* Prion-Like Element	*Podospora anserina* Prion-Like Element
Normal form	PrP^{c}	Ure2p	Sup35p	pHET-s*
Prion form	PrP^{Sc}	Ure2p*	Sup35p*	pHET-s
Genetic element of the normal form	—	[ure3]	[psi-]	[Het-s*]
Genetic element of the prion form	—	[URE3]	[PSI+]	[Het-s]
Phenotype of the normal form	PrP^{c} could play roles in • sleep behavior/circadian activities • synaptic processes • survival of purkinje neurons	block of ureidosuccinate uptake	termination of translation	heterocaryon formation
Phenotype of the prion form	amyloid formation vacuolization astrocytosis gliosis causes TSEs	ureidosuccinate uptake	no termination of translation—readthrough	heterocaryon incompatibility
Benefit for the affected organism	no	no	no	yes, prevents transmission of viruses and other micro-organisms damaging the fungi
Consequence	death	cell damage	cell damage	

9. THE PRION-LIKE ELEMENT Het-s IN *PODOSPORA ANSERINA*

Podospora anserina represents a filamentous fungus with a network of multinucleate cells (hyphae). These hyphae are able to fuse to heterokaryotic filaments. Heterokaryon formation is dependent on the genotype of the fusing cells. *Het* are the genes in *P. anserina,* which are responsible for heterokaryon incompatibility. The gene locus *het*-s has two alleles: *het*-s and *het*-S. If both strains (*het*-s/*het*-S) are growing close to each other, heterokaryons from both allelic cells can be formed but fail to survive (Rizet, 1952). Both proteins expressed from the *het*-s and *het*-S alleles are 289 amino acids in length and differ in 14 amino acids. If both proteins differ in just one amino acid, heterokaryon incompatibility exists (Deleu et al., 1993). The incompatibility is due to the formation of a poisonous complex consisting of the *het*-s and *het*-S encoded proteins (Begueret et al., 1994). Homodimeric as well as heterodimeric complexes between the het proteins have recently been confirmed in a yeast two-hybrid system (Coustou et al., 1997).

P. anserina includes two non-Mendelian elements [Het-s] and [Het-s*]. The corresponding proteins are pHET-s and pHET-s*. They behave like the prion-like elements in yeast sup35p and ure2p and the mammalian prions PrP^{c} and PrP^{Sc}. The prion form pHET-s is more proteinase K resistant than the normal form pHET-s* (Coustou et al., 1997) and is able to convert pHET-s* to pHET-s. The spontaneous conversion of pHET-s* to pHET-s was achieved by simple overexpression of the *het*-s gene (Coustou et al., 1997). In the case of *Podospora anserina,* the prion-like element is denoted pHET-s whereas in yeast the prion-like elements are denoted Sup35p* and Ure2p*.

Table 16.4 summarizes the mammalian prions and prion-like factors in *S. cerevisiae* and *P. anserina.* The only prion-like factor that is advantageous for the organism is the pHET-s protein. Heterokaryon incompatibility prevents the transmission of viruses and other harmful microorganism from one hyphae to the other. We are convinced that the mechanism of prions represents a general mechanism, and that we will find many more examples of this in nature.

10. THREE DIFFERENT FACTORS LEADING TO TSE: SPORADIC, FAMILIAL, TRANSMITTED (IATROGENIC)

Three different possibilities exist for an organism to develop a TSE: (1) sporadic in the absence of any mutation within the *Prn*-p gene, (2) familial due to defined mutations within the *Prn*-p locus and (3) transmitted. Iatrogenic cases describe humans who were infected by surgery or treatment with products such as the human growth hormone (HGH) extracted from humans suffering from CJD. About 55 iatrogenic CJD cases have been reported in France.

10.1. THE GENOMIC STRUCTURE AND LOCATION OF *Prn*-p GENES

Prn-p represents a single-copy gene, which is located on chromosome 20 in humans (Sparkes et al., 1986), on chromosome 2 in mice (Sparkes et al., 1986), on chromosome 3 in rats (Kuramoto et al., 1994) and on chromosome 11 in minks (Khlebodarova et al., 1995). The *Prn*-p gene of humans contains two exons (Puckett et al., 1991) (for review, see Schätzl et al., 1995). The *Prn*-p genes of hamsters and mice contain two and three exons, but the complete *Prn*-p open reading frame encoding PrP^c is located on exon two in hamsters and humans and exon three in mice (Prusiner, 1991). A detailed comparison of the primary structures of PrP^c molecules of different species has been performed by Schätzl (Schätzl et al., 1995).

10.2. THE OCTA(PEPTIDE)REPEAT (GLYCINE-PROLINE) REGION

Characteristic for PrP are repetitive sequences between amino acids 51 and 93 (in the case of human PrP), which encode a Glycine-Proline-rich sequence encompassing eight amino acids (octa-repeats) and which are identical in all species examined so far. Whereas most of the investigated species show alleles with five octa-repeats, cows contain alleles with five and six repeats (additional octa-repeats at position 86; Figure 16.12,A). A recent report claims that the octa-repeat region specifically binds copper in vivo (Brown et al., 1997). Moreover, additional octa-repeats (more than 7) in the hamster prion protein led to a higher aggregated form of the protein concomitant with an increased proteinase K resistance (Priola and Chesebro, 1998). Additional octa-repeats in human PrP causes Creutzfeldt-Jakob disease (Owen et al., 1990), suggesting that the octa-repeat region is of high pathogenic relevance for the development of a TSE.

10.3. POLYMORPHISMS AND MUTATIONS WITHIN THE *Prn*-p LOCUS OF DIFFERENT SPECIES

In the case of sheep, polymorphisms of PrP^c at positions 136, 156 and 171 are known. These influence incidence rate and incubation time after experimental infection with scrapie prions (Figure 16.12, A) (Goldmann et al., 1994). Mice encompass two alleles of the *Prn*-p gene, which differ in codons 108 and 189 (Figure 16.12, A). Mice encompassing the allele *Prn*-p^a show after inoculation with RML-prions a short incubation time whereas mice with the *Prn*-p^b allele reveal a longer incubation time (for review, see (Prusiner, 1991).

In humans 85% of CJD cases occur spontaneously. However, 15% of the CJD cases, are due to mutations within the *Prn*-p gene. In these patients, mutations within the *Prn*-p gene have been identified, which led to amino

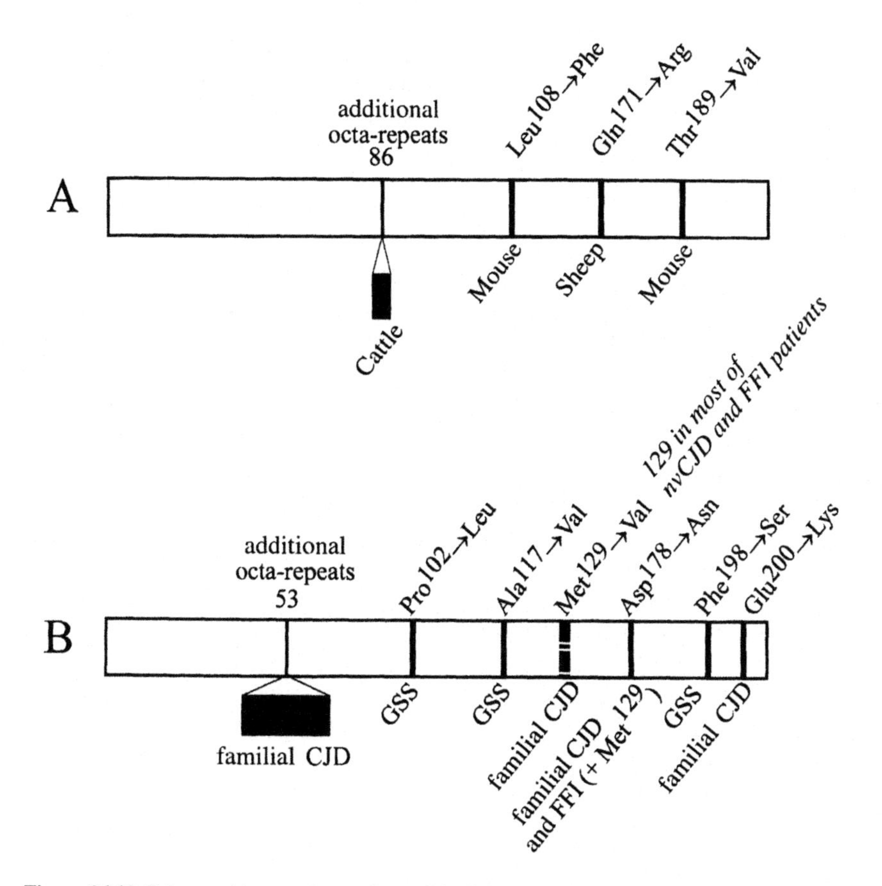

Figure 16.12 Polymorphisms and mutations of the PrP^c primary structure in various animals (A) and in humans (B).

acid substitutions at positions 178 and 200 as well as to additional octa-repeats at position 53 (Owen et al., 1990) (Figure 16.12, B). On the contrary, GSS and FFI are in any case due to mutations and, therefore, represent exclusively familial TSEs (Figure 16.12, B). Mutations within the *Prn*-p gene resulting in amino acid substitutions at positions 102, 117, and 198 lead to GSS. The amino acid substitution at position 117 (Alanine to Valine) is associated with the recently discovered ^{Ctm}PrP isoform (Hegde et al., 1998). The *Prn*-P genotype 178^{Asn} together with the polymorphism 129^{Met} causes fatal familial insomnia (FFI; Tateishi et al., 1995).

11. TRANSMISSION STUDIES AND SPECIES BARRIERS—CJD—NEW VARIANT CJD

Inoculation of a TSE to a new host species leads to low transmission rates, if any at all, and prolonged incubation times. This phenomenon is referred to as the *species barrier.* Its force relies upon several parameters: the strain of the agent, the new host species and, to a lesser extent, the donor host. Therefore, interspecies transmission of a particular agent to a given host cannot be predicted and, until recently, transmission of BSE to humans remained a matter of speculation.

11.1. TRANSMISSION RISKS OF BSE TO HUMANS: CJD EPIDEMIOLOGY AND APPEARANCE OF A NEW CJD SUBTYPE IN ENGLAND (NEW VARIANT CJD)

Classical CJD epidemiological studies have not answered the question whether BSE is transmissible to humans. However, a new variant form of CJD appeared suddenly in the U.K. 10 years after the first histological description of BSE in cattle. It was immediately suspected that the appearance of nvCJD was likely a consequence of BSE and that it was caused by the consumption of products originating from cattle suffering from BSE.

11.2. CLASSICAL CJD EPIDEMIOLOGY IN EUROPE

Mortality rates for classical CJD excluding iatrogenic cases in Europe are listed in Table 16.5. The data have been provided by the EU Collaborative Group for CJD with R. G. Will, CJD Surveillance Unit, U.K., as the responsible scientist.

The rates in individual countries vary between 0.20 cases/million (Spain) and 1.44 (Switzerland) and give rise to an overall annual mortality rate for CJD in Europe of 0.84 cases/million. The extremely high mortality rates in the Netherlands and the low rate in Slovakia and Spain may reflect the small population sizes in these countries and/or underreporting. In summary, there has been a relative consistency in mortality rates, both with time and between countries, until 1997.

11.3. NEW VARIANT (nv) CJD IN ENGLAND AND FRANCE

In March 1996 a new variant of CJD (nv)CJD was reported (Will et al., 1996). Until September 1999, 46 cases of nvCJD have appeared in England accompanied by one case in France (Deslys et al., 1997). These patients differ in the following respects from classical CJD patients: (1) All patients are extremely young, between 16 and 50 years of age (mean of 29 years versus a mean of about 65 years in sporadic CJD). (2) The time between occurrence of the first symptoms and death (course of disease) is 15 months (sporadic CJD, six months). (3) All patients have florid plaques in the brain, which has never been reported in classical CJD patients and show high amounts of

TABLE 16.5. Total Cases of Sporadic CJD Worldwide: 1993–1997.[a] Annual Mortality Rates (Cases/Million/Year) Excluding Iatrogenic Cases.

	Australia	Austria	France	Germany	Italy	Netherlands	Slovakia	Spain	Switzerland	UK
1993	1.02	0.77	0.60	0.44	0.46	0.72	0	0.35	1.44	0.65
1994	0.62	1.28	0.77	0.80	0.58	1.18	0.80	0.20	1.42	0.87
1995	1.05	1.15	1.02	1.00	0.49	0.52	0.40	0.41	1.27	0.60
1996	1.32	1.15	1.19	0.89	0.83	0.83	0.40	0.51	1.41	0.70
1997	1.13	0.77	1.38	1.22	0.74	1.21	0.40	0.43	1.41	0.99
Mean Rate	1.03	1.02	0.99	0.87	0.62	0.89	0.4	0.38	1.39	0.76

[a] *Source:* The European and allied Countries Collaborative Study Group of CJD (EUROCJD).

proteinase K resistant material in the basal ganglia and cerebellum. (4) Pseudo-periodic EEG waves observed in classical CJD are absent in nvCJD patients. (5) First analysis revealed that all nvCJD patients were homozygous for methionine at position 129 (Zeidler et al., 1997). However, it cannot be stated that heterozygocity at position 129 leads to nvCJD immunity. It is supposed that nvCJD patients have consumed BSE-infected material during the late eighties or the early nineties where high amounts of BSE-contaminated material has been present in the food chain.

11.4. EXPERIMENTAL DATA IN FAVOR OF THE TRANSMISSIBILITY OF BSE TO HUMANS RESULTING IN nvCJD

(1) Macaques inoculated with prions originating from BSE-infected cattle developed the same florid plaques as nvCJD patients (Lasmézas et al., 1996b). (2) Experiments employing transgenic animals carrying the *Prn*-P gene coding for human PrP developed the disease after inoculation with BSE prions (Hill et al., 1997). (3) Wild-type mice inoculated with prions from humans suffering from nvCJD developed the disease after the same period of time (incubation time) as animals inoculated with BSE prions and showed the same lesion profile (Bruce et al., 1997); and (4) human PrP can be converted by bovine PrP^{BSE} into the proteinase K resistant state during in vitro conversion assays (Raymond et al., 1997). From these data we conclude that BSE is transmissible to humans.

11.5. TRANSMISSION OF BSE TO HUMANS, SPECIES BARRIERS: ORAL, PARENTERAL AND INTRACEREBRAL TRANSMISSION

BSE transmission to humans cannot be tested by oral, parenteral or intracerebral routes for obvious ethical reasons. However, data are available from oral transmission attempts of BSE to several mammalian species. The agent has been transmitted by oral route from cows to sheep (Bradley and Wilesmith, 1993), to goats (Bradley and Wilesmith, 1993), to minks (Robinson et al., 1994), to mice (Barlow and Middleton, 1990) and to cows (Roy Bradley, Paris Symposium, 1996, Personal Communication) but not to chickens (Bradley and Wilesmith, 1993) or pigs (Bradley and Wilesmith, 1993). The BSE agent, however, has been transmitted to pigs by parenteral routes (Dawson et al., 1990). Recently, an immunohistological study of PrP in the digestive tracts of 200 apparently healthy lemurs and experimentally orally contaminated lemurs revealed labeling indicative of a BSE infection in these animals (Bons et al., 1999).

The BSE agent has been transmitted intracerebrally to cows (Barlow and Middleton, 1990), minks (Robinson et al., 1994), mice (Fraser et al., 1988);

(Barlow and Middleton, 1990), marmosets (Baker et al., 1993) and macaques (Lasmézas et al., 1996b). The scrapie agent from sheep has also been transmitted intracerebrally to mice (Fraser et al., 1988; Barlow and Middleton, 1990; Fraser et al., 1992) and cows (Cutlip et al., 1994). The agent causing feline spongiform encephalopathies in cats (FSE) has been transmitted to mice, which revealed its identity with the BSE agent (Fraser et al., 1994); indeed, the lesion profile characterization of an agent strain for a given host was the same in syngenic mice.

11.6. INFECTION RISKS BY BODY FLUIDS [MILK, CEREBROSPINAL FLUID (CSF), BLOOD, URINE]

The agent has been proven to be present in the CSF. There are some reports that TSE agents could be present in the blood, but only in very small amounts. The finding that B-lymphocytes play an essential role in scrapie pathogenesis (Brown, 1997; Klein et al., 1997) gave rise to new safety concerns regarding blood and blood products. In milk and milk products TSE agents have never been identified so far (Middleton and Barlow, 1993). It is also highly speculative if the agent could be present in urine, and some diagnostic assays focusing on this body fluid are based on molecular markers different from the agent itself.

12. DIAGNOSTICS FOR TSE

12.1. BRAIN AUTOPSY/BIOPSY

The classical diagnostic method for detection of a TSE is an autopsy of the brain post mortem. The neuropathologist confirms the disease by an observation of vacuolization, PrP^{res} accumulation, astrocytosis, gliosis and other parameters of TSE infection. A brain biopsy is also possible but is sometimes ethically problematic.

12.2. TONSIL BIOPSY

Tonsil tissues are analyzed from the vital organism. After proteinase K treatment and incubation with a PrP antibody, PrP can be detected in the tissue sections or by western blotting. The assay works in the case of sheep (Schreuder et al., 1996) and also in humans infected with nvCJD (Hill et al., 1999).

12.3. PROTEINASE K ASSAY FOLLOWED BY PrP^{res} DETECTION WITH PrP ANTIBODIES

In the classical proteinase K assay, homogenates from the brain of potentially

TSE-infected individuals or animals are digested with proteinase K. The cellular PrP^c becomes completely degraded whereas the scrapie isoform PrP^{Sc} is processed to PrP27–30, which can be detected with a PrP-specific antibody. This methodology is highly reliable and has led recently to the development of powerful post-mortem diagnostic tests.

12.4. PROOF OF PROTEIN MARKERS SPECIFIC TO TSE IN BODY FLUIDS

The 14.3.3 protein (Hsich et al., 1996; Zerr et al., 1998) and the neuron-specific enolase (NSE) (Zerr et al., 1995) have been detected in the cerebrospinal fluid (CSF) of patients suffering from TSE. Unfortunately, both protein markers are not 100% specific for TSEs. The 14.3.3 protein has also been detected in cases of stroke and encephalitis. A further disadvantage of these protein markers is the time of detection. Both markers become visible only when the first symptoms of the disease occur.

12.5. ELECTROCHEMICALLY ACTIVE METABOLITES AS TSE-SPECIFIC MARKERS IN URINE

R. Jackman, U.K. is developing an assay initiated in France (Brugère et al., 1991) for detection of TSE-induced electro potential changes that occur by oxidation of catechol and tryptophan-like metabolites in the urine of BSE-infected cattle and scrapie-infected sheep (Jackman and Everest, 1993). The sensitivity of this assay and the strict correlation of this oxidation reaction with the occurrence of TSE symptoms has still to be evaluated.

12.6. PrP^{Sc}-SPECIFIC ANTIBODIES

For a long time antibodies failed to distinguish between PrP^c and PrP^{Sc} (Groschup et al., 1997). Recently, a PrP^{Sc}-specific antibody was developed, which recognizes specifically PrP^{Sc}, failing to interact with PrP^c by immunoprecipitation (Korth et al., 1997). This antibody (15 B3) might be a very suitable tool for detection of TSEs. The sensitivity of this antibody and its practicability have to be demonstrated.

12.7. PrP^c- AND PrP^{Sc}-SPECIFIC RNA APTAMERS

Employing in vitro selection or SELEX (systematic evolution of ligands by exponential enrichment), we identified RNA aptamers (Lat. aptus = to fit), which recognized specifically PrP in brain homogenates of mice, hamster and cattle (Weiss et al., 1997). Those aptamers could be used for the development of a diagnostic assay for the detection of TSEs.

13. THERAPEUTICS FOR TSE

13.1. MOLECULAR CHAPERONES

To date, no effective therapy is available for TSE. The research in this field is heavily hampered by the uncertainties remaining about the replication mechanisms of the etiological agent. However, one central event is the conversion of PrP^{c} into PrP^{Sc}. Molecules such as chaperones may support the conversion process or, on the contrary, stabilize the normal form of the protein. It is reasonable to expect that the blockage of PrP^{c} conversion into PrP^{Sc} would impede prion propagation. Thus, research on molecular chaperones involvement in PrP conversion constitutes one facet of TSE therapeutic studies.

13.2. THERAPEUTIC DRUGS

Most of the classical drugs used in human therapy have been tested. Among them, only a few delay the appearance of the disease in rodents experimentally infected with TSE agents; these are presented below. Presently, their mode of action remains unknown.

13.2.1. Polyanions

Polyanions such as heteropolyanion 23 [HPA-23], dextran sulfate 500 [DS-500] and pentosan sulfate [SP-54] are efficient in experimental scrapie only when the treatment occurs at the time of infection (Kimberlin and Walker, 1986; Ladogana et al., 1992). Pentosane polysulfate is even able to prevent the disease under certain conditions (Farquhar et al., 1999). These drugs probably act on the primary replication phase of the agent in the LRS. They are taken up by the lymphoid cells and may inhibit entrance or replication of the agent, thereby slowing down the disease process.

13.2.2. Congo Red

Congo red inhibits PrP^{Sc} accumulation and prion replication in chronically, scrapie-infected, mouse neuroblastoma cells (ScN2a cells) (Caughey et al., 1993). In vivo, it increases the survival time of hamsters inoculated intraperitoneally and treated by the time of infection (Ingrosso et al., 1995). Congo red may also act during the early phases of infection; its ability to insert into amyloid proteins suggests that it interferes directly with the formation of PrP^{Sc}.

13.2.3. The Anthracycline 4′-Iodo-4′-deoxy-doxorubicin (IDX)

IDX is a derivative of the drug doxorubicin, which has proven efficacy in a large number of malignancies (Tagliavini et al., 1997). IDX binds to amyloid fibrils and even induces amyloid resorption. When coincubated with the inoculum, IDX is able to delay the onset of clinical disease in hamsters infected by intracerebral route with the 263K strain of agent. At a molecular level, a significant decrease of PrP^{Sc} accumulation was observed in the brains of IDX-treated hamsters as compared with the infected control animals. The authors suggested that the delay in PrP^{Sc} accumulation may have been the consequence of IDX's binding to abnormal PrP, diminishing its ability to act as a template for the conversion of the normal protein (Tagliavini et al., 1997).

13.2.4. Amphotericin B (AmB)

AmB is a polyene macrolide antibiotic widely used for the treatment of systemic fungal infections such as candidiasis, histoplasmosis and aspergillosis (Medoff et al., 1983). The beneficial effects of AmB in scrapie have been evidenced mainly in the 263K/hamster model; they are dose-dependent and correlate with the duration of the treatment (Pocchiari et al., 1989). AmB prolongs the incubation time of intracerebrally and intraperitoneally infected rodents. Treatment during the early stages of infection appears again to be the best regimen, and no significant effect could be observed when the animals were treated after the appearance of symptoms (Pocchiari et al., 1989). The retardation of the clinical signs is accompanied at a molecular level by a delay in the accumulation of PrP^{Sc} and GFAP in the brain. At terminal stage of disease, however, the level of these marker proteins is similar in treated and untreated animals. The development of this drug as a therapeutical tool for TSEs is restricted, however, by its acute nephrotoxicity and its inefficiency in TSE agents other than 263K. It has constituted the backbone for other studies on AmB derivatives.

13.2.5. The AmB Derivative MS-8209

M-8209 is derived from a modification of the mycosamine of AmB. It is at least five times less toxic than AmB, allowing the use of doses that could never be attained with the parental molecule. This permitted widening of the spectrum of TSE strains that were sensitive to treatment with polyene macrolide antibiotics (in particular to experimental mouse scrapie and BSE, Adjou et al., 1996) and to show that even a treatment beginning late during infection, after half of the incubation period had elapsed (that is, when neuroinvasion had been primed), was able to increase the survival time of the animals (Demaimay et al., 1997). When MS-8209 was administered as a long-term treatment in hamster scapie at high doses (10 and 25 mg/kg of body weight), a doubling of the survival time was observed compared with infected, untreated animals (Adjou et al., 1995). This represents the most important delay ever reported for TSEs.

The effect of MS-8209 is also observed at the molecular level with a delay in PrP^{Sc} and GFAP accumulation. Moreover, the therapeutic efficacy of the molecule upon the different strains of agent tested seems to be coincidental with the extent to which PrP^{Sc} accumulates in the brain of terminally ill animals, with the following gradient of efficiency and PrP^{Sc} levels: 263K/hamster >scrapie C506M3/mouse>BSE 4PB1/mouse. These data favor the hypothesis that polyene antibiotics act by interfering with the conversion of PrP into its abnormal isoform either by interacting directly with one of the isoforms or with a third factor of the conversion process, or by intercalating in the cell membranes (which is the mechanism leading to the death of fungi).

14. CONCLUDING REMARKS

TSEs are exceptional lethal diseases caused by a completely new class of infectious agents named TSE agents or prions. A total of 47 young people (status September 1999) suffering or died from new variant CJD in Europe are more than enough to justify the search for powerful diagnostics and therapeutics for detection and treatment of these harmful diseases. Common efforts worldwide will be necessary to investigate the complex life cycle of prions which might involve cell surface receptors, molecular chaperones, completely new PrP isoforms and other types of molecules.

15. ACKNOWLEDGMENTS

We thank Sabine Gauczynski, Christoph Hundt, Christoph Leucht and Roman Rieger for critical reading of the manuscript. Stefan Weiss thanks the Bundesministerium für Bildung, Forschung, Wissenschaft und Technologie (BMBF) grant # KI 01–9760 and the European Union (EU) grants # FAIR5-CT97-3314, # FAIR6-CT98-7020 and BIOMED PL976054 for financial support, and Ernst-Ludwig Winnacker for valuable advice and continous support.

16. REFERENCES

Adjou, K. T., Demaimay, R., Lasmezas, C., Deslys, J. P., Seman, M., and Dormont, D. 1995. "MS-8209, a new amphotericin B derivative, provides enhanced efficacy in delaying hamster scrapie," *Antimicrob. Agents Chemother.* 39:2810–2812.

Adjou, K. T., Demaimay, R., Lasmézas, C. I., Seman, M., Deslys, J. P., and Dormont, D. 1996, "Differential effects of a new amphotericin B derivative, MS-8209, on mouse BSE and scrapie: implications for the mechanism of action of polyene antibiotics," *Res. Virol.* 147:213–218.

Alper, T., Haig, D. A., and Clarke, M. C. 1968. "The exceptionally small size of the scrapie agent," *Biochem. Biophys. Res. Commun.* 22:278–284.

Alper, T., Cramp, W. A., Haig, D. A., and Clarke, M. C. 1967. "Does the agent of scrapie replicate without nucleic acid?" *Nature* 214:764–766.

Baker, H. F., Ridley, R. M., and Wells, G. A. H. 1993. ''Experimental transmission of BSE and scraple to the common marmoset,'' *Vet. Rec.* 132:403–406.

Barlow, R. M. and Middleton, D. J. 1990. ''Diatery transmission of bovine spongiform encephalopathy to mice,'' *Vet. Rec.* 126:111–112.

Begueret, J., Turcq, B., and Clave, C. 1994. ''Vegetative incompatibility in filamentous fungi: het genes begin to talk,'' *Trends Genet.* 10:441–446.

Bessen, R. A., Kocisko, D. A., Raymond, G. J., Nandan, S., Lansbury, P. T., and Caughey, B. 1995. ''Non-genetic propagation of strain-specific properties of scrapie prion protein,'' *Nature* 375:698–700.

Bons, N., Nestre-Frances, N., Belli, R., Cathela, F., Gajdusek, D. C., and Brown, P. 1999. ''Natural and experimental oral infection of non-human primates by bovine spongiform encephalopathy agents,'' *Proc. Nat. Acad. Sci. USA* 96:6046–6051.

Borchelt, D. R., Scott, M., Taraboulos, A., Stahl, N., and Prusiner, S. B. 1990. ''Scrapie and cellular prion proteins differ in their kinetics of synthesis and topology in cultured cells,'' *J. Cell. Biol.* 110:743–752.

Bradley, R. and Wilesmith, J. W. 1993. ''Epidemiology and control of bovine spongiform encephalopathy (BSE),'' *British Med. Bull.* 49:932–959.

Brandner, S., Isenmann, S., Raeber, A., Fischer, M., Sailer, A., Kobayashi, Y., Marino, S., Weissmann, C., and Aguzzi, A. 1996. ''Normal host prion protein necessary for scrapie-induced neurotoxicity,'' *Nature* 379:339–343.

Brown, D. R., Qin, K., Herms, J. W., Madlung, A., Manson, J., Strome, R., Fraser, P. E., Kruck, T., von, B. A., Schulz, S. W., Giese, A., Westaway, D., and Kretzschmar, H. 1997. ''The cellular prion protein binds copper in vivo,'' *Nature* 390:684–687.

Brown, P. 1997. ''B lymphocytes and neuroinvasion,'' *Nature* 390:662–663.

Bruce, M. E., Will, R. G., Ironside, J. W., McConnell, I., Drummond, D., Suttie, A., McCardle, L., Chree, A., Hope, J., Birkett, C., Cousens, S., Fraser, H., and Bostock, C. J. 1997. ''Transmissions to mice indicate that 'new variant' CJD is caused by the BSE agent,'' *Nature London* 388:498–501.

Brugère, H., Banissi, C., Brugère-Picoux, J., Chatelain, J., and Buvet, R. 1991. ''Recherche d'un témoin biochemique urinaire de l'infection du mouton par la tremblante,'' *Bull. Acad. Vét. de France* 64:139–145.

Bueler, H., Aguzzi, A., Sailer, A., Greiner, R. A., Autenried, P., Aguet, M., and Weissmann, C. 1993. Mice devoid of PrP are resistant to scrapie,'' *Cell* 73:1339–1347.

Bueler, H., Fischer, M., Lang, Y., Bluethmann, H., Lipp, H. P., DeArmond, S. J., Prusiner, S. B., Aguet, M., and Weissmann, C. 1992. ''Normal development and behaviour of mice lacking the neuronal cell-surface PrP protein,'' *Nature* 356:577–582.

Caughey, B., Ernst, D., and Race, R. 1993. ''Congo red inhibition of scrapie agent replication,'' *J. Virol.* 67:6210–6272.

Caughey, B., Race, R. E., Ernst, D., Buchmeier, M. J., and Chesebro, B. 1989. ''Prion protein biosynthesis in scrapie-infected and uninfected neuroblastoma cells,'' *J. Virol.* 63:175–181.

Caughey, B. and Raymond, G. J. 1991. ''The scrapie-associated form of PrP is made from a cell surface precursor that is both protease and phospholipase-sensitive,'' *J. Biol. Chem.* 266:18217–18223.

Chernoff, Y. O., Lindquist, S. L., Ono, B., Inge-Vechtomov, S. G., and Liebman, S. W. 1995. ''Role of the chaperone protein Hsp104 in propagation of the yeast prion-like factor [psi+],'' *Science* 268:880–884.

Collinge, J., Whittington, M. A., Sidle, K. C. L., Smith, C. J., Palmer, M. S., Clarke, A. R., and

Jefferys, J. G. R. 1994. "Prion protein is necessary for normal synaptic function," *Nature* 370:295–297.

Coustou, V., Deleu, C., Saupe, S., and Begueret, J. 1997. "The protein product of the het-s heterokaryon incompatibility gene of the fungus *Podospora anserina* behaves as a prion analog, *Proc. Natl. Acad. Sci. USA* 94:9773–9778.

Creutzfeldt, H. G. 1920. "Uber eine eigenartige Erkrankung des Zentralnervensystems," *Z. f. d. g. Neur. u. Psych. O.* LVII:1–18.

Cutlip, R. C., Miller, J. M., Race, R. E., Jenny, A. L., Katz, J. B., Lehmkuhl, H. D., DeBey, B. M., and Robinson, M. M. 1994. "Untracerebral transmission of scrapie to cattle," *J. Infect. Dis.* 169:814–820.

Dawson, M., Wells, G., Parker, B., and Scott, A. C. 1990. "Primary parenteral transmission of bovine spongiform encephalopathy to the pig," *Veterinary Record* 127:338.

Deleu, C., Clave, C., and Begueret, J. 1993. "A single amino acid difference is sufficient to elicit vegetative incompatibility in the fungus *Podospora anserina,*" *Genetics* 135:45–52.

Demaimay, R., Adjou, K. T., Beringue, V., Demart, S., Lasmézas, C. I., Deslys, J.-P., Seman, M., and Dormont, D. 1997. "Late treatment with polyene antibiotics can prolong the survival time of scrapie-infected animals," *J. Virol.* 71:9685–9689.

Deslys, J. P., Lasmézas, C. I., Streichenberger, N., Hill, A., Collinge, J., Dormont, D., and Kopp, N. 1997. "New variant Creutzfeldt-Jakob disease in France," *Lancet British edition* 349:30–31.

Dickinson, A. G. and Outram, G. W. 1979. "The scrapie replication-site hypothesis and its implication for pathogenesis," in Prusiner, S. B. and Hadlow, W. J., *Slow Transmissible Diseases of the Nervous System, eds., Academic Press.* New York: pp. 13–31.

Dickinson, A. G. and Outram, G. W. 1988. "Genetic aspects of unconventional virus infections: the basis of the virino hypothesis," Ciba Foundation Symposium. *Novel Infectious Agents and the Central Nervous System.* Chichester: John Wiley & Sons. pp. 63–83.

Dickinson, A. G., Outram, G. W., Taylor, D. M., and Foster, J. D. 1989. "Further evidence that scrapie agent has an independant genome," in eds., Court, L. A., Dormont, D., Brown, P., and Kingsbury, D. T. Unconventional Virus Diseases of the Central Nervous System, Fontenay-aux Roses, France: CEA Diffusion. pp. 446–459.

Donne, D. G., Viles, J. H., Groth, D., Mehlhorn, I., James, T. L., Cohen, F. E., Prusiner, S. B., Wright, P. E., and Dyson, H. J. 1997. "Structure of the recombinant full-length hamster prion protein PrP(29-231): the N terminus is highly flexible," *Proc. Natl. Acad. Sci. USA.* 94:13452–13457.

Edenhofer, F., Weiss, S., Winnacker, E.-L., and Famulok, M. 1997. "Chemistry and molecular biology of transmissible spongiform encephalopathies," *Angew. Chem. Int. Ed. Engl.* 36:1674–1694.

Edenhofer, F., Rieger, R., Famulok, M., Wendler, W., Weiss, S., and Winnacker, E. L. 1996. "Prion protein PrPc interacts with molecular chaperones of the Hsp60 family," *J. Virol* 70:4724–4728.

Farquhar, C., Dickinson, A., and Bruce, M. 1999. "Prophylactic potential of pentosane polysulphate in transmissible spongiform encephalopathies," *Lancet* 353:117.

Fischer, M., Rulicke, T., Raeber, A., Sailer, A., Moser, M., Oesch, B., Brandner, S., Aguzzi, A., and Weissmann, C. 1996. "Prion protein (PrP) with amino-proximal deletions restoring susceptibility of PrP knockout mice to scrapie," *EMBO J.* 15:1255–1264.

Fraser, H. and Farquhar, C. F. 1987. "Ionising radiation has no influence on scrapie incubation period in mice," *Vet. Microbiol.* 13:211–223.

Fraser, H., McConnell, I., Wells, G. A., and Dawson, M. 1988. "Transmission of bovine spongiform encephalopathy to mice," *Vet. Rec.* 123:472.

Fraser, H., Bruce, M. E., Chree, A., McConnell, I., and Wells, G. A. 1992. "Transmission of bovine spongiform encephalopathy and scrapie to mice, *J. Gen. Virol.* 73:1891–1897.

Fraser, H., Pearson, G. R., McConnell, I., Bruce, M. E., Wyatt, J. M., and Gruffydd, J. T. 1994. "Transmission of feline spongiform encephalopathy to mice," *Vet. Rec.* 134:449.

Gajdusek, D. C. and Zigas, V. 1957. "Degenerative disease of the central nervous system in New Guinea," *New Eng. J. Med.* 257:974–978.

Gerstmann, J. 1928. "Über ein noch nicht beschriebenes Reflexphänomen bei einer Erkrankung des zerebellaren Systems," *Wiener Med. Wschr.* 906–908.

Goldmann, W., Hunter, N., Smith, G., Foster, J., and Hope, J. 1994. "PrP genotype and agent effects in scrapie: change in allelic interaction with different isolates of agent in sheep, a natural host of scrapie," *J. Gen. Virol.* 75:1984–1995.

Griffith, J. S. 1967. "Self-replication and scrapie," *Nature* 215:1043–1044.

Groschup, M. H., Harmeyer, S., and Pfaff, E. 1997. "Antigenic features of prion proteins of sheep and of other mammalian species," *J. Immunol. Methods* 207:89–101.

Harris, D. A. and Lehmann, S. 1996. "A cell culture model of familial prion disease," in. Court, L. and Dodet, B. *Transmissible Subacute Spongiform Encephalopathies: Prion Diseases,* eds., Paris: Elsevier. pp. 339–346.

Hegde, R. S., Mastrianni, J. A., Scott, M. R., DeFea, K. A., Tremblay, P., Torchia, M., DeArmond, S. J., Prusiner, S. B., and Lingappa, V. R. 1998. "A transmembrane form of the prion protein in neurodegenerative disease," *Science* 279:827–834.

Hill, A. F., Desbruslais, M., Joiner, S., Sidle, K. C., Gowland, I., Collinge, J., Doey, L. J., and Lantos, P. 1994. "The same prion strain causes vCJD and BSE," *Nature* 389:448–450.

Hill, A. F., Biutterworth, R. J., Joiner, S., Jackson, G., Rosser, M. N., Thomas, D. J., Frosh, A., Tolley, N., Bell, J. E., Spencer, M., King, A., All-Sarrey, S., Ironside, J. W., Lantos, P. L., and Collinge, J. 1999. "Investigation of Creutzfeldt-Jakob disease and other human prion diseases with tonsil biopsy samples," *Lancet* 353:183–189.

Hsich, G., Kenney, K., Gibbs, C. J., Lee, K. H., and Harrington, M. G. 1996. "The 14–3–3 brain protein in cerebrospinal fluid as a marker for transmissible spongiform encephalopathies [see comments]," *New England J. Med.* 335:924–930.

Ingrosso, L., Ladogana, A., and Pocchiari, M. 1995. "Congo red prolongs the incubation period in scrapie-infected hamsters," *J. Virol.* 69:506–508.

Jackman, R. and Everest, S. 1993. "Further development of the electric chemical analysis of urine from cows with BSE," *Proceedings of a Consultation on BSE with the Scientific Veterinary Committee of the commission of the European community.* pp. 369–376.

Kellings, K., Meyer, N., Mirenda, C., Prusiner, S. B., and Riesner, D. 1992. "Further analysis of nucleic acids in purified scrapie prion preparations by improved return refocusing gel electrophoresis," *J. Gen. Virol.* 73:1025–1029.

Khlebodarova, T. M., Malchenko, S. N., Matveeva, N. M., Pack, S. D., Sokolova, O. V., Alabiev, B. Y., Belousov, E. S., Peremislov, V. V., Nayakshin, A. M., Brusgaard, K., and et al. 1995. "Chromosomal and regional localization of the loci for IGKC, IGGC, ALDB, HOXB, GPT, and PRNP in the American mink (*Mustela* vison): comparisons with human and mouse," *Mamm. Genome* 6:705–709.

Kimberlin, R. H. and Walker, C. A. 1980. "Pathogenesis of mouse scrapie: evidence for neural spread of infection to the CNS," *J. Gen. Virol.* 51:183–187.

Kimberlin, R. H. and Walker, C. A. 1986. "Suppression of scrapie infection in mice by heteropolyanion 23, dextran sulfate, and some other polyanions." *Antimicrob. Agents Chemother.* 30:409–413.

Kimberlin, R. H. and Walker, C. A. 1988. "Incubation periods in six models of intraperitoneally injected scrapie depend mainly on the dynamics of agent replication within the nervous system and not the lymphoreticular system,'' *J. Gen. Virol.* 69:2953–2960.

Kitamoto, T., Muramoto, T., Mohri, S., Doh-Ura, K., and Tateishi, J. 1991. ''Abnormal isoform of prion protein accumulates in follicular dendritic cells in mice with Creutzfeldt-Jakob disease,'' *J. Virol.* 65:6292–6295.

Klein, M. A., Frigg, R., Raeber, A. J., Flechsig, E., Hegyi, I., Zinkernagel, R. M., Weissmann, C., and Aguzzi, A. 1998. ''PrP expression in B-lymphocytes is not required for prion neuroinvasion,'' *Nat. Med.* 4:1429–1433.

Klein, M. A., Frigg, R., Flechsig, E., Raeber, A. J., Kalinke, U., Bluethmann, H., Bootz, F., Suter, M., Zinkernagel, R. M., and Aguzzi, A. 1994. ''A crucial role for B cells in neuroinvasive scrapie [see comments]'' *Nature* 390:687–690.

Kocisko, D. A., Priola, S. A., Raymond, G. J., Chesebro, B., Lansbury, P. J., and Caughey, B. 1995. ''Species specificity in the cell-free conversion of prion protein to protease-resistant forms: a model for the scrapie species barrier,'' *Proc. Natl. Acad. Sci. USA* 92:3923–3927.

Kocisko, D. A., Come, J. H., Priola, S. A., Chesebro, B., Raymond, G. J., Lansbury, P. T., Jr., and Caughey, B. 1994. ''Cell-free formation of protease-resistant prion protein,'' *Nature* 370:471–474.

Korth, C., Stierli, B., Streit, P., Moser, M., Schaller, O., Fischer, R., Schulz, S. W., Kretzschmar, H., Raeber, A., Braun, U., Ehrensperger, F., Hornemann, S., Glockshuber, R., Riek, R., Billeter, M., Wuthrich, K., and Oesch, B. 1994. ''Prion (PrPSc)-specific epitope defined by a monoclonal antibody,'' *Nature* 390:74–77.

Kozak, R. W., Golker, C. F., and Stadler, P. 1996. ''Transmissible spongiform encephalopathies (TSE): minimizing the risk of transmission by biological/biopharmaceutical products: an industry perspective,'' *Dev. Biol. Stand.* 88:257–264.

Kuramoto, T., Mori, M., Yamada, J., and Serikawa, T. 1994. ''Tremor and zitter, causative mutant genes for epilepsy with spongiform encephalopathy in spontaneously epileptic rat (SER), are tightly linked to synaptobrevin-2 and prion protein genes, respectively,'' *Biochem. Biophys. Res. Commun.* 200:1161–1168.

Ladogana, A., Casaccia, P., and Ingrosso, L. 1992. ''Sulphate polyanions prolong the incubation period of scrapie-infected hamsters,'' *J. Gen. Virol.* 73:661–665.

Lansbury, P. T. J. and Caughey, B. 1995. ''The chemistry of scrapie infection: implications of the 'ice 9' metaphor,'' *Chemistry & Biology* 2:1–5.

Lasmézas, C. I., Cesbron, J. Y., Deslys, J. P., Demaimay, R., Adjou, K. T., Rioux, R., Lemaire, C., Locht, C., and Dormont, D. 1996a. ''Immune system-dependent and -independent replication of the scrapie agent,'' *J. Virol.* 70:1292–1295.

Lasmézas, C. I., Deslys, J. P., Demalmay, R., Adjou, K. T., Lamoury, F., Dormont, D., Robain, O., Ironside, J., and Hauw, J. J. 1996b. ''BSE transmission to macaques,'' *Nature* 381:743–744.

Lasmézas, C. I., Deslys, J. P., Robain, O., Jaegly, A., Beringue, V., Peyrin, J. M., Fournier, J. G., Hauw, J. J., Rossier, J., and Dormont, D. 1997. ''Transmission of the BSE agent to mice in the absence of detectable abnormal prion protein,'' *Science* 275:402–405.

Lehmann, S. and Harris, D. A. 1996. ''Mutant and infectious prion proteins display common biochemical properties in cultured cells,'' *J. Biol. Chem.* 271:1633–1637.

Liautard, J. P. 1991. ''Are prions misfolded molecular chaperones?'' *FEBS Lett.* 294:155–157.

Lindquist, S. 1996. ''Mad cows meet mad yeast: the prion hypothesis,'' *Mol Psychiatry* 1:376–379.

Lledo, P. M., Tremblay, P., DeArmond, S. J., Prusiner, S. B., and Nicoll, R. A. 1996. ''Mice deficient for prion protein exhibit normal neuronal excitability and synaptic transmission in the hippocampus,'' *Proc. Natl. Acad. Sci.* 93:2403–2407.

Manson, J. C., Clarke, A. R., Hooper, M. L., Aitchison, L., McConnell, I., and Hope, J. 1994. ''129/

Ola mice carrying a null mutation in PrP that abolishes mRNA production are developmentally normal,'' *Mol. Neurobiol.* 8:121–127.

Manuelidis, L., Murdoch, G., and Manuelidis, E. E. 1988. ''Potential involvement of retroviral elements in human dementias,'' Ciba Foundation Symposium. *Novel Infectious Agents and the Central Nervous System.* Chichester: John Wiley & Sons, pp. 117–129.

Martins, V. R., Graner, E., Garcia, A. J., de, S. S., Mercadante, A. F., Veiga, S. S., Zanata, S. M., Neto, V. M., and Brentani, R. R. 1997. ''Complementary hydropathy identifies a cellular prion protein receptor [see comments],'' *Nat Med* 3:1376–1382.

Masison, D. C. and Wickner, R. B. 1995. ''Prion-inducing domain of yeast Ure2p and protease resistance of Ure2p in prion-containing cells,'' *Science* 270:93–95.

McBride, P. A., Eikelenboom, P., Kraal, G., Fraser, H., and Bruce, M. E. 1992. ''PrP protein is associated with follicular dendritic cells of spleens and lymph nodes in uninfected and scrapie-infected mice,'' *J. Pathol.* 168:413–418.

Medoff, G., Brajtburg, J., Kobayashi, G. S., and Bolard, J. 1983. ''Antifungal agents useful in therapy of systemic fungal infections,'' *Annu. Rev. Pharmacol. Toxicol.* 23:303–330.

Meyer, R. K., McKinley, M. P., Bowman, K. A., Braunfeld, M. B., Barry, R. A., and Prusiner, S. B. 1986. ''Separation and properties of cellular and scrapie prion proteins,'' *Proc. Natl. Acad. Sci. USA* 83:2310–2314.

Middleton, D. J. and Barlow, R. M. 1993. ''Failure to transmit bovine spongiform encephalopathy to mice by feeding them with extraneural tissues of affected cattle,'' *Vet. Rec.* 132:545–547.

Muramoto, T., Kitamoto, T., Tateishi, J., and Goto, I. 1992. ''The sequential development of abnormal prion protein accumulation in mice with Creutzfeldt-Jakob disease,'' *Am. J. Pathol.* 140:1411–1420.

Owen, F., Poulter, M., Shah, T., Collinge, J., Lofthouse, R., Baker, H., Ridley, R., McVey, J., and Crow, T. J. 1990. ''An in-frame insertion in the prion protein gene in familial Creutzfeldt-Jakob disease,'' *Mol. Brain Res.* 7:273–276.

Ozel, M. and Diringer, H. 1994. ''Small virus-like structure in fractions from scrapie hamster brain,'' *Lancet* 343:894–895.

Ozel, M., Xi, Y. G., Baldauf, E., Diringer, H., and Pocchiari, M. 1994. ''Small virus-like structure in brains from cases of sporadic and familial Creutzfeldt-Jakob disease,'' *Lancet* 344:923–924.

Pan, K.-M., Baldwin, M., Nguyen, J., Gasset, M., Serban, A., Groth, D., Mehlhorn, I., Huang, Z., Fletterick, R. J., Cohen, F. E., and Prusiner, S. B. 1993. ''Conversion of α-helices into β-sheets features in the formation of the scrapie prion proteins,'' *Proc. Natl. Acad. Sci. USA* 90:10962–10966.

Patino, M. M., Liu, J. J., Glover, J. R., and Lindquist, S. 1996. ''Support for the prion hypothesis for inheritance of a phenotypic trait in yeast,'' *Science* 273:622–626.

Paushkin, S. V., Kushnirov, V. V., Smirnov, V. N., and Ter, A. M. 1996. ''Propagation of the yeast prion-like [psi+] determinant is mediated by oligomerization of the SUP35-encoded polypeptide chain release factor,'' *EMBO J.* 15:3127–3134.

Pocchiari, M., Schmittinger, S., Ladogana, A., and Masullo, C. 1989. ''Effects of amphotericin B in intracerebrally scrapie inoculated hamster,'' in Unconventional Virus Diseases of the Central Nervous System, eds., Court, L. A., Dormont, D., Brown, P. and Kingsbury, D. T. Paris. pp. 314–323.

Priola, S. A. and Chesebro, B. 1998. ''Abnormal properties of prion protein with insertional mutations in different cell types,'' *J. Biol. Chem.* 273:11980–11985.

Priola, S. A., Caughey, B., Wehrly, K., and Chesebro, B. 1995. ''A 60-kDa prion protein (PrP) with properties of both the normal and scrapie-associated forms of PrP,'' *J. Biol. Chem.* 270:3299–3305.

Prusiner, S. B. 1982. "Novel proteinaceous infectious particles cause Scrapie," *Science* 216:136–144.

Prusiner, S. B. 1991. "Molecular biology of prion diseases," *Science* 252:1515–1522.

Prusiner, S. B., Groth, D. F., Bolton, D. C., Kent, S. B., and Hood, L. E. 1984. "Purification and structural studies of a major scrapie prion protein." *Cell* 38:127–134.

Prusiner, S. B., McKinley, M. P., Bowman, K. A., Bolton, D. C., Bendheim, P. E., Groth, D. F., and Glenner, G. G. 1983. "Scrapie prions aggregate to form amyloid-like birefringent rods," *Cell* 35:349–358.

Prusiner, S. B., McKinley, M. P., Groth, D. F., Bowman, K. A., Mack, N. I., Cochran, S. P., and Masiarz, F. R. 1981. "Scrapie agent contains a hydrophobic protein," *Proc. Natl. Acad. Sci. USA* 78:6675–6679.

Puckett, C., Concannon, P., Casey, C., and Hood, L. 1991. "Genomic structure of the human prion protein gene," *Am. J. Hum. Genet.* 49:320–9.

Race, R. and Chesebro, B. 1998. "Scrapie infectivity found in resistant species," *Nature* 392:770.

Raymond, G. J., Hope, J., Kocisko, D. A., Priola, S. A., Raymond, L. D., Bossers, A., Ironside, J., Will, R. G., Chen, S. G., Petersen, R. B., Gambetti, P., Rubenstein, R., Smits, M. A., Lansbury, P. J., and Caughey, B. 1997. "Molecular assessment of the potential transmissibilities of BSE and scrapie to humans [see comments]," *Nature* 388:285–288.

Rieger, R., Edenhofer, F., Lasmézas, C. I., and Weiss, S. 1997. "The human 37-kDa laminin receptor precursor interacts with the prion protein in eukaryotic cells," *Nat. Med.* 3:1383–1388.

Riek, R., Homemann, S., Wider, G., Billeter, M., Glockshuber, R., and Wuthrich, K. 1996. "NMR structure of the mouse prion protein domain PrP(121–321)," *Nature* 382:180–182.

Rizet, G. 1952. "Les phenomenes de barrage chez *Podospora anserina:* analyse genetique des barrages entre les souches s et. S.," *Rev. Cytol. Biol. Veg.* 13:51–92.

Robinson, M. M., Hadlow, W. J., Huff, T. P., Wells, G. A., Dawson, M., Marsh, R. F., and Gorham, J. R. 1994. "Experimental infection of mink with bovine spongiform encephalopathy," *J. Gen. Virol.* 75:2151–2155.

Rogers, M., Serban, D., Gyuris, T., Scott, M., Torchia, T., and Prusiner, S. B. 1991. "Epitope mapping of the Syrian hamster prion protein utilizing chimeric and mutant genes in a vaccinia virus expression system," *J. Immunol.* 147:3568–3574.

Sakaguchi, S., Katamine, S., Nishida, N., Moriuchi, R., Shigematsu, K., Sugimoto, T., Nakatani, A., Kataoka, Y., Houtani, T., Shirabe, S., Okada, H., Hasegawa, S., Miyamot, O. T., and T. N. 1996. "Loss of cerebellar Purkinje cells in aged mice homozygous for a disrupted PrP gene," *Nature* 380:528–531.

Schätzl, H. M., Da-Costa, M., Taylor, L., Cohen, F. E., and Prusiner, S. B. 1995. "Prion protein gene variation among primates," *J. Mol. Biol.* 245:362–374.

Schreuder, B., Keulen, L. V., Vromans, M., Langeveld, J., Smits, M. A., and Van, K. L. 1996. "Preclinical test for prion diseases," *Nature London* 381:563.

Shmerling, D., Hegyi, I., Fischer, M., Blaettler, T., Brandner, S., Goetz, J., Ruelicke, T., Flechsig, E., Cozzio, A., von Mering, C., Hangartner, C., Aguzzi, A. and Weissmann, C. 1998. "Expression of amino-terminally truncated PrP in the mouse leading to ataxia and specific cerebellar lesions," *Cell* 93:203–214.

Shyng, S. L., Heuser, J. E., and Harris, D. A. 1994. "A glycolipid-anchored prion protein is endocytosed via clathrin-coated pits," *J. Cell Biol.* 125:1239–1250.

Sparkes, R. S., Simon, M., Cohn, V. H., Fournier, R. E., Lem, J., Klisak, I., Heinzmann, C., Blatt, C., Lucero, M., Mohandas, T., and et al. 1986. "Assignment of the human and mouse prion protein genes to homologous chromosomes," *Proc. Natl. Acad. Sci. USA* 83:7358–7362.

Stahl, N., Baldwin, M. A., Teplow, D. B., Hood, L., Gibson, B. W., Burlingame, A. L., and

Prusiner, S. B. 1993. "Structural studies of the scrapie prion protein using mass spectrometry and amino acid sequencing," *Biochemistry* 32:1991–2002.

Supattapone, S., Bosque, P., Muramoto, T., Wille, H., Aagaard, C., Peretz, D., Nguyen, H.-O. B., Heinrich, C., Torchia, M., Safar, J., Cohen, F., DeArmond, S. J., Prusiner, S. B., and Scott, M. 1999. "Prion protein of 106 residues creates an artificial transmission barrier for prion replication in transgenic mice," *Cell* 96:869–878.

Tagliavini, F., McArthur, R. A., Canciani, B., Giaccone, G., Porro, M., Bugiani, M., Lievens, P. M.-J., Bugiani, O., Peri, E., Dall'Ara, P., Rocchi, M., Poli, G., Forloni, G., Bandiera, T., Varasi, M., Suarato, A., Cassutti, P., Cervini, M. A., Lansen, J., Salmona, M. and Post, C. 1997. "Effectiveness of anthracycline against experimental prion disease in Syrian hamsters," *Science* 276:1119–1122.

Tateishi, J., Brown, P., Kitamoto, T., Hoque, Z. M., Roos, R., Wollman, R., Cervenakova, L., and Gajdusek, D. C. 1995. "First experimental transmission of fatal familial insomnia," *Nature* 376:434–435.

Tatzelt, J., Prusiner, S. B., and Welch, W. J. 1996. "Chemical chaperones interfere with the formation of scrapie prion protein," *EMBO J.* 15:6363–6373.

Tatzelt, J., Zuo, J., Voellmy, R., Scott, M., Hartl, U., Prusiner, S. B., and Welch, W. J. 1995. "Scrapie prions selectively modify the stress response in neuroblastoma cells," *Proc. Natl. Acad. Sci. USA* 92:2944–2948.

Telling, G. C., Scott, M., Mastrianni, J., Gabizon, R., Torchia, M., Cohen, F. E., DeArmond, S. J., and Prusiner, S. B. 1995. "Prion propagation in mice expressing human and chimeric PrP transgenes implicates the interaction of cellular PrP with another protein," *Cell* 83:79–90.

Tobler, I., Gaus, S. E., Deboer, T., Achermann, P., Fischer, M., Rülicke, T., Moser, M., Oesch, B., McBride, P., and Manson, J. C. 1996. "Altered circadian activity rhythms and sleep in mice devoid of prion protein," *Nature* 380:639–642.

Tuite, M. F., Mundy, C. R., and Cox, B. S. 1981. "Agents that cause a high frequency of genetic change from [psi+] to [psi−] in Saccharomyces cerevisiae," *Genetics* 98:691–711.

Vey, M., Pilkuhn, S., Wille, H., Nixon, R., DeArmond, S. J., Smart, E. J., Anderson, R. G., Taraboulos, A., and Prusiner, S. B. 1996. "Subcellular colocalization of the cellular and scrapie prion proteins in caveolae-like membranous domains," *Proc. Natl. Acad. Sci. USA* 93:14945–14949.

Weiss, S., Rieger, R., Edenhofer, F., Fisch, E., and Winnacker, E.-L. 1996. "Recombinant prion protein rPrP27–30 from Syrian Golden Hamster reveals proteinase K sensitivity," *Biochem. Biophys. Res. Commun.* 219:173–179.

Weiss, S., Famulok, M., Edenhofer, F., Wang, Y. H., Jones, I. M., Groschup, M., and Winnacker, E. L. 1995. "Overexpression of active Syrian golden hamster prion protein PrPc as a glutathione S-transferase fusion in heterologous systems," *J. Virol.* 69:4776–4783.

Weiss, S., Proske, D., Neumann, M., Groschup, M. H., Kretzschmar, H. A., Famulok, M., and Winnacker, E. L. 1997. "RNA aptamers specifically interact with the prion protein PrP," *J. Virol.* 71:8790–8797.

Weissman, C. 1991. "A 'unified theory' of prion propagation." *Nature* 352:679–683.

Weissman, C. 1996. "PrP effects clarified," *Curr. Biol.* 6:1359.

Wickner, R. B. 1995. "Prions of yeast and heat-shock protein 104: 'coprion' and cure," *Trends Microbiol.* 3:367–369.

Wickner, R. B., Masison, D. C., and Edskes, H. K. 1995. "[PSI] and [URE3] as yeast prions," *Yeast* 11:1671–1685.

Will, R. G., Ironside, J. W., Zeidler, M., Cousens, S. N., Estibeiro, K., Alperovitch, A., Poser, S., Pocchiari, M., Hofman, A., and Smith, P. G. 1996. "A new variant of Creutzfeldt-Jakob disease in the UK [see comments]," *Lancet* 347:921–925.

Zeidler, M., Stewart, G., Cousens, S. N., Estibeiro, K. and Will, R. G. 1997. "Codon 129 genotype and new variant CJD," *Lancet* 350:908–910.

Zerr, I., Bodemer, M., Racker, S., Grosche, S., Poser, S., Kretzschmar, H. A., and Weber, T. 1995. "Cerebrospinal fluid concentration of neuron-specific enolase in diagnosis of Creutzfeldt-Jakob disease," *Lancet* 345:1609–1610.

Zerr, I., Bodemer, M., Gefeller, O., Otto, M., Poser, S., Wiltfang, J., Windl, O., Kretzschmar, H. A., and Weber, T. 1998. "Detection of 14–3–3 protein in the cerebrospinal fluid supports the diagnosis of Creutzfeldt-Jakob disease," *Ann. Neurol.* 43:32–40.

Index

About the Editors

Jeffrey Cary earned his Ph.D. in Microbiology from Louisiana State University in 1986 under the guidance of Dr. V. R. Srinivasan. He performed postdoctoral research on the molecular biology of the *Clostridium acetobutylicum*, acetone/butanol fermentation in the Department of Biochemistry and Cell Biology at Rice University under the guidance of Dr. G. N. Bennett. He subsequently joined the UDSA, ARS, Southern Regional Research Center at New Orleans in 1989. Dr. Cary currently serves as a molecular biologist in the Food and Feed Safety Research Unit in New Orleans which has received international recognition for its work in the elucidation of the genes involved in aflatoxin biosynthesis and their regulation. In addition, he is involved in research developing transgenic plants that express novel antimicrobial peptides/proteins as a means to control aflatoxin contamination in food and feed crops. He has served on the editorial board of *Applied and Environmental Microbiology* since 1999 and has served as a technical reviewer for several granting agencies. Dr. Cary has appeared as an author in 45 publications.

John Linz received his Ph.D. in Microbiology from Louisiana State University in 1983 under the guidance of Dr. Michael Orlowski. He performed postdoctoral research in Microbiology and Molecular Genetics at the University of California, Irvine under the guidance of Dr. Paul Sypherd from 1983 to 1986. He joined the faculty of Michigan State University in 1986 as an Assistant Professor and is currently a Professor and Director of Graduate Studies in Food Science and Human Nutrition. Dr. Linz's research has three areas of focus: (1) molecular mechanisms which regulate cell differentiation and gene expression in eukaryotes with an emphasis on filamentous fungi; (2) elucidation of the molecular mechanisms which regulate mycotoxin biosynthesis for the development, of novel strategies to reduce mycotoxin contamination in food; (3) mechanisms of pathogenesis and acquisition and transfer of antibiotic resistance in *Campylobacter jejuni.* The work has generated over 40 publications. Dr. Linz has received awards in recognition for teaching (MSU Teacher Scholar, 1992; "Gerber G. Smith Award," 1996) and research (MSU Sigma Xi Junior Meritorious Research Award, 1994).

Deepak Bhatnagar earned his Ph.D. in Biophysics from the Indian Agricultural Research Institute, New Delhi, in 1977. After postdoctoral research in chloroplast bioenergetics at Purdue University and the study of the catalytic mechanism of phosphotransferases at Louisiana State University Medical Center, New Orleans, he joined the Southern Regional Research Center of the USDA's Agricultural Research Service at New Orleans in 1985. Dr. Bhatnagar has achieved international recognition for his scientific accomplishments in elucidation of molecular events in aflatoxin biosynthesis and in the application of biotechnology for producing novel solutions to problems associated with toxin contamination of food and feed. He has been the recipient of numerous ARS and USDA Secretary's awards including the ARS Outstanding Senior Scientist of the Year Award (1998). He has served on the editorial boards of *Mycopathologia* and *Applied and Environmental Microbiology* since 1990 and served on technical panels of the National Research Council of the U.S. National Academy of Sciences. He is a technical reviewer for several granting agencies and has edited three books on mycotoxins and applied mycology.